D1190979

BOSTON STUDIES IN THE PHILOSOPHY OF SCIENCE

VOLUME XXXIX

ESSAYS IN MEMORY OF IMRE LAKATOS

SYNTHESE LIBRARY

MONOGRAPHS ON EPISTEMOLOGY,

LOGIC, METHODOLOGY, PHILOSOPHY OF SCIENCE,

SOCIOLOGY OF SCIENCE AND OF KNOWLEDGE,

AND ON THE MATHEMATICAL METHODS OF

SOCIAL AND BEHAVIORAL SCIENCES

Managing Editor:

JAAKKO HINTIKKA, *Academy of Finland and Stanford University*

Editors:

ROBERT S. COHEN, *Boston University*

DONALD DAVIDSON, *Rockefeller University and Princeton University*

GABRIËL NUCHELMANS, *University of Leyden*

WESLEY C. SALMON, *University of Arizona*

VOLUME 99

IMRE LAKATOS 1922–1974

BOSTON STUDIES IN THE PHILOSOPHY OF SCIENCE

EDITED BY R. S. COHEN AND M. W. WARTOFSKY

VOLUME XXXIX

ESSAYS IN MEMORY OF IMRE LAKATOS

Edited by

R. S. COHEN, P. K. FEYERABEND AND M. W. WARTOFSKY

D. REIDEL PUBLISHING COMPANY

DORDRECHT-HOLLAND / BOSTON-U.S.A.

Library of Congress Cataloging in Publication Data

Main entry under title:

Essays in memory of Imre Lakatos.

(Boston studies in the philosophy of science ; v. 39)
(Synthese library ; v. 99)
Bibliography: p.
Includes index.
1. Science—Philosophy—Addresses, essays, lectures. 2. Logic, Symbolic and mathematical—Addresses, essays, lectures. 3. Lakatos, Imre. I. Lakatos, Imre. II. Cohen, Robert Sonne. III. Feyerabend, Paul K., 1924– IV. Wartofsky, Marx W., V. Series. Q174.B67 vol.39 [Q175.3] 501s [501]
76-16770
ISBN 90-277-0654-9
ISBN 90-277-0655-7 pbk.

Published by D. Reidel Publishing Company,
P. O. Box 17, Dordrecht, Holland

Sold and distributed in the U.S.A., Canada, and Mexico
by D. Reidel Publishing Company, Inc.
Lincoln Building, 160 Old Derby Street, Hingham,
Mass. 02043, U.S.A.

Q175.3
.E8
1976x
0 0260898 8

All Rights Reserved
Copyright © 1976 by D. Reidel Publishing Company, Dordrecht, Holland
and copyrightholders as specified on appropriate pages within
No part of the material protected by this copyright notice may be reproduced or
utilized in any form or by any means, electronic or mechanical,
including photocopying, recording or by any informational storage and
retrieval system, without written permission from the copyright owner

Printed in The Netherlands

PREFACE

The death of Imre Lakatos on February 2, 1974 was a personal and philosophical loss to the worldwide circle of his friends, colleagues and students. This volume reflects the range of his interests in mathematics, logic, politics and especially in the history and methodology of the sciences. Indeed, Lakatos was a man in search of rationality in all of its forms. He thought he had found it in the historical development of scientific knowledge, yet he also saw rationality endangered everywhere.

To honor Lakatos is to honor his sharp and aggressive criticism as well as his humane warmth and his quick wit. He was a person to love and to struggle with.

PAUL K. FEYERABEND

ROBERT S. COHEN

MARX W. WARTOFSKY

TABLE OF CONTENTS

JOHN WORRALL

IMRE LAKATOS (1922–1974): PHILOSOPHER OF MATHEMATICS AND PHILOSOPHER OF SCIENCE

Through the sudden death of Imre Lakatos on 2 February 1974 the intellectual world has lost not only an important and influential philosopher but also an exceptional human being.

His life reflects in one way or another many of the major events in recent European history. He was born in Hungary in 1922. He was a member of the anti-Nazi resistance, fortunately evading arrest, unlike his mother and grandmother both of whom were killed in Auschwitz. (During the Nazi occupation of Hungary he changed his name from the patently Jewish Imre Lipschitz to the safer Imre Molnár. After the war he was, however, reunited with a set of his shirts monogrammed 'I.L.'. Faced with this major problem (shirts like most other things were in short supply) and now a devoted communist, he again changed his name to the more working class Imre Lakatos.) In 1947 he became a high-ranking official in the Hungarian Ministry of Education, but, never a man to bow to authority, his 'revisionist' tendencies soon got him into trouble. In 1950 he was arrested and spent over three years in a Stalinist jail. After the Hungarian uprising in 1956 he was informed of the likelihood of his re-arrest and he fled to Vienna. From there he went eventually to Cambridge where his academic career began in earnest.

As well as his intellectual legacy, he left behind him at the London School of Economics (where he taught from 1960 until his death) fond memories and a fund of well-remembered stories and jokes. He embellished the English language (at least as it is spoken in the Philosophy Department at the L.S.E.): he turned 'thinking aloud' into 'thinking loudly' and the body of accepted scientific theories into the 'body scientific'. He also, in one of his seminar papers, accused a prominent Wittgensteinian, who had recently produced an enormous tome, of committing an unforgivable 'book act'. The one lesson above all others that his students learned (by example) from him was that serious scholarship can be fun.

Lakatos made important contributions to philosophy. His first love,

R. S. Cohen et al. (eds.), Essays in Memory of Imre Lakatos, 1–8. *All Rights Reserved*
Copyright © 1976 by D. Reidel Publishing Company, Dordrecht–Holland

once he had turned to the West and to academic life, was the philosophy of mathematics. His Cambridge Ph.D. Thesis: 'Essays in the Logic of Mathematical Discovery' became the basis of his 'Proofs and Refutations'.[1] This paper takes the form of an imaginary discussion between a teacher and a group of his students which reconstructs the history of the attempts to prove the Descartes-Euler conjecture about polyhedra (that the number of vertices *minus* the number of edges *plus* the number of faces is equal to two for any polyhedron). The real history is told in the many footnotes. This paper (which as well as having great philosophical and historical value is a superb literary piece) was circulated in offprint in enormous numbers, but although he had a long-standing contract for its publication in book form, Lakatos characteristically withheld it in the hope of improving it still further.[2]

The thesis of 'Proofs and Refutations' is that the development of mathematics does not consist (as conventional philosophy of mathematics tells us it does) in the steady accumulation of eternal and undeniable truths. Mathematics develops, according to Lakatos, in a much more dramatic and exciting way – by a process of conjecture, followed by attempts to 'prove' the conjecture (i.e. to reduce it to other conjectures) followed by criticism *via* attempts to produce counter-examples both to the conjectured theorem and to the various steps in the proof.

An important theme of this work is the claim that by criticising proofs ('proof analysis') mathematics very soon progresses beyond the naive trial-and-error stage of fortunate conjecture followed by undirected search for a counter example. Lakatos in fact argues that there is such a thing as mathematical *heuristic*, which is susceptible of rational analysis; in other words that the process of mathematical discovery is not simply a non-objectively analysable affair to be studied by trying to delve into the psyches of the great mathematicians. Both Popper (whom Lakatos joined at the L.S.E. and who influenced Lakatos considerably) and the logical positivists had accepted the distinction between questions about the *discovery* of scientific theories and questions about the *justification* of ready-articulated scientific theories. These philosophers claimed that philosophy was concerned solely with questions of the latter kind. Questions of the former kind were alleged to be purely psychological questions about individual scientists' thought-processes and so were claimed 'neither to call for logical analysis nor to be susceptible of it'.[3]

Lakatos (inspired here by Polya) argued that there exists a realm of rational mathematical heuristic between these two realms. This became an enduring theme of his work and was, as we shall see, carried over into his philosophy of the natural sciences.

Implicit in 'Proofs and Refutations' is a new approach to the philosophy of mathematics completely transcending the three 'foundational' schools of logicism, intuitionism and formalism, which despite known difficulties have so far dominated 20th century philosophy of mathematics. In a 1962 paper called 'Infinite Regress and the Foundations of Mathematics',[4] Lakatos placed Russell's logicism and Hilbert's formalism in a more general epistemological framework and extended the sceptics' arguments against foundations of knowledge into the sphere of mathematics.

Lakatos showed in this paper how one of the traditional ways of attempting to justify some branch of knowledge has been to try to find some indubitably true 'first principles', containing only 'crystal clear' terms, from which the whole of that branch of knowledge is derivable *via* the infallibly truth preserving rules of deductive logic. Any such enterprise Lakatos called 'Euclidean'. He showed how the logicist programme of Frege and Russell is a supreme example of such a Euclidean enterprise and how Hilbert's formalist programme falls essentially in the same category. He traced the development of these two programmes and showed how the response to difficulties within both has been one he calls 'Rubber Euclideanism'. This consists of stretching the notions of a 'crystal clear term' and of an 'indubitable *a priori* truth' so as to include precisely those terms and those 'truths' required to get round the difficulties. Lakatos argued that all such Euclidean programmes are doomed to failure even within mathematics – Euclideanism's last stronghold; and that mathematics is in need of foundations no more than are the physical sciences.

This paper together with 'Proofs and Refutations' and another important paper called 'A Renaissance of Empiricism in the Recent Philosophy of Mathematics?' (only a part of which has so far been published[5]) sets out a new philosophy of mathematics. This philosophy recognizes Euclideanism as utopian, but does not on this account embrace the despairing claim that there are no objective standards of acceptability in mathematics or that which proofs and which axioms are accepted is at best a question for aesthetics. Rather this philosophy proposes that growth in mathematics is controlled by objective standards no less than is growth in the physical sciences, and that there can, therefore, be good growth, or progress, and bad growth, or degeneration, in mathematics just as there is in physics.

Lakatos still felt that his philosophy of mathematics needed further improvement and development, but before he could supply them, circumstances turned his attention to the philosophy of science. He was asked to write a paper on Popper's philosophy. Lakatos had so far regarded himself as extending Popperian fallibilism into the domain of mathematics but he now critically scrutinized Popper's philosophy itself and found within it some open problems. Lakatos also agreed to organize an international colloquium on philosophy of science which was held in London in the summer of 1965. Having organised this conference with great success, he edited its proceedings in four volumes.[6]

The major intellectual outcome of all this activity was a series of four important papers. The first of these is called 'Changes in the Problem of Inductive Logic'.[7] It critically analyses the debate between Carnap and Popper concerning the relations between scientific theories and evidence. It charts the development of the two approaches and argues that Carnap's approach solved no philosophical problems except ones of its own creation. It also argues the importance of the switch in the Popperian programme from concentrating simply on a theory's testability to requiring that a theory have *independent* or *excess* testability over its rival theories. Lakatos argued that scientific theories can *only* be corroborated by successfully predicting the outcomes of *independent tests* (i.e. tests the outcome of which is not also predicted by rival theories). He pointed out that this makes the question of whether or not a theory is corroborated by a piece of evidence depend on what rival theories are around when the corroboration appraisal is made and hence gives corroboration a historical character.

Lakatos also showed in this paper that while Popperians and others were correct in their attribution of metaphysical synthetic *a priori* assumptions to Carnap and other inductive logicians (for example ascriptions of a specific value to Carnap's λ parameter reflect metaphysical assumptions about the degree to which nature is uniform), the postulation of a weak metaphysical 'inductive principle' is necessary also within the Popperian system. A statement ascribing a degree of corroboration to a scientific theory given certain evidence is, for Popper, analytic. It simply records the extent to which the theory has *in the past* stood up to 'severe' testing. Lakatos persuasively argued that the choice of the best corroborated theories for use in technological applications hence involves an assumption that a theory's past performance is a guide to its future performance.[8]

The outline of Lakatos' major contribution to the philosophy of science appear already in this 1968 paper, with its emphasis on scientific growth rather than on falsification and its use of the notions of progressive and degenerating problem-shifts. But his methodology of scientific research programmes is developed in detail only in his 1970 paper on 'Falsification and the Methodology of Scientific Research Programmes'.[9]

Several philosophers and historians of science had pointed to certain typical kinds of development in science which do not cohere well with the falsificationist model of scientific rationality. It was pointed out (for example by Kuhn) that the typical response of a theoretician to an experimental 'refutation' was *not* to reject the theory but to retain it whilst trying to modify the auxiliary and observational assumptions involved in the 'refutation' in the hope of explaining it away as merely 'apparent'. Lakatos himself documented several cases where a theory which is alleged to have been defeated in a 'crucial' experiment was in fact developed by its protagonists so as to keep up with its rival for some time after the supposedly crucial result.

The methodology of research programmes is the result of Lakatos's creative development of two discoveries. First the discovery that falsificationism could be developed so as to deal with these apparently anomalous aspects of scientific development; and secondly the discovery of the critical role played in science by the heuristic principles whose importance for mathematics Lakatos had already stressed.

According to this methodology the basic unit of scientific discovery is not an isolated theory but rather a research programme. Such a programme, developing under the guidance of its heuristic, issues in a series of theories. Each such theory though it may contain an irrefutable ('metaphysical') part, will be refutable, but the typical response of the proponent of the programme to an experimental refutation will be to amend his theory – leaving certain assumptions (the 'hard core' of the programme[10]) unchanged, whilst replacing other ('auxiliary') assumptions.[11] (But the auxiliary assumptions are changed not only under the pressure of anomalies but also (and more importantly in the case of the best research programmes) under the guidance of the heuristic.) There will in general be rival research programmes in any field and it may be (and generally is) the case that the latest theory produced by each of these programmes is inconsistent with accepted experimental reports (i.e. 'refuted').

But if even the best scientific programmes are always in experimental difficulties and if their proponents are allowed to elaborate and amend their assumptions rather than give them up in the face of experimental difficulties, what distinguishes one programme from a better one? And what distinguishes the 'best' science from apparently pseudo-scientific programmes, like the Freudian and Marxist ones, whose proponents seem to defend them in precisely this way? Lakatos showed that the distinction is one between 'progressive' and 'degenerating' research programmes. Given any finite set of anomalies to, or refutations of, a theory it will in general be trivially easy to construct modified auxiliary assumptions which deal with these anomalies. Many such modifications will, however, be *ad hoc*; only a few will, on the contrary, have extra predictive power over the original theory, and even fewer will have their extra predictions empirically confirmed. Newton's programme, for example, produced theories which not only dealt with some of their predecessors' anomalies but also correctly predicted new facts; it was therefore progressive. The Cartesians, on the other hand, managed to incorporate the Newtonians' successes within their programme but only in a *post hoc* way and without at the same time predicting anything new (and hence this programme degenerated).

Lakatos developed his views on how historical case-studies can be used as a source of criticisms of philosophies of science in his paper on 'History of Science and its Rational Reconstructions',[12] where he attempts to give substance to his (increasingly famous) paraphrase of Kant: 'Philosophy of science without history of science is empty; history of science without philosophy of science is blind.' He had always been eager (as befits an ex-Hegelian) to bring the philosophy both of mathematics and science closer to their histories. He now became in this paper the first to propose a general method for the evaluation of rival methodologies in terms of the 'rational reconstructions' of the history of science they provide and of the historical accuracy of these reconstructions. This meta-methodological criterion synthesized, he argued, the *a prioristic* approach to methodology (which claims that there are immutable *a priori* general rules for scientific appraisal) and the antitheoretical approach to methodology (which claims that there are no general standards of appraisal and that all we have to go on is the scientific élite's instinctive decisions in individual cases).

Lakatos was a master of the methodologically motivated study of specific historical cases (a subject which, according to Feyerabend, he turned into 'an art form'). His last publication, a joint paper with his colleague Elie Zahar, was such a case-study. It argues that amongst available methodologies only the methodology of scientific research programmes can explain the Copernican revolution as consisting of the replacement of one theory by an objectively better one without distorting the historical facts.[13]

Imre Lakatos leaves behind him a mass of so far unpublished material and a set of thwarted plans to reply to some of his critics (like Kuhn, Feyerabend and Toulmin, for Lakatos's methodology had become one of the focal points of debate in philosophy of science) and eventually to apply his methodological ideas to other fields. Fortunately he also leaves behind him (and it was of this achievement that he was most proud) a thriving research programme manned, at the London School of Economics and elsewhere, by young scholars engaged in developing and criticising his stimulating ideas and applying them in new areas.

London School of Economics

NOTES

[1] This was published in four parts in *The British Journal for the Philosophy of Science*, in 1963–4.

[2] The piece has also occasionally been performed by groups of mathematics students in the U.S.A. It was also 'pirated' in the Soviet Union where it is apparently a best-seller. (*Proofs and Refutations* is finally to appear in bookform in 1976, published by Cambridge University Press.)

[3] Popper, *Logic of Scientific Discovery*, p. 2.

[4] Published in *Proceedings of the Aristotelian Society, Supplementary Volume* **34** (1962).

[5] This part is in Lakatos (ed.), *Problems in the Philosophy of Mathematics*, 1967.

[6] *Problems in the Philosophy of Mathematics*, 1967; *The Problem of Inductive Logic*, 1968;

Problems in the Philosophy of Science, 1968; and *Criticism and the Growth of Knowledge*, 1970. (The last two works were edited jointly with Alan Musgrave.)

[7] This paper appeared in *the Problem of Inductive Logic* volume.

[8] This argument is also further developed in Lakatos's paper 'Popper on Demarcation and Induction', in P. A. Schilpp (ed.), *The Philosophy of Karl Popper*, earlier published in German as 'Popper zum Abgrenzungs– und Induktionsproblem' in H. Lenk (ed.), *Neue Aspekte der Wissenschaftstheorie*, 1971.

[9] This was published in the *Criticism and the Growth of Knowledge* volume; a briefer and less fully argued account of the methodology had already appeared in the *Proceedings of the Aristotelian Society* **69** (1968).

[10] The 'hard core' of the Newtonian programme, for example, consisted of the three laws of motion and the law of universal gravitation.

[11] Lakatos called the set of auxiliary assumptions (which included in the Newtonian case theories of optics and in particular of atmospheric refraction) a programme's 'protective belt' since it protects the hard core from refutation.

[12] This paper forms part of Buck and Cohen (eds.), *Boston Studies in the Philosophy of Science*, vol. VIII, 1971. See also his 'Replies to Critics' in the same volume.

[13] "Why did Copernicus's programme supersede Ptolemy's?" in R. Westerman (ed.) *The Copernican Achievement*, 1976.

JOSEPH AGASSI

THE LAKATOSIAN REVOLUTION*

1. Lakatos as a Teacher

Lakatos' classical 'Proofs and Refutations' reports the ongoings in a classroom in Utopia. Lakatos himself tried out the Utopian experiment in a real class early in the day – it was in Popper's seminar, and while he was writing his doctoral dissertation which includes an early draft of his masterpiece. Not surprisingly, then, he was acidly critical of some aspects of the accepted modes of mathematical teaching (to be discussed below). Nonetheless, I know of no discussion of his educational philosophy – printed, manuscript, or orally presented.

Perhaps this reticence, thus far, relates to the fact that Lakatos took active part in the student revolt affair, and the definitely wrong part. There was, I am of the opinion, no right part, at least no obviously right part, to the students' revolts of the sixties anywhere in evidence (except, I think, for the initial demands of the French students, the rejection of which sparked off their revolt); but there were more or less degrees of wrong. The students have demanded at times the wrong things; and they usually did so in the wrong way. But this is not to condemn them offhand, much less to take a reactionary stand in the name of the preservation of all that deserves preserving, refuse to yield on any point, and recommend non-negotiation. Lakatos did take such a stand.

Perhaps the reticence, thus far, is due to the fact that Lakatos evolved into a highly successful university lecturer of the old style and was highly censorious of those colleagues of his who, he felt, did not live up to the obligations of a university lecturer as he understood them, failed to prepare for each lecture elaborate lecture-notes, cover much informative material in each lecture, etc. etc. I am too averse to the oldfashioned view to do it justice, but I think the reader may be familiar enough with it, even if he is not familiar with its Central European rich and thick and learned and emotionally charged and witty and sweeping variants. I assure the reader that all these epithets fit Professor Lakatos' later perfor-

R. S. Cohen et al. (eds.), Essays in Memory of Imre Lakatos, 9–21. *All Rights Reserved*
Copyright © 1976 by D. Reidel Publishing Company, Dordrecht-Holland

mances as a university lecturer in his lecture courses in the London School of Economics; that he would have taken this ascription of them to him as a high compliment; that I for one do not see anything to praise in the tradition or mold he was so proud to belong to; and that his Utopian lectures, in his classic 'Proofs and Refutations' as well as in his earliest performances in the London School of Economics as a guest speaker in Professor Popper's seminar on scientific method, which I so admire, had as little to do with that tradition as is conceivable at all.

We once used to wonder about the fact that makers and starters of revolutions so often stayed behind. T. S. Eliot has changed this with his deep insight in his *Journey of the Magi*.

2. Schools and scholarship

Inasmuch as schools supposedly convey in a condensed manner all that is worth preserving in our heritage, it may just as well be an excellent representation of it. Even when what schools pass on to the vast population that passes through them is not the very best, nevertheless it often is representative. I found from experience that some puzzlements about a foreign culture may be solved by even a superficial perusal of the curriculum, the set texts, and similar school articles.

I suppose it is very customary still, in the curricula and syllabi of many university departments around the globe, to claim that students are offered there not only materials but also the means of acquisition of more materials. These means are soft-wear and hard-wear, with instruments and libraries as hard-wear and techniques of using them as soft-wear. The soft-wear are often carried in the student's – or researcher's – head or notebooks, and are known by the embracing name of methods and/or methodology. Schools, to repeat, often claim to be teaching methods and/or methodology. They do not.

An exception, noted by Lakatos any number of times, is this. Whereas most teachers claim that they teach methods, some of the most high-powered mathematics teachers openly deny that they teach methods. They think that the acquisition of large doses of mathematical knowledge in itself develops in the student a high level of mathematical sophistication which enables him to resolve old puzzles his teacher has chosen to leave unsolved and to develop the necessary skills to continue in his predecessor's lives of investigation into new areas of knowledge.

What Lakatos briefly suggested is that this is a mere excuse; that, in other words, the puzzles a teacher leaves unsolved he cannot solve – at the early stage in which they arise or even at any other state – and that the methods he does not teach he does not know. A fish, Lakatos was fond of saying, may well be able to swim with not the slightest knowledge of hydrodynamics. Lakatos never explained.

Now, let us be open-minded about this. Popper's philosophy, which influenced Lakatos at the time he was writing his drafts, starts with the assertion that there is no scientific method. And, Lakatos is right in claiming that the view that there exists scientific method in the sense in which Popper combats it – the Baconian sense of a sausage-making machine with a sure output as long as input keeps flowing – thanks to Popper "at least among philosophers of science Baconian method is now only taken seriously by the most provincial and illiterate" (Schilpp's Popper Volumes, p. 259). Inductivism, Lakatos thought, was definitely out. I am not even quite happy about that since I read Konrad Lorenz's and Nikolaas Tinbergen's Nobel Lectures that are fairly Baconian and considering that philosophers of science are still not ready to dismiss as insignificant views of such big fish. Anyway, at least in the present essay we can take it for granted that no Baconian method exists. It is clear that Popper – and Lakatos too – assumes that in other senses methods do exist, however vague this statement is; for example, it is left open whether the more general method of testing in science or more specific methods, more peculiar to given ages and fields, like the mathematical methods in theoretical physics, the empirical methods common in ever so many contemporary laboratories, etc. Can these be taught? Are laboratory manuals of much use, and, even if they are, should one read them rather than consult them on occasion?

I speak of the teaching of mathematical methods because, I think, Lakatos was right: mathematicians did not teach them because they were ignorant of them. I shall claim that the only excuse teachers have for teaching is ignorance – their own; their students' is taken for granted, of course. If so, then teachers will do better to study rather than teach, since the more they will know the less they will wish to teach: which is all to the good. This is particularly true concerning methods: we all know that foreign aid in the form of consumers' goods is only good for emergencies and as a stand-by, that the important goods foreign aid can transmit are means of producing consumers' goods – both hard-wear and soft. What is true of underdeveloped countries is true of underdeveloped individuals.

3. The Problem of the Magic of Mathematics and Lakatos' Solution

Consider the observation of a magician pulling a rabbit out of a top-hat. He can do so spontaneously, or after he is requested to do so by members of his audience. It is in a sense less surprising – namely, expected – in a sense more surprising – that he could do so upon a request. The second sense of surprise, not the first, concerns a mathematics student. For, in the first sense surprise is momentary and can be cushioned by preparation of an anticipation; in the second sense surprise can be retained until alleviated by explanation, and so is a kind of puzzlement. (This is why some magicians like it this way, some that way. Those who like it this way keep talking to avoid interruption, those who like it that way plant people in their audience who make the proper request.) Now observing a magician we are puzzled and are meant to be puzzled; not to understand. But when a mathematics teacher pulls rabbits out of a top-hat we are meant to comprehend, yet we are puzzled. How can we both comprehend and remain puzzled?

There is no doubt that mathematics teachers do pull rabbits of exciting theorems from top-hats of all sorts of axioms and definitions. Whether they do so apologetically – sorry to be unable to relieve the puzzlement as any magician does to his apprentices – or with a vengeance – defying his students as any magician defies his audience, especially if the audience is rather presumptuous – is not a matter I shall go into. I do accept Lakatos' observation: they cannot resolve the puzzlement: at least not as yet. Perhaps later on they can explain the puzzlement to a budding colleague – after he has acquired much more insight to the working of the mathematical method. All right, says Lakatos, let us spell out to ourselves what is this working of mathematical method and how the puzzlement is to be resolved by it. It turns out to be no mean matter: very few studies exist on mathematical methods, whether written by working mathematicians or not, and their whole product is quite meager.

Before Lakatos came to make his mark, a few ideas were extant on the matter. The simplest idea, one which is still not rare in mathematics departments, one which he felt compelled to fight time and time again though now it is somewhat less popular and so towards the end of his life he felt he could ignore, is the sausage-machine idea. Essentially, the view goes, you choose a set of axioms, try to avoid inconsistency, and feed them into any old deduction-machine which will start churning out theorems, mostly worthless and uninteresting, but some useful, some interesting, some both. If so, then mathematicians are simply finding needles in haystacks and have no idea which haystack – axioms – to choose or how to increase the frequency of finding needles – interesting or useful theorems – in the hay-stack. If so, the puzzle is unresolvable, and all a budding mathematician can hope for is better luck and a better intuition for short-cuts. But that's all. The promise which the mathematics professor makes to his students, of a better understanding in maturity, then, is quite pointless: he does not have it himself.

Another theory, which I have heard from Popper in his lectures but I think is not unknown, looks the very opposite of the above idea, but in a surprising manner turns out to be a variant of it. We start not with axioms but with problems, starts the variant, where problems are theorems to be proved. The rest is a matter of intuition and luck: you choose your axioms or you have them given, and you wade through innumerably possible deductive routes, looking for the one which leads you from your axioms to your theorem. If you succeed, you have to tell your students that you can deduce the theorem (or its negation, or its independence), but not why this route was successful, whereas other routes lead not to the same destination. (Nor can you tell whether an open case is decidable, or which way.) This method is a variant, since it only adds lucky shortcuts.

There is a lot to this variant: the student is told what he is expected to understand – namely, each single step in the deduction – and what remains a puzzle – namely, how a specific deductive route was chosen; he can see how the few attempts which have failed have failed, and that the successful one succeeds. The strong evidence in support of this explanation, of the success of the short-cut as lucky, is also mentioned by Popper (in Schilpp's Popper volumes, p. 1077): once a proof is discovered, "almost invariably it can be simplified." That is to say, even when the needle is found, it is part needle part hay to be removed; yet the very removal of the hay shows the existence of a proof idea, an idea to be hit upon by trial and error. It is this proof idea, incidentally, that allegedly the mathematical students learns to appreciate when he acquires an ever increasing number of proofs. If so, says George Polya, we can teach him from the start both proof ideas and the method of trial and error by which he may learn to hit upon a successful proof idea himself – first a known one, when he is a true apprentice, rather than a passive student, and then a new one, when he is a novice.

Suppose all this is true. Can we, then, explain the method of mathematics, so as to make a success more understandable than drawing a rabbit out of a hat? Not quite. We may, perhaps, explain one proof-idea but each proof has a different one. And so, Polya's idea of letting students discover proofs for themselves is too hard to execute, though his intentions are perfect.

Such was the state of the understanding of methods of mathematics when Lakatos came on to the stage, with a bow and an expression of gratitude to Polya, while adopting Popper's philosophy to his end; he achieved his end and threw things into a state of havoc.

The relation between Polya and Lakatos is the same as between Whewell and Popper (even though none of the other three knew of Whewell; likewise Lakatos learned about Popper only after his work was begun). All four believe in trial and error, in deductive explanation, in starting with problems, with the search for explanations. Yet, whereas Whewell and Polya believe in verification, and see the corpus of (scientific or mathematical) knowledge as the set of successful trials, of verified explanations, Popper and Lakatos are fallibilists and view refuted theories as part and parcel of our heritage.[1]

Lakatos' masterpiece deals with mathematics, not physics. So his refutations are not physical; they are potential counterexamples; at times, but rarely and gratuitously, they are even actual. The task is to look for them; proofs facilitate refutations as they specify conditions which we may violate in our search for counter-examples. This makes us able to stretch and shrink concepts, as well as to criticize theorems as at times too narrow – not covering all cases of the theorem – and at other times as too wide – covering counter-examples (whereas actual counter-examples in physics only prove a theorem too wide; its being too narrow only leads to a quest for a further or a better explanation). In a manner not quite clear to readers of Lakatos, a theorem's transformation and transfiguration may end up in formalization. In a sense formal systems are end-products. But only in a sense; formalization does not end the process of proofs and refutations since we can always ask, is the formal system the same as the preformal one, or does it have an unintended model?

So much for the view of Lakatos. He claims that he is the first to give a theory about the role of proof in mathematical research. Before him people saw a proof as closing an issue; he saw it as a part of the process of the growth of mathematics. Anyone who reads his masterpiece must notice that the very fact that the proof procedure is a long process of fumbling, of trying again and again, makes the growth of a proof idea reasonable, especially intriguing and human – as opposed to being an act of magic – in that it is far from perfect and we poke holes in it and see in its strong points and in its weak points matters of great insight and an increased interest. It is the claim that proof-ideas are essentially correct, even if they need a correction or a simplification, that leads to the attempt to offer a simplified condensed version, and thus, further, according to Lakatos, to an appearance of magic. Only if we appreciate errors enough to incorporate some obvious errors into the dynamic growth of a proof-idea do we offer our audiences those details of the making of the act which makes a magician utterly unable to be surprised when he sees a colleague use tricks he himself knows so well and draws rabbits out of top-hats galore.

4. Can Lakatos' Method Be Applied?

If I were a mathematics teacher asked whether I should emulate the Lakatos method of Utopian teaching, I should hesitate. I would not hesitate if I were asked to teach mathematics to non-mathematicians, or if I were to teach whatever I do teach (since in fact I exhort my students to break the equestrian and asinine habit of hard-work and passivity in preference for intelligent work – and the two only seldom overlap). As a mathematics teacher teaching, say, topology, I should have to cover the material which Lakatos discusses in his 'Proofs and Refutations' in the space of a fraction of one lecture. I would have no time at all for his details. Oh, I would gladly advise my students to read Lakatos on a weekend or during a vacation or as relaxing bedtime reading. Still better, if my mathematics students can take credits in either the history of mathematics or in the philosophy of mathematics, then I would gladly propose Lakatos as their major text. But the cruel fact is that Lakatos' very lengthy discussion ends more or less where the modern textbook of topology begins, to wit with Poincaré's algebraic version of the Descartes-Euler theorem. What shall we do about advanced topology? Can I present it *à la* Lakatos? Do I have the time to do it within a prescribed semester course? Do I know how?

Let us not be finicky. Let us assume that in a small seminar of advanced mathematics, where we are going over raw new material, we can make better use of the deliberation and fumbling of recent students of the field in order to see how they work and not be puzzled to see them pull rabbits out of hats. Let us also agree that if we do not have to move quickly and cover a lot of ground, we can, indeed, use the Lakatos method at leisure. Let us also observe, as a matter of fact, that Lakatos did succeed to make non-mathematicians partake in his Utopian mathematical discussions and reproduce historical cases – with the aid of the teacher, of course.

This last point signifies much. It shows that students' interruptions need not be any impediment; that, erroneous as they are, they help dispel the impression that mathematicians draw rabbits out of top-hats.

Let me stress all this. The general view of the matter is that students' participation is of low quality and so at best a necessary evil. And the minority view – the view known as the discovery-method of teaching – is that students' participation is or should be of high quality, the highest indeed, if only certain conditions are met. But, no doubt, the discovery method is an unrealizable dream: students cannot possibly be systematically so good as to emulate the best minds and the best results in a given field, no matter what that field is. In any case, I insist, whether students are of high quality or low quality matters little, since we want to help them raise their quality such as it is – indeed we should worry more about the low quality ones. The fallibilist view that the students interruptions, however low quality, can be used in class – this view is thrilling because it makes the question of quality superfluous. It assumes that students do fail, and do need help to learn what to do about their failure, how to improve performance, where to find some solution which they have overlooked, and how to try to assess these and perhaps even transcend them. There is one hard question: can we assume that we can pursue such a line of activity for a whole university course? I do not think so.

Let me also concede that in certain junctions even small doses of the Lakatos technique may be added to traditional courses with some excellent results. Let me take an unpublished example of Lakatos. The foundation of the calculus taught in a rough and ready manner to uncomprehending students is very hard. It is much better to tell them something about the early calculus, say Newton's; Berkeley's criticisms; attempts by diverse writers to answer him, culminating with Weierstrass. This can be even read into Bell's most conventional and wrong headed history. But, says Lakatos, there is much more to it. Let us look closely at Cauchy. You take a Cauchy series, and you prove that it converges. You have a criticism: that it converges assumes a convergence point to converge into: is there one for every Cauchy series? Comes Weierstrass and says, yes if we identify the series with the point. I need not explain how revealing this example is, especially of the historical proximity of Cauchy and Weierstrass.

Can we, however, do this for a whole course of mathematics; present a body of mathematical knowledge as series of proofs and refutations? Is Lakatos right? In other words, is Lakatos' view of mathematics comprehensive?

I think not.

5. Mathematical systems and sub-systems

Lakatos himself was aware of the fact that his own researches presented not a total view of mathematics and its development but first and foremost a criticism of all prevalent views – since these were invariably verificationists of one kind or another – and second a tentative view of mathematics which, when viewed as comprehensive is found wanting, exactly in the way that Popper's view of physics is.

I do not think Lakatos ever developed a comprehensive view of mathematics and of its history. I do not think he even had an idea that satisfied him about the way comprehensive views of mathematics have interacted with the growth of mathematics. He wanted to rewrite his 'Proofs and Refutations' in a manner that would include a comprehensive view. He never did.

Let me take an example. Lakatos enthusiastically endorsed Russell's thesis *(Foundations of Geometry)* that nineteenth century mathematics is largely the outcome of a response to Kant's comprehensive view of mathematics. I have myself commented on this point elsewhere and shall not go into it now, except to say that Lakatos wanted to incorporate such facts as significant factors in his view of mathematics as a whole, yet he never did. In some place he even declared that formalism, his pet enemy, had a role to play in the history of mathematics. But he never put this into a comprehensive framework.

But let me touch upon less comprehensive instances. I am loath to take cases of any established mathematical truths – seemingly or in truth – as I remember how Lakatos fought like a lion when presented with these. He either disproved them, or claimed that his interlocutor was offering the latest modification of a theorem, designed to overcome a recent refutation. But one example, I think, he did concede and had to concede.

The example is the field of ordinary differential equations. You can, of course, say that it is not so much pure mathematics as applied mathematics. This will raise a host of important questions – more questions that Lakatos' early demise prevented him from taking up as he intended to do. So let us not go into that.

As a student I found this field particularly irksome, as I suppose – though I could not articulate it then – a book will be found irksome if each of its paragraphs makes sense but as a whole it makes none (e.g. a book by L. Wittgenstein). But putting it into its context, offering its problems and methods, makes the field of differential equations eminently lucid and sensible.

Differential equations are puzzles or riddles. We cannot find answers to the puzzles, but we can guess them, try each guess and refute or, on occasion, verify it. This makes the field eminently in accord with Whewell and with Polya. It does not accord with Lakatos, since refutations of blind guesses in this field lead nowhere, but verifications are true successes. To prove this perhaps we need proofs of existence theorems and of uniqueness or generality of solutions or of forms of solutions (uniqueness up to a constant, or up to the product or sum or such of some function or another). Yet, whether existence and uniqueness theorems are above criticism or not, clearly, when we solve a problem set by a differential equation by guessing what the function may be, differentiating and substituting, and arrive at the set equation, then we have solved the problem even though perhaps not uniquely or generally enough. That much, but not much more, Lakatos did concede. Moreover, we are taught a few useful tricks such as guessing a transformation of the variable(s) which just might transform an unsolved equation to a solved one; and of such tricks we can say we do not know why they work but many people tried many tricks and some are indeed successful. All this is as plain as your nose and quite outside Lakatos' concept of mathematics; his concept is too narrow. Moreover, the idea of transforming differential equations has very wide extensions in modern mathematics, which have axiomatic systems based on similar general ideas though within them the Lakatos method may well be very useful and enlightening.

This includes category theory. Indeed, the natural way to introduce both category theory and a specific category (with respect to a specific composition rule) is axiomatic. On the whole, since mathematicians these days all too often introduce abstract entities axiomatically prior to investigating them, the role of axiomatics has radically changed. It is well-known, of course, that in the nineteenth century axioms ceased to be self-evident and competing axiom systems for geometry and the different geometries studied. But these geometries were seen as having their own characteristics, depicted axiomatically or otherwise – especially since Klein showed embedding to be a general way of presenting geometries. When Hilbert introduced his metamathematics and theorems became objects of a different kind, and deduction became an operation that took its own quality rather than the way to show – in a way developed by Hilbert himself! – that the characteristics of a geometry are indeed successfully depicted by the axioms. Nowadays the axioms do not depict the characteristics of a geometry (a category, an algebra, a space with a topology, etc.) but, in abstract cases of some sorts, the axioms generate the system. We have here, in this short outline quite a few views of axiomatics and these need much further study. Lakatos merely repudiated the Hilbertian one.

What Lakatos had to say about the choice of axioms, of such systems or of geometry, was largely negative. We start not with axioms but with problems, theorems, proofs, refutations, he said. We axiomatize a system only after its concepts were quite sufficiently knocked into shape – he gave no criterion – and often systems are not axiomatized or only quasi-axiomatized. Yet the fact is that often a new field springs into being almost axiomatized almost from the start, not struggling towards its axioms as the calculus or as the theory of probability did.

Let this be. How do we get axioms (choose axioms or move towards them)? By trial and error, of course. But what do we aim at? There proofs and refutations systems do not make. We have peculiarities of systems, and Lakatos says nothing about them.

Take a trite example that also troubled me as a student. In the calculus we postulate the existence of divergence of functions, and we allow our variable(s) to vary from minus to plus infinity. Yet we postulate one point at infinity. In affine geometry we postulate a line at infinity. This is puzzling. It is more puzzling if we know that measure theory postulates two points of infinity for each variable, but I did not know this then. I asked my math professor why the difference between the calculus and affine geometry. He said, the one projects a sphere on the plane, the other a semi-sphere. I asked why and never got an answer.

I asked a few mathematicians the same question since. They all know the answer but many could not articulate it. The answer is this. The calculus is concerned with well-behaved functions and tends to lump together unpleasant and bothersome exceptions; in affine geometry we are concerned with directions (i.e. complete sets of parallel lines) and each point at infinity corresponds exactly to one direction.

Speaking as conventionalists, mathematicians are bound to view their choices as arbitrary, and so not explain them; speaking as naturalists they feel an urge to prove the correctness of their choices. Lakatos quite rightly rejected both of these philosophies and wanted problem orientation or the dialectical view to take over. But he showed no way to explain overall concerns and overall or global problems of mathematical subfields.

Work in this direction should well suit his general attitude, and it has already started – but is not yet at a stage where we can fully apply the Lakatos method.

6. The ongoing revolution

The feeling that mathematicians, and teachers of mathematics, are pulling rabbits out of hats is rightly disturbing – at least to an apprentice who wishes to know how the master does it and how he can emulate him.

We cannot fully explain this, as there is no systematic way for inventing mathematical ideas – there is no algorism of discovery. Yet much can be done. First, the aim of a given exercise can be explained. Second, attempts to accomplish the aim that look fairly obvious may be presented and criticized. These two kinds of steps take much of the mystery away and explain to some extent given discovery, thereby it also offers partial algorisms of sorts, as I have explained elsewhere when examining the parallel situation in empirical science. In mathematics, but not in empirical science, often the partial algorisms play a significant role in that they become subject to investigation in attempts to complete them into algorisms proper.

Lakatos has shown that the formalist view is erroneous, in that what sets of problems we do have algorisms for the solution of we do not much care about, that where the action is matters are fluid and criticism is the daily routine. Nevertheless, here is a complete and unbridgeable break within the system; whereas in physics we have nothing but conjecture and test and no algorism, in mathematics algorisms are an important and an ever increasing sedimentation.

All this, I feel, require much more study. And the more we cover in the spirit of Lakatos, the more we can teach in class in a dialectical method, as practiced in Lakatos' lovely utopian class.

Boston University and Tel Aviv University

NOTES

* I am grateful to Professor Peggy Marchi for her comments on earlier drafts.

[1] In his contribution to the Schilpp Popper volume Lakatos ridicules this view, and opts for a more inductivist philosophy of science. I have expressed my view on his philosophy of science elsewhere; here we come to praise him, not to bury him. Let me just mention one fact that sharply exhibits the overall change in Lakatos' philosophy when he permitted a bit of inductivism in. His later works repeatedly – 3 times – attack the Popperian view (he quotes me on this) that empirical learning from experience is the discovery of counter-examples. This very idea, implicit in Popper, was explicitly stated in Lakatos' own classic 'Proofs and Refutations' (IV, pp. 303–4) as an anti-inductivist stand.

D. M. ARMSTRONG

IMMEDIATE PERCEPTION

I wish that I had a paper for this volume which was closer to Imre Lakatos' own philosophical interests. He worked on questions which are central to our scientific and intellectual culture. His contributions to these questions conveyed a profound sense of excitement, adventure and progress. What I offer here is something on a very much smaller scale. Nevertheless, any philosophy of science must not merely make use of, but must investigate, the notion of observation and so the notion of perception. In what follows I try to sort out what is living and what is dead in the notion of *immediate* perception.

1. The Concept of Immediate Perception

The philosopher who, above all others, we associate with the notion of immediate perception is Berkeley. He says in the *First Dialogue*:

> ...in truth the senses perceive nothing which they do not perceive *immediately*: for they make no inferences.

It is clear that the central notion which Berkeley has in mind is that of the *completely uninferred* element in perception. Perceptions may involve inference, as in the case when, seated in the study, I automatically take it for granted, on the basis of what I hear, that a coach has gone past in the street. But, in default of infinite regress, whatever is inferred must be based upon what is not inferred. But since our knowledge and/or belief about our current environment *begins* with perception, there must in every case be perceptions which involve no inference. These are our 'immediate perceptions'.

This line of argument seems to be sound. Unfortunately, however, a number of other much more dubious notions cluster around this central notion of a perception involving no inference, both in Berkeley and in later philosophers. Some of these notions receive little countenance among contemporary philosophers. I devote the rest of this section to

R. S. Cohen et al. (eds.), Essays in Memory of Imre Lakatos, 23–35. *All Rights Reserved*
Copyright © 1976 by D. Reidel Publishing Company, Dordrecht-Holland

noticing and dismissing these notions briefly. This leaves me free to attend to the main objective of this paper: criticism of the notion that there are such things as immediate *objects* of perception.

The *first* correction we need to make is to reject a notion that is present in the quotation from Berkeley. This is the notion that immediate perception is the only true perception and that mediate perception, because it involves inference, is not really perception at all. If a man in his study takes it for granted that he hears a coach go by outside then it seems clear that this should be said to be a perception. It would be awkward to call it anything else. At the same time, the perception clearly involves inference, even if inference of a quite automatic, instantaneous and unselfconscious sort. Berkeley seems wrong to oppose perception and inference so sharply.

A *second* notion which Berkeley and others have associated with immediate perception is that it is an awareness of a certain special class of objects: 'ideas', sense-data or the perceiver's own sensory states. I reject this notion, first on the ground that there are no such things as 'ideas' or sense-data, second on the ground that, although we do have sensory states, we do not *perceive* them and so cannot immediately perceive them. We have, or are in, sensory states and we may, or we may not be, introspectively aware of them. I excuse myself from arguing for this now fairly orthodox view.

The *third* notion, also to be found in Berkeley, is that this completely uninferred element in perception constitutes logically indubitable knowledge. Here again I believe that this notion must be rejected, first on the ground that there is no such knowledge (see Armstrong, 1968, Chapter 6, Section 10), second that there is no reason to think that what is completely uninferred, in perception or any other field of cognitive activity, must even be knowledge, let alone indubitable knowledge. It may simply be belief, and false belief at that. The notion that immediate perception involves logically indubitable knowledge is, of course, linked with the idea that it is perception of a special class of mental, or at least private, objects. But it has even deeper roots in the 'search for super-certainty' which has bedevilled philosophy from the time of Descartes almost up to the present day.

But even when we have rejected the notions that immediate perception is the only true perception, that it is perception of private objects or states, and that it furnishes us with logically indubitable knowledge, still the notion of immediate perception requires further purging before it can be accepted.

2. Immediate perception is not perception of objects

Immediate perception has traditionally been thought to be the perception of objects. As usual, the notion is found in Berkeley. Even those who, like Armstrong (1961), have rejected the idea that immediate perception is of private objects, have simply substituted the notion of immediate perception of certain public objects. To try to make it clear that immediate perception is not perception of *objects* at all, I will now say something about the vocabulary of perception. I will distinguish between:

(1) 'Object-perception' statements
(2) 'Propositional-perception' statements
(3) 'Object-appearance' statements
(4) 'Pure-appearance' statements.

I believe that most, if not all, ordinary statements of perception can be brought under these four headings.

2.1. *Object-Perception Statements*

'Object-perception' statements are statements like 'Jones saw a (the) fox'. Their general form is '*A* perceives *O*.' But the object perceived need not be an object in any narrow sense of the word. 'The child saw the march past' counts as an object-perception statement, although the march past would be normally classed as an event or happening, not as an object.

It is an essential feature of object-perception statements that they entail that the object perceived has existence. If Jones saw a fox, then, of necessity, there was a fox there to see. '*A* perceived *O*' entails '*O* existed'. The statements have what may be called 'existence-grammar' (Armstrong, 1968, p. 215).

It must be admitted that it is possible, even if uncommon, to have *sentences* of the form '*A* perceives *O*' where what is stated by the sentence carries no commitment to *O*'s existence. An oculist may say to a patient 'Tell me what you see' and the patient may say that he sees something or other without being committed to asserting the existence of that thing. In the case of sentences of this sort I simply stipulate that they do not express 'Object-perception' statements. They express statements of type (3) or (4).

But although 'Object-perception' statements entail that the thing perceived exists, they do not entail that the perceiver knows, or truly believes or even has the thought that the object perceived exists. Nor need the object perceptually appear to him to be the object that it in fact is. This point was first emphasized, to my knowledge, by Warnock in an important paper (Warnock, 1954–5). It can be true that Jones saw a fox, yet also true that Jones neither knew nor believed that he saw a fox, that the thought of a fox did not cross his mind, and that his perceptual experiences were not fox-like. Suppose, for instance, there looked to him to be some sort of object in the far distance whose nature he was quite unable to make out but which was in fact a fox. It might still truly be said of him that he saw a fox.

The fact is that object-perception statements are hardly *cognitive* idioms at all. This is connected with the fact that the argument-form:

(1) A perceives O
(2) $O = P$
∴(3) A perceives P

is a *valid* form. If A sees a table, and the table is a cloud of fundamental particles, then A sees a cloud of fundamental particles, although A be the most ignorant peasant in the world. As Quine would put it, in 'Jones saw a table' the phrase 'a table' occurs in a referentially transparent context.

But once we see the non-cognitive nature of object-perception statements, we see further that we cannot draw the immediate/mediate distinction in respect of such statements. For the distinction between the uninferred and the inferred is a *cognitive* distinction. The uninferred is something known or believed, while the inferred is something known or believed on the basis of what is uninferred.

Berkeley gives as an instance of immediate perception hearing a sound while seated indoors and as an instance of mediate perception hearing the coach which made that sound. But if we take the *object-perception statement* 'B heard a coach' then this does not entail that B made any inference from the occurrence of the sound to the presence of a coach. A man who was quite ignorant of the noise coaches make could still truly be said to have heard the coach provided only that he heard the noise. Even a man who knew perfectly well how to recognize the presence of coaches by their sound might truly be said to hear the sound but fail to realize that a coach was passing. For he might hear the sound, but, because of some disturbance of auditory perception, it might not sound to him like the sound of a coach.

2.2. *Propositional Perception Statements*

By contrast with object-perception statements, propositional-perception statements, which have the form '*A* perceives that *p*', are cognitive statements *par excellence*. If Jones sees that there is a fox before him then it is entailed that at that time Jones *knows* that there is a fox before him. This entailment is pointed out by Unger (1972). To see that it holds consider the unacceptable sentence 'Jones sees that there is a fox over there but he does not know that there is a fox over there.' There seems no way to explain the unacceptability of this third-person sentence without postulating the entailment. With knowledge, referential opacity is at once introduced. If a peasant sees that there is a table before him, then it by no means follows from the additional fact that a table is a collection of fundamental particles that the peasant sees that there is a collection of fundamental particles before him.

Knowledge may be inferred or it may be uninferred and, on pain of vicious regress, all inferred knowledge must ultimately rest upon uninferred knowledge. The distinction between immediate and mediate perception can therefore be very easily drawn in the case of propositional-perception statements. 'He heard that there was a coach going past in the street' seems very clearly mediate perception, involving inferential knowledge. But 'He heard that a certain sort of sound was sounding in his environment' is a good candidate for a statement of immediate perception, unless one holds the view, which I am taking to be incorrect, that the presence of sounds in the environment is only inferred from the awareness of having *sensations* of sound. (If the latter view is correct, it *would* seem reasonable to speak of the "immediate perception" of sensations of sound.)

2.3. *Object-Appearance Statements*

'Object-appearance' statements have the form '*O* perceptually appears $\emptyset$ to *A*.' 'The rose looks red to him' is such a statement. Such statements entail the truth of a corresponding object-perception statement. Thus, 'The rose looks red to him' entails 'He sees the rose.' It follows that the word 'rose' occurs in a referentially transparent context, permitting free substitution. From 'He sees the rose', however, all that can be deduced is that 'The rose looks *some way* to him.' Some more concrete object-appearance statement must be true, but it cannot be deduced what one. The word 'red', of course, occurs in an opaque context. The rose need not look to him to have the colour associated with the red wave-lengths.

A distinction between immediate and mediate perception can be drawn in the case of object-appearance statements. But needless repetition will be avoided if we pass straight on to pure-appearance statements where the same distinction appears in a more disentangled way.

2.4. *Pure-Appearance Statements*

'Pure-appearance' statements have as their form 'To *A*, there perceptually appears to be an *x* which is $\emptyset$.' To Macbeth, there looks to be a dagger before him. Pure-appearance statements are characteristically *used* to describe hallucinations, such as the one Macbeth suffered from, but their *meaning* goes beyond such situations. Indeed, for every perception which a man has, a pure-appearance statement might be found to render it, lack of vocabulary only excepted. Such statements create purely opaque contexts (after the verb 'appears').

It would seem possible to draw the distinction between immediate and mediate perception within pure-appearance statements. Thus, 'There looks to him to be a dagger before him' would seem to be a *mediate* perception, based upon an immediate perception. 'There looks to him to be a thing having a certain shape and colour before him' (the visual properties associated with a dagger) would be a plausible candidate for the statement of the immediate perception involved.

It is, presumably, an empirical question just what is or is not the completely uninferred element in a perception. It will therefore be a matter for psychological theory rather than philosophy to decide just where the line is to be drawn. Philosophy simply discusses the nature of the line.

In the case of appearance-statements of both sorts, however, we must reckon with a difficulty which does not arise in the case of propositional-perception statements. It is possible that, although there looks to somebody to be a dagger before him, he does not actually believe that there is a dagger before him. Now, in inferring, presumably we move from belief to belief, and so it might seem that there is no room for the immediate/mediate distinction in belief-free perception.

The point is well taken, but I think that even in belief-free perception an analogue of the original immediate/mediate distinction can be found. Suppose that there looks to somebody to be a dagger or a table before him, but he believes that there is no dagger or table before him (whether truly or falsely is immaterial). We can surely ask on what basis the world looks to him to be this way. The natural answer would seem to be that there looks to him to be things having certain shapes, colours and perhaps other visual properties before him: things which, if they exist, would naturally be identified as daggers and tables. These are his immediate perceptions. The immediate/mediate distinction is here drawn in terms of the beliefs which would have been held if the perceptions had not been 'belief-free'.

3. Why was the mistake made?

It seems, then, that the immediate/mediate distinction can be drawn in the case of propositional perception, and again in the realm of perceptual appearance, but not in the case of object-perception. It may be wondered why philosophers from Berkeley to Armstrong persisted in trying to force a cognitive distinction upon a non-cognitive idiom.

One reason may be the influence of the doctrine that it is sense-impressions which are the immediate objects of perception. They are very peculiar objects. They might be described as propositions which have been congealed into objects, and they retain a propositional flavour. As a result, it is relatively easy to treat them as what is immediately perceived.

A second reason may have been the fatal influence of *first-person present-tense* object-perception statements. If I say that I see a fox, then I imply that I know or I believe there is a fox there. 'I can see a fox but I do not believe there is one there' is a paradoxical statement, although not self-contradictory. It thus became easy, by concentrating on this particular case, to overlook that object-perception statements are non-cognitive.

In this connection, therefore, it is worth noticing that 'I can see a fox but I cannot see *that* it is a fox' is not a paradoxical statement. For suppose that the fox looks to me to be but an indistinguishable object in the distance, but a friend who is nearer shouts out that it is a fox. I can then make that statement with propriety and truth.

It may be, however, that there is a third, dimly felt, reason for persisting in trying to force the immediate/mediate distinction upon objects of perception. For there may be another distinction to be drawn in the realm of object-perception which is not a cognitive one but which gets confused with the immediate/mediate distinction. In the fourth section of this paper I try to draw this new distinction.

4. Direct and non-direct perception

It is necessary first to consider more deeply the nature of object-perception statements. It is widely, and I believe correctly, accepted among contemporary philosophers that such statements entail that *O* is the *cause* of those perceptions of *A* which constitute his perceiving of *O*. (The perceptions themselves can, in principle at least, be exhaustively described by pure-appearance statements. In the same way that the term 'father' attaches to a certain man, but attaches to him in virtue of his relation to a certain further thing, a child, so the phrase 'perceptions of *O*' attaches to certain sensory states of *A* – his perceptions – but attaches to them in virtue of their relation to the object *O*.) The entailment was canvassed for a number of years by C. B. Martin, but to my knowledge was first argued for in print by H. P. Grice (Grice, 1961).

The entailment may best be demonstrated by using what I call "the method of subtraction" (Armstrong, 1972, p. 81). If we are interested in the question whether or not *C* is a logically necessary condition for *D* we consider a situation which is *D*-like but where it is explicitly stipulated that *C* does not hold. It may then become obvious that, lacking *C*, we also lack *D*. The technique has obvious analogies with *reductio ad absurdum* in logical and mathematical proof and with the 'method of difference' in experimental investigation. It has considerable utility in conceptual analysis.

Consider, then, a case where (a) there is an orange on the table; (b) *A* has perceptions as of an orange on the table, but (c) it is not the case that the orange brings about these perceptions of *A*'s. We can imagine a story. Suppose that the light-waves travelling from the orange towards *A*'s eyes are in some way diverted from their path, but that, at the same time, a probe introduced into *A*'s brain at that moment happens, by an incredible coincidence, to produce in *A* just those perceptions of an orange which the light waves would have produced if they had not been diverted. Given that conditions (a) to (c) obtain, I do not think that anybody would wish to maintain that *A* saw the orange on the table. The best which could be said is that he had a 'veridical hallucination' as of an orange on the table.

Now, however, change the case and allow that the orange acts upon *A*'s eyes in the ordinary way and is responsible for *A*'s perceptions of an orange on the table. We will at once be prepared to say that *A* sees the orange. Indeed, we will be prepared to concede that condition (b) was much too strong. *A*'s perceptions might not have been orange-like, yet, given the causal condition, we would still be prepared to say that *A* saw the orange.[1]

So it seems that the 'existence-grammar' of '*A* perceives *O*' is explained by its *causal* grammar. If *O* is to cause *A*'s perceptions, *O* had better exist.

Given this preliminary, I proceed to draw a distinction in the field of object-perception which bears a shadow-resemblance to the epistemological distinction between immediate and mediate perception.

Consider the following puzzle cases. (1) *A* sees an apple. (2) *A* sees a half-apple, but the outer skin of the half-apple is turned towards *A* so that his eyes are affected just as in case (1). Now consider case (1) again. We would be happy to say, ordinary language would support us in saying, that *A* cannot see the back half of the apple. But this jostles with (1). If *A* cannot see the back of the apple, then he cannot see the whole apple.

Perhaps what he sees is only the front half of the apple? But it is clear that there can be no stopping there. *A* cannot see the back half of the apple, and he cannot see most of the front half of the apple. It seems that we will end up having to say that he can see no more than the *surface* of the front half of the apple. And the surface is not even a physical object, although it belongs to a physical object.

Faced with this puzzle, it is natural to wonder whether we might not re-introduce a distinction somewhat like that between immediate and mediate perception in the sphere of object-perception. I will use the words 'direct' and 'indirect'. Might we not say that the front surface of the apple is the *directly* perceived 'object', while the rest of the object is not directly perceived? One might wonder if this is not the epistemological distinction once again. But I do not think it can be. For, first, the surface could be utterly misperceived (a mere blur on the table), so that *A* could draw no inference at all from the content of the perception, but *A* could still be said to be seeing the apple, and, it would seem, it would still be the front surface only which he perceived 'directly'. Second, even where the surface was correctly perceived, *A* might still fail to infer the presence of the whole apple, yet still be said truly to be seeing the whole apple. Suppose, for instance, that *A* lived in a world where half-apples were the rule, but that what he was looking at was, for once, a whole apple. He might think that it was a half-apple with its skin turned towards him. But we would be prepared to say that he actually saw a whole apple.

But if we remember that '*A* perceives *O*' entails '*O* is the cause of *A*'s perception of *O*', a suggestion for drawing the direct/indirect distinction comes to mind. When *A* sees a (whole) apple, the bulk of the apple is doing little to bring about *A*'s perception of the apple. *But the front surface is playing an absolutely vital causal role.* So perhaps we can say that the object which is 'directly' perceived is that portion of the perceived object which plays the causal role in the perception of the object.

Even this, perhaps, is too liberal. When a thing acts, it acts in virtue of certain characteristics it has. What our philosophy of characteristics should be is immense question: it is the problem of universals. But there is *some* clear sense in which things have characteristics or properties, and ordinary thought assumes that, when things act, they act in virtue of some and only some of these characteristics: in virtue of the sort of things which the things are.

Hence, given that *A* sees an apple, we can first distinguish between those portions of the apple which do, and those which do not, act upon *A* to bring about those visual perceptions of *A*'s which are his seeing of the apple. Furthermore, even when we restrict ourselves to the causally effective portion of the apple, we can distinguish between those characteristics of the effective portion in virtue of which it so acts upon *A*, and those characteristics which are causally irrelevant to that effect. Hence, although we can say both with propriety and truth that *A* sees an apple, we could also say that what *A directly* sees is *a portion of the apple considered as having certain characteristics.*

Let me emphasize again that this introduced notion of direct perception is distinct from the epistemological notion of immediate perception. *A* may misperceive what he 'directly' perceives, and even when he perceives it veridically he may draw incorrect or no inferences from his veridical perceptions. Statements of direct perception create a referentially transparent context. It matters not how that portion of the object perceived which actually works upon *A*'s sense-organs is referred to or described. It is still that portion which is at work upon *A*'s sense-organs, and so is directly perceived. It matters not how the characteristics of the portion in virtue of which that portion acts be referred to or described. It is still those characteristics in virtue of which the portion acts, and so what is directly perceived is still that portion as having those characteristics. All this is quite different from the epistemological notion of immediate perception. But perhaps 'direct' perception has a shadow-resemblance to immediate perception, a resemblance which has helped to encourage the confused idea that the immediate/mediate distinction can be applied to object-perception.

5. THE UNIMPORTANCE OF NON-DIRECT PERCEPTION

We noted at the beginning of this paper that there has been a tendency for those who have drawn the immediate/mediate perception distinction to take mediate perception to be only perception by courtesy. Yet the only thing which gives support to this notion is the rather dubious contention that perception and inference are completely opposed.

But suppose that we now consider the distinction between direct and non-direct perception as we have drawn it in connection with object-perception. There really does seem to be some reason to denigrate and play down *non-direct* perception. An important point here is the great arbitrariness there is in the rules which govern what objects we are said to see. Warnock has pointed out (Warnock, 1954–5, p. 56) that if all I see of Lloyd George is his left leg, then I can hardly be said to have seen Lloyd George. Yet a mariner gaining a distant and indistinct glimpse of a small piece of shoreline might be said to have been the first Westerner to have seen Australia. Of course, it is possible to see why there should be these differences in the two cases, but the reasons do not have much to do with perception. In the Lloyd George case, there is the psychological and social importance of seeing a man's face, or at least a fair portion of him (although the portions seen may be clothed). In the Australia case, there is the fact that we treat the continent as a unitary thing for many purposes. Particularly in the context of discovery, we are naturally led from the thought of the portion sighted to the thought of the whole. In fact, however, the continent of Australia really played little part in the mariner's perception. It is easy to say that 'in truth and strictness' the mariner saw but a tiny portion of Australia.

Hence I am led, not too unwillingly, to say that when *A* is said to see an apple, in truth and strictness he really sees a good deal less. Ordinary speech confesses this because, as we have noted, not just the philosopher but everybody will admit that, when *A* is said to see an apple, it is not the case that *A* can see the rear of the apple or, indeed, see any portion of the apple except that portion of its skin which faces *A*.

Some ordinary language is more ordinary than other portions of language. It can be a very ordinary matter indeed what objects we are said to perceive. Within the sphere of object-perception we can draw a distinction between what in this paper was called direct and non-direct object-perception. Only *direct* object-perception seems of much theoretical importance. In the case of immediate and mediate perception, however, both sorts of perception are important theoretical notions.

University of Sydney

NOTE

[1] *A*'s perceptions cannot be anything at all, and it still be said that he saw the orange. At the same time, they can be extraordinarily unrepresentative of the nature of the object seen. If a man looks at Alpha Centauri, then the cause of his perception, and the thing which he can be said to be seeing, is that huge object Alpha Centauri as it existed four years ago. But his perception is a perception as of a tiny light source which exists contemporaneously. If, however, a poke in the eye with a stick makes *A* 'see stars', we would not say that his perceptions were perceptions of the stick. How is the line between these two sorts of cases to be drawn? For the best attempt I know to draw it, see Grice (1961) Section V. But I feel sure that Grice would agree that it is desirable to say more than he says there.

BIBLIOGRAPHY

Armstrong, D. M., *Perception and the Physical World*, Routledge, London, 1961.
Armstrong, D. M., *A Materialist Theory of the Mind*, Routledge, London, 1968.
Armstrong, D. M., *Belief, Truth and Knowledge*, Cambridge University Press, Cambridge, 1972.
Grice, H. P., 'The Causal Theory of Perception', *Aristotelian Society Proceedings*. Suppl. Vol. **35** (1961).
Unger, P., 'Propositional Verbs and Knowledge', *Journal of Philosophy* **59** (1972).
Warnock, G. J., 'Seeing', *Aristotelian Society Proceedings* **55** (1954–5).

W. W. BARTLEY, III

ON IMRE LAKATOS

I knew Imre Lakatos intimately between the autumn of 1958 and the summer of 1965. During this period we met almost every day and corresponded on those days we did not meet. In the summer of 1965 we had an extraordinary and widely discussed quarrel, after which we did not speak. Since the implications of this quarrel shattered me and altered permanently my attitude to my fellow professionals in philosophy, Lakatos must count among those who have influenced my life most deeply.

There was a time, both before and after our quarrel, when I regarded Lakatos as the most immoral man I had ever met. I later came to think this judgement naive. Lakatos merely talked openly and appreciatively – with a certain connoisseurship – of the sort of behaviour which is widespread and almost universally *covert*. I can now appreciate the merits of his practice: although I often saw Lakatos lie when it suited his purposes, he was never a hypocrite. He was remarkably without *self*-deception and quite without cant. In this regard he was morally my superiour and moral light years ahead of some of our friends and colleagues at the London School of Economics. I remember vividly the scolding he gave me in the autumn of 1964: "Beel, you moralize too much. It doesn't get you anywhere. I used to do that in Hungary, and I ended up in prison."

Although Lakatos was of course a highly talented thinker who contributed many valuable ideas, his chief importance both in Hungary and in England was as an educator promulgating ideas. His fascination with research programmes in the history of science reflected his interest in the strategies whereby a good idea might come to power. In this connexion he explained to me in 1961 – against my considerable scepticism and resistance – that Popper was quite wrong to say that words do not matter. Quite the contrary, Lakatos insisted, ideas are of secondary importance compared to the *names* one gives to them: if you give your ideas good names, they will be accepted – and you will be named the father. It was a double-edged strategy, he explained, which brought one fame without demand at the same time it advanced the ideas one favoured. Thus this

R. S. Cohen et al. (eds.), Essays in Memory of Imre Lakatos, 37–38. *All Rights Reserved*
Copyright © 1976 by D. Reidel Publishing Company, Dordrecht-Holland

second Adam explained and defended his term 'monster-barring'.

It is then appropriate that Lakatos should have acquired his chief fame during the final decade of his life for his 'scientific research programmes.' This was an idea that he took over completely developed from the accounts by Popper, Agassi, and Watkins of 'metaphysical research programmes.' Lakatos had the good sense to see that the word 'metaphysics' presented an insuperable public-relations obstacle to the professional philosophers of scientific bent who lacked his own sense of humour. So he calmly changed the word 'metaphysical' to the word 'scientific' and won the acclaim that he had intended for the notion.

What Popper *meant* of course is that words *shouldn't* matter. But Lakatos knew better than to moralize.

California State University Hayward

WILLIAM BERKSON

LAKATOS ONE AND LAKATOS TWO: AN APPRECIATION

1. Lakatos enters into purgatory

When I was pondering what I should write about Lakatos in this article, I ran into what seemed an insuperable difficulty. On one hand I thought that part of my job should be to give a portrait of Lakatos' personal manner. This would be informative from the point of view of history of thought, as it would help people read his works with greater understanding. And it would be interesting, if I succeeded at all: everyone who knew Lakatos would agree that he was one of the most remarkable people they had met. On the other hand memorial essays tend to give portraits which are all sweetness and light. But such a portrait of Lakatos would not be a portrait of Lakatos. Everyone has his dark and his light side; sometimes we see one, occasionally the other. But in Lakatos both were simultaneously and brilliantly displayed at every instant. It would be trying to describe bitter sweet chocolate with the vocabulary for milk chocolate. On the other hand, to give no description at all would be a real loss.

Last night I fell asleep while troubling over this problem, and in my dreams I saw Imre:

Lakatos: Where am I?

Angel: These are the pearly gates.

Lakatos: Terrible, I must be in hell.

Angel: But I just said these are the *pearly* gates.

Lakatos: Then why are you speaking English? The only axiom of theology is that in heaven they speak Hungarian.

Angel: You started speaking English.

Lakatos: An unfortunate habit I picked up in my later years. Give me an office next to Archimedes.

Angel: But on my paper you are listed under 'philosophy.'

Lakatos: Well, I did teach philosophy, but that was one of my little lapses. Actually methodology of mathematical discovery is my...

Angel: Speaking of little lapses, you have to be judged before you are

R. S. Cohen et al. (eds.), Essays in Memory of Imre Lakatos, 39–54. *All Rights Reserved*
Copyright © 1976 by D. Reidel Publishing Company, Dordrecht–Holland

sent anywhere. We have to go over this list before any decision is made.

Lakatos: Is that list for $Lakatos_1$, $Lakatos_2$, or $Lakatos_3$?

Angel: What?

Lakatos: You mean you don't know about my methodology of research programs? Individual actions cannot be accepted or rejected, only series of actions. You have to know what research program they are part of, and I have been identified with fourteen research programs at least.

Angel: I don't understand.

Lakatos: Well you can't judge an action out of context. You have to know whether it's a progressive or a degenerating character shift; to judge this in turn you need to know whether the action is $acceptable_1$ or $acceptable_2$. Otherwise you will botch up the job. Just come along to my office – it's next to Archimedes – and I'll explain the whole thing.

Angel: (alarmed) Botch the whole job! Well you may be right. Frankly I'm new at this job. My last job was as administrator of purgatory and I botched that one up. Plato and Aristotle were arguing as usual, and Archimedes sided with Plato. It got a little heated, and they told Aristotle he didn't understand mathematics. He could take that from Plato, but with Archimedes too, he took offence and refused to talk to them for a hundred years. Socrates could not do a thing with them this time, and I was called in. I couldn't do anything either. On earth I was a baker, and even after a thousand years in purgatory I cannot handle these intellectuals. Well, I was called up before the Lord. He told me that in purgatory he wouldn't have people quarreling like hell.

Lakatos (grinning): It's a relief to know the Lord has a sense of humor; his seminars must be wonderful.

Angel: Well, frankly, I didn't find it too funny. But he told me I still lacked the sense of humor necessary in dealing with vain men, and sent me here.

Lakatos: Well, you must lead me over to Archimedes and the others right away. I am an expert in these situations at pouring oil over troubled fires.

Angel: You mean troubled waters.

Lakatos: These English metaphors never make sense.

Angel: I already told you Archimedes is in purgatory; are you sure you want to go?

Lakatos: Don't bother me with details, I am a busy man.

Angel: Well, I must admit, you'll fit right in; Plato still hasn't realized he's not in heaven – too preoccupied. And to tell you the truth, purgatory is a very 'in' spot ever since Socrates came. We have constant visitors from heaven. Of course I think it's not only the conversation, but my coffee and strudel. Even the heavenly hosts admit it's divine. Say, do you think the Lord would find that one funny?

Lakatos: Well, it was a try. But I think you'd better concentrate on your present job. Of course, you'll have to learn the rudiments of the methodology of research programs to have any hope of getting back with us in purgatory. My seminars will be on Monday afternoons from 2–4; and my office is open to you at any time. I'll do my best to get you back with us. I can't stand the thought of refectory food for an eternity!

As I woke up, a vision of Imre and the Angel walking away toward purgatory faded from my eyes. Though my dream was no doubt a figment of my own imagination, the vision of Lakatos *sub specie aeternitatis* gave me the resolve to try to portray the real Lakatos.

2. Lakatos Gossip and Lakatos' Gossip

2.1. *Lakatos Gossip*

Lakatos was charming, infuriating and brilliant. I did not know Lakatos well enough or long enough to comment on his character. But I did see him often enough to get a clear impression of his manner, which was both unvarying and remarkable. Lakatos spoke in witty remarks which were often sparkling, sometimes brilliant, and always so contorted that you never knew exactly how to take them.

Let me give an example of a typical Lakatos remark. I remember once, Lakatos was engaged in his usual banter with some visitor from 'foreign' (non-Popperian) or possible 'enemy' (anti-Popperian) territory. Perhaps later the 'enemy' would have been pro-Popperian! In response to some remark of the visitor on Popper's ideas, Lakatos said, "But don't you know that the *Zeitgeist* has descended upon the LSE Philosophy Department?" LSE means, of course, the London School of Economics. The phrase was always pronounced 'zie LSE' by Lakatos, so that it rhymed. His Hungarian accent made many of his stock phrases seem like they had a special significance, only fully understood by Lakatos himself. His 'series of theories' is one that comes to mind. A tongue-twister at best, this always came out a 'series of see-o-ries' with the guttural 'r'. But I digress.

Many aspects of this remark make it typical; one of them being that it has *many* aspects. First, it is a disarming, unanswerable question. Second, it is an intellectual joke, but at the same time a serious remark. The *Zeitgeist* or 'spirit of the times' is of course an idea of Hegel, victim of a scathing attack in Popper's book *The Open Society and its Enemies*. In fact the idea of *Zeitgeist* and the notions of historical inevitability it involves are one of the main targets of that book. So to say that the *Zeitgeist* belongs to Popper and his followers is some sort of contradiction which makes fun of Hegel and Popper at the same time. On the other hand, no one denies that there are currents of opinion, and schools of thought which turn out to have been most important. So Lakatos might have been claiming that Popper and followers are the only ones who count philosophically. Could Lakatos be so audacious as to believe that? Well, yes; but how much 'yes', neither I nor the visitor nor, I suspect, Lakatos knew.

The conflicting forces of this remark become even stronger when we consider the social situation. The remark is both an invitation (join us and do important work) and a threat (don't join us and you will waste your life). It is both a compliment (you are able enough to take the advantage of this information, you are on the way in) and an insult (you dummy, why haven't you seen the importance of our work; you are on the way out). But it is only when you add the presence of Lakatos that you get the full impression. First, he has just made four or five similar remarks. Second, the remark is delivered with his usual expression. He has an infectious grin and a twinkle in his eyes and he is constantly in motion. His expression is at the same time affectionate and calculating. That is, you can't tell whether the affection is that of a fox for a chicken, or for a fellow fox. The affection is genuine and the smile even vulnerable; but the twinkling eyes seem fully in control and warn you not to get on his wrong side. That so many conflicting emotions and intentions could be packed into an offhand remark may seem incredible. But such was the bread and butter of Lakatos' conversation.

2.2. *Lakatos' Gossip*

The previous remark is also typical of Lakatos in that it is an evaluation of someone's place in intellectual history. Lakatos was more than anything preoccupied with the history of ideas, and particularly the patterns of growth and change in ideas. His gossip was applied to the living and the dead with equal vivacity, and was usually concerned with a point of intellectual history. I arrived at the LSE in the fall of 1965 as a beginning graduate student, and attended his course in mathematical logic. I was constantly amazed and delighted to find that a subject I thought might be dry was actually – in Lakatos' version – roughly two-thirds gossip. There was always a personal element in the gossip: *x* was actually a much more important thinker than *y*, and *z*, who is thought to be so great is nothing at all. But the gossip always had a much richer side to it. The pathways of thought were being traced and re-evaluated with a fresh eye.

In the logic course of '65–'66, the gossip was influenced by Lakatos' recent visit to California and his talk with the logicians there. Lakatos thought that 'foundational' studies were basically misguided in their philosophy, and that the real progress that had been made was not to give mathematics any better 'foundations', but rather to give some deeper explanations of mathematical systems and in turn the improvement of the systems. Furthermore he thought that the formalists were misguided in thinking that any absolute standard of rigor could ever be established. The standards of rigor were something which grew by proofs and refutations, like the rest of mathematics, and would continue to grow. So he went to California to catch up on the developments in logic in the last fifteen years, hoping to find reinforcement for his own views.

This he found in the work of Leon Henkin and Abraham Robinson. Henkin had changed the notion of logical consequence ('every model of the premises is a model of the conclusion') by pointing out that by 'model', one could either mean a 'standard' or 'non-standard' model. If one included the latter in one's concept of model, then the higher orders of the predicate calculus were actually complete; only if one was restricted to standard models did Gödel's proof apply. This delighted Lakatos as it was a perfect example of the 'concept stretching' involved in the growth of mathematics which he had pointed out. Furthermore it changed the notion of what was a logical consequence, and so changed the standards of rigor. The French school of Bourbaki, he thought, had tried to be the modern Euclid, and settle the standards of rigor and proof once and for all. But Henkin's work had turned much of their work into 'sloppy talk'. Furthermore, Robinson had used the new ideas of non-standard models to produce new methods of proof in mathematics proper, such as the calculus. Thus the (so-far) end result of 'foundational' studies was not to give foundations, but to enrich mathematics itself.

This story I have told for two reasons. One is that Lakatos never published these ideas. He intended these to be part of a book expanded from 'Proofs and Refutations', showing how the ideas applied equally well to present day mathematics as to 19th-century mathematics. Hopefully there exists a manuscript which will be published when Lakatos' collected essays come out. The second reason is to form a backdrop for the typical pieces of Lakatos' gossip, which he related during the course on mathematical logic.

A bit of gossip, number 1: Lakatos said he asked Tarski, (a name mentioned reverentially) "Don't you think that the work of Russell would to-day be regarded as a waste, had it not been for Hilbert, whose ideas were really productive?" I forget what Tarski was supposed to have answered. This remark if typical of Lakatos' gossip in that it is both personal – not necessarily generous! – and thought provoking on issues which are really quite deep.

A bit of gossip, number 2: Lakatos had been trying for some time to figure out exactly what had happened in mathematical logic when he had a talk with Abraham Robinson. He asked Robinson for the gossip: what had been the most important contributions; what important problems were people working on now; what generally was the problem situation. Robinson answered his questions in a few minutes, and suddenly Lakatos saw clearly the whole development of logic over the past twenty years and the present situation. He then asked Robinson: "why don't you publish this?" Robinson replied, "Everybody knows it." Lakatos asked, "How many people are 'everybody'?". Robinson replied, "Five".

One of the points of this story was that mathematics properly speaking, is two-thirds gossip and it should be published. Of course, if you look into almost any work of mathematics, and if you look into *any* text-book, you will not find such gossip, but rather axioms, theorems, proofs. Lakatos' view then, seems peculiar, if not crazy. However to continue to tell the truth, on this point it seems to me that Lakatos was right and the rest of the world wrong. Which brings me to the next section.

3. Lakatos One

By 'Lakatos One', I mean the Lakatos of 'Proofs and Refutations'. The main point of this and the next section is that 'Proofs and Refutations' is a work which should be of fundamental importance for history of mathematics, the teaching of mathematics, the social community of mathematicians, philosophy of mathematics and perhaps even directly to mathematics itself.

These claims are extravagant, but I think justified. Let me first briefly explain Lakatos' revolution, and then its consequences. 'Proofs and Refutations' is a work which reflects the personality of the creator. It is a scintilating dialogue in which it is never entirely clear where Lakatos stands on many issues. It is an unfamiliar territory which we are led through. All the scenery is dazzling and new; but in what way it is new is not clear. We ask our guide, Lakatos, a question about a pretty piece of scenery; he responds with three new questions we have never heard before. The first effect of 'Proofs and Refutations' is to dazzle and to numb.

Margaret Marchi, who came to the LSE at the same time as I, has spent time from then till now trying to sort out what exactly is going on in that great work and to improve on it. This with the help and opposition of Lakatos. Her latest results are elsewhere in this volume. Briefly, what she regards as fundamental is Lakatos' discovery that *new mathematical discoveries modify old mathematical discoveries*. That is, what was regarded as true is now seen to be *generally* speaking false, and only true in some special cases. For example, Euclidean geometry seen as true only in 'Euclidean Space'.

This idea can be seen most clearly as an analogy to one of Popper's fundamental discoveries. Whewell, and others following him saw that there was a stage of trial and error leading to a new discovery. This new discovery, though, once confirmed in an independent test, became an established truth, never to be changed in the future growth of science. Duhem and others saw that the new science seemed to modify the old one, but they denied this by saying it was only the metaphysics that changed; considered as a system of classification, there was no need to see anything once established in science as changing. Popper faced up the real changes, but denied that this meant the 'bankruptcy of science', to use Poincaré's phrase. The old theories had been indeed refuted, and the new theories may be in the future; but the new theories explained the successes and the failures of the old theory and added new successes in explaining the facts. Thus the new theories constitute progress over the old.

The relation of Lakatos to his predecessors is somewhat similar. Most historians and philosophers of mathematics are like Whewell. They see a period of trial and error, then establishment of the truth. But Lakatos showed by his historical examples – which can be continued up till to-day – that this is a delusion. The delusion is better hidden than in the history of science by the fact that many theorems appear in up-to-date texts in exactly their old form. But in fact they have changed greatly, because *the meaning of the terms in the theorem has changed.* The concepts have been 'stretched' and altered into a meaning quite different from that originally intended. These are changes no less substantial than the changes in the actual form of the equations which appear in science. These changes show themselves in the very unintuitive definitions in modern mathematics. The changes in definitions can be traced through the development of mathematics from article to article and book to book.

The first reaction of most people who understand Lakatos' point is either that it is not true or that it was true but is no longer now. However, in my opinion Lakatos' discovery cannot be brushed off. Mathematics consists not only of formalism but also of interpretations and the many concepts of to-day can and no doubt will be changed to-morrow. Lakatos' discovery is, I believe, in one respect like that of universal gravitation; our theories and explanations of it may undergo important change, but its existence is a fact.

In one very important aspect the parallel between Lakatos' and Popper's discovery breaks down. That is, Lakatos does not give us a clear idea in 'Proofs and Refutations' of how to measure progress. It seems to me that probably the question of what it means to say that a mathematical theory is *true* – the old problem – will have to be faced to give a full answer. But rather than consider any of the many philosophical questions raised by Lakatos, I will take on the lesser task of pointing out that some of the consequences of what seems to me a fundamental breakthrough.

One of the important consequences of Lakatos' work is the point I mentioned before, namely that mathematics, is properly speaking, two-thirds gossip. Let me put this point in a straight forward way. As long as we believe that old established truths will not be modified by further developments, the history of a subject is not an essential part of the subject. But once we acknowledge that the results of past discoveries may be modified by future discoveries, the situation changes: *There can be no sharp boundary between the history of a subject and the subject itself.* The only difference can be of emphasis, or of focus of interest. The reason is simple: the existing theories or theorems are not only statements of fact, but they are also *solutions to problems.* If we wish to apply the theory intelligently, or specially if we wish to improve the theory, to advance the subject, then we must also understand the problems. For the new improved theories must be able to solve the old problems as well as the old theories did. The old problems are as much a guide to our research as the theories which solved them. But to understand the problem is just to see what was the situation in which they arose, and what were the solutions alternative to the latest one.

Of course all that I have said applies also to Popper's theory of science, and accounts for Popper's insistence that to understand a theory we have to understand the problems it came to solve. How much we concentrate on old problems and solutions and how much on recent ones, depends on our judgement and interests. In the study of history, the priorities will be different than in contemporary efforts to improve the subject. But the material of interest may at times be exactly the same. Every case I have looked into in the history of science, it has turned out that the discoverer began with a critical survey of the recent history (at least) of his subject.

The effects of the realization of the near identity of a subject and its history are great in science, but even greater in mathematics. Let me begin with the consequences for the community of mathematicians. Lakatos' story above indicates that mathematicians do not very often publish surveys of the problem situation in their field, even for experts, much less for those who want to enter the field. And my spies in the world of mathematics, (I am many miles away from being a mathematician) tell me that generally speaking, mathematicians do not even talk to each other. Either you are a genius, in which case there is no use explaining; or you are not in which case there is no use explaining. But if Lakatos is right, then this ideology is only true because it is self-fulfilling; because each thinker has to reconstruct the problem situation from hints and guesses. The publication of the other two-thirds suppressed part of the content of mathematics would greatly promote progress in mathematics.

In history of mathematics, the consequences of Lakatos' work are more obvious. The new territory discovered by Lakatos, which is indeed most of the history of mathematics, lies waiting to be explored. To trace the real *changes* in mathematical concepts is the first obvious question which can be asked in each incidence of a breakthrough. Second, Lakatos has pointed out that the interaction between formal and informal mathematics, between formalism and its interpretations, has been the real engine of growth in mathematics. This side, again, has yet to be mined. And there are the detailed ideas of 'concept stretching' 'monster-barring' and so on, which can be used as a guide for research into other historical examples.

In philosophy of mathematics, if Lakatos is right, there must be a new look at the whole enterprise of mathematics. On one hand, there have been those who have seen mathematics as being essentially concerned only with formal structures; and on the other hand, those who believe that it is concerned with the development of *a priori* true intuitions. However, philosophy of mathematics must explain how interpretations are both an essential part of mathematics, and yet may be modified and improved. And they must answer the question of how we gage mathematical improvements even without the establishment of truths which are absolutely certain. There are many detailed questions involved with the problems of how 'proofs and refutations' operate, how standards of rigor can change and so on. Again new territory is opened up.

Perhaps the field which *should* be most changed by Lakatos' discovery is the teaching of mathematics. If Lakatos is right, mathematics text books would be a hundred times easier to understand if they were three times longer – if they contained the other two-thirds of the subjects they discuss. Now the student is forced to reinvent the other two-thirds for himself. This closes the subject to all but those who have both this particular gift, and the knowledge that it has to be done, *and* the interest to do it. Now mathematics is usually taught either by the rote method, driving in somehow mechanical skills with a minimum of understanding, or taught by 'new' methods which try to give the latest up-to-date formal version, whose concepts have been stretched beyond the understanding of most who do not already know the earlier version of the subject and the connection of its mechanics to the recent modifications.

If we accept Lakatos' basic idea, then the teaching of mathematics should have a good deal of history in it. The instruction should begin with intuitive concepts, and show step by step why and how they have been modified to solve the problems which modern theories solve. Thus the student would be shown the path from his recent intuitive notions of mathematical operations and objects, to the present, less intuitive more powerful ideas. The combination of such an approach with Polya's ideas on methods of problem solving would constitute a revolution in the teaching of mathematics – a revolution from either traditional or 'the new mathematics' methods. And in my opinion it would be all to the good. It would be a joy to students and teachers, and a boon to mathematics.

Finally, there are problems in mathematics itself which might be raised by Lakatos' work. These questions would probably concern the relation between formalism and interpretations, and the limits of rigor. But unfortunately I am no mathematician and have no ideas on how to make these problems clearer. Whatever the direct effect Lakatos' work might have on mathematics, I think that the effect it *could* and should have on the history, the philosophy and the teaching of mathematics would eventually pay great dividends in the progress of the subject itself.

4. Lakatos Two

By 'Lakatos Two', I mean the Lakatos of 'Criticism and the Methodology of Scientific Research Programmes' and of the second version of that article, 'Falsification and the Methodology of Scientific Research Programmes'. If you read the previous section, you may wonder how this section can clarify Lakatos' contribution to the philosophy of mathematics. The answer is, by clearing the air of the now fairly famous fog which constitutes the above two articles. Lakatos' notoriety seems to rest mainly on the latter of these two articles, which has a *few* good points, but which is nothing compared to his 'Proofs and Refutations', and obscures the importance of 'Proofs...'.

Briefly my view of this latter work of Lakatos is, that it has one good idea, that of a *heuristically* degenerating research program, and the rest of the work *is* a degenerating research program. To make clear my position let me begin by explaining Lakatos' problem situation, on the issue of research programs. Basically, Lakatos was trying to reconstruct Popper's theory of the growth of scientific knowledge while accepting the criticisms of Agassi and Feyerabend, and some points of Kuhn's theory. Feyerabend pointed out that historically speaking a new theory was sometimes contradicted by old evidence, but it was not rejected. Instead it was developed and eventually the evidence taken account of. This was indeed what Popper himself had pointed out in his paper 'The Aim of Science'; Kepler's description of the path of planets strictly speaking contradicted Newton's theory, which included perturbations. But this means that Popper's method of theory choice has to go. Agassi objected to the same point on somewhat different methodological grounds. He found Popper's rule of choice too conservative, and unnecessary. He pointed out that it was unnecessary to adopt it in order to measure progress in science, for, he noted, it is possible to explain past data as *false*, as well as true. And such explanations can be independently tested. Neither Agassi nor Feyerabend, though, provided a new rule for the acceptance and rejection of theories. Lakatos felt this was necessary to rescue 'scientific rationality'.

In constructing a new theory of theory choice, Lakatos was guided by two other ideas; one coming from Kuhn and the other again from Agassi. The idea from Kuhn was that somehow many scientists are held together in their research by agreement on some fundamental assumptions. Agassi, on the other hand had noted that not only were scientists guided by 'metaphysical research programs', an old idea, but what gave the community a certain unity is that they concentrated on problems connected with the *competition* between different metaphysics. In particular, scientists tried to render a metaphysics testable by developing a theory within its framework; and this theory was generally intended to be one which could not be fitted, or easily fitted, within the competing metaphysical frameworks. (This process, Agassi noted, also contradicted Popper's advice to choose the most testable theory first.) Lakatos saw that Agassi was more historically accurate than Kuhn, with his idea of *competition* between metaphysics; yet he thought that something like Kuhn's agreement on a common 'paradigm' was also necessary.

Out of Lakatos' effort at compromising these conflicting views, his own 'methodology of research programmes' was born. His first critical point, a strengthening of the criticism against Popper was that choice could not be between different *theories.* This was for two reasons. First was Duhem's argument that if the conclusion of a deduction from theory was wrong, we could not see what premise was responsible. Second there were an infinite number of possible 'hidden' premises. This meant that nothing resembling 'refutation' could take place.

Second, Lakatos gave a new rule of choice between *research programs* rather than theories. The new rule involved the idea of a 'degenerating' and a 'progressive' 'problem-shift' within a research program. The idea was basically that a research program was *heuristically* successful if it were able to produce new independent tests with each modification of it, and to pass these tests. To produce a clear-cut (?) method of choice, Lakatos had to pin down what was essential in the research program. This he did with his idea of the 'hard core' of the program. The 'hard core' resembled Kuhn's paradigm in that scientists engaged in the research program had to agree on its fundamental tenets. The difference was that there were many different 'hard cores' competing at any stage of history. The 'hard core' is a set of statements which are protected from refutation, by the attempt to modify other assumption when apparent refutation arises. This is the 'negative heuristic'. And subsidiary assumptions are introduced and improved which, together with 'hard core' explain old facts and predict new ones. This is the 'positive heuristic'. When successive modifications by the 'positive heuristic' fail to produce any new independent tests which are passed, the research program is 'degenerating'. If the research program is successful in guiding us to new discoveries, though, then it is 'progressive'. *In the long run, progressive are chosen over degenerating research programs.* Thus we arrive at Lakatos' solution to his problem; a new rule of choice which rescues 'scientific rationality'.

Lakatos' system, considered as a modification of Popper's, is a compound of error, confusion, and a good idea. These I will deal with separately beginning with 'error'. There are two fundamental errors in Lakatos' system. The first is in the new criticism of Popper. As he puts it in his first paper on the subject (p. 162), according to Lakatos, *"No theory forbids some state of affairs specifiable in advance."* If true, this would indeed be damning to Popper's system in a fundamental way. But it is false.

Newton's law '$F = ma$' forbids the occurrence of an object with a unit mass and a unit force on it which accelerates at the other at other than a unit rate. Of course to *determine* that such a case has actually happened in an experimental set up is difficult, and never fully conclusive. But this is quite a different matter from the exclusion of certain possibilities, in principle. (For a discussion of this matter, see my book *Fields of Force*, pp. 241ff. and 354–355. See also my review of *Criticism and the Growth of Knowledge*.)

What is involved is a simple logical principle, and I still don't understand how Lakatos could make the mistake: If a conclusion is *validly* derived from a set of premises, and the conclusion is false, then at least one of the premises *explicitly* listed must be false. My only explanation for this blunder is that Lakatos had been so immersed in the changes in ideas of rigor in mathematics, and the subtleties of the notion of validity, that he overlooked the fact that he was dealing with the most unproblematic of deductions: That if $F = ma$, and $F = 1$ and $m = 1$, then $a = 1$. If Lakatos were right, then we couldn't compare the explanatory success and failure of individual theories, and Popper's measure of progress would indeed be in trouble (a point I will return to.) But he is wrong.

The second mistake is a historical one. The notion of a 'hard core' does not in general find any substantiation in the history of science, just as Kuhn's 'paradigm' on which all scientists agree, does not historically exist. Its non-existence I have shown in detail for the development of field theory in the 19th century. Every scientist had his own 'hard core', and some times more than one, which shows that the 'cores' are not exactly hard in that they were protected from change at almost all cost. Scientists actually, contra Kuhn and Lakatos, have had enough intellectual independence to make up their own minds about what to take as fundamental, and enough independence to *change* their minds also, or to keep an open mind and try different alternative ideas as fundamental. (See my book, especially the table on p. 254 for an elucidation of this point.)

Now let me progress from Lakatos' errors to his confusion. As Maurice Finocchario has pointed out in his review of *Criticism and the Growth of Knowledge*, Lakatos confuses a variety of problems, and gets himself into a tangle as a result. The two problems I have in mind which he confuses are: (1) How can we measure the *progress* of science? (2) How should scientists choose between theories? Lakatos, I think was looking for an improved solution (over Popper) to the second question. What he actually produced was a bad answer to the first question. – It is bad because the measure of progress Popper gives is better, and thus Lakatos' solution a degeneration, not an improvement. Lakatos partly realized (under criticism by me – who was mainly ignored – and Kuhn and Feyerabend) that he had no solution to the second problem, and thus no theory of the 'rationality of scientific change', under his own conception. I say this because he at the same time refused to say that his evaluations were anything except in long retrospect, and advocated ('History of Science and its Rational Reconstructions', p. 105) that editors should refuse to publish the papers which are part of a 'degenerating research program', and that scientists writing them should not be funded by research foundations. It is such offhand completely illiberal statements which I must admit I found infuriating in Lakatos, as well as contrary to the true spirit of science.

However, with credit to Lakatos, I must admit that Agassi and Feyerabend's criticisms of Popper do destroy his theory of theory choice, and do not replace it. Agassi provided an improved method of rational *discussion* of theories, though. In my paper 'Rationality and Science' I have tried to show how this method could be used to guide rational *choice.* Lakatos' own efforts in this direction seem to me authoritarian and repugnant, just as do Kuhn's.

Finally, let me turn from Lakatos' confusions to his good idea in this matter. The good idea is, I think, that a research program is in fact adopted by an individual not only because it may further the debate over what world picture is correct, but also for its power to guide the researcher to new problems and ideas, that is, its *heuristic* power. If it has lost the power to guide to new ideas, then it cannot even help its partisans much. Research programs do progress and degenerate in this way, even though they usually have no 'hard core.' And this idea can be used to give us a better idea of the paths along which science has developed.

But though I do think this is a good idea, I would like to conclude by repeating that it is a very minor one compared to the revolution initiated by Lakatos in 'Proofs and Refutations'. I remember several years after he had turned away from his work on philosophy of mathematics, a student gave a paper in his seminar on quantum mechanics. He had explained the idea of a 'vector' as a certain type of invariant, according to the modern definition, and as usual the definition left those who didn't know the subject matter baffled. Then Lakatos got up in his usual state of smiling high agitation and said. "Wait a minute. Don't give us this mystification. If you want to explain what a vector is, you should begin by saying that a vector is an arrow. Then explain what problems this idea solves, and then what it fails to solve, and why it had to be changed and generalized successively to the present idea." My heart leapt with delight and regret. Here was the old giant come back to life; why was he ever abandonned? Lakatos' ideas on mathematics were so rich; he had just begun to pluck their fruits. John Aubrey said (The line is in the Roy Dotrice play): "When a man of great learning dies, what a lot of learning goes." Whenever I think of Lakatos somehow this line echoes in my mind as I see in my mind's eye Lakatos, so energetic, so contrary, so full of life.

Bridgewater State University

BIBLIOGRAPHY

Agassi, J., 'Sensationalism', *Mind* (1966); 'The Nature of Scientific Problems and Their Roots in Metaphysics', in M. Bunge (ed.), *The Critical Approach to Science and Philosophy*.

Berkson, W., *Fields of Force: The Development of a World View from Faraday To Einstein*; 'Review of *Criticism and the Growth of Knowledge*', in *The Journal of Critical Analysis* **4**, No. 3; 'Rationality and Science', talk given before the 1974 Philosophy of Science Association.

Feyerabend, P., 'Explanation, Reduction, and Empiricism', in *Minnesota Studies in the Philosophy of Science*, vol. 3.

Finocchiaro, M., 'Review of *Criticism and the Growth of Knowledge*', in *Studies in the History and Philosophy of Science* **3**, No. 4.

Lakatos, Imre, 'Proofs and Refutations', *British Journal for the Philosophy of Science* (1963–64). 'Criticism and the Methodology of Scientific Research Programmes', *Proceedings of the Aristotelian Society* **69**; 'Falsification and the Methodology of Scientific Research Programs' in I. Lakatos and A. Musgrave (eds.), *Criticism and the Growth of Knowledge*, and 'History of Science and Its Rational Reconstructions', in *PSA 1970*, vol. VIII of *Boston Studies in the Philosophy of Science* (ed. by R. Buck and R. Cohen), D. Reidel, Dordrecht.

Popper, Karl, *Logic of Scientific Discovery, Conjectures and Refutations, Objective Knowledge* (especially 'The Aim of Science); and 'Replies to Lakatos' in P. A. Schilpp (ed.), *The Philosophy of Karl Popper*.

I. BERNARD COHEN

WILLIAM WHEWELL AND THE CONCEPT OF SCIENTIFIC REVOLUTION[1]

Discussions of scientific change have, in recent years, come to center more and more on the concept of scientific revolution. No doubt, one of the reasons for the attention given to scientific revolutions has been the challenging thesis of Thomas S. Kuhn, in his widely-read book, *The Structure of Scientific Revolutions.*[2] The critical response to Kuhn's book[3] has largely been of three kinds: criticising his fundamental thesis or some aspect of it,[4] challenging the propriety of using the concept of revolution in the analysis of scientific change,[5] and substituting an alternative schema for our understanding of the development of science.[6] But no one thus far has seriously taken on the assignment of tracing the origin and vicissitudes of development of the concept and name 'revolution' as it has been applied to specific changes in the sciences (as in the Copernican, Newtonian, and Darwinian revolutions)[7] or to large-scale changes that have affected science as a whole and man's views of himself, his universe, and nature. Many historians appear to believe that it is only in very recent years that the concept of revolution has become imposed harshly and anachronistically on the scientific events of the past. Accordingly, there is a special relevance to exploring the history of the successive attempts to explain scientific change by revolutions.

I shall not enter here into the ways in which the innocent mathematical and astronomical term 'revolution' became a heavily loaded concept in the realm of civil affairs.[8] But it may be noted that in this transfer, the original meaning of *revolutio* (a late mediaeval scientific word[9]) was altered from the cyclical sense of return or revolve, roll back, unroll or unwind; so that 'revolution' came to signify a secular change, often accomplished by violence, that created a rather complete break with the social or political order of the past. These two very different meanings may have some points in common: for a revolution is an 'overturning', which has implications of at least part of a cycle. Furthermore, a revolutionary change in the secular, rather than the cyclical, sense may be conceived as a return to a better and purer state that existed in the past.

R. S. Cohen et al. (eds.), Essays in Memory of Imre Lakatos, 55–63. *All Rights Reserved*
Copyright © 1976 by D. Reidel Publishing Company, Dordrecht-Holland

Not unexpectedly, then, the writers of the eighteenth century are often far from clear in their discussions of revolutions; and great care must be used when reading such men as Voltaire, Rousseau, or Montesquieu, to be sure as to which of these two senses of the word 'revolution' they may have had in mind at each occurrence. And it is equally the case that care must be exercised in interpreting those writers who used the word 'revolution' in regard to science in the eighteenth century.[10] An exception to this rule occurs whenever the word 'epoch' occurs in conjunction with 'revolution'; for then there can usually be no ambiguity whatsoever about the author's intent. An example is Clairaut's statement that Newton's great book, the *Philosophiae naturalis principia mathematica* "a été l'époque d'une grande révolution dans la Physique".[11] Another is Fontenelle's remark, a propos of the enormous advances (made with "pas de géant") attendant on the discovery (or invention) of the infinitesimal calculus ("le calcul de l'infini"):

> L'infini. . .aména tout à une facilité, dont on n'eût pas osé auparavant concevoir l'espérance; et c'est là l'époque d'une révolution presque totale arrivée dans la géométrie.[12]

Fontenelle, by the way, is the earliest writer I have encountered, who wrote *expressis verbis* about a revolution in the sciences.

By the nineteenth century, the effects of the French Revolution had become so wide-spread that in general discourse, as well as in political and social discussions, the word 'revolution' came to have the exclusive meaning of a secular change of great magnitude. Yet, the obvious etymological meaning could still trouble a critical mind. An example may be seen in John Stuart Mill's *Principles of Political Economy* (1848), in the statement:

> The opening of foreign trade ... sometimes works a complete industrial revolution in a country whose resources were previously undeveloped for want of energy and ambition in the people.

As G. N. Clark shrewdly observed, "Mill was accustomed to weigh his words". When, therefore, he saw the phrase "a complete industrial revolution" on the printed page, this combination must surely "have appeared to him infelicitous". The word "complete" is applicable only in the original or literal sense of the word, "like the revolution of a wheel which takes it back to the starting point", whereas in this instance Mill was "concerned with breaches of continuity, not with recurrent cycles". Accordingly, Clark concludes, "Mill crossed out the word 'complete' and in the later editions we read: 'a sort of industrial revolution'".[13]

All philosophers and historians of science did not introduce the concept of scientific revolution into their presentations; and of those who did so, some made more prominant use of 'revolution' than others.[14] William Whewell is a major figure in the group who did employ this concept and name in a very significant way. And although it has recently been asserted that Whewell "made virtually no use of the concept of scientific revolution",[15] I find him to be the most notable example turned up by my search for nineteenth-century analysts of science who based their discussions in a primary way on the notion of revolutions in science. His *History of the Inductive Sciences*, in fact, displays a veritable philosophy of scientific development in relation to revolutions. Very near the opening of volume one, it is announced, under the heading, 'Terms record Discoveries':

> We shall frequently have to notice the manner in which great discoveries ... stamp their impress upon the terms of a science; and, like great political revolutions, are recorded by the change of the current coin which has accompanied them.[16]

Here Whewell has only made a comparison between "great discoveries" and "great political revolutions"; he has not actually said that such discoveries are or may be revolutions. The unwary reader who goes no further than the opening pages of the Introduction to volume one might thence conclude that Whewell did not approve of the concept of revolution in application to the sciences. And this impression might easily be strengthened by an admonition given by Whewell on the facing page, to the effect that "the history of each science, which may... appear like a succession of revolutions, is, in reality, a series of developments"; the reason is that "earlier truths are not expelled but absorbed, not contradicted but extended".[17] But shortly afterwards, Whewell refers to the "great changes which... take place in the history of science" as "the revolutions of the intellectual world".[18]

Whewell introduced the work of Lavoisier with this observation: "Few revolutions in science have immediately excited so much general notice as the introduction of the theory of oxygen".[19] He contrasted Lavoisier with those "who had, in earlier ages, produced revolutions in science", noting that Lavoisier "saw his theory accepted by all the most eminent men of his time, and established over a great part of Europe within a few years from its first promulgation".[20] By this time (1837), such references to Lavoisier's chemical 'revolution' were not uncommon.[21] Of more interest is the fact that in his discussion of Lavoisier, Whewell seems to have lost sight of the caution he had advocated earlier with respect to 'revolutions' in science.[22] And he introduced 'revolution' in an even more striking way in his account of the *Genera plantarum* (1789) of the younger Jussieu, published after he "had employed himself for nineteen years upon botany". This work, said Whewell, "was not received with any enthusiasm; indeed, at that time, the revolutions of states absorbed the thoughts of all Europe, and left men little leisure to attend to the revolutions of science".[23]

Almost a decade later, in his *Philosophy of the Inductive Sciences, Founded Upon Their History* (first published in 1845), Whewell particularly evoked the metaphor of revolution in relation to the sciences, especially in Book 12, 'Review and Opinions on the Nature of Knowledge, and the Means of Seeking It'. There, in discussing "the revolutions" in opinions as to the methods whereby the physical sciences might progress, he referred to two "movements" – "*the Insurrection Against Authority* and the *Appeal to Experience*". In language more befitting the French Revolution than the rise of science, Whewell wrote about the "submission of the mind" that had been demanded by "the Scholastic System" and related that "the natural love of freedom in man's bosom, and the speculative tendencies of his intellect, rose in rebellion... against the ruling oppression".[24] His aim was to trace "historically the views which have prevailed respecting such methods, at various periods of man's intellectual progress". He found, "Among the most conspicuous of the revolutions which opinions on this subject have undergone, is the transition from an implicit trust in the internal powers of man's mind to a professed dependence upon external observation; and from an unbounded reverence for the wisdom of the past, to a fervid expectation of change and improvement".[25]

Following a presentation of certain presages of a "philosophical revolution",[26] Whewell turned his attention to "the practical reformers of science" who had a "greater share in bringing about the change from stationary to progressive knowledge, than those writers who so loudly announced the revolution".[27] The list includes Leonardo da Vinci, Copernicus, Fabricius, Maurolycus, Benedetti, Tycho, Gilbert, Galileo, Kepler. Introducing Copernicus, Whewell noted that "even in those practical discoverers to whom, in reality, the revolution in science, and consequently in the philosophy of science, was due", we do not always find a "prompt and vigorous recognition of the supreme authority of observation" and a "bold estimate of the probable worthlessness of traditional knowledge".[28] Some chapters later, after describing the work of these practioners of science, Whewell announced the coming on the scene of Francis Bacon in these words:

A revolution was not only at hand, but had really taken place, in the great body of real cultivators of science. The occasion now required that this revolution should be formally recognized; – that the new intellectual power should be clothed with the forms of government; – that the new philosophical republic should be acknowledged as a sister state by the ancient dynasties of Aristotle and Plato. There was needed some great Theoretical Reformer, to speak in the name of the Experimental Philosophy; to lay before the world a declaration of its rights and a scheme of its laws. And thus our eyes are turned to Francis Bacon, and others who like him attempted this great office.[29]

Political rhetoric abounds in Whewell's presentation of Bacon's philosophy of science. He is "not only one of the Founders, but the supreme Legislator of the modern Republic of Science"; he is "the Hercules who slew the monsters that obstructed the earlier traveller"; and he is "the Solon who established a constitution fitted for all future time".[30] But above all, he was properly hailed "as the leader of the revolution",[31] that is, the leader of the "revolution in the method of scientific research" that had been going on in the sixteenth century, and of which he became the articulator and chief spokesman.[32] "If we must select some one philosopher as the Hero of the revolution in scientific method, beyond all doubt Francis Bacon must occupy the place of honour".[33]

Over and again in the *Philosophy of the Inductive Sciences*, Whewell invokes the concept of revolution. Thus Harvey's "reflections on the method of pursuing science" are said to be "strongly marked with the character of the revolution that was taking place".[34] The Cartesian philosophy – held to be "an endeavour to revive the method of obtaining knowledge by reasoning from our own ideas only" and to erect a "philosophical system" in "opposition to the method of observation and experiment" – is called "an attempt at a counter-revolution".[35] And so on.

Thus far, I have indicated the ways in which Whewell referred to revolutions in the sciences.[36] I shall conclude with a few words on Whewell's conception of two revolutions in philosophy. In a 'Review of

Opinions on ... Knowledge' (Book 12), he remarked that the "mode of research by experiment and observation, which had, a little time ago, been a strange, and to many, an unwelcome innovation, was now become the habitual course of philosophers". And this led him to conclude, "The revolution from the philosophy of tradition to the philosophy of experience was completed". Chapter 14 of Book 12 is devoted to "Locke and his French followers". Here Whewell observed that Locke "came at a period when the reign of Ideas was tottering to its fall". Locke "by putting himself at the head of the assault, became the hero of his day: and his name has been used as the watchword of those who adhere to the philosophy of the senses up to our own times".[37] This led Whewell to the following discussion of revolutions and revolutionaries:

Locke himself did not assert the exclusive authority of the senses in the extreme unmitigated manner in which some who call themselves his disciples have done. But this is the common lot of the leaders of revolutions, for they are usually bound by some ties of affection and habit to the previous state of things, and would not destroy all traces of that condition: while their followers attend, not to their inconsistent wishes, but to the meaning of the revolution itself; and carry out, to their genuine and complete results, the principles which won the victory, and which have been brought out more sharp from the conflict.[38]

This analysis leaves no doubt that Whewell was not merely using 'revolution' as metaphor, but had a rather fully developed theory of revolutions in thought.

Book 12 ends with Chapter 15, on 'The Reaction Against the Sensational School'. Here there is a discussion of Kant. According to Whewell,

... The revolution in the customary mode of contemplating human knowledge which Kant's opinions involved, was most complete. He himself, with no small justice, compares it with the change produced by Copernicus's theory of the solar system.[39]

Then Whewell quotes at length from the preface to Kant's *Kritik der reinen Vernunft*, (ed. 2, 18) including the sentences;

Let us then try whether we cannot make out something more in the problems of metaphysics, by assuming that objects must be regulated by our knowledge, since this agrees better with that supposition, which we are prompted to make, that we can know something of them *à priori*. This thought is like that of Copernicus, who, when he found that nothing was to be made of the phenomena of the heavens so long as everything was supposed to turn about the spectator, tried whether the matter might not be better explained if he made the spectator turn, and left the stars at rest.[40]

It is to be observed that neither in Whewell's own statement, nor in the quotation from Kant, is there any indication that Kant had specifically referred to a 'Copernican revolution' or had compared his own aims in philosophy to such a 'Copernican revolution' – as Kant is (incorrectly) said to have done by so many writers.[41] Perhaps this error, a feature chiefly of American and British philosophers and historians of philosophy, derives from a careless and hasty reading of Whewell's page. But a careful reading shows that it is Whewell himself, and not Kant, who is making the judgment concerning the Kantian 'revolution'. And such a judgment is wholly in keeping with the theory of scientific and philosophical revolutions which we have seen to characterize Whewell's historical thought.

Harvard University

NOTES

1 The present article is based on research supported by the Spencer Foundation. Some related studies of mine are: (i) 'The Eighteenth-century Origins of the Concept of Scientific Revolution', *Journ. of the History of Ideas*, Apr. 1976; (ii) 'The Copernican Revolution from an Eighteenth-century Perspective: with Notes on Jean-Sylvain Bailly's Views on Revolutions in Science', to appear in the *Festschrift* for Prof. Willy Hartner of Frankfurt; (iii) *Scientific Revolution: the History of a Concept and a Name*, (Science History Publications, 1976, New York, in press).

2 T. S. Kuhn, *The Structure of Scientific Revolutions* (The University of Chicago Press, Chicago, 1962 – also issued as Vol. 2, of the *International Encyclopedia of Unified Science*; second edition, enlarged, 1970).

3 The number of scientists, philosophers, and social scientists who have applied Kuhn's analysis to one problem or another is too large for me to consider here.

4 For this kind of response to Kuhn's analysis, see Imre Lakatos and Alan Musgrave (eds.), *Criticism and the Growth of Knowledge* (at the University Press, Cambridge, 1970), comprising a primary paper by T. S. Kuhn, followed by critical discussions by J. W. M. Watkins, S. E. Toulmin, L. Pearce Williams, K. R. Popper, Margaret Masterman, I. Lakatos, P. K. Feyerabend, plus a final 'Reflections on my Critics' by Kuhn. Among many reviews and review articles, particular attention may be called to those by Gerd Buchdahl, Dudley Shapere, and Israel Scheffler.

5 The propriety of using the word and concept of 'revolution' in relation to science is discussed by Stephen E. Toulmin, in the course of a lengthy historical narrative of, and critique upon, Kuhn's views, in *Human Understanding*, Vol. 1 (Princeton University Press, Princeton, 1972), pp. 100–117, and esp. 117–118.

6 An example is Imre Lakatos's notion of "methodology of scientific research programmes" – for which see his contribution to the volume cited in Note 4 *supra*.

7 On the Copernican Revolution, see the work cited as (ii) in Note 1 *supra*. I am currently completing a study of *The Newtonian Revolution in Science*, to be published by Cambridge University Press.

8 This topic is explored at length in the work cited as (iii) in Note 1 *supra*.

[9] *Revolutio* was primarily a technical term in astronomy and astrology, introduced in the late Middle Ages; it encompassed what we today know as revolution (in an orbit) and as rotation (about an axis). In the early Middle Ages, there are a few occurrences (by Augustine and Martianus Capella) of *revolutio* in the meaning of *conversion.*

[10] These topics are explored in the work cited as (iii) in Note 1 *supra.*

[11] Alexis-Claude Clairaut, "Du système du monde, dans les principes de la gravitation universelle". *Suite des mémoires de mathématique, et de physique, tiré des registres de l'Académie Royale des Sciences de l'année M.DCCXLV* (chez J. Schreuder & Pierre Mortier le jeune, Amsterdam, 1754), p. 465; Clairaut's paper was read "à l'Assemblé publique du 15 Nov. 1747". Clairaut's statement was repeated almost verbatim by Joseph-Louis Lagrange, John Playfair, and Thomas Henry Huxley.

[12] On Fontenelle's uses of 'revolution' to characterize scientic progress, see the works cited as (i) and (iii) in Note 1 *supra.*

[13] On Mill's use of 'revolution' in this context, see George Norman Clark, *The Idea of the Industrial Revolution* (Jackson, Son & Company, Glasgow, 1953): "Being the twentieth Lecture on the David Murray Foundation in the University of Glasgow, delivered on 15th October 1952", p. 12.

[14] Among those who made use of 'revolution', but not in a very significant way, are J. S. Mill (in his *Logic*), John Herschel, Hans Christian Oersted. Those for whom revolutions in science were of capital inportance include Claude Henri de Saint-Simon, Auguste Comte, and Antoine Augustin Cournot. This topic is explored further in the work referred to under (iii) in Note 1 *supra.*

[15] By Lewis F. Feuer, *Einstein and the Generations of Science* (Basic Books, New York, 1974), p. 245. This work, which was published after my basic research on this topic had been completed, contains the only attempt (with which I am familiar) to trace out the history of the concept of revolution in science. Unfortunately, there are many errors and omissions; see Notes 20 and 23 *infra.*

[16] William Whewell: *History of the Inductive Sciences, from the Earliest to the Present Times* (John W. Parker, London, J. and J. J. Deighton, Cambridge, 1837), Vol. 1, p. 11.

[17] *Ibid.*, p. 10.

[18] *Ibid.*, p. 11.

[19] *Ibid.*, Vol. 3, p. 128: Book 14, Ch. 6, Sect. 1.

[20] *Ibid.*, p. 136. Cf. Feuer, *op. cit.* (Note 15 *supra*), p. 245, it is said: "In his classical *History of the Inductive Sciences* Whewell used the concept of revolution only once, and that was in reference to the philosophical work of Francis Bacon: 'He announced a New Method, not merely a correction of special current errors; he thus converted the Insurrection into a Revolution, and established a new philosophy Dynasty'", This quotation is taken from "the third edition, with additions", of Whewell's *History* (D. Appleton & Co., New York, 1865), p. 339, where it occurs in a lengthy extract, printed within square brackets, of a supplement added by Whewell in the '2nd Ed.'. It appears at the very end of Book 6, Ch. 2, Sect. 5. Between the first and second editions of the *History*, Whewell published his *Philosophy of the Inductive Sciences*, discussed below, in which he discussed more fully the rôle of Bacon as a revolutionary. Despite very specific references by Whewell to the 'revolution', produced by Lavoisier and Jussieu, Feuer, *op. cit.*, p. 246, states that in the *History*, Whewell "did not use it [i.e., the analogy of revolution] to characterize any of the specific landmarks of scientific discovery".

[21] The question of an individual scientist referring to his own achievement as revolutionary is discussed in the works listed under (i) and (iii) in Note 1 *supra*; Lavoisier appears to have been the first scientist to have so described his own scientific work.

[22] See the previous quotation, corresponding to Note 17.
[23] *Op. cit.*, Vol. 3, p. 337: Book 16, Ch. 5.
[24] *The Philosophy of the Inductive Sciences, Founded Upon Their History* (John W. Parker, London; J. and J. J. Dighton, Cambridge, 1840), Vol. 2, pp. 319–320: Book 12 ('... of Opinions on the Nature of Knowledge and the Methods of Seeking It'), Ch. 7. A facsimile reprint of the second edition (London, 1847) was issued, with an introduction by John Herivel, by Johnson Reprint Corporation (New York and London) in 1967.
[25] *Ibid.*, p. 318.
[26] *Ibid.*, p. 366: Ch. 9. In introducing the work of some sixteenth-century philosophers, Whewell again employed the language of political revolution, writing that "these insurrections against the authority of the established dogmas, although they did not directly substitute a better positive system in the place of that which they assailed, shook the authority of the Aristotelian system, and led to its overthrow".
[27] *Ibid.*, p. 366: beginning of Ch. 10.
[28] *Ibid.*, p. 370: Ch. 10.
[29] *Ibid.*, p. 388: Ch. 10.
[30] *Ibid.*, p. 389.
[31] *Ibid.*, p. 392.
[32] *Ibid.*, p. 391.
[33] *Ibid.*, p. 392. But Whewell also said (p. 414: beginning of Ch. 12) that "Bacon was by no means the first mover or principal author of the revolution in the method of philosophizing which took place in his time".
[34] *Ibid.*, p. 414: Ch. 12.
[35] *Ibid.*, p. 418.
[36] These include, of course, references to revolutions in the methods or philosophy of science. I do not claim to have listed all of the references to 'revolution' in Whewell's writings.
[37] *Op. cit.* (Note 24 *supra*), pp. 424–425, 458. He also observed, in an unexpected fashion for one who had rung the clarion so vehemently for Bacon, that: "Perhaps too, as was natural in so great a revolution, the writers of this time, especially the second-rate ones, were somewhat too prone to disparage the labours and talents of Aristotle and the ancients in general, and to overlook the ideal element of our knowledge, in their zealous study of phenomena. They urged, sometimes in an exaggerated manner, the superiority of modern times in all that regards science, and the supreme and sole importance of facts in scientific investigations" (*ibid.*, p. 426).
[38] *Ibid.*, p. 458.
[39] *Ibid.*, p. 479.
[40] *Idem.*
[41] The currency of this expression has apparently been greatly increased by Thomas S. Kuhn's book, *The Copernican Revolution: Planetary Astronomy in the Development of Western Thought* (Harvard University Press, Cambridge, 1957). Two double units of the second-level course of the Open University (of Great Britain) on 'The Age of Revolutions' are entitled *Kant's Copernican Revolution: Speculative Philosophy* and *Kant's Copernican Revolution: Moral Philosophy* (these volumes have been published in 1972 by The Open University Press, A202, units 15–16, 17–18). Despite the prevalence of the opinion that Kant referred to his own work as a 'Copernican revolution', a close scrutiny of Kant's text shows that he never in fact did make this alleged reference. On this subject, see Norwood Russell Hanson: 'Copernicus's Rôle in Kant's Revolution', *Journal of the History of Ideas*, 1959, Vol. 20, pp. 275–281.

L. JONATHAN COHEN

HOW CAN ONE TESTIMONY CORROBORATE ANOTHER?

Imre Lakatos gave new life to every problem he touched. There are many issues that I regret never having discussed with him. Among them is the one I examine below.

1. The mutual corroboration[1] of different testimonies and the convergence of different items of circumstantial evidence are two very familiar patterns of argument for conclusions about individual facts, whether in a law-court, a laboratory or elsewhere. These two patterns of argument seem to share a common logical structure. But how is their validity to be elucidated in terms of a concept of probability that conforms to the familiar principles of the mathematical calculus of chance? How can it be proved that probability is raised by corroboration or convergence?

At its simplest testimonial corroboration occurs when two witnesses both testify, independently of one another, to the truth of the same proposition. On its own the admitted fact that *A* testifies that **p** would do something to raise the probability that **p**, and so would the fact that *B* testifies that **p**. But both facts together raise this probability substantially more than either would alone, provided that one important condition holds good. This condition is roughly that neither fact may be connected with the other (other than through the truth of what is testified). If one witness has been told what to say by the other, or is influenced by what he hears him testify, or is involved in any other kind of collusion, there may be no real corroboration. But if there was no special reason to suppose that the two testimonies would be identical, they do give each other some degree of corroboration.

Two items of circumstantial evidence converge when both facts, independently of one another, probabilify the same conclusion. Suppose it has been established beyond reasonable doubt that *A* was due to inherit under *B*'s will, so that *A* had a motive for murdering *B*. Then this fact on its own would do something to raise the probability that *A* was

R. S. Cohen et al. (eds.), Essays in Memory of Imre Lakatos, 65–78. *All Rights Reserved*
Copyright © 1976 by D. Reidel Publishing Company, Dordrecht-Holland

B's killer. Similarly, if it were established that *A* was a guest of *B*'s at the time, so that he had an opportunity to commit the murder, this too, on its own, would do something to raise the probability that *A* was the killer. But both facts together raise this probability substantially more than either would alone, provided that a certain important condition holds. This condition is that neither fact is dependent on the other except through the guilt of the accused. If everyone who had a motive for the crime is known to have been deliberately invited by *B* to his house at the same time (as would appear to happen in some detective stories!), there is no genuine convergence. But if there was no special reason to infer *A*'s presence in the house from his having had a motive, or *vice versa*, the two facts do converge to increase the probability of his having been the killer.

It is easy to describe in general terms the logical structure that is common to corroboration and convergence. If a conclusion, S, has its probability raised by each of two premisses, R_1 and R_2, when these are considered separately, and R_2 is unconnected with R_1, unless through the truth of S, then the conjunction of R_1 and R_2 makes S more probable than does R_1 alone. But how is it possible to elucidate this familiar principle in terms of mathematical probability?

2. The elucidation that has been most commonly proposed is that of Bernoulli (1713, part IV, chapter iii). But this analysis is more easily discussed in the admirably perspicuous form in which it was expounded by Boole (1857), who acknowledged his debt here to Cournot and de Morgan. The analysis has undergone a good deal of adverse criticism: for example, by Keynes (1921, p. 180ff.). But it is still occasionally adopted as if it were unexceptionable: for example, by Schreiber (1968, p. 41). And an investigation of why the analysis goes wrong will help to lead us in the right direction.

Boole wrote

Let **p** be the general probability that **A** speaks truth, **q** the general probability that **B** speaks truth; it is required to find the probability that, if they agree in a statement they both speak truth. Now agreement in the same statement implies that they either both speak truth, the probability of which beforehand is **pq**, or that they both speak falsehood, the probability of which beforehand is $(1-\mathbf{p})(1-\mathbf{q})$. Hence the probability beforehand that they will agree is $\mathbf{pq}+(1-\mathbf{p})(1-\mathbf{q})$ and the probability that if they agree they will agree in speaking the truth is accordingly expressed by the formula

$$\mathbf{w}=\frac{\mathbf{pq}}{\mathbf{pq}+(1-\mathbf{p})(1-\mathbf{q})}.$$

This formula can obviously be applied also to the convergence of circumstantial evidence, with as much plausibility as to testimonial corroboration. Let **p** be the probability that fact *A* incriminates a certain person correctly, and **q** the probability that fact *B* does so. It would then be required to find the probability that, if they independently agree in whom they incriminate, they both incriminate correctly. Now, if they agree independently in whom they incriminate, this convergence implies that they either both incriminate correctly with a probability of **pq**, or both incriminate incorrectly with a probability of $(1-\mathbf{p})(1-\mathbf{q})$. So it looks as though just the same formula should express the probability that, if facts *A* and *B* do converge, the person whom they incriminate is the killer.

The formula is obviously plausible enough when **p** and **q** are independent of one another and both greater than .5, and nothing is at issue but the truth or falsehood of testimony, or the correctness or incorrectness of an incrimination. The formula then always assigns **w** a greater value than either **p** or **q**, and so it looks as though corroboration, or convergence, as the case may be, has taken place. But unfortunately the formula also tells us that if the mathematical probability of *A*'s telling the truth (or of *A*'s being correctly incriminatory) is less than .5, and ditto for *B*, then the mathematical probability that they are both telling the truth (or incriminating correctly) is even less. So one common type of testimonial corroboration, or convergence of circumstantial evidence, is not elucidatable in this way. One witness, for example, may seem rather unreliable because of his shifty demeanour, and another may seem rather unreliable because of his bad eyesight. Yet perhaps, quite independently, they both testify to precisely the same set of propositions even though each could have told any number of other stories. In such a case Boole's formula produces a lower probability for their joint veracity, whereas normal juries would assign a higher one. Boole's formula makes the two testimonies undermine one another, whereas our problem is to elucidate why in fact they corroborate one another. Similarly, the fact that the accused had a motive for murdering the victim may be only mildly incriminatory, since perhaps four other people also are known to have had motives, and the fact that he had an opportunity to commit the murder may also be only mildly incriminatory, since perhaps four other people are also known to have had opportunities. But the combination of motive and opportunity in one person out of the nine is more seriously incriminatory, and Boole's formula cannot represent this. If the mathematical probability of the accused's guilt on the evidence of motive were .2, and that of his guilt on the evidence of opportunity were also .2, Boole's formula gives a mathematical probability of approximately .06 for the accused's guilt on the combined and converging evidence.

3. The specific trouble with Boole's formula is that it envisages a situation in which the domain of possibilities is a binary one. The choice has to be between a witness's speaking the truth and his speaking falsehood. Yet, if the evidence of one unreliable witness is understood to corroborate that of another in terms of mathematical probability, this corroboration seems to occur just because there are so many other stories that each could have told. The domain of possibilities is a multiple one. Similarly when two pieces of circumstantial evidence converge to incriminate the same man, the mathematical probability of his guilt seems to be increased just because there are so many other men that either piece of evidence might have incriminated.

But what this means is that any adequate representation of corroboration in terms of mathematical probability must take into account the prior probability that a certain story – which is one out of so many competing possibilities – is true, as well as the probability of its truth on the evidence of this or that witness's testimony. Where agreement is relatively improbable (because so many different things might be said), what is agreed is more probably true. Similarly any adequate representation of convergence in terms of mathematical probability must take into account the prior probability that, say, the accused – who is one out of so many possible criminals – committed the crime, as well as the probability of this on the facts before the court. Where the guilt of any arbitrarily selected man is relatively improbable, a particular man's possession of both motive and opportunity is more probably significant. Boole's formula could hardly be expected to give us what we want here because it fails to take account of a vital element in the situation. The underlying trouble with the formula is that it takes no account of the relevant prior probabilities.

Indeed these prior probabilities *must* be taken into account if each of two corroborating or converging premisses may itself give a less than .5

probability to the desired conclusion. In such a case it cannot be the absolute level of the conclusion's probability on the premiss that makes the premiss eligible to corroborate, or converge towards, the conclusion, since by the usual complementational principle for negation the denial of the conclusion then has a greater absolute probability. So instead what is vital is that the premiss should raise the probability of the conclusion. However low $p(S, R)$ may be, if it is greater than $p(S)$, $p(\text{not-}S, R)$ will be less than $p(\text{not-}S)$.

4. More recently another principle of convergence or corroboration has been suggested by Ekelöf (1964). At first sight this new principle seems to avoid the difficulties that the Bernoulli-Boole principle encounters.

Let us suppose that in an action concerning a highway accident there are two facts tending to prove that one of the cars concerned had a speed exceeding 60 m.p.h.; length of the braking marks, and a witness who observed the collision. We further make the unrealistic assumption that by examining a great number of similar situations it has been possible to ascertain that each of these evidentiary facts implies in 3 cases out of 4 a faithful description of reality, whereas in the fourth case it has no value whatever as evidence of the speed of the car. At least if the value of each evidentiary fact is independent of that of the other, the value [*sc.* of the combined evidence] must be greater than $\frac{3}{4}$. But how much greater? The length of the braking marks proves that the speed exceeded 60 m.p.h. in 12 out of 16 similar cases; at the same time this is proved by the witness-statement in 3 out 4 remaining cases. The convincing force of the combined evidentiary facts would thus be $\frac{15}{16}$.

According to Edman[2] (1973) the principle that Ekelöf has in mind is

$$\mathbf{w} = \mathbf{p} + \mathbf{q} - (\mathbf{pq})$$

where **p** and **q** are the separate forces of two converging but independent evidential facts and **w** is their combined force. This principle has the advantage that, unlike the Bernoulli-Boole principle, it seems to work just as well where **p** or **q**, or both, are less than .5. So long as both **p** and **q** are less than 1 and greater than 0, the combined force **w** is greater than either **p** or **q** alone.

But this is only a superficial merit. Ekelöf's principle does not escape the underlying difficulty about prior probabilities that wrecked the Bernoulli-Boole solution of the convergence problem. Consider a case where **p** is .25 and is the mathematical probability that the criminal was a male on the evidence that he had long hair, and **q** is .25 and is the probability that the criminal was a male on the evidence of testimony to that effect by a supporter of the women's liberation movement. By Ekelöf's principle these two evidential facts converge to give an increased value to the combined evidence of .44. Yet if the mathematical probability that the criminal was a male, on each separate piece of evidence, is .25, the probability that the criminal was a female is .75, and so the combined evidence has a force of .94 in favour of the conclusion that the criminal was a female. We thus have the paradox that, according to Ekelöf's principle, the two pieces of evidence converge in opposite directions at the same time. Or – to put the point in other words – the evidence of the witness, on Ekelöf's view, corroborates whichever conclusion you prefer to draw from the fact that the criminal had long hair; and evidence that purports to corroborate opposite conclusions does not in fact corroborate either. Moreover, if the force of the combined evidence is to be conceived of as a mathematical probability, we have a straightforward contradiction between the calculation that the probability of the criminal's being a male is .44 and the calculation that the probability of the criminal's being a female is .94, since presumably the probability of being a female is in fact the complement of the probability of being a male.

What has gone wrong here? Obviously the trouble is that some of the posterior probabilities involved – viz. those of .25 – are lower than the corresponding prior ones, if the prior probability of the criminal's being a male is, say, .5. In fact therefore the evidence decreases the mathematical probability that the criminal is a male and increases the mathematical probability that the criminal is a female. We should normally infer in such a situation that the evidence converges in the direction in which it increases the probability. But to represent such an inference we need some other principle than Ekelöf's. We need a principle that takes some appropriate account of prior probabilities.

To try and avoid this conclusion it might seem plausible to argue that in assessing corroboration or convergence we should consider the improbability of the testimony or evidence on the assumption that the desired conclusion is false, rather than the probability of the conclusion on the given testimony or evidence. For, if in the circumstances of a particular case motive and opportunity were relatively independent of one another, the multiplicative law for the probability of a conjunction would ensure its being substantially more improbable that the accused had both motive and opportunity on the assumption that he was not guilty than that he had either of these alone on that assumption. I.e. we should have $p[\text{not}(R_1 \,\&\, R_2), \text{not-}S] > p[\text{not-}R_1, \text{not-}S]$, because $p[R_1, \text{not-}S] > p[R_1 \& R_2, \text{not-}S]$ and therefore $1 - p[R_1 \& R_2, \text{not-}S] > 1 - p[R_1, \text{not-}S]$.

But this move also fails because it has no room to take account of prior probabilities. Consider the case where both R_1 and R_2 are intrinsically irrelevant to S, i.e. where R_1 and R_2 are about matters unconnected with S and so $p[S, R_1] = p[S] = p[S, R_2]$. In such a case, if R_1 and R_2 are independent of one another, we shall still have $p[\text{not}(R_1 \,\&\, R_2), \text{not-}S] > p[\text{not-}R_1, \text{not-}S]$. So apparently corroboration or convergence can occur even when the testimony or evidence is intrinsically irrelevant!

5. It is therefore tempting to try and contrive an elucidation of the issue in terms of some familiar theory of probability that admits a concept of relevance. But we can obtain no help either from Carnap's relevance-measure or from Reichenbach's principle of the common cause.

Carnap (1950, p. 360ff.) introduced a relevance measure r, where a relevance function $r(\mathbf{i}, \mathbf{h}, \mathbf{e})$ represents the change in the probability of **h** on **e** by the addition of new evidence **i**. Presumably the corroboration of **i** by **j**, in relation to a proposed conclusion **h** and other evidence **e**, is then to be construed as an increase in the relevance of $\mathbf{i} \cdot \mathbf{j}$ over that of **i** alone. But one theorem for Carnap's r is

$$r(\mathbf{i} \cdot \mathbf{j}, \mathbf{h}, \mathbf{e}) = r(\mathbf{i}, \mathbf{h}, \mathbf{e}) + r(\mathbf{j}, \mathbf{h}, \mathbf{e}) - r(\mathbf{i} \vee \mathbf{j}, \mathbf{h}, \mathbf{e}).$$

Now, suppose **i** states that witness A's testimony incriminates C, and **j** states that witness B's testimony incriminates C, while **h** states that C is guilty. Then presumably $\mathbf{i} \vee \mathbf{j}$ may be at least as relevant to **h**, on the other evidence **e**, as is **i**, or **j**, on this evidence. So if the corroboration of **i** by **j** were construed as an increase in the relevance of $\mathbf{i} \cdot \mathbf{j}$ over that of **i** alone, no testimonial corroboration would take place, according to Carnap's theorem, even when two witnesses spoke quite independently of one another.

It is perhaps tempting instead to think of two mutually corroborating testimonies as outcomes of what Reichenbach (1956, p. 157ff.), called a common cause. The existence of the cause may then be inferred, with increased probability, from the fact that corroboration has occurred. According to Reichenbach a common cause S is indicated for two events R_1 and R_2, when the following four conditions are satisfied[3].

$$p[R_1 \,\&\, R_2, S] = p[R_1, S] \times p[R_2, S]$$
$$p[R_1 \,\&\, R_2, \text{not-}S] = p[R_1, \text{not-}S] \times p[R_2, \text{not-}S]$$
$$p[R_1, S] > p[R_1, \text{not-}S]$$
$$p[R_2, S] > p[R_2, \text{not-}S].$$

If these four conditions are satisfied, Reichenbach shows, it is possible to infer such conclusions as

$$p[R_1, R_2] > p[R_1]$$

and

$$p[S, R_1] > p[S] < p[S, R_2].$$

But the trouble is that these conditions do not always cover the crucial factors in the corroboration-convergence situation. For example, let R_1 be C's being one out of just five men who had the opportunity to commit the crime, while S is C's committing the crime. Then $p[S, R_1] > p[S]$. But suppose that it is extremely rare for just five men to have had the opportunity to commit this kind of crime, when it has been committed, and much commoner for such quintets of opportunity-possessors to occur when the crime has not actually been committed. (Perhaps prospective delinquents are put off by the thought that only four others could be suspect on ground of opportunity.) Then $p[R_1, \text{not-}S] > p[R_1, S]$. So here a typically incriminatory piece of circumstantial evidence R_1 is known. But the common cause analysis, because it is sensitive to some rather irrelevant facts of criminological statistics, fails to represent the incriminatoriness of the evidence.

6. To determine a correct analysis, a closer look at typical situations will be necessary.

Typically we have a situation in which, if S is the proposition to be established and R_1 and R_2 are the two premisses, then both R_1 and R_2 raise the probability of S. That is to say, $p[S, R_1] > p[S]$ and $p[S, R_2] > p[S]$. Hence, if the prior probabilities were evenly distributed over a binary domain, like the truth or falsity of S, they would each be equal to .5, and we should have each of the two posterior probabilities,

$p[S, R_1]$ and $p[S, R_2]$, greater than .5, as Boole's formula requires. But with a larger domain for evenly distributed prior probabilities, each of the prior probabilities would be less than .5, and the posterior probabilities could then be less than .5.

There is also another important requirement. The premisses, in a case of corroboration or convergence, are supposed to be unconnected with each other unless through the truth of the conclusion. This involves two independence conditions.

First, if one witness is to corroborate another, the former must not be particularly inclined to give false testimony when the latter does. For example, there must be no conspiracy to deceive. Similarly, if one piece of evidence is to converge with another, the former must not be particularly prone to indicate an incorrect conclusion when the latter does. For example, the fact that a man is scared of the police and the fact that he has recently committed a similar crime do not converge to raise the probability of his having committed the present crime, since even if he is innocent of the present crime his other misdeed is likely to make him scared of the police. So the first independence condition is that $p[R_2, R_1$ & & not-$S]$ must not be greater than $p[R_2$, not-$S]$. That is, R_1 must not be favourably relevant to R_2 on the assumption that S is false.

Secondly, if one witness is to corroborate another, the former's inclination to give true testimony must not be less, but may be greater when the latter's testimony is true. For example, the former must not be more inclined to contradict the latter on those occasions when the latter's testimony happens to be true, since agreement of the two testimonies would then indicate their incorrectness. So $p[R_2, R_1 \& S]$ must not be less than $p[R_2, S]$. But the other witness's testimony may make the corroborating witness more inclined to testify to what he believes to be true. True testimony is genuinely corroborative whether it springs from the witness's belief about the fact at issue, or is influenced by what the other witness has said. Hence $p[R_2, R_1 \& S]$ could be greater than $p[R_2, S]$. Similarly, if one piece of evidence is to converge with another, the former must not reduce but may increase the probability of the latter when the indicated conclusion is correct. For example, a man's consumption of both barbiturates and whisky before going to bed do not converge to raise the probability that he slept well, even though either fact indicates this on its own. For, if he really consumed both, the combination might well make him ill. And it is easy to see what blocks convergence here. If the man slept well, and therefore was not ill, he is less likely to have consumed both barbiturates and whisky beforehand. That is, $p[R_2, R_1 \& S]$ is less than $p[R_2, S]$, and so convergence fails. But a man's having a motive for killing the victim and his lack of grief at the victim's death could converge to raise the probability of his being the killer, even though either fact would increase the probability of the other. For, if he was the killer, his lack of grief would confirm the strength of his motive and his motive would confirm that his apparent lack of grief was not due to concealment of his feelings. I.e., even if $p[R_2, R_1 \& S]$ exceeds $p[R_2, S]$, convergence may still occur. So the second independence condition is just that $p[R_2, R_1 \& S]$ must be greater than or equal to $p[R_2, S]$. R_1 must not be unfavourably relevant to R_2 on the assumption that S is true.

Schematically, therefore, when all the premisses in a simple case of corroboration or convergence are made explicit, and expressed in terms of mathematical probability, we have two premisses about probability-raising, viz. $p[S, R_1]>[pS]$ and $p[S, R_2]>p[S]$, and two premisses about independence, viz. $p[R_2, \text{not-}S]\geqslant p[R_2, R_1 \& \text{not-}S]$ and $p[R_2, S] \leqslant p[R_2, R_1 \& S]$. From these four premisses, plus the obviously appropriate assumption that $R_1 \& R_2$ is not self-contradictory and therefore does not have zero probability, and the equally appropriate assumption that $p[S, R_1]$ has room for increase, it is demonstrable, with the help of nothing but standard mathematical principles, that $p[S, R_2 \& \& R_1]>p[S, R_1]$. In other words, however the mathematical probabilities in question are measured, the formal structure of the mathematical calculus of chance ensures that corroboration or convergence takes place under appropriate conditions.

Indeed, if this were not demonstrable, the title of the mathematical calculus of chance to be interpreted as a theory of probability would be very substantially weakened, since corroboration or convergence is so important a pattern in everyday human reasonings about probabilities. Hence the demonstration deserves to be set out in detail.

7. The premisses are:

1. $p[S, R_1]>p[S]$.
2. $p[S, R_2]>p[S]$.

3. $p[R_2 \,\&\, R_1]>0$.
4. $p[R_2, S] \leqslant p[R_2, R_1 \,\&\, S]$.
5. $p[R_2, -S]\geqslant p[R_2, R_1 \,\&\, -S]$.
6. $1>p[S, R_1]$.

From these premisses it is required to prove that

$$p[S, R_2 \,\&\, R_1]>p[S, R_1].$$

It will be convenient to set out the proof in algebraic form. So let us put

7. $p[S]=\mathbf{s}$.
8. $p[R_1]=\mathbf{r}_1$.
9. $p[R_2]=\mathbf{r}_2$.
10. $p[R_1 \,\&\, S]=\mathbf{a}$.
11. $p[R_2 \,\&\, S]=\mathbf{b}$.
12. $p[R_2 \,\&\, R_1]=\mathbf{c}$.
13. $p[R_2 \,\&\, R_1 \,\&\, S]=\mathbf{d}$.

The proof will use certain standard principles of the calculus of chance, viz. 14–19:

14. $p[A]=1-p[-A]$.
15. $p[A, B]=1-p[-A, B]$.
16. $p[A \,\&\, B]=p[A]\times p[B, A]=p[B \,\&\, A]$.
17. $p[A, B]=\dfrac{p[A \,\&\, B]}{p[B]}$ where $p[B]>0$.
18. $p[A, B]=\dfrac{p[B, A]\times p[A]}{p[B]}$ where $p[B]>0$.
19. $p[A \,\&\, -B]=p[A]-p[A \,\&\, B]$.

We first prove

20.	$p[R_1]>0<\mathbf{r}_1$	3, 8, 16.
21.	$p[R_2]>0<\mathbf{r}_2$	3, 9, 16.
22.	$p[S]>0<\mathbf{s}$	1, 7, 18, 20.
23.	$1-p[S]>0$	6, 14, 15, 18, 20.
24.	$p[R_1 \,\&\, S]>0<\mathbf{a}$	1, 10, 16, 20, 22.
25.	$p[R_1 \,\&\, -S]>0$	6, 15, 16, 20.

We next obtain algebraic expressions, along with 7–13, for all the probabilities mentioned in the premisses 1–6, as follows (26–31):

26. $p[S, R_1]=\frac{\mathbf{a}}{\mathbf{r}_1}$ 17, 20; 8, 10, 16.

27. $p[S, R_2]=\frac{\mathbf{b}}{\mathbf{r}_2}$ 17, 21; 9, 11, 16.

28. $p[R_2, S]=\frac{\mathbf{b}}{\mathbf{s}}$ From 18, 22, and 7, 9, 27, we obtain $p[R_2, S]=\frac{\frac{\mathbf{b}}{\mathbf{r}_2}\times\mathbf{r}_2}{\mathbf{s}}$.

29. $p[R_2, R_1 \,\&\, S]=\frac{\mathbf{d}}{\mathbf{a}}$ 17, 24; 10, 13.

30. $p[R_2, -S]=\frac{\mathbf{r}_2-\mathbf{b}}{1-\mathbf{s}}$.

i. $p[R_2, -S]=\frac{p[R_2]-p[R_2 \,\&\, S]}{1-p[S]}$ 17, 19; 14, 23.

ii. $p[R_2, -S]=\frac{\mathbf{r}_2-\mathbf{b}}{1-\mathbf{s}}$ i, 7, 9, 11.

31. $p[R_2, R_1 \,\&\, -S]=\frac{\mathbf{c}-\mathbf{d}}{\mathbf{r}_1-\mathbf{a}}$

i. $p[R_2, R_1 \,\&\, -S]=\frac{p[R_2 \,\&\, R_1 \,\&\, -S]}{p[R_1 \,\&\, -S]}$ 17, 25.

ii. $p[R_2, R_2 \,\&\, -S]=\frac{\mathbf{c}-\mathbf{d}}{\mathbf{r}_1-\mathbf{a}}$ i, 19; 8, 10, 12, 13.

The first five premisses are reformulated in algebraic terms (32–36):

32. $\frac{\mathbf{a}}{\mathbf{r}_1}>\mathbf{s}$ 1, 7, 26.

33. $\frac{\mathbf{b}}{\mathbf{r}_2}>\mathbf{s}$ 2, 7, 27.

34. $\mathbf{c}>0$ 3, 12.

35. $\frac{\mathbf{b}}{\mathbf{s}}\leqslant\frac{\mathbf{d}}{\mathbf{a}}$ 4, 28, 29.

36. $\frac{r_2-b}{1-s} \geqslant \frac{c-d}{r_1-a}$ 5, 30, 31.

The proof then proceeds algebraically as follows:

37. $\frac{\frac{b}{s}-b}{1-s} > \frac{r_2-b}{1-s}$ 7, 23; 22, 33.

38. $\frac{\frac{b}{s}-b}{1-s} = \frac{b-bs}{s(1-s)} = \frac{b}{s}$

39. $\frac{d}{a} > \frac{r_2-b}{1-s}$ 35, 37, 38.

40. $\frac{d}{a} > \frac{c-d}{r_1-a}$ 36, 39.

41. $\frac{d}{a} > \frac{\frac{c}{a}-\frac{d}{a}}{\frac{r_1}{a}-1}$ 24, 40.

42. $\left(\frac{r_1}{a}-1\right) \times \frac{d}{a} > \frac{c}{a}-\frac{d}{a}$ 41.

43. $\frac{r_1}{a} \times \frac{d}{a} > \frac{c}{a}$ 42.

44. $\frac{d}{c} > \frac{a}{r_1}$ 20, 34, 43.

Finally, we translate 44 back into the language of probability in order to obtain what we set out to prove: –

45. $\frac{d}{c} = p[S, R_2 \,\&\, R_1]$ 12, 13; 16, 17, 34.

46. $p[S, R_2 \,\&\, R_1] > p[S, R_1]$ 44; 26, 45.

In fact line 1 above is a rather stronger premiss than is needed and could be replaced by $p[S, R_1] > 0$, because line 32 is otiose. We then prove that additional evidence, if favourable (premiss 2), normally increases the mathematical probability of a conclusion irrespectively of whether the previous evidence was favourable or unfavourable, provided that the two pieces of evidence are compossible (premiss 3) and appropriately independent (premisses 4 and 5) and provided that the previous evidence did not wholly rule out either the truth or the falsity of the conclusion (premisses 1 and 6). The extent of this increase in probability, and how exactly it is determined, need not concern us here.

It is, however, quite another matter to determine whether the above elucidation of correboration and convergence can apply to proofs of fact in Anglo-American courts of law or in any legal system in which an accused person is presumed absolutely innocent until proved guilty. In such a system it looks as though no positive prior probability of guilt is admissible. But a zero prior probability of guilt will, in accordance with 16 and 17 above, make the posterior probability also zero. The problem is discussed in Cohen (1977).

The Queen's College, Oxford

BIBLIOGRAPHY

Bernoulli, James, 1713, *Ars Conjectandi.*

Boole, George, 1857, 'On the Application of the Theory of Probabilities to the Question of the Combination of Testimonies or Judgments', in *Transactions of the Royal Society of Edinburgh* xxi. Reprinted in G. Boole, *Studies in Logic and Probability*, 1952, p. 364ff.

Carnap, Rudolf, 1950, *Logical Foundations of Probability.*

Cohen, L. Jonathan, 1977, *The Probable and the Provable* (forthcoming).

Edman, Martin, 1973, 'Adding Independent Pieces of Evidence', in *Modality, Morality and Other Problems of Sense and Nonsense: Essays dedicated to Sören Halldén* (ed. by J. Hintikka *et al.*).

Ekelöf, Per Olof, 1964, 'Free Evaluation of Evidence', in *Scandinavian Studies in Law* (ed. by Folke Schmidt).

Keynes, J. M., 1921, *A Treatise on Probability.*

Reichenbach, Hans, 1956, *The Direction of Time.*

Schreiber, Rupert, 1968, *Theorie des Beweiswertes für Beweismittel im Zivil prozess.*

NOTES

[1] The term 'corroboration' is used throughout in its everyday sense and is not intended to have any Popperian echo. I have profited considerably from discussion of several of the issues involved with Jonathan Adler, Peter Neumann and Wesley Salmon.

[2] Edman claims to report a proof by Hallden of Ekelöf's principle. But the significance of his symbols is neither defined clearly nor illustrated by any example, so that it is not possible to say exactly where the proof goes wrong.

[3] '$p[S, R]$' is here to be read 'the probability of S on R': i.e. Reichenbach's ordering has been reversed.

ROBERT S. COHEN

CONSTRAINTS ON SCIENCE

I

(1) Truths may be used for personal or group interests, as may false beliefs and illusions. So also truth-seeking may be ideological in social function, just as may the concealment of truth. Thus the scope of ideologies is widened, to reach every hope, action and institution. Whatever serves a special interest has its ideological aspect in addition to any other aspect. Not only astrology but also astronomy; not only *kitsch* but also Shostakovitch; not only religion but also rationalist atheism; not only dogmas but also free thought. One of the Christian faith's earliest claims was that the truth shall make the believer free; and the spirit of the 18th century was supported as much by belief in liberation through the truths of science as by belief in liberation through political change. Truth serves as the ideology of liberation; or it may; or it is claimed as such.

(2) At all times, to identify science wholly with the search for truth was too strong a claim, even for those most devoted to science. But those who reject traditional doctrines and find speculative metaphysics untrustworthy, unpersuasive, or without clear and helpful meaning, usually turn to science as a guide for dealing with the problems of individual and social life. Typically, they are plainly critical of religion, but they also value investigators who explain the social facts and beliefs of different religions with the help of scientific theories from sociology and psychology. Just as modern natural science replaced the cosmologies of ancient myths and classical philosophies, so social science has promised to replace mythical views of man and society. So the sciences appear to provide the most substantial part of whatever truth mankind can attain in the modern period. They are incomplete always, but this defect is also an opening toward change, development, and correction of error. Incompleteness may stimulate simply more of the same, and thereupon miss all the more radical gaps; or incompleteness may be

R. S. Cohen et al. (eds.), Essays in Memory of Imre Lakatos, 79–86. *All Rights Reserved*
Copyright © 1976 by D. Reidel Publishing Company, Dordrecht–Holland

discovered by sociological, philosophical (as well as technical) criticism, and thereby stimulate radically new thought and practice. But typically, the sciences may be identified with both technical practice and abstract reason, in the separate historical developments of technology and reason, and in their interaction. Indeed, modern science may be understood as a fusion of the advanced technological arts with a geometrized and newly dynamic mathematics.

(3) What gives science its appeal to good and bad men alike is its help in their making, doing and transforming; first and fundamentally transforming the natural surroundings, but soon the social environment of other human beings as well. The mark of scientifically known truths is that they are established through such deliberate action, practical or 'pure', as it may be. Science does not necessarily contradict ancient wisdom, or religious hopes, or philosophical theories of man. It supplements them with a power that technology alone was finally unable to provide, despite its magnificent achievements in the great civilizations of China, India, Africa and Islam, and the traditional European development to Leonardo.

Neither were philosophy despite its most abstract logical power, nor mystical nature-religions and number-mysticisms, able to do what science finally makes possible. Science makes it possible to transform philosophy of man from mere philosophy into practicality. This achievement is itself not yet real but only possible, and the question still remains: does science go beyond philosophies? As physics replaced speculative philosophy of nature, and chemistry replaced mystical alchemy, and experimental and physiological psychology develop to displace unscientific philosophies of the mind, so may we say that modern science generally replaces philosophy?

Indeed philosophies vary greatly from time to time, epoch to epoch, serving to articulate the values and world-views of those who hold them. While not relativized to individual preferences and hopes, philosophy is nevertheless the historically situated testimony to social and cultural relativisms of viewpoint, morality, hopes and fears with accompanying harmonized conceptions of the world. Reality is seen differently from different social places; and it is unclear whether the different conceptions of the world shall each be counted as knowledge, if only partial knowledge. Now, even when pure philosophy is sophisticated and logically self-consistent, internally structured and dialectically evolved, it provides no workable instruments for transforming nature. With respect to nature, philosophy classically must project its concepts upon nature by metaphors drawn from social structure and technical practice. With respect to human nature, philosophy oscillates between proto-science, empirically based, and plain reflective justification either of existing relations of dominance and submission, or of existing desires for change.

And yet beyond these limitations due to partiality of projection and to ideological teaching, philosophers repeatedly have returned to the genuinely humane ideal of discovering what is reason, what is a reasonable way of life, and what is the way, if there be one, toward establishing reasonable relations among human beings in society. That this effort occurs, and recurs, poses a problem: What converts the philosophical endeavor from articulation and investigation of class and self interests to an inquiry into selfless classless all-human interests? What provokes self-criticism? Under what conditions does species-interest prevail?

(4) To know what is reasonable is also to recognize what is irrational; the philosopher as rational critic has a social function quite different from that of the philosopher as spokesman for any particular group interest. Since concepts of rationality have their history too, what is rational for one era may no longer serve at another time. And yet, what is self-defeating, even contradictory, in the processes of a society (just as what is self-frustrating, even neurotic, in the behavior of a person), will be understood as irrational as judged by differing political and cultural times. The productive link between science and philosophy, then, is provided by their common belief in rationalism, which means genuine potentialities, and hence the real possibilities of a change from whatever are the present observed facts of life to new ones, equally natural but possibly humanly more fulfilling. There are different links between science and philosophy, but the productive link is genuine: hope made possible, and technique judged. Where science goes further than technology is in generalization, beyond any particular or any 'typical' concrete local reality. Science, taken as a whole, is distinguished from technology as much as from philosophy, by virtue of this universalizing understanding. Traditional technology rested upon practical success of trial and error, rules of thumb, while science trans-

cends both common-sense perceptions and even the craftsman's most subtle technological knowledge. Using its intellectual main strength of hypothetical thinking (with experiments or not), science locates realistic potentials and choices within nature as known by mankind. And thereby science brings the objective material power of transforming natural objects, so dearly achieved by previous technology, into a new stage of transformation not only of specific objects but of nature. Purposeful action was always characteristic of mankind, but with the modern age of science and industry, of geographical exploration and theoretical inventiveness, human actions became deliberate, consciously so, became choices among objective alternatives. Science both transforms nature and remains objective. Here, to be objective is to be faithful to the material characteristics of nature changing forms and structures within objectively general laws. This is modern science, a novel and advanced way of living with nature, with new materials, processes, behavior, animals, plants, a new human environment.

(5) Previously, technology showed the historically changing relations of man with nature. And likewise technological divisions of labor entailed the class divisions of social life, at least in fundamental outline. Now, after nearly two centuries of increasingly close mutual support of science and technology, it is science which leads, and now there is the different relation of modern industrial man with nature. But science changes too. Although still consciously devoted to learning the processes of nature, still cognitive and curious, as well as practical, science has become a new productive force of society, now the mechanized, energized, late industrial society of the 20th century. It is modern industry, as economic historians have demonstrated, which vastly amplified the utility of science so that science now leads a split existence as practical and as visionary. Philosophy has been expected to serve two functions: propaganda which educates and codifies how things are, and prophecy which uncovers what is hidden and inspires to what might be. It remains to ask whether science can any better serve two social functions than did philosophy?

II

(6) Simply put, science is limited in several ways by social factors,

partly constrained, partly focussed, partly distorted. To be sure, it is also socially liberated, enabled to be partly autonomous. Just as simply and surely, the socially imposed constraints and freedoms have been there all along, from classical antiquity through the ages of Galileo, Leibniz, Helmholtz and Bohr. The history of science is a social history, with respect to problems attacked, tools, techniques and instruments, metaphors and symbols, criteria of explanation, and the specificity of purposive practice which tests, checks, rejects or corroborates. But its social-historical genesis and continued social support need not compel us to identify scientific truth with scientific usefulness. Only for science, among the several truthseeking cultural enterprises, must the test of the truth of a judgment be practically (as well as theoretically) distinguished from the evaluative test of its importance for human purpose. If we say that science is ideological, it is not because scientific judgments are wilfully or intrinsically delusive, only that science serves powers and interests aside from truth alone, and here too not always wilfully. Scientifically established truths serve certain purposes; scientific methods have been developed to establish those truths of nature; and the purposes, since European feudalism, have been the mastery of nature so as to make of the natural order not a super-tool but a bodily organ of what Marx called the metabolism (*Stoffwechsel*) of man with nature. Whether the new metabolic system has been a success is problematic, but this new place of man in nature which produces a partly new nature-for-man seems unique to the scientific era.

(7) Whether any specific scientifically achieved truths are in fact of human use is a matter of empirical inquiry as well as philosophical judgment, and the use will change as social needs do. Nor is the question only one of whether any given truth is useful, for beyond mere utility is the human issue of repression and liberation. For nature is immensely flexible, with alternative possibilities of development. Benevolent fulfillment of individual life is consistent with what is known of human and social nature; but so is sadistic satisfaction of some individuals along with repressive and murderous degradation of others. It is an ideological half-truth that truth is intrinsically liberating. Even the mastery of natural laws does not assure power to fulfill every human purpose. Nor is the mastery clear-cut even when it has occurred. A century ago Engels wrote: "Let us not be very hopeful about our human

conquest of nature. For each such victory, nature manages to take her revenge".

(8) Nor is this all. The mastery of nature through the collaboration of science with modern technology is not yet mastery by all the people, but only by a few. Judged by the specific situations of daily life, the individual has access to power over nature only within narrowly circumscribed conditions, themselves determined primarily by the authority structures of property relations and administrative powers. Decisions to use technological mastery or to initiate scientific inquiries into their own problems are not made by ordinary individuals, and in fact individuals are more likely to have been victims, or at most passive recipients, than conscious agents or actual masters. As a Marxist might, so the Christian theologian C. S. Lewis concluded that "what we call Man's power over Nature turns out to be a power exercised by some men over other men with Nature as its instrument". Nature-for-man turns from the organic metaphor of metabolism to the craft and weapon metaphor of an instrument; no surprise that critics of science speak scornfully of 'instrumental reason'.

On balance, science is a servant of power, an intrinsic part of the intellectual culture of the dominant forces of modern societies, and integrated within the productive apparatus in both the superstructure and the base, as *both* cognition and instrumentality. Being an aspect of power, science must be judged along with that power. When those in power need truth, they will sponsor science; when they need partial truth, then they sponsor incomplete science; when self-delusion, then pseudoscience; when deception of others, then the half-truths of manipulative social-psychological science. And when the social structure is divided among interests and classes, scientific inquiries may be differential. Linked as it has been to differing human interests, science may be conservative or revolutionary. Scientists are a product of this society and not of previous societies, as surely as factory workers are; and yet the potentially classless truths of literature and art have their analogue in the achievements of science. Science is thereby doubly analogous: first, to the class positions of its sponsors and its workers, whether rulers or ruled; and second, to the universal human, or potentially classless, position of its discoveries. In the long run, these two might combine in the lovely prospect of an all-human democratic society without rigid distinctions of race or economic class role. But even to state this at this time seems utopian, and delusive. The democratic prospect is thereby ideological too, but its interest is not wholly pernicious, even while the utopian spirit may have the practical outcome of blunting consciousness of real and present conflict.

(9) Science has triumphed in field after field, with immense practical results, and with its own internal dynamic of theoretical crises and some striking cognitive achievements. But the general crisis of this century is one of political economy, not of science. In this crisis, science is one of several factors which have not, and could not, carry out their apparent functions. Despite materials, skills, mechanisms, the world is not benefited as it might be; the wealth of scientific achievement is beset and then overwhelmed by the fate of other forces of production, for it is applied in gross disharmony with the real needs of human beings. Nor is it at all certain that scientists themselves are able to diagnose the causes and resolutions of this discord, unless and until they are bound to different social interests or find themselves developing a new consciousness of their own. It is more likely that the competition between critical scientific self-awareness and the ideology of establishment servitude will continue along with liberal statements of conscientious concern, such as the honorable guilt of Snow's *Two Cultures*; allocation of intelligence and material resources will be dominated by the established order. A change in that allocation will only come with a change in the established order.

The ideological character of our scientific civilization consists far less in error or lies than in obscurity about deeply important problems side by side with clarity about whatever is pragmatically directed or contracted. Ideology within science "comes to light in what science closes its eyes to" (Horkheimer, 1931). The human truths to which scientific eyes have mainly been closed might be accessible just as the rest of nature so far has been, for mankind is as natural an object for study as every species even though changed by the self-transcending effects of being the only species with a qualitative history. But in any effort to perceive the sources of its own unfulfilled application to the liberation of individual lives, science is still deeply ensnared by historical roots and contemporary masters. The basic ideology of science is mastery of nature, a basic component of modern society. But that mastery must be mastered, by philosophy and politics, which is to say, by insight, activity and change.

Boston University *

NOTE

* These brief notes are based on my lecture at the 5th International Conference 'Science and Society', Dubrovnik 13 July 1973, as published in *Analysen und Prognosen*, September 1973. The last phrase reflects a theme of Bernard S. Robbins.

GENE D'AMOUR

RESEARCH PROGRAMS, RATIONALITY, AND ETHICS

In footnote 122 of his 'History of Science and its Rational Reconstructions', Imre Lakatos states, "Let me point out here that the methodology of research programmes may be applied not only to norm-impregnated historical knowledge but to any normative knowledge, including even ethics and aesthetics." He goes on to suggest that such an application would supersede the naive falsificationist 'quasi-empirical' approach as carried out by J. W. N. Watkins in his 'Negative Utilitarianism'. In this article I shall attempt to show that a rational reconstruction of at least some theories in normative ethics is possible.[1] Professor Lakatos' criterion of rationality is based on his criterion of demarcation between science and pseudoscience. Hence, I shall begin my argument by exposing the latter criterion.

Lakatos' criterion is neither inductivist nor justificationist: he does not advise the scientist to make observations and infer theories or to put forth theories and attempt to confirm or 'justify' them. Rather, it is deductivist and critical: he sees the scientist as one who proliferates theories in the form of a research program and severely tests these research programs by deducing risky statements from them.

A research program is composed of a 'hard core' theory, or theory which is protected from falsification, and a series of 'auxiliary' theories or hypotheses which protect the hard core theory from falsification. Lakatos entitles such a series of auxiliary theories 'The Protective Belt.' Theories never confront reality alone but in the context of a research program. Thus, one does not accept or reject theories but research programs. Research programs are not just any 'patch-work-quilt' of theories but have a certain continuity brought about by what Lakatos calls a "negative and positive heuristic." The negative heuristic involves the prescription not to direct the *modus tollens* in an encounter between a research program and a basic statement toward the hard core but rather to an auxiliary hypothesis. The positive heuristic, on the other hand, protects the research program from getting bogged down in a sea of

R. S. Cohen et al. (eds.), Essays in Memory of Imre Lakatos, 87–98. *All Rights Reserved*
Copyright © 1976 by D. Reidel Publishing Company, Dordrecht-Holland

anomalies by suggesting how one might go about changing and sophisticating the protective belt.

The risky statements deduced from a research program in order to test it are entitled 'basic statements.' A basic statement is a statement which meets the formal requirements that it be a singular statement (a statement which could not be derived from a theory without initial conditions, and whose negation is derivable from some theory) and the material requirement that it be generally accepted as true by the scientific élite (professional scientists) after simple, controlled experiments (experiments which are simple enough that the risk of being mistaken is minimal). Lakatos believes there is no hard and fast line between theories and basic statements; basic statements are always 'theory laden.' Lakatos is a conventionalist with respect to basic statements: Basic statements are not provable.[2]

The heart of Lakatos' demarcation criterion is best exposed in his own words.

> Let us take a series of theories, T_1, T_2, T_3, where each subsequent theory results from adding auxiliary clauses to ... the previous theory in order to accommodate some anomaly, each theory having at least as much content as the unrefuted content of its predecessor. Let us say that such a series of theories is *theoretically progressive* (or 'constitutes a *theoretically progressive problem shift*') if each new theory has some excess empirical content over its predecessor, that is, if it predicts some novel hitherto unexpected fact. Let us say that a theoretically progressive series of theories is also *empirically progressive* (or '*constitutes an empirically progressive problem shift*') if some of the empirical content is also corroborated, that is, if each new theory leads us to the actual discovery of some *new fact*. Finally, let us call a problem shift progressive if it is both theoretically and empirically progressive, and degenerating if it is not. We 'accept' problem shifts as 'scientific' only if they are at least theoretically progressive; if they are not, we 'reject' them as 'pseudo-scientific.' Progress is measured by the degree to which a *problem shift* is progressive, by the degree to which the series of theories leads us to the discovery of novel facts.[3]

It is important to note that the above quote expresses not only Lakatos' demarcation criterion but also the broader notion of scientific progress. To be a science, it is *only* necessary that a research program be *theoretically* progressive. To make scientific 'progress' a research program must be both theoretically *and* empirically progressive. Progress is *not* a necessary, though it is a desirable, characteristic of science.

Lakatos is also a conventionalist with respect to research programs.[4] A research program can be tested (though, like a theory, it never confronts reality alone but only against a competing research program), but can never be justified or falsified for once and for all. It can however be 'overtaken' by another research program, and in such a case, it would be rational to give up or 'shelve' the overtaken research program. Research program *A* would 'overtake' research program *B* if research program *A* is progressing empirically while research program *B* is degenerating empirically, or more specifically, if research program *A* could explain all the unrefuted content of research program *B* while predicting novel facts, facts improbable in the light of *B*, some of which were corroborated.

Lakatos' broader criterion of rationality includes all that his criterion which demarcates between science and pseudo-science includes as science, plus those research programs which meet all the conditions of being a science except for the fact that their basic statements are quasi-empirical, that is, do not refer to observables. Therefore, I shall argue next that some ethical theories meet Lakatos' criterion for being rational, that is, that some ethical theories are part of a quasi-empirical, *theoretically* progressive research program. I shall support this claim in two parts: I shall argue first that there are statements in ethics, namely, the conclusions of ethical counter examples, which meet Lakatos' criterion for quasi-empirical basic statements, and second that at least one ethical research program, namely, act utilitarianism, has gone through theoretically progressive problem shifts.

An oft cited counter example to utilitarianism is the desert island counter example.[5] This is an idealized, imaginary case in which two men are stranded on a desert island and one of them is dying. The healthy man, let us call him Mr. *X*, promises the dying man, Mr. *Y*, that he will do something for Mr. *Y* after Mr. *Y* dies, for example, put up a tombstone for Mr. *Y* in his hometown if Mr. *X* ever returns to civilization. It is further posited that Mr. *Y* has no relatives or friends living, that Mr. *Y* does die, and that Mr. *X* does get back to civilization. Now, since no happy consequences would occur if Mr. *X* keeps his promise and even though the only unhappy consequence which would follow from putting up a tombstone is the small amount of trouble that Mr. *X* would encounter in erecting it, Mr. *X*, according to the utilitarian *theory*, ought not to keep his promise. For, this act would not lead to the greatest happiness for the greatest number.

On the other hand, both utilitarians (when not looking at the example through their 'theoretical glasses' but facing it 'head on') and other researchers in the field of normative ethics (intuitionists, egoists, etc.) will generally agree that the promise *ought* to be kept.[6] Hence, given that it follows from the utilitarian principle in conjunction with some auxiliary hypotheses and initial conditions that this promise ought not be kept, and that the theorists in the normative ethical research area generally agree that the promise ought to be kept, the claim 'Mr. *X*'s promise ought to be kept' provides a counter example or 'anomaly' for the utilitarian theory.

The statement 'Mr. *X*'s promise ought to be kept' satisfies the formal criteria that basic statements must meet. It is a singular statement – it could not be derived from a universal statement without initial conditions. Also, its negation (in conjunction with certain initial conditions) can be derived from a theory, namely, the utilitarian theory. Indeed, an ethical counter example must supply a statement whose negation is derivable from the ethical theory to which it is a counter example.

Before answering the question whether the conclusions of ethical counter examples meet the material requirement of acceptance, it must be realized that in ethics, as in science, those of whom we must require acceptance if a statement is to be deemed 'basic' are those who are involved in the testing of theories, in this case, normative ethical theorists. Given this constraint, the conclusions of ethical counter examples do meet the material requirement. In fact, if professional normative ethical theorists did not agree with the conclusion of a counter example it could not even be called a 'counter example.' Historically, a purported ethical counter example, usually a description of some simple, ideal type situation, is presented to professional normative ethical theorists, and if, after a simple, imaginary experiment, these professionals agree with its conclusion it becomes accepted as a counter example and is often published in the professional literature.[7] Such a counter example will present a problem, an anomaly, for some theory and ethical theorists may work for years trying to patch up, revise, or possibly adopt an entirely new theory in order to take care of this anomaly.[8]

Of course, it might be countered here that not everyone *outside* the professional research area will agree with the conclusion of an ethical counter example. This point, which might be supported by citing anthropological data about the relativity of moral beliefs, is, however, no more relevant in determining whether the conclusions of the counter examples of ethics are truly basic statements than it would be relevant in determining whether the statements which describe the observations of an astronomer are really basic statements.[9] Consider the pandemonium which would occur if researchers in astronomy could not accept statements describing data acquired through telescopic observation as basic unless there was wide-spread agreement among those outside the profession as to what was being seen. The typical non-astronomer would have no idea as to how to differentiate between, say, a planetary nebula and a Strömgren sphere.

The clearest way to see if there are any ethical research programs and if they do go through theoretically progressive problem shifts is to see how ethical theorists handle anomalies, in this case, ethical counter examples. Let us consider, for example, how utilitarians have handled the counter example previously presented. A. K. Stout aptly describes one of the defenses most commonly put forth by utilitarians in response to such a counter example.

> If it is proper to speak of man's 'setting himself an example' when he begins to act in a way which develops into a habit, we could then say that by breaking a rule he sets an example which may be followed by himself or by others, and which may also be 'stretched' by himself or others. We need not deny that a man can set himself an example. The point is that there are psychological laws under which the original action that breaks the rule can contribute to produce further breaches by the agent himself or by others, either in the same or in a wide range of circumstances.
>
> The conscientious rule-breaker may be quite clear about why the particular case is exceptional, and he may be strong-minded enough to be in no danger of relaxing his principles and stretching the exception on a future occasion. But (as with ordinary habitforming) he takes the risk of believing himself stronger or clearer-minded than he really is. In any case, he may by his example influence others who are not at all clear about it. They will notice that he is breaking the rule, rather than *why* he is breaking it. His right action will thus help to cause wrong actions to others.[10]

Thus, Mr. *X*'s breaking his promise would result in unhappy consequences and the utilitarian theory is saved.

In this reply, the utilitarian is setting forth a psychological hypothesis concerning the effect upon oneself of an instance of breaking a rule and a socio-psychological hypothesis concerning rule-following behavior. Such hypotheses result in novel and risky predictions. A. K. Stout, among others, has noted, "it is easy to exaggerate the influence of example, especially of a single breach of rule."[11]

This type of response to a counter example is quite typical not only among act utilitarians but among *normative* ethical theorists generally. Because the handling of the desert island counter example reflects the typical way act utilitarians have handled counter examples, act utilitarianism in conjunction with its auxiliary psychological and sociological hypotheses may be seen as an example of a theoretically progressive research program in which the hard core is occupied by an ethical theory and the implicit negative heuristic prescribes that we direct the *modus tollens* away from it. The guiding belief of the positive heuristic seems to be that such things as dispositions to obey rules are 'threshold phenomena,' i.e., the disposition to break a rule is not broken down little by little but all at once. Such a heuristic has wide-ranging effects on one's view of such psychological factors as guilt and such sociological factors as institutional behavior. The act utilitarian research program is theoretically progressive because it deals with anomalies in a non *ad hoc* manner; that is, by proposing auxiliary hypotheses which result in novel and risky predictions. According to Lakatos' criterion then, act utilitarianism provides one example of an ethical research program which is rational.[12]

One might reply here that the fact that these sociological and psychological predictions are made by an ethical theorist does not really have much bearing on the status of ethical theories. Even if the auxiliary hypotheses turn out to be true, this surely does not prove the ethical theory to be true, let alone rational. With this I agree. The proposing of auxiliary hypotheses which lead to novel and risky predictions does not make an ethical theory rational. Given the Lakatosian criterion, it is inappropriate even to speak of a theory being 'rational'; only the kinds of moves that take place in the context of a research program can be appropriately labeled 'rational.' The foregoing reply, then, does not create problems for the claim that some ethical *research programs* are rational, that is, that some ethical theories are part of a quasi-empirical, theoretically progressive research program. I also agree that even if these auxiliary hypotheses turn out to be true this would not prove the hard core ethical theory to be true. This is not, however, a situation peculiar to ethical research programs. For, in *any* kind of research program there is always the possibility that it will progress even though one or more of the theories it incorporates are false. Indeed, this is one of the insights of conventionalism: true predictions may follow from false theories. The best we can do is work within a progressive research program and hope that none of the theories it incorporates are false. Of course, an ethical research program *E* could be considered falsified (for the time being) if it is overtaken by a research program which has excess empirical content, some of which is corroborated, and explains the unrefuted content of research program *E* (including the normative content).

One might still reply that there is no real continuity in such a program since, given the fact-value gap, the program incorporates two logically different types of statements. To this my reply is that even if psychophysical dualism is true, it would surely not be the case that a research program in astronomy which utilizes psychological theories of vision would therefore be called "irrational."[13] Further, the psychological and sociological auxiliary hypotheses of the act utilitarian research program are not put forth in an arbitrary or hodgepodge manner, but, as mentioned above, are guided by an implicit negative and positive heuristic. My account, though not denying the naturalistic fallacy, shows how it is possible to have fact and value statements related in an evidential context. Though there may be a logical gulf between the two types of statements, there is no reason why the two cannot complement each other, with pay off for both, in a progressive research program.

Finally, one might still have reservations concerning my account since no theoretically progressive ethical research program, including act utilitarianism, has also been empirically progressive. That is, the novel predictions of ethical research programs have not been corroborated and hence such research programs could not be said to be involved in progressive problem shifts. Although this is definitely a minus for the field of ethics, it is not a minus for my claim. For, given our commitment to the Lakatosian criterion, we can accept this lack of progress without having to conclude that these ethical research programs are irrational. On the other hand, there is the possibility that the predictions of an ethical research program such as act utilitarianism might be corroborated, that, as C. West Churchman states, "Ethics has been a backward science only because it has lacked the kind of empirical evidence it requires for its advancement."[14]

West Virginia University, Morgantown

NOTES

[1] An earlier version of this paper was read at the eastern division meeting of the American Philosophical Association in December 1973. The generous help in the preparation of this paper given me by Imre Lakatos before his regrettable and untimely death, as well as that of J. W. N. Watkins, J. Howard Sobel, Keith Gunderson, Theodore Drange and Ralph Clark is most gratefully acknowledged. The article by Lakatos quoted above appeared in Buck, Roger C. and Cohen, Robert S. (eds.), *Boston Studies in the Philosophy of Science*, Vol. III, D. Reidel Publishing Company, Dordrecht, Holland, 1971, p. 132. Watkins' article appeared in the *Aristotelian Society Supplementary* **37** (1963), 95–114.

[2] This is not to say, however, that basic statements are neither true nor false. This would conflate conventionalism and instrumentalism.

[3] Imre Lakatos, 'Falsification and Methodology of Scientific Research Programmes', in Imre Lakatos and Alan Musgrave (eds.), *Criticism and Growth of Knowledge*, Cambridge University Press, 1970, p. 118.

[4] This does not, however, imply that a research program is neither true nor false in any sense. It might be further mentioned that Lakatos differs from traditional conventionalism in basically three ways: first, he is a radical conventionalist in that he not only rejects the possibility of theories being proven true or false, but also of basic statements being proven true or false. Second, the decision to accept or reject a theory is not left up to common sense or some type of subjective taste, but is sternly regulated by clear-cut, objective rules. Finally, Lakatos does posit an extra-methodological inductive principle in an attempt to tie up acceptance and rejection with verisimilitude. [For more on this final point see Imre Lakatos, 'Popper on Demarcation and Induction', in H. Lenk (ed.), *Neue Aspekte der Wissenschaftstheorie*, Vieweg 1971, Section 2, b.]

[5] Variations of this counter example have appeared many times in philosophical literature. I believe the first appearance was in W. D. Ross's, *The Right and The Good*, Oxford, 1930, pp. 34–39. The first reaction to it came from W. A. Pickard-Cambridge, 'Two Problems about Duty', *Mind*, **41** (1932), 145–172. In more recent times it can be found in J. J. C. Smart, 'Extreme and Restricted Utilitarianism', *Philosophical Quarterly* **6**; D. H. Hodgson, *Consequences of Utilitarianism*, Oxford 1967; Jan Narveson, 'The Desert Island Problem', *Analysis* **23**, 63; and James Cargile, 'Utilitarianism and the Desert Island Problem', *Analysis* **25**, 23–24.

[6] For an interesting discussion of the counter example move by normative ethical theorists see Richard W. Eggerman, 'Moral Theory and Practicality', *Ethics* **84**, 78–85. This is an extremely interesting move. It suggests that there are certain cases in which no theory is needed to make an ethical judgment and that, in these cases, an ethical theorist is actually willing to consider the possibility that his theory is false. It is similar to the move made by the scientist who accepts the fact that there are certain cases which are so simple that we can judge as to the truth or falsity of statements describing these cases without appealing to a high level theory and who is willing to agree that such a case may provide an anomaly to his theory. It is the above described move by the ethical theorist which, I believe, more than anything else suggests the objectivity of ethical data. It calls to mind the words of Popper that:

> "It is only the idea of truth which allows us to speak sensibly of mistakes and of rational criticism and which makes rational discussion possible – that is, to say, critical discussion in search of mistakes with the serious purpose of eliminating as many of these mistakes as we can, in order to get nearer to the truth. Thus the very idea of error – and of fallibility – involves the idea of an objective truth as the standard of

which we may fall short." Karl Popper, *Conjectures and Refutations*, Harper & Row, 1963, p. 229.

[7] Using imaginary experiments with ideal type cases for the purpose of testing a theory is by no means foreign to either the social or physical sciences. For a thoughtful discussion of the use of imaginary experiments in science see Thomas Kuhn, 'A Function for Thought Experiments', in *Melanges Alexandre Koyre, L'Aventure De L'esprit*, Vol. II (ed. by Rene Taton and M. Cohen), Hermann, 1964, pp. 307–323.

It is interesting to note the similarity between the basic statements of ethics and the basic statements of linguistics as described by Noam Chomsky in Chapter I, Part 7 of his *Aspects of the Theory of Syntax*, M.I.T. Press, 1965.

[8] Some Utilitarians who used to take counter examples seriously appear now to be changing their minds. J. J. C. Smart, for example, in 'An Outline of a System of Utilitarian Ethics' (J. J. C. Smart and Bernard Williams, *Utilitarianism: For and Against*, Cambridge University Press, 1973) states: "It is also necessary to remember that we are considering utilitarianism as a normative system. The fact that it has consequences which conflict with some of our particular moral judgments need not be decisive against it. In science general principles must be tested by reference to particular facts of observation. In ethics we may well take the opposite attitude, and test our particular moral attitudes by reference to more general ones." Smart is somewhat ambivalent about this claim, however, as he later states: "The utilitarian, then, will test his particular feelings by reference to his general principle, and not the general principle by reference to his particular feelings. Now while I have some tendency to take this point of view I have also some tendency to feel the opposite, that we should sometimes test our general principles by how we feel about particular applications of them." Further, he does give serious consideration to a number of counter examples that historically have plagued utilitarianism. I believe Smart's real worry is not about what I have called "ethical basic statements" but rather about what he calls "the common moral consciousness". In his 'Extreme and Restricted Utilitarianism' (Phillipa Foot, *Theories of Ethics*, Oxford University Press, 1967) Smart states: "How are we to decide the issue between extreme and restricted utilitarianism? I wish to repudiate at the outset that milk and water approach which describes itself sometimes as 'investigating what is implicit in the common moral consciousness' and sometimes as 'investigating how people ordinarily talk about morality'. We have only to read the newspaper correspondence about capital punishment or about what should be done with Formosa to realize that the common moral consciousness is in part made up of superstitious elements, of morally bad elements, and of logically confused elements. I address myself to goodhearted and benevolent people and so I hope that if we rid ourselves of the logical confusion the superstitious and morally bad elements will largely fall away. For even among goodhearted and benevolent people it is possible to find superstitious and morally bad reasons for moral beliefs. These superstitious and morally bad reasons hide behind the protective screen of logical confusion. With people who are not logically confused but who are openly superstitious or morally bad I can of course do nothing. That is, our ultimate pro-attitudes may be different. Nevertheless I propose to rely on *my own* moral consciousness and to appeal to *your* moral consciousness and to forget about what people ordinarily say." The idea stated in the last sentence of this quote is reiterated in Smart's 'An Outline of a System of Utilitarian Ethics': "In setting up a system of normative ethics the utilitarian must appeal to some ultimate attitudes which he holds in common with those people to whom he is addressing himself." What I have called "ethical basic statements" have little to do with the 'common moral consciousness' but, I suspect, a good deal to do with Smart's readers who are primarily normative ethical theorists. Smart's position does not, therefore, appear to be inconsistent with the position I am suggesting.

[9] Of course, the ethical subjectivist might claim that even if one could call the conclusions of ethical counter examples "basic," they still differ from the basic statements of science in that the agreement on ethical counter examples is not objective. This position is usually stated in one of two ways:

A. Whereas non-scientists may disagree concerning the basic statements of science, they could be brought to agreement if they were apprised of all the facts and trained to be competent observers, while it is *logically possible* for two people to be apprised of all of the facts and still disagree about a basic value judgment.
B. Whereas non-scientists may disagree concerning the basic statements of science, they could be brought to agreement if they were apprised of all the facts and trained to be competent observers, while it is *physically possible* for two people to be apprised of all the facts and still disagree about basic value judgments.

Both of these subjectivist claims are far too strong as we could not even have objective basic statements in science if we could only admit statements which pass these tests. Concerning claim *A*, it is by no means difficult to conceive of a case in which two irrational, stubborn people disagree over some simple basic statement in, for example, physics which everyone in the research area readily accepts. With respect to claim *B*, there is no evidence which would rule out the physical possibility of two observers who meet the conditions specified in *B* disagreeing over a basic statement in science. Further, even if such an occurrence did take place (and, I would suggest, it does take place a lot more than most people realize) we surely would not say that the basic statements of science are not objective.

One might respond here by weakening the ethical subjectivist position in the following way: whereas with respect to the basic statements of science it is generally the case that *most* reasonable, medically sound, trained observers, when well informed about all relevant information and after careful observation, will agree on their truth or falsity, with respect to the basic statements of ethics it is *often* the case that two reasonable, medically sound, trained observers who are well informed of all the relevant facts and have carefully considered the case, will disagree concerning their truth or falsity. Put in this way, the oft camouflaged empirical content of the subjectivist claim becomes transparent and the question readily leaps to mind "How could such a claim ever be empirically established?" Of course the answer to this question also comes quickly: "By the evidence provided by Cultural Value Relativism."

This answer, however, will not do. For, cultural value relativists have come nowhere near to proving a claim such as the one set forth above. First, although there is much data on moral disagreement, there is none concerning disagreement over simple moral judgments in ideal cases such as those found in ethical counter examples. Rather, cultural value relativists look for moral agreement with respect to such complex and sophisticated concepts as happiness, justice, killing, etc. Consider what would happen in an analogous situation where a physiologist was attempting to show that all 'normal' people agree with respect to color judgments. He would immediately run into insurmountable problems if he began his study by asking questions about such sophisticated color difference as that between puce and maroon. [For further elucidation of just such problems see B. Berlin and P. Kay, *Basic Color Terms: Their Universality and Evolution*, University of California, 1971.]

Second, in the revised subjectivist hypothesis we have the occurrence of such concepts as 'reasonable,' 'medically sound,' 'trained observer,' 'well informed,' 'carefully considered,' etc., and, I would ask, when has any program of cultural value relativism ever incorporated any concepts such as these (except possibly "being well informed concerning all the relevant data") into a research program? Although such criteria are never used, they are surely important. Consider again our case of the physiological research program in color agreement. Surely such a program would never get off the ground if the researcher could not 'select out' those who were irrational-bordering-on-insane, or those who were color-blind, or those who did not know what to look at when asked to look at a color or those who had such a short attention span that they couldn't look at a color carefully. All of these factors would be essential for such research.

One might ask at this point "But what could ever count as selecting criteria in such a research program?" Possible answers to this question have been proposed repeatedly throughout the history of ethics. An ethical 'observer' must: be clear about the concepts involved in the case under consideration (E. F. Carritt, *Ethical and Political Thinking*, Clarendon Press, 1947, pp. 5–6.), have knowledge of all the relevant facts (W. D. Ross, *Foundations of Ethics*, Oxford University Press, 1939, p. 17.), be impartial (E. F. Carritt, for example, declares that most men are prejudiced by their personal moral beliefs in his *Ethical and Political Thinking, op. cit.*, p. 8. Also see Roderick Firth, 'Ethical Absolutism and the Ideal Observer', *Philosophy and Phenomenological Research* **12**, 338.), be disinterested and unemotional (R. Firth, *op. cit.*, p. 342.), and/or have sufficient moral maturity (W. D. Ross, *The Right and the Good*, Oxford University Press, 1930, p. 78; Aristotle, *Nicomachean Ethics*, Martin Oswald, Trans., Bobbs Merrill, 1962, pp. 160 and 167.). Some of these criteria have even been worked out in scientific research programs. (See for example the work of Lawrence Kohlberg, 'Stages and Sequence: The Cognitive-Developmental Approach to Socialization', in D. Goslein (ed.), *Handbook of Socialization Theory and Research*, Rand McNally, 1969, p. 352; 'The Hierarchical Nature of Stages of Moral Judgment', in L. Kohlberg and E. Ruriel (eds.), *Recent Research in Moral Judgment*, Holt, Rineholt and Winston, 1971; Jean Piaget, *The Moral Judgment of a Child*, Free Press, 1948.)

Of course, one might further ask "How could such theories of 'observation' ever be independently tested without presupposing some ethical theory?" This question requires a two-part answer. First, it is true that such theories could never be independently tested. This is not, however, a situation peculiar to ethics. For, as has been pointed out by Lakatos, no theory ever confronts reality alone but only in the context of a research program. Ethical theories of 'observation' would have to be embedded in a research program of cultural value absolutism. Such a research program might present simple, ideal type cases (such as those put forth in ethical counter examples), disqualify those in a culture (possibly a whole culture) who do not meet the 'observational' requirements, and then see if uniform responses are obtained. Of course, it must test its disqualification judgments from time to time to show that it has disqualified the right people for the right reasons. That is, if an 'observational' theory declares that Mr. *X* disagrees with judgments concerning ideal case *Y* because he is in imperfect 'observational' state *Z*, it must be the case that if Mr. *X* is brought out of 'observational' state *Z* into a normal 'observational' state, Mr. *X* will come to agree about the status of ideal case *Y*.

Such a research program of cultural value absolutism would naturally meet anomalies, that is, cases in which it could not account for ethical disagreements over basic statement type examples. Its success or failure, as with other scientific research programs, would depend on how it deals with these anomalies; that is, if in the course of its testing it generates new auxiliary hypotheses which account for the anomalies and result in the prediction and discovery of novel anthropological, sociological or psychological facts.

It should now be noted that in the above description of a possible research program which has embedded in it theories of ethical 'observation,' at no point is the truth of a normative ethical theory presupposed. This then, is the second part of the answer to the question mentioned above; the testing of ethical 'observation' theories in no way presupposes the truth or falsity of a normative ethical theory.

[10] A. K. Stout, 'But Suppose Everybody Did the Same', *The Australasian Journal of Philosophy* **32**, 24.

[11] *Ibid.*, pp. 34–35. For a more thorough discussion of some of the psychological and sociological theories that underlie utilitarianism see Rolf Sartorius, 'Individual Conduct and Social Norms: A Utilitarian Account', *Ethics* **82**, 200–218.

It is interesting to note that whereas act utilitarianism takes care of a counter example by putting forth novel empirical hypotheses concerning rule following behavior, rule utilitarianism simply gives a semantical reinterpretation of the utilitarian principle which decreases the empirical content of the utilitarian program and hence results in a theoretically degenerating problem shift. For an obvious example of this '*ad hoc*' strategy by some rule utilitarians see R. F. Harrod, 'Utilitarianism Revised', *Mind* **45**, 148.

[12] Of course, this is not to say that act utilitarianism is the only rational normative ethical research program; in fact, I think egoism also nicely meets this test. Nor do I mean to imply that all ethical research programs are rational. Just as we have certain research programs put forth by scientists which end up in theoretically degenerating problem shifts, so too we have ethical research programs which are degenerating and hence 'metaphysical.'

[13] I might mention here that I believe there is a great deal to be learned about the fact-value gap by looking at arguments about the mindbody gap.

[14] *Prediction and Optimal Decision*, Prentice-Hall, Inc., 1961, p. viii.

YEHUDA ELKANA

INTRODUCTION: CULTURE, CULTURAL SYSTEM AND SCIENCE

Author's note: Imre was a deep thinker and a wonderful friend. He was not an anarchist and he did not toy with ideas. While often acting the buffoon in everyday life, thus misleading many a superficial observer, he was deeply earnest when ideas came up for discussion. In my last few meetings with him I tried to 'sell' anthropological thinking to him and he wanted to show me how wrong I was. I am now working on a slim volume called 'Science as a Cultural System' intended as an Introduction to a more extensive study of Intellectual History which I hope to pursue. This introductory volume will be dedicated to Imre. Here is the first chapter of it: 'Introduction: Culture, Cultural System and Science'.

The slim volume presented here is, as the subtitle suggests, an attempt at developing analytical tools for a general intellectual history of Europe, i.e., of Western culture from the vantage point of science. There are many intellectual histories from different points of view: the most easily acceptable are those of political and social ideas and institutions – some emphasize the central importance of religion, others the deep social role of the arts and especially of visual art; but ambitious histories of Western music also exist (for example, Paul Henry Lang's). Most of them try only superficially to cover their real aim of being a 'total intellectual history'. Some histories of philosophy have the same ambition. Yet, depending on the author's image of knowledge, science is almost always upgraded or downgraded to be in a class by itself – different. Science is rarely considered as much a totality of human culture as art or religion because it is seen as different, unique, all apart. In those rare cases, when science is considered as the whole of human culture it represents an arrogant technocratic view stemming not from viewing science as inclusive of art, religion, etc., but rather by eliminating the other dimensions of culture by treating them as obsolete and irrelevant for modern man.

The fundamental presupposition of this introductory volume and the subsequently planned 'Intellectual history' is that the various

R. S. Cohen et al. (eds.), Essays in Memory of Imre Lakatos, 99–107. *All Rights Reserved*
Copyright © 1976 by D. Reidel Publishing Company, Dordrecht–Holland

dimensions of culture: religion, art, science, ideology, common-sense, music, are all on a par – they are all cultural systems. Moreover, that approach to culture according to which it can be viewed as a sum of the dimensions like a cake called Culture cut into slices of Religion, Art, Science, is rejected here. Even allowing for minor distortions I do not believe that such a summation is meaningful in any sense. On the contrary, it seems to me that an interpretative venture can only be aimed at totality by selecting a focus, a vantage-point and from this view the whole culture. Thus, one can consider the whole of Western Culture primarily as Religion which does, however, have aspects relating to art, science, ideology; just as the few really good books on religious history tried to see their subject. One can see the whole of culture through the spectacles of philosophy, or art, or music. For some primitive culture it even sometimes seemed that Art would be one of the few legitimate spectacles; magic used to be the other. I claim this for Science. Any culture, and especially Western or European Culture, can be viewed as the Culture of Science with mystical, religious, artistic, musical, ideological aspects connected with it. Or again, in some cases illumination is gained by seeing the whole of culture as an area of class-conflict or of ideation conflict: take Marxism or T. D. White's 'The Warfare of Science and Christianity'. Naturally no such interpretive essay is complete or total. But, in my opinion, any one such attempt is more illuminating than the summation-of-dimensions-view. Moreover, this approach is an open-ended one: while a cake can be cut up only into a finite number of slices, and their size diminishes as their number grows, there is no limitation to piling one upon the other, layers upon layers of 'total view', each from a given chosen perspective.

The title of the book echoes the well-known papers of Clifford Geertz.[1] Geertz's concept of culture

> ... is essentially a semiotic one. Believing with Max Weber that man is an animal suspended in webs of significance he himself has spun, I take culture to be those webs, and the analysis to be therefore not an experimental science in search of law but an interpretive one in search of meaning.[2]

The word 'semiotic' is either misleading, or the analogy of the word is not to be taken seriously: for though the web is spun by man himself, yet it has a reality beyond what we tend to denote as 'semiotic'. We should stick to this reality of the web. If we do, we can now go from the concept of culture to that of a 'cultural system' with greater ease. In his 'Common Sense as a Cultural System' Geertz says:

> ...if common sense is as much an interpretation of the immediacies of experience, a gloss on them, as are myths, paintings, epistemology, or whatever, then it is like them historically constructed, and like them subjected to historically defined standards of judgement. It can be questioned, disputed, affirmed, developed, formalized, contemplated, even taught, and it can vary dramatically from one people to the next. It is, in short, a cultural system...

(p. 5 – the numbering of pages refers to the preprint).

The first strong thesis here with which I fully agree as a basis of any explication of science, is that there is no general theory of culture or of a cultural system: it does 'vary dramatically from one people to the next'; it also varies continuously (whether continuous changes can be considered dramatic or not is a question of taste – I think it can) with time; more specifically it varies as the various forces in the system realign periodically: in science it changes with the periodic redistribution of the whole system into disciplines. *Thus, there is no theory of culture.* What does that mean? What would it imply to affirm the existence of a theory of culture? It would "permit one to apply a template to the varieties of human behaviour and arrive at an 'analysis'".[3] It would be the quest for a better and more general template which ideally would be true for all cultures and all times. It would be then subdivided into sub-theories: a theory of Art, a theory of Science, a theory of Religion, a theory of Ideology, all being dimensions of Culture. Above all, it would include firm criteria of progress in the progressive dimensions of culture, like Science and perhaps Art, and it would supply rules of demarcation between the various dimensions of culture. In short, a theory of culture would in general be what philosophers of science have been after for decades as relates to science. Our thesis thus affirms that no such theory can exist.

The objection could be raised – and judging by his other views, Geertz would welcome that objection – that the above argument applies to culture in general, or to most of its several dimension, but not to Science. The reason being – or so the objection could run – that while religions are many, art is versatile, ideologies vary, values are a multitude, science is only one kind. Naturally there can be pseudo-science, pre-science, non-science, etc., but whatever the real science is, it is only one. It is exactly this view of science which I hope to have convincingly disestablished by the end of this essay. Science is a cultural system like all others.

What aspects of culture – or, better, which of the social institutions of man can be treated as a cultural system? Geertz, as is well-known, wrote an 'Ideology as a Cultural System', 'Religion as a Cultural System', and recently, the above-mentioned 'Common Sense as a Cultural System' (1973). I do not have much doubt that in the above sense, epistemology, myth, magic, art in general, and science too, are cultural systems. All of these are interprative attempts in search of meaning. An interpretive science is not an experimental one; neither is it a science of pure theory, it is something in between. If we accept the usual philosophical dichotomy between theory and experiment, between observation language and theoretical language, then there is no in-between. Geertz rejects this dichotomy for ethnography. Instead there is an interpretive effort going on – a thick description. My thesis is that the same is true for science.

The analytical tool of an interpretive science is 'thick description'. The term is Ryle's[4] – the novel broad usage of it is Geertz's. Thick description is the most elementary daily work of the ethnographer: he interviews informants, observes rituals, elicits kin terms, traces property lines, censuses households... writes his journals.

Thick description for Ryle is a way of describing the complexity of thinking: he starts from the most elementary one-layer activity, like for example, counting the number of cars on the street. Describing this activity involves a very thick description. Then, layer by layer (or step-by-step on a ladder) the activity becomes more complex and its description thicker. Thus, the kind of description we have to give when describing what a person is doing, is somewhere on a continuum between the very thin to the very thick and, the thickness depends on the kind of activity we are describing. In ethnography the problem is always one of translation so whatever the ethnographer describes cannot be a thin description any more. That is Geertz's point.

All this involves "a multiplicity of complex conceptual structures many of them superimposed upon or knotted into one another..." (*ibid.*, p. 10). But this is just what the scientist is doing: formulating problems, choosing phenomena, i.e., demarcating the seemingly self-evident from the seemingly puzzling, observing selected relevant motions, changes, processes, sizes, colours, etc., tracing introductions and connections between previously firmly or vaguely admitted non-connected phenomena; reducing one phenomenon to another and then changing frameworks going vice versa; counting population of a newly determined independent unit... describing his experiments. No doubt the scientist has to deal with a multiplicity of complex conceptual structures mostly superimposed on one another. Kepler, looking at the retrograde movement of Mars, observed the changes of the position of the star in relation to what tradition and authority had taught him were fixed stars; accordingly he believed God's universe to be harmonious as revealed to man in the 'Book of Nature' and compared this with his calculations which he knew to be correct, logic being logic, and calculations being based on it. He also knew from common-sense (albeit a 17th century common-sense) that one should expect terrestrial bodies to behave fundamentally in the same way as do far-away celestial ones and superimposed on this common-sense knowledge (not shared by Galileo's Aristotelian critics) is his theory of optics according to which it is reasonable to expect to see what is out there better with a telescope than without one. This mixture of different sources of knowledge is a kind of thick description. Which source of knowledge is considered legitimate in a given culture, at a given place and time, depends again on the socially defined images of knowledge and have little to do with the body of optics or astronomy, or physics. (This will be outlined in detail in the next chapter.) Again, which of the sources will be primary is context-dependent: if 'celestial harmony' as a metaphysical principle clashes with 'observation' or with 'calculation', or both, it is only multiplicity of superimposed complex conceptual structures in the metaphysics of science and the images of knowledge which can decide which is primary; thus a decision can be made only for a given, narrowly specified case.

Before continuing, let me, in accordance with the hermeneutic circle which this approach represents, focus on the mere outline of a much-discussed case study in history of science: Galileo's use of the telescopic findings for supporting his Copernican conception of the universe. The case will involve a thick description (thus illustrating this chapter) and show that the thick description involves body of knowledge, images of knowledge, ideology and propaganda (thus illustrating the next chapter).

Galileo had a set of legitimate sources of knowledge other than his opponents and even among these the various kinds of knowledge had different priorities than for his opponents, or even for Kepler. He had no theory of optics (Kepler did) and thus his firm commitment to Copernicanism came first and supplied the certainty that what are seen through his tube must be Jupiter's moons. He shared Kepler's anti-Aristotelian non-distinction between terrestrial and celestial phenomena. The problems connected with the work of Galileo are fascinating: what did he know, what did he merely intuit, what experimental evidence was accessible to him? Did Galileo know enough optics to construct a telescope? What were the common-sense expectations of what could be seen through a telescope? What auxiliary optical theories and theories of vision were necessary so as to be able to consider the phases of Venus and the moons of Jupiter – seen with the help of the telescope – as confirmation of the Copernican system of the world? Or, was Copernicanism the primary commitment (in spite of much known empirical difficulties connected with it)? Then, the fact that the telescope showed things which could be interpreted as supporting Copernicus served to convince Galileo of the non-distinction between terrestrial and celestial phenomena or of the truthfulness of optical devices; or, perhaps, of the validity of Keplerian optical theory (which he may even not have known then!).[5]

Is science, then, not at all different from other human artifacts? Is it just like art, religion, magic, ideology of common-sense? A priori, yes. Historically, no. It happens to be that in our time in Western culture, it is the most important single human creation which has heavily been influencing every other aspect of life during the last four hundred years or so. Indeed, in my opinion, it is so central that it justifies an attempt to write an intellectual history as the story of the culture of science. But this non-uniqueness does not rest on a relativistic approach to the various sources of knowledge: according to the socially defined images of knowledge the legitimate sources of knowledge can be: revelation; ratiocination; evidence of the senses; authority; intuition; tradition or values. There is a tendency to identify social institutions with one or at most two legitimate sources of knowledge: religion with revelation; science with evidence of the senses; politics and ideology with tradition and authority and values; interpersonal relationships with intuition; old science' or scholastic philosophy with ratiocination. If one is a relativist then, with respect to the legitimacy of the sources of knowledge, science becomes just 'one of these', I do not agree with that tendency: there is no social institution – or, to use our language – there is no cultural system which relies on a single source of knowledge although, admittedly, at some stages and in some parts of the system, some source of knowledge predominates in being considered the primary 'legitimate source of knowledge'. Thus it is irrelevant whether one is a relativist or not with respect to the primacy of desirability of these sources of knowledge. If religion is, as in Geertz's definition,

'a system of symbols which act to establish powerful, pervasive and long-lasting moods and motivations in men by formulating conceptions of a general order of existence and clothing these conceptions with such an aura of factuality that the moods and motivations seem uniquely realistic[6]

then it involves revelation, ratiocination, reliance on the senses, tradition, authority, all in a mixture. Revelation is important in the choice of symbols; tradition is the source of our ability to decide what is a general order of existence, and we use our sense to ascribe an aura of factuality to some conceptions. That art relies on a combination of sense-impressions, tradition, authority, intuition and perhaps sometimes even revelation is trivially true. According to Parsons if "the problem of ideology arises when there is a discrepancy between what is believed and what can be (established as) scientifically correct"[7] then it is clear that such a discrepancy is neither directly 'observed' nor simply 'thought out' nor independently of tradition just concluded. It is all of them together. Geertz (I disagree with his conception of science) shows, when comparing science and ideology, that ideology too has diverse sources of knowledge.

When science is the diagnostic, the critical dimension of culture, ideology is the justificatory, the apologetic one – it refers to that part of culture which is actively concerned with the establishment and defense of patterns of belief and value...[8]

Before discussing the different sources of knowledge and their legitimization in the various contexts in greater detail, let me close this section with another Geertzian quotation on thick descriptions:

'Hopping back and forth between the whole conceived through the parts which actualize it and the parts conceived through the whole which motivates them we seek to turn them by a sort of intellectual perpetual motion, into explications of one another...[9]

Caveat: The approach which I shall employ here throughout the book is a methodological thick description and it entails again the seemingly circular argument: in order to explain sources of knowledge; thick descriptions; observation/theory distinction; justification/discovery distinction, I already used the concept of images of knowledge and the related view of growth of knowledge which I had. Now I shall try to elaborate this view and the concept of images by relying on thick descriptions, on the non-singularity of the sources of knowledge, and on the non-distinguishability between the contexts of justification and the context of discovery.

NOTES

[1] These are: 'Ideology as a Cultural System' (1964, reprinted in his 'The Interpretation of Cultures', Basic Books, 1973, pp. 193–233) and 'Religion as a Cultural System' (1966, *ibid.*, pp. 78–125), and the more recent 'Common Science as a Cultural System', (Antioch Review, 1972).

While my approach preceded my acquaintance with the work of Clifford Geertz, it was on reading Geertz that my dissatisfaction with other approaches and my groping in this direction, crystallized into the kind of analysis attempted here. It is also when reading Geertz that I felt courageous enough to advocate publicly a 'thick description' which recognized no absolutes. For earlier attempts, see my 'The Discovery of the Conservation of Energy' (Harvard University Press, 1974); 'Boltzmann's Scientific Research Programme and Its Alternatives' in Y. Elkana (ed.): *The Interaction Between Science and Philosophy*, Humanities Press, New York, 1974; and especially, 'The Problem of Knowledge in Historical Perspective', Proc. of the 2nd International Humanistic Symposium, Greece, 1972, pp. 191–247.

[2] C. Geertz: 'Thick Description: Toward an Interpretive Theory of Culture'; Chapter 1 of 'The Interpretation of Cultures', Basic Books, 1973, p. 5.

[3] One of Geertz's reviewers, R. N. Adams in *Reviews of Anthropology* **1**, November, 1974, p. 583.

November, 1974, p. 583.

[4] Who has the right of authorship on terminology? This is a thorny problem in sociology of science. My attitude is that any novel usage of a term which makes it useful or applicable to a broad field is rightly identified with the person who introduced the novel usage: Ryle introduced 'thick description' in his 'Thinking and Reflecting', (1966, reprinted in his Collected Papers, Hutchinson, 1971, Vol. II, pp. 465–479), and in 'The Thinking of Thoughts: What is the 'Penseur' Doing?' (1968, reprinted, *ibid.*, pp. 480–486). Geertz uses Ryle's 'thick description' as a novel tool for the whole of ethnography. In my opinion it is also the correct tool for history and philosophy of science. Thus, I shall refer to Geertz's 'thick description'.

A similar fate has befallen the much-abused concept of paradigm. Wittgenstein used it in his Cambridge lectures. Merton wrote a 'Paradigm for the Sociology of Knowledge' in 1945 (reprinted in R. K. Merton: 'The Sociology of Science' edited by N. Storer, University of Chicago Press, 1973, pp. 7–40). Toulmin used in in his 'Foresight and Understanding' (Bloomington, Indiana University Press, 1961) meaning 'ideals of natural order' (p. 56). Thomas Kuhn made it the core concept of his 'The Structure of Scientific Revolutions' (University of Chicago Press, 1962). Yet, only Merton's and Kuhn's usage became fundamental analytical tools for a whole discipline. Thus, there is a Mertonian paradigm and a Kuhnian paradigm. For more on the history of the usage of this concept, see S. Toulmin: 'Human Understanding' (Clarendon Press, Oxford, 1972, pp. 106–7. On what constitutes novelty in knowledge, see J. Agassi: 'The Novelty of Popper's Philosophy of Science' (*Internat. Phil. Quart.* **8** (1968), 442–463).

[5] See P. Feyerabend: 'Problems of Empiricism, II' in R. Colodny (ed.), 'The Nature and Function of Scientific Theory', (University of Pittsburgh, 1970, pp. 275–353); and criticism thereof: P. K. Machamer: 'Feyerabend and Galileo: The Interaction of Theories and the Reinterpretation of Experience' (*Stud. Hist. Phil. Sci.* **4** (1973), 1–46); and again: P. Feyerabend: 'Against Method' (NLB 1974) where in Chapters 6–11 the argument is repeated and enlarged and there is an attempt to deal with Machamer's criticism in Appendix 2 (pp. 112–128).

In order to be a Copernican, to construct a telescope, and to use the evidence of the telescope as support for Copernicanism, Galileo not only had to disregard some empirical evidence and to proceed counterinductively, but he had to introduce new images of knowledge and new scientific metaphysics – these two are not distinguished by Feyerabend but are called new *natural interpretations* (e.g. circular inertia) and new sensations (e.g. the telescopic data). Actually this is acknowledged by Machamer too: "If Feyerabend's talk about the 'changed sensory core of everyday experience' and the 'redrawing of conceptual lines' is taken in the sense of alternative or new descriptions as I have suggested, I think his insight is of great importance". (*ibid.*, p. 32). Both 'natural interpretations' and 'sensations' involve views about the world and views about knowledge intertwined. More about that elsewhere.

[6] 'Religion as a Cultural System', *ibid.*, p. 90.

[7] T. Parsons: 'An Approach to Sociology of Knowledge', Transl. of the 4th World Congress of Sociology (Milan and Stresa, 1959), pp. 25–49, reprinted in Curtis James E. and John W. Petras (eds.): 'The Sociology of Knowledge', London, 1970, pp. 282–306.

[8] The whole quotation is from C. Geertz: 'Ideology and Cultural System', *ibid.*, p. 231; the quotation in quotation is from Fallers: 'Ideology and Culture in Ugando Nationalism', *Amer. Anthrop.* **63** (1961), 677–686.

[9] C. Geertz, 'From the Native's Point of View', p. 23. This procedure is very similar to what Dilthey called a 'hermeneutic circle', and I have referred to it above.

P. K. FEYERABEND

ON THE CRITIQUE OF SCIENTIFIC REASON

SUMMARY. The historical studies which have been carried out and are being carried out with the help of the methodology of research programmes define two types of relation between research programmes and the evidence. Let me call these types type A and type L respectively. They examine episodes where one research programme R'' replaces another research programme, R' (or fails to be replaced by it), i.e. R' is made the basis of research, argument, metaphysical speculation by the great majority of competent scientists. The authors find that the relation of R'' to the evidence is usually of type L while that of R' to the evidence is of type A (other circumstances being present when this is not the case). Assuming their historical analysis to be correct this is an interesting *sociological law*. The authors do not present their results in such terms, however. Making A and L part of a *normative methodology* they claims to have shown that the acceptance of R″ was *rational* while the continued defence of R' would have been *irrational*, and they express this belief of theirs by calling research programmes exhibiting relation L to the evidence *progressive* while research programmes which stand in relation A to the evidence are called *degenerating*. They also claim that such judgements are *objective*, independent of the whims and subjective convictions of the thinkers who make them. Using such a normative interpretation of their sociological results they also claim to possess *arguments* for and against well known research programmes. For example, they would say that today most versions of environmentalism degenerate and that it is irrational to continue working on them. Fortunately this puritanical superstructure of the otherwise excellent sociological studies need not be taken seriously. The reason is that the superstructure is arbitrary, or 'subjective' in at least five different ways. (1) the basic philosophy behind the normative apprisals makes modern science the source of the standards without giving reasons for the choice; (2) despite all its praise for modern science the methodology of research programmes uses a streamlined version of it without (a) making the principles of streamlining explicit and without (b) arguing for them; (3) the standards that are obtained via the arbitrary steps (1) and (2) are not strong enough to praise any action as 'rational' or condemn it as 'irrational' which means that such judgements are without support from the arbitrarily selected standards; (4) in some of the studies research programmes are selected in an idiosyncratic manner, the purpose being to make the general philosophy appear true (not that such truth would be of much use – see (3)); (5) the attempt to show that competent scientists always acted 'rationally' is not applied to all scientists but only to those whose actions seem to fit into the general methodology (for 'seem' see again item (3)). The superstructure of rationality which is subjective in the five ways just enumerated is supposed to guide scientists while the case studies are to show that the guide has substance – he is not merely a philosopher indulging in abstract dreams of law and order. The alleged substantiality is moonshine and one can reject the standards just as arbitrarily as they have been introduced. In sum: in the case studies we have (a) the discussion of certain sociological regularities; (b) the proposal of arbitrary standards which have no practical force; (c) the insinuation that the regularities are not merely factual, but are features of rationality, that they lend support to the standards and are justified by them. (a) may

R. S. Cohen et al. (eds.), Essays in Memory of Imre Lakatos, 109–143. *All Rights Reserved*
Copyright © 1976 by D. Reidel Publishing Company, Dordrecht–Holland

be accepted with the caution we extend to any new 'discovery' in sociology, (c) must be rejected (and with it the tendentious terminology used in all the case studies), (b) may be accepted, or rejected, depending on mood, the weather etc. Environmentalists, however, may continue on their path, for no argument has been raised against their enterprise.

(1) *Two fundamental questions; only one of them examined by the methodology of research programmes.*

There are two questions that arise in the course of any critique of scientific reason. They are:

(a) *What is science* – how does it proceed, what are its results, how do its procedures, standards and results differ from the procedures, standards and results of other enterprises?

(b) *What's so great about science* – what makes sciences preferable to other forms of life, using different standards and getting different kinds of results as a consequence? What makes modern science preferable to the science of the Aristotelians, or to the ideology of the Azande?

Note that in trying to answer question (b) we are not permitted to judge the alternatives to science by scientific standards. We are now *examining* these standards, we are *comparing* them with other standards rather than making them the *basis* of our judgements. Azande results must be judged by Azande standards and the question is: are these results and these standards preferable to science, or are they not. And if they are not – then, what are the reasons for the deficiency?

Questions (a) and (b) arise with all abstract concepts. We can ask them about truth, knowlegde, beauty, goodness, and so on. In the history of thought answers to question (b) are often taken for granted. For example, it is taken for granted that Truth is someting quite excellent and that all we need to know are the detailed features of this Excellent Thing. This means that one starts with a *word* and uses the enthusiasm created by its *sound* for the support of questionable ideologies (cf. the Nazis on Freedom). A rare exception is the "long standing quarrel between poetry and philosophy" (Plato, *Rep.* 607b6f). The philosophers' case was stated as follows: "If you ... admit the honeyed muse in epic or in lyrical verse, then pleasure and pain will usurp the sovereignty of law and of the principles always recognised by common consent as the best" – i.e. poetry will drive out knowledge. The case of the poets was stated, in a very weak form, in the following story in Plutarch's *Life of Solon*: "When the company of Thespis began to exhibit tragedy, and its novelty was attracting the populace but had not yet gone as far as public competition, Solon being fond of listening and learning and being rather given in his old age to leisure and amusement, and indeed to drinking parties and music, went to see Thespis act in his own play, as was the practice in ancient times. Solon approached him after the performance and asked him if he was not ashamed to tell so many lies to so many people. When Thespis said *there was nothing dreadful in representing such works and actions in fun*, Solon struck the ground violently with his walking stick: 'If we applaud these things in fun', he said 'we shall soon find ourselves honouring them in earnest.'" The italicised passage gives the weak defence of Thespis. (The story seems historically impossible, yet it elucidates a widespread attitude in early Greece; cf. also Chapter 8 of John Forsdyke, *Greece before Homer*, New York 1964). Platon challenges the champions of poetry "to plead for her ...that she is no mere source of pleasure but a benefit to society and to human life" (607d5). The challenge is taken up by Aristotle in his *Poetics* where it is shown that "poetry tends to give general truths while history gives particular facts": poetry, dealing with the *nature* of things, is more philosophical than history (1451b2ff). Note that the champions of poetry are already on the defensive: they defend poetry by trying to show that it *aids* knowledge. They do not defend it as an *alternative* to the kind of knowledge sold by the Presocratics and their followers. In the course of their defence they emphazise historical *tendencies* over individual idiosyncrasies (primacy of the plot in Aristotle) and thus inhibit the advance of freedom. Cf. my *Einführung in die Naturphilosophie*, vol. i, Braunschweig 1977.

There are at least two ways of dealing with question (a). We may use the method of *anthropologist* who examines the behaviour and the ideology of an interesting and peculiar tribe. In this case statements such as 'science proceeds by induction' are *factual* statements of the same kind as statements describing how a particular tribe builds houses, how the foundation is laid, how the walls are erected, what rites accompany the procedure, and so on. On the other hand we may consider *ideal demands* and examine their consequences. Such a procedure is only loosely connected with actual (scientific) practice, and it may be entirely divorced from it. This applies to many investigations dealing with the 'logic' of science. Occasionally the difference is noticed, but emphasised as an advantage: actual science has not yet achieved the purity of an enterprise that agrees with the demands, of a socalled 'rational' enterprise; it must be 'reconstructed' and the reconstruction, obviously, will be different from the real thing. Of course, nobody can say whether a reconstruction, *when inserted into the historical surrounding that gave birth to actual science* will produce comparable results. It may not give any results at all. (Who would expect that one can climb Mount Everest using the 'rational' steps of classical ballet?) We need the anthropological method to find out whether reconstruction improves science, or whether it turns it into a useless though perfect adornment of logic books. The procedure of the anthropologist therefore takes precedence over the procedure of the logician.

There is a third way in which science and, for that matter, any practice can be examined. Considering the standards and the aims of a certain form of life we may ask whether the practice agrees with the standards and whether it leads up to the aim. In this case we compare the results of an anthropological inquiry (the practice, the aims, the standards which have been found to constitute the form of life) with what we know, or think we know about the laws of nature and man's relation to them. For example, we may point out that a certain way of building houses is not very efficient and that houses built in this way cannot last very long. Or we may point out that induction does not get us very far, and does not provide certainty. The first criticism assumes (a) that the builder *wants* to build a solid house and (b) that our knowledge of building houses is at least as good as the knowledge of the culture we are examining. That (a) is not always satisfied can be seen from our own civilization which relies on obsolescence. And as regards (b) it suffices to remind the reader of recent discoveries which show that 'primitive' procedures are often superior to their scientific rivals.[1] The second criticism, the criticism of induction assumes that its users want to get very far (they do not) and that the classes corresponding to universal properties have misleading subclasses (if there are no misleading subclasses then induction will succeed despite the alleged invalidity of the inference from the particular to the general). We see: a true Critique of Scientific Reason cannot take anything for granted. It must examine the most obvious assumptions.

(2) *The excellence of science is* assumed, *it is not argued for. The same is true of the standards proposed by the methodology of research programmes. These standards are obtained by an analysis of modern science. Their excellence is therefore again assumed, it is not argued for. There is not a single argument to show that they are better than the standards that underlie the practice of magic.*

Such a critical attitude is only rarely found among philosophers of science. Scientists have by now gone very far in the revision of basic cosmological ideas and they have come up with some amazing suggestions (subject-dependence of the physical world; synchronicity in addition to causality; telekinesis; non-sensory information-gathering by plants and ability to recognize individuals; non-causal reaction of deep sea organisms to the position of the sun and of the moon; artificial character of the first satellite of Mars; existence of an international astronomy at 30 000 B.C. – and so on). There is no longer any antagonism between the most advanced parts of science and ancient points of view which have degenerated because of scientific warfare. Ancient myths are reconsidered, brought into testable form, examined. Surprising and revolutionary results have been obtained, in the Soviet Union [1a], in China [1b], in the United States [1c]. Speculation on the frontiers of knowledge is often indistinguishable from mythmaking and does not follow any easily recognizable method. There may be law and order in some domains, there is absolutely no law and order in others. It is true that the great majority of scientists is still quite hostile towards such mobility (the National Science Foundation, for example, refuses to support the most interesting research on plant-communication that is being carried out by some members of the Stanford Research Institute in Menlo Park). This is a familiar phenomenon, explainable by prejudice and an anxious commitment to the status quo. On the other hand, one would have thought that philosophers of science, being aware of such developments and being less impressed by scientific orthodoxy than their specialist colleagues might have developed a suitable philosophy, providing additional stimuli for speculation. This has not been the case. Quite the contrary – most philosophers of science still seem to be living in Victorian times when only a few clouds were dimly perceived on a distant horizon. Their craving for orderliness easily exceeds that of the most systematic scientist and approaches that of a catatonic. They have a strong faith in the basic orderliness of science, they have a strong faith in the excellence of (non-dialectical) logic (and this despite the many open problems one finds in this discipline) and they spend their lives trying to find a point of view that enables them to uphold both kinds of faith. In this they often succeed, for 'science' is for them a particular logical system, or set of systems rather than the historical process usually designated by that name, and 'logic' is a very simple and dull part of that discipline, a kind of pidgin logic. 'Problems of Science' however are the internal problems of the chosen system, or set of systems, illustrated with the help of bowdlerised examples from science itself.[2] Kuhn has shown the dreamlike character of the whole enterprise. The work of Lakatos and his followers and the methodology of research programmes that forms their background are an attempt to move from dream to reality *without* any loss of logic and reconstruction. Let us inquire to what extent the attempt succeeds.

We see at once that question (b) remains unanswered. It is of course *assumed* that science is vastly better than any other research programme of comparable scope and generality. But we do not find a single reason in favour of this assumption. Occasionally the assumption enters a detailed argument concerning some different matter, apparently lending it additional force when all we have is a dogmatic and ritual reassertion of the greatness of science.[3]

Thus John Worrall in a position paper on critical rationalism[4] compares the measures which Marxists use to get rid of prima facie refuting instances with the measures used by scientists and he asserts that the former do not lead to any increase of content while the latter do. Had he examined the matter with the care he has spent on his story of Young and Fresnel he would have come to a different conclusion.[5] But let us assume he is right – what follows? We can infer that Marxism does not agree with the standards of science as reconstructed by critical rationalists. We cannot infer that Marxism is *inferior* to science *unless* we have reasons for the standards which are independent of the fact that they are parts of science. No such reasons are found in the philosophy of research programmes. Quite the contrary: it is explicitly stated that science is the measure of method and that good method is the method practiced by the "scientific elite".[6] This, at least, is the theory defended by Lakatos.

According to Lakatos methodologies are tested, i.e. either defended or attacked by reference to historical data. The historical data which Lakatos uses are "'basic' appraisals of the scientific elite"[7] or "basic value judgements"[8] which are *value* judgements about *specific* achievements of science. Example: "Einstein's theory of relativity of 1919 is superior to Newton's celestial mechanics in the form in which it occurs in Laplace". For Lakatos such value judgements (which constitute what he calls a "common scientific wisdom") are a suitable basis for methodological discussions because they are accepted by the great majority of scientists: "While there has been little agreement concerning a *universal* criterion of the scientific character of theories, there has been considerable agreement over the last two centuries concerning *single* achievements."[9] Basic value judgements can therefore be used for checking theories about science or *rational reconstructions* of science much in the same way is which 'basic' *statements* are used for checking theories about the world. The ways of checking depend on the particular methodology one has chosen to adopt: a falsificationist will reject methodological rules *inconsistent* with basic value judgements, as follower of Lakatos will accept methodological research programmes which "represent a *progressive shift* in the sequence of research programmes of rational reconstructions".[10] The standard of methodological criticism thus turns out to be the best methodological research programme that is available at a particular time. So far a first approximation of the procedure of Lakatos.

The approximation has omitted two important features of science.

On the one side basic value judgements are not as uniform as has been assumed. 'Science' is split into numerous disciplines each of which may adopt a different attitude towards a given theory and single disciplines are further split into schools, heresies, and so forth. The basic value judgements of an experimentalist will differ from those of a theoretican (cf. Rutherford, or Michelson, or Ehrenhaft on Einstein), a biologist will look at a theory differently from a cosmologist, the faithful Bohrian will regard modifications of the quantum theory with a different eye than will the faithful Einsteinian. Whatever unity remains is dissolved during revolutions, when no principle remains unchallenged, no method unviolated. In addition there are individual differences: Lorentz, Poincaré, Ehrenfest thought that Kaufmann's experiments had refuted the special theory of relativity and were prepared to abandon relativity in the form in which it had been introduced by Einstein, while Einstein himself retained it because of its comprehensiveness.

On the other hand, basic value judgements are only rarely made for good reasons. Everyone agrees now that Copernicus' hypothesis was a big step forward but hardly anyone can give a halfway decent account of it, let alone enumerate the reasons for its excellence. Newton's theory of gravitation was "highly regarded by the greatest scientists"[11] most of whom were unaware of its difficulties and some of whom believed that it could be derived from Kepler's laws. The quantum theory which suffers quantitative and qualitative disagreements with the evidence and which is also quite clumsy in places is accepted not *despite* its difficulties, in a *conscious violation* of naive falsificationism, but because "all evidence point with merciless definiteness in the ... direction ... [that] all processes involving ... unknown interactions conform to the fundamental quantum law".[12] And so on. *These* are the reasons which produce the basic value judgements whose "common scientific wisdom" Lakatos occasionally gives such great weight.[13] Add to this the fact that most scientists accept basic value judgements on trust, they do not examine them, they simply bow to the authority of their specialist colleagues and one will see that *common scientific wisdom is not very common and it certainly is not very wise.*

Lakatos is aware of the difficulty. He realises that basic value judgements are not always reasonable[14] and he admits that "the scientists' judgement [occasionally] fails".[15] In such cases, he says, it is to be balanced and perhaps even overruled, by the "philosophers statute law".[16] The 'rational reconstruction of science' which Lakatos uses as a measure of method is therefore not just the sum total of all basic value judgements; nor is it the best research programme trying to absorb (or to produce) them. It is a "pluralistic system of authorities"[17] in which basic value judgements are a dominating influence as long as they are uniform *and* reasonable. But when the uniformity disappears, or when "a tradition degenerates"[18], then general philosophical constraints come to the fore and enforce (restore) reason and uniformity.

Now I have the suspicion that Lakatos vastly underestimates the number of occasions when this is going to be the case. He believes that uniformity of basic value judgements prevailed "over the last two centuries"[19] when it was actually a very rare event (here he is in the same predicament as Kuhn who assumes that a particular normal science may have lasted for decades when in fact it was a very rare event). But if that is the case, then his rational reconstructions are dominated either by commonsense,[20] or by the abstract standards of the "philosopher's statute law". Moreover, he accepts a uniformity only if it does not stray too much from his standards: "When a scientific school degenerates into pseudoscience, it may be worthwhile to force a methodological debate".[21] This means, of course, that the judgements which Lakatos passes so freely are ultimately neither the results of research, nor parts of scientific practice; they are part of an *ideology* which he tries to lay on us in the guise of a 'common scientific wisdom'. We discover here a most interesting difference between the *wording* of Lakatos' proposals and their *cash value.* The methodology of research programmes is introduced with the purpose of *aiding* rationalism. It is supposed to find historical support for methodological standards. Such standards are to be grounded in history, not in the abstract discussion of abstract possibilities. But the reconstructions which are to provide the historical support are very close to the abstract methodologies supposedly aided and they merge with them at times if crisis. Despite the difference in rhetorics ("Is it not *hybris* to try to impose some a priori philosophy of science on the most advanced sciences? ... I think it is"[22] – a sentiment that is forgotten the moment Lakatos enters "the most advanced" parts of atomic physics[23]) despite the decision to keep things concrete ("there has been considerable agreement ...concerning single achievements"[24]) Lakatos does not really differ from the traditional epistemologists *except that they argue for their abstract principles while he does not but uses propaganda instead*: he announces that he is going to support his principles by historical research when the results of this research are overruled the moment they conflict with what he thinks a 'rationalist' should do. Here I prefer the procedure of Watkins who in the position paper I have already mentioned simply says, "in a letter to Santa Claus" that the science described by critical rationalists is the science he "would like to have".[25] This is not exactly the most sophisticated answer to question (b) but it is the answer that emerges whenever we examine the procedure of our most recent critical rationalists in somewhat greater detail. It is the answer implicit in the work of the followers of Lakatos. Any charge of irrationalism or praise of rationalism which these papers contain is therefore purely subjective, unsupported by either abstract or historical reasons. This will become even clearer as our analysis proceeds.

Let us look at the matter from a slightly different point of view. A 'rational reconstruction' as described by Lakatos comprises concrete judgements about results in a certain domain as well as general standards (we have seen that it is the general standards that really run the reconstruction, and in an arbitrary manner, but let us forget this for the time being). A rational reconstruction as described by Lakatos is rational in the sense that it reflects *what is believed to be a valuable achievement* in the domain. It reflects what one might call the *professional ideology* of the domain. Now even if this professional ideology consisted of a uniform bulk of basic value judgements only, even if it had no abstract ingredients whatsoever, even then it would not guarantee that the corresponding field has worthwhile results, or that the results are not illusory. Every medicine man proceeds in accordance with complex rules, he compares his tricks and his results with the tricks and the results of other medicine men of the same tribe, he has a rich and coherent professional ideology – and yet no rationalist would be inclined to take him seriously. Astrological medicine employs strict standards and contains fairly uniform basic value judgements, and yet critical rationalists reject its entire professional ideology as 'irrational'. For example, they are not prepared even to consider the 'basic value judgement' that the tropical method of preparing a chart is preferable to the sidereal method or vice versa (the latter opinion being that of Kepler). This possibility to reject professional standards *tout court* shows that 'rational reconstructions' *alone* cannot solve the problem of method. To find the right method one must reconstruct the *right discipline*. But what is the right discipline?

Lakatos does not consider the question, and he need not consider it as long as his aim is to find out how post-17th-century science proceeds and as long as he can take it for granted that this enterprise rests on a coherent and uniform professional ideology (we have seen that it does not). But Lakatos and his followers go much further. Having finished their 'reconstruction' of modern science they turn it against other fields *as if it had already been established* that modern science is superior to magic, or to Aristotelian science, and that it has no illusory results. *It is assumed* that question (b) has already been answered, and that it has been answered in the affirmative. However, there is not a shred of an argument to support this assumption.[26] 'Rational reconstructions' take 'basic scientific wisdom' for granted, they do not show that it is better than the 'basic wisdom' of witches and warlocks. Nobody has shown that only science (of "the last two centuries"[27]) has results that conform to its 'wisdom' while other fields have no corresponding results that conform to their 'wisdom'. What *has* been shown, by more recent anthropological studies, is that *all* sorts of ideologies and associated institutions produce, and have produced, results that conform to their standards and other results that do not conform to their standards. For example, Aristotelian science has been able to accommodate numerous facts without changing its basic notions and its basic principles, thus conforming to its own standard of *stability*. We obviously need further considerations for deciding what field to accept as a measure of method.[28]

Exactly the same problem arises when we consider *particular* methodological rules. It is hardly satisfactory to reject naive falsificationism because it conflicts with the basic value judgements of eminent scientists. Most of these eminent scientists retain refuted theories not because they have some insight into the limits of naive falsificationism, but because they do not realise that the theories are refuted. Besides, even a more rational practice would not be sufficient to reject the rule: universal leniency towards refuted theories may be nothing but a mistake. It certainly is a mistake in a world that contains well defined species whose properties are only rarely misread by the senses. In such a world the basic laws are manifest and recalcitrant observations are rightly regarded as indicating an error in our *theories* rather than in our *methodologies.*[29] The situation changes when the disturbances become more insistent and assume the character of an everyday affair. A cosmological discovery of this kind forces us to make a choice: shall we retain naive falsificationism and conclude that knowledge is impossible; or shall we opt for a more abstract and recondite idea of knowledge and a correspndingly more liberal (and less 'empirical') type of methodology? Most scientists, unaware of the nomological-cosmological background of the problem, and even of the problem itself, retain theories that are incompatible with established observations and experiments and praise them for their excellence. One might say that they make the right choice *by instinct*[30] – but one will hardly regard the resulting behaviour as a measure of method, especially in view of the fact that the 'instinct' has gone wrong on more occasions than one. The *cosmological criticism* just outlined (omnipresence of disturbances) is to be preferred (and was, as a matter of fact, prefered by Aristotle: see his criticism of the Presocratics).[30a]

To sum up: the methodology of research programmes does not argue for the superiority of science ("of the last two centuries"), it takes this superiority for granted and pretends to use it as a basis for the standards it employs. It doesn't use it as such a basis either because it implies "a pluralistic system of authority" in which "the philosopher's statute law"[31] plays an important role, side by side with "common scientific wisdom" ("the philosophers' statute law" – these are the abstract principles the methodology of research programmes was supposed to support by appeal to historical facts). Now: what is the content of this "philosopher's statute law", what are the reasons for it, and when does it come to the fore and overrule "common scientific wisdom"? It comes to the fore "when the scientists' judgement fails"[32] and *that* occurs whenever there is massive support of degenerating research programmes.[32a] Thus the standards, instead of being supported by history are the *criterion* by which we decide when to accept historical trends and when to reject them. Moreover, the methodology of research programmes does not offer any abstract (philosophical) arguments in their favour (or against alternative standards). The standards are therefore arbitrary, subjective, and 'irrational'. They do not provide *objective* reasons for eliminating Marxism, or Aristotelianism, or Hermeticism, or for attacking new developments in the sciences. They merely indicate what critical rationalists would "like to have" at this stage of the development of their ideology.[25]

But the situation is even worse. So far I have argued that Imre Lakatos has *not* provided *any objective reasons for accepting the standards*, he has not shown them to be rational in any sense of the word he is prepared to accept. I shall now argue that *the standards have no force either*, they are too weak to condemn any action as 'irrational'. It follows that an author who uses Puritanical language of this kind – and the followers of Lakatos use it rather frequently – either subscribes to a rationality theory different from Lakatos or else he is content with rhetorical flourishes, unconnected with any argument. In the latter case he gives us an interesting *sociological* study and uses it as a club for forcing people to accept standards which are very different from those he pretends to defend. Let us now examine the assertion that the standards which Lakatos recommends have no force to condemn any action as irrational.

(3) *Nor are the standards strong enough to praise individual actions as 'rational' or condemn them as 'irrational'. All that can be said is that the actions have taken place, and that they have certain features.*

When a theory enters the scene, it is usually somewhat inarticulate, it contains contradictions, the relation to the facts is unclear, ambiguities abound, the theory is full of faults. However it can be developed, and it may improve. The natural unit of methodological appraisals is therefore not a single theory, but a succession of theories, or a *research programme*; and we do not judge the *state* in which a research programme finds itself at a particular moment, we judge its *history*, preferably in comparison with the history of rival programmes.

According to Lakatos the judgements are of the following kind: "A research programme is said to be *progressing* as long as its theoretical growth anticipates its empirical growth, that is as long as it keeps predicting novel facts with some success . . .; it is *stagnating* if its theoretical growth lags behind its empirical growth, that is, as long as it gives only *post hoc* explanations of either chance discoveries or of facts anticipated by, and discovered in a rival programme".[33] A stagnating programme may *degenerate* further until it contains nothing but "solemn reassertions" of the original position coupled with a repetition, in its own terms, of (the successes of) rival programmes.[34] Judgements of this kind are central to the methodology Lakatos wishes to defend. They *describe* the situation in which a scientist may find itself. *They do not yet advise him how to proceed.*

Considering a research programme in an advanced state of degeneration one will feel the urge to abandon it, and to replace it by a more progressive rival. This is an entirely legitimate move. *But it is also legitimate to do the opposite* and to retain the programme. For any attempt to demand its removal on the basis of a *rule* can be criticised by arguments almost identical with the arguments that eliminate say, naive falsificationism: if it is unwise to reject faulty theories the moment they are born because they might grow and improve, then it is also unwise to reject research programmes on a downward trend because they might recover and attain unforeseen splendour (the butterfly emerges when the caterpillar has reached its lowest state of degeneration.[35] Hence, one cannot *rationally* criticise a scientist who sticks to a degenerating programme and there is no *rational* way of showing that his actions are unreasonable.

Lakatos agrees with this. He emphasizes that one "may rationally stick to a degenerating programme until it is overtaken by a rival *and even after*"[36] – "programmes may get out of degenerating troughs".[37] It is true that his rhetoric frequently carries him much further, showing that he has not yet become accustomed to his own liberal proposals.[38] But when the issue arises in explicit form, then the answer is clear: the methodology of research programmes provides standards that aid the scientist in *evaluating* the historical situation in which he makes his decisions; it does not contain *rules* that tell him what to do.

However, even this very modest formulation still goes much too far. Speaking of *risks* (cf. the quotation in note 38) it assumes that the progress initiated by progressive phases will be greater than the progress that follows a degenerating phase – after all, it is quite possible that progress is always followed by longlasting degeneration while a short degeneration (say, 50 or 100 years) precedes overwhelming and longlasting progress.[39] Speaking of *evaluation* and using evaluative terms such as "progressive" and "degenerating" it assumes that "progress" is preferable to "degeneration" both *intrinsically*, and as regards *consequences*. The second case has just been dealt with. The first case (intrinsic advantage of "progress") leads back to question (b). Question (b) is unanswered, hence the question of the intrinsic advantage of progress is unanswered, too.[40]

Alan Musgrave has written an interesting paper in which he agrees with some of the criticisms voiced above.[41] Lakatos, he writes (p. 15 of the draft) "develops an elaborate account of what is good science and what bad, but he refuses (apart from 'Thou shalt not lie') to give advice to the scientists". "'Anything goes' is the position which Lakatos finally adopts" (footnote 53). But a methodology must "provide advice or directives" (p. 22). This advice and these directives are to be addressed "to science, ... to the community of scientists, as a whole" (p. 22). They would "forbid wholesale persistence with degenerating programmes, or premature mass conversion to a budding one" but permit the individual scientist to go his way: "we cannot condemn Priestley for his die-hard adherence to phlogistonism; but we could condemn the community of late nineteenth century chemists had they all done the same". Musgrave thinks he "can provide a purely deductive argument" for such directives (p. 23). The argument proceeds from the premise that "science ought to devote energy to investigating unsolved scientific problems". Now, a "progressive research programme throws up more unsolved problems than a degenerating one" hence, "science ought to devote more energy to" progressive programmes than to degenerating ones (p. 24). In reply one can point out, first, and still in accordance with the methodology of research programmes that every success of a progressive programme is a problem for its degenerating rival, so in the end it will be the degenerating programme that "throws up more unsolved problems". Secondly, it is not only the *number* of problems that counts, but also their *quality*.

Now it is certainly more difficult to find the right questions than to answer "problems" that are already spelled out in detail. Again the directive advises us to pay attention to degenerating programmes. Thirdly, letting the individual scientist doing the dirty work of improving a degenerating programme prejudges the issue in a very unfair way. Today an individual can only rarely attack, let alone solve the problems that arise in the course of research. Without computers, without expensive equipment, without the help of colleagues and assistants he is doomed from the start (just consider the expenses involved in experiments such as those of Reines, or Weber; and where would general relativity be today had Einstein had to carry the expenses of all the experiments carried out to test it?). Musgrave's directive, and that is my fourth point, is also uncomfortably close to the directive of some politicians who advise us to spend educational funds only on those who are already well educated and to let the less educated fight for themselves the difference being that the advantages of an educated person are much more obvious than the advantages of a 'progressive' research programme. And this is my fifth point (question): why should one prefer programmes which have successfully anticipated experimental discoveries to programmes which have no such record? Does such a preference not indicate an inductivistic prejudice? Musgrave thinks it should not and this is why he speaks of problems rather than of the successful anticipation of novel facts. But why should a research programme that creates lots of problems be preferable to a research programme that creates none? For an Aristotelian absence of problems is a sign that certainty and agreement with facts has been achieved. Popperians do not like certainty and they reject the moves that help us achieve it. They do not like certainty and they think they have also found arguments to support their dislike: certainty is not part of science, therefore it should be rejected. This is first not true (cf. the arguments of early Newtonians and Cartesians) and secondly not sufficient: why should we accept science as a measure of excellence? We have to conclude that Musgrave's rescue manoevre does not succeed. Its principles are arbitrary, and they lead to results very different from those envisaged by him.

To sum up this part of the argument: The standards which Lakatos has chosen neither issue abstract orders (such as 'abandon degenerating research programmes') nor do they support general judgements concerning the rationality, or irrationality of a certain course of action (such as 'it is irrational to support degenerating research programmes'). Such orders and such judgements give way to concrete decisions in complex historical situations. Hence, if the enterprise that contains the standards is to be different from the 'chaos' of anarchism, *then such decisions must be made to occur with a certain regularity*. Taken by themselves the standards cannot achieve the regularity, as we have seen. But psychological or sociological *pressures* may do the trick.

Thus assume that the institutions which publicize the work and the results of the individual scientist, which provide him with an intellectual home where he can feel safe and wanted and which because of their eminence and their (intellectual, financial, political) pull can make him seem important adopt a *conservative attitude* towards the standards, they refuse to support degenerating research programmes, they withdraw money from them, they ridicule their defenders, they do not publish their results, they make them feel bad in any possible way. The outcome can be easily foreseen: scientists who are as much in need of emotional and financial security as anyone else, especially today, when science has ceased to be a philosophical adventure and has become a business, will revise their 'decisions', and they will tend to reject research programmes on a downward trend.

This conservative attitude adopted by the institutions is not irrational, for it does not conflict with the standards. It is the result of collective policies of the kind encouraged by the standards. The attitude of the individual scientist who adapts so readily to the pressures is not irrational either, for he again decides in a way that is condoned by the standards. We have thus achieved law and order without reducing the liberalism of our methodology. And even the complex nature of the standards now receives a function. For while the standards do not prescribe, or forbid, any particular action, while they are perfectly compatible with the "anything goes" of the anarchist who is therefore right in regarding them as mere embroideries, they yet give content to the actions of individuals and institutions who have decided to adopt a conservative attitude towards them. *Taken by themselves* the standards are incapable of forbidding the most outrageous behaviour.[42] *Taken in conjunction with* the kind of conservatism just described they have a subtle but firm influence on the scientist. And this is precisely how Lakatos wants them to be used. Considering a degenerating programme he suggests that "editors of scientific journals should refuse to publish... papers [by scientists pursuing the programme].... Research foundations, too should refuse money."[43] The suggestion is not in conflict with the standards, as we have seen. Nor can it be used to raise the charge of irrationality against alternative suggestions: measured by the standards of the methodology of research programmes the conservative attitude expressed by the suggestion is neither rational nor irrational. It is an interesting sociological fact – nothing more. *But it is eminently rational according to other standards*, for example, according to the standards of common sense.[44] This wealth of meanings of the word 'rational' is used by Lakatos to maximum effect. In his arguments against naive falsificationism he emphazises the new 'rationalism' (p. 22) of his standards which permits science to survive. In his arguments against Kuhn and against anarchism he emphasises the entirely different 'rationality' of common sense but without informing his audience of the switch and so he can have his cake – have more liberal standards – and eat it too – have them used conservatively, and he can even expect to be regarded as a rationalist in both cases. Indeed, there is a great similarity between Lakatos and the early Church Fathers who introduced conservative doctrines in the guise of familiar prayers (which formed the common sense of the time) and who thereby gradually transformed common sense itself.[45]

(4) *Using the methodology of research programmes as our theory of rationality we must therefore regard the case studies of Lakatos and his school as* sociological studies *and we must* disregard *the frequent judgements of rationality or irrationality that occur in them (on the other hand, these judgements may be retained if we adopt a different theory of rationality, for example Hegel's). But though the aim of rationalising history is never reached, the attempt to reach it has produced a history that is richer in content and more conceptual than its predecessors.*

Taking all these things into consideration it is clear that Lakatos has not succeeded in showing "rational change" where "Kuhn and Feyerabend see irrational change".[46] A revolution occurs when a new research programme has accumulated a sufficient number of successes and the orthodox programme suffered a sufficient number of failures for both to be regarded as serious rivals, and when the protagonists of the new programme proclaim the demise of the orthodox scheme. Seen from the point of view of the methodology of research programmes they do this not just because of their standards – the standards are not strong enough for making such a judgement – but because they have adopted a conservative attitude towards the standards (all this assumes of course, that both sides use the methodology of research programmes in their deliberations – a matter that is open to considerable doubt: see below see 5). Their orthodox opponents have what one might call a 'liberal' attitude; they are prepared to tolerate a lot more degeneration than the conservatives. The standards permit both attitudes. They have nothing to say about the 'rationality' or 'irrationality' of these attitudes and of the developments initiated by them. It follows that the fight between the conservatives and the liberals and the final victory of the conservatives is not a 'rational change"[47] but a "power struggle" pure and simple, full of "sordid personal controversy".[48] It is a topic neither for methodology, nor for the theory of rationality, but for "mob psychology",[49] or, to use a more traditional term, *it is a topic for the sociology of knowledge*.

Exactly the same is true of the case studies that have been carried out by Lakatos and his followers. Each of these essays describes a battle between alternative research programmes and the victory of one of them. Each essay reconstructs the battle in terms of the methodology of research programmes, using the standards of this methodology as a basis for the proper application of its evaluative terms. The standards are not strong enough to guide such application. Basing our judgements on them we can only say that one programme was *accepted* while the other *receded into the background*, we cannot add that the acceptance was *rational*, or that a rational development took place. Thus the essays are interesting contributions to the *history*, or the *sociology* of science no matter how hard the authors try to present them as something different. The reader must not be misled by the frequent and rather assured use of terms such as 'rational' or 'irrational' which suggests that the authors have some deeper insight into the historical process. Using the methodology of research programmes they cannot have such insight, as I have tried to show. Of course, they have their *preferences*, they know what kinds of things they "would like to have" and they defend them with religious fervour. But neither they nor the authorities on whom they rely succeeded in turning this fervour into a rational procedure.

Let us therefore from now on regard such case studies as historical studies and let us evaluate them on that basis. We see at once that they are superior to earlier studies of the same kind. The procedure is always the same. First, a certain historical episode is identified, whether it be the Copernican Revolution; the Einsteinian Revolution; the chemical revolution of Lavoisier and its altercation with the phlogiston theory; the rejection of Young's version of the wave theory of light; or the battle between phenomenological thermodynamics and the kinetic theory of heat. Then follow explanations of the episode that have played a role in the literature. Some of these explanations are mere narratives, without conceptual connexions, others are psychological, still others are methodological, they try to show how the events arose, in a "rational" manner, as a result of the determined use of methodological rules. Among the methodologies surveyed are inductivism, naive falsificationism, conventionalism, and the views of Kuhn. Next comes the demolition of the traditional explanations. This demolition is almost always historical: the explanations omit important facts, they conflict with others (inductivism alone is removed by logical arguments). Finally, we have an account in terms of the methodology of research programmes. Guided by a complex methodology this account is richer and more sophisticated that the alternative accounts. It is history of ideas in the best sense of the word. It is *history* because it deals with facts. It is history of *ideas* because is shows conceptual connexions between these facts. It is *sophisticated* history of ideas because it uses a rich inventory of conceptual tools (hard core; protective belt; heuristic; progressiveness; degeneration; monsteradjustment; recovery of hidden lemmas; and so on) rather than relying on intuition in all cases.[25] The researcher is equipped with instruments that aid him on his way and are open to inspection so that he can criticise them and replace them by better instruments. It is true – the wish of the writers to arrive at some "objective" judgement has made these instruments overly intellectual: a researcher who wants to inquire into motivation, or sociological causes hardly receives any help. Even worse, he is discouraged from putting too much weight on causes of this kind. This explains the rather primitive sociology and psychology of some of the papers and the complete absence of any inquiry into when, how, and why certain standards were accepted. But there are definite advantages when one compares the history with what other methodologists have to offer. An inductivist, for example, will consider a theory (or, rather, a "logical reconstruction" of it), the "evidence" (which again is a reconstruction of the complex experimental results that govern science), and the relation between the two. He has two abstract elements and he examines them irrespective of the historical surroundings in which they arose. Writing history he is interested in the "rational" parts of science only which are again the elements and their relation. This is why inductivist history is so arid or, if richer, so lacking in conceptual penetration. A naive falsificationist is not much better for all *he* wants is *some* evidence – no matter in what historical surroundings – that contradicts the theory. The methodology of research programmes, on the other hand, does not examine theories, it examines sequences of theories connected by hard cores, heuristics and intuitive attitudes not all of which need to be formulated explicitly. Already at this point it goes much deeper into the structure of a theory than do the rivals. Secondly, the methodology of research programmes does not examine research programmes by themselves, it examines them in comparison with other research programmes. So the investigation spreads and must ultimately reach every research programme at the period in question. The whole intellectual scene must be taken into account. In the sixteenth and seventeenth centuries this includes theology, Aristotelian physics and metaphysics, magic (neoplatonism), the philosophy of Paracelsus (which centered around medicine and chemistry), alchemy – all subjects which were studied with care by the great Newton himself. Then we must examine whether the predictions made were novel predictions, or whether they were repetitions of things already known. This means we must examine the way in which the research programme was originally introduced,[50] the expectations of the age, the 'accepted facts' and the relation of these facts to current theory. We must know a great amount of material that belongs to the history of ideas and that is often missing even from rather detailed *historical* accounts. Nor is this material just *aufgerafft*. It is collected with an aim in mind, described in terms adapted to this aim and thus essentially ideational. It is true that the basic scheme lends a certain uniformity to the procedure which in the hand of less gifted writers can introduce an element of boredom. It is also ironical to realise that the aim – to give a 'rational' account of developments – is never reached and that we are left with a historical narrative only.[50a]

But on the way to the aim history has been transformed to such an extent that a slight change in our standards say, from research programme standards to Hegelian standards enables us to read it as a history of reason itself.[51]

(5) *Remaining lacunae can be explained by the rationalising tendency of the authors and their blind acceptance of modern science.*

Still, the rationalising tendencies of the followers of Lakatos and their assumption of the excellence of science make themselves felt and are responsible for some lacunae even in the historical account. Let me mention two of them.

All the authors assume that the defenders of rival research programmes (and perhaps even the onlookers) are influenced by, and proceed in accordance with the methodology of research programmes. They use this methodology as a measure of good science and they behave in accordance with the advice given in the standards. It has emerged that the standards are incapable of giving any advice – but let us now disregard this drawback. Let us assume that the standards do indeed favour progressive programmes (they don't just *state* their progressiveness) and condemn degenerating programmes. It is then assumed by the authors that all good scientists distribute praise and blame in accordance with the methodology of research programmes. They may not be able to give an account of the *principles* of this methodology, but they still proceed as if they held such principles to be the basis of research.

Now this assumption may have some plausibility in periods of peace and uniformity which are most likely ruled by a single methodological framework. It becomes improbable when we are dealing with developments such as the Copernican Revolution, or the rise of 20th century science, especially quantum theory. Lakatos has emphasized that his standards are based on outstanding science of "the last two centuries"[52] and he has thereby conceded the possibility of different standards *before* this period. But in his joint paper with Zahar on the Copernican Revolution he contends that "Copernicus' and Kepler's and Galileo's adoption of the heliocentric theory" are "rationally explainable".[53] "The Copernican Revolution" the authors write[54] "became a great scientific revolution... simply because it was scientifically superior" where "scientifically superior" means, of course, superior in terms of the methodology of research programmes. Even Aristarchus' claim should have been taken seriously because "the goecentric programme had already heuristically degenerated".[55] Similarly Peter Clark in summarising the result of his paper says "It was the degeneration of these attempts to provide a foundation for thermodynamics in terms of some deeper theory, compared with the empirical progress of the phenomenological theory which led to the evolution of thermodynamics into a 'paradigm' of great science".[56] This again assumes that the moves of contemporary scientists were motivated by an implicit understanding of the methodology of scientific research programmes. Is this assumption correct?

I do not think it is.

In the case of the Copernican Revolution we know that the methodology of research programmes, if it was used at all, was not the only methodology in existence. One influential methodology was connected with the slogan of *saving the phenomena*. The slogan presupposes a distinction between *basic physics* and *auxiliary assumptions*. Basic physics describes the processes that are expected to occur in this world, auxiliary assumptions link the processes to the phenomena. In Aristotle basic physics is decidedly empirical, much more so than modern science could ever aspire to be. It starts from 'phenomena' which are either observed facts or assumptions of common sense,[57] develops terminology for their description and principles for their explanation[58] and incorporates 'new facts' in a 'degenerating' manner. *This is regarded as an advantage*: the fact that theory and phenomena are related in a manner described as 'degenerating' by Lakatos shows the truth of the principles and the adequacy of the terminology used. The auxiliary assumptions are eventually separated from basic theory and assembled into various disciplines (astronomy; optics; mechanics; and so on). The task of these disciplines is to save the phenomena, not to give a physical account of the processes (motions) that created them.[59] Attempts to *reabsorb* them into basic theory in a 'degenerating' way continue throughout the Middle Ages.[60] They continue with Copernicus who tries to find an arrangement of circular movements that agrees with the basic Aristotelian philosophy[61] to such an extent that it can again be regarded as an account of real motions leading to a coherent *system* of the world rather than as a set of independent *hypotheses* for the calculation of planetary phenomena.

Neither the tradition of saving the phenomena nor the attempt to absorb the mathematical devices of this tradition into basic theory look for novel facts or novel explanations of existing facts in the sense of the methodology of research programmes. The former cannot be found for all facts are related to the same basic principles and formulated in their terms, the latter cannot be admitted for that would deny the existence of a stable basic theory. To a certain extent this is true even of Copernicus himself (though not of Rheticus, and certainly not of Kepler.[62]) But if that is the case then the revolution that was started by Copernicus' hypothesis was much more dramatic than a transition, *within the methodology of research programmes*, from one research programme to another. It brought in new standards and thus constituted a true paradigm change in the sense of Kuhn.[63]

A second historical lacuna is created by the authors' insistence on the 'objective' character of their evaluations: 'objectively' Copernicus (Einstein, the phenomenologists etc.) progressed, hence it was 'rational' to follow their lead. (I must again repeat that the standards of the methodology of research programmes are not strong enough to permit us to make such a judgement of rationality; let us forget this drawback now that we are dealing with another difficulty.) But we have seen above[64] that 'correct' actions ('correct' in the sense of the methodology of research programmes) are only rarely carried out for the 'correct' reasons. It is therefore quite possible that the Copernicans, wanting to do one thing, ended up doing another. (Newton wanted to achieve certainty; what he did achieve was unceasing progress.) In this case they acted rationally (in the sense of the methodology of research programmes) for irrational (again in the sense of the methodology of research programmes) reasons and their rationality was a matter of luck, of accident, of the propitious collaboration of irrational causes. (Hegel called this the *List der Vernunft*.) The authors, concentrating on the internal features of a perhaps quite fortuitous result leave such possibilities unexamined and thereby add to the false impression of the overwhelming 'rationality' of modern science.

(6) *Judgements of progress and of degeneration are often arbitrary, for they depend on an arbitrary selection of the research programmes to be compared. Althogether the appraisals made by the writers of case studies*

are arbitrary in at least five different ways. Once more it becomes clear that science needs and uses *a plurality of standards and that scientists work best without any authority, the authority of 'reason' included.*

In Section 3 I argued that the standards of the methodology of research programmes are not strong enough to recommend particular actions or to characterize such actions as being rational, or irrational. We may describe the heuristics of a research programme, we may say that the programme progresses, or degenerates, we cannot infer that it may be retained, or that it must be abandoned, or that continued work on it is rational, or irrational. It was then assumed that at least the judgements of progress and degeneration themselves are made in an unambiguous and objective manner. This is not always the case.

To see this, consider two rival research programmes, R and R' According to the methodology of research programmes both R and R' are plagued by anomalies. Let r and r' be corresponding sub-programmes of R and R' and let the anomalies of R and R' be so distributed that r progress while r' degenerates. A follower of the methodology of research programmes will then support r over r'. It seems that some followers of Lakatos are in this situation prepared to support R over R' as well. But in doing so they may go against their own principles for it is quite possible that R' progresses while r' degenerates and r progresses.

As an example, let R' be the Aristotelian cosmology, r' Prolemaic astronomy, r the astronomy of Copernicus and R a dynamical research programme consisting of r and suitable dynamical principles. Lakatos and Zahar have shown that r progresses while r' degenerates. Now they are either instrumentalists, or realists. In the first case they will remain content with what have shown. In the second case they will regard the motions of r as real motions and infer that R' must be abandoned. But R' continued to progress long after Copernicus as can de seen from the work of Harvey.[65]

Similar remarks apply to Zahar's comparison of the research programmes of Lorentz and Einstein. We have to distinguish L and E, the research programmes of Lorentz and Einstein, L' and E', the programmes that are compared by Zahar, L^t and E^t, the *theories* that are usually regarded as the decisive rivals in 1905 and L'' and E'', those parts of the programmes that deal with inertial space-time only. L^t and E^t are often asserted to be equivalent, and Zahar repeats that judgement.[66] He is mistaken,[67] but the mistake does not matter. Not theories, but research programmes are decisive. Lorenz' programme is said to consist of Maxwell's equations, Newton's laws of motion (with Galilean transformations) and the Lorentz force.[68] Zahar never mentions the hard core of Einstein's programme though he mentions a programme, E'', that contains the relativity principle together with the principle of the constancy of c.[69] In 1905 this programme started degenerating while Lorentz' programme L' was advancing,[70] and had been advancing for quite some time. Now Zahar wants to explain, in a rational way "why... brilliant methematicians and physicists like Minkowski and Planck abandon[ed] the classical programme in order to work on Special Relativity"[71] and he also wants to show that the success of general relativity in the explanation of the perihelion of Mercury was "a success for the *whole* relativistic programme".[72] To achieve both aims he replaces E'' by E' in the following manner:[73] c is used not for 'empirical' reasons (it is not used in a degenerating way), but because of Einstein's belief in (a) the basic nature of Maxwell's equations[74] and (b) their limited validity.[75] The principle of the constancy of light velocity is all that can be salvaged of (a) in view of (b) which means that it is a fundamental principle not only because of its position in the theory, but because of the nature of things. Thus E' no longer degenerates, but it does not seem to advance either. To explain its acceptance by Planck, Minkowski and others in a 'internal' way, i.e. without recourse to psycho-sociology Zahar turns to heuristics. To preserve continuity with the general theory of relativity and to refute Whittacker's conjecture that the ether programme was "developed into the Relativity Programme"[76] he concentrates on this heuristics to the exclusion of any hard core.[77] This brings him very close to E, the always unknown research programme underlying all of Einstein's activities and thus he seems to move in the right direction. But E is still compared with L', the *truncated and frozen* programme of Lorentz and L, which contains atomism as well as the possibility of a more fundamental explanation of electromagnetic phenomena is never considered. But L can take care of all the facts which E'' obtains from the relativistic formulation of the electrodynamics of media.[78] It yields the constancy of c as a contingent fact and is in this respect closer to the general theory of relativity than E'' and E' where the constancy of c is a basic law.[79] And its heuristics is at last as powerful as the heuristics of E for every law produced by a research programme can of course be used in the heuristics of that research programme.[80] We see: the choice of research programmes and rivals is pretty arbitrary, and so are the judgements that rest on them. But these judgements are the basis for Zahar's "objective" or "internal" appraisal of the actions of Planck, Minkowski and others.

Such appraisals (and the corresponding appraisals in all case studies that have been carried out in accordance with the methodology of research programmes) therefore turn out to be arbitrary, or subjective, or "irrational" in at least four different ways. They are arbitrary, because they proceed from an arbitrarily chosen authority: science of "the last two centuries" (see above, Section 2). Science is chosen not because its excellence has been shown by argument, but because everybody is impressed by it.[81] They are arbitrary because it is not really science that decides the issue – science is much too chaotic for that – but a streamlined image of it, and there are no independent arguments for the principles of streamlining chosen. Thirdly, the appraisals are arbitrary because the standards that are obtained via steps one and two are not strong enough to support judgements of rationality, or irrationality. Any such judgement is independent of the standards, it receives no authority from the standards, there are no other arguments in its favour, and so it is arbitrary, or subjective in a very strong sense of the word. Fourthly, the appraisals are arbitrary because they rest on an arbitrary choice of rival research programmes. In the case of Zahar we find still another source of arbitrariness, and it is rather amusing: Planck and Minkowski start working on relativity, Planck and Minkowski are great scientists, hence their actions must be explained in an 'objective' manner. But there were many great scientists who either rejected the theory, or did not pay any attention to it. As a matter of fact, "It was only in Germany that the theory was elaborated upon".[82] How are the actions of the dissenters to be explained? Are they to be explained in the same way as the acceptance of the theory by Planck and Minkowski? Not likely. Are they to be explained 'internally'? This would mean that a theory has advantages as well as disadvantages, that different people look at it in different ways and come to different results though they use the same set of standards. It does not seem that Zahar would accept such an explanation. But then the only way out is to admit that the dissenters acted 'irrationally', and for external reasons. But if *they* can act irrationally, then why not Planck and Minkowski? This is the fifth type of arbitrariness found in the case studies. It is surprising to see that a philosophy that makes such a fuss about 'rationality' and 'objectivity' should possess so abjectly little of either.

Zahar tries to show that Planck and Minkowski acted rationally when deciding to work on Einstein's programme[5] and is thus faced with the problem of the rationality of those who stayed either with Lorentz, or with aether models. He neither states the problem, nor does he indicate how he would solve it. Peter Clark, in his essay on the kinetic theory which I read in manuscript perceives an analogous problem. Towards the turn of the century the kinetic theory "was subject to severe attacks from some of the leading scientists of the day".[83] This is not supposed to be due to purely philosophical preferences, such as a preference for positivism. All these are "external explanations".[84] The correct explanation, according to Clark, is that "in the last decades of the 19th century [the kinetic research programme] was *degenerating* [while] from the two laws of thermodynamics a number of startling novel facts were deduced".[85] Was it therefore irrational to work on the kinetic theory? Not at all! The kinetic theory has a heuristic, there are means "of systematically *improving* the theory"[86] while the phenomenological theory "lacks a heuristic".[87] It is therefore also "rational"[88] to try to make the latter supersede the former.

Now it would seem that such an attempt is rational only if a heuristic is *needed*. The kinetic theory which ran into one difficulty after another definitely needed a heuristic. The phenomenological theory which was applied to an ever increasing domain of problems without ever failing to live up to its promise (this is Clark's description of the situation, not mine!) did not. It was too good to be in need of becoming a research programme. Besides, it was this very universality of the theory, this absence of models that made it so attractive to Einstein. Einstein[89] distinguished between "constructive theories" which move through various stages and gradually conquer one problem after another and "theories of principle" which remain valid no matter how far they are extended, and he preferred the latter, mentioning thermodynamics as an outstanding example. He viewed his own theory of relativity as a theory of principle, not as a constructive theory. We see: Clark does not succeed to explain the rationality *both* of those who objected to the kinetic theory and of those who continued working on it. *At least one party must be criticised as being irrational.* Given the methodology of research programmes it is arbitrary which party we choose.[90]

This result can be generalised. Research programmes are superseded by other research programmes only because people are not invariably impressed by success and progressiveness. They are not impressed by 'rationality' either because they have standards different from those accepted by the proponents of the 'successful' programmes, or because they are moved by external considerations. In the first case we have a plurality of *standards*, in the second case a plurality of *motives*. Science *as described by Lakatos and his followers* cannot exist without such a plurality and where it seems to succeed it does so only because certain problems have been overlooked (the problem of the opponents of relativity in the case of Zahar) or because of the biased terminology (talk of 'rationality' or 'objectivity' when no rational grounds for such talk have been given). The methodology of research programmes most certainly has led to some interesting historical discoveries. This is not surprising. Any hypothesis, however implausible, can widen our horizon. It has not led to a better understanding of science and it is even a hindrance to such a better understanding because of its habit of beclouding facts with sermons and moralizing phrases.

University of California, Berkeley

NOTES

[1] Cf. *Against Method*, London 1975, Chapter 4 and the literature given there.

[1a] Ostrander and Schroeder, *Psychic Discoveries behind the Iron Curtain*, Prentice Hall, 1970, as well as the literature in Thelma Moss' contribution to E. D. Mitchell (ed.), *Psychic Exploration, A Challenge for Science*, New York 1974.

[1b] Cf. the immense literature on traditional medicine in China as well as my brief account in Chapter 4 of *Against Method*, London 1975.

[1c] Cf. the reports and literature in E. D. Mitchell *op. cit.*

[2] According to Nelson Goodman 'A World of Individuals' quoted from Benacerraf and Putnam (eds.), *Philosophy of Mathematics*, Prentice Hall, 1964, p. 207, the inventions of the scientists "become raw material for the philosopher whose task is to make sense of all this... in understandable terms". Considering that logicians are only rarely capable of following scientists on their flights of fancy this would indeed seem to be 'the task' of 'the philosopher' – only Goodman is not that modest. If *he* does not comprehend a thing, then the thing uncomprehended is *intrinsically obscure* and must be 'clarified', i.e. it must

be translated into a language which he understands (pidgin logic, in most cases). If *he* understands a language, then the language is intrinsically clear and must be spoken by everyone. This is also how the demand for reconstruction arises. Logicians cannot make sense of science – but they can make sense of logic and so they stipulate that science must be presented in terms of their favourite logical system. This would be excellent comedy material were it not the case that by now almost everyone has started taking the logicians seriously.

[3] It is surprising to realise how difficult it is to see science in perspective. Carl Sagan, surely one of the most imaginative scientists alive warns us not to unduly restrict the possibilities of life, and he mentions various types of 'chauvinism' (oxygen-chauvinism: if a planet has no oxygen, then it is uninhabitable; temperature chauvinism: low temperatures such as those on Jupiter and high temperatures such as those on Venus make life impossible; carbon chauvinism: all biological systems are constructed of carbon compounds) which he regards as unwarranted (*The Cosmic Connection*, New York 1975, Ch. 6). He writes (page 179): "It is not a question of whether we are emotionally prepared in the long run to confront a message from the stars. It is whether we can develop a sense that beings with quite different evolutionary histories, beings who may look far different from us, even 'monstrous' may, nevertheless, be worthy of friendship and reverence, brotherhood and trust". Still, in discussing the question whether the message on the plaque of *Pioneer 10* will be comprehensible to extraterrestrial beings he says that "it is written in the only language we share with the recipients: science" (18; cf. p. 217: messages to extraterrestrial beings "will be based upon commonalities between the transmitting and the receiving civilization. Those commonalities are, of course, not any spoken or written language or any common, instinctual encoding in our genetic materials, but rather what we truly share in common – the universe around us, science and mathematics.") In times of stress this belief in science and its temporary results may become a veritable maniac-making people disregard their lives for what they think to be the truth. Cf. Medvedev's account of the Lysenko case.

[4] *Workshop No. 1 – Position Paper. Criteria of Scientific Progress, A Critical Rationalist View*, pp. 2/21f.

[5] Critical Rationalists take great care to show that prima facie disreputable procedures in science, when looked at in detail, turn out to be quite acceptable (cf. Zahar on the Lorentz-Fitzgerald contraction, or Worrall on the fate of Young's version of the wave theory). They also know that there are good scientists and bad scientists and that the procedures of the former are not discredited by the errors of the latter: no one would abandon science because it contains complementarity. The attitude towards Marxism, or astrology, or other traditional heresies is very different. Here the most superficial examination and the most shoddy arguments are deemed sufficient. Worral uses some Marxist interpretations of events in Hungary to discredit the whole approach but without saying what the interpretations are, who has put them forth, and where they can be found. Popper (*Open Society*[5], Vol. ii, 187ff) mentions the hypothesis of colonial exploitation as a perfect example of an ad hoc hypothesis although it is accompanied by a wealth of novel predictions (the arrival and structure of monopolies being one of them). And whoever has read Rosa Luxemburg's reply to Bernstein's criticism of Marx or Trotsky's account of why the Russian Revolution took place in a backward county (cf. also Lenin 'Backward Europe and Advanced Asia', *Collected Works*, Vol. 19, pp. 99ff) will see that Marxists are pretty close to what Lakatos would like any upstanding rationalist to do, though there is absolutely no need for them to accept his rules. After all, all he can say in favour of these rules is that the elite of some enterprise he loves sometimes sticks to the rules (see below).

[6] 'History of Science and its Rational Reconstruction', p. 111.
[7] *ibid.*, p. 117.
[8] p. 117.
[9] p. 111.
[10] p. 117f.
[11] p. 112.
[12] Leon Rosenfeld in *Observation and Interpretation*, London 1957, p. 44.
[13] "Is it not... *hybris* to try to impose some *a priori* philosophy of science on the most advanced sciences?... I think it is." p. 121.
[14] Footnote 80.
[15] p. 121.
[16] p. 121.
[17] p. 121.
[18] p. 122.
[19] p. 111.
[20] Footnote 58.
[21] p. 122.
[22] p. 121.
[23] Footnote 130.
[24] p. 111.
[25] p. 1/3.
[26] At this point critical rationalists and followers of the methodology of research programmes usually introduce the criterion of content increase: Aristotle was defeated, and justly so, because he did not conform to this criterion. This assumes (a) that Aristotelians *wanted* to conform to the criterion (they did not – see below, p. 130) and (b) that the criterion is preferable to, say, the criterion of stability, or to the criterion that the best explanations are posthoc explanations. But (b) is the assumption under examination.
[27] Lakatos, p. 111.
[28] Watkins, in his 'letter to Santa Claus' points out that the ideal preferred by him is "really an amalgam of ideas" found in Bacon and Descartes (p. 1/4). That may be so but does not establish its superiority over, say, the ideology of the Aristotelians or of John Dee. Thus the argument always becomes circular at the most decisive point. Among the followers of Lakatos it is only Elie Zahar who approaches the problem in a more rational manner. In his account of the Copernican Revolution he assumes that all the competitors *shared the same standards* and that the Aristotelians lost because their theories did not comform to these shared standards. This still does not give us an answer to question (b) – all it does is to extend the domain from which the basic value judgements are taken. But it gives a rational account of the victory of the Copernicans *provided* the shared standards had some *force*. This problem will be discussed in the next section.
[29] In such a world the demand for depth (Watkins, pp. 1/4ff. of the Position paper) is unrealistic and cannot be satisfied.
[30] "Up to the present day it has been the scientific standards, as applied 'instinctively' by the scientific elite in *particular cases* which have constituted the main – although not the exclusive – yardstick of the philosopher's *universal* laws" – Lakatos, p. 121.
[30a] This possibility of *choosing* a methodology on the basis of cosmological considerations shows that there can be different types of science: given fairly clear species with not too many disturbances we may decide to remain naive falsificationists and absorb the exceptions by methods such as monster barring, or various means of adaptation, but we may also decide to use basic laws for the explanation of *all* events and so becomes research program-

mists. Aristotle made the first decision, Galileo as seen by some thinkers made the second. Cf. Kurt Lewin's most interesting essay 'Aristotelische und Galileische Betrachtungsweise in Biologie und Physik', *Erkenntnis*, vol. ii. Thus the idea that there can be only one science – one physics, one biology, one chemistry – which is found even among socalled dialectical materialists (Cf. Zhores A. Medvedev, *The Rise and Fall of T. D. Lysenko*, New York and London 1969, pp. 133, 247) is again but a result of insufficient analysis.

[31] Lakatos, p. 121.

[32] p. 121.

[32a] According to Lakatos it seems that modern particle physics represents a degeneration: footnote 130 of his essay. He also speaks of the development of "new bad traditions" (p. 122) such as modern sociology, psychology, social psychology. These traditions are indeed *new*. But they are *bad* only if the standards of science "of the last two centuries" have been shown to be good. They are *assumed* to be good – this much is sure. But there is no argument to support this assumption and, besides, they are overruled whenever they seem to conflict with the house philosophy of the methodology of research programmes.

[33] p. 100.

[34] p. 105; cf. also 'Falsification and the Methodology of Scientific Research Programmes' in Lakatos and Musgrave (eds.), *Criticism and the Growth of Knowledge*, Cambridge 1970, pp. 116ff.

[35] This remark shows that the methodology of research programmes, too, makes certain *cosmological* assumptions concerning the relation between research programmes, and the world.

[36] p. 104.

[37] 'Falsification', p. 164.

[38] "I give rules for the 'elimination' of whole research programmes"; p. 100 – note the ambiguity introduced by the quotation marks. Occasionally the restrictions are formulated in a different way, by denying the "rationality" of certain procedures. "It is perfectly rational to play a risky game", says Lakatos, p. 104 – "what is irrational is to deceive oneself about the risk": one can do whatever one wants to do if occasionally one remembers (or merely recites?) the standards *which, incidentally, say nothing about risks or the size of risks*. Speaking about risks either involves a *cosmological* assumption (nature rarely permits research programmes to behave like caterpillars), or a *sociological* assumption (*institutions* only rarely permit degenerating research programmes to survive) and thus lead to exactly the same conundrum which Hume explained so nicely more than 200 years ago. Lakatos in passing (p. 101) admits the need for such additional assumptions: only they "can turn science from a mere game into an epistemologically rational exercise". But he does not *discuss* them in detail and those he takes for granted are very doubtful, to say the least. Take the cosmological assumption I have just mentioned. It is interesting, and it certainly deserves to be studied in greater detail. Such a study, I venture to suggest, would reveal that the research programme corresponding to it is now in a degenerating phase (to see this, one needs only to consider anomalies such as the Copernican Revolution, the revival of the atomic theory, the revival of assumptions concerning celestial influences; and so on). The sociological assumption, on the other hand, is certainly true – which means that given a world in which the cosmological assumption is false we shall forever be prevented from finding the truth.

[39] This is, of course, again a version of Hume's problem. Hume turns up in all methodologies because all methodologies make cosmological assumptions. The naive falsificationist assumes that there are no oceans of anomalies. The conventionalist assumes that the world is built in a simple way. The research programmist assumes that progress, once realised,

does not put an end to further progress and that it leads to the truth (after all, a progressive theory may lead us further and further away from the truth; cf. the life of Paphnutius as presented by Anatole France) and so on.

[40] John Worrall writes (p. 2/29, footnote 3 of the position paper already mentioned): "A scientist *would* be pronounced 'irrational' (or rather mistaken) by the methodology of research programmes if he stuck to the old programme denying that the new programme had any merits not shared by the old one and thus denying that his own programme needed improvement in order to catch up with the new one. It is in such circumstances that we shall begin to suspect the operation of extra-rational motives". The arguments in the last section, in the text above, and in note 38 show that it is rather this judgement of Worrall's which makes us "suspect the operation of extra-rational motives" where by "extrarational" we mean motives either not in accordance with the standards, or not dictated by them. Assume, I have a research programme which degenerates and I am told so by a research programmist. My reply might well be that I am interested in certainty and not in novelty and that I prefer a programme that can incorporate newly found facts without revision to a programme that constantly upsets basic convictions. When being told that this means I am not being "scientific" I can reply that the excellence of science is still a matter of debate, that it is *assumed* by my opponent (though discarded by him when it goes counter to his own pet ideology – see above) that it is not supported by argument (nor is there any argument to show that non-scientific ideologies are worse than science in addition to being different from it; of course, there is a general *belief* that this is so, and this belief may even be quite reasonable but what I am now talking about is the ability of the methodology of research programmes to give a *reason* for the belief). Adopting the point of view of science I can add that degeneration when taken seriously may be followed by bigger progress than progress and that progress may lead away from the truth. Or is a scientist supposed to be satisfied with temporary spectacles only? Is it enough for him to impress everyone by first predicting, and then discovering, a new planet (Neptune, e.g.) *without any implications for the quality of future research*? And, finally, one might comment on the futility of a point of view where a thief can steal as much as he wants, is praised as an honest man by the police and by the common folk alike provided he tells everyone that he is a thief. If *that* is the sense in which the methodology of research programmes differs from anarchism (Worrall, p. 2/30, footnote 1 of the position paper), then I am ready to become a research programmist. For who does not prefer being praised to being criticised if all he has to do is to describe his actions in the lingo of a particular school?

[41] 'Method or Madness? – Can The Methodology of research programmes be rescued from epistemological anarchism?' this volume.

[42] For a minor exception which by now seems to have become the only point of resistance of the methodology of research programmes cf. note 40.

[43] Lakatos, p. 105.

[44] "In such decisions" says Lakatos, referring to decisions such as those leading to a conservative use of standards, "one has to use ones *common sense*" – note 58. Right on – as long as we recognize that in doing so we *leave* the domain of rationality as defined by the standards and move to an 'external' medium, or to other standards. Lakatos does not always make the change clear. Quite the contrary. In his attack upon opponents he makes full use of our inclination to regard common sense as inherently rational and to use the word 'rational' in accordance with *its* standards. He accuses his opponents of 'irrationality'. We instinctively agree with him, quite forgetting that the methodology under debate does not support the judgement and does not provide any reasons for making it.

[45] Using the *psychological* hold which the baptismal confession had over the members of

the early Christian Church and taking the non-Gnostic interpretation "as its selfevident content" (A. Harnack, *History of Dogma*, vol. ii, New York 1961, p. 26) Irenaeus succeeded in defeating the Gnostic heresy. Using the psychological hold which common sense has over philosophers of science and taking the conservative interpretation of his standards as *its* self evident content, Lakatos has almost succeeded in convincing *us* of the rationality of his own law-and-order philosophy and the non-ornamental character of his standards: now as before the best propagandists are found in the Church, and in conservative politics.

[46] Lakatos, p. 118.

[47] p. 118.

[48] p. 120.

[49] 'Falsification', p. 178 – italics in the original.

[50] E. Zahar, 'Why did Einstein's Programme supersede Lorentz?', *British Journal for the Philosophy of Science* 24 (1973), 103 writes: "My redefinition of novelty amounts to the claim that *in order to assess the relation between theories and empirical data, one has to take into account the way in which a theory is built and the problems it was designed to solve.* This new criterion of novelty of facts also implies that the traditional methods of historical research are even more vital for evaluating experimental support than Lakatos had already suggested. The historian has to read the private correspondence of the scientist whose ideas he is studying; his purpose will not be to delve into the psyche of the scientist, but to disentangle the heuristic reasoning which the latter used in order to arrive at the new theory".

[50a] Environmentalists therefore need not be intimidated by Peter Urbach's 'Progress and Degeneration in the I.Q. Debate' *BJPS* 1974 (two parts). What is shown in these papers is that the relation of environmentalism to the evidence is different from the relation of some versions of geneticism to the evidence. *This is all that is shown* though both terminology ('degenerating'; 'progressive') and philosophical insinuations *create the impression* that the one type of relation is better, more 'scientific' than the other. My remarks in the text above make it clear that this is not so. Value judgements of this kind are completely arbitrary and subjective and nobody needs to be intimidated by them. But alas! The propagandistic genius of Lakatos has concocted a mixture of propaganda and sham argument that is only difficult to unravel and so he will have his way, because of the power of his rhetorics. One might well call this the *List der Unvernunft*.

[51] Cf. The introduction to Hegel's *Geschichte der Philosophie*.

[52] p. 111.

[53] 'Why did Copernicus' programme supersede Ptolemy's?' in R. S. Westman (ed.), *The Copernican Achievement*, p. 17.

[54] p. 31.

[55] p. 22, footnote 1.

[56] p. 5.

[57] G. E. L. Owen, 'Tithenai ta phainomena', *Aristote et les problemes de methode*, Louvain 1961.

[58] For the role of principles in Aristotle cf. W. Wieland, *Die Aristotelische Physik*, Göttingen 1970. Wieland makes it clear how principles in Aristotle are designed with the explicit purpose of achieving 'degenerating' adaptations of facts (of course, he does not use this terminology).

[59] Originally, in Aristotle and in his immediate successors the task is to *give an account of phenomena* in terms of basic physics which in turn must be constructed in such a way that the phenomena can be accounted for. Later on basic physics is taken for granted and phenomena must be explained in its terms. This is how the idea of *saving* the phenomena

(rather than giving an account of them) arises. Cf. Fritz Krafft, 'Der Mathematikos und der Physikos – Bemerkungen zu der angeblichen Platonischen Aufgabe, die Phaenomene zu retten' in *Alte Probleme – Neue Ansätze Drei Vorträge von Fritz Krafft*, Wiesbaden 1965. An account of phenomena deals with the nature of things. The auxiliary assumptions that are used to save the phenomena have no such pretensions. The distinction is prepared by *Physics* B 2.

[60] The attempts start with Ptolemy, *Planetary Hypotheses*, are continued by Arab astronomers of the 11th and 13th centuries who demand a realistic account of planetary motions and last until the 16th century when the system of Alpetragius is taken up by Purbach. Cf. Duhem, *Le Systeme du Monde*, vol. ii, 130ff.

[61] Cf. *De Revolutionibus* i 5–8 with Birkenmaijer's comments in footnote 82ff of G. Klaus (ed.), *Copernicus über Kreisbewegung*, Berlin 1959. Cf. also Zahar, Einstein, II, p. 241.

[62] "Copernicus was not aware of his own riches" writes Kepler, *Mysterioum Cosmographicum*, Ch. i, fn. 4 by which he means that Copernicus was not aware (a) that he was dealing with a research programme rather than with a single theory and (b) that this research programme was capable of producing novel predictions. One novel prediction, mentioned by Kepler, is that the synodic anomaly of the planets depends on the *true* motion of the sun not on its mean motion as had been assumed by Ptolemy.

[63] The Aristotelian methodology of finding a point of view that could accommodate facts by degenerating absorption was not the only alternative in existence. Neoplatonists and reformers of magic such as Agrippa von Nettesheim emphasised the *hidden virtues* of objects that were not accessible to normal observation and had to be brought forth by special methods. In our period they may well have been the only thinkers to come close to the methodology of scientific research programmes. Cf. E. H. Kocher, *Science and Religion in Elisabethan England*, New York 1951.

[64] Cf. text to notes 11 to 13.

[65] W. Pagel, *William Harvey's Biological Ideas*, New York 1967; 'William Harvey Revisited., *History of Science* **8** (1969). Cf. also Charles B. Schmitt, 'Towards a Reassessment of Renaissance Aristotelianism', *History of Science*, 1974. Schmitt's problem is: what is the reason for the "dogged persistence of the Aristotelian tradition in the sixteenth and seventeenth centuries"? (p. 171). His answer: the ability of the systen "to adopt itself and to absorb within itself many novel elements" (p. 178). The success of such absorption in the eyes of the contemporary indicates that the methodology of research programmes was not universally taken as a basis of evaluation. In addition there were progressive developments.

[66] *loc. cit.*, p. 237.

[67] The Lorentz contraction involves real forces and should therefore lead to oscillations. No oscillations are to be expected on Einstein's account. No oscillations were found in the experiment of Wood, Tomlinson and Essen, *Proc. Roy. Soc.* **158** (1937), 606.

[68] Zahar, p. 100.

[69] p. 232.

[70] The light principle "is thrown out with no justification whatever" (p. 232) while Lorentz "explained Michelson's result in a non ad hoc way ... and he explained the invariance of c" (P. 122). Also "there was no build up – of unsolved anomalies which Einstein's theory dissolved better than Lorentz'" (p. 238).

[71] p. 237.

[72] p. 249.

[73] pp. 233, 235, footnote 2.

[74] p. 233.

[75] p. 235 footnote 2.
[76] p. 238.
[77] pp. 252, 224f.
[78] $E = mc^2$ as, well as the one-sided nature of electromagnetic emission was obtained by Poincaré in 1900 without invoking the relativistic point of view. Cf. *Archives Neérland.* **5** (1900), 252. Hasenöhrl arrived at a more restricted result four years later. It is quite true that Lorentz himself gives no indication "that the rest mass is a variable quantity" (Zahar, p. 249) – but we are not talking about Lorentz, we are talking about his research programme.
[79] Cf. Einstein's comparison between "constructive theories" such as the theory of Lorentz and "theories of principle" such as the special theory of relativity in his autobiographical notes. *Albert Einstein: Philosopher-Scientist* Evanston 1951, p. 53.
[80] Thus the derivation on pp. 246ff might have been carried out by Lorentzian, though he would have given a very different interpretation to its result.
[81] "Is it not hybris to try to impose some a priori philosophy of science on the most advanced sciences? ... I think it is", Lakatos, p. 121.
[82] Stanley Goldberg 'In Defense of Ether', *Historical Studies in the Physical Sciences*, Vol. ii, Russell Mc Cormach (ed.), Philadelphia 1970, p. 97.
[83] Clark, p. 1.
[84] p. 3. This definition of 'external' makes of course many of Einstein's reasons external.
[85] p. 3.
[86] p. 4.
[87] p. 4.
[88] p. 5.
[89] Cf. footnote 79.
[90] Peter Clark no longer believes the phenomenological theory to be without any heuristic. But, he says "that heuristic was a weak one in a very specific sense, namely that it was *fact dependent* in much the same way that the Ptolemaic heuristic was." (Letter of March, 26, 1975).

ARTHUR FINE

THE YOUNG EINSTEIN AND THE OLD EINSTEIN

> Fashionable 'sociologists of knowledge' – or 'psychologists of knowledge' – tend to explain positions in purely social or psychological terms when, as a matter of fact, they are determined by rationality principles. A typical example is the explanation of Einstein's opposition to Bohr's complementarity principle on the ground that 'in 1926 Einstein was forty-seven years old. Forty-seven may be the prime of life, but not for physicists.' (I. LAKATOS in 'Falsification and the Methodology of Scientific Research Programmes'.)

In 1911 Einstein, then thirty-two years old, held the post of Professor of Theoretical Physics in the German University at Prague. He had indicated to his friend and former classmate Marcel Grossmann his interest in returning to Zürich to teach at the Polytechnic (The Federal Institute of Technology). The dossier in support of his appointment there contained a letter from Henri Poincaré who wrote, in part,

> What I admire in him in particular is the facility with which he adapts himself to new concepts and knows how to draw from them every conclusion. He has not remained attached to classical principles and when presented with a problem in physics he is prompt to envisage all its possibilities. ([12], p. 134)

This same liberated quality of Einstein's thought was emphasized in the letter written two years later by Max Planck, Walter Nernst, Heinrich Rubens and Emile Warburg in proposing Einstein for membership in the prestigious Royal Prussian Academy of Science. After summarizing Einstein's accomplishments in physics they wrote,

> Apart from his own productivity Einstein has a peculiar talent for probing alien, original views and premises and from his experience judging their inter-relationship with uncanny certainty. ([12], p. 145)

These words were written about the young Einstein, the Einstein who had yet to create the general theory of relativity. The old Einstein, however, was to receive a very different testimonial. From Werner Heisenberg,

> Most scientists are willing to accept new empirical data and to recognize new results, provided they fit into their philosophical framework. But in the course of scientific progress

R. S. Cohen et al. (eds.), Essays in Memory of Imre Lakatos, 145–159. *All Rights Reserved*
Copyright © 1976 by Arthur Fine.

it can happen that a new range of empirical data can be completely understood only when the enormous effort is made to enlarge this framework and to change the very structure of the thought process. In the case of quantum mechanics, Einstein was apparently no longer willing to take this step, or perhaps no longer able to do so. ([2], p. x)

And from his good friend Max Born,

At first there were quite a number of serious scientists who did not want to know anything about the theory of relativity; conservative individuals, who were unable to free their minds from the prevailing philosophical principles.... Einstein himself belonged to this group in later years; he could no longer take in certain new ideas in physics which contradicted his own firmly held philosophical convictions. ([2], p. 72)

This is the tale told of the old Einstein.

In one form or another, the biographies have repeated this tale suggesting the contrast between the young man free in his thought whose special talent was to wrest the truth from 'alien' points of view and the old man whose mind traveled only in the circle of ideas of his youth no longer able to cope with the new ideas of the developing quantum physics. To understand the transition between this young Einstein and this old one we might attend to Einstein's own words of 1934 written in memorial to Paul Ehrenfest and about Ehrenfest's struggle with the quantum theory. Einstein wrote,

Added to this was the increasing difficulty of adaptation to new thoughts which always confronts the man past fifty. I do not know how many readers of these lines will be capable of fully grasping that tragedy. ([4], p. 238)

These are words from the heart and there is a strong temptation to see them as reflecting Einstein's own struggle with quantum theory for, as he wrote to Born in 1944,

Even the great initial success of the quantum theory does not make me believe in the fundamental dice-game, although I am well aware that our younger colleagues interpret this as a consequence of senility. ([2], p. 149)

In the same vein, reflecting on the permanence (or otherwise) of his work in physics, after an operation in 1949, he wrote to his lifelong friend Maurice Solovine,

The current generation sees in me at the same time a heretic and a reactionary, one who has, so to speak, outlived himself. ([3], p. 94)

Nevertheless I think we should resist the temptation of applying Einstein's diagnosis of Ehrenfest's tragedy to himself. For not only did Einstein's thinking about the quantum theory keep pace with the developing physics, his thinking all along was at least as radical in its break with the conceptions of classical physics as were the interpretations promulgated by Niels Bohr and the visitors to the Institute in Copenhagen.[1]

Einstein's own contributions to the development of quantum theory are well-known (and have recently been studied in detail by Martin Klein; see [7] and references in the 'Bibliography' that follows on pp. 332–333). The background to the mature theory lies in the work on radiation of Planck and Einstein. It was this work, particularly in the hands of Einstein, that set the problem of the dual nature of light. For despite his interest in a physics based on the continuum, Einstein emphasized the necessity for assuming a grainy structure to light and with it the problem of reconciling the wave and particle aspects. I do not want to rehearse Einstein's other contributions to quantum theory here. I do, however, want to emphasize the way in which Einstein exercised his special talent for probing alien modes of thought during the period of development of quantum theory, a period ending around 1930 when the current quantum formalism was well in hand. For, as we shall see, Einstein was less bound by classical preconceptions than were most of his contemporaries and, in many instances, it was Einstein who first piloted the flights of fancy that marked the rise of the theory.

The first of these flights was the already mentioned introduction of light quanta in 1905. This move was, of course, a radical break with the continuum physics of the time and was not to become an acceptable option in physics for many years. Equally important, however, is to notice the method of treating physical concepts that Einstein used to find room for the light quantum hypothesis. In the famous 1905 paper 'On a Heuristic Viewpoint Concerning the Production and Transformation of Light' Einstein wrote,

> However, it should be kept in mind that optical observations refer to values averaged over time and not to instantaneous values. Despite the complete experimental verification of the theory of diffraction... it is conceivable that a theory of light operating with continuous three-dimensional functions will lead to conflicts with experience if it is applied to the phenomena of light generation and conversion. (Quoted in [5], p. 52)

In an unpublished manuscript written around 1910 Einstein reiterated this theme, saying that Maxwell's electrodynamics provided only time-averaged values and hence could be viewed as merely an intermediate stage in the development of a theory that would deal with discrete but instantaneous values of quantities like energy. (See [5], p. 53.) One cannot help noticing the similarity between this line of thought and the well-known analysis of simultaneity that Einstein proposed in the 1905 paper on special relativity. In both cases Einstein examined the way physical concepts are actually applied in order to show that the full range of application was not completely determined. He then went on to develop new rules that extended the concepts into this not-determined area. Although these rules were different from the extrapolations one might make on the basis of then accepted theories, they were nevertheless constrained by requiring that the refined concepts be co-extensive with the originals inside the already determined area of application, or at least approximately so. Thus in the background to relativity Einstein noticed options about distant simultaneity, other than the use of infinitely fast signals, which yield the same locally simultaneous events. And behind the quantum postulate Einstein found options about, say, when a particular energy value occurs which yield, in good approximation, the apparent continuity of energy in classical electrodynamics. I would characterize this method as that of examining the limits of application of a concept in order to make room for a constructive theoretical refinement. No doubt Einstein learned the first part of this method from the analysis of concepts in Hume and Mach but the constructive use of the analysis is characteristically his own. (See, for example, [13], p. 391, where Einstein acknowledges the influence of Mach and Hume, especially that of Hume whose *Treatise* he studied during the early years in Bern.)

In his 1916 obituary for Mach, Einstein described this method as follows:

> ... concepts which have proved useful for ordering things easily assume so great an authority over us, that we forget their terrestrial origin and accept them as unalterable facts. They then become labeled as 'conceptual necessities,' '*a priori* situations,' etc. The road of scientific progress is frequently blocked for long periods by such errors. It is therefore not just an idle game to exercise our ability to analyse familiar concepts, and to demonstrate the conditions on which their justification and usefulness depend, and the way in which these developed, little by little, from the data of experience. In this way they are deprived of their excessive authority. Concepts which cannot be shown valid are removed. Those which had not been coordinated with the accepted order of things with sufficient care are corrected, or they are replaced by new concepts when a new system is produced which, for some reason or other, seems preferable. (Quoted in [2], p. 159)

The analysis of space, time and motion represented by the four-dimensional manifold of general relativity is an application of this method. Thus in the shift from special to general relativity, a shift that brought Einstein away from Mach's sensation-based positivism toward Planck's realism (see Holton, [6]) this method of conceptual analysis remained firm. Most of Einstein's later writings on scientific method refer to this process of conceptual refinement as a basic tool in the kit of the theoretical physicist.

Thus from the beginnings of his work on quantum phenomena Einstein employed a method of investigation that was bound not to respect classical theories and presuppositions. His method looked forward to a series of revisions of the classical concepts by first seeking the limits of their application in experimental situations and then by building a theory to refine them. Contrary to the tale of the old man locked into a circle of fixed ideas, it was precisely this open-ended and questioning attitude towards the concepts of classical physics that brought Einstein into conflict over the quantum theory.

This aspect of Einstein's relationship to the quantum theory is brought out in an exchange of letters between Einstein and Erwin Schrödinger. Einstein had been corresponding with Schrödinger in 1926 about problems of gas degeneracy, the so-called 'Einstein-Bose' statistics in the quantum theory of gases. Apparently Planck pointed out Schrödinger's development of the famous 'wave equation' to Einstein, who seems to have gotten it garbled. Thus on April 16, 1926 Einstein wrote to Schrödinger arguing that the additivity of energy values necessitated an equation different from the one he (mistakenly) attributed to Schrödinger. Einstein then made the 'correction' (deriving, in fact, the time-independent Schrödinger equation!). Within a few days Einstein realized his mistaken attribution and wrote again to Schrödinger to acknowledge the error. Before receipt of this acknowledgement, however, Schrödinger had written back to point out Einstein's mistake in memory. He wrote,

... the whole thing would certainly not have originated yet, and perhaps never would have (I mean not from me), if I had not had the importance of de Broglie's ideas really brought home to me by your second paper on gas degeneracy. The objection in your last letter makes me even happier. It is based on an error in memory. The equation [which Einstein had attributed to Schrödinger] is *not* mine, as a matter of fact, but my equation really runs *exactly* like the one that you constructed free hand.... I am, moreover, very grateful for this error in memory because it was through your remark that I first became consciously aware of an important property of the formal apparatus. [Presumably, it is the superposition principle to which Schrödinger alludes.] Besides, one's confidence in a formulation always increases if one – and especially if *you* – construct the same thing afresh from a few fundamental requirements. ([9], pp. 26–27)

Here then we see Einstein open to the ideas of Schrödinger and so much on top of the problems of the quanta that he is actually able to reconstruct the Schrödinger equation from fundamentals, one of which is a restricted version of the principle of superposition, a principle that has turned out to be basic to the whole of quantum theory.

In the next letter (April 26, 1926) Einstein displays his knowledge of the development of matrix mechanics by Heisenberg and Born. He thinks this formalism does not satisfy superposition and for this reason he remarks to Schrödinger that "the Heisenberg-Born route is off the track." ([9], p. 28) Earlier (March 7, 1926) he had written to Hedi Born, "The Heisenberg-Born concepts leave us all breathless, and have made a deep impression on all theoretically oriented people. Instead of dull resignation, there is now a singular tension in us sluggish people." ([2], p. 88) By December 4, however, he wrote to Max Born, "Quantum mechanics is certainly imposing. But an inner voice tells me that it is not yet the real thing." ([2], p. 91) If that inner voice included the conviction that the superposition principle must hold but that, as he had written to Schrödinger, it actually fails in the matrix mechanics, then Einstein was right in principle but wrong in fact. For, as Schrödinger was the first to show (in March of 1926), the wave and matrix mechanics are formally equivalent and hence superposition holds in both representations. Yet here too we see Einstein openminded, receptive to new developments and quite properly subjecting them to his keen sense, born of his own work on quantum phenomena, as to what principles might be validly generalizable. How then does Einstein's open-ended method of conceptual refinement bring him into conflict with the developing quantum theory?

The conflict comes out in the letter of May 30, 1928, from Schrödinger to Einstein. Schrödinger enclosed a copy of a letter from Niels Bohr, a letter that Bohr had wanted Einstein to be aware of. Commenting on a recent paper by Bohr, Schrödinger had written him about the limitation on the applicability of the concepts of position and momentum embodied in the Heisenberg indeterminacy relations. Schrödinger expressed the view that these concepts will have eventually to be replaced by new ones that apply not approximately but exactly. He went on to say that these new concepts will be hard to find, so deep is our feeling for the classical concepts of space, time and causality. (See the editor's note 14, [9], p. 29. The original letter is in the Einstein Archives, microfilm idex I B 1, number 22.) Bohr responded in a way that was to become characteristic of his view of quantum theory. He wrote that he strongly disagreed with Schrödinger's emphasis on the necessity for the development of new concepts. He contended that we have only the old concepts with which to organize experience for only these seem to be linked with the human capacity for conceptualization. (I paraphrase here from the copy of Bohr's letter in the Einstein Archives, microfilm index I B 1, number 22. For a similar argument see also the 1929 address by Bohr reprinted in [1], especially pp. 94–96.) This theme, that we must stick with the classical concepts and can only attend carefully to their limitations, was a perennial one in Bohr's later discussions of the quantum theory. Later, however, he sometimes shifted his argument for it to considerations involving the necessity for an unambiguous, intersubjective language in which to communicate laboratory experience. Einstein's response to this controversy came in his letter to Schrödinger of May 31, 1928,

> Your claim that the concepts p, q [momentum, position] will have to be given up, if they can only claim such 'shaky' meaning seems to me to be fully justified. The Heisenberg-Bohr tranquilizing philosophy – or religion? – is so delicately contrived that, for the time being, it provides a gentle pillow for the true believer from which he cannot very easily be aroused. So let him lie there. ([9], p. 31)

Here then is the root of the conflict.

Einstein, with his method of conceptual refinement, was quite ready to scrutinize the concepts of classical dynamics and kinematics with great care, he was prepared to acknowledge their limitations and he was even willing to take seriously Schrödinger's suggestion that they will eventually have to be replaced. It was Bohr who balked at the idea that one might give up the classical concepts and it was then Bohr who worked out the method of complementary descriptions in order to save these very concepts. This is the method that Einstein castigates as a "tranquilizing philosophy." Thus the tale of Einstein grown conservative in his later years is here seen to embody a truth dramatically reversed. For it is Bohr who emerges the conservative, unwilling (or unable?) to contemplate the overthrow of the system of classical concepts and defending it by recourse to those very conceptual necessities and *a priori* arguments that Einstein had warned about in his memorial to Mach. Whereas, with regard to the use of classical concepts, Einstein's analytical method kept him ever open-minded, always the gadfly who would not be tranquilized.

These considerations should help dispel the myth of the Einstein grown conservative (reactionary, senile, etc.) when faced with quantum theory. But they do not yet display the interesting fine structure of Einstein's dispute. To do that one must draw out the deep methodological similarity between Einstein's approach to the theory and that of the Copenhagen school. Both parties to the dispute shared the methodological heritage of Mach, the method of conceptual analysis that seeks to find the limits of application of a concept (or a system of concepts). Mach (as earlier Hume and Berkeley and others) had used this method only as a critical tool; as a tool, that is, for eliminating or severely limiting the use of a concept (as in Mach's attack on Newton's concept of absolute space). Einstein well understood this aspect of Mach's method. He expressed it graphically to Michele Besso in 1917, "I do not inveigh against Mach's little horse; but you know what I think about it. It cannot give birth to anything living; it can only exterminate harmful vermin." (Quoted in [6], p. 240. See [13], p. 114 for the original letter.) In Einstein's hands the search for the limits of application of a concept became only the preliminary step in the constructive method of conceptual refinement. His way of making Mach's horse fertile was to supplement the analysis of concepts with a good theory that employs new concepts constrained to relate to the old ones in the following two ways. The first constraint is that the new concepts must coincide with (i.e., be co-extensional with) the old ones, at least approximately, in the central region where the application of the old ones is fully determinate. The second is that the new concepts themselves must apply in a determinate way beyond the region central to the old. The first of these constraints merely expresses the requirement that the new concepts generalize the old ones. The second condition represents the scientist's hope that a good theory will penetrate into nature more deeply than its predecessors; it should advance our understanding by allowing us to put to nature certain questions which were not clearly expressible on the basis of the old concepts and it should lead us to expect that such questions have determinate answers.

Bohr, too, tried to utilize Mach's method in a constructive way. Like Einstein he recognized that the construction of a good theory calls for a refinement of the old concepts. And the first constraint, pioneered by Einstein's early work on relativity and the light quantum hypothesis, was one that Bohr employed in his 1913 atomic theory. The formal statement of this constraint became known as 'the correspondence principle' and this principle was one of the major guidelines in the development of Heisenberg's version of quantum theory. For Bohr the correspondence principle was the means to insuring that quantum mechanics is a 'rational generalization' of the classical theory. He expressed these considerations in 1929 as follows,

> A method for making such concepts [mass and charge] useful in other fields than that in which the classical theories are valid has been found, however, in the demand of a direct concurrence of the quantum-mechanical description with the customary description in the border region where the quantum of action may be neglected. The endeavors to utilize in the quantum theory every classical concept in a reinterpretation which fulfills this demand... found their expression in the so-called correspondence principle. However,... only in recent years has it been possible to formulate a coherent quantum mechanics which may be regarded as a natural generalization of the classical mechanics,... ([1], p. 110)

So far, then, are Einstein and Bohr in agreement as to what constitutes a constructive use of Machian analyses. They diverge, however, at the next step. For, unlike Einstein, Bohr did not feel that the new concepts must extend the range of application of the old in a uniform way.

Bohr's position over this second stage in the constructive refinement of a system of concepts is not easy to state, for it embodies Bohr's notoriously obscure doctrine of complementarity. The underlying metaphor of complementarity is that we must view the world through the old concepts, that when we attempt to look into the quantum domain in this manner we must select which of these concepts to employ but that what we see there will depend on our point of view (i.e., on the chosen concepts) in such a way that different perspectives cannot be pieced together into one unitary picture of the quantum world. In terms of the methodology for conceptual refinement that I have been exploring above, one can shift the metaphor this way. According to Bohr the system of classical concepts splits into mutually exclusive packages if one attempts to use this system outside the region of macroscopic physics, the region where all the concepts of the system have a determinate, simultaneous application. To get beyond this central core one must select which package of concepts to use. Different selections will enable one to get beyond the core in different ways. But the results of these different explorations do not combine into some unified picture of a region beyond the core. Bohr thus views the product of conceptual refinement as a wheel-like structure: a central hub from which there extends a number of disjoint spokes. Different explorers can move out separately along different spokes but, according to Bohr, the reports they send back will not enable one to piece together an account of some region between the spokes or of a rim that connects them. Thus the new conceptual structure for Bohr looks like the steering wheel of an old-fashioned ship. The beauty of this steering mechanism and the aspect that Einstein saw as a 'gentle pillow' is that it enables one to navigate into the quantum domain using only a classical chart in any given direction.

Einstein's dispute with Bohr (and others) is a dispute over this wheel-like structure. Einstein asks whether the spokes must really be disconnected, could there not at least be a rim? This is the question as to whether the quantum theory allows for a realist interpretation, a picture of the world as a single entity with a rich set of simultaneously determinate properties that are observer-independent. And Einstein asks whether the spokes must really be made of the same material as the hub. Must we, that is, stick with just the classical concepts? I do not want to tell here the story of how these questions were asked and what attempts were made to answer them. It is sufficient to recognize these three points about the dispute. First, it was Bohr not Einstein who felt bound by the classical concepts. Second, that both Bohr and Einstein shared Mach's critical method of searching for the limits of application of concepts and that they both wanted to go beyond this purely critical analysis to attempt some sort of constructive extension that would generalize the limited concepts. Third, that the dispute between them was a dispute about the possibilities and nature of this constructive refinement in the context of quantum theory. This last point is a crucial one for it shows that the dispute was not over accommodating experimental facts, or understanding the new mathematics of the quantum formalism, or taking in new physical ideas, or the like. It was a dispute at the level of methodology over what processes of concept formation are progressive; i.e., constitute a fundamental scientific advance.

There is one further aspect of Einstein's approach to quantum problems that I want to emphasize, for it too represents an early insight of his that later became part of the accepted outlook of quantum theory, which point of view was then used to suggest Einstein's supposed recidivism. (Unfortunately, the criminal overtones of this word, recidivism, are not too strong. In a letter of March 23, 1936 Schrödinger recounted to Einstein a conversation in which Bohr called the attitude towards quantum theory of Schrödinger, von Laue and Einstein 'appalling' and 'high treason.' See the Einstein Archives microfilm index I B 1, number 22.) What I have in mind is the easy-going pragmatism of the quantum theory. For the theory is most often seen as merely a coherent framework for the deduction (or 'prediction') of experimental consequences and a framework whose principle justification lies in its uncanny success in this enterprise. Einstein's persistent refusal to accept the quantum framework as a part of fundamental physics might seem to run counter to this pragmatic point of view. Because of this, no doubt, many of the brightest students of physics have been directed by their mentors away from examining Einstein's actual position. Alfred Landé, for example, was interested in Einstein's thought but he writes

> The more pragmatic Sommerfeld... warned his students, one of them this writer, not to spend too much time on the hopeless task of 'explaining' the quantum but rather to accept it as fundamental and help work out its consequences,... ([8], p. 460)

It is Einstein, then, who is seen as engaged in the hopeless task of explaining the quantum rather than accepting it and working out its consequences. Curiously, or at least it ought to seem curious to those who see things that way, Einstein's procedure in the development of relativity theory was just the opposite of this. In the case of relativity Einstein looked for what he called "generalizable facts" (such as the constancy of the velocity of light) which could be elevated to the status of principles. The theory itself was to be merely a framework for drawing out the consequences of such principles. (See Einstein's account in [11], Vol. 1, esp. pp. 51–53.) Einstein called a framework of this sort a "principle-theory" and contrasted it with a "constructive" theory, one that "attempts to build up a picture of the more complex phenomena out of the materials of a relatively simple formal scheme from which they start out." (Quoted in [10], p. 62, which contains an illuminating discussion of the role of Einstein's option for a principle-theory in the genesis of special relativity.) The charge against Einstein then is that whereas in the case of relativity he was content with the pragmatism of a principle-theory, in the quantum case he made the unpragmatic demand that one build a constructive theory.

But if this were the case then why was Einstein receptive and so enthusiastic to Bohr's 1913 atomic theory (a principle-theory *par excellence*), as he was later to be excited by the 1925 matrix mechanics of Heisenberg and Born? The answer is that Einstein's own intensive work on the quantum problems once again led him, quite early, in the direction of the mature quantum theory, in this instance towards the feeling that a pragmatic framework (or principle-theory) was the best hope. After a period of concentrated work on quantum problems Einstein wrote to Michele Besso on May 13, 1911,

> I no longer ask whether the quanta really exist. Also I no longer try to construct them because I now understand that my brain is incapable of proceeding in this way. Rather I look most carefully at the consequences [of this quantum representation]... ([13], pp. 19–20)

Here then we find Einstein expressing in 1911 the attitude that Sommerfeld later tried to impress on Landé and that is characteristic of the pragmatic quantum theory: look for the consequences. That Einstein maintained this attitude constantly, is evident from a letter to Born in 1953 where he wrote, "I am quite satisfied if we have the machinery for making predictions, even if we are unable to understand it clearly." ([2], p. 209) It is true that Einstein never felt that merely looking at the empirical consequences was sufficient justification for accepting a physical theory. He always emphasized that one must also look for inner simplicity and perfection among the constellation of theoretical principles. But it is nevertheless clear that from 1911 on he was open to accepting the sort of theory that eventually emerged – a framework or 'principle-theory' – as the fundamental theory of the quantum domain.

I think our story is nearly done. If indeed the thought of the old Einstein was fixed in the circle of ideas of his youth then it travelled among the following stations well-known in the current versions of quantum theory: a firm belief in the existence of quanta, a constant attention to what are the limits of the concepts of classical physics, an openness to the revision or replacement of these concepts, and the sense that a theory construed as a framework for drawing empirical consequences is the right sort of theory for the quantum domain. In travelling this route we have seen how Einstein hit upon the notions central to the technical development of quantum theory (the correspondence principle and the principle of superposition) and how quickly, in fact, he was able to assimilate this technical development. Einstein's dispute with the quantum theorists was not a dispute due to hardening of the channels of thought. It was a genuine dispute about the conceptual adequacy of the developed theory. In pressing his side of the story Einstein was no doubt frustrated for the reasons he expressed to Solovine in a different context; namely, "because physicists have little understanding for logical and philosophical arguments." ([3], p. 106)

In the end Einstein was more radical in his thinking than were the defenders of the orthodox view of quantum theory, for Einstein was convinced that the concepts of classical physics will have to be replaced and not merely segregated in the manner of Bohr's complementarity. In 1936 he described the quantum theory which results from the decision to retain these concepts as "an incomplete representation of real things, although it is the only one which can be built out of the fundamental concepts of force and material points (quantum corrections to classical mechanics)". ([4], p. 88) He took the advice of the quantum theorists, which was his own to begin with, and attended carefully to the consequences of the theory, especially for macroscopic bodies. He argued that the theory was unable to account for (=predict) even the simplest of these phenomena unless we understand the theory never to treat of individual systems but only of statistical aggregates of such systems. So understood he had a deep respect for the theory and for its predictive success, which he expressed in 1936 in this way,

> There is no doubt that quantum mechanics has seized hold of a beautiful element of truth and that it will be a test stone for any future theoretical basis, in that it must be deducible as a limiting case from that basis,... ([4], p. 92)

Because, as he saw it, quantum theory is essentially statistical he felt that it could not be fundamental in the sense of providing a framework for all of physics. He hoped, instead, that the framework of general relativity – the space/time manifold and the analytical methods appropriate thereto – would enable one to establish the new concepts and the theoretical basis from which the quantum theory would emerge as a statistical approximation. For Einstein this recourse to the space/time manifold was neither a reversion to classical concepts nor a working out of entrenched philosophical commitments. (For example, in responding sympathetically to a suggestion by Karl Menger that one might be able to develop a probabilistic geometry to replace the space/time manifold Einstein wrote in 1948, "Adhering to the continuum originates with me not in a prejudice, but arises out of the fact that I have been unable to think up anything organic to take its place." ([11], Vol. 2, p. 686)) Einstein hoped that the space/time framework of general relativity would contain possibilities for the systematic treatment of individual quantum systems. He devoted his last years to exploring the possibility for thus finding an account that would penetrate the quantum domain more deeply than does the present theory. He did not succeed. But I think he was genuinely content to have made the effort and that he would be pleased to have this effort seen in the context of his own remark to Besso in 1950,

> I have learned one thing in my long life: It is devilishly difficult to get nearer to Him if one is not willing to remain on the surface. ([13], p. 439)

ACKNOWLEDGMENTS

Some of the work on this paper was supported by National Science Foundation Grant #GS-37820. The paper was written during the tenure of a Senior Fellowship from The National Endowment for The Humanities. I want to acknowledge the cooperation of the trustees of the Einstein estate in permitting me access to the Einstein microfilms and the pleasant assistance of the staff of the Rare Book Room of Princeton University's Firestone Library, where the microfilms are kept. I want to thank Judith Weinstein and Dr. Theodore Weinstein for their hospitality and kindness while I was gathering materials for this study. I owe Helene Fine credit for leading me to the metaphor describing Bohr's complementarity, and I owe thanks to Sharon Fine and to Dana Fine for help in clarifying the writing. Others have helped too and I hope that the paper itself will be a sufficient expression of my thanks.

University of Illinois at Chicago Circle

NOTES

[1] In the essay which follows I take Bohr to be Einstein's main protagonist in the debate over the quantum theory. I do this in part for stylistic ease but also because Bohr's discussions of the theory as it developed are the most challenging in the literature, just as historically they were the most influential.

BIBLIOGRAPHY

[1] Bohr, N., *Atomic Theory And The Description of Nature*, Cambridge University Press, Cambridge, England, 1961.

[2] Born, M., *The Born-Einstein Letters*, Walker and Co., New York, 1971.

[3] Einstein, A., *Lettres à Maurice Solovine*, Gauthier-Villars, Paris, 1956.

[4] Einstein, A., *Out of My Later Years*, The Philosophical Library, New York, 1950.

[5] Hermann, A., *The Genesis of Quantum Theory*, The M.I.T. Press, Cambridge, 1971.

[6] Holton, G., 'Mach, Einstein and the Search for Reality' in his *Thematic Origins of Scientific Thought*, Harvard University Press, Cambridge, 1973, pp. 219–259.

[7] Klein, M., 'Einstein, Albert' in C. C. Gillespie (ed.), *Dictionary of Scientific Biography*, Charles Scribner's Sons, New York, 1971, Vol. 4, pp. 312–319.

[8] Landé, A., 'Albert Einstein and the Quantum Riddle', *American Journal of Physics* **42** (1974), 459–464.

[9] Przibram, K. (ed.), *Letters On Wave Mechanics*, The Philosophical Library, New York, 1967.

[10] Schaffner, K., 'Einstein versus Lorentz: Research Programmes and the Logic of Comparative Theory Evaluation', *British Journal for the Philosophy of Science* **25** (1974), 45–78.

[11] Schilpp, A. (ed.), *Albert Einstein: Philosopher-Scientist*, Open Court, La Salle, 1969, third edition.

[12] Seelig, C., *Albert Einstein, A Documentary Biography*, Staples Press Ltd., London, 1956.

[13] Speziali, P. (ed. and transl.), *Albert Einstein-Michele Besso Correspondence, 1903–1955*, Hermann, Paris, 1972.

ERNEST GELLNER

AN ETHIC OF COGNITION

The following argument can be put against all and any methodology: presumably a methodology contains prescriptions which are more than just the requirement to observe the formal commands of logic. No-one who writes a book on method thinks he is merely replicating the precepts of consistency, non-contradiction and so forth. So, a methodology must have some meat which is not merely logic. But if it asserts, or presupposes, something over and above the formal requirements of logic, will not that *something else*, whatever it may be, have some implications concerning *the world*? And if so, can one not imagine or construct a possible world within which those implications are false, and within which consequently those recommendations are misguided? And if such a world is conceivable, obviously we cannot say, in advance of all inquiry, that such a world is not the *real* world. But what use is a methodology which prejudges the nature of the world we are in, before we have investigated the matter, and before we have any right to an opinion about it? So we cannot use it *before* we inquire, as a tool of investigation, or as a guide to what tools to use. And we certainly do not wish to use it *after* our inquiry is over. For one thing, it is too late by then; for another, by then we can presumably enjoy some much more meaty conclusions, and will hardly have much time for the relatively thin and abstract doctrines of methodology, even if, from another viewpoint, they evidently were not thin and abstract enough.[1]

This argument can be used and I think has been used as an attack on the very idea of method: method, purporting to be more than mere logic, must have some substance. But method, being method, must also be neutral, and hence must have no substance. Enterprise which is simultaneously required both to have some substance and to have none, cannot satisfy the demands made of it. Hence, no method is possible.

I wish to accept the argument, but use it, not as an attack on the very idea of method, not as a reduction ad absurdum of the very notion of method (which is the spirit in which I think it was used), but on the

R. S. Cohen et al. (eds.), Essays in Memory of Imre Lakatos, 161–177. *All Rights Reserved*
Copyright © 1976 by D. Reidel Publishing Company, Dordrecht–Holland

contrary as a means of highlighting the true nature of methodology. Yes, method is more than logic; consequently it does indeed prejudge some results of inquiry prior to the very start of the inquiry, eliminating certain possibilities, excluding certain imaginable worlds; *and rightly so.* Some worlds are indeed eliminated, in an a priori fashion, by philosophical methodology. The aprioristic exclusion of those worlds is of the essence of the thing. But, though method is aprioristic and on points of substance at that, this admission should not be worn as a badge of shame. It is, on the contrary, a badge of honour. Certain worlds are *out.* We exclude them, not because we can know in advance (or, perhaps, ever) that the world we actually inhabit is not one of these excluded ones, but for some other reason or reasons. I for one think that those reasons happen to be good ones.

Which are the excluded worlds?

I am not sure I can give a general characterisation of *all* such excluded worlds; but it is I think possible to offer a general outline of *some* of them, of an important sub-class of the excluded worlds. Let us call them, without prejudice, the Ideologist's Worlds.

The world of an ideologist, the world satisfying the requirement of some well-constructed Ideology, has certain general traits. It is always a non-alienated world, in the important sense that in it the Real and the Rational sustain and support each other; the realm of fact and the realm of value are not separated from each other. The True and Good and the Beautiful converge and ratify each other. Such a world has a basic plot, in which the Good faces its opposite. Within it, this cannot be doubted, for this is central to the belief system in question, and were it to be in doubt, all else would tumble. But it need not be doubted, either: for the plot of our cognitive adventure neatly dovetails with the story of moral, political and other confrontation. In essence, within it, the story of our cognitive and exploratory adventures in this: the good men have cognitive equipment or attitudes or whatnot which make them perceive the real plot, whereas the evil and accursed ones are blinded, by forces or mechanisms specified in the overall story itself. So, the sheer distribution of faith and disbelief itself confirms the truth of faith. Only those devoid of Grace doubt the existence of God; only neurotic resisters doubt the insights of depth psychology; only class enemies fail to see the cogency of scientific historical materialism; only victims of rigid models of how language works spurn the mature Wittgenstein, etc.

You may say that there is a certain circularity in such forms of reasoning, and no doubt this is so. But note that this in no way constitutes a conclusive refutation of any vision of this kind. It isn't only that, from the inside, to a true believer, the circularity does not in the very least seem like a *circularity*; one has to be a nasty minded outsider-sceptic to see it as such. From the inside, it does not look like circularity at all, it looks like the consensus and convergence of so many manifest signs and indications, which one would have to be blind or wicked to ignore. Indeed it is precisely those wicked ones who insist perversely on casting doubt, and supporting their cussedness by specious reasoning about circularity!

But the important point is not that, from the inside, it does not look or feel circular at all. What *is* important is the fact that we have *no* logical or independent way of proving that such a 'circular' world, or strictly speaking a world sustained by reasoning which seems circular to an outside and hostile critic (or outside and *hence* hostile), cannot exist. On the contrary: it could well exist. A world so constructed as to make its most important features manifest to the good, and obscured from the wicked, might well exist. Perhaps, indeed, it does exist; and perhaps *this* world is just such a world. There is nothing in the very least logically self-contradictory in such a supposition. It cannot be excluded by logic. And it cannot be excluded by fact either, for it is constructed precisely in a way such that all facts can be accommodated. It is only to the outsider or, in its own terms, to the wicked, that facts are 'accommodated', an expression which suggests wilful manipulation. From the inside, for a healthy undisturbed vision, all facts support and endorse the vision.

So a world of this kind *is* possible, and there is no purely logical reason for excluding its a priori. There is nothing self-contradictory about the idea of such a world. It is true that there is a certain suspicion-enhancing consideration, which runs as follows: if the real world is *not* like that, nonetheless a movement which possesses a well-constructed belief-system of this kind, will never have to admit that it is in error, even though, on our supposition, it *is* in error. But this merely shows that *if* the world is indeed different, then nevertheless believers in an ideologically neat world need never admit defeat. But that in no way proves that the world is not as they claim it to be.

Furthermore, there is a certain consideration on their side, which I shall call the Argument from Elementary Decency. Decent people live decent orderly lives in decent homes and good districts. People who live in filthy houses in areas of ill repute can hardly be surprised if some of that ill repute rubs off on them. What goes for houses and districts also goes for universes. Decent people live in decent universes. A decent universe is one with some measure of propriety and order. Knowledge and morality are not, in such an universe, strangers to each other. They dovetail. Even people not too sure that the universe is like that often feel that it ought to be, or that society ought to be such, or that, if things are not like that, philosophy really ought to do something about it.

I am not saying that this is a cogent argument. It is open to the objection that we did not make the universe. I for instance personally disclaim all responsibility either for the existence of the universe or for its general character. I had nothing whatever to do with it. I was not even consulted. Things are as they are. I did not make them. I just tell it like what it is, but that does not make me responsible. Still, though this would seem cogent, people are curiously sensitive to the view that you can tell whether a man is a bastard by the kind of universe he inhabits. The quality of badness rubs off onto bearers of bad news. So, there is a certain moral sentiment on the side of the assumption of an ideological universe.

When people speak of 'alienation' one of the things they nebulously mean is the overcoming of a world in which life and meaning diverge, in which there is a lack of fittingness and justice. The Hegelian-Marxist tradition for instance, sees the plot of human history in the overcoming of such alienation. Marx's transcendence of philosophy is not a form of positivism but means, if I understand it, that henceforth alienation will no longer be overcome by mere words and in a fictitious world, but in real life and in this world.

The idea or feeling of a 'meaningful' world seems rooted deep in our minds and moral expectations, perhaps in virtue of our earliest family experiences, perhaps through some cultural traditions, or perhaps for other reasons. Whatever the basis of these anticipations, they provide a powerful impulse towards the acceptance of an ideological world; for only such a world, in the sense here defined, can satisfy this craving.

The substantive content of certain methodologies in fact excludes (whatever else it may exclude as well) such worlds. Consider, in the simplest possible terms, what these methodologies actually prescribe. Take the Baconian version. It commands us to read the book of nature with attentive care and to find what is actually there, rather than what some other consideration may lead us to wish, expect, or deem appropriate. The Popperian variant allows us, indeed encourages us, to anticipate whatever we may wish, for any reason whatever; this doctrine professes total indifference to our motives and our sources of inspiration. All that matters is that, once we have been inspired, never mind how and never mind for what reason, that then our anticipation should be subjected to trial by the book of nature. Generically, it would be fair to class the inductivist and the Popperian philosophies together as species of *empiricism*, and no doubt there are further kinds.

What *is* empiricism, in general?

The rough but correct definition, i.e. the one which brings out what really matters about the doctrine, is also highly paradoxical. It runs approximately as follows: empiricism is the a priori exclusion of a certain class of possible worlds, namely those worlds which satisfy some very deep general moral yearnings, roughly indicated above.

The definition is paradoxical in as far as it underscores, as seems to me entirely proper, that empiricism is indeed itself an a priori doctrine. To attempt to establish it empirically, as Hume did (as a kind of finding about how the human mind happens to work), is not merely circular but more seriously objectionable. As a piece of genetic psychology it is blatantly false, for reasons which ought to have been obvious though they have only become familiar thanks to Noam Chomsky; as a schema for the construction of psychological theories it is quite outstandingly sterile. The merits of empiricism, which are very considerable, lie entirely in the field of normative epistemology. It is a cognitive ethic, a code of intellectual comportment. But even within this sphere, there are great and well-known difficulties about defining it correctly or in a satisfactory manner.

The most natural way of defining it is by invoking the notion of *experience*. I remember, at the time of the height of confidence of the linguistic school in philosophy and of rampant 'common sense', a very prominent linguistic philosopher ironising empiricism and its philosophical systematization in roughly the following words: are these thinkers saying any more than that we learn by experience? So what is new? (The answer to this question is of course that empiricism does not say merely that we learn from experience; it says he can learn *in no other way*. And that is very new, radical, and contentious.) Another prominent thinker of this persuasion thought he had solved or dissolved the problem of induction as follows: of course we learn from experience. (No folk saying could be false.) But learning from experience is not using some funny kind of argument. So there is no induction (for its principles would have been the principles of that unusual type of argument, and there is no such type). But if there is no such thing, there cannot be a problem of how to justify it. Problem solved. (In those days, problems were generally solved with such breath-taking simplicity and speed, though all this was sometimes spun out to article or book length.)

The key idea in empiricism is the sovereignty of experience, and above all the exclusive and sole sovereignty of experience, in cognitive matters. What seems a trite truism (we learn by experience) becomes an unbelievably daring, radical, destructive, and difficult doctrine if reformulated more strongly so as to say – we learn in no other way. Let us leave aside for a moment the problem of how such a doctrine, or rather, such an ethic, can be justified or supported. There are grave enough problems concerning the definition of this view. These problems hinge on the definition of the notion of 'experience', which is used by our preliminary account of 'empiricism'. Intuitively and naturally, 'experience' is what we actually 'encounter', come across, sense, etc., that which we can't stop coming, before we start 'adding' to it. The trouble with this natural and intuitive notion of experience is well known.

The trouble is, to put it in a simple way, that what is simply there and in front of us, 'experienced' by us, is absolutely saturated by our expectations, our concepts, our culture. It is not wholly determined by these things, of course; we cannot altogether think away or transform reality by our thoughts. But though events are given, the *kind* of events they are, the way they are classified and interpreted, is not given, but made by men, their language, their culture. This is a familiar and well-supported theme and there is no need to expand it.

Empiricists who wish to articulate and refine their position, either in pursuit of precision for its own sake, or as a step on the way towards defending their philosophy, have one obvious strategy open to them: they can try to isolate define, characterise that *given* element which plays such an important role in their philosophy, which is the ultimate cognitive sovereign, the intellectual arbiter. I do not think they have been successful in this. It is easy enough to convey, in a rough and ready way, what we mean by experience; but it is difficult to do so in a non-metaphorical, and more than sketchy way. Perhaps empiricist philosophers will be more successful in the future in this endeavour.

But I doubt it. I suspect the boot is on the other foot. It is not so much that we need to define 'experience' or 'the given', so as to identify that which we have made into our cognitive sovereign, so as to make all theoretical structures subject to this master, and ultimately accountable to his judgment. It is the other way round. The essence of empiricism is that all, but *all*, theoretical structures are accountable; that none can claim such an awful majesty as to be exempt from the indignity of inquiry and judgment; and that substantive theoretical systems so constructed as to elude and evade this indignity, are out. *Out*.

What we can define and exclude are the world pictures which are, by an a priori decree, *excluded*. We can so exclude a whole class of them, by indicating certain formal features which they possess – for instance that kind of dovetailing of moral and cognitive ideas which evades testing by morally disqualifying the sceptic. There are of course other devices such as the characteristic one of demanding assent with menaces. No attempt is made here to list them all. 'Experience' of 'the given' is then defined or identified only negatively, as that which is not under the control of any theoretical system, but independent of them all. This is metaphorical and formally unsatisfactory, but I doubt whether we can do much better. Professor H. H. Price's celebrated account of the meaning of 'sense data', in his *Perception*, really amounts to this. The term or notion of 'sense data' is of course an attempt to give a philosophic account of experience, breaking up into its alleged constituents, and using a term which has both a singular and a plural.

Thus empiricism is normally present as an account of how in fact we know (though experience), or sometimes as a metaphysic (the world is made up of experiences), such that, on either interpretation, certain worlds which defy or evade experience are, as a corollary or consequence of the initial position, excluded. On my account, all this is back-to-front.

The essence or real starting point of the position is *the exclusion of certain kinds of world,* in a quite a priori manner. A certain possible misunderstanding should perhaps here by prevented: people may assume that describing a theory as quite an aprioristic one, is an attempt to denigrate it. This is no way intended. Empiricism is an a priori doctrine; and it is also a *good* doctrine.

The question now arises how one can know that a doctrine normally read one way, should 'really' be read in another and reversed way. The answer is roughly that it is this reading of the theory which really brings us the reasons why the theory is as persuasive as it is.

Let us return to the general consideration of ideologies or belief systems. It is they who create the problem situation within which empiricism offers a putative solution. We have stressed one trait of ideologies, a trait in terms of which we have in fact defined them: they are so constructed that they are not refutable from inside. They generate, each of them, a world so neatly rounded off that, though it is well fed by confirmations, no refutation can occur within it. But, though this is an important characteristic of ideologies, it is not the only one. Another one of the first importance is: they are all bi-lingual. Though they speak a language which is closed and generates a full-circle world, they must invariably – if only for the purposes of proselytising, defence, and so forth – also entertain diplomatic-conceptual relations with other worlds, and meet on more or neutral ground, or at any rate ground not wholly controlled by either side.

For practical reasons, they have no choice but to do this, though it inevitably makes them look naive, childish and contradictory. Inter-ideological détente inevitably involves double-think. A belief system within which certain principles are sacred, and define the very criterion of reality or truth, so that their recognition is the very touchstone of the integrity of thought, an entry qualification to any debate, will at the same time, when meeting outsiders on relatively neutral ground, argue for or defend those principles ... But *arguing* in their favour of course presupposes that there are some more authoritative principles, some more authoritative sources of evidence, in terms of which the sacred axioms of the system are justified; which automatically makes those previously sacred principles less than ultimate and hence less than sacred. Some theologians like Barth have noticed this and tried to turn it to good account: to defend the faith would be to contradict it; hence it is true without any defence. Neat and simple. Unfortunately, (a) this itself is an argument, however brief, and thus sins against its own premiss; and if you sin, you might as well sin strongly, and give us some better arguments, and (b) qua argument, it is easily and automatically applicable, if applicable at all, to *any* faith, and not exclusively to *the* faith.

In fact, most belief systems are not so summary and Barthian. They generally do argue, often at great length, sometimes with enormous scholastic elaboration. When they notice, as sometimes they do, the contradiction between the total and ultimate nature of their claims on the one hand, and the willingness to argue at all and thus appeal to more ultimate standards on the other, they develop elaborate and sophisiticated second-level theories to explain it all. Only Grace can give you faith but reason can help in this way or that. (This doesn't really solve the problem: either this help is not essential, in which case it is irrelevant, or it is essential, in which case the problem reappears with all its initial force.)

However, the point which concerns us here is not so much just what ideologists say when they speak in the outside or neutral language, but the fact that they speak it at all. The language has certain norms and conventions. Whilst conversing in Neutral-Speak, the ideologists must at least seem to observe its rules and procedures. Prominent amongst these, perhaps paramount, is the requirement not to beg the question, not to proceed in a circular manner. Your premisses must not, at least blatantly, contain that which you wish to establish. The premisses must be acceptable to both sides to the dispute. The game then proceeds in accordance with a rule which grants victory to him who establishes his own point from shared premisses. If the shared premisses do not already contain the conclusion, at least in a covert way, how can you ever get there? In fact, the contest is rather like those card games in which players, in addition to being dealt a certain number of cards at the start, also receive further ones in the course of the game. The subsequently dealt cards are known as 'evidence'. It turns up, and can favour one side or the other, notwithstanding their shared general premisses.

One way of looking at empiricism is as an attempt to codify and make general, absolute, the conventions of Neutral-speak. Hence that need to define 'experience', i.e. what counts as a legitimate card which can be dealt in the course of the game. But leaving aside the technical difficulties of properly codifying empiricism, by adequately defining its key terms, what are the reasons for holding it valid? It says, on this interpretation, that Neutral-speak is good speak, the best speak, not just an interim lingua franca for use in buffer zones; and that all Closed-speaks are bad speaks, and should be eschewed even when the power and diplomatic situation does not actually compel us to do so. We should give up Closed-speak even when we can get away with not doing so.

Why should we obey such an injunction? It *is* an injunction rather than a theory. It cannot really be translated into a positive world-picture (though of course the doctrine of the world-made-of-sense-data is an attempt to do so), for its substantive descriptive content is negative – it excludes certain kinds of worlds, as indicated. But anyway: what is the authority of this injunction?

Here one gets into those final stages of the regress of validation where things get very difficult and elusive. Above all it becomes hard to maintain a distinction between description and prescription or endorsement. I am inclined to use descriptive language and simply indicate the various factors which, in my view, incline us in a certain direction (i.e. towards the acceptance of the empiricist injunction); but I should be the first to admit that this is a bit spurious, that the descriptive account is accompanied by a sotto voce 'and rightly so'. Anyway, which are those considerations which so incline us (and rightly so) towards empiricism?

(1) The plausibility of the empiricist model.
(2) The correct (sociological) argument from illusion.
(3) The lessons of the Big Divide.
(4) The extension of Neutral-speak.
(5) Kantian-protestant ethics.

Each of these needs to be spelt out. I should find it difficult to assess their relative importance of persuasiveness. No significance attaches to the way in which they are ordered in this list.

(1) The key image in the empiricist model of cognition is very simple. It sees knowledge as built up from bits, 'data', which come and arrive 'from the outside', or are 'given'. The validity of knowledge hinges on this 'given' status. Truth is a kind of fidelity to the facts as supplied. Now any non-empiricist, apriorist theory faces a very difficult problem, whatever other reasons may support it: it must claim that there is some

kind of pre-established harmony between mind and reality. There are occasionally ways out of this: it may resign itself to a kind of autism and refrain from claiming external reference for our convictions at all, or it may claim that the harmony is ensured not by any constraint on the fact, but only by constraint on how *we* handle the facts. (In areas in which I hold apriorism to be valid, this is how I justify it.) But these recherché cases apart, apriorism must explain or justify this surprising harmony. It can only do so as a corollary of some wider theory – and what reasons can we have for holding such a wider theory to be true? (Some such theories seem attractive. The benevolence of God constituted such a theory for Descartes; natural selection and evolutionary adaptation provided a similar background premiss for various kinds of pragmatist.)

But on the whole, it is difficult to find, let alone justify, such background warrants for pre-established harmony.

No pre-established harmony; we are given what we are given, and cannot go beyond that. These are the persuasive insights, which supply the empiricist model with its strong, albeit not irresistible, appeal.

(2) The correct formulation of argument from illusion runs as follows: we observe fervently held belief systems which contain blatant and pervasive error. In fact we see many such, denouncing each other, and indeed often disavowing their own past. Why should *we* be exempt? (The incorrect formulation of the argument from illusion is about straight sticks which look bent in water, or is it the other way round?) This being so, should we not subject belief systems to scrutiny? – and how can they be tested, other than by the 'given', i.e. that which, whatever it be, is not controlled by *them*?

(3) The biggest, most conspicuous single fact about the human world is the Big Divide between what may roughly be called the industrial-scientific societies and the Rest. The former possess, for good and ill, enormous manipulative and predictive powers over nature (though not over social processes), endowing them with the means both of mass destruction and of mass affluence and leisure. The latter miserably scrape a precarious living by agriculture or even cruder methods. Their techniques for either feeding or killing people are slow, inefficient, and labour-intensive.

It would be idle to pretend that the Big Divide is properly understood; but it is widely and plausibly held to depend on the growth of 'science';

and science in turn is widely and plausibly held to depend on the systematic application of the 'experimental method' (and no doubt other factors, such as standardised and operationalisable notation and measurement).

But the experimental method and its rationale can best be expressed in terms of, or as a corollary of, the empiricist model. The two appear but aspects of each other, and they seem to confirm each other. Thus the cognitive history of mankind, the radical growth in knowledge and control of nature since the 17th century, and the contrast with other periods and regions of cognitive stagnation, seem to confirm a model which in any case has a great deal of intrinsic plausibility.

The importance of original, Comtian positivism was that it combined an articulation of the model with an historico-sociological awareness of the Big Divide. The trite and scholastic nature of many 20th century formulations of positivism are a consequence of an interest restricted to the model alone. The empiricist story of how an individual accumulates information about the world, is only useful if treated as an account of how some societies (but some only) have learned to investigate nature, and so as a parable of the Big Divide.

(4) As stated earlier, ideologists are bi-lingual and meet in neutral territory and converse in Neutral-speak. This on its own does not necessarily convict them of hypocrisy. They often possess quite elaborate theories explaining why all this is allowed, why a language permitting or imposing doubt may on occasion be employed notwithstanding the fact that in the True Language, doubt is inarticulable. They possess conceptual exit permits, rather the way in which authoritarian regimes grant temporary exit visas to their more trusted citizens.

But, in the modern world, the area in which the conventions of Neutral-speak apply has been steadily growing, and the spheres in which the Closed languages are spoken have been contracting. In fact, one more often hears apologetic and tortuous, devious justifications for the occasional, Sunday employment of faith-language, than for the daily employment of Neutral-speak. Modern ideologies are a bit like central banks in dictatorships which employ quite different exchange rates in different markets and diverse circumstances. (Modernist theology, for instance, does not believe in anything, but merely debases the conceptual currency.) But the occasions on which they dare use the 'real' exchange rate as they officially see it, use their 'internal' concepts, have been steadily contracting.

Looking at their comportment can only strengthen the doubts one already has in virtue of the Argument from Illusion.

(5) The appeal underlying the closed vision is, as stressed, the moral gratification of seeing a coherent, un-'alienated' world. But this is not the only moral vision operative amongst us. There is also what might be called the protestant-Kantian ethic, which spurns extraneous confirmations and sanctions. We hold certain values, it says, and we do not need nature to confirm them for us. Let the facts be what they will; our norms are not at their mercy. We take no bribes – not even from the Top Management.

Thus there is an important ethic which actually encourages an expectation that facts will not fit in with a closed, integrated world, but will form their own patterns which need not be, and in all probability will not be, such as would somehow dovetail with human aims. Whether it is a sour grapes doctrine, which merely accommodates itself to this disagreeable world, or an independently held and persuasive morality (as I hold), we need not here discuss. But I am sure that either way it is, by now, an additional prop of the empiricist vision.

I have now sketched out what I think empiricism 'truly' is, and why we hold it or at least are attracted by it. Now for a certain admission, and some comments on the implications of this view.

The admission is of course that in a sense, empiricism is an ideology amongst others, though it is an ethic of cognition, and its substantive prejudgements are indirect and negative – a *differentia* which is all we can invoke to save empiricism from falling inder its own axe. It is an ideology, though, in my view, a very very good one. As already conceded, it cannot refute its rivals from neutral, shared premisses. Like the hero of the fairy tale who cannot be harmed by the ferocious evil spirits, as long as he does not step out of the chalk circle sanctified by holy water, they cannot be demolished as long as they remain within their own circle of ideas. Secondly, the empiricist model is itself a kind of complete circle. Data, whatever they be, are sovereign, and cognitive contentions are either about nothing at all (autistic in some way or other), or about the data; and only the data can decide. This simple and persuasive model, underlying empiricist epistemologies and metaphysics, can hardly be destroyed on its own terms. Furthermore – and this is perhaps the most crucial part of the admission – empiricism is, in practice, fluently bi-lingual. Its bi-lingual fluency is so great that it is almost a bit worrying.

What basically happens is this. As a 'prior philosophy', in Quine's sense – an independently established self-sustaining principle, prior to any specific inquiry, and one conferring legitimacy on more specific procedures – empiricism derives its authority from consideration (1), from the plausible and persuasive abstract model, the repudiation of either autism or pre-established harmony. But it would be dishonest to pretend that those of us who are highly sensitive to the claims of empiricism, are persuaded by this philosophically pure consideration *alone*. We are swayed by it in part; but we are also significantly influenced by considerations of what we know *in* and *about* the world, by specific information, notably the historic record of various belief systems, cognitive styles, and diverse societies and traditions. It is the argument from illusion, so copiously fed and encouraged by the record of human credulity and its institutionalisations, and the conspicuous difference between successful and abortive styles of knowing, which encourage our empiricist faith.

Philosophically, this is untidy and messy. In as far as empiricism is a recipe for constructing a world, a touchstone for what bricks of evidence may be used in erecting a universe, it seems a bit circular and illogical also to be swayed by evidence *inside* the world. But, circular and inelegant or not, this is how in fact we think. I do not for one moment accept Quine's oft-repeated belief that we can do without prior philosophy altogether, and trust the world which we seem to inhabit, the 'aggregate of existing science'; this may describe a possible course of conduct in periods of civil peace in the realm of our ideas, but it does not describe any possible course whatever in cases of dramatic revolutionary discontinuity, and these do happen. The Quinian view corresponds to a political philosopher who would believe that history records continuous change, that the legitimate line of succession is always reasonably manifest! At times it is nothing of the kind, and some crises are quite unique and are not illuminated by the previous ones. In such situations of conceptual anarchy or the dissolution of cognitive norms, we absolutely need a 'prior philosophy'. (Empiricism happened to be a very good one.) But at the same time, man cannot live by prior philosophy alone. For better or for worse, we do shuttle-cock, in choosing our position, between independent and prior considerations on the one hand, and evidence drawn from within the world, about the comportment of human societies and ideologies. The appeal of empiricism derives, inelegantly, from such a double, strictly-speaking incommensurate, type of support.

So much by way of concession. What are the implications of this account of empiricism? There may be many; but some are of immediate interest.

The central element in empiricism, on this account, is a substantive, a priori and negative principle – the exclusion of certain logically possible and historically upheld visions. In the more elaborate account of what empiricism is, which admits our shuttle-cocking between prior-philosophic and intra-mundane historical considerations, this exclusion is seen to be both aprioristic (in as far as derived from the ultimate model of what knowledge must be), and in part based on evidence which, admittedly, is strong but never conclusive. (The fact that many closed visions are, when characterised in Neutral-speak, meretricious frauds, cannot establish that *all* such visions must be humbugs. Perhaps, who knows, the world is a Big Cosy Meaningful Unity after all, only we haven't hit on it yet; or perhaps even – excuse my shudder – one of the existing faiths is the true one.)

But if this is so – if the central idea in empiricism is paradoxically the a priori substantive exclusion of certain worlds – then certain more narrowly philosophical implications follow. If it is of the essence of a very important philosophy that it excludes certain visions, then the view that philosophy is, could be or ought to be 'neutral', is absurd. Another one is that falsificationists are wrong to spurn, as they tend to do, verificationism and Baconian inductivism. The difference between the two is not so very significant. The really important element in both forms is the deprivation of all positive, concrete visions, of the status of final norms of truth. (In their non-bilingual state, this is what ideologies do claim, even though they half-suspend the claim when talking to outsiders.) This end is achieved equally by a positive insistence on the sovereignty of experience (verificationism) as by a negative insistence on the right of experience to dethrone any particular theory (falsificationism). Historically and psychologically, it may well have been inevitable that the verificationist formulation should have been earlier, more influential and persuasive than the falsificationist one, whatever technical advantages the latter may possess; a revolutionary movement which promises a positive alternative and a deliverence is more attractive than a merely negative one.

This leads to another and closely related question. As W. W. Bartley pointed out, there is a certain development in Popper's thought from the problem of demarcation to the problem of rationality, from the cult of falsification to the cult of criticism. There are indeed certain passages in the later Popper which read as if the notion of criticism has replaced that of falsification, which seem to suggest that what matters is the critical spirit, and not any particular method it employs, and which seems to lose interest in the delimitation of rightful methods of criticism. (The earlier position still seems articulated in something akin to the empiricist metaphysic, in terms of a vision of the world made up of facts which eliminate the theories floating above them; the latter position is spelt out rather in a non-prior-philosophical idiom of the ordinary *Lebenswelt*. This evolution is curiously parallel to that which other philosophers went through, when they moved from the pure prior philosophy of logical positivism, to the comforting and joyful return to the ordinary world of daily speech.)

Now this difference corresponds to a question which pervades this paper – namely, is the real meat of empiricism contained in its *exclusion* of closed worlds, with the talk of 'experience' merely a picturesque way of achieving that end, or is the actual stress on experience as that which is now sovereign, of the very essence of the position? I find myself vacillating on this. The grave difficulties facing any attempt to define or even indicate what 'experience' is, other than metaphorically or in a circular manner, incline me to think that what is at the heart of the matter is the exclusion of closed worlds. (In other words that it is the critical spirit is what counts). On the other hand, it is also difficult to give a good account of 'criticism'. One can imagine conditions of mere instability and rapid change, which are yet cognitively most unsatisfactory. The positive identification of the ultimate arbiter *does* seem to matter: it is *experience*, and not the last speaker, or the most powerful or eloquent one. This consideration inclines me to favour the formulation of empiricism in terms which do stress the independent and extraneous nature of the source of confirmations or refutations. It also suggests that it is historically ungrateful to spurn that empiricist metaphysics which habituated us to this idea and to its appeal.

University of London

NOTE

[1] Dr E. Zahar of the LSE drew my attention to the argument and to roughly this formulation of it, and cited Paul Feyerabend as its source.

JERZY GIEDYMIN

INSTRUMENTALISM AND ITS CRITIQUE: A REAPPRAISAL

The main objection against various empiricist and – in particular – positivist doctrines in 19th- and 20th-century philosophy of science has been that they unduly restrict theorizing: as descriptive accounts of natural and social science they refuse scientific status to various theories which many would normally count as science; as programmes for research they impose on theorizing limitations which would hamper the development of certain theories and in this way – so some argue – stifle the progress of science. In addition, they branded as nonsense or – at least – as empirically meaningless various metaphysical speculations. Since apart from philosophers also many scientists have indulged in such speculations, it is felt that not only vital components of traditional philosophy but also of science have thus been ostracised. To be sure, very few of those empirically minded scientists and philosophers of science wanted to dispense with or ban theories altogether and to restrict science to the 'empirical basis' and empirical generalizations. Either because they felt that few scientists would want to be deprived of the 'theoretician's paradise', or because they appreciated (with Poincaré) that "theoretical physics is a fact to be explained" or else because they saw the philosopher's role as descriptive rather than prescriptive, they have usually ruled against the ban on theories invoking an ancient clause in the epistemologist's legislature known in various forms as instrumentalism (formalism, anti-realism, etc.). The intended effect of that clause is that those theoretical terms and sentences which do not satisfy certain definite requirements in terms of relations to observables (translatibility into observables in the extreme case of phenomenalism, descriptivism or positivism; increase in predictive power of the theory in the case of more liberal empiricist doctrines) are acknowledged as components of scientific theories with the status of formal symbols, metaphors, useful fictions, instruments, etc. In contemporary social sciences (economics, in particular) the instrumentalist clause would allow so-called 'models' apart from proper empirical theories as well as theories based on "un-

R. S. Cohen et al. (eds.), Essays in Memory of Imre Lakatos, 179–207. *All Rights Reserved*
Copyright © 1976 by D. Reidel Publishing Company, Dordrecht-Holland

realistic assumptions".[1] This piece of liberal legislation has not been appreciated at all by recent 'antipositivist' critics of those empiricist doctrines, for it seems to those critics that it does not take theorizing sufficiently seriously and, consequently, grants theoreticians too great freedom blurring the boundary between science and fiction. Thus modern empiricism has been blamed by its various critics for being intolerant and at the same time too permissive.

The aim of the present essay is to re-examine some recent accounts and critical assessments of the instrumentalist tradition in natural science and its philosophy. I shall claim that those accounts are descriptively inadequate, that the instrumentalist tradition comprised many doctrines and that its leading representatives, usually criticised, held much more moderate instrumentalist views than attributed to them and that the unqualified condemnation of instrumentalism is, in this light, unjustified. I shall also argue that the view of science (or its programmatic counterpart) formed within the instrumentalist tradition has in many respects anticipated the views of science advocated by contemporary critics of instrumentalism as 'progressive', 'critical', 'tolerant'.

I shall first report two accounts and assessments of instrumentalism given by Karl Popper and Paul Feyerabend, respectively. Then I shall try to show their inadequacy with respect to the instrumentalist tradition in astronomy and its philosophy and with respect to 19th-century conventionalism.

1. Two views concerning instrumentalism and its effect on natural science

Perhaps the earliest and strongest condemnation of instrumentalism in contemporary philosophy of science came from Sir Karl Popper. It was made in the context of the philosophy of the natural sciences, primarily of physics and astronomy.

Popper contrasted three views concerning the status of physical theories, essentialism, instrumentalism and 'the third view', realism combined with fallibilism (hypotheticism), which is also Popper's own view.[2]

Essentialism, which originated in Aristotle's philosophy and shared its age-old influence on science and philosophy, asserts that (a) the aim of science is to give a true description of the world, (b) it is possible for science and philosophy to demonstrate which theories about the world are true, (c) scientific and philosophic theories about the world concern the hidden nature of things, essential properties, true causes by contrast to observable phenomena, (d) explanations of phenomena in terms of latent causes, natures or essences are ultimate, i.e. in no need of further explanation.

From the third point of view, i.e. Popper's own, claims (a) and (c) are descriptively correct and programmatically healthy. Claims (b) and (d), on the other hand, are mistaken and unacceptable. It is, however, the last of the claims which Popper blamed as obscurantist, since by preventing scientists and philosophers from asking fruitful questions concerning deeper and deeper explanations, it hampered the progress of science.

Instrumentalism was identified by Popper with the view concerning the status of scientific theories "... founded by Osiander, Cardinal Bellarmino and Bishop Berkeley..." but also advocated "in various ways" by Mach, Kirchhoff, Hertz, Duhem, Poincaré, Bridgman, Eddington as well as by Niels Bohr and those who have adopted the so-called Copenhagen interpretation of quantum physics.[3] The following tenets are, according to Popper, characteristic of instrumentalism:

Scientific laws and theories are not proper, descriptive statements; they are nothing but instruments to derive observational predictions from other observational statements. Dispositional and theoretical terms (e.g. magnetic, force, field of forces), unlike observational terms, do not refer to real entities but are merely symbols which facilitate deductions. When a theory is found not to 'work' in a given area, its field of application is suitably modified.

Since scientific theories are nothing but computational rules, of the same character as the computational rules of 'applied science', there is no difference between 'pure' and 'applied' science; all science is applied.

By contrast, realism combined with fallibilism claims that: Scientific laws and theories are not only instruments for making predictions but are also genuine, descriptive (true or false) statements; they are empirical hypotheses which have explanatory power. Dispositional and theoretical terms refer to real entities just as so-called observational terms do. All terms are theoretical to some degree, though some are more so than others; the distinction, therefore, between observational and theoretical terms, assumed by instrumentalists, is mistaken.

Scientific theories, though not verifiable by observation, can clash with reality, provided they are falsifiable; they do clash, if – when subjected to genuine tests, i.e. attempts at falsification – they are shown to be falsified by experimental outcomes; crucial experiments, though unable to establish any theory, can eliminate some as false.

The fundamental aim of theoretical (pure) science is to provide a (hopefully) true description of the world, explanations of increasing depth (which means, of increasing abstractness, referring to unobservables, more and more removed from everyday experience) and predictions some of which at least are predictions of new kind of events, i.e. are real discoveries.[4]

There are several reasons why, according to Popper, instrumentalism gained popularity among scientists and philosophers. One of those reasons was the fact that it enabled them "to deal with inconvenient scientific hypotheses". So, for example, it enabled Osiander and Bellarmino to deal with the Copernican hypothesis in such a way that it no longer appeared to contradict the Scripture. Another reason, related to the previous one, was the concern that if science is credited with the ability to discover the truth about the world unaided by divine revelation, religion would suffer. This was, for example, Berkeley's concern. There was a third reason. Many scientists and philosophers rejected the essentialist view that the aim of science was to discover hidden causes or essences of things, either because they did not believe that such hidden causes or essences existed (e.g. Mach) or because they did not believe science was able with its methods to discover them (e.g. Berkeley, Duhem, Poincaré). Consequently they denied that theoretical concepts referred to latent causes or properties and that the aim of science was to give explanations which transcend our knowledge of the phenomenal world. Similarly, from the correct belief that science is unable to demonstrate the truth of its theories and that so-called crucial experiments cannot establish any theory as true, instrumentalists (e.g. Osiander, Bellarmino, Duhem, Poincaré) drew the wrong conclusion that scientific theories are never descriptive, possibly true statements but are merely mathematical tools for making predictions. Similarly, from the (right) premiss that the method of science cannot yield but tentative, revisable hypotheses and from the further premiss that scientific laws and theories (e.g. Newton's laws of motion, conservation laws) cannot be contradicted by experimental findings nor can a crucial experiment be instituted to refute, e.g. one (physical) geometry in competition with another, some instrumentalists (e.g. Poincaré) drew the wrong conclusion that scientific laws and theories are nothing but terminological conventions, useful for systematizing observational data in the way a catalogue systematizes books in a library, but having no descriptive, empirical content.[5]

Paul Feyerabend gives two related characterizations of instrumentalism and realism. In an earlier article[6] he defined 'instrumentalism' in a way similar to Popper's, except that – in conformity with the traditional usage established in the history of astronomy – he does not mention fallibilism (hypotheticism) in formulating the position of realism. In a later article[7] he associates instrumentalism with the claim that there is a theory-independent observational language available for making comparisons between rival theories and realism with the denial of this claim (Cf. the same denial on Popper's list of the tenets of the 'third view') as well as with the practice of using "the most abstract terms" of relevant theories to give theoretical meaning to all observables. In effect, an instrumentalist "makes commensurable all those theories which are related to the same observation language and are interpreted on its basis". Not so the realist: theoretical re-interpretation of observables in the light of a new theory may be so radical that they disappear from the class of consequences shared with earlier theories; the result is incommensurability.[8]

In his earlier treatment of instrumentalism Feyerabend pointed out that in Copernicus' theory three types of arguments were used in support of instrumentalism: religious (the inconsistency of the realist version of the theory with the Scripture), epistemological (only theories whose truth has been conclusively demonstrated can be descriptive of physical reality) and physical (inconsistency with Aristotelian physics – well-confirmed, given available evidence). In a somewhat "naive falsificationist" mood, Feyerabend$_1$, to borrow Imre Lakatos's phrase and slicing razor, contended that a "definitive refutation" of the epistemological arguments (of Bellarmino and Duhem) may be found in Popper's article on the three views. He therefore concentrated on the physical arguments – in order to show that even if these (according to him, weightier) arguments are taken into account, realism is preferable to instrumentalism. Now, the examination of the physical arguments in support of instrumentalism show – so Feyerabend continued – that one has to revise drastically traditional methodology which demanded factual support of new theories as a condition of their acceptability. Since certain defects of established theories cannot be detected without the help of alternative theories, one must not wait for the established theory to be directly refuted by experimental findings but proliferate rival theories, inconsistent with the established one, take them in the strongest form, i.e. in the realist fashion and – if any of them is more successful in solving problems or yielding new predictions – then the old theory, though unrefuted directly, ought to be replaced by the new, more successful one.[9] This methodological principle of proliferation of mutually inconsistent and incommensurable theories (Cf. Feyerabend's second characterization of realism) is offered by Feyerabend as a "methodological justification for realism" and claimed by him to be demanded by "the principle of testability, according to which it is the task of the scientist relentlessly to test whatever theory he possesses."[10]

I have two interrelated objections to both outlined accounts of instrumentalism. I think they are descriptively inadequate and the unqualified condemnation of instrumentalism which we find in them seems to me unjustified. They are historically inadequate since Popper's (and the first of Feyerabend's) characterizations of instrumentalism does not fit well enough either the position of Osiander as we know it from his uninvited preface to *De Revolutionibus* and from his letters to Copernicus and Rheticus, nor the views of Roberto Bellarmino expressed in his celebrated letter to Foscarini. Duhem's and Poincaré's views on the epistemological status of scientific theories are not exactly instrumentalist in Popper's sense either. Instrumentalism in Popper's sense is a rather extreme doctrine which was held by few scientists and philosophers. Osiander, Bellarmino, Duhem, Poincaré and many others were much less extreme in rejecting realism. It would be more appropriate to refer to their views on the status of theories as *moderate instrumentalism* but one could equally well label them *moderate (or critical) realism.* As to Feyerabend's second characterization of instrumentalism (and realism) it seems to me that it fails to draw the boundary line where it was intended to be drawn. Again, both Duhem and Poincaré may reasonably claim to be realists in this sense. More importantly, however, any thoroughgoing fictionalist seems to satisfy Feyerabend's second criterion of realism.

As regards Popper's and Feyerabend's condemnation of instrumentalism, I think that it ought to be reappraised in the light of the foregoing comments on the factual inadequacy of their accounts. Moreover, their negative assessment of the role of instrumentalism was based, I believe, on a comparison with what Popper called "the third view", i.e. realism combined with hypotheticism (fallibilism). The third view was hardly held by any scientist engaged in the controversy with instrumentalists. Although such an external assessment is, I believe, not only perfectly legitimate but also interesting, a fairer appraisal of instrumentalism would be obtained by comparing it also with the epistemological views held by actual realists and with the influence those views had on the intellectual life. Finally, I think that in their unqualified appraisal of instrumentalism both Popper and Feyerabend have disregarded certain methodological features associated with that doctrine which they themselves have considered as valuable in other contexts.

2. THE INSTRUMENTALISM OF OSIANDER AND BELLARMINO

Without attempting to solve the historical problem of the genesis of instrumentalism (or of anti-realism), I shall assume here as uncontentious that the views concerning the aims and status of astronomical theories expressed by Osiander and Bellarmino did not originate with those two authors but were the result of a long tradition going through the Middle Ages back to ancient Greek astronomy. They originated directly from the distinction established by ancient Greeks between mathematical and physical astronomy. As Geminus put it (in the 1st century B.C.), the business of a physicist is "to consider the substance of the heavens and stars, their force and quality, their coming into being and destruction... The physicist will prove each fact by considerations of essence and substance...(he)... will in many cases reach the cause by looking to creative force". Not so the astronomer who is interested in "the shapes and sizes and distances of the earth, sun and moon, and of eclipses and conjunction of the stars"; accordingly he makes use of "arithmetic and geometry" and "is not qualified to judge of the cause...

sometimes he does not even desire to ascertain the cause, as when he discourses about an eclipse; at other times he invents by way of hypothesis and states certain expedients by assumption of which the phenomena will be saved". The task of mathematical astronomy, accordingly, is to invent kinematic systems capable of saving the phenomena "... without being concerned with the question whether such systems are realized in the physical structure of the heavens, and if so, in what way...".[11] The distinction, apparently influenced by Aristotle's conception of physics, does not imply instrumentalism in Popper's sense, i.e. the claim that astronomical hypotheses are not descriptive, true or false sentences. All that is implied by it is that *mathematical astronomers are entitled to leave the question of the physical realizations of their mathematical systems open and that – as astronomers – they are not qualified to do otherwise.* This was repeated almost verbatim by Osiander in his preface to Copernicus's *De Revolutionibus:*

> ... For it is the duty of an astronomer to compose the history of the celestial motions through careful and skillful observation. Then turning to the causes of these motions or hypotheses about them, he must conceive and devise, since he cannot in any way attain to the true causes, such hypotheses as, being assumed, enable the motions to be calculated correctly from the principles of geometry, for the future as well as for the past. The present author has performed both these duties excellently. For these hypotheses *need not be true* nor even probable; if they provide a calculus consistent with the observations, that alone is sufficient....[12]

The point that although astronomical hypotheses *may be true or false* the astronomer does not commit himself to the belief in either, is made even more clearly in Osiander's letter to Copernicus:

> ... I have always felt about hypotheses that they are not articles of faith but the basis of computation; so that *even if they are false* it does not matter, provided that they reproduce exactly the phenomena of the motion....[13]

Similarly, and even more cogently, Bellarmino wrote in his letter to Foscarini:

> ...To say that on the supposition that the earth moves and the sun stands still all the appearances are saved better than on the assumption of eccentrics and epicycles, is to say very well – there is no danger in that, and it is *sufficient for the mathematician:* but to wish to *affirm* that in reality the sun stands still in the center of the world, and only revolves upon itself without travelling from east to west, and that the earth is located in the third heaven and revolves with greater velocity about the sun, is a thing in which there is much danger not only of irritating all the scholastic philosophers and theologians, but also of injuring the Holy Faith by rendering false the Sacred Scriptures... I say that *once there is a real demonstration* that the sun stands in the center of the world and the earth in the

third heaven, and that the sun does not go round the earth, but the earth around the sun, then we must go to work with much thoughtful consideration to explain the passages of Scripture that seem to oppose this view, and better to say that we have not understood these passages, than to say that that which has been demonstrated is false. But I shall not believe that there is any such demonstration until it has been shown to me: nor is it the same thing to demonstrate that supposing the sun to stand in the center and the earth to move in the heaven will save the appearances, and to demonstrate that in truth the sun does stand in the center and the earth moves in the heaven; for I believe that the first demonstration can be given, but concerning the second I have the gravest doubt, and *in case of doubt* one should not abandon the Sacred Scriptures as they have been expounded by the Holy Fathers.[14]

This seems to be the passage in Bellarmino's letter which Popper construes as concluding the non-descriptive nature of the hypothesis of the earth's motion (instrumentalism in Popper's sense) from the fact that the truth of the hypothesis has not been demonstrated and it is this alleged epistemological argument for instrumentalism, along with a similar Duhemian one, which Feyerabend believes to have been definitely refuted by Popper. But what Bellarmino explicitly says in this passage are the following points (a) the truth of the hypothesis of the earth's rotation and revolution has not been demonstrated, for (b) to show that the hypothesis explains (saves) all relevant phenomena is not (the same as) to demonstrate the truth of the hypothesis, therefore (c) until such demonstration is presented, one should *abstain from affirming its truth* and be satisfied with using it *ex suppositione (ex hypothesi)*, (d) this is, in fact, sufficient for the mathematical astronomer and (e) resolves for the time being (i.e. until a demonstration of the hypothesis is given) the clash with the Scripture and with peripatetic physics.[15]

Clearly, to advise not to assert a hypothesis as true but merely entertain it – which is what Bellarmino does – is not the same as to advise to divest it of its physical interpretation. Logical inconsistency with the Scripture and with peripatetic physics may be avoided not only through desinterpretation but also by *suspending assertion* of one of the mutually inconsistent statements. But suspension of assertion and mere entertainment were essential from the anti-realist viewpoint in astronomy since proliferation of "factually adequate and mutually inconsistent theories" or "invention of alternatives in addition to the view that stands in the centre of discussion"[16] were no less strongly emphasized by its methodology than by Feyerabend's, although for somewhat different reasons. Here is Osiander's plea for tolerance in matters astronomical:

... The peripatetics and theologians will be readily placated if they hear that there can be different hypotheses for the same apparent motion; that the present hypotheses are brought forward, not because they are in reality true, but because they regulate the computation of the apparent and combined motion as conveniently as may be; that it is possible for someone else to devise different hypotheses; that one man may conceive a suitable system, and another a more suitable, while both systems produce the same phenomena of motion; that each and every man is at liberty to devise more convenient hypotheses; and that if he succeeds, he is to be congratulated. In this way they will be diverted from stern defense and attracted by the charm of the inquiry; first their antagonism will disappear, then they will seek the truth in vain by their own devices, and go over to the opinion of the author.[17]

In his own way, Osiander argued that progress in astronomy may be achieved better through tolerance by letting everyone invent his alternative to the existing theories, through greater simplicity (given, however, the same empirical adequacy of alternatives), by avoiding antagonism and attracting opponents through "the charm of the inquiry", in other words – replacing the old theory by better ones without any of the theories being either conclusively refuted or demonstrated. In effect, this is Osiander's methodological justification of anti-realism.

Duhem's most often discussed and criticised comments on the controversy between realists and antirealists in Renaissance astronomy were made in connection with Galileo's indirect retort to Bellarmino. In a statement written towards the end of 1615 and addressed to the consultants of the Holy Office, Galileo first appeared to grant Bellarmino that one must not expect anyone to believe in the mobility of the earth without a demonstration; then he also conceded that "... it is not the same thing to show that on the assumption of the sun's fixity and the earth's mobility the appearances are saved and to demonstrate that such hypotheses are really true in nature..."; finally, Galileo postulated:

... But it should also be granted, and is much more true, that on the commonly accepted system there is no accounting for these appearances, whence this system is indubitably false; just so should it be granted that a system that agrees very closely with appearances may be true; and one neither can nor should look for other or greater truth in a theory than this, that it answers to all the particular appearances.[18]

Were it not for the last sentence, one might have thought that Galileo's only intention was to stop opponents from sneering at the Copernican system and regarding it as "... so paradoxical that no clear demonstration could conceivably ever be given of it." The last sentence, however, went farther than that: it claimed, in effect, that the Copernican system had been demonstrated in the only way a scientific theory can be demonstrated, viz. by showing that it saves the phenomena while its rival, the Ptolemaic system, does not. In effect, Galileo argued as follows: (a) One must not expect anyone to believe in the mobility of the earth (or in any other hypothesis) without a demonstration, so Bellarmino was right to demand one; (b) to show that a hypothesis saves the phenomena is not the same as to demonstrate that it is true in nature; again, Bellarmino was right to claim this; (c) to show that a hypothesis saves the phenomena while its rival does not is a demonstration and indeed the only one possible in science; therefore, Bellarmino (or anyone else) must not demand more than that.

Rightly or wrongly, Duhem attributed to Galileo the claim that the Copernican system has been demonstrated conclusively in competition with the Ptolemaic system by a crucial experiment. He characterized Galileo's position in the following way:

> ... Galileo's notions of the validity of the experimental method and the art of using it are nearly those that Bacon was later to formulate. Galileo conceives of the proof of a hypothesis in imitation of the *reductio ad absurdum* proofs that are used in geometry. Experience, by convicting one system of error, confers certainty on its opposite. Experimental science advances by a series of dilemmas, each resolved by an *experimentum crucis.*[19]

To show that Galileo's conception of experimental method was wrong Duhem formulated his celebrated argument against the possibility of crucial experiments in the strict (Baconian) sense:

> ... Grant that the phenomena are no longer saved by Ptolemy's system; the *falsity* of that system must then be acknowledged. But from this it does not by any means follow that the system of Copernicus is true; the latter is, after all, not purely and simply the contradictory of the Ptolemaic system. Grant that the hypotheses of Copernicus manage to save all the known phenomena; that these hypotheses *may be true* is a warranted conclusion, not that they are *assuredly true*. Justification of this last proposition would require that one prove that no other set of hypotheses could possibly be conjured up that would do as well at saving the phenomena. The latter proof has never been given. Indeed, was it not possible, in Galileo's own time, to save all the appearances that could be mustered in favour of the Copernican system by the system of Tycho Brahe?[20]

As Popper rightly points out, Duhem's argument merely shows that there are no conclusive crucial experiments, i.e. experiments which by refuting one of two rival (contrary but not mutually contradictory) hypotheses would conclusively prove the truth of the other. Popper, Feyerabend, Rosen and others were, of course, also right in claiming that the impossibility of conclusive crucial experiments does not imply at all the instrumentalist status of all scientific hypotheses, i.e. the claim that scientific hypotheses have no descriptive content. But how relevant are these critical comments to Duhem's own position?

As already indicated, there is no doubt that Duhem's argument was directed against the view – which Duhem attributed to Galileo – that there are conclusive crucial experiments and that such an experiment has decided against the Ptolemaic system and conclusively established the Copernican system. It is possible that the Galilean conception of the experimental method was more along inductivist lines and that by "demonstration" he meant "a method capable of establishing a hypothesis with very high probability." Galileo's account of experimental method, quoted by Duhem, was similar to that given later by Huygens ("... It is possible... to establish a probability which is little short of certainty. This is the case when the consequences of the assumed principles are in perfect accord with the observed phenomena, and especially when these verifications are numerous...") and by Newton ("... although the arguing from experiments and observations by induction be no demonstration of general conclusions; yet it is the best way of arguing which the nature of things admits of...")[21]. If the Galilean account of experimental method is taken in this inductivist sense, then Duhem's argument may be understood accordingly as a criticism of the probabilistic idea of induction and of the concept of inductive probability (confirmation).

As regards Popper's, Feyerabend's and Rosen's[22] claims that the Duhemian argument is a misconceived, erroneous inference for instrumentalism (in Popper's sense) or fictionalism, one should point out that those claims do not take into account and, therefore, do not explain the following fact: in the very formulation of the argument Duhem explicitly acknowledges that the Ptolemaic and Copernican theories are true or false, or rather that given suitable evidence one can show that one of them is definitely false and the other may be true; this seems more like a straightforward rejection of instrumentalism (in Popper's sense) rather than an argument in its favour. Why did Duhem's critics disregard this fact and draw the opposite conclusion? Perhaps because in the last chapter of his work *To Save The Phenomena* Duhem wrote:

The physicists of our day, having gauged the worth of the hypotheses employed in astronomy and physics more minutely than did their predecessors, having seen so many illusions dissipated that previously passed for certainties, have been compelled to acknowledge and proclaim that logic sides with Osiander, Bellarmino and Urban VIII, not with Kepler and Galileo – that the former had understood the exact scope of the experimental method and that, in this respect, Kepler and Galileo were mistaken.[23]

But, of course, on the interpretation of Osiander's and Bellarmino's (relevant) texts proposed in the present article, neither of them was an instrumentalist in the strict Popperian sense. Moreover, this was admitted by Popper, albeit only in a footnote.[24] So to say that logic sides with Osiander and Bellarmino does not amount, after all, to saying that it sides with instrumentalism as understood by Popper. However, one would be mistaken to think that Duhem could be saved from the charge of instrumentalism so easily. To make things more difficult he also wrote in the same concluding chapter:

Despite Kepler and Galileo, we believe today, with Osiander and Bellarmino, that the hypotheses of physics are mere mathematical contrivances devised for the purpose of saving the phenomena. But thanks to Kepler and Galileo, we now require that they save all the phenomena of the inanimate universe together.[25]

It appears that Duhem both admitted the descriptive (true or false) nature of astronomical theories and claimed that the hypotheses of astronomy and physics are "mere mathematical contrivances." Has he thus contradicted himself? It will take some hermeneutic effort to argue, in the next section, that – perhaps – he has not.

3. THE INSTRUMENTALISM OF DUHEM AND POINCARÉ

The title of one of the sections of Duhem's article 'Physical law' (in *The aim and structure of physical theory*) announces that "a law of physics is, properly speaking, neither true nor false...".[26] What clearer evidence could there be of Duhem's instrumentalism? Is this not enough to disregard, as his critics have, Duhem's apparent admission in *To Save The Phenomena* of the descriptive (true or false) nature of the rival astronomical systems? I shall consider now these questions in relation to another version of the Duhemian argument against the possibility of crucial experiments, viz the argument from the existence of experimentally indistinguishable hypotheses. Both arguments rely on the ancient claim, associated with the antirealist tradition in astronomy, that any given set of observed facts (phenomena) can be accounted for ('saved') by more than one mutually incompatible hypothesis, so that the process of attempted elimination can never terminate and is, thus, ineffective. Both may be regarded as criticisms of induction by elimination, the first of them ante-dating Hume, since it had been known to ancient and Renaissance astronomers.

Of course, Duhem's statement concerning the laws of physics appears to be a clear indication of his instrumentalism only as long as it remains – as quoted above – in incomplete form. When completed it goes as follows: "A law of physics is, properly speaking, neither true nor false but approximate."

One does not refer to mere instruments, which cannot perform a descriptive function as "approximate". And indeed from Duhem's elucidation of what he means by "approximate" it becomes obvious that he regards physical laws as descriptive, although not in the same, simple way as "common-sense laws".[27] For common-sense laws, which are simply generalizations of everyday experiences, e.g. "In Paris, the sun rises every day in the east, goes up in the heavens, then comes down and sets in the west", the question whether they are true or false is determinate and can be answered with a "yes" or "no". Not so for scientific laws which are symbolic representations picturing "... reality in a more or less precise and detailed manner". *The question whether such symbolic representations of reality are true or false is indeterminate and must not be asked.* The reason is this: A physical law is always underdetermined by the evidence we possess, in this sense it may be said to be indeterminate; a given 'group of facts' (evidence statements) is compatible with an infinity of formulae which may express our law. As an example let us consider a physical law corresponding to the aforementioned common-sense law. Such a physical law would be a function which assigns to each moment of time the position of the sun in the sky as seen from Paris. Of course the law would be concerned not with the sun as we see it but with the positions of the centre of the sun conceived as a geometrically perfect sphere disregarding atmospheric refraction and annual aberrations; this is one of the senses in which a physical law is symbolic, i.e. it is a re-description of facts in the idealized and precise language of mathematics. To make this symbol 'correspond to reality' we have to effect complicated measurements, using a lens equipped with a micrometer, make many readings, subject these to corrections and calculations, the legitimacy of which depends on various theories, e.g. on the theory of aberrations and of atmospheric refraction. Our formulae tell us the coordinates of the point we call the centre of the sun; the meaning of these cannot be understood without knowing the laws of cosmography and their values designate points in the sky only given various previous determinations (e.g. of the meridian of the place, etc.). Now, owing to *limited sensitivity of our optical instruments*, we shall be able to determine for each instant the longitude and the latitude of the centre of the sun only with a *limited precision*, e.g. not greater than 1′; we shall not be able to discriminate between two positions which differ from one another by less than that value. Hence, in spite of the fact that the sun's centre occupies at each instant of time only one position, we shall be able to give for each instant an infinity of values for the longitude and an infinity for the latitude. Consequently "in order to represent the path of the longitude as a function of the time, we shall be able to adopt not a single formula, but an infinity of different formulas, provided that for a given instant all these formulas give us values for the longitude differing by less than 1′." And the same applies, of course, to the latitude. We shall, then, be able to represent our observations on the path of the sun by an infinity of different laws, expressed by equations which are mutually incompatible, i.e. if one of them is true, no other is. Each of these equations will trace a different curve on the celestial sphere:

> ... It would be absurd to say that the same point describes two of these curves at the same time, yet, to the physicist all these laws are equally acceptable, for all determine the position of the sun with a closer approximation than can be observed with our instruments. The physicist does not have the right to say that any of these laws is true to the exclusion of the others ....[28]

The physicist will eventually choose one of the laws but his considerations will not be necessarily related to the idea of truth and they may vary from case to case depending, for example, on the use to which the law is to be put. Among the considerations which guide his choice may be simplicity but especially the deducibility of the given law from a theory which he had accepted; however "... physical theories are only a means of classifying and bringing together the approximate laws to which experiments are subject; theories, therefore, cannot modify the nature of these experimental laws and cannot confer absolute truth on them..."[29].

The degree of the indetermination of abstract symbols (theoretical concepts, as we would say today) depends on the degree of approximation with which quantities can be measured. However, the latter varies with time and increases with the improvements of instruments. Accordingly, what was acceptable as sufficiently precise at one time, will not be so acceptable at a later time. On the other hand, the degree of precision required varies with purpose. Moreover the same physical law is often simultaneously accepted and rejected by the same physicist in the course of the same work; this would result in a formal contradiction "if a law of physics could be said to be true or false". Finally, a physical law is in the process of constant modifications and improvements; factors previously disregarded are being taken into account, exceptions are explained which had only been enumerated before, etc. As a result, it is more a series of statements than one complete statement.

From my report of Duhem's views it seems obvious that he regarded laws of physics as descriptive statements about reality, though not in the same straightforward way as 'common-sense laws'. It seems equally obvious that he did not regard physics as having the same characteristics as applied science. On these counts he did not qualify as an instrumentalist in Popper's sense. However, he believed that, from the viewpoint of "a strict logician" as he put it, presumably to emphasize that it would be irrelevant to scientific practice but relevant to an epistemological understanding of physics, the laws of physics cannot be regarded as true or false.

Now, Duhem was certainly right that idealizations (i.e. sentences in which terms such as 'perfectly elastic body', 'point-mass', etc. occur) are not descriptive in the same way as simple sentences we use in everyday life to refer to tables, chairs, stars, etc. If idealized sentences are construed as so-called formal implications, i.e. sentences of the form 'For every x: if P_1x then P_2x', they are sometimes said to be 'vacuously true', since their antecedents are false (there are no perfectly elastic bodies, no point-masses, no perfect competition) and an implication with a false antecedent is, on the conventions of classical two-valued logic, true. Such a construal of idealizations as true sentences was not known – or, at least, not known outside logical circles at the time Duhem wrote. It is not surprising, therefore, that he did not make use of it, but the issue between instrumentalism and realism had nothing to do with this fact. Idealizations are descriptive in a sense for which the concept of approximation is essential.[30] Attempts have been made recently to introduce for that reason some concepts of 'approximate truth".[31]

In my view, the most essential points in Duhem's arguments concerning the cognitive status of scientific laws are as follows:

(a) For common-sense laws, which are simply generalizations of everyday experiences, the question whether they are true or false is determinate; this is because the language of common-sense laws is the same as that of our everyday experiences; common-sense laws may, therefore, be regarded as true or false;

(b) Physical laws, unlike common-sense laws, are formulated in the highly idealized and precise language of mathematics; even those which originate from common-sense laws are re-descriptions of the latter in mathematical language; predictions derived from scientific laws are in the same precise and idealized mathematical language;

(c) Owing to the limited precision of our measuring instruments and measuring procedures the outcomes of measurements of observed quantities are always imprecise and so compatible with an infinity of mathematical functions which may express our scientific laws; in other words, for any set of measurements of relevant quantities there is an infinity of experimentally indistinguishable laws (expressed as mathematical functions), i.e. laws which agree (account for, etc.) with those measurements; the scientist has thus no right to say that one of those laws is true to the exclusion of others; the question whether any of those laws is true or false is, therefore, indeterminate and must not be asked (is meaningless).

Is the Duhemian argument from the existence of experimentally indistinguishable laws for any set of measurements to the indeterminacy of the question of the truth-value of those laws valid or is it the mistaken epistemological argument for instrumentalism, which Feyerabend claimed had been destroyed by Popper? Because of the ambiguities of the Duhemian text, only a conditional and tentative answer may be attempted. If 'true' in (c) is used in the semantical sense (as originally defined by A. Tarski), then the argument is invalid: from the (physical) impossibility of deciding at a given time which, if any, of an infinite set of mutually incompatible sentences is true, it certainly does not follow that none of them is true or false (in the semantic sense). A sentence may express 'the true value of a parameter' and thus be true in the semantical sense, without our being able ever to find what the true value of the parameter is. If Duhem used here 'true' in the classical or semantical sense, then he should have said – as he did in his comments on Galileo v. Bellarmino controversy – that each of these laws equally compatible with evidence (taken singly) may be true, though they cannot be true simultaneously. If, however, 'true' in (c) is used not in the classical but in some constructivist (relative and time-dependent) sense – familiar from the writings of pre-intuitionists (Kronecker, Poincaré), from intuitionistic literature on logic and mathematics, from the writings of some logical empiricists, from the way 'true' is often used by scientists (as 'confirmed') and by Duhem himself in the second chapter of his *The Aim and Structure of Physical Theory*[32] – then the argument may be valid. "Diminution of the excess of accepted questions over possible answers" – to use Quine's phrase – was clearly as much Duhem's motive as it was Brouwer's. It is also tempting to read into Duhem's statement "a law of physics is neither true nor false but approximate" an implicit rejection of "the law of bivalance" (the meta-law of excluded middle) which would result in a three-valued logic. His claim that "theories... cannot confer *absolute* truth on (laws)" seems to show that he did distinguish between the absolute, classical conception of truth and the relative, non-classical, constructive ones. Laws which are neither true nor false are, presumably, not asserted or 'asserted' only with the third truth-value, which would avoid inconsistency in using several alternative ones at the same time. On various occasions Popper claimed that he has been using the concept of truth in the classical or semantical sense. From this point of view, the inference expressed in (c) is invalid. However, the decision not to use 'truth' in the semantical but rather in some constructivist sense – which may be sufficient (depending on how the given constructivist concept is characterized in detail) to make the inference under discussion valid – cannot be regarded in any clear sense as an error or mistake which might be exposed by philosophical or other arguments. Like any decision it will have some consequences which may be evaluated differently depending on one's problems and aims. To conclude, if – as is quite likely – Duhem used 'true' not in the semantical sense, unlike Popper but in some constructivist sense and 'neither true nor false' in the sense of 'experimentally indistinguishable' or 'equally acceptable', then his argument expressed in (c) not only may be valid but cannot possibly be construed as an argument in support of the main instrumentalist thesis in Popper's sense, since in the latter 'true or false' is intended to be understood in the semantical sense. We have seen, moreover, that Duhem regards scientific laws as descriptive statements though 'descriptive' is not characterized in his philosophy as 'true or false', since there may be descriptive but indeterminate statements.

Now it is possible, of course, that in 'Physical Laws' (1906) Duhem contradicted what he was to say (1908) in *To Save The Phenomena* about the possibility of the Copernican system being true. This could have happened either because he changed his mind or because he did not remember or did not quite realize the implications of his comments on Galileo's retort to Bellarmino. On the other hand, there is no inconsistency between the two texts if we accept one of the following interpretative hypotheses: (a) 'true' was used in two different senses in the two texts, viz. in the classical (semantical) sense in the argument against crucial experiments in *To Save The Phenomena* and in a non-classical, constructivist sense in the chapter on 'Physical Laws'. There would be nothing unusual about such usage: 'true' in the classical sense is very much part of our everyday language and authors using 'true' in some non-classical sense (e.g. William James, Brouwer and others) sometimes shift from one sense to the other, without – apparently – realizing it. As I pointed out before, there are indications that Duhem was aware of the difference between the classical (absolute) and non-classical conceptions of truth. (b) 'true' was used in both texts in the same constructive, non-classical sense; on the assumption, made by Galileo, that there was evidence contradicting the Ptolemaic system, the falsity of that system was effectively established (relative to the evidence). If the Copernican system explained the same evidence and was not contradicted by any, a constructivist was allowed to say, if only for argument's sake, that on that assumption it might be true provided that truth could be in principle effectively established at some time; it was only in the last step of the argument against the possibility of crucial experiments that Duhem showed – by claiming existence of alternative hypotheses able to explain any given evidence – that there was no such effective procedure. So, the possibility of one of the rival theories being true was granted in the earlier premiss for argument's sake, to be denied in the last step of the argument.

There remains Duhem's claim that although laws are usually selected from the point of view of theories, from which they are deducible, theories "...are only a means of classifying and bringing together the approximate laws...". Compare this with Poincaré's equally celebrated claim that science may be likened to 'a library' which must go on increasing indefinitely; experimental physics alone can enrich this library, "...as for mathematical physics, her duty is to draw up the catalogue. If the catalogue is well done the library is none the richer for it...".[33] Claims and similes like these have been used all too often to attribute to conventionalists, like Duhem and Poincaré, the instrumentalist view of physical theories in both Popper's sense and in the second of the two senses distinguished by Feyerabend. For one argues that neither a classificatory system nor a catalogue may be said to be true or false or provide explanations, though they may be simple, convenient to use, etc. Similarly, one might think that those who regard theories as merely performing a humble though useful role of catalogues will not join the club of Feyerabend's realists who "use the most abstract terms of whatever theory (they) contemplate" in order to give a unified account of both observable and unobservable matters, i.e. to give theoretical meaning even to observational terms and sentences. However, similes and analogies are at least as deceptive as they are illuminating. So are 'isms' unless they are used with appropriate qualifications. Contrary to the widespread belief, neither of the two founding fathers of 19th century conventionalism, Poincaré or Duhem, claimed that all scientific laws and theories are nothing but terminological conventions, though both believed – quite rightly – that conventions play an important role in science.[34] Poincaré's simile notwithstanding, physical theories were regarded by both Poincaré and Duhem as languages, partly formal (mathematical), partly metaphorical or both. ("The sun is a perfect geometrical sphere" is a mathematical metaphor, which taken literally is false, but may be used to express the same as the phrase "Assume, contrary to fact, that the sun is a perfect sphere").[35] As such they are conceptual, therefore, classificatory systems, in a way similar to library catalogues. They can and do perform a descriptive function: they are redescriptions of laws in theoretical language and they are capable of expressing, usually with the help of equations, certain relations (true or false) which remain invariant under many changes of metaphorical expressions used to phrase successive theories in the given area.[36] These metaphorical, theoretical expressions, of course, must not be understood in a literal sense unless independent, direct evidence allows one to transform an analogy or metaphor into a literal description. There is something we call atom and, to help our imagination, its structure is likened to the structure of the planetary system; there is something we call molecule and – again – it is in a way likened to a billiard ball, etc. But theories describe only relations between those things, provided, of course, that our theories are not mistaken.

Duhem's statement that physical theories are merely means of classifying and bringing together approximate laws must not be understood as in any way depreciating the importance of theories. For in the same book he also claimed that abstract scientific concepts (like mass, temperature, pressure, etc. "assume meaning only by grace of physical theory", that "according to whether we adopt one theory or another, the very words which figure in a physical law change their meaning, so that the law may be accepted by one physicist who admits a certain theory and rejected by another physicist who admits some other theory".[37] In a similar vein, Poincaré emphatically denied that it was possible to make experiments "without preconceived ideas"; one has to use language and "our language is necessarily steeped in preconceived ideas."[38] In his criticism of the extreme conventionalism of E. LeRoy, Poincaré declared that "the scientific fact is only the crude fact translated into a convenient language", i.e. the language of a theory and that "... the enunciation of our laws may vary with the conventions we adopt..." and that these theoretical conventions may even modify the relations between these laws.[39]

My conclusions are as follows:

Firstly, neither Poincaré nor Duhem deny that theories play a descriptive role though both insist that theoretical description in science is not as straightforward as descriptions in everyday life and has to be sieved through critically, which is not an easy task; they would both say that in this case they reject not realism but its naive form which amounts to taking all theoretical expressions literally and hypothesizing objects with all properties and structures suggested by scientific metaphors, models, images, etc. Moreover, Duhem's idea of the descriptive function of laws and theories cannot be adequately expressed in terms of 'true or false', if 'true' is used in the classical sense (hence his idea of 'natural classification').

Secondly, neither Poincaré nor Duhem were instrumentalists in Feyerabend's second sense, according to which instrumentalists leave the meanings of observables unaffected by theories whereas realists do not; they both believed that by subsuming an empirical law under a theory one 'interprets' or re-describes the observational terms in the law from the viewpoint of the given theory; they did not believe, however, that this results in logical or empirical incommensurability of alternative theories considered for the given area.[40]

Thirdly, the belief that all terms, including observables, are – or should be – "theory impregnated" or "steeped in theoretical meaning", to use the nowadays fashionable, picturesque slogans, has little – if anything – to do with realism (instrumentalism) in the semantical or ontological sense of implying the existence (non-existence) of so-called theoretical entities, i.e. referents of theoretical terms. After all, both 'critical positivism', i.e. the rejection of so-called naive realism and the theory – or 'conceptual framework' – mystique originated from Kant's critical philosophy and were both inherited by various post-Kantian doctrines, including Lange's neo-Kantianism, Poincaré's conventionalism, Nietzsche's perspectivism and Vaihinger's fictionalism. These philosophies show that one can regard all observables as not only tinted by but part of theories, conceptual frameworks, perspectives, etc., however metaphorical, fictional and distorting all these linguistic structures may be claimed to be. In a fairy-tale a uniform fairy-account may be given of both 'observable' and 'unobservable' fairy-tale matters, without involving any serious ontological commitment so dear to the heart of traditional realists.

The intermediate position of all these post-Kantian doctrines with respect to semantical (ontological) realism and instrumentalism (similar to that of Bellarmino) can be seen from the following passage in which Vaihinger distinguishes criticism (also called by him "logical positivism" or "critical positivism") from logical (epistemological) pessimism and optimism:

... True criticism or logical positivism proceeds to examine the thought instrument dispassionately and objectively. With logical pessimism it frees itself from childish beliefs in the power and unlimited validity of thought, and with optimism it holds firmly to the fact of the ultimate practical coincidence of thought and existence. The valuable outcome of pessimism is the habit of seeing in these conceptual constructs primarily nothing more than subjective products. Instead, therefore, of demanding with the dogmatist that we accept their reality until their unreality is proved – a thesis that from a practical point of view is the only useful one – it reverses the process and *mutatis mutandis* applies the juristic formula '*Quisque praesumatur malus, donec probetur bonus*', demanding that every logical function be taken for what it actually is, a mere logical construct; and insisting on a special proof before the reality of any given mental construct or logical form is assumed...[41].

4. Logical reconstruction of indeterminate languages

Popper's (and Feyerabend's first) characterization of instrumentalism and realism (cum fallibilism) leaves the impression that it is somehow up to the scientist (or philosopher) to decide the status of theoretical statements, i.e. to make them either descriptive, true or false (in the semantical sense) or non-descriptive, neither true nor false. For some philosophical or opportunistic reasons instrumentalists deprive theoretical sentences of the descriptive function whereas realists-fallibilists preserve that function to make strong, falsifiable theoretical claims. However, if the semantical concept of truth (and falsity) is to be applicable to the language of science, at least in principle, that language has to satisfy certain conditions, syntactical and semantical. On the semantical side, it is required, for example that the non-logical (so-called descriptive) expressions of the language be assigned interpretations or denotations, i.e. sets of objects (sets of ordered pairs, triples, etc. of objects) in a domain, called the realization or semi-model of the language. In an empirical language this assignment of denotations cannot be made by purely verbal means, i.e. through the axioms serving as terminological conventions alone. Some non-verbal methods are essential, e.g. pointing out, using standard objects (as exemplars), measuring procedures, etc. Obviously, not all scientific terms can be given interpretations in this way. Typically, so-called theoretical or abstract terms, cannot. Before, therefore, we are entitled to apply the terms 'true', 'false' to theoretical statements, we have to decide how their interpretations are to be fixed. One way of doing this is by relating theoretical terms with observables with the help of sentences usually called "correspondence rules", "bridge principles", etc. If we choose this way, then we are faced with the problem of deciding what logical form these correspondence rules are allowed to have. If we allow only explicit definitions as correspondence rules, then all theoretical terms become definable in terms of observables and, consequently, eliminable in principle. A language with this characteristic is purely phenomenalistic. If, on the other hand, we decide to have a language which is closer to the language of actual science, then we shall allow as correspondence rules sentences which do not have the form of equivalences; in effect we shall allow various forms of conditional sentences. As a result of such a decision, the theoretical terms of our language will be interpreted only in a partial way, they will become so-called "open concepts". Partially interpreted theoretical terms will have no interpretation, at least in some contexts. Moreover, an examination of observational terms in the language of science shows that they are all vague terms. If the vagueness of observational terms is to be reflected in the semantics of our language, we have to decide how this is to be done. One way consists in interpreting the observational part of the language not with the help of one, single (semi)model but with the help of a family of models, i.e. a set of structures in each of which a given vague term is assigned a slightly different set of objects as its denotation. If this method is chosen, then the interpretations of the theoretical terms will become even vaguer so that at least some of the theoretical sentences are bound to be indeterminate, i.e. neither true nor false, though they are well-formed. Duhem's discussion of the imprecision of measurements which give only approximate values (e.g. of the position of a physical object) and of the idealized nature of some theoretical concepts and laws which will, presumably, yield predictions in terms of precise values (e.g. of the position of the centre of a perfect sphere), might be rephrased as concerned with the interpretations of scientific terms, with the possible result that predictions in terms of precise values would be indeterminate sentences, i.e. neither true nor false. In this way Duhem's argument – or, rather, certain hints therein – may be turned into a semantical argument in support of a weak thesis of instrumentalism which claims that some theoretical sentences are neither true nor false in the given reconstruction of the language of physics, even though the reconstruction in question may be based on the two-valued classical logic.[42]

As we see, decisions are involved in these considerations. However, they are not decisions whether to regard all theoretical sentences as either true or false or neither. They are decisions (or clarifications) concerning the way scientific terms are given physical interpretations.

5. Conclusions

Are unqualified condemnation of instrumentalism and praise for realism, which we found in the writings of Popper and Feyerabend, justified?

Their assessment is based on the claims concerning the effects these two epistemological views have had and are likely to have on the progress of science. It seems to me that the influence of epistemological views of scientists on actual science, if there is such influence, is much more difficult to ascertain than critics seem to believe. Each epistemological doctrine allows not one but several methodological strategies. No set of methodological rules, as the critics admit, is able to ensure success and it seems reasonable to suggest that the effects of following certain methodological prescriptions will vary with historical circumstances. For example, the requirement of consistency transcending one scientific discipline and demanding in effect that the theories of one discipline (e.g. astronomy in natural science, economics in social science) ought to be compatible with the theories of another discipline (e.g. physics in natural science, sociology, psychology in social science) may be either progress inducing or be an obstacle to progress depending on the state of the other discipline at the given moment. In ancient astronomy instrumentalism (antirealism) was associated with the division of labour between the mathematical and physical astronomy and, in effect, meant relative independence of mathematical astronomy of Aristotelian physics and ontology. Historians of science agree that ancient mathematical astronomy progressed much better than ancient physical astronomy and that the latter "retarded rather than promoted" the development of science.[43] However, the division was not always strictly observed which is why there was a variety of views concerning the nature of astronomy and its methods, from realism through moderate instrumentalism to fictionalism. Ptolemy adopted the antirealist view in his *Almagest* and the realist view in *Hypothese Planetarum*. Even in the former, however, he rejects the hypothesis of the mobility of the earth on the grounds that it contradicted Aristotelian physics, i.e. he rejects it on physical grounds; on consistently antirealist grounds Ptolemy need not have rejected any hypotheses. What effect did such realist observance of the consistency condition have in this case? "... The rejection on physical grounds of any astronomical theory which conceived the earth to be moving, prevented at the same time its elaboration as a purely mathematical description; at least, there is not a single indication that this was ever attempted...".[44]

The instrumentalism of Osiander and Bellarmino resulted from this ancient tradition of distinguishing between mathematical and physical astronomy. It was no more opportunist than had been the ancient division of labour: in the latter case its point was to allow the practising of astronomy in relative independence of Aristotelian physics and ontology, in the former of Aristotelian physics and of Christian theology. Its main idea was to suspend assertion of certain hypotheses (mainly of the earth's mobility) in conflict with peripatetic physics and with Church dogmas, rather than the claim that those hypotheses were non-descriptive, neither true nor false statements. Suspension of the assertion of hypotheses does not prevent their deductive elaboration and use in research. Contrary to Feyerabend's claim, instrumentalism – at least in this sense – does not make the content of a given theory weaker than it would have been under the realist interpretation, since suspension of assertion (a pragmatic factor) is irrelevant to the consequence class of the theory, usually identified with its content.

Ancient and Renaissance instrumentalism was associated with a pluralist view of scientific method. Its descriptive account of the method of science was based on a poly-theoretic model: one invents freely imaginative hypotheses to account for available observed data; one eliminates those which do not save the phenomena; however, this process is never conclusive and never terminates: there are always alternative hypotheses explaining facts equally well and predicting new facts with equal precision; the invention of new, mutually incompatible though observationally equivalent hypotheses is encouraged and their critical assessment is based on criteria transcending factual support; since incompatible, rival hypotheses are, of course, not asserted, one is free to use several of them in research and see what results they yield. From the normative-programmatic viewpoint, therefore, instrumentalism has been advocating "the proliferation of hypotheses". By contrast, realism as we know it from the history of science (not to be confused with Popper's third view, which is realism combined with fallibilism and the critical approach) was usually associated with the mono-theoretic conception of scientific method according to which scientific method is an (inductive) inference from facts to a theory; in one of its variants (the Baconian) rivals are conclusively eliminated by crucial experiments; from the normative viewpoint it recommends concentration on one hypothesis or theory (e.g. Newton's rule IV).

Conventionalists (Poincaré, Duhem) added to the poly-theoretic model of scientific method certain new ideas: scientific facts are re-descriptions of common-sense facts in theoretical-mathematical language; there is, thus, change of language (and meaning) in the transition from everyday common-sense knowledge to scientific knowledge and scientific observations are theory-dependent. Scientific laws and theories are descriptive but in a much more roundabout and complicated way than common-sense observations and laws; owing to the imprecision of measurements there are always infinitely many experimentally indistinguishable laws. Radical conventionalists (LeRoy, Ajdukiewicz), not discussed in the present article, claimed that many rival theories, usually regarded as mutually incompatible, are – in fact – not inter-translatable and have no expressions in common so that neither the method of crucial experiments nor one common concept of truth (in the semantical sense) are applicable to them.

In their articles referred to in my analysis both Popper and Feyerabend evaluate epistemological and methodological doctrines from the point of view of whether and how much these doctrines encourage criticism. They also regard the existence and development of rival theories as essential to criticism, to the emergence of new problems and thus to the progress of science. So did instrumentalists, except that they also advocated suspension of assertion and greater tolerance towards alternative theories. Are we to believe that suspension of assertion (a phenomenon of the 'second world') may be detrimental to the progress of science? But this seems only plausible on the view of science as 'personal knowledge' which develops through passions rather than through the dialectic of problems and answers in the 'third world'. Or is tolerance for rival theories and the plea for relative independence of research in one area from others incompatible with the progress of science? This can hardly be claimed by those who, like Feyerabend, advocate proliferation of rival theories and an anarchistic theory of knowledge. Have the differences between instrumentalism and their own philosophy not been overstated by both Popper and Feyerabend? Now, the classical concept of truth plays a paramount role in Popper's philosophy. This *is* a major difference. The same does not apply, however, to Feyerabend's views most of which – their anarchist colouring excepting – had been anticipated by the instrumentalist tradition.

University of Sussex

NOTES

[1] Cf. the dispute over the paper by Friedman, M., 'The Methodology of Positive Economics' in *American Economic Review, Papers and Proceedings*, 1963.
[2] Popper, Karl, 'Three Views Concerning Human Knowledge', *Conjectures and Refutations*, London 1963, pp. 97–120.
[3] *ibidem*, p. 99.
[4] *ibidem*, pp. 107–110, 114–119.
[5] *ibidem*, p. 98.
[6] Feyerabend, Paul, 'Realism and Instrumentalism: Comments on the Logic of Factual Support', M. Bunge (ed.), *The Critical Approach to Science and Philosophy*, London 1964, pp. 280–309.
[7] Feyerabend, Paul, 'Consolations for the Specialist', in I. Lakatos and A. Musgrave (eds.), *Criticism and the Growth of Knowledge*, Cambridge 1970, p. 220.
[8] *ibidem*, p. 220.
[9] Feyerabend, Paul, 'Realism and Instrumentalism', p. 307.
[10] *ibidem*, p. 305.
[11] Cohen, M. R. and Drabkin, I. E. (eds.), *A Source Book in Greek Science*, Harvard 1969, pp. 90–91; 'Geminus: The Scope of Astronomy Contrasted with that of Physics'.
[12] Rosen, Edward (ed.), *Three Copernican Treatises*, New York (2nd. ed.), 1959, Introduction, pp. 24–5.
[13] *Ibidem*, p. 22.
[14] Duhem, Pierre, *To Save The Phenomena*, Chicago and London 1969, p. 107.
[15] By the time Bellarmino wrote his letter to Foscarini (1615), leading Jesuit astronomers, on whose opinion Bellarmino relied, were convinced by Galileo's and their own discoveries, that the Ptolemaic system did not save the phenomena. Since there were both religious and physical arguments (e.g. failure to discover stellar parallax) against the Copernican hypotheses, they were in favour of the Tychonic system.
[16] Feyerabend, Paul, 'How to be a Good Empiricist', in P. H. Nidditch (ed.), *Philosophy of Science*, O.U.P., 1968, p. 29.
[17] Rosen, E., *op. cit.*, p. 23.
[18] Duhem, P., *op. cit.*, p. 108.
[19] *ibidem*, p. 109.
[20] *ibidem*, pp. 109–10.
[21] Madden, E. (ed.), *Theories of Scientific Method*, Seattle and London 1966, pp. 120, 123.
[22] Rosen, E., *op. cit.*, p. 33.
[23] Duhem, P., *op. cit.*, p. 113.
[24] Popper, K., *op. cit.*, p. 98.
[25] Duhem, P., *op. cit.*, p. 117.
[26] Duhem, P., *The Aim and Structure of Physical Theory*, 1954. 'Physical Law' (reprinted

in J. Kockelmans (ed.), *Philosophy of Science: The Historical Background*, New York 1968, p. 302).

[27] *ibidem*, pp. 299–306.

[28] *ibidem*, p. 305.

[29] *ibidem*, pp. 305–6.

[30] *ibidem*, p. 302; Körner, S., *Experience and Theory*, London 1966.

[31] Wojcicki, R., *Metodologia Formalna Nauk Empirycznych* [Formal Methodology of Empirical Sciences], 1974, Wrocław. Zakład im. Ossolinskich. pp. 173–186.

[32] Duhem, P., *The Aim and Structure of Physical Theory*, Princeton 1954, Chapter II, 1 (Cf. also W. James's rejection of the classical, absolute concept of truth.) It is interesting to note that in the disputes over intuitionist logic in the nineteen-twenties the members of the Paris School refer to that logic as 'empiricist logic' by contrast to the usual (classical two-valued) 'formal logic'. Cf. R. Wavre 'Logique formale et logique empiriste', *Revue de Métaphysique et de morale* (Janvier 1926), also articles in same journal by P. Levy, M. Barzin, A. Errera and E. Borel.

Poincaré, Henri, *Dernières Pensées*, Paris 1920, pp. 161–2. The analogy between the anti-metaphysical intolerance of logical empiricists and of intuitionists was pointed out by Karl Menger in his article 'The New Logic', *Krise and Neuaufbau in den exakten Wissenschaften*, Wien 1933.

[33] Poincaré, H., *Science and Hypothesis*, New York 1952, p. 144.

[34] H. Poincaré clearly rejected extreme conventionalism in his polemic with E. LeRoy in *The Value of Science*, New York 1958, Part III, 'The Objective Value of Science', pp. 112–128. For his view of the descriptive function of physical theories, cf. *Science and Hypothesis*, Chapter X, particularly pp. 160–2.

[35] On the role of metaphors in explanations, Cf. Hesse, M., 'The Explanatory Function of Metaphor', in Y. Bar-Hillel (ed.), *Logic, Methodology and Philosophy of Science*, Amsterdam 1965; also Kmita, J., 'Wyjasnianie naukowe i metafora', *Studia Filozoficzne* **3** (1967), 143–159.

[36] Poincaré, H., *Science and Hypothesis*, Chapter X.

[37] Duhem, P. 'Physical law', p. 301, *The Aim and Structure of Physical Theory*, 1954. Cf. however, *ibidem*, Appendix: 'Physics of a believer' (1905), p. 285. Some statements here appear to contradict the doctrine of theory-dependence of observables and to express extreme instrumentalist views.

[38] Poincaré, H., *Science and Hypothesis*, p. 143.

[39] Poincaré, H., *The Value of Science*, p. 120.

[40] For Poincaré's criticism of LeRoy's 'incommensurability thesis', see *The Value of Science*, part III, Chapter X, 4, pp. 122–128.

[41] Vaihinger, H., *The Philosophy of 'As if'*, New York 1925, p. 163.

[42] In an empirical theory so reconstructed there may be three types of theoretical statements, analytic, empirical and indeterminate. Carnap, R., *Philosophical Foundations of Physics*, New York 1966; Przełecki, M., *The Logic of Empirical Theories*, London 1968; Williams, P. M., 'On the Conservative Extensions of Semantical Systems: A Contribution to the Problems of Analyticity', *Synthese* **25** (1973), 398–416. For a different view of the semantics of empirical theories Cf. e.g. Tuomela, R., *Theoretical Concepts*, Wien, New York 1973.

[43] Dijksterhuis, E. J., *The Mechanization of the World Picture*, London 1961.

[44] Dijksterhuis, E. J., *op. cit.*

MARJORIE GRENE

IMRE LAKATOS: SOME RECOLLECTIONS

Imre read a paper at Belfast one of the years I was a lecturer there – part of 'Proofs and Refutations.' It was, it seemed to me then and still does, an immensely illuminating object lesson in the complexities of the life of science, or rather of mathematics. When some one raised a question about the application of this historico-philosophical theme to a case in physics, Imre betrayed a surprising naiveté about empirical science, in contrast to his expertise in a mathematical subject matter. All that was clear to us was that, come hell or high water, physics, in his view, had to be treated with Popperian falsificationist dogmatism. Yet the lesson of 'Proofs and Refutations' is surely anti-Popperian. What it demonstrates, among other things, is how tenaciously, and even acrobatically, one can hold onto a thesis when the evidence runs against it.

As every one knows, the publication of 'Proofs and Refutations' in book form was delayed for the sake of Imre's ever expanding excursion into his 'Research Programme' program. What that vast enterprise meant in terms of refutation of Popper or modified support of Popper I could never make out. But there were, in the course of my occasional encounters with Imre, two occasions which shed some light on his interests and objectives, and these I should like to report.

At one stage, the research programme program appeared to Imre to bring his interests close to Polanyi's, as indeed, his earlier work on mathematics had done, although he didn't recognize the convergence then. At his request, I tried a number of times to explain what, in my view, Polyani's philosophy of science comes to. Somehow, I was never able to put this over, until on one occasion I mentioned Polanyi's motivation in relation to the movement for scientific planning in the early thirties, and in particular the impression made on him by a visit with Bukharin in 1935. Told that the notion of 'pure' science was a capitalist, and politically motivated, delusion, as distinct from the dialectical fusion of theory and practice in the U.S.S.R., Polanyi became concerned about the problems of scientific independence, scientific objectivity, and basically, as he put it,

R. S. Cohen et al. (eds.), Essays in Memory of Imre Lakatos, 209–212. *All Rights Reserved*
Copyright © 1976 by D. Reidel Publishing Company, Dordrecht–Holland

"the justification of dubitable beliefs." "That explains a great deal," Imre commented on this report. "Polanyi, I now see, was concerned to save free investigation in a time when planning loomed as a threat. So he could emphasize the search for understanding, and play down prediction as the hallmark of science. But now we *have* planning, whether we like it or not. Science depends on public or industrial support; so we have to develop a conceptual apparatus that will show which research programme is likely to pay off." In other words (and of course the words I have put in quotes are also my reconstruction), now that we're condemned to mission-oriented research, we have to think in mission-oriented terms: which research programme is 'advancing' in the sense of promising predictable results in empirical observations and publishable reports? While admitting, of course, this fact of the politics of science, I found, I must confess, Imre's comparison irrelevant to the philosophical problem of scientific explanation. Prediction of observables is one *test* (not the only one) of a good theory, but the *aim* of a theory is understanding: not in any mystical sense, but rational understanding of how some real, and formerly puzzling, aspect of nature works. The need to justify funding, at least semi-hypocritically, by relating one's research to some practical need and to some shortly and explicitly foreseeable outcome is a regrettable public reality, to be studied by sociologists of knowledge, not, in the first instance, by philosophers. Should the pressures of funding agencies altogether suppress 'pure' research, it would still be, epistemologically, a different kind of conceptual enterprise from the attempt to cure cancer or put a man on Mars. I can, I believe, give a fair outline account of the differences between Aristotelian and modern science, even though the former is extinct, as the latter may well become. True, a given style of scientific inquiry can arise and flourish only within a given historical – social and political – framework. It would be foolish to condemn Aristotle for failing to practice Galilean measurement, let alone use differential equations. Yet Galileo and his successors were right to condemn him for the sake of the new enterprise they could initiate only through the overthrow of Aristotelian science. There is a tension here, admittedly, between the historical and the conceptual, as there is in any analysis of style, but to respond soleley to the immediate demands of contemporary history is to miss the very existence of philosophy.

Unfortunately (in my view), Imre persisted stubbornly along this path.

The scope and profundity of our disagreement became clear to both of us in the last conversation we were to have, a long talk over and after breakfast at the Harvard Faculty Club when I was teaching at Boston University in the Fall of '72 and Imre came through en route to perform one of his famous demolitions at the Philosophy of Science Association meeting in East Lansing, Michigan. He proclaimed quite plainly, on that occasion, his adherence to a classical predictivism (a position which he seems to have modified slightly, but only trivially, before his death). And he gave me two examples in support of his view. No one but a couple of Englishmen, he insisted, paid any attention to Newton till after the return of Halley's comet in 1751. Possessing only the ordinary philosopher's third-hand knowledge of the history of physics, I could do no more than register silent astonishment. But soon thereafter I stumbled on an elegant early (posthumously published) essay by Adam Smith, which celebrates the systematic triumph of Newton – and which at least adds a Scotsman (in addition to David Hume!) to Imre's "couple of Englishmen." More important, Voltaire's *Métaphysique de Newton* was published in 1740 and his *Eléments de la Philosophie de Newton mis à la Portée de tout le Monde* in 1741. This should add, if not twenty million Frenchmen, at least quite a few.

The other instance Imre put forward was just as shocking, indeed, more so, since I know a little bit about it. What about Darwin, I asked, whose overview of the transmutation of species made so much sense that, despite gaps in the fossil record, apparent counterexamples like 'altruism' in the social insects (assimilated to selectionist thinking only since 1964), and the disagreement of many contemporary scientists, many others of his contemporaries reacted with a "Why hadn't I thought of that?". And of course, quite apart from the fact that evolutionary theory is 'retrodictive' rather than predictive, it still rests on sweeping extrapolations from the data, both field and laboratory data, extrapolations which carry conviction because of their comprehensive scope, the elegance with which the genetical calculations that support them can be manipulated, and the ease with which they fit into the metaphysics of a naturalistic view of living things. Imre's answer to my question about Darwin was simple. He replied, "I have a student whose dissertation will show that Darwin was a lousy scientist." A characteristically Lakatosian parrying move. This time I was silent, not out of ignorance, but out of horror. I'm happy to say he did not name the student in question, and I trust that that person has abandoned his (her) misguided quest. Of course Darwin didn't know genetics, any more than Galileo knew calculus; but the thinking of evolutionists today is still fundamentally Darwinian, and, despite the vast increase in their empirical knowledge and in the sophistication of their methods, it remains Darwinian not so much for the experimental predictions it mediates as for the sweep and coherence of its governing principles. It is tragic that Imre postponed – as it turned out, too long – the elaboration of his brilliant incursion into the history of mathematics and the philosophy of discovery and confirmation that it appeared to promise, for the sake of so outworn and even, if I may say so without speaking too unkindly of a regretted friend, so silly a position as he had come to take.

Imre would not, I think, have minded the derogatory adjective. For the point of these anecdotes has been, partly to capture, by reporting them in writing, my recollection of some of our conversations, and partly to express the hope that *Proofs and Refutations*, the solid contribution that Lakatos made to the philosophy of science, will, on publication, receive the serious attention it merits. Others will have to try to interpret its philosophic import, since its author was not afforded time to do so.

University of California, Davis

ADOLF GRÜNBAUM

IS FALSIFIABILITY THE TOUCHSTONE OF SCIENTIFIC RATIONALITY? KARL POPPER VERSUS INDUCTIVISM*

1. INTRODUCTION

There is already a sizeable literature in which the question posed in the title of this essay is answered in the negative. That literature includes the writings of Imre Lakatos.

My justification for now undertaking to add to that literature is as follows: Popper's very influential version of falsificationism asserts the primacy of falsifiability in at least *four* distinct major epistemological or methodological theses. I claim that each of these four central theses is beset by fundamental and seemingly irremediable difficulties of its own. But so far as I know, most of my cardinal objections to these four Popperian doctrines have not been raised by others. Since present-day exponents of these doctrines have found the prior critiques by others unconvincing, I hope that such exponents will be challenged by my critical scrutiny of Popper's views in the present essay.

To avoid misunderstanding of my intent, let me say that the philosophical moral I draw from my critique of Popper's falsificationism is *not* the *wholesale* agnosticism espoused by Lakatos, who wrote: "Can we ... learn from experiment that some theories are false? ... We cannot learn from experience the falsehood of any theory."[1] A statement of my reasons for not embracing this universal *falsificational* agnosticism must await another occasion. On the other hand, Lakatos did espouse some of the major anti-inductivist tenets of Popper's which I shall contest below.

Let me outline the four major groups of Popperian falsificationist theses which I shall challenge:

(1) In his quest for a criterion of demarcation between science and non-science, Popper invokes his own historiography of inductivism and concludes the following: In 1919 the account of testability furnished by all forms of inductivism was such that "there clearly was a need for a different criterion of demarcation."[2] Popper then enunciates falsifiability

R. S. Cohen et al. (eds.), Essays in Memory of Imre Lakatos, 213–252. *All Rights Reserved*
Copyright © 1976 by D. Reidel Publishing Company, Dordrecht–Holland

as the linchpin of the *scientific entertainability* of a hypothesis. Thus, falsifiability by potential negative instances is claimed to play a distinguished role to the exclusion of inductive supportability or probabilistic confirmability by positive instances. Says he: "Testability is falsifiability" (C & R, p. 36). I shall speak of Popper's 'demarcation asymmetry' in order to allude to the asymmetry in the roles which he assigns to the falsifiability and the supportability of a hypothesis vis-à-vis its scientific status.

According to Popper, Freudian psychoanalytic theory does not meet his standards of falsifiability and should therefore be regarded as unscientific. Moreover, this unfavorable appraisal of psychoanalysis is claimed to illustrate the superiority of Popper's falsificationist criterion of demarcation over the inductivist requirements for scientific status.

(2) In his theory of *quantitative* verisimilitude and quantitative content of a hypothesis, Popper invokes *a priori* or 'logical' probabilities of universal hypotheses which are *non*-zero. But he rejects *a posteriori* probabilities of specifically *such* hypotheses as absurd. Hence he rejects Bayesian probabilistic inductivism. And he offers an account of corroboration in which the corroboration of a theory T by an observable fact F which was successfully predicted by T is *parasitic* on the logical relations of F to T's *known* rivals $R_1, R_2 \ldots, R_n$ as follows: For each known rival R, the accepted fact F *either* deductively falsifies R *or* F is *novel* with respect to R in the sense of being logically independent of R. In other words, a successfully predicted positive instance F of T corroborates T, if for each R, the fact F is either contrary to R or not predicted by R but without being contrary to R. Such knowledge as we have is therefore held to rest probatively solely either on refutations of rival theories by accepted negative instances or on novel facts. This is the cornerstone of Popper's proposed solution of the problem of induction. As against Pierre Duhem, Popper asserts an asymmetry of fallibility between the falsification and the corroboration or confirmation of a hypothesis.

(3) According to Popper, the comparative degrees of falsifiability of a theory and of a modified form of it containing a new auxiliary hypothesis are the yardstick for the admissibility of the auxiliary hypothesis and for whether it is *ad hoc*.

(4) In *Objective Knowledge*, falsifiability is claimed to play a fundamental role in the quest for theories of greater verisimilitude: The method of bold conjectures and *refutations* is held to be THE method of science,

because of its purported conduciveness to theories of greater verisimilitude.

Due to limitations of space, I shall now deal in turn with only the first two of these four distinct assertions of the epistemological and methodological primacy of falsifiability. My critique of the remaining two is given elsewhere.[3]

2. Popper's Historiography of Inductivism, and the Test Case of Freudian Psychoanalytic Theory

Philosophers whom Popper disapprovingly calls "inductivists" try to use supportive instances of hypotheses or theories to make either absolute or relative *credibility assessments* of them. An example of an absolute credibility judgment would be to say that a given hypothesis is more likely to be true than false. A relative credibility judgment might take the form of saying that a theory is more credible than a certain rival. *Some* inductivists propose to 'probabilify' hypotheses on given evidence by holding that, in principle, hypotheses can be assigned numerical degrees of confirmation which are mathematical probabilities in the sense of satisfying the mathematical calculus of probability. But other inductivists deny that either absolute or relative degrees of credibility of hypotheses *must* be construed as mathematical probabilities.[4]

Popper's historiography of inductivism is unfortunately oblivious of a paramount fact. It is the fact that there are very important differences among inductivist epistemologies concerning the requirements which must be met by an observational finding, if this finding is to count as a *bona fide supportive* instance of a hypothesis. And, as we shall see, just these crucial differences render quite untenable Popper's declaration that "The fundamental doctrine which underlies all theories of induction is *the doctrine of the primacy of repetitions.* ... According to this doctrine, repeated instances furnish a kind of *justification* for the acceptance of a universal law."[5] Indeed, the test case of psychoanalysis will illustrate the following conclusion, which will emerge from my impending critique of Popper's historiography of inductivism: The inductive use of supportive instances to 'credibilify' hypotheses in one way or another does *not* automatically commit inductivism as such to grant *credible scientific status* or even to lend at least *some* credence to a hypothesis *H merely* because there are *numerous* observationally true consequences of H, and no known negative instances. For, as will become clear, the mere quest for credibilifying a hypothesis by supporting evidence does NOT require at all that *every* positive instance of a hypothesis be automatically held to be supportive of the hypothesis to some *non*-zero degree or other! *A fortiori*, the program of inductive credibilification *as such* in no way requires that all positive instances count as *equally* supportive to some *non*-zero degree. Furthermore, I shall argue that whereas Popper's own criterion of demarcation does not entitle him to indict psychoanalysis as unscientific, certain classical inductivist canons do seriously impugn the scientific credentials of Freud's therapeutic theory and also have the capability of similarly discrediting Freud's psychodynamics.

In the first chapter of Popper's C & R, he adduces astrology, Freudian psychoanalysis, the Marxist theory of history and Adlerian psychology in an endeavor to show the following: (1) all forms of inductivism are logically committed to grant scientific status to these four theories in the face of the numerous positive instances marshalled by them, and (2) whereas the requirements of inductive supportability are hopelessly incompetent to derogate the theories in question as pseudo-scientific, an alternative demarcation criterion based on falsifiability succeeds in excluding them from the body of science.[6]

Let us begin our appraisal of Popper's historiography of inductivism by considering the views which the arch-inductivist Francis Bacon (1561–1626) expressed on astrology three centuries before Popper.

Popper (C & R, pp. 255–256 and 112) illuminatingly calls attention to major weaknesses in Bacon's methodological prescriptions, qua purported methods for *discovering* theories and for certifying reliably that theories thus discovered must be *true*. Nonetheless, Popper credits Bacon to the extent of noting that "the problem of *drawing a line of demarcation*" between science and pseudo-science "has agitated many philosophers since the time of Bacon" (C & R, p. 255). But then Popper tries to discredit Bacon's inductivism as an *answer* to the problem of demarcation by writing as follows:

> ... many superstitious beliefs ... have no doubt often been based on something like induction. Astrologers, more especially, have always claimed that their 'science' was based upon a great wealth of inductive material. This claim is, perhaps, unfounded; but I have never heard of any attempt to discredit astrology by a critical investigation of its alleged inductive material (C & R, p. 256).

Yet when we turn to Bacon's writings, we find that he explicitly refers to astrology when discussing the 'idols of the tribe' in his *Novum Organum* (Book I, Section 46). Speaking of the relative importance of attention to negative and positive instances of a theory, Bacon condemns astrology as superstitious because, among other things, it neglects evidence *unfavorable* to it. The Baconian aphorism dealing with astrology merits quotation in its entirety because it is delicious. Bacon wrote:

> The human understanding when it has once adopted an opinion (either as being the received opinion or as being agreeable to itself) draws all things else to support and agree with it. And though there be a greater number and weight of instances to be found on the other side, yet these it either neglects and despises, or else by some distinction sets aside and rejects; in order that by this great and pernicious predetermination the authority of its former conclusions may remain inviolate. And therefore it was a good answer that was made by one who when they showed him hanging in a temple a picture of those who had paid their vows as having escaped shipwreck, and would have him say whether he did not now acknowledge the power of the gods, – "Aye," asked he again, "but where are they painted that were drowned after their vows?" And such is the way of all superstition, whether in astrology, dreams, omens, divine judgments, or the like; wherein men, having a delight in such vanities, mark the events where they are fulfilled, but where they fail, though this happen much oftener, neglect and pass them by. But with far more sublety does this mischief insinuate itself into philosophy and the sciences; in which the first conclusion colors and brings into conformity with itself all that come after, though far sounder and better. Besides, independently of that delight and vanity which I have described, it is the peculiar and perpetual error of the human intellect to be more moved and excited by affirmatives than negatives; whereas it ought properly to hold itself indifferently disposed towards both alike. Indeed in the establishment of any true axiom, the negative instance is the more forcible of the two (*Novum Organum* I, 46).[7]

Long before Popper's injunction that "Confirmations should count only if they are the result of *risky predictions*" (C & R, p. 36), Francis Bacon made a vital contribution toward distinguishing *merely positive* from *supportive* instances of a theory by emphasizing that some kinds of positive instances can differ radically from others in *evidential value* and treating the evidentially significant ones under the heading of *'Prerogative Instances'* (*Novum Organum* II, Section 21ff.). I shall say in more modern parlance that an instance is a 'positive' one with respect to a *non*-statistical theory T, if its occurrence or being the case can be deduced from T in conjunction with suitable initial conditions. But an instance is supportive of T, if it is positive *and* has the probative significance of conferring a stronger truth presumption on T than T has without that instance. As will be shown in our impending discussion of the very important case of causal hypotheses, post-Baconian inductivists appealed to the distinction between merely positive and supportive instances of a theory: they did so in order to guard against unsound causal inferences such as *post hoc ergo propter hoc* and against other ravages of the unbridled use of the hypothetico-deductive method in inductive inference. The very odd one-sidedness of Popper's account of the Baconian inductive method of comparative instances becomes further evident from the exposition given, for example, in the 1916 standard book on *Logic* by the English philosopher Joseph.[8]

As Joseph stresses, Bacon was grievously mistaken in supposing that scientists could devise an *exhaustive* finite list of all logically possible alternative theories relevant to a given phenomenon. Having made that unsound assumption, Bacon felt justified in maintaining that scientists could then irrevocably establish the truth of *one* of these specified theories by refuting all of its rivals in the purportedly exhaustive disjunctive class of relevant theories. But while rejecting this Baconian conclusion as ill-founded, Joseph also points out that, according to Bacon, the confident affirmation of the truth of the one remaining theory "would rest *not* on the positive testimony" of its positive instances, "but upon the fact that we had disproved all possible rival theories."[9] In short, for Bacon these positive instances were prerogative or *supportive* of the one theory, because these same instances were *also* negative ones for its rival theories. Says Joseph:

> We must proceed then by *exclusions*. Where a hundred instances will not *prove* an universal connexion, one will *disprove* it. This is the corner-stone of his [Bacon's] method: *maior est vis instantiae negativae* [negative instances have more force].[10]

Not only Bacon's tables of comparative instances but also J. S. Mill's "inductive methods" for appraising *causal* hypotheses had long ago led inductivists to demand *controlled experiments* or so-called *controls* as an indispensable check on whether positive instances do have the probative significance of being supportive instances! But in fairness to Popper, it should be pointed out that there has been one important school of inductivists among both philosophers and eminent scientists who championed the doctrine that *any* positive instance of a hypothesis also necessarily qualifies as a *supportive* instance of the hypothesis. This doctrine that any positive case of a hypothesis is automatically also supportive to *some* degree is sometimes called "the instantiation condition" (Nicod's criterion).[11] In an *unbridled* use of the instantiation condition, any positive instance of a hypothesis – e.g. one resulting from the hundredth repetition of the same experiment – will *increase* the credibility of the hypothesis *as much as any other*. And hence in *such* an invocation of the instantiation condition, one considers the mere *number* rather than the relative weight of positive instances when assessing the credibility of a hypothesis. On the latter *stronger* version of the instantiation condition, degrees of credibility could not be construed as mathematical probabilities, since a sufficiently large finite number of positive instances would then yield a 'probability' *greater* than unity.

But it was none other than the confirmed inductivist J. S. Mill who emphatically rejected the instantiation condition after Bacon had rejected induction by simple enumeration of positive instances as puerile. Thus, when discussing the case of a *spurious* plurality of causes in his *Logic*, Mill *denies* the probative value of *mere* repetitions of positive instances just as much as Popper does (LSD, p. 269), and indeed for much the same reasons.[12] Mill stresses the need to find positive instances of a specified causal hypothesis such that these instances *also* refute one or more *rival* hypotheses as to the cause. And he writes:

> The tendency of unscientific inquirers is to rely too much on number, without analyzing the instances; without looking closely enough into their nature to ascertain what circumstances are or are not eliminated by means of them. Most people hold their conclusions with a degree of assurance proportioned to the mere *mass* of the experience on which they appear to rest; not considering that by the addition of instances to instances, all of the same kind, that is, differing from one another only in points already recognized as immaterial, nothing whatever is added to the evidence of the conclusion. A single instance eliminating some antecedent which existed in all the other cases [i.e., an instance which refutes a *rival* hypothesis as to the cause] is of more value than the greatest multitude of instances which are reckoned by their number alone.[13]

Clearly, there is at least one historically influential inductivist conception of testability which makes three assertions as follows: (i) it calls attention to the difference between the genus of positive instances of a hypothesis *H* and the species *S* of those special positive instances which serve to eliminate one or more *specified* rivals of *H*, (ii) it rejects the instantiation condition by denying that every member of the genus of positive instances automatically qualifies as supportive, and (iii) it regards the species *S* of positive instances as supportive in the sense of conferring some degree of credibility on *H*, even if the amount of available supportive evidence is *insufficient* to make it more likely that *H* is true rather than false. Note that among these three assertions, only the third endorses inductive inference. Yet Popper is so preoccupied with dissociating himself from the third of these assertions (LSD, p. 419) that he ignores the first two. Thus, he is prompted to misportray *all* post-Baconian theories of induction as assigning *supportive* probative significance to all positive instances *alike*. In this way, he is *misled* into charging that all such versions of inductivism are unable to indict psychoanalysis, for example, as unscientific in the face of its many positive instances. He gives a concise statement of this misportrayal by making the following previously cited statement (LSD, p. 420, italics in original): "The fundamental doctrine which underlies all theories of induction is *the doctrine of the primacy of repetitions*.... According to this doctrine, repeated instances furnish a kind of *justification* for the acceptance of a universal law." But this claim is untenable in the face of the espousal of assertions (i) and (ii) by the eliminative inductivism of Bacon and Mill. And, as Section 3 below will show, Popper's portrayal of all forms of inductivism is likewise refuted by the fact that Bayesian probabilistic inductivism disavows the instantiation condition. Popper's preoccupation with opposing assertion (iii), to the exclusion of appreciating assertions (i) and (ii), is made evident by his appraisal of the eliminative inductivism of Bacon, Whewell and Mill. Speaking of the latter on the immediately preceding page of LSD (p. 419), he says:

> ... the sole purpose of the elimination advocated by all these inductivists was to *establish as firmly as possible the surviving theory* which, they thought, must be the *true* one (or perhaps only a *highly probable one*, in so far as we may not have fully succeeded in eliminating every theory except the true one).
>
> As against this, I do not think that we can ever seriously reduce, by elimination, the number of the competing theories, since this number remains always infinite.

When we discuss Popper's theory of corroboration in Section 3, the last sentence of this quotation from him will return to haunt his own theory!

For brevity, I shall use the term 'instantionist inductivism' to refer to the very special version of inductivism which does espouse the instantiation condition. Then I can say that Popper's characterization of all theories of induction as being instantionist also runs counter to the fact that Mill demanded a check by the Joint Method of Agreement and Difference on the *probative* significance of the positive instances which had been accumulated by the merely *heuristic* Method of Agreement. Hence we reach the following important conclusion: At least some post-Baconian inductivists *deny* that every positive instance of a hypothesis *H* is automatically supportive of *H* to some non-zero degree or other. And *a fortiori* these inductivists – as distinct from the exponents of *unbridled instantionist* inductivism! – deny that all positive instances count as *equally* supportive. In view of the importance of this conclusion, let us restate it by saying that all *anti*-instantionist versions of inductivism espouse the following cardinal epistemic principle: *The ability of a theory T to 'explain' and/or predict certain phenomena deductively* (with the aid of suitable initial conditions) *is generally only a necessary and NOT a sufficient condition for qualifying these phenomena to count as T-supporting instances over and above being merely positive instances of T!*

Note here that within the genus of positive instances, the term 'confirming instance' is disastrously ambiguous as between a supportive and a non-supportive species of positive instances. By the same token, logical mischief has been wrought by the weasel word 'verification,' and by the equivocal verb 'verify.' Popper's misportrayal of inductivism as universally instantionist is abetted by the ambiguity of the terms 'confirmation' and 'verification,' when he says misleadingly: "It is easy to obtain confirmations, or verifications, for nearly every theory – if we look for confirmations" (C & R, p. 36). Contrary to Popper, we shall see very soon that whereas it is easy to find positive instances for the therapeutic claims made by psychoanalysts, it is at best moot whether there exists any *bona fide* or *significant amount* of genuine inductive support for these claims, let alone for the theory undergirding psychoanalytic treatment.

So much for my examination of Popper's historiography of inductivism, which prompted his claim that "there clearly was a need for a different criterion of demarcation" (C & R, p. 256). Lakatos echoes Popper's grossly one-sided concentration on the least sophisticated parts of Bacon's inductivism. Bacon was surely oblivious of the fact, later noted by Kant, that "percepts without concepts are blind," and yet he stressed rightly that theoretically-aseptic fact collecting is either a delusion or sterile (*Novum Organum*, Book I, Aphorism 95). But this patent deficiency does not justify the imbalance in Lakatos's account of Bacon's conception of scientific inference. When speaking of Popper's role as "the scourge of induction" (PDI, p. 258), Lakatos distinguishes three strands in Popper's anti-inductivist campaign and says: "First, there is the campaign against the *inductivist logic of discovery*. This is the Baconian doctrine according to which a discovery is scientific only if it is *guided* by facts and not *misguided* by theory." Having thus one-sidedly dealt with Baconian doctrine, Lakatos can then declare (PDI, p. 259) that "at least among philosophers of science Baconian method is now only taken seriously by the most provincial and illiterate."

The Illustrative Test Case of Psychoanalysis

Turning now to Freudian psychoanalysis, we note that psychoanalysis offers not only a psychodynamic theory to explain behavior but also a type of treatment – prejudicially dubbed 'therapy' – to be practiced for the alleviation of psychological disturbances. I shall focus on the *therapeutic* theses of psychoanalysis as a test case for those of Popper's allegations against inductivism on which he bases his demarcation asymmetry. In so doing, I make due allowance for the following: If it were shown that psychoanalytic therapeutics is not entailed by Freud's psychodynamics, then Freud's theory *may* conceivably have a correct account of the etiology and pathogenesis of certain disorders while failing to provide successful therapeutic prescriptions for them. Generally speaking, it is not incumbent upon a psychodynamic theory to produce effective recommendations for psychotherapy in order to qualify as etiologically explanatory, any more than a theory which *explains* carcinogenesis can be expected to *guarantee* that cancer is curable.

Using his falsificationist criterion of demarcation, Popper felt entitled to indict psychoanalysis as unscientific after having addressed the following challenge to Freudians:

But what kind of clinical responses would refute to the satisfaction of the analyst not merely a particular analytic diagnosis but psycho-analysis itself? And have such criteria ever been discussed or agreed upon by analysts (C & R, p. 38)?

Whereas Popper is concerned with the scientific entertainability of a hypothesis, the inductivist is concerned with its scientific *credibility*. Hence let us now see quite specifically how the Bacon-Mill inductivist can impugn the scientific credentials of psychoanalytic therapeutics by his particular requirements for *supporting* evidence. It will then be clear how, *mutatis mutandis*, inductivist checks on the probative significance of positive instances can prevent Freudian *psychodynamics* from adducing such instances uncritically with impunity in favor of its etiological claims concerning, say, hysteria or fetishistic sexual behavior. Thus, it will emerge, for example, how the inductivist can challenge Freudian claims about the homosexual etiology of paranoia. In this way, the inductivist will be seen to match Popper's ability to disparage the scientific status of, say, the Freudian Oedipus hypothesis, which Popper does by pointing to the failure of psychoanalysis to specify what kind of behavior (of sons toward their fathers) would warrant its abandonment.

Just how do the time-honored inductivist canons of Bacon and Mill enjoin us to employ *experimental controls* as a curb on giving *undeserved* causal credit to psychiatric treatment for such improvement as is shown by patients during or after treatment? As is well known, though apparently overlooked by Popper, their inductivist canons discount instances of subsequently improved patients as non-supportive, *unless* the incidence of improvement among treated patients *exceeds the spontaneous remission rate!* Existing sets of cases of *improved* psychoanalytic patients may indeed qualify as *positive* instances of the hypothesis that Freudian treatment is effective in a specified diagnostic category. But this improvement does *not* inductively redound to the credit of this treatment, unless its incidence exceeds the spontaneous remission rate, which is practically always positive. In particular, the *mere* fact that *at least* some psychoanalytically-treated neurotics in a certain diagnostic category get better *after* treatment does *not* show inductively that this treatment is effective, unless *other evidence* shows the spontaneous remission rate for that diagnostic category to be *zero.* Thus, in the absence of further information, positive instances of improvement can fail to be supportive. Indeed, it is conceivable that patients who undergo treatment have a remission rate which is *lower* than the spontaneous one, thereby giving rise to the possibility that the treatment made at least some of them worse! In psychiatry no less than in somatic medicine, there can be iatrogenic or doctor-induced disease.

Furthermore, by *Popper's* standards, the claim that Freudian treatment is effective is falsifiable. For Popper could hardly deny that the failure to surpass the spontaneous remission rate would serve to refute the hypothesis of therapeutic effectiveness under the assumption of suitable auxiliary claims. And according to Popper's demarcation criterion, falsifiability is *sufficient* for scientific status though *not*, of course, for being a *corroborated* scientific hypothesis. Therefore, by *Popper's* standards the thesis that psychoanalytic treatment is therapeutic does qualify for scientific status without prejudice to whether this therapeutic claim will turn out to be a *corroborated* scientific claim by Popper's requirements.

As for the inductivist, I have been concerned to note so far that the question as to the scientific *credentials* of psychoanalytic therapeutics is tantamount to the following question: What are the actual empirical findings as to the success of psychoanalytic treatment in *surpassing* the spontaneous remission rate in any given diagnostic category? But I must now emphasize that there is a host of difficult subsidiary questions here which must be answered before this question becomes susceptible of a *meaningful* and empirically-supported reply. For example, there is much imprecision in the diagnostic categories of neuroses and in certifying membership of patients in one of these. Also, it is unclear what criteria are to be employed in assessing *improvement* in a given diagnostic category such as depressives. Is there improvement in a previously depressed man who is full of joy after treatment, if he also now beats his wife and children when they cannot keep up with his energetic demands?

Whatever these various subsidiary questions, which are well-known to those who are familiar with treatment-outcome research, the inductivist places the logical burden of formulating a *meaningful* claim of therapeutic success for Freudian treatment on the *advocates* and *dispensers* of psychoanalytic or psychiatric treatment. And I would add that the *moral* burden of doing so likewise belongs squarely on the shoulders of these advocates and dispensers. In sum, if we are inductivist sceptics, it is emphatically *not* incumbent on the rest of us to show that psychoanalytic treatment is *in*effective! In this vein, psychologists such as Eysenck,[14] Rachman[15] and others have argued that there is no satisfactory evidence showing that psychoanalytic treatment succeeds in surpassing the spontaneous remission rate, although others have sharply challenged their analysis and consider the question moot.[16]

In any case, one can only be baffled and distressed by the immense epistemological crudity encountered in *some* of the literature of psychoanalysis in this connection. Thus, in his 1970 book *The Crisis in Psychoanalysis* Erich Fromm acknowledges the justification for a crisis of confidence in psychoanalysis. But when Fromm turns to the issue of the causal efficacy of psychoanalytic treatment, he gives not even a hint of being aware of the need for a control group in assessing that efficacy. This, even though anyone who ever studied medicine could hardly be unaware of the need for a control group if the therapeutic efficacy of a drug or chemical agent were at issue, for example. Indeed, Fromm reasons in accord with *post hoc ergo propter hoc* that Freudian psychotherapy is causally efficacious, because psychoanalysts have 'observed' improvement in their patients. One might as well argue that coffee consumption cures colds, because we can all observe *with our own eyes* that when we have a cold, we get better every time after we drink coffee for a sufficient number of days! Fromm writes:

Indeed, the facile denial of the therapeutic success of psychoanalysis says more about the difficulty of some fashionable authors to grasp the complex data with which psychoanalysis deals than about psychoanalysis itself. Criticism by people with little or no experience in this field cannot stand up against the testimony of analysts who have observed a considerable number of people relieved of troubles they complained about. Many patients have experienced a new sense of vitality and capacity for joy, and no other method than psychoanalysis could have produced these changes.[17]

Extra ecclesiam nulla salus!

The exponent of anti-instantionist inductivism says to the psychoanalyst: Suppose that Eysenck is right and psychoanalytic treatment does not improve on the spontaneous remission rate. Then how, if at all, can you nonetheless hope to adduce such improvement as is shown by your patients in support of your claim that Freudian psychotherapy *is* effective? Bergin has argued that the studies used by Eysenck and others in reaching their conclusion dealt with *average* effects of treatment and thus overlooked evidence indicating the following: After treatment, the improvement of *some* of the treated patients does exceed the gains exhibited by the untreated patients in the control group, but other treated patients get *worse* after treatment than the untreated controls. Says Bergin:

When these contrary phenomena are lumped together in an experimental group, they cancel each other out to some extent, and the overall yield in terms of improvement (in these particular studies) is no greater than the change occurring in a control group via 'spontaneous remission factors.'[18]

Significantly, however, Bergin goes on to concede that, on his view, it is not the treatment *as prescribed by the therapeutic theory* which is effective in a certain subclass of cases, but only certain *dispensers* of the treatment having as yet unknown special characteristics or knacks whose efficacy is not understood *and cannot be credited to the therapeutic theory*!

Let me suggest a different putative reply which might be offered by the psychoanalyst and then appraise it according to 'neo-Baconian' inductivist standards. Very loosely put, he might conjecture that, in at least some diagnostic categories, patients who seek and receive Freudian treatment are, in some unspecified way, *more sick* than those neurotics who do not seek such treatment and remain untreated by a psychoanalyst. Thus he might go on to venture the counterfactual claim that if the presumably sicker subclass of treated neurotics had been left *untreated*, then they would have exhibited a remission rate which is LOWER than the spontaneous one among the actually untreated neurotics. In short, the putative conjecture is that the treated and untreated neurotics do *not* have the same capacity for spontaneous remission, the treated neurotics having been initially more sick.

But to confer credibility on this counterfactual claim by the stated inductivist standards, the psychoanalyst would *at least* have to produce evidence showing the following: Neurotics in a given diagnostic category who seek but are deliberately denied psychoanalytic treatment – say by being put on a waiting list – do exhibit the allegedly lower remission rate. Until and unless this is done, the claim of therapeutic effectiveness stands indicted, according to the specified inductivist canons, as gratuitous and hence as devoid of scientific credibility or credentials. Moreover, even if the putative psychoanalyst did furnish the required evidence of *lower spontaneous* remission in the treated subclass, the Baconian inductivist would go on to issue the following challenge to him: In the absence of further information, the positive instances provided by this evidence cannot *redound to the credit of the theory* undergirding psychoanalytic therapeutics. Before such credit can be given, at least the *rival hypothesis* of autosuggestive 'faith-healing' or *attention-placebo-effect* would need to be refuted. Note that it is *not* incumbent on the inductivist critic or sceptic to *establish* the rival hypothesis of placebo effect. But it *is* incumbent upon the proponent of psychoanalytic theory to *refute* it! The latter refutation would presumably require showing that in the allegedly sicker subclass, various quite different kinds of psychotherapy *fail* to produce as high a remission rate as psychoanalysis does. To my knowledge, there is no such evidence of any superiority of psychoanalytic treatment. On the contrary, a careful 1970 book by Meltzoff and Kornreich [19] concludes that there is no difference among a whole gamut of different schools of traditional psychotherapy in regard to outcome success. And this important finding refutes the aforecited claim by Erich Fromm that "no other method than psychoanalysis could have produced these changes" [remissions].[20]

Clearly, contrary to Popper, post-Baconian inductivism has at least the capability of challenging the scientific credentials of psychoanalytic therapeutics by issuing the stated two-fold challenge to the above putative defender of its therapeutics. Thus, post-Baconian inductivism does command the logical resources to derogate the scientific credentials of psychoanalysis in the face of the positive instances adduced by Freudians. Inductivism has that capacity in regard to psychoanalysis just as it has the capacity to question, for example, whether the relatively low incidence of coronary heart disease among *athletic* adult males shows that physical exercise contributes to cardiovascular health, unless one can rule out the possibility that precisely those who are otherwise healthy cardiovascularly are the ones who engage in physical exercise to begin with.

Let me merely mention, but not rehearse, the criticism in the literature that Popper's deductive falsifiability of a hypothesis is much too strong a requirement for scientific entertainability, as illustrated by the *mere example* of the hypothesis "All men are mortal." For the latter says that for each man, *there is some time or other* at which he dies, i.e., that no man lives forever. To *falsify* this deductively by a test statement, we must produce at least one man who NEVER dies. For no matter how long any Methusalah lives, he may still die later on. Hence no basic observation statement can deductively falsify the old saw that we are all mortal. Other examples are furnished by assertions of non-existence concerning *perpetual* motion machines of the first and second kinds, as in the first two laws of thermodynamics.

Surely I am not being captious if I conclude from my analysis that Popper's demarcation asymmetry is either unsound or too strong: It is unsound if it is claimed to have demarcational capabilities with respect to psychoanalysis and astrology, for example, which are not also possessed by the requirements of anti-instantionist inductivism, or it is too strong as just explained. As shown by my examples, the mere fact that anti-instantionist inductivists try to use supportive instances to 'probabilify' or credibilify hypotheses does NOT commit them to granting credible scientific status to a hypothesis *solely* on the strength of existing positive instances, however numerous. Indeed as shown by thousands of cases of people who are improved *after* psychiatric treatment, Mill's inductivism can discount even such a multitude of positive instances as *non*-supportive of the hypothesis of therapeutic effectiveness. But I now need to forestall a possible misunderstanding of the moral I draw from my comparison of Popper's demarcation asymmetry with the conception of scientificality advocated by an inductivist who rejects the instantiation condition.

As already noted, Popper is primarily concerned with scientific *entertainability*, whereas the inductivist is intent upon the scientific *credibility* of a hypothesis. For Popper, the mere falsifiability of a hypothesis suffices for according scientific status to it, but the inductivist may be prepared to grant no more than *potential* scientific status to it in virtue of its inductive support*ability* in principle. And Popper's *corroborated* scientific hypothesis is the *counterpart* of the inductivist's *actually* scientific hypothesis. Hence, if one wished, one could treat Popper's falsifiability and the inductivist's genuine support*ability* as counterpart criteria in their respective endeavors to effect a demarcation between non-science, on the one hand, and those theories which are at least scientifically *entertainable*, on the other.

Can it be held that the inductivist's criterion avoids any defect corresponding to the weakness of Popper's falsifiability requirement, which fails by being too strong? I think not, because the inductivist's genuine supportability likewise fails to the extent that there is no satisfactory general theory of evidential support which states unambiguous necessary and sufficient conditions that are successfully applicable to *every* concrete case. We must be mindful in this connection of the fact that science covers a whole gamut of kinds of claims which pose problems of empirical validation: Comprehensive theories like neo-Darwinism or general relativity, causal hypotheses like 'Asbestos is carcinogenic,' statistical hypotheses, and even simple generalizations like 'All ravens are black.' Consider just the last of these examples, which lends itself to the statement of Hempel's paradox of confirmation. If the inductivist avoids that paradox by rejecting the instantiation condition, then it is still incumbent upon him to tell us under what conditions a positive instance of 'All ravens are black' *does* count as supportive. Yet no satisfactory general answer is available, as far as I know. Nor is it clear that inductivism can furnish a generally satisfactory method for assigning *non-zero* posterior mathematical probabilities to even the empirically most successful *universal* law hypotheses (cf. note 21 and Section 3, items I and II below). Hence I think that the inductivists have no more succeeded than Popper in stating *general* criteria for effecting a *neat* demarcation of science from non-science.

Thus, the upshot of my comparison of inductivist conceptions of scientificality with Popper's is *not* the claim that there is a viable inductivist counterpart to Popper's defective demarcation criterion for scientific entertainability. Instead the moral I draw is the following: Popper was seriously mistaken in claiming that IN THE ABSENCE OF NEGATIVE INSTANCES, all forms of inductivism are necessarily committed to the (probabilified) scientific credibility of a theory, merely because that theory can adduce numerous positive instances.

As for psychoanalysis in particular, contrary to Popper, Freudian therapeutic theory is no less falsifiable by Popper's own standards than it is in principle genuinely supportible inductivistically. But in regard to actual scientific credibility, the inductivist can indict that theory is at best gratuitous, because its positive instances do not have the probative force of being sufficiently supportive and perhaps are not even supportive at all. Thus, the inductivist's willingness to either probabilify or somehow credibilify theories which *can* marshall genuinely supportive positive instances does *not* render the inductivist helpless to dismiss the positive instances adduced by psychoanalysts as non-probative. And, as we shall soon see, the inductivist can justly complain that Popper's pure deductivism has no non-trivial answer to the question: "What does it mean to say that a successful risky prediction COUNTS in favor of the theory that made it?"

3. Popper's Theory of Corroboration versus Bayesian Probabilistic Inductivism and Duhemian Holism

I. *Popper on the Probability of Universal Laws*

One of Popper's major reasons for alleging the 'impossibility of an inductive probability' is as follows: In a universe containing infinitely many distinguishable things or spatio-temporal regions, *"the probability of any (non-tautological) universal law will be zero"* (LSD, Appendix vii,

p. 363; italics in original). Commenting on the meaning of the term 'probability' in this sentence, he says (LSD, p. 364):

By 'probability,' I mean here either the *absolute* logical probability of the universal law, or its probability *relative to some evidence*; that is to say, relative to a singular statement, or to a finite conjunction of singular statements. Thus if a is our law, and b any empirical evidence, I assert that

(1) $p(a)=0$

and also that

(2) $p(a, b)=0$.

These formulae will be discussed in the present appendix. The two formulae, (1) and (2), are equivalent.

And for the case in which the universal statement a is interpreted as entailing an infinite conjunction of singular statements, he then goes on to give several arguments for concluding that $p(a)=0$. He regards one of these arguments to be "incontestable" (LSD, p. 366). But Colin Howson has offered rebuttals to these arguments.[21]

The same conclusion is reiterated and amplified in C & R, where Popper writes:

... in view of the high content of universal laws, it is neither surprising to find that their probability is zero, nor that those philosophers who believe that science must aim at high probabilities cannot do justice to facts such as these: that the formulation (and testing) of *universal laws* is considered their most important aim by most scientists (C & R, p. 286).[22]

Furthermore, he contrasts the inevitably vanishing inductive probabilities of universal law statements with the *non*-vanishing degrees of corroboration of which such statements are capable on appropriate evidence in his theory of corroboration, saying:

But it can be shown by purely mathematical means that *degree of corroboration can never be equated with mathematical probability*. It can even be shown that all theories, including the best, have the same probability, namely zero. But the degree to which they are corroborated (which, in theory at least, can be found out with the help of the calculus of probability) may approach very closely to unity, i.e. its maximum, though the probability of the theory is zero (C & R, pp. 192–193).

But Popper's conclusion that, for any universal statement a, $p(a)=0$ can now be shown to be incompatible with his *quantitative* theory of the content and verisimilitude of a hypothesis. For when we turn to the latter theory (C & R, pp. 390–397, 234, 218), we find the following:

(1) Both in C & R and in his *Objective Knowledge*,[23] Popper defines the *measure* $\mathrm{ct}(a)$ of the content of a theory a as $\mathrm{ct}(a)=1-\mathrm{p}(a)$, where

'p(a)' is 'the logical probability' of a. He further characterizes the latter probability very meagerly (OK, p. 51) as "the logical probability that it [a] is true (accidentally, as it were)."

(2) When speaking of the logically *incompatible* theories of Einstein and Newton – respectively denoted by 'E' and 'N' – Popper tells us without any explanation that "the content measures $ct(N)$ and $ct(E)$" bear out the intuition that, as between these two theories, "Einstein's has the greater content" (OK, p. 53). But if $ct(E) > ct(N)$, then $p(N) > p(E)$, and hence $p(N) > 0$ even though N is a conjunction containing universal statements and should therefore have *zero* probability according to Popper's cited claim. Let us now see how this inconsistency carries over into his theory of quantitative verisimilitude.

(3) Popper states his view as to what outcome of attempted falsifications of E would epistemically warrant the conjecture that the quantitative falsity-content of E does not exceed the corresponding falsity-content of N (OK, p. 53). And he then deduces (OK, p. 53) that E has greater quantitative verisimilitude than N from the conjunction of the following three assertions: (i) the aforementioned claim that $ct(E) > ct(N)$, (ii) the stated conjecture as to the comparative falsity-contents of E and N, and (iii) the contention that "the stronger theory, the theory with the greater content, will also be the one with the greater verisimilitude *unless its falsity content is also greater*" (OK, p. 53). I should mention incidentally that, in the case of *logically incompatible* theories such as E and N, Popper's justification for this third assertion breaks down: His 1966 theorem on truth-content, which serves as a premise in his derivation of the third assertion (OK, pp. 52–53), is *not* proven for the purported content-relation $ct(E) > ct(N)$ of *incompatible* theories such as E and N but only for the corresponding proper inclusion relation among the *Tarskian* logical contents of a pair of theories one of which unilaterally entails the other.[24] In any case, his cardinal example of greater verisimilitude makes the claim that $ct(E) > ct(N)$.

But since N and E are each replete with universal statements, consistency with his claims in LSD (Appendix vii) requires Popper to say that both $p(N) = 0$ and $p(E) = 0$. And since $ct(a) = 1 - p(a)$, this has the consequence that $ct(E)$ and $ct(N)$ have the *same* value unity. Yet, as we just noted, he told us in OK (p. 53) that $ct(E) > ct(N)$. Furthermore, Popper asserts (C & R, p. 218) that (1) "with increasing content, probability decreases, and *vice versa*," and that (2) if one (universal) theory B *unilaterally entails* another (universal) theory A, then ct(B) > ct(A) (OK, pp. 51 and 53). These assertions (1) and (2) require that if B unilaterally entails A, then p(A) > p(B), and hence that at least one of the two probabilities p(A) and p(B) be *non*-zero. But for universal non-tautological theories A and B, the latter requirement contradicts the cited LSD claim that *both* of them must be zero. Moreover, even for the case in which B *unilaterally* entails A, the probability calculus requires only that p(A) ⩾ p(B) rather than that p(A) > p(B), as Popper does here.

Thus, Popper hoists himself on his own petard: If he is to furnish an indictment of inductive probabilities by his LSD claim that $p(a)=0$ and $p(a, b)=0$ for any non-tautological universal a and any empirical evidence b, then his theory of quantitative content and quantitative verisimilitude is untenable; conversely, if the latter is affirmed, the claim $p(a)=0$ is no longer available as a premiss for a *reductio ad absurdum* argument against inductive probabilities.

II. *Popper on Probability vis-à-vis the Aims of Science*

One of the citations in the preceding subsection I contains Popper's complaint that the aim to achieve high probabilities as in Bayesian inductive inference *fails* to vindicate the quest for *universal* laws: As Lakatos has summarized it (PDI, p. 259): "...*if* inductive logic was possible, *then* the virtue of a theory was its improbability rather than its probability, given the evidence." Concerning this complaint, Ronald Giere has aptly made a personal remark to the following effect: The probabilistic inductivist does *not* say that the aim of science is or ought to be to play it safe. Instead, this inductivist tells us (i) what risks we take, *if* we choose *not* to play it safe, and (ii) that *if* we wish to play it safe, then we should act on the *less* daring of two theories whenever possible. On this understanding of probabilistic inductivism, Popper's complaint no longer applies.

III. *Comparison between Popper's Methodology and Bayesian Inductivism*

It has been or ought to have been well-known since the nineteenth century that the prior probabilities in Bayesian inference pose formidable if not insoluble difficulties, at least if they are to be construed *non*-subjectively. Hence, if I now proceed nonetheless to assess the capabilities of Bayesian inductivism to implement Popper's methodological prescriptions, I do *not* do so in the spirit of d'Alembert, who is said to have declared: "Allez en avant, et la foi vous viendra." Rather I do so, because I hope to show that this assessment is instructive, although it is moot whether the status of Bayes's prior probabilities will be satisfactorily clarified in the future. I shall inquire: Can Popper do any better?

A. *The Epistemic Capabilities of Popper's Methodology*

The following two epistemic contentions of Popper's are pertinent here:

(i) "Confirmations should count only if they are the result of *risky predictions*; that is to say, if, unenlightened by the theory in question, we should have expected an event which was incompatible with the theory – an event which would have refuted the theory" (C & R, p. 36), and

(ii) "When trying to appraise the degree of corroboration of a theory we may reason somewhat as follows. Its degree of corroboration will increase with the number of its corroborating instances. Here we usually accord to the first corroborating instances far greater importance than to later ones: once a theory is well corroborated, further instances raise its degree of corroboration only very little. This rule however does not hold good if these new instances are very different from the earlier ones, that is if they corroborate the theory in a *new field of application*. In this case, they may increase the degree of corroboration very considerably" (LSD, p. 269).

Popper clarifies the latter of these two demands by writing elsewhere (C & R, p. 240) as follows:

(iii) "A serious empirical test always consists in the attempt to find a refutation, a counter example. In the search for a counter example, we have to use our background knowledge; for we always try to refute first the *most risky* predictions, the '*most unlikely* ... consequences' (as Peirce already saw [footnote omitted here]); which means that we always look in the *most probable kinds* of places for the *most probable* kinds of counter examples – most probable in the sense that we should expect to find them in the light of our background knowledge. Now if a theory stands up to many such tests, then, owing to the incorporation of the results of our tests into our background knowledge, there may be, after a time, no places left where (in the light of our new background knowledge) counter examples can with a high probability be expected to occur. But this means that the degree of severity of our test declines. This is also the reason why an often repeated test will no longer be considered as significant or as severe: there is something like a law of diminishing returns from repeated tests (as opposed to tests which, in the light of our background knowledge, are of a *new kind*, and which therefore may still be felt to be significant). These are facts which are inherent in the knowledge-situation; and they have often been described – especially by John Maynard Keynes and by Ernest Nagel – as difficult to explain by an inductivist theory of science. But for us it is all very easy. And we can even explain, by a similar analysis of the knowledge-situation, why the empirical character of a very successful theory always grows stale, after a time" (C & R, p. 240).

Before commenting on the capabilities of the Bayesian inductivist to implement these methodological prescriptions of Popper's, let me point out why I think that Popper himself cannot justify these demands within his own deductivistic framework.

(i) As for Popper's demand (i) that positive instances should count only if they are the results of successful *risky* predictions, let us ask: What does the word '*count*' mean in his deductivist framework, when Popper declares that "Confirmations should count only if they are the result of *risky predictions*"? Count toward or for what? Qua *pure* deductivist, can Popper possibly maintain without serious inconsistency, as he does, that *successful* results of initially risky predictions should 'count' in favor of the theory making the risky prediction in the sense that in these "crucial cases ... we should expect the theory to fail if it is not true" (C & R, p. 112)? The latter statement is part of his characterization of 'crucial cases' in which he says:

> A theory is tested not merely by applying it, or by trying it out, but by applying it to very special cases – cases for which it yields results different from those we should have expected without that theory, or in the light of other theories. In other words we try to select for our tests those crucial cases in which we should expect the theory to fail if it is not true (C & R, p. 112).

Note Popper's claim that – on the basis of background theories – we should expect the [new] theory to fail [a 'crucial' test] *if it is not TRUE*. As against this assertion, I say that what is warranted *deductivistically* is the following:

(a) The older theories (*cum* initial conditions) predict an outcome contrary to the result *C* predicted by the new theory H, or the older theories are at least deductively non-committal with respect to *C*, thereby making the prediction of the new theory 'risky' in *that special sense*. Popper is here concerned with the more highly risky kind of prediction *C*, which is *contrary* to each of the older theories rather than logically independent of them. For this reason as well as in order to simplify the discussion, I shall sometimes concern myself with just the more highly risky kind of *C*.

(b) If the new theory H does *not* fail the crucial test but passes it, clearly nothing follows deductively about *its* truth. What does follow from the observation statement *C* is the truth of the 'infinitely' weaker disjunction D of ALL and only those hypotheses which individually entail C. For *C* itself is one of the disjuncts in the infinite disjunction D. And thus C entails D in addition to being entailed by D.[25]

Note that for any true disjunct in D, the latter has an infinitude of *false* disjuncts which are pairwise *incompatible* with that true disjunct in D and, of course, also incompatible with the background knowledge. Yet – contrary to the background knowledge – each of this infinitude of false theories predicts C no less than H does, and indeed H may be false as well. Hence of what avail is it to Popper, *qua deductivist*, that by predicting C, H is one of an infinitude of theories in D incompatible with those *particular* theories with which scientists had been working by way of historical accident, and that scientists happened to have thought of the particular disjunct H in D?

According to Popper's definition of the term 'severe test,' the experiment E which yielded the riskily predicted *C* does qualify as a 'severe test' of H. But surely the fact that H makes a prediction C which is incompatible with the prior theories constituting the so-called 'background knowledge' B does *not* justify the following contention of Popper's: A *deductivist* is entitled to expect the experiment E to yield a result *contrary* to C, *unless H is true.*

Indeed, Popper's reasoning here is of a piece (as I showed in *Brit. J. Phil. Sci.* **27**, 108–110) with the reasoning of the Bayesian proponent of *inductive* probabilities which has been stated by Salmon as follows:

A hypothesis risks falsification by yielding a prediction that is very improbable unless that hypothesis is true. It makes a daring prediction, for it is not likely to come out right unless we have hit upon the correct hypothesis. Confirming instances are not likely to be forthcoming by sheer chance.[26]

It would seem, therefore, that when Popper assessed the epistemic significance of severe tests in the quoted passage, he was unmindful of his own admonition against Baconian eliminative inductivism. As we recall from our citation of LSD in Section 2, this admonition reads as follows:

> ... the sole purpose of the elimination advocated by all these inductivists was to *establish as firmly as possible the surviving theory* which, they thought, must be the *true* one (or perhaps only a *highly probable one*, in so far as we may not have fully succeeded in eliminating every theory except the true one).
>
> As against this, I do not think that we can ever seriously reduce, by elimination, the number of the competing theories, since this number remains always infinite (LSD, p. 419).

What then would be a *deductivistically* sound construal of Popper's injunction that "Confirmations should count only if they are the result of *risky predictions*"? It would seem that deductively speaking, *C* can be held to *count* FOR H only in the following Pickwickian sense: *C* deductively refutes or *counts* AGAINST those *particular* rivals of H which belong to *B* while leaving fully intact an infinitude of other rivals in the disjunction D. Given the latter infinitude of relevant competing theories rightly stressed by Popper, I see no rational basis on which a *pure deductivist* can even hope for a good chance that *severe* tests will weed out *false* theories any better than *non*-severe tests. And hence *I cannot see any purely deductivistic rationale for advocating severe tests as being conducive to the discovery of true theories.* What then is the purportedly superior deductivistic rationale for Popper's corroborations as contrasted with neo-Baconian eliminative inductions? I shall return to this point in IV below.

So much for the first of Popper's demands, stated under (i) above.

(ii) Is it 'all very easy' for him in keeping with his major anti-inductivist tenets to espouse the *reasons* offered by him for *not* repeating the same corroborating experiment *ad nauseam* and for claiming that "there is something like a law of diminishing returns from repeated tests" (C & R, p. 240)? Alan Musgrave has argued cogently that it is not. And Musgrave makes the following telling points:

(a) Let *p* qualify as a *very risky* kind of prediction with respect to the initial background knowledge B_0, and let *p* be tested successfully, say, ten times, thereby yielding ten corroborating instances for the new theory T entailing *p*. Musgrave says:

> ... after each performance of the test, the particular results are incorporated into our

background knowledge, changing it successively from B_0 to B_1, to B_2, and so on, up to B_{10}. Popper's claim seems to be that the probability of the prediction in the next instance *gradually rises* in the light of our successively augmented background knowledges. Hence the severity of the successive repetitions of the same test *gradually falls*, and so does the degree of corroboration afforded by them to T.

But this clearly involves a straightforward *inductive argument*. It involves the inductive argument that each positive instance of the universal test-implication *p* increases its probability. (Alternatively, if *p* is construed as a singular prediction, it involves the inductive argument that the accumulation of past instances renders the next instance of the same kind more and more probable.) It would seem, therefore, that if we eschew inductive arguments, then the incorporation of past results of a severe test into background knowledge can do nothing to reduce the severity of future performances of that test. It would seem, in other words, that if Popper's theory of corroboration really is a non-inductivist one, then it cannot provide us with diminishing returns from repeated tests.[27]

(b) Now suppose that Popper were to attempt to *escape* the charge of inductivism as follows: On grounds whose specification remains urgent, he takes the aforementioned ten corroborations of *p* to be 'sufficiently many' repetitions of the experiment to accept *p*. Commenting on this putative approach, Musgrave writes:

Suppose that we test *p* ten times, that we always get the same result, and that ten repetitions is 'sufficiently many' for us to accept the universal test-implication *p* and reject the falsifying hypothesis. Suppose that we incorporate into background knowledge, not the *particular* results of our ten experiments, but the *universal* statement *p*. Now we need no inductive argument to make the severity of future tests of *p* decline. Indeed, it will decline sharply to *zero*, because *p* follows trivially from our augmented background knowledge as well as from the theory T.

This account is perfectly consistent with Popper's claim to have a non-inductivist theory of corroboration. But it can hardly be said to provide us with *diminishing* returns from repeated tests, if this means *steadily diminishing* returns. Instead, we have a 'one-step function': before 'sufficiently many' repetitions of the same test have been performed, each one has the same severity; after 'sufficiently many' repetitions, the severity of all future ones is zero. Perhaps this is what Popper had in mind all along: for he only claimed that he could provide *something like* a law of diminishing returns, and one step down is, I suppose, something like a gradual slide.[28]

(c) But the latter 'saltation'-version of the law of diminishing returns leaves unsolved for Popper the problem of '*how* often is sufficiently often' when repeated performances of a corroborating test are to be held adequate for incorporating the *universal* hypothesis *p* into the background knowledge.

The arguments given so far against Popper's ability to accommodate his own demands (i) and (ii) might be no more than *tu quoque* arguments, unless the Bayesian inductivist can do better in regard to justifying the gradual diminution of the epistemic returns from repeated trials. I shall now try to show in what sense the latter can indeed do better. We already saw earlier (Section 2) that the kind of inductivism which espouses the *unbridled* invocation of the instantiation condition can indeed *not* do better. For in *this* particular version of inductivism, all positive instances alike, including those resulting from repetitions of the same kind of test, will raise the credibility of the relevant hypothesis by the *same* amount. In this *latter* sense, there is warrant for the concessions by Keynes and Nagel which are mentioned by Popper.

Hence let us now look at the capabilities of the Bayesian framework.

B. *Bayesian Inference: The Rejection of the Instantiation Condition, and the Law of Diminishing Epistemic Returns from Repeated Tests*

In comparing Bayesian inference with Popper's epistemology, it behooves us to see first that the Bayesian schema rejects even the weaker version of the instantiation condition. After furnishing the specifics of that rejection, we shall see how the Bayesian obtains diminishing epistemic returns from repetitions of the same kind of experiment.

We are about to consider a form of inductivism which regards degrees of credibility to be mathematical probabilities and thus holds Bayes' theorem to be applicable to the probability of hypotheses. But I disregard here whether a subjectivistic or an objectivistic construal is given to these probabilities and will use the term "Bayesian inductivism" in a sense neutral to the difference between them.

Let B assert an initial condition relevant to the universal statement or deterministic (causal) law-hypothesis H, and let B also assert some other as yet unspecified background knowledge. Let C be the phenomenon which is predicted deductively by B and H. Assume that C occurs.

Then in the Reichenbach notation, where P(X, Y) is the probability *from* X *to* Y, Bayes' theorem (division theorem) can be written for this case as

$$P(B \& C, H) = \frac{P(B, H)}{P(B, C)} \times P(B \& H, C),$$

where $P(B \& H, C) = 1$.

Now assume that $P(B, C) \neq 0$: Unless this assumption is made, the right-hand-side ('r.h.s.') of the equation and hence the posterior probability would be undefined, and it would be incoherent to have gone ahead with a test for C. And be mindful of the fact that, in our case of a universal (nonstatistical) hypothesis H, P(B & H, C) = 1. Then the necessary and sufficient condition that the occurrence of the positive instance C will yield a *posterior* probability which is *equal* to H's *prior* probability is as follows: Either P(B, H) = 0 or P(B, C) = 1. Given our assumption that P(B, C) ≠ 0, the first of these two disjuncts assures that both the prior and the posterior probability of H vanish.

Clearly, *if* one can find a rationale for rejecting the view that universal law statements H necessarily have a vanishing probability, the first disjunct P(B, H) = 0 is clearly a very extreme and hence uninteresting example of the violation of the instantiation condition in the Bayesian version of inductivism.[29]

Is the second disjunct P(B, C) = 1 a similarly extreme and dull example? At first glance, it might be thought that it *clearly* is *not*. For suppose that the background knowledge *B* is now held to comprise an older hypothesis H_0 which is incompatible with H but such that under the initial condition asserted in B, the older H_0 entails C no less than the newly proposed H. In that case, it would be non-trivial that P(B, C) = 1. Moreover, since B is now presumed to comprise H_0, the conjunction B & H is an inconsistent system, and we also have P(B, H) = 0, while P(B & H, C) = 1 as before. Thus, in this putative way of assuring P(B, C) = 1, we would obtain P(B & C, H) = 0. And it would seem that the Bayesian version of inductivism would then implement with a vengeance the Popperian requirement (C & R, p. 36, item (1)) that a positive instance C of H should count in H's favor only if B entails a prediction which is *contrary* to C or if C is novel for B (C & R, p. 390).

But this way of non-trivializing the case of P(B, C) = 1 – i.e., by the incorporation of H_0 in B – must meet the following challenge: It is at least unclear how the old hypothesis H_0 can ever be supplanted by the new one H within the framework of the Bayesian formalism, once the old hypothesis H_0 has been made an integral part of the background knowledge that is brought to bear on the evaluation of H. William Harper[30] has proposed a way of meeting this challenge of subsequent exportation of H_0 from B. If Harper's proposal is viable, then the case of P(B, C) = 1 comprises repudiations of the instantiation condition by Bayesian inductivism which are at once non-trivial and splendidly meet

Popper's stated requirement. Once H_0 has become dubious, it should be exported from B.

In any event, given $P(B, C) \neq 0$, we know that in the case of a hypothesis H for which $P(B \& H, C) = 1$, the positive instance C will always be somewhat supportive of H except when either $P(B, C) = 1$ or $P(B, H) = 0$. For in all other cases in the stated category, the ratio of the posterior and prior probabilities of H will always exceed 1 even if only slightly, so that the occurrence of C will assure that the posterior probability of H is greater than its prior probability.

When $P(B \& H, C)$ is *less* than 1, we must get clear in what sense C is or is not a *positive* instance of H. We are interested in the case when C does so qualify as positive. Then we can say the following: Given both $P(B, C) \neq 0$ and $P(B, H) \neq 0$, we have

$$\frac{P(B \& C, H)}{P(B, H)} = \frac{P(B \& H, C)}{P(B, C)}.$$

Thus in this case a sufficient condition for the equality of the posterior and prior probabilities of H is that the ratio on the r.h.s. have the value 1, i.e. that the evidence C be equally probable whether or not H is true.[31]

Let us suppose for argument's sake that the previously discussed way of assuring *non*-trivially that both $P(B, C) = 1$ and $P(B \& C, H) = 0$ fails after all, say, because Harper's proposed method of 'exporting' H_0 from *B* is beset by an unexpected difficulty. And return to the case $P(B \& H, C) = 1$. But now suppose that $P(B \& C, H) > P(B, H)$.

Then we are concerned with supportive rather than non-supportive positive instances, and we ask: To what extent, if any, can the Bayesian inductivist comply with Popper's two stated demands? As we recall, these were that positive instances should count just if they are furnished by successful risky predictions and that repeated trials yield diminishing epistemic returns.

Thus as before, let B assert an initial condition relevant to the law-hypothesis H *and* also assert other background knowledge. Let C be the phenomenon which is predicted deductively by B and H. Assume that C occurs.

Then in the Reichenbach notation, Bayes' theorem (division theorem) can be written for this case as before:

$$P(B \,\&\, C, H) = \frac{P(B, H)}{P(B, C)} \times P(B \,\&\, H, C),$$

where $P(B \,\&\, H, C) = 1$.

Assume that neither of the two prior probabilities in the fraction on the r.h.s. vanishes. Then we can write

$$\frac{P(B \,\&\, C, H)}{P(B, H)} = \frac{1}{P(B, C)}.$$

We are now concerned with the case of a successful *risky* prediction in which the probability of C on the background knowledge and given initial condition *without* H is *very* low – say 10^{-6} – so that P(B, C) is *near* zero. Then the r.h.s. is *very* large, say 1,000,000. But in that case the *ratio* of the posterior probability of H to its prior probability is likewise huge, viz. 1,000,000.

But this means that according to this theorem of the probability calculus, the following is the case: If a hypothesis H predicts a phenomenon C whose occurrence is very unlikely without that hypothesis on the basis of the background knowledge alone, then the occurrence of C confers strong support on H in the sense that the *factor* by which C *increases* the probability of H is enormous, say a *factor* of 1,000,000.

It is important to note here that a huge *factor* of probability increase need not be tantamount to a large *amount* of increase in the probability of H. To see this, note first from an expanded equivalent of the term P(B, C) in an alternate form of Bayes theorem, that in our putative case of P(B & H, C) = 1, the value of P(B, H) will typically be even smaller than P(B, C). This is also intuitive because hypotheses *other than* H whose prior probability on *B* is likewise very low will also entail the prediction C. Hence for a sufficiently small value of P(B, H) – say 10^{-8} – even a *million-fold* increase in the probability of H will yield an *amount* of increase of even less than 1/100 in the probability of H.[32] For the *difference* between the prior and posterior probabilities of H will be only 10^{-2}–10^{-8}.

This *caveat* concerning the distinction between the *factor* by which the probability of H changes and the *amount* of such change should now be borne in mind when considering the Bayesian's ability to implement Popper's demand for diminishing epistemic returns from repeated trials. We now ask: What is the Bayesian *supportive* significance of *repeated* occurrences of the positive kind of event instantiated by C (hereafter 'C-like' event)? Can the Bayesian succeed where unbridled instantianist inductivism failed, and show why the same confirming experiment should *not* be repeated *ad nauseam*?

In the Bayesian inductivist framework, each occurrence of a C-like event can be held to change the *content* of the background knowledge B such that the prior probability P(B, C) of the next C-like event will *increase*, so that its reciprocal will *decrease*. Therefore, the *ratio* of the posterior and prior probabilities of H will *decrease* with such repeated positive occurrences. Hence the successive *factors* by which repeated occurrences of C-like events will increase the probability of H will get ever smaller. But this decrease in the successive *factors* of probability-increase does *not* itself suffice at all to assure a *monotonic* diminution in the *amounts* of probability-increase yielded by repeated occurrences after the first occurrence C! For suppose as before that prior to the first such occurrence $P(B_0, H) = 10^{-8}$, so that just before the *second* such occurrence the revised prior probability $P(B_1, H)$ becomes 10^{-2} after the above putative million-fold increase. Suppose that before the second occurrence, $P(B_1, C)$ were, say, 1/50. Then the second occurrence would increase the probability of H by a factor of only fifty as contrasted with a million. But the *amount* of such probability-increase effected by the first occurrence would be less than 1/100, whereas the amount of increase yielded by the second occurrence would have the much greater value 49/100. It is true that the posterior probability can become at most 1, so that the successive amounts of probability-increase can neither keep growing on and on nor even indefinitely remain above some non-zero lower bound. But the latter restriction by itself is a far cry from *assuring* a *monotonic* diminution in the successive amounts of probability-increase. Such a diminution might be assured, however, if the decrease in the successive values of the *ratios* of the posterior and prior probabilities of the hypothesis were sufficiently drastic each time.

Clark Glymour (*Journal of Philosophy*, August, 1975) has argued that there are cases in scientific practice in which the successful outcome of a 'severe' test nonetheless justifiably does *not* redound to the credibility of the hypothesis in the scientific community. For example, measurements of the gravitational red shift predicted by the general theory of relativity ('GTR') would seem to qualify as a 'severe' test of the latter's field equations in the sense that such a red shift is not predicted by the earlier Newtonian rival theory of gravitation. But Glymour adduces Eddington's analysis of this case to exhibit a justification for the view of physicists that the gravitational red shift is a very weak test of the GTR, while other phenomena are rather better tests, although the latter do not qualify as equally 'severe.'

The foregoing comparison between Popper's methodology and Bayesian inductivism in regard to the capability of *implementing* Popper's avowed "law of diminishing returns from repeated tests" shows the following: Vis-à-vis Bayesian inductivism, Popper is hardly entitled to claim that "it is all very easy" for him to show that his methodology has a superior capability of justifying that 'law.'

One important point remains to be considered in our comparative appraisal of Bayesian inductivism: Does Bayesianism sanction the credibilification of a new hypothesis by a sufficient number of positive instances which are the results of NON-risky predictions? For example, would a sufficient number of cases of people afflicted by colds who drink coffee daily for, say, two weeks and recover not confer *a posterior* probability greater than 1/2 on the new hypothesis that such coffee consumption *cures* colds? And how, if at all, does the Bayesian conception of inductive support enjoin scientists to employ a *control group* in the case of this causal hypothesis K with a view to testing the rival hypothesis that coffee consumption is causally irrelevant to the remission of colds?

As for the first question, note that in the case of a test of H by a *non*-risky prediction, $P(B, C)$ is even initially quite *high*, i.e., $P(B, C) > 1/2$, as in our example of the recoveries from colds. And as we accumulate positive instances C_i $(i = 1, 2, 3 \ldots)$, the content of B changes after each new positive instance C_i such that $P(B, C)$ increases further. Hence the latter's reciprocal will decrease. But that reciprocal is the previously discussed *factor* $1/P(B, C)$ of probability-increase of the hypothesis under test. In our example of the hypothesis K concerning the therapeutic benefits of coffee drinking, the *factor* of its probability-increase will therefore itself decrease. The *general* question is whether the probability increases in the hypothesis which ensue from the accumulation of indefinitely many positive instances will ultimately raise its probability to nearly 1 or whether these increases will be such as to yield a posterior probability which is *only asymptotic to very much less than 1/2.*

Unfortunately, apart from some rather restricted conditions which Allan Gibbard has investigated (unpublished), it would seem that more typically the Bayesian formalism does indeed permit the following: In the case of those successful but *non*-risky predictions which *are* supportive, a sufficient number of them will credibilify a hypothesis to be more likely true than false. And it would seem that one must likewise *supplement* the formalism by the injunction to employ experimental controls in testing causal hypotheses like K: Such an injunction could be based on Hilpinnen's version of the principle of *total* evidence which enjoins us to *seek out* as much potentially relevant evidence as possible.[33]

In any case, the Bayesian probabilistic form of inductivism, no less than the Baconian version, is able to *deny* that a mere mass of positive instances of a hypothesis must automatically be held to confer a substantial measure of credibility on the hypothesis. And in Popper's theory of corroboration, the corroboration of a theory T by an observable fact F which was successfully predicted by T is *parasitic* on the logical relations of F to T's *known* rivals R_1, R_2..., R_n as follows: For each known rival R, the accepted fact F *either* deductively falsifies R *or* F is *novel* with respect to R in the sense of being logically independent of R. In other words, a successfully predicted positive instance F of T corroborates T, if for each R, the fact F is either contrary to R or not predicted by R but without being contrary to R (C & R, pp. 36 and 390).

On the basis of our comparative scrutiny above of Popper's methodology vis-à-vis two major versions of inductivism, it will be apparent why I conclude the following: Popper is not justified if he claims to have advanced over all forms of post-Baconian inductivism "in explaining satisfactorily what is ... a 'supporting instance' of a law" (OK, p. 21; see also item (5) on p. 20). In saying this, I am also being mindful of my arguments above that Popper has given us no *purely deductivistic* rationale for advocating *severe* tests as being at all conducive to the discovery of true theories. Indeed, we saw that his critique of Baconians boomerangs massively.

IV. *The Inductivist Capstone of Corroboration in 'Objective Knowledge'*

I now need to call attention to an important *tacit* inductivist commitment in Popper's assignment of *epistemic* significance to his concept of truth preference for corroborated theories over refuted ones.

He tells us that "since we are searching for a true theory, we shall prefer those whose falsity has not been established" (OK, p. 8). And he articulates his concept of corroboration as well as the latter's relation to his notion of preference-with-respect-to-truth in the following passages:

By the degree of corroboration of a theory I mean a concise report evaluating the state (at a certain time *t*) of the critical discussion of a theory, with respect to the way it solves its problems; its degree of testability; the severity of tests it has undergone; and the way it has stood up to these tests. Corroboration (or degree of corroboration) is thus an evaluating *report of past performance*. Like preference, it is essentially comparative: in general, one can only say that the theory A has a higher (or lower) degree of corroboration than a competing theory B, in the light of the critical discussion, which includes testing, *up to some time t*. Being a report of past performance only, it has to do with a situation which may lead to preferring some theories to others. *But it says nothing whatever about future performance, or about the 'reliability' of a theory* (OK, p. 18).

... It may perhaps be useful to add here a point about the degree of corroboration of a statement *s* which belongs to a theory *T*, or follows logically from it, but is logically much weaker than the theory *T*.

Such a statement *s* will have less informative content than the theory *T*. This means that *s*, and the deductive system *S* of all those statements which follow from *s*, will be less testable and less corroborable than *T*. But if *T* has been well tested, then we can say that its high degree of corroboration applies to all the statements which are entailed by it, and therefore to *s* and *S*, even though *s*, because of its low corroborability, could never on its own attain as high a degree of corroboration.

This rule may be supported by the simple consideration that the degree of corroboration is a means of stating *preference with respect to truth*. But if we prefer *T* with respect to its claim to truth, then we have to prefer with it all its consequences, since if *T* is true, so must be all its consequences, even though they can be less well tested separately.

Thus I assert that with the corroboration of Newton's theory, and the description of the earth as a rotating planet, the degree of corroboration of the statement *s* 'The sun rises in Rome once in every 24 hours' has greatly increased. For on its own, *s* is not very well testable; but Newton's theory, and the theory of the rotation of the earth are well testable. And if these are true, *s* will be true also.

A statement *s* which is derivable from a well tested theory *T* will, *so far as it is regarded as part of T*, have the degree of corroboration of *T*... (OK, pp. 19–20).

Consider the statement *s* that 'The sun rises in Rome once in every 24 hours' and its rival statement *u* that 'The sun rises in Rome once *biennially*.' It would seem that according to Popper's stated criteria, the augmented Newtonian theory T which entails *s* now has a very much higher degree of corroboration than some equally well tested rival theory H that entails *u*. Given a particular time *t* at which we assess the past performance of the theories T and H, the statements *s* and *u* pertain to the *future* of *t* no less than to its past. Thus both *s* and *u* have entailments involving times which are in the future of *t*. Therefore we can say that, if we identify the time *t* as 'now,' *s* and *u*, and hence T and H, *respectively* entail the following two rival future-tensed statements *p* and *r*, made at time *t*: 'In the future, the sun will rise in Rome once every 24 hours,' and 'In the future, the sun will rise in Rome once biennially.' Since Popper holds that the relative degree of corroboration is preserved under logical entailment, the future-tensed statement *p* 'inherits' a very much higher degree of corroboration from T than the likewise future-tensed statement *r* inherits from H. Thus, while alike being about the future, the incompatible statements *p* and *r* differ greatly in degree of corroboration and *therefore* in preferability-with-respect-to-truth (empirical verisimilitude). Yet Popper told us above that relative degree of corroboration and hence difference in empirical verisimilitude *"says nothing whatever about future performance, or about the 'reliability' of a theory"* (OK, p. 18, italics in the original).

How then are we to understand as a *compatible* trio of statements the following claims made by Popper, if we are mindful of the important fact that many theoretical law-statements are *universally quantified with respect to the time*?

(i) "Obviously, the degree of corroboration at the time *t* (which is a statement about preferability at the time *t*) says nothing about the future.... It is just a report about the state of discussion at the time *t*, concerning the logical and empirical preferability of the competing theories" (OK, p. 19).

(ii) The relative degree of corroboration of a hypothesis is preserved under logical entailment.

(iii) "...the degree of corroboration is a means of stating *preference with respect to truth*" (OK, p. 20).

It would seem to follow on the strength of (ii) and (iii) that for the case of two rival theories, both of which do have logical entailments with respect to the future but possess very different degrees of corroboration, there can be *future*-tensed entailments, such as our statements *p* and *r*, which differ greatly in corroboration *and* hence in truth preferability or empirical verisimilitude. If that is so, *what does it mean* to say with Popper that because of their relatively different degrees of corroboration, we prefer the future-tensed statement *p* to the future-tensed statement *r with respect to truth*, while *also* saying the following: Despite their differing strengths as candidates for being true, according to (i) there is complete parity of 'reliability' between *p* and *r* in the sense that for each one alike, its degree of corroboration at the time *t* 'says nothing about the future'?

Can Popper tell us how his claims (i), (ii) and (iii) can all be both meaningful and compatible? I cannot see how, in the stated context, they can be meaningful without an incompatibility between (i), on the one hand, and the conjunction of (ii) and (iii), on the other.

Let us suppose that, for the sake of securing compatibility, Popper were willing to weaken the very strong denial (i) in order to preserve his much more justified claims (ii) and (iii). Vis-à-vis Hume's challenge, Popper's epistemic position would then become very similar to that of the (Bayesian) inductivist. Indeed, in the context of law statements each of which pertain to all times but differ in corroboration, Popper's criteria (ii) and (iii) do seem to commit him to some form of inductivism in different terminological garb. For in the stated context, these criteria require that some *future-tensed* statements are truth-preferable to others or have greater Popperian empirical verisimilitude – 'reliability' – than others.

V. *Duhem versus Popper: Epistemological Comparison of Refutations and Corroborations*

Shortly after the 1959 English publication of LSD, I noted that Popper had misportrayed Duhem's position in that book.[34] There Popper wrote (LSD, p. 78, fn. 1): "Duhem denies (English translation p. 300) the possibility of crucial experiments, because he thinks of them as verifications, while I assert the possibility of crucial *falsifying* experiments." Thus, here Popper erroneously depicts Duhem as *allowing* that, *for the kind of hypothesis typically at issue in science*, there is an *asymmetry of fallibility* between its verification and its falsification such that falsifications have logical credentials which verifications fail to possess. Having made this imputation in LSD, Popper asserted the compatibility of Duhem's position with his own championship of falsifications as preeminently logically reliable vis-à-vis verifications as tests of a hypothesis (theory).

Before the 1962 publication of his C & R, Popper had been made aware of the need to correct his exegesis of Duhem. Thus in 1962, Popper acknowledged Duhem's espousal of symmetry of fallibility, saying (C & R, p. 112. fn. 26): "Duhem, in his famous criticism of crucial experiments (in his *Aim and Structure of Physical Theory*), succeeds in showing that crucial experiments can never *establish* a theory. He fails to show that they cannot *refute* it."

How are we to construe Popper's concurrence with Duhem in regard to verifiability and his 1962 avowal of disagreement with him in regard to falsifiability? Surely Popper is *not* charging Duhem with unawareness that *modus tollens* is a deductively valid form of inference while commending him for appreciating the deductive fallaciousness of affirming the consequent! If someone were to insist on documentation that Duhem is cognizant of the validity of *modus tollens* no less than the rest of us, we would only need to point to his statement that "when the experiment is in disagreement with his {the physicist's} predictions, what he learns is that at least one of the hypotheses constituting this group {i.e., the comprehensive theory being tested experimentally as a whole} is unacceptable and ought to be modified."[35] Thus, it is clearly agreed on all sides that there is an asymmetry of deductive validity between affirming the consequent and *modus tollens*. I shall speak of this asymmetry as "the deductive asymmetry."

How then are we to interpret Popper's 1962 divergence from Duhem? The 1962 statement of disagreement with Duhem which we quoted from Popper above is appended as a footnote to the following much fuller text in which Popper declares (C & R, p. 112):

But while Bacon believed that a crucial experiment may establish or verify a theory, we shall have to say that it can at most refute or falsify a theory. [At this point, Popper states his previously quoted disagreement with Duhem in a footnote.] It is an attempt to refute it; and if it does not succeed in refuting the theory in question – if, rather, the theory is successful with its unexpected prediction – then we can say that it is corroborated by the experiment....

Against the view here developed one might be tempted to object (following Duhem) that in every test it is not only the theory under investigation which is involved, but also the whole system of our theories and assumptions – in fact, more or less the whole of our knowledge – so that we can never be certain which of all these assumptions is refuted. But this criticism overlooks the fact that if we take each of the two theories (between which the crucial experiment is to decide) *together* with all this background knowledge, as indeed we must, then we decide between two systems which differ *only* over the two theories which are at stake. It further overlooks the fact that we do not assert the refutation of the theory as such, but of the theory *together* with that background knowledge; parts of which, if other crucial experiments can be designed, may indeed one day be rejected as responsible for the failure. (Thus we may even characterize a *theory under investigation* as that part of a vast system for which we have, if vaguely, an alternative in mind, and for which we try to design crucial tests.)

Here Popper alleges that Duhem has overlooked two points. To state and evaluate Popper's rebuttal, suppose that a theory T_1 composed of a major hypothesis H_1 and an auxiliary A_1 entails a consequence C_1, while the conjunction $H_2 \cdot A_2$ constituting a *rival* theory T_2 entails a consequence C_2 *incompatible* with C_1. And suppose further that a purportedly crucial experiment yields evidence which is taken to be favorable to the truth of C_1 but adverse to the truth of C_2. Popper has us consider two such comprehensive theoretical systems T_1 and T_2 which respectively yield correct and incorrect predictions and whose respective auxiliaries A_1 and A_2 are identical. Then his first point is that the common auxiliary A will not be at issue between T_1 and T_2 and that the latter systems differ only with respect to the disputed component hypotheses H_1 and H_2. I cannot see that Duhem 'overlooked' Popper's first point. For the latter is unavailing against Duhem's important claims that (i) the experimental falsification of T_2 via *modus tollens* does *not* have the logical force of a like falsification of H_2 itself, and yet it is the pair H_2 and H_1 rather than T_2 and T_1 which is at stake in the given context of scientific controversy, and (ii) there is *symmetry* of fallibility between the falsification of H_2 itself and the verification of H_1.

Plainly, Duhem did not deny that *if* the (common) auxiliary A is *not* at issue *and* is presumed to be true, then the false prediction made by T_2 will permit the valid deduction of the falsity of H_2 via *modus tollens*. Instead, Duhem maintained that precisely because we cannot have any guarantee of the truth of A, we cannot construe the experimental decision against T_2 and in favor of T_1 as logically tantamount to a crucial refutation of H_2 itself. Far from being gainsaid by Popper's first point, the soundness of this Duhemian contention is explicitly conceded and even endorsed in the second point that Popper goes on to make. For there Popper tells us that "parts" of the background knowledge A "may indeed one day be rejected as responsible for the [experimental] failure" of T_2.

As for Popper's second point, Duhem recognized explicitly, as we saw, that under the posited conditions, at least one of the component hypotheses H_2 and A of the comprehensive theory T_2 "is unacceptable and ought to be modified." Hence Duhem has no quarrel with Popper, if Popper is indeed content to assert the crucial falsifiability or refutability of the total system T_2 vis-à-vis T_1, in contra-distinction to claiming this kind of refutability for such component hypotheses as H_2 which are typically the foci of scientific controversy. Oddly enough, Popper does come close to contenting himself with only *global* falsifiability of a *"whole system"* (LSD, p. 76 and fn. 2 there) without mentioning Duhem at this point.

Yet only two pages later (LSD, p. 78, fn. 1), he refers to Duhem and says: "I assert the possibility of crucial *falsifying* experiments," where he presumably means that *component* hypotheses are crucially falsifiable.

4. Conclusion

In the light of the several groups of considerations presented in this paper, how can Popper be warranted in maintaining that falsifiability is the touchstone of scientific rationality, to the exclusion of inductive supportability? And how can Popperians justify adhering to Popper's indictment of inductivism while maintaining that he has given us a viable epistemological alternative on genuinely deductivist foundations?

University of Pittsburgh

NOTES

* This essay is based on research supported by the National Science Foundation.

1 Lakatos, I., 'The Role of Crucial Experiments in Science', *Studies in History and Philosophy of Science* **4** (1974), 310; hereafter this paper will be cited within the text as 'RCES'.

2 Popper, K. R., *Conjectures and Refutations*, Basic Books, New York and London, 1962, p. 256. Hereafter this work will be cited as 'C & R' within the text.

3 See Grünbaum, A., 'Can a Theory Answer more Questions than one of its Rivals?' *British Journal for the Philosophy of Science* **27** (1976), 1–24 and the author's subsequent articles in that volume in the June and December issues.

4 Cf. Russell, B., *Human Knowledge*, Simon and Schuster, New York, 1948, p. 381.

5 Popper, K. R., *The Logic of Scientific Discovery*, Hutchinson, London, 1959, p. 420 (italics in original). Hereafter this work will be cited within the text as 'LSD'.

6 According to Lakatos [RCES, p. 315 and 'Popper on Demarcation and Induction' (hereafter 'PDI'), *The Philosophy of Karl Popper*, The Library of Living Philosophers (ed. by P. A. Schilpp), Open Court, LaSalle, 1974, Book I, pp. 245–246], Popper *tailored* his demarcation criterion to the requirement of *not* according scientific status to the aforementioned four theories.

7 Burtt, E. A. (ed.), *The English Philosophers From Bacon to Mill*, Random House (Modern Library Series), New York, 1939, p. 36. Any quotations from Bacon are taken from Burtt's edition. I am grateful to Ernan McMullin for calling my attention to Aphorism 95 in Book I of Bacon's *Novum Organum*, where Bacon seems to stress the sterility of such fact-collecting as is *not* guided by theory. We shall recall this view of Bacon's when noting below that it runs counter to Popper's portrayal of Bacon as echoed by Lakatos.

8 Joseph, H. W. B., *An Introduction to Logic*, 2nd revised edition, Oxford University Press, Oxford, 1916.

9 Ibid., p. 393.

[10] Ibid; cf. also p. 565 n. 1.

[11] Cf. R. Giere's illuminating discussion ['An Orthodox Statistical Resolution of the Paradox of Confirmation', *Philosophy of Science* **37** (1970), 354–362] of the resolution of Hempel's paradox of confirmation in statistical theory by means of *rejecting* a certain version of the instantiation condition. But note the *caveat* in fn. 31 below concerning the difference between my concept of 'positive instance' and the corresponding concept relevant to Giere's analysis.

[12] In Section 3 below, we shall consider to what extent, if any, Popper's own methodology entitles him to claim that *mere* repetitions of an initially corroborating type of instance increase the corroboration of a hypothesis only very little, if at all.

[13] Mill, J. S., *A System of Logic*, 8th edition, Harper and Brothers, New York, 1887, p. 313. Thus, in § 2 of chapter X, Mill explains how his inductive methods can invalidate actually false claims of a plurality of causes. But in §§ 1 and 3 of ch. X, he contends that there are genuine cases of multiple causation. And in § 3, he maintains that his inductive methods can handle the latter as well.

[14] Eysenck, H. J., *The Effects of Psychotherapy*, International Science Press, New York, 1966, and 'The Effects of Psychotherapy', *International Journal of Psychiatry* **1** (1965), 97–178.

[15] Rachman, S., *The Effects of Psychotherapy*, Pergamon Press, New York, 1971.

[16] Cf. Bergin, A. E., 'The Evaluation of Therapeutic Outcomes', A. E. Bergin and S. L. Garfield (eds.), *Handbook of Psychotherapy and Behavior Change*, John Wiley and Sons, New York, 1970, pp. 217–270; see also the important additional references given on pp. 217–218. H. J. Eysenck and G. D. Wilson [*The Experimental Study of Freudian Theories*, Methuen, London, 1973, pp. 378–379] point out that whereas Freud's theory emphatically denies the existence of spontaneous remission, the latter is one of the best attested findings of psychiatry.

[17] Fromm, E., *The Crisis of Psychoanalysis*, Holt, Rinehart and Winston, New York, 1970, pp. 3–4. Emanuel Peterfreund's discussion of 'The Nature of Therapeutic Changes in Psychoanalysis' [*Information Systems, And Psychoanalysis*, International Universities Press, New York, 1971, pp. 351–358] does not improve on Fromm, despite Peterfreund's strongly *revisionist*, learning-theoretic stance toward Freud.

[18] Bergin, A. E., op. cit., pp. 246–247. For a criticism of this claim, see Rachman, S., op. cit., ch. 3 and p. 16.

[19] Meltzoff, J. and Kornreich, M., *Research in Psychotherapy*, Atherton Press, New York, 1970, p. 200. See pp. 258, 113–4 and 190 for *placebo* effect.

[20] Fromm, E., op. cit., p. 4.

[21] Howson, C., 'Must the Logical Probability of Laws be Zero?', *British Journal for the Philosophy of Science* **24** (1973), 153–160. See also Hintikka, J., 'Carnap and Essler Versus Inductive Generalization', *Erkenntnis* **9** (1975), 235–244.

[22] I am indebted to Noretta Koertge for some of these Popper references.

[23] Popper, K. R., *Objective Knowledge*, Oxford University Press, Oxford, 1973. p. 51. Hereafter this work will be cited as 'OK'.

[24] Popper, K. R., 'A Theorem on Truth-Content', in P. Feyerabend and G. Maxwell (eds.), *Mind, Matter and Method*, University of Minnesota Press, Minneapolis, 1966, Theorem 1, p. 350. For a discussion of relevant details, see Section 2 (i) of the first of my articles cited in note 3 and p. 134 of the *second* of these articles.

[25] That D is indeed equivalent to C becomes intuitive by reference to Tarskian logical contents as follows: The logical content of a disjunction is the *intersection* of the respective contents of the disjuncts. And the intersection of the contents of the infinitude of disjuncts

in D will be just the content of C. To illustrate, take the *entire* infinitude of pairwise different curves, all of which go through or contain each of a finite set S of points in the *xy* plane. This totality T of curves or point sets will have exactly the set S as its set-theoretical intersection. For no point outside of S can belong to ALL of the curves in T. The points in S play the role of 'data points' in analogy to the observation statement C. And the curves in T play the role of the hypotheses in D each one of which entails C.

[26] Salmon, W. C., *The Foundations of Scientific Inference*, University of Pittsburgh Press, Pittsburgh, 1966, p. 119.

[27] Musgrave, A., 'Popper and "Diminishing Returns from Repeated Tests"', *Australasian Journal of Philosophy* **53** (1975), 250.

[28] Ibid., p. 251.

[29] In the paper by Howson cited in note 21 above, he has offered a counterexample on pp. 161–162 to the claim that $p(a)=0$ for every universal a. See also Hintikka's paper in note 21.

[30] Harper, H. W., 'Rational Belief Change, Popper Functions and Counterfactuals', *Synthèse* **30** (1975), 221.

[31] Good, I. J., *British Journal for the Philosophy of Science* **17** (1966–7), 322) has given a perhaps far-fetched example having the following features: C states that a randomly selected bird is a black raven, H states that 'all ravens are black', and $P(B \,\&\, H, C) \ll 1$ (i.e., $10^2/10^6$). Good takes a black raven to be a 'case' (positive instance) of H *in a sense DIFFERENT from our above sense*, since his B does *not* assume the initial condition that the randomly selected bird is a raven! And despite the universality of H, the special feature of Good's example then is that not only is $P(B \,\&\, H, C) \ll 1$, but also $P(B \,\&\, H, C) < P(B, C)$, so that the ratio on the r.h.s. is *less than* 1. But this means that the 'case' C of the hypothesis yields a *posterior* probability of H which is *smaller* than its prior probability: A perhaps somewhat far-fetched but even more resounding repudiation of the instantiation condition than the case of *equal* prior and posterior probabilities of H.

I am indebted to Wesley Salmon and William Harper for helpful comments relating to the violation of the instantiation condition in Bayesian inference and I thank Laurens Laudan for stimulating criticisms of Bayesian inference.

[32] The need to distinguish the *factor* of probability increase from the *amount* was overlooked in this context in Salmon's *The Foundations of Scientific Inference*, op. cit., pp. 118–120.

[33] Hilpinnen, R., 'On the Information Provided by Observations', Hintikka, J., and Suppes, P. (eds.), *Information and Inference*, Reidel, Dordrecht, 1970, Section II, esp. pp. 100–101. I am indebted to Teddy Seidenfeld not only for this reference but also for very clarifying comments on Allan Gibbard's results, which I mentioned above.

[34] Grünbaum, A., 'The Duhemian Argument', *Philosophy of Science* **27** (1960), 76, fn. 1. Cf. also A. Grünbaum, *Philosophical Problems of Space and Time*, Alfred Knopf, New York, 1963, p. 109 n. 4. The second enlarged edition of the latter work (Reidel, Dordrecht and Boston 1973) has the same comment in the same place.

[35] Duhem, P., *The Aim and Structure of Physical Theory*, Princeton University Press, Princeton, 1954, p. 187.

JAAKKO HINTIKKA AND UNTO REMES

ANCIENT GEOMETRICAL ANALYSIS AND MODERN LOGIC

1. Ancient analysis and modern logic

The old heuristic method known as analysis (geometrical analysis) was famous in Antiquity, and in the course of the history of Western thought its generalizations have played an important and varied role. It is nevertheless far from obvious what this renowned method of the ancient geometers really was. One reason for this difficulty of understanding the method is the scarcity of ancient descriptions of the procedure of analysis. Another is the relative failure of these descriptions to do justice to the practice of analysis among ancient mathematicians.

An attempt to spell out the import of the method has recently been made by Hintikka and Remes.[1] Much of their effort is devoted to the interpretation of the only extensive extant description of analytical method in the ancient mathematical literature, viz. the one found in Pappus (see Section 2 below).

It is even unclearer what the heuristic rationale of the method of analysis is. One admittedly partial attempt to elucidate this heuristic usefulness of the method is made here. This attempt takes the form of a comparison between the old method of analysis and certain relatively new techniques in symbolic logic which may be called *natural deduction methods*. In fact, more than a mere comparison is involved here, for the thesis put forward here is that the method of analysis is almost a special case of these natural deduction methods. Perhaps the most profound link between the two is a direct corollary to our interpretation of the ancient method as an analysis of a definite figure (i.e., configuration of geometrical objects). This implies, as pointed out below in Section 9, that the logic of the method satisfies the so-called subformula property, which is the characteristic feature of natural deduction methods.

A similar thesis was put forward by E. W. Beth already in 1955 when he suggested that his approach to logical deduction by means of the so-called semantical *tableaux* (for these, see below, Section 4) "realizes to

R. S. Cohen et al. (eds.), Essays in Memory of Imre Lakatos, 253–276. *All Rights Reserved. Copyright © 1976 by D. Reidel Publishing Company, Dordrecht–Holland*

a considerable extent the conception of a purely *analytical method* which has played such an important role in the history of logic and philosophy."[2] However, Beth neither connected this logical idea in so many words with the ancient geometrical analysis nor applied his approach to the elucidation of historical problems.

Beth's silence notwithstanding, it seems that this relationship between the two methods, the geometrical and the logical, can be used to discuss certain interesting questions concerning the methods of analysis which are found exemplified in the historical material. (See below, especially Sections 5–7.) This comparison was the tacit – and occasionally more than tacit – methodological backbone of the Hintikka-Remes study. Here it will be treated somewhat more systematically. This comparison is not vitiated by the fact that there are obvious differences between the method of analysis and natural deduction methods. These discrepancies can be understood as being due to the exigencies of the task of a practicing mathematician. Even if ancient geometers had known natural deduction methods and tried to use them strictly, these practical difficulties would have pushed them to the familiar form of the traditional method, we are tempted to say. (Cf. Section 8 below.)

2. Pappus on Analysis and Synthesis

The only extensive general description of the method of analysis and synthesis to be found in the surviving ancient literature is given by Pappus of Alexandria (*floruit c.* 300 A.D.). Its central part is as follows:

Now analysis is the way from what is sought – as if it were admitted – through its concomitants [τὰ ακόλουθα, usually translated 'consequences'] in order to something admitted in synthesis. For in analysis we suppose that which is sought to be already done, and we inquire from what it results, and again what is the antecedent of the latter, until we on our backward way light upon something already known and being first in order. And we call such a method analysis, as being a solution backwards. In synthesis, on the other hand, we suppose that what was reached last in the analysis to be already done, and arranging in their natural order as consequents the former antecedents and linking them one with another, we in the end arrive at the construction of the thing sought. And this we call synthesis. Now analysis is of two kinds. One seeks the truth, being called theoretical. The other serves to carry out what was desired to do, and this is called problematical. In the theoretical kind we suppose the thing sought as being and as being true, and then we pass through its concomitants [consequences] in order, as though they were true and existent by hypothesis, to something admitted; then, if that which is admitted be true, the thing sought is true, too, and the proof will be the reverse of analysis. But if we come upon something false to admit, the thing sought will be false, too. In the problematical kind we suppose the desired thing to be known, and then we pass through its concomitants [consequences] in order, as though they were true, up to something admitted. If the thing admitted is possible or can be done, that is, if it is what the mathematicians call given, the desired thing will also be possible. The proof will again be the reverse of analysis. But if we come upon something impossible to admit, the problem will also be impossible. (The translation is from Hintikka and Remes, Chapter III.)[3]

Just before and after the quoted passage, Pappus connects his conception of the method with so-called 'Treasury of Analysis', a collection of books by earlier mathematicians. In fact, it can be shown that he at least in practice followed rather closely the example of his famous predecessors. This entitles us to consider Pappus' description representative of ancient geometrical analysis at large.

From Pappus' description it is seen that among the important features of the method of analysis there are at least the following.

(i) An analyst starts by assuming what he wants to prove or to construct and argues 'backwards', that is, uses this desired end in his analytical reasoning calculated to uncover a suitable proof or construction.

(ii) Analysis is complemented by a synthesis in which the same steps as were taken in analysis are traversed in the opposite direction.

It is argued in Hintikka and Remes, Chapter II, that Pappus' description of analysis and synthesis is, appearances notwithstanding, almost completely consistent. (The only exception to this claim is discussed below in Section 7.) The contrary appearance is due to the fact that Pappus seems to be describing both the analysis and the synthesis as a process of drawing of logical conclusions (consequences). Since they proceed in opposite directions, Pappus' account seems to be consistent only if analyses and syntheses consist entirely of equivalences. This is ruled out by collateral evidence concerning the ancient method of analysis, however.[4]

It is shown in Hintikka and Remes that the terms (τὰ ἀκόλουθα, and its cognates) which Pappus uses to describe the progress of analysis do not mean anywhere in his writings logical consequence but are used rather loosely to express almost any sort of 'going together'. (Cf. the parenthetical explanations in the quotation from Pappus above.) Hence Pappus' general description of analysis depicts it consistently as a search for premisses, not as drawing of consequences.

This makes fairly good sense of Pappus' text, but it does not suffice to reconcile his general description of analysis with his own mathematical practice, or with the practice of other ancient mathematicians. In fact, the two seem to be in a rather serious conflict with each other. In his general description of analysis, Pappus depicts it (if we are right) as a search for premisses from which the desired conclusion could be drawn, whereas his actual analyses proceed consist in a series of conclusions from premisses which include the desired conclusion and hence proceed in a direction diametrically opposed to that envisaged in the general description quoted above. No wonder that most commentators have been left puzzled by such violent discrepancies. We shall return to the resulting problems later. (See especially Section 8 below.)

One more general point is perhaps worth making here. Although Pappus distinguishes theoretical and problematical analysis in his general description of the method, it can be shown that this distinction is conventional and is not essential for our appreciation of the method. However, the significance of the distinction has often been overestimated by modern students of the method.

3. Example

In order to illustrate the actual mathematical practice, we reproduce an analysis of a problem from Pappus' *Collectio* in T. L. Heath's translation.

Given a circle *ABC* and two points *D*, *E* external to it, to draw straight lines *DB*, *EB* from *D*, *E* to a point *B* on the circle such that, if *DB*, *EB* produced meet the circle again in *C*, *A*, *AC* shall be parallel to *DE*.

Analysis

Suppose the problem solved and the tangent at *A* drawn, meeting *ED* produced in *F*.

(Part I. Transformation.) Then, since *AC* is parallel to *DE*, the angle at *C* is equal to the angle *CDE*. But, since *FA* is a tangent, the angle at *C* is equal to the angle *FAE*. Therefore the angle *FAE* is equal to the angle *CDE*, whence *A*, *B*, *D*, *F* are concyclic. Therefore the rectangle *AE*, *EB* is equal to the rectangle *FE*, *ED*.

(Part II. Resolution.) But the rectangle *AE*, *EB* is given, because it is equal to the square on the tangent from *E*. Therefore the rectangle *FE*, *ED* is given; and, since *ED* is given, *FE* is given (in length). [Data, 57.] But *FE* is given in position also, so that *F* is also given. [Data, 27.] Now *FA* is the tangent from a given point *F* to a circle *ABC* given in position; therefore *FA* is given in position and magnitude. [Data, 90.] And *F* is given; therefore *A* is given. But *E* is also given; therefore the straight line *AE* is given in position. [Data, 26.] And the circle *ABC* is given in position; therefore the point *B* is also given. [Data, 25.] But the points *D*, *E* are also given; therefore the straight lines *DB*, *BE* are also given in position.

Synthesis

(Part I. Construction.) Suppose the circle *ABC* and the points *D*, *E* given. Take a rectangle contained by *ED* and by a certain straight line *EF* equal to the square on the tangent to the circle from *E*. From *F* draw *FA* touching the circle in *A*; join *ABE* and then *DB*, producing *DB* to meet the circle at *C*. Join *AC*. I say then that *AC* is parallel to *DE*.

(Part II. Demonstration.) Since, by hypothesis, the rectangle *FE*, *ED* is equal to the square on the tangent from *E*, which again is equal to the rectangle *AE*, *EB*, the rectangle *AE*, *EB* is equal to the rectangle *FE*, *ED*. Therefore *A*, *B*, *D*, *F* are concyclic, whence the angle *FAE* is equal to the angle *BDE*. But the *FAE* is equal to the angle *ACB* in the alternate segment; therefore the angle *ACB* is equal to the angle *BDE*. Therefore *AC* is parallel to *DE*.[5]

The example shows what the different parts of the total method of analysis and synthesis are. The names of the parts are not found in Pappus, however, and the 'transformation' is sometimes called 'analysis' in the narrower sense of the word. In its wider sense, 'analysis' also comprises the 'resolution'.

Special attention is worth paying to the role of constructions in the argument as distinguished from deductive steps. (For instance, the drawing of the tangent at *A* is a construction in the example above while the rest of transformation is deductive. The 'construction' part is entirely devoted to what it is said to be devoted to, whereas no constructions are carried out in the 'demonstration'.) Constructions are also found in theoretical analyses, for good reasons which will be commented on later in this paper. (See Section 10 below.)

4. Natural deduction methods

How is modern logic supposed to turn light on this venerable method of analysis? In order to find an answer, we must have a look at the rules of what is known as the natural deduction method. These rules may be considered either in the form given to them in Beth's method of semantical *tableaux* or in the form given to them in a Gentzen-type sequent calculus (in German, *Sequenzenkalkül*). The former was first presented in Beth's 1955 paper *Semantic Entailment and Formal Derivability*. The latter is found for instance in Kleene's *Mathematical Logic* p. 289ff.[6] The tableau method proceeds as follows. When one wants to prove the disjunction of V_1, V_2,... from U_1, U_2,..., one begins by inserting U_1, U_2,... in the left column of the tableau and V_1, V_2,... into the conjugate (right) column. (Rule (i)) The rest proceeds according to the following rules:

(ij) If $\sim X$ appears in some column, then we insert X in the conjugate column.

(iij) If $(x)\ X(x)$ appears in a left column or $(Ex)\ X(x)$ in a right column, then we insert in the same column $X(p)$ for each individual p which has been or will be introduced.

(iv) If $X \& Y$ appears in a left column or $X \vee Y$ in a right column, then we insert both X and Y in the same column.

(v) If $X \supset Y$ appears in a right column, then we insert Y in the same column and X in the conjugate left column.

(vi) If (a) $X \vee Y$ appears in a left column, (b) $X \supset Y$ in a left column, (c) $X \& Y$ in a right column, (d) $X \equiv Y$ in a left column, or (e) $X \equiv Y$ in a right column, than the tableau must be split up, and we insert:

(a) X in one left subcolumn and Y in the other.

(b) Y in one left column and X in the right column *not* conjugate with it.

(c) X in one right subcolumn and Y in the other.

(d) Both X and Y in one left column and once more both X and Y in the right column *not* conjugate with it.

(e) X in one left subcolumn, Y in the right subcolumn conjugate with it, Y in the other left subcolumn, and X in the remaining right subcolumn.

(vij) If $(Ex)\ X(x)$ appears in a left column or $(x)\ X(x)$ in a right column, then we introduce a new individual p and we insert $X(p)$ in the same column (it is convenient not to apply this rule until all possibilities of applying other rules have been exhausted; the part of the construction which starts with the introduction of the kth individual and which terminates at the introduction of the $(k+1)^{st}$ one is denoted as its kth *stage*).

(viij) If one and the same formula appears in two conjugate columns, then the corresponding tableau (or subtableau) is closed.

(ix) If all subtableaux subordinate to some tableau (or subtableau) are closed, then that tableau (or subtableau) itself is also closed.

(x) If all possibilities of applying rules (i)–(vij) have been exhausted, whereas the tableau is not closed, then the tableau provides us with a suitable counter-example and is terminated.

We have in the original tableau two (conjugate) *columns*, the left one and the right one. If the tableau must be split up, we will have two left subcolumns, and likewise two right subcolumns. These subcolumns are pairwise combined to form a subtableau. A left and a right column which together form a tableau or a subtableau are said to be *conjugate*. A subtableau may later be split up again.

The rules for the Sequent Calculus are as follows:

$$\frac{A, \Gamma \rightarrow \Theta, B}{\Gamma \rightarrow \Theta, A \supset B.} \rightarrow\supset \qquad \frac{\Gamma \rightarrow \Theta, A \quad B, \Gamma \rightarrow \Theta}{A \supset B, \Gamma \rightarrow \Theta.} \supset\rightarrow$$

$$\frac{\Gamma \rightarrow \Theta, A \quad \Gamma \rightarrow \Theta, B}{\Gamma \rightarrow \Theta, A \& B.} \rightarrow\& \qquad \frac{A, B, \Gamma \rightarrow \Theta}{A \& B, \Gamma \rightarrow \Theta.} \&\rightarrow$$

$$\frac{\Gamma \to \Theta, A, B}{\Gamma \to \Theta, A \vee B.} \to \vee \qquad \frac{A, \Gamma \to \Theta \quad B, \Gamma \to \Theta}{A \vee B, \Gamma \to \Theta.} \vee \to$$

$$\frac{A, \Gamma \to \Theta}{\Gamma \to \Theta, \neg A.} \to \neg \qquad \frac{\Gamma \to \Theta, A}{\neg A, \Gamma \to \Theta.} \neg \to$$

$$\frac{A, \Gamma \to \Theta, B \quad B, \Gamma \to \Theta, A}{\Gamma \to \Theta, A \sim B.} \to \sim \qquad \frac{A, B, \Gamma \to \Theta \quad \Gamma \to \Theta, A, B}{A \sim B, \Gamma \to \Theta.} \sim \to$$

$$\frac{\Gamma \to \Theta, A(b)}{\Gamma \to \Theta, \forall x\, A(x)} \to \forall \qquad \frac{A(r), \forall x\, A(x), \Gamma \to \Theta}{\forall x\, A(x), \Gamma \to \Theta.} \forall \to$$

where b does not occur free in $\Gamma \to \Theta, \forall x\, A(x)$.

$$\frac{\Gamma \to \Theta, \exists x A(x), A(r)}{\Gamma \to \Theta, \exists x A(x).} \to \exists \qquad \frac{A(b), \Gamma \to \Theta}{\exists x A(x), \Gamma \to \Theta} \exists \to$$

where b does not occur free in $\exists x A(x), \Gamma \to \Theta$.

As can be seen, these are all of the nature of rules of inference. As axioms we may use sequents of the form

$$C, \Gamma \to \Theta, C$$

The two methods look *prima facie* quite different. The tableau method can be thought of as proceeding from the desired deductive connection between $U_1, U_2, \ldots$ and $V_1, V_2, \ldots$ and examining how it can be obtained, while the Sequent Calculus is set up like our ordinary proof method. In spite of those superficial differences, the virtual identity of the two sets of rules is obvious. One sees readily that a closed tableau can be construed with $U_1, U_2, \ldots$ as its initial left-hand formulas and $V_1, V_2, \ldots$ as its initial right-hand formulas if and only if a proof can be obtained for the sequent

$$U_1, U_2, \ldots \to V_1, V_2, \ldots$$

by the rules of the Gentzen method. This can be seen by comparing pairwise the two sets of rules and by observing that they are precisely identical except for the direction in which one is proceeding.

Quite a few proof theoretical results have been established for the Gentzen method. Some of these results are not without relevance for the interpretation of the ancient method of analysis. However, already the broad outlines of the Beth method or of the Gentzen method are useful here.

5. Analysis as a Natural Deduction Method

The intuitive strategy underlying the method of semantical *tableaux* should be clear.[7] As was already indicated, when we are trying to prove the disjunction of V_1, V_2,... from given premisses U_1, U_2,..., we place the former into the right-hand column of a tableau and the latter into its left-hand column. Then what we do in building up the left-hand column downwards can be thought of as drawing conclusions from U_1, U_2,..., while in the right-hand column we look for premisses from which the disjunction of V_1, V_2,... can be concluded. This is not quite an exhaustive description, for it does not yet explain the possibility of transferring a formula from the one column to the other. However, this does not affect the possibility of looking at the tableau method as a stepwise procedure which starts from the desired deductive connection and tries to utilize its two ends to find a way to establish it.

This procedure as it were works its way from both ends to the middle where a link is hopefully eventually found. The possibility of transferring formulas from a column to the conjugate column means that this procedure is not quite linear, however.

From this point of view, most of the tableau rules can readily be understood. For instance, the import of (vi) (c) from this point is simply that in order to prove X & Y (from the given premisses) we must prove both X and Y.

The interpretation of the instantiation rule (vij) is perhaps not quite as clear. However, we can think of it as effecting simply a transition from a bound-variable formula to the corresponding free-variable expression. For instance, the second part of (vij) may be taken to say that in order to prove $(x)\ X(x)$ it suffices to prove $X(p)$ for a *new* free variable p which has nothing to do with the variables so far used.

The naturalness of this way of looking at Beth's *tableaux* is perhaps seen even better if we introduce derived rules into our treatment. Thus we can move to Y from X and $X \supset Y$ in a left-hand column of Beth's tableau, etc.

Each pair of conjugate columns must of course be thought of as a separate line of proof which must be completed before the desired overall deduction is accomplished.

The rule for closing a pair of conjugate columns also receives a very natural meaning in this way. It simply means that when the same formula appears in both of the conjugate columns, the right connection has been established between the different consequences of the given premisses and of different potential premisses from which at least one of the desired consequences could be drawn: one of the former matches one of the latter.

This procedure is strongly reminiscent of Pappus' description of the ancient method of analysis. In the tableau method, described in slightly different terms (and in slightly inaccurate terms), we start (especially in the right-hand column) from the desired consequence and proceed from it through its potential premisses to something that follows from the premisses through normal 'synthetic' deduction. In other words, in the tableau method we assume the desired consequence as if it were already accomplished, and ask from what it results, where the latter could come from, and so on, until we come either to one of the premisses or to one of their known consequences (established in the left-hand column).

This description of the tableau method agrees obviously very closely with Pappus' description of analysis as given *in extenso* in Section 2 above (and as interpreted in the way done there). Thus there obtains a strong *prima facie* similarity between ancient analysis and suitably formulated natural deduction technique.

6. The Direction of Analysis

This kinship can be explored further. The tableau method can be given another twist which is reminiscent of what we find in the historical material. If we simply replace all formulas in the right-hand column by their negations, we obtain a method in which we are drawing conclusions when we proceed downwards along either column (taking each pair of conjugate columns as an alternative line of argument which must be concluded for a conclusion concerning the original formulas to be drawn). In fact, any pair of two conjugate columns can now be merged into one. The whole procedure becomes very much like Hintikka's method of model set construction in which the distinction between columns is unnecessary.[8]

This technique is of course equivalent with the original one. It involves little more than a notational change. However, the intuitive interpretation of the purpose of the whole manoeuvre is now different. Instead of trying to derive a formula or one of a set of formulae $\{V_1, V_2, \ldots\}$ from the premisses $U_1, U_2, \ldots$, we are now trying to reduce *ad absurdum* the joint assumption of $U_1, U_2, \ldots$ and $\sim V_1, \sim V_2, \ldots$ Now the closure condition for a pair of conjugate columns is even more intuitive than before. A line of argument comes to a close when it leads to a contradiction, i.e., to the joint assertion of a formula and its negation.

Assuming classical logic, essentially *the same 'analytical' method* can thus be used both in constructing direct proofs backwards, by starting partly from the conclusion to be proved, and proofs by *reductio ad absurdum*, that is, in proofs which reduce the counterexample to absurdity. This identity is strongly reminiscent of what we find in historical material.

Note, first, that in the original tableau method we proceeded on the right 'upstream', against the direction of relations of logical consequence, looking for intermediate premisses from which the top right-hand formula (or one of a number of formulae) could be inferred. In the inverted method, conclusions were drawn on the right from the top right-hand formula (or formulae), which means proceeding 'downstream', along the direction of relations of logical implication. Since the two methods are nevertheless basically identical with one another, it is to be expected that there will easily be some confusion in the actual historical discussions of the analytical method as to what its direction really is as compared with the direction of relations of logical entailment. Every student of Greek mathematics knows that this expectation is amply fulfilled. There is an extensive, albeit inconclusive literature on the very problem of the direction of geometrical analysis. Now we can see that there are good topical reasons why this literature has remained inconclusive, and why the primary sources likewise exhibit a certain amount of confusion. A case in point is the discrepancy between Pappus' general description of analysis and his own analytical practice mentioned above in Section 2. (Perhaps neither the Greek geometers nor their commentators should be censored very severely, for we have seen that the subject matter itself exhibits a considerable subtlety.) In one sense, analysis can even be said to proceed in either direction. The whole problem of the direction of analysis is also superficial in the sense that it is not connected with the heuristic usefulness of the method of analysis.

Notice, however, that the direction makes a difference as to what we may be expected to conclude from a successfully completed analysis.

7. Analysis and Reduction

Another historical comparison is possible here. It has been surmised that the analytical method originated from the method of proof by *reductio ad absurdum*. The basic identity of suitable versions of the two which we have noted points to the systematical naturalness of such a historical thesis. (Notice that assuming classical logic the difference which is sometimes made between reductive proofs which prove a (negative) result, say $\sim P$, by assuming P (classically, $\sim\sim P$), and which prove $P(=\sim\sim P$, classically) by assuming $\sim P$ is not essential for the details of the historical thesis.)[9]

Moreover, in the historical sources we find clear indications of the idea that the analytical method can provide both kinds of proofs, that is, both direct proofs and reductive proofs. For instance Pappus writes:

> If that which is admitted be true, the thing sought is true, too. But if we come upon something false to admit, the thing sought will be false, too. (Hultsch 636, 2–6)

This is said of the theoretical analysis, and the same claim concerning the effectiveness of the problematical analysis is also made by Pappus (cf. Section 2 above).

Here we see at the same time that Pappus must be said to be somewhat confused, when interpreted in the light of our observations. For one cannot hope to a direct proof of a hypothesis H and a disproof of H (i.e., proof of $\sim H$) from given premisses $P_1, P_2,\ldots$ through one and the same analytical procedure, interpreted consistently in our way, as Pappus seems to be assuming. In the former case one must start either from $P_1, P_2,\ldots$ (on the left) and H (on the right), and proceed in accordance with the original rules, or from P_1, P_2,... (on the left), and $\sim H$ (on the right), and use the new (mirror-image) rules. In the latter case, the roles of the two sets of rules must be reversed. This is what is meant above when it was said that the direction of analysis makes a difference to what one may conclude from it when it comes to an end.

Usually, Pappus' confusion has been attributed to a confusion in his mind between the two possible directions of analysis. Now we can see that it may in the last analysis be due to a subtler misunderstanding, viz. to a confusion as to how precisely the direction of analysis hangs together with its expected (or, rather, hoped for) conclusion. Notice, further, that this is not incompatible with saying, as Gulley has done, that several different traditions are present in Pappus' confused account.[10] If the reductive idea was older, we might even find different historical strata in Pappus' account.

8. Why synthesis?

An important *prima facie* difference between the ancient method of analysis and modern natural deduction methods (such as the tableau method) is that the latter one provides us with a definitive result as soon as the analysis is finished (i.e., as soon as all pairs of conjugate columns are closed) whereas the old geometrical procedure of analysis went together with a complementary synthesis in all its typical uses in Antiquity. Only in conjunction with synthesis did analysis provide the mathematician with an acceptable proof. The whole method thus perhaps might be called, as it in fact was often called, the method of analysis and synthesis.

Why this discrepancy between geometrical analysis and natural deduction? Two different explanations can be given which jointly serve to show that this apparent discrepancy scarcely tells against the parallelism of the two and in fact reinforces the parallelism.

First, what a tableau construction gives us does not look like an ordinary deductive argument. Above, it was indicated how a tableau construction can be viewed as a search for a deductive connection. An even more natural interpretation is possible, however. It can be further enhanced by turning the tableau construction into a model set construction à la Hintikka. It consists in understanding a tableau construction as an attempted model construction, not as an attempted proof. On this interpretation, in building up a tableau one is trying to describe (in a sense to construct) a counterexample to the desired deductive connection, that is to say, a model in which $U_1, U_2, \ldots$ are true but $V_1, V_2, \ldots$ all false. The tableau rules are all very naturally understood from this vantage point. A line of construction (pair of conjugate columns) is closed when it leads to a dead end, and the whole construction succeeds when all the different ways of constructing a counterexample are thus aborted.

Although some kind of logical proof can always be obtained from such a successful construction, as a comparison with the Gentzen method easily shows, the transformation will require a certain amount of re-adjustment, especially if a conversion to a form of argument more familiar than the Gentzen technique is what is aimed at. This readjustment can be compared with the synthesis part of the double-barrelled method of analysis and synthesis.

This comparison cannot be pushed very far, however. An even better interpretation is obtained by pointing out the practical inconvenience of distinguishing between the two conjugate columns in a Beth tableau construction. Such a distinction presupposes a double bookkeeping which is quite cumbersome in the case of an informal argument. For that reason, the methods of ancient geometers differed from the usual forms of modern natural deduction methods. If their procedure is compared (by hindsight) with present-day logical methods we must say that they, in effect, proceeded merrily downwards in a Beth-tableau also in looking for direct proofs, that is, in effect drew logical conclusions in both columns (whose difference thus disappeared). When such a procedure comes to halt, only an undesired negative conclusion can *ipso facto* be drawn. In the case of a negative outcome, the premisses turn out to imply, not the desired conclusion, but its negation. This happens when (as it were) contradictory sentences turn up in each pair of conjugate columns (i.e., in each line of argument). However, in a positive case (the same formula popping up in each pair of conjugate columns) no definitive conclusion can be drawn. Such a conclusion can only be ascertained by means of a separate examination calculated to ascertain whether the rules used in their analysis can be inverted so as to obtain an actual direct proof.

Thus the two-barrelled character of the ancient method of analysis *and* synthesis becomes understandable as a result of the inconvenience of the (self-sufficient) double-column procedure of simultaneously drawing inferences from the premisses and looking out for intermediate premisses from which the desired conclusion could be drawn. It is much more convenient to proceed in one direction only and subsequently (in the light of the insights obtained through this useful but hazardous stratagem) re-examine the steps taken in analysis to establish their soundness.

This explains partly one major puzzle in Pappus. In his general description of analysis, Pappus speaks of a stepwise search of premisses from which the desired conclusion could be drawn, assuming of course the Hintikka-Remes interpretation mentioned above in Section 2. If this were strictly true, however, synthesis would merely be empty formality. And in his practice Pappus proceeds to draw inferences from this very desired conclusion quite as happily as the next geometer. This discrepancy between theory and practice was caused by the features of the logical situation which we saw described and which Pappus apparently never quite saw through. His inconclusiveness is also shown by the fact that he treats positive and negative outcomes of analysis on a par even when analysis is conceived of as a search for a direct proof. This occurs in Pappus' general descriptions of analysis (see Section 7 above).

9. Why is analysis analytical?

The heuristic usefulness of the analytical method requires an explanation different from all questions of direction. It seems to us that the heuristic value of the analytical method depends to a considerable extent on the fact that the proofs obtained by this method can be thought of as dealing with one specific kind of constellation of individuals (members of our universe of discourse). In the original case of elementary geometry, the proof will deal with one particular kind of geometrical configuration. As a consequence, the search for this proof can be conceived of as a study of the interdependencies within this configuration of individuals. Since we usually have a pretty good idea of these interdependencies, our tacit knowledge is easier to bring to bear here than in the case of many other types of proof procedure.

This thesis may be dubbed analysis as an analysis of configurations. (In the special case of geometry, analysis becomes an analysis of figures.)

Our thesis here is closely related to what is said in Chapter IV of Hintikka and Remes.

Once one has (so to speak) a figure before one's eyes one can trace series of interrelations in it both forwards (in the direction of logical consequence) and backwards (looking for suitable premisses for given conclusions) with essentially equal facility. It is this 'configurative' nature of analytical arguments that is heuristically essential here, not the direction of analysis.

Our thesis of analysis as an analysis of configurations will turn out to require an important qualification. (See Section 10 below.) It might also seem that the possibility of so conceiving of one's attempt to find a proof has little to do with the proof procedure one is envisaging. However, the two are in reality closely connected.

As the first approximation, we can say that a search for a proof by some such technique as the tableau method can be viewed as an analysis of configurations only if this proof technique satisfies what is known as the *subformula property*. This property is present if in the process of looking for a proof formulas are all the time chopped into simpler ones. The reason why it is needed in order for us to be able to think of the resulting proof in terms of a given figure or configuration is not difficult to fathom. If we may introduce any old complex formula in the midst of the proof, there is no hope of always being able to interpret them as speaking of the same figure or configuration. The use of such complex formulae may make the proof shorter, but it may also make it more difficult to interpret.

In order to see that this is not an idle point, let us consider the logical situation. In the study of Gentzen-type systems, rules additional to those listed above and unlike them (or rules derivable from them) are often considered. A typical (and important) case in point is the so-called cut rule (see Kleene p. 331.):

$$\frac{\Delta \to \Lambda, C \quad C, \Gamma \to \Theta}{\Delta, \Gamma \to \Lambda, \Theta}$$

Since by omitting Γ, Λ, and Δ we can obtain from the cut rule the rule for *modus ponens*:

$$\frac{\to C \quad C \to \Theta}{\to \Theta}$$

we may consider the cut rule as a kind of strengthened principle of *modus ponens.*[11] (By this time, even an imperceptive reader will have noticed that the rules for the Gentzen-type Sequent Calculus given above do not contain any form of *modus ponens.*)

The analogue of the cut rule in Beth's *tableaux* method is obvious. We can formulate it as follows.

(xi) If X appears in the left column and Y in the right column of a tableau (or a subtableau), the tableau can be split. Put a new formula C in one right column, and also in the left column not conjugate with it.

From this formulation it is seen that the unrestricted use of the cut rule destroys every hope of thinking of the resulting proofs as somehow dealing with the configurations specified in the original premisses of the proof (top left-hand formulas in Beth's method). It imports a new formula C (the cut formula) into the argument. Since there is no restriction to the complexity of C, this cut formula can transcend any and every situation described in U_1, U_2,... in all kinds of ways. Unless the cut rule is dispensed with, there is no hope of understanding a logical argument à la Beth as an analysis (in the ordinary sense of the word) of a definite configuration of individuals. The same obviously goes for other proof techniques.

Now the study of a Gentzen-type approach to logic shows that what distinguishes the simple and transparent proof methods which it yields is precisely the possibility of dispensing with the cut rule and with other comparable variants of rules of the *modus ponens* type. What makes natural deduction methods natural, we might say, is that they preserve the subformula property (read of course from down up in the case of Gentzen proofs). This is what gives Gentzen-type proofs (without the cut rule) their simplicity, and attempts to find such proofs their heuristic ease. In looking for proofs, we can in virtue of the subformula property always (with certain qualifications to be noted later) study simpler parts of the total situation considered in the given premisses and/or in the conclusion.

The same goes of course for the tableau method. What makes all these methods interesting and important logically is thus essentially the same thing as makes it possible to look for searches of the resulting proofs (such as the traditional geometrical analyses are) as studies ('analyses') of a given configuration or situation. It is our thesis that this is an important part of the heuristic usefulness of the traditional method of analysis.

It is also worth noting how the possibility of using rules that violate the subformula property illustrate the difference between different conceptions of analysis. When such rules are for instance incorporated in the tableau method, nothing changes in the *direction* of our procedure. Hence the result is still an analytical method (as much as it was before the change) in any purely directional sense of analysis and synthesis. From what we have found it nevertheless follows that the result cannot always be analytical in the sense of analysis of configurations.

The obvious fact that no mathematician would ever recognize the heuristic method of analysis in a procedure which fails to satisfy the subformula property and which therefore can change the configuration of individuals being examined supports further our emphasis on the nature of heuristically relevant variants of the method of analysis as so many analyses of configurations.

10. The role of constructions

This idea of analysis as an analysis of configurations or figures nevertheless requires a major qualification. In its briefest form, the qualification may perhaps be expressed by saying that the figure analysed in analysis is not the figure of which the theorem to be proved speaks, but this figure *supplemented by suitable auxiliary constructions.* For instance, in the example given above of Pappus' practice, such auxiliary constructions are carried out in the beginning of analysis. (See Section 3 of this paper.)

The need of auxiliary constructions was a well known fact of life for the practicing geometers of Antiquity, although its deeper reasons were not fully understood.[12] In the tableau method, the role of auxiliary constructions is played by the rule (vij). In its both uses, it introduces a new 'arbitrarily chosen' individual to the configuration so far studied. This rule is, as Beth already pointed out, closely related to the ancient concept of *ecthesis* which is a kind of intuitive counterpart to the instantiation rules of modern logic.[13] In Kleene's formulation of the Gentzen method, the same role is played by the rules $\rightarrow \forall$ and $\exists \rightarrow$.

We may also put the point as follows. In order for a logical argument to be interpretable as an analysis of a fixed configuration, it is necessary (but not sufficient) that the argument satisfy the subformula property. It is also necessary that the number of individuals considered together in their relation to each other in the argument is not increased. The precise sense of this number is spelled out in Hintikka's concept of the degree of a (first-order) sentence.[14] In simple intuitive terms, it is basically the different instantiation rules that introduce new individuals (however 'arbitrarily chosen') into a logical argument and thus raise its degree.

In ancient geometry, the introduction of new geometrical objects in an auxiliary construction or *kataskeue* took place by means of the so-called postulates. The difference between postulates and axioms is on one interpretation just what this use illustrates: the application of postulates introduces new individuals, whereas the application of axioms deals with the geometrical objects already at hand. In any case, reliance on auxiliary constructions does not take us outside the axiomatic framework of geometry.[15] Auxiliary constructions are in fact little more than ancient counterparts to applications of modern instantiation rules.

It is nevertheless remarkable in several other respects. For one thing, it can be shown that in so far as the general logical situation is concerned, the auxiliary constructions needed to prove a putative theorem are not predictable in general (in the sense that they cannot be effectively found on the basis of Gödel number of the putative theorem). Even the number of auxiliary individuals needed is in general unpredictable in the same sense.[16]

Admittedly these auxiliary constructions may be recursively predictable in sufficiently elementary parts of geometry.[17] However, this special case is not representative of the general situation. We may note here as an illustration that one of the first curves the ancients came to consider in geometry already put them face to face with non-elementary difficulties. The periodic sine function which is implicit in the concept of *quadratrix*, a curve discovered by the sophist Hippias, is not 'elementary'.[18] To add it to an elementary mathematical system already effects the unpredictability of instantations we have been discussing.[19]

Intuitively, the situation is in geometrical terms such that we cannot in general know when we have carried out enough auxiliary constructions for an analysis to succeed. We cannot usually restrict ourselves to analysing the original figure envisaged in the theorem to be proved, and we cannot in principle be always sure whether we have amplified it sufficiently.

Heuristically, this means that the discovery of suitable auxiliary constructions will be a crucial feature in an attempted analysis. Again this phenomenon is but a special case of the problems of effective theorem-proving in first-order logic, where the choice of the right instantiations is often the most vital consideration.[20]

At the same time, the need of auxiliary constructions, and their unpredictability, shows that the analytical method cannot amount to a general foolproof discovery procedure, notwithstanding its projected use as such a general method by Descartes.[21] Even in elementary geometry, the idea of proceeding analytically cannot alone be a full-fledged discovery procedure, but requires the support of a method of determining what Leibniz called "the best constructions".[22] (See *Nouveaux Essais*, Book IV, Chapter 3, § 6 and Chapter 17, § 3.) In this direction logical analysis is thus apt to throw particularly sharp light on the heuristic situation.

11. The relative order of different rules

Ancient geometers appreciated keenly the role of constructions in geometrical proofs. Their terminology is proof enough of this appreciation: proofs were often referred to as 'constructions' or 'figures' (διάγραμμα), and proving was likewise referred to as 'drawing' (γράφειν).[23] In the traditional division of an Euclidean proposition into several parts, the auxiliary construction or *kataskeue* (κατασκευή) had a niche of its own, separate from the proof proper or *apodeixis* (ἀπόδειξις) during which no new individuals were introduced into the figure, and preceding this *apodeixis*.[24]

This possibility of so segregating steps of instantation from other steps in a logical proof is not quite trivial. In principle, a geometer could mix applications of postulates and axioms. The fact that instantiations (constructions) can be gathered together so as to precede *apodeixis* is in effect a result concerning the possibility of permuting applications of different kinds of rules in a logical argument. (In the geometrical case, the premisses and the conclusion can be thought of as being in the so-called prenex form.) Logicians know that such permutation results constitute one of the cornerstones of modern proof theory. They are used in establishing many nontrivial methodological results in proof theory.[25]

The Euclidean separation of the auxiliary constructions from the proof proper is a kind of modest early anticipation of a special case of such permutability results.

The same segregation of constructions (synthesis in the narrower sense) from the *apodeixis* was practiced in the synthesis part of the double method of analysis and synthesis. In fact, synthesis (in the narrower sense) and *apodeixis* were (in this order) the main parts of the synthesis (in the wider sense). (See our example in Section 2.)

This led the ancient analyst into a problem which they never mastered in abstract logical terms, even though this did not mar their successful analytical practice. For in analysis the relative order of steps of construction and steps of deduction could scarcely be the same as in synthesis nor yet its mirror image. Pappus says in so many words that in the synthesis we go through the same steps as in the analysis, but in the reverse order. This simply cannot be true of the relative order of steps of construction and steps of deduction; we cannot take 'the former antecedents' as 'consequents' in the 'reverse' order. (Cf. Section 2). If the analytical order were in this respect the opposite to the order of synthesis, all auxiliary constructions would in an analysis come after the non-constructive ones. However, it was pointed out that analysis cannot usually succeed unless enough auxiliary constructions have been carried out, i.e., carried out before or during the analysis. Hence in recasting an analysis into an ordinary synthetic (deductive) form, the relative order of different kinds of steps could not simply be reversed but had to change in other ways, too, as it can be seen to change in our example above in Section 3. (There the construction of the tangent at A was the first step of analysis, but it did not by any means become the last step of synthesis. Rather, it was one of the earlier steps in the first half of synthesis.) The logical subtlety of the principles governing such permutations was beyond the ken of ancient geometers as a general logical problem, no matter how successful (and strict) they were in practice.

Moreover, changes of order in the application of different kinds of principles were often needed for another purpose, too. If analysis proceeded 'downwards', i.e., if it consisted (inter alia) in drawing logical conclusions from the premisses, then some of the auxiliary constructions carried out in analysis may have depended on the desired conclusion, too. In justifying this procedure, the geometer therefore also had to demonstrate the independence of the auxiliary constructions in the analysis from the desired conclusion. Often this, too, implied a reshuffle among the different kinds of argument steps.[26]

The considerable hesitancy and uncertainty which there seems to have prevailed among ancient geometers concerning the precise nature of the justification of analysis has thus a solid systematical reason in the difficulty and subtlety of the permutation rules of modern proof theory.

12. THE 'RESOLUTION'

An analysis contains yet another part, called conventionally the 'resolution', as shown by our example above in Section 3. Its interpretation is not a simple matter, but its main function can be gathered from the contrast between the steps of construction and the steps of deduction.[27] This 'resolution' shows, typically, that the steps of the former kind can be carried out on the basis of the given elements. In algebraic language, we could say that the 'resolution' points out that the values of the unknowns are determined as functions of the known arguments.

This is forcefully illustrated by our example in Section 3 above. It serves to establish that the desired lines *DB*, *BE* are 'given'. And the force of this characteristic term is precisely what was indicated. A geometrical object is 'given' in so far it is determined by the data of the theorem or problem, independently of the heuristic assumption that the problem has already been solved which forms the starting-point of analysis. This is just what our sample resolution accomplishes for the lines searched is the problem in question.

There are no constructions (or arithmetical operations) in the 'resolution' itself. It relies on already known theorems (typically on Euclid's *Data*), using them to establish the independence of the constructions needed in analysis of its outcome, but it does not yet show how the actual constructions can be carried out. This may prompt a question concerning its function in the overall analysis-synthesis scheme. A partial answer is obtained from our discussion in Section 11 above where it was shown that the transition from analysis to synthesis often involves difficult changes of order among different kinds of steps. Seeing that the 'resolution', with its exclusive concentration on what is 'given', can stay closer to the analysis proper than the synthesis can and that it therefore offers a kind of shortcut through these difficulties, we perhaps begin to understands its role better.[28]

Academy of Finland

NOTES

[1] Hintikka, Jaakko and Remes, Unto, *The Method of Analysis: Its Geometrical Origin and Its General Significance* (*Boston Studies in the Philosophy of Science*, Vol. 25), D. Reidel Publishing Company, Dordrecht, 1974.

[2] Beth, E. W., 'Semantic Entailment and Formal Derivability', *Mededelingen van de Koninklijke Nederlandse Akademie van Wetenschappen*, Afdeling Letterkunde, N.R., **18**, n:*of* **13**, Amsterdam 1955, reprinted in Jaakko Hintikka (ed.), *The Philosophy of Mathematics*, Oxford University Press, Oxford, 1969. – Beth refers to Plato's *Philebus* 18 B-D, and to Aristotle's *Metaphysics* **IV**, 3, 1005b2, and to Leibniz. Cf. *The Philosophy of Mathematics*, p. 19, note 8.

[3] Cf. *Pappi Alexandrini Collectionis Quae Supersunt*, Fr. Hultsch (ed.), Weidmann, Berlin, Vols. I–III, 1876–77, 634–36. Cf. Hintikka, Jaakko, *Logic, Language-Games, and Information* (= *LLGI*), Clarendon Press, Oxford, 1973, Chapter IX, and Hintikka-Remes, Chapter II, for secondary literature on Pappus' description.

[4] Hintikka-Remes, Chapter II.

[5] Hultsch, pp. 830–32. The translation is from Heath, T. L., *The Thirteen Books of Euclid's Elements*, Cambridge University Press, Cambridge, 1926, pp. 141–142. *Pace* Heath, we speak of construction in analysis as a part of the transformation (Part I of analysis).

[6] See Kleene, Stephen C., *Mathematical Logic*, John Wiley & Sons. Inc., New York, 1968, p. 289.

[7] Usually the intuitive strategy is described as a search of counterexamples, of course.

[8] For the rules of Hintikka type model set construction, see *LLGI*, Chapter I, p. 24. There we also see especially clearly the other intuitive way of looking at the tableau construction, viz. an abortive attempt to describe step by step a counterexample to the desired conclusion (i.e., to describe a model of U_1 & U_2 & ... & $\sim V_1$ & $\sim V_2$ & ...).

[9] Cf. Hintikka-Remes, Appendix 1. This Appendix is by Prof. Arpad Szabó. For the two types of reductive arguments, see Lakatos, I. (ed.), *Problems in the Philosophy of Mathematics*, North-Holland Publishing Company, Amsterdam, 1967, p. 10 (comment by Prof. Kalmar), and p. 25 (reply by Prof Szabó). For the early history of the reductive arguments, see Knorr, W., *The Evolution of the Euclidean Elements*, D. Reidel Publishing Company, Dordrecht, 1975, Chapter II.

[10] See Gulley, Norman, 'Greek Geometrical Analysis', *Phronesis* **3** (1958), 1–14.

[11] For the original treatment of the cut formula in the sequent calculus, see Gentzen, G., 'Untersuchungen über das logische Schliessen', *Mathematische Zeitschrift* **39** (1934), 176–210 and 405–431; for an earlier idea of eliminating *modus ponens*, see Herbrand, J., in Warren D. Goldfarb and J. van Heijenoort (eds.), *Logical Writings*, D. Reidel Publishing Company, Dordrecht-Holland, 1971, pp. 40ff.

[12] For the observations concerning the necessity of dealing with auxiliary constructions in geometry, see *Proclus In. Pr. Eucl. Comm.* (ed. Friedlein), p. 78, line 12ff., and *Euclides: Suppl. Anaritii Comm.* (ed. by Curtze), pp. 88 and 106.

[13] Cf. Beth's article in *The Philosophy of Mathematics*, p. 37, and Beth, *Aspects of Modern Logic*, D. Reidel Publishing Company, Dordrecht-Holland, 1970, p. 44; Hintikka, *LLGI*, p. 215.

[14] Cf. *LLGI*, pp. 141 and 178–185.
[15] Auxiliary constructions can be introduced also by assuming – in the midst of proof – a problem which we know to have been solved earlier as a premise. This clearly amounts to an exception of the subformula principle. However, this can be always avoided by using postulates and axioms.
[16] For if we could predict this number, we would have a decision procedure.
[17] Cf. Tarski, A., 'What Is Elementary Geometry?' in *The Philosophy of Mathematics*, pp. 165–175, reprinted from *The Axiomatic Method* (ed. by L. Henkin, P. Suppes, and A. Tarski), North-Holland Publishing Company, Amsterdam, 1959, pp. 16–29. We refer the reader to Tarski's paper also for the sense of 'elementary' which we are presupposing here.
[18] Tannery, Paul, *Mémoires Scientifiques* **II**, p. 1 for the *quadratrix*.
[19] Cf. Tarski, A., *A Decision Method for Elementary Algebra and Geometry*, second edition, University of California Press, Berkeley and Los Angeles, 1951, p. 45.
[20] Cf., e.g., Prawitz, Dag, 'Advances and Problems in Mechanical Proof Procedures', *Machine Intelligence* **4**, 59–71.
[21] See *La Géométrie*, p. 299 of the original.
[22] See Prawitz, Dag, *op. cit*.
[23] Cf. Hintikka-Remes, Chapter VII.
[24] Cf. *LLGI*, Chapter IX.
[25] Such as Gentzen's Extended *Hauptsatz*.
[26] For such a possible dependency, see Hintikka-Remes, Chapter III.
[27] Cf. Chapter VI of Hintikka-Remes.
[28] This paper represents independent work by the Junior Author, based on suggestions by the Senior Author. – We have profited from the comments of an anonymous referee.

POSTSCRIPT

This paper seems to us peculiarly appropriate as a contribution to a volume dedicated to the memory of Imre Lakatos, for it owes its inception to him. In this paper, we are examining the deeper logical and systematical foundation of our predominantly historical monograph, *The Method of Analysis* (note 1 above). An early version of the monograph was presented at the International Colloquium in the History and Philosophy of Science at Jyväskylä, Finland, in June 1973. One of our two commentators on that occasion was Imre Lakatos. Imre's untimely death prevented the written version of his comments ever reaching the print. But his verbal comments already touched a subject on which Imre's long-time interest in the method of analysis had led him to an emphasis altogether different from ours, to wit, to a strong positive emphasis on the directional peculiarities of the method and to a strong negative emphasis on the role of logic in this celebrated method. While agreeing with Imre on the essentially heuristic nature of the analytical method, we nevertheless feel no less strongly that a true understanding of the method, including the reasons for its heuristic value, can only be reached through a careful analysis of its logic. Heuristics does not preclude logic here, but rather presupposes it. Having failed to persuade Imre in Jyväskylä, we were moved to formulate our own position more fully and more explicitly. We feel especially deprived by not having the prospect of learning Imre's rejoinder to our attempt which we expected to be delivered with that spirit and outspokenness which all his friends learned to cherish.

Added in proof by Jaakko Hintikka: It is my sad duty to record here the untimely and tragic death of my co-author, Unto Remes, on June 2, 1975. In him, we have lost not only a gifted young scholar but a friend and a man with exceptionally high principles.

COLIN HOWSON

THE DEVELOPMENT OF LOGICAL PROBABILITY*

1. Introduction

I shall in the course of this paper investigate some notable attempts to provide a theory of a weak consequence relation ostensibly subsisting between pairs of arbitrarily chosen sentences. At various points in their careers James Bernoulli and Leibniz, Keynes and Carnap were preoccupied with this problem. Leibniz was the first, to my knowledge, to articulate it; the remaining three have been preeminent in their attempts to provide a solution. Part of the significance of a solution lies in its furnishing a criterion of verisimilitude; for we know that if $\Sigma \vdash \sigma$ holds, and all the sentences in Σ are true, then so is σ. A measure of the strength of a weak relation of consequence between arbitrary Σ and σ could therefore be regarded as a measure of the verisimilitude of σ, given the truth of all the sentences in Σ. The exigencies of life in an uncertain world appear to require, moreover, for their mitigation, such a criterion by which to discriminate between conjectures which transcend experience at some given time. (I am not, by the way, claiming that this is the only sense that can be given to the vague term 'verisimilitude' – Popper, for example, has given an entirely distinct explication[1] –; but that term is indeed so vague that it will surely support the gloss I have given it, the history of which reaches back to Leibniz if not before.)

Bernoulli, as we shall see, entertained also the possibility of another use for the theory we shall be discussing: namely, as furnishing the foundation of an explanatory theory of a realm of phenomena which he thought predicted by his theory. These phenomena are what we should today call statistical phenomena. Whether Bernoulli was justified in his belief is a question I shall try to answer in due course.

2. Keynes and Carnap

In his (1921) Keynes advanced the thesis that

R. S. Cohen et al. (eds.), Essays in Memory of Imre Lakatos, 277–298. *All Rights Reserved*
Copyright © 1976 by D. Reidel Publishing Company, Dordrecht-Holland

we believe that there is some real objective relation between Darwin's evidence and his conclusions which is independent of the mere fact of our belief, and which is just as real and objective, though of a different degree, as that which would exist if the argument were as demonstrative as a syllogism. We are claiming, in fact, to cognise correctly a logical connection between one set of propositions which we call our evidence... and another set which we call our conclusions... (p. 6).

In Chapter VIII of his book we find the statement of a position very much reminiscent of Russell's (it is well known that Keynes was very strongly influenced by Russell, and it is difficult to exaggerate the contribution of Russell's ideas to the formal middle sections of Keynes's *Treatise*):

That part of our knowledge that we obtain directly supplies the premisses of that part which we obtain by argument. From these premisses we seek to justify some degree of rational belief about... conclusions... The kind of rational belief which we infer in this manner is termed probable... and the logical relations, by the perception of which it is obtained, we term relations of probability (p. 111).

Not even Russell, however, is the progenitor of these ideas: that honour seems to be due to Leibniz. The epigraph to Keynes's first chapter is from Leibniz ("J'ai dit plus d'une fois qu'il faudrait une nouvelle éspèce de logique, qui traiteroit des degrés de probabilité"). Leibniz, it is well known, received early training in jurisprudence, where his attention was drawn to the canons of juridical proof, and to the criterion of probability, as a standard not only of value in jurisprudence, but also in judging the force of all non-metaphysical (i.e. non demonstrative) reasoning: "La philosophie pratique est fondée", he once wrote, "sur l'art d'estimer les degrés des probations, qui ne se trouve pas encore dans les auteurs logiciens, mais dont les jurisconsults ont donné des échantillons qui ne sont pas à mépriser et peuvent servir de commencement pour former la science des preuves". Kindred views appear to be taken by James Bernoulli, in the splendid part IV of his *Ars Conjectandi* (though Bernoulli and Leibniz were celebrated correspondents and Leibniz claimed to have aroused Bernoulli the mathematician's interest in probability):

we are said to know or to understand those things which are certain and beyond doubt; and other things we are said merely to conjecture or to guess about... to conjecture about something is to measure its probability; and therefore the art of conjecturing or the stochastic art is defined by us as the art of measuring as exactly as possible the probabilities of things with this end in mind: that in our decisions or actions we may be able always to follow what has been perceived as being superior, more advantageous, safer or better considered: in this alone lies all the wisdom of the philosopher and all the discretion of the statesman (p. 13).

I shall show later that Bernoulli's development of these ideas is significantly distinct from that of Keynes (and Carnap).

Let us return to Keynes. What justification was there, in 1921, for Keynes's contention that the formal theory of probability furnished an extended theory of deduction? At this point it is worth noting that by the end of the nineteenth century the intellectual feat of distinguishing the formal laws of probability from some particular *measure* e.g. as the ratio of favourable to possible cases, had been accomplished. That the generalised logic of which he spoke satisfied the calculus of probability was according to Keynes discerned by logical intuition (Carnap provides a similar justification in his (1950)), though that the probability calculus generalises deductive logic in one sense is of course actually provable. For consider a set Σ of sentences in some language L and let σ, τ be any sentences of L. Suppose Σ is consistent, and let $\mu(\sigma/\Sigma)$ be the truth-value of σ on the assumption that all the members of Σ are true (with respect to some fixed interpretation), where $T=1$, $F=0$. Then:

(i) $\mu(\sigma \vee \tau|\Sigma)=\mu(\sigma|\Sigma)+\mu(\tau|\Sigma)$ if $\Sigma \vdash (\sigma \rightarrow \neg\tau)$
(ii) $\mu(\sigma \vee \neg\sigma|\Sigma)=1$
(iii) $\mu(\sigma \wedge \neg\sigma|\Sigma)=0$
(iv) $\mu(\sigma|\Sigma)\geqslant 0$

If we identify σ with the equivalence class generated by logical equivalence of which it is the canonical representative, then it follows that the classical valuation μ is a probability measure. Hence, where $\mu(\cdot|\Sigma)$ is any real valued function satisfying (i)–(iv), μ is a probability function. It follows that the probability calculus, which for fixed Σ is just (i)–(iv), can be regarded as defining a class of measures of the extent to which the joint satisfaction of Σ determines the truth-values of sentences of L. Thus the probability calculus can be mapped into the theory of classical valuations of sentences from a language L. To this extent the probability calculus is a generalisation of deduction. But what of the apprehension of particular values by our logical intuition? Ramsey, in reviewing Keynes's *Treatise*, wrote that

there really do not seem to be any such things as the probability relations he describes. He supposes that, at any rate in certain cases, they can be perceived; but speaking for myself I feel confident that this is not true. I do not perceive them, and if I am to be persuaded that they exist it must be by argument; moreover I shrewdly suspect that others do not perceive them either, because they are able to come to so little agreement as to which of them relates any two given propositions. All we appear to know about them are certain general propositions, the laws of addition and multiplication; it is as if everyone knew the laws of geometry but no one could tell whether any given object were round or square; and I find it hard to imagine how so large a body of general knowledge can be combined with so slender a stock of particular facts. (Ramsey, 1931, p. 162).

Ramsey's innuendo is surely correct: Keynes's faculty of logical intuition apprehended no such relation. Where did the general laws of this relation come from? – it might well be asked. Before I answer this, and the answer seems to me to be rather obvious, I should point out that the Carnap of the *Logical Foundations* evades partly the thrust of Ramsey's attack by holding intuition responsible only for the general principles, and attempting to use these to define a suitably small class of logical measures. However, how intuition can tell one, in default of more particular information, that the probability of two mutually disjoint statements, granted the truth of some other statement or set thereof, is precisely the sum of their separate probabilities, I do not know.[2] But then, this law is "generally accepted in practically all modern theories of probability_1 [logical probability]" (Carnap, 1950, p. 285). To return to the question asked above, I should answer it by pointing out that both Keynes and Carnap were looking for a suitable theory of partial proof: probability theory, with its historically extended association with induction, was conveniently at hand, and formally provided a generalisation of the consequence relation.

So much for the genesis of this theory. Attempts by Carnap to mitigate the *embarras de richesses* represented by the existence of uncountably many probability measures, by invoking certain symmetry principles, have been unsuccessful to the extent that they still leave as many admissible measures as there are real numbers, and the programme has run into a veritable desert of sand, from which it is unlikely that it will ever emerge.

3. The Work of James Bernoulli

Mathematical probability commenced life as the principal application of the newly developed mathematics of permutations and combinations. The rather mundane origins of the subject are well known, consisting of a series of requests by an assiduous gambler, the Chevalier de Méré, to

Pascal for information about the correct computation of odds, and division of stakes, in simple games of chance. One of the problems posed by de Méré was discussed by Pascal in a letter to Fermat, and is as follows: de Méré claimed to have located an inconsistency in the fact that the odds on throwing a six in four throws of one die differ from those on throwing two sixes in twenty four throws of two dice. "This" wrote Pascal to Fermat, "is his 'great scandal' which makes him say loftily that the propositions are not constant and that arithmetic is self-contradictory".[3] de Méré was of course right that the odds differ and wrong in that he supposed this to indicate inconsistency.

Mathematically, the form of these early problems was as follows. A set of repeatable conditions (e.g. throwing m dice), when instantiated, yields one of a set S of n possible outcomes. An event A, extensionally, is an r membered subset of S. The odds on A are computed according to the formula:

$$\text{odds}\,(A) = \frac{r}{n-r} = \frac{r/n}{1-r/n}$$

and the quantity r/n was isolated as the fundamental notion of the new theory, and called the probability of A, or the chance of A.

The mathematical theory was thus very simple. At first, up to 1713 (actually, before: this is the date of the posthumous publication of the *Ars Conjectandi*) its development consisted in determining the probabilities of events defined relative to increasingly complex games of chance: the conditions of these problems, for example, that of the 'fair' division of stakes at an arbitrary termination of the game, or that of the probability of the gambler's ruin, guaranteed solubility, and the problems were essentially a test of the mathematician's expertise. The situation was pretty much one of puzzle-solving. No explicit interpretation was assigned to the probability function by any of the mathematicians who developed the theory until the publication of the *Ars Conjectandi* – I shall argue later than the 'equal ease' with which the outcomes must occur for them to be rendered equiprobable, refers not to any physical quality but is a defined term within the mathematical theory. The theory was widely used in computing odds in games of chance, but does not seem to have been granted more than a classificatory significance in defining a 'fair' game as one in which expected and actual winnings are approximately equal.

Bernoulli was the first of the mathematicians to consider using the theory in an explanatory role, precisely in relation to empirical statistics. Indeed, the work that he did within the mathematical theory made its interpretation, for the first time, a matter of some importance. Bernoulli proved (and it took him twenty years to do so) his 'golden theorem', the first of the celebrated limit theorems of the theory. In its conditional form, using the modern notation that did not appear until the late nineteenth century, the theorem is expressed as follows:

$$P\left(\left|\sum^{n} \frac{x_i}{n} - p\right| < \varepsilon \,\middle|\, S^n\right) \to 1$$

where S is the set of m possible outcomes of some stochastic experiment, A is a particular r membered subset of S, $P(A|S)=p=r/m$, S^n is the nth Cartesian product of S with itself, and $x_i=1$ if at the ith trial on S an outcome in A occurs – if not, $x_i=0$. ε is an arbitrarily small positive constant. Now in this theory the extreme numbers 1 and 0 have a special significance, for both clearly correspond to deductive certainty: if in the formula above the arrow could be replaced by the equality sign, then the event

$$\left|\sum^{n} \frac{x_i}{n} - p\right| < \varepsilon$$

would be logically certain given S^n; and the event

$$\left|\sum^{n} \frac{x_i}{n} - p\right| \geqslant \varepsilon$$

logically impossible. The theorem seems to make, in short, a qualified prediction: if n is permitted to increase without bound, the event

$$\left|\sum^{n} \frac{x_i}{n} - p\right| < \varepsilon$$

approaches arbitrarily close to being entailed by the description of the space S^n. This result extended the scope of the mathematical theory by apparently bringing it into a (quasi) logical relation with fact – statistical fact; though the fact, which now we denominate by such terms as 'the stability of statistical sequences', did not have its factual nature established until the experiments of Buffon and others, when it was shown to be a property of the sample means defined relative to repeated binomial trials, like the repeated tossing of a coin, that after a sufficient number of such trials these means were confined within a fixed interval whose end-points could be taken arbitrarily close to each other (note that the other familiar statistical fact – of convergence of binomial sample means – is not, even asymptotically, explained by this theorem). This result of Bernoulli's, in bringing the theory of probability into coincidence with empirical fact, even though only in a limiting sense, elevates him at once from the ranks of the ingenious puzzle-solvers like Huyghens and Montmort, to that of a creative developer of what is now the science of mathematical statistics.

Bernoulli's theorem made sense only for rational p, relative to a set S generated by some experiment or other, where the event whose probability is p can be decomposed into a finite number of equiprobable elementary events. But a mathematician of Bernoulli's stature would certainly have realised that the theorem was valid for all real p in the closed unit interval, being a consequence just of the additivity property of probability functions and the assumptions of probabilistic independence and constant probability from trial to trial. Clearly, these two last assumptions are, as it were, built into the early mathematical theory; for

$$P(x_1=x_1', x_2=x_2', \ldots, x_n=x_n' \mid S^n)= \frac{r^k(n-r)^{n-k}}{m^n}$$

where $x_i'=0$ or 1, $i=1, 2, \ldots, n$, and k values of the x_i' are 1.

$$=\left(\frac{r}{m}\right)^k\left(\frac{n-r}{m}\right)^{n-k}$$

$$=p^k(1-p)^{n-k}$$

$$=\prod_{i=1}^{n} P(x_i=x_i').$$

This result suggests the possibility of an explanatory theory of very much greater scope than the 'classical' theory based on ultimate equiprobable partitions with which Bernoulli worked; frequencies arising from the repetition of an experiment which were not, in the long run, approximately equal to the classical magnitudes r/n, were apparently inexplicable according to the classical theory. Yet they become explicable 'with high probability' if the assumptions of independence and constant probability are assumed to be satisfied, and to the appropriate event in the outcome space of the experiment is attributed an appropriate value to the parameter p. If, subsequently, the variance, for example, of the sample is found not to be explicable on these hypotheses, it is an obvious step to weaken the hypothesis of constant probability to the hypothesis that the probability varies from trial to trial; this was in fact Poisson's procedure on examining a set of birth statistics, and he proved a law of large numbers relative to this weaker assumption. Clearly, to take Bernoulli's theorem as a theorem about a magnitude p capable of assuming any value in $[0, 1]$ signals a striking departure from the contemporary theory towards a new mathematical theory whose axioms are those of what we refer to today as the probability calculus, and – it would seem – to the interpretation of the class of probability *functions* as representing a class of possible physical magnitudes characteristic of the experimental conditions being instantiated. The adoption of such an explanatory theory nevertheless had to await Laplace, where it appears very sketchily in the *Essai philosophique sur les probabilités*. It was not adopted by Bernoulli; and in fact Bernoulli thought his own theory would suffice in such an explanatory rôle.

Bernoulli's theory was, I claim, a theory of a weak consequence relation, measured by the classical ratios r/n. He defined, as is well known, the classical probability function as the measure of 'degree of certainty'. He wrote that

> if, for example, the whole and absolute certainty – which we designate by the letter *a* or the unity symbol 1 – is supposed to consist of five probabilities or parts, three of which stand for the existence or future existence of some event, the remaining two standing against its existence or future existence, this event is said to have 3/5 *a* or 3/5 certainty. (Bernoulli, 1713, p. 8)

A little later we come to the condition that the parts, or cases, be equally possible, or that they can all happen with equal ease. The phrases 'equally possible' and "with equal ease" are not glossed explicitly by Bernoulli, but they are familiar among the early writers and are invariably applied to cases whose degree of composition out of a given class of elementary events is the same; Galileo, who also used these terms, does give an explicit criterion which is nothing but 'having equal degree of composition':

Some [numbers in a dice game] are more easily and more frequently made than others, which depends on their being able to be made up with more variety of numbers. Thus a 3 and an 18, which are throws which can only be made in one way with 3 numbers... are more difficult to make than, for example, 6 or 7, which can be made in several ways.[4]

The outcome space S of an experiment or trial is clearly the set of all the elementary cases permitted by the description of the set-up; it is the set of all structurally distinct worlds satisfying a sentence from some appropriately simple language. Thus, if the set up consists of n balls in an urn, white and black in an unknown ratio, then the space of elementary outcomes or 'cases' is the set of the 2^n structurally distinct models satisfying the sentence

$$(\alpha) \qquad \bigwedge_{1 \leqslant i < j \leqslant n} a_i \neq a_j \wedge \forall x [\bigvee_{1 \leqslant i \leqslant n} x = a_i]$$

from the language characterised by one one-place predicate P and n individual constants $a_1, \ldots, a_n$. If to (α) is conjoined the sentence that r individuals have the property P then the proportion of the appropriately factored models satisfying each of the sentences Pa_j, $j = 1, \ldots, n$, is, of course, equal to the classical ratio r/n. The classical mathematical theory of probability is therefore interpretable as a theory of weak consequence, relative to premiss sets satisfiable in domains of at most n elements, constructed with a monadic language. Bolzano, in his (1837), gave formally the same definition to his notion of the satisfiability of a sentence relative to a set of sentences satisfiable only in finite domains, and later Sigwart was to call the classical mathematical theory of probability "nothing but a mathematically formulated chapter of logic".[5] It is clear, indeed, that by 'certainty' Bernoulli intended deductive and not subjective certainty; and his use of the contemporary mathematical theory of probability to define the notion 'degree of certainty', so understood, bears out Jevons's dictum that "in the study of probabilities mathematicians have unconsciously employed logical processes far in advance of the logicians" – in this instance, in the anticipation by Bernoulli of the semantic analysis of inference.

Of course, Bernoulli adumbrated his theory without the advantage of an adequate logical notation. This rendered most applications difficult if not impossible to evaluate, because of the inability of the pioneers to characterise the class of indivisible elementary cases satisfying the relevant data. A rule of thumb was used by Bernoulli and his great successor Laplace (who adopted Bernoulli's theory[6] and in whose analysis of what he called "the probability of causes" attempted to extend it to continuous spaces), which was that if one could discern no reason why any given event was no more favoured, so to speak, by some stock of information than any other within a definite class, then every event within that class has the same degree of certainty relative to that information. This rule, christened the Principle of Insufficient Reason by von Kries and the Principle of Indifference by Keynes, is notoriously inconsistent.[7]

Clearly, there are cases where such a rule of thumb does indeed become essential: namely, where it is in principle impossible to decompose the data or premisses of an inference into a finite set of elementary 'cases'. One of the earliest objections, made by Leibniz, to Bernoulli's theory was that the data might indeed admit an infinity of primitive possibilities. Bernoulli's reply to Leibniz was that having a quotient whose terms were both infinite did not of itself mean that the ratio was impossible to evaluate, because numerator and denominator may tend to infinities of different orders. But this is a specious point: the classical ratio is, after all, not a limit; it is a simple proportion, without ordinal significance. It might be countered that Bernoulli had merely failed to anticipate the notions of a random variable and (Lebesgue) measure; for if a variate may exhaust some finite interval $[a, d]$ then an intuitively natural extension of the classical theory is to take the probability that the variate is contained in the subinterval (b, c) to be the ratio $c-b/d-a$, $0<a<b\leqslant \leqslant c<d$. In general the extension of the mathematical theory to Euclidean space of n dimensions consists in defining the probability of an n-dimensional variate being contained in a Lebesgue-measurable set A, given that all its possible values fill up some measurable region B of finite measure, to be $L(A)/L(B)$. Called 'geometrical probability', this is how the theory with which Bernoulli worked did in fact develop (while the modern frequency theory due to Kolmogorov is an application of the theory of normed measures); among its recent advocates are numbered Emile Borel and Jerzy Neyman.

The trouble with this apparently obvious extension is that it is vulnerable to the 'paradoxes of geometrical probability' which caused Bertrand, to whom the paradoxes are due, to believe – I think rightly – that the theory could not be extended to infinite domains (i.e. it cannot be extended to the case where the data – or conditions defining a stochastic experiment – do not break down into a finite number of distinct cases, or, in a variant vocabulary, where their class of models is not finite). Bertrand considered the following problem: a chord of a circle is selected at random; required the probability that its length exceeds that of the side of the inscribed equilateral triangle. Bertrand, as is well known, found that it was possible to give more than one apparently correct solution to this problem. I shall follow Neyman's presentation of the problem (Neyman, 1952, pp. 15–17). We have to determine the probability that a chord randomly drawn to a circle of radius r should have a length ($2h$) in excess of a fixed quantity $2k$ ($2k<2r$). Now every chord of this circle is determined by a pair of coordinates (x, y) taken relative to the circle, where y is the perpendicular distance from the midpoint of the chord to the centre of the circle, and x the angle between a fixed radius and the straight line joining the midpoint of the chord with the centre of the circle. Now the probability that an arbitrary chord will have a length greater than $2k$ is obviously the probability that its y coordinate is less than $\sqrt[+]{r^2-k^2}$. The probability is then equal to the area of the rectangle of side lengths 2π, $\sqrt[+]{r^2-k^2}$, divided by the area of the rectangle of side lengths 2π, r i.e. $\sqrt[+]{1-(k/r)^2}$. But a given chord is also determined uniquely by the angles x and y between the fixed direction and the end-points of the chord. Here the variates x and y satisfy the conditions $(0\leqslant x<\pi)$, $(x\leqslant y\leqslant x+\pi)$ and $(\pi\leqslant x<2\pi)$, $(x\leqslant y<x+\pi)$. Clearly $2h>2k$ just when $r\sin((y-x)/2)>k$ i.e. when $y>n+2\sin^{-1}(k/r)$. The area of the parallelogram defined by $0\leqslant x\leqslant 2\pi$, $y>x+2\sin^{-1}(k/r)$ is $2\pi(\pi-2\sin^{-1}(k/r))$ and so the required probability is $1-(2/\pi)\sin^{-1}(k/r)$. These two solutions are clearly not identical. Formally, what has happened is that we have transformed our coordinate system by a non-linear transformation in passing from one coordinate representation of a chord to another. If we retain the letters x and y for our second set of coordinates, and call our first u and v, these two systems are related by the following transformation.

$$u=r\cos\left(\frac{y-x}{2}\right) \qquad 0<y-x<\pi$$

$$v=\frac{y+x}{2}$$

$$-u = r\cos\left(\frac{y-x}{2}\right) \qquad \pi \leqslant y-x < 2\pi$$

$$v = \frac{x+y}{2}$$

Putting it another way, we have selected two distinct two-dimensional random variables to represent the same phenomenon, which are related by a non-linear transformation. Now being uniformly distributed is not a property of variables that is invariant under non-linear transformations in general; and this means that when we try to extend the mathematical theory to possibility spaces that can be mapped into R^n in such a way that the space fills up some finite interval, we can do this in more than one way. Thus, for example, we may measure some empirical magnitude, like heights of individual human beings, on a standard (linear) scale, or on a logarithmic scale, and neither is 'more correct' than the other (though one may be more convenient). But the extended definition of probability will yield distinct results, depending on which scale is selected. Consequently, it seems impossible to measure the strength of the postulated logical relation between sentences concerning such a magnitude in which no scale is stipulated. But even if a scale *is* specified, we are still not out of the wood. For let α, α' be two sentences (or sets of sentences) which specify distinct non-linearly related scales of measurement for the same magnitude, β, β' a corresponding pair; then in general

$$P(\beta|\alpha) \neq P(\beta'|\alpha')$$

although $\vdash \alpha \leftrightarrow \alpha'$, $\vdash \beta \leftrightarrow \beta'$. The problem of infinite possibility spaces seems indeed, *pace* James Bernoulli, an intractable one for this theory.

Bernoulli nevertheless thought his theory, weak though it would seem to be, of practical value in the 'art of conjecture', and the reason he did so explains to some extent the fact that he declined to take the step I have indicated above of adopting for the explanation of statistical phenomena the non-classical theory of probability whose laws are those of the modern calculus of probabilities, and the hypothesis of independence. Bernoulli's great successor, Laplace, did as a matter of fact take this step, while also adopting Bernoulli's own theory to define and discriminate between degrees of certitude, relative to a body of information; indeed, Laplace was the first to regard the calculus of probabilities simultaneously as a theory to explain frequency data, and also as a theory of a logical relation (whose strength or intensity is measured by the classical ratios). Of course, the interpretation of the probability function in the former theory will be determined by the requirement that probabilities close to 1 or 0 can be understood as endowing the respective events with a tendency to necessary occurrence or necessary non-occurrence, given the conditions of the experiment. Bernoulli, as we shall see, chose always to interpret the necessity as being of a purely logical or formal character; Laplace, on the other hand, when using the calculus in its explanatory role, opted for a physical interpretation of this tendency and construed his generalised probability functions as measuring the strength of constant 'causes' present in the physical set-ups, anticipating the modern propensity theory of Sir Karl Popper. Using and extending some results obtained earlier by the English clergyman Thomas Bayes, Laplace defined within the 'degree of certainty' theory a conditional probability distribution for these 'causes', whose condition simply described frequencies observed in a sequence, of arbitrary length, of previous trials. This posterior distribution, from which can be obtained by integration over the range [0, 1] of variation of the 'cause', or propensity, variable the famous, or infamous, Rule of Succession, represents the first attempt to extend the logical theory to continuous possibility spaces. Laplace's use of two notions of probability, one explanatory of statistical phenomena, in which what he called the 'secret causes' of these phenomena[8] are measured by an appropriate real valued, non-negative, additive normed function defined on an algebra of events, the other in which 'degree of certainty' is defined as the classical ratio of favourable to possible cases, was endorsed by Poisson.[9] By this time the existence of almost constant long run relative frequencies generated by the repetition of a given type of stochastic experiment had become regarded as established fact: it was, wrote Poisson, "un fait général, résultant d'observations de toutes natures".[10]

Bernoulli thought that frequency data could be explained within his logical theory. This is a virtue not usually advanced as a feature of a logical interpretation of probability, though it is the case that according to Bernoulli's own theory, if an experiment has an outcome space *S*, of *m* elements, then

$$P\left(\left|\sum^{n}\frac{x_i}{n}-\frac{r}{m}\right|<\varepsilon \mid S^n\right)\to 1$$

where x_i takes the value 1 on the event whose (classical) probability is equal to r/m given S, and 0 elsewhere. Hence, it seems, we have only to determine the outcome space of the experiment to obtain a weak prediction about long-run frequencies, 'derived' from the description of that space. But this is clearly an unsound procedure; a biassed coin will appear with a relative frequency *ex hypothesi* outside a small interval $(\frac{1}{2}-\delta, \frac{1}{2}+\delta)$ after sufficiently many trials; yet the outcome space of the experiment is apparently just the pair $S=\{H, T\}$, giving, on Bernoulli's theory, the probability $P(H|S)=\frac{1}{2}$.

Bernoulli's response to the empirical phenomenon of biassedness away from the value of the long-run relative frequency predicted by his own theory was to deny that in such cases the apparent outcome space of the experiment was in fact the real space of irreducible elementary possibilities. Bernoulli considered the probability of death at various ages, and sought to explain the frequencies that would be recorded in a mortality table by postulating an occult possibility space, relative to each age considered, such that the outcomes 'life' and 'death' within an interval of the one year determined by that age are themselves merely phenotypes, as it were, determined by sets of genotypic possibilities, whose degree of composition out of the elementary possibilities would provide a good fit to the data available in the table. In other words Bernoulli proposed to use such frequency data to estimate the 'degree of certainty' of death in the first, second, etc. years after birth. Bernoulli proposed his 'hidden variable' theory in a famous letter to Leibniz (who pronounced himself sceptical of Bernoulli's claim to be able to estimate probabilities *ex datis*):

I place in an urn several hidden pebbles, black and white ones, and the number of white ones is twice the number of black ones; but you do not know this ratio, and you wish to determine it by experiment. And so, you draw one pebble out after another (replacing the pebble which you drew out in each single choice before you draw the next one, so that the number of pebbles in the urn is not diminished) and you note whether you have picked a black or a white one. Now, I claim (assuming that you have two estimates of the two-to-one ratio which are, though quite close to one another, different, one being larger, the other being smaller – say, 201:100 and 199:100) that I can determine scientifically the necessary number of observations so that with ten, a hundred, a thousand, etc. times more probability, the ratio of the number of drawings in which you choose a white pebble to the number of drawings in which you choose a black pebble will fall within, rather than outside of, the limits of the two-to-one ratio. But if now in place of the urn you substitute the human body of an old man or a young man, the human body which contains the tinder of sicknesses within itself as the urn contains pebbles, you can determine in the same way through observations how much nearer to death the one is than the other.

Each person, then, is represented by an urn; each age corresponds to a fixed composition of black and white balls (death and life) and the statistical data correspond to a series of drawings of one ball each from a large number of urns with the same constitution of black and white balls.

Over two hundred years later Emile Borel, the eminent French mathematician, wrote that

we can now formulate the general problem of mathematical statistics as follows: determine a system of drawings made from urns having a fixed composition, so that the results of a series of drawings, interpreted with the aid of coefficients conveniently selected, lead with a very great likelihood to a table which is identical with the table of observations (Borel, 1950, p. 138.).

The proposal to explain statistical results by reference to combinations of urns, using the data to fix values for the relevant parameters, is, of course, analogous to the Ptolemaic proposal to use combinations of uniform circular motions to explain planetary phenomena, again using the results of observations to evaluate parameters. Just as any planetary motion can be approximated with arbitrary accuracy by this method, so data like bills of mortality and birth statistics can be represented as the results of drawings from a suitable combination of urns. The urn, or system of urns, is a concrete model only, of course, though its structure is supposed to reflect the underlying structure of the physical experiment. Bernoulli's programme, then, consists in examining the conditions productive of the observable characters for which data have been collected and seeking at the fundamental level a finite set of possibilities the proportion of which are favourable to a given phenotypic character of an outcome is approximately equal to the probability of the outcome estimated from the gross data. What is rather surprising in view of the ad hoc character of Bernoulli's hypothesis, that with regard to each set of data recording the results of a series of stochastic experiments there is a space of primitive possibilities with the required property, is the fact that in the domains of genetics and physics such spaces have been physically identified (perhaps this is too bold a description of what has actually happened during research on the kinetic theory; but Mendel's postulated discrete factors of inheritance can, of course, actually be located as regions of chromosomes).

Formally, what Bernoulli has done is to take the report of an n-fold relative frequency ρA, of some event A where n is a suitably large integer, and to conjecture that there exists, relative to the experiment whose repetition is under consideration, a finite space S' of elementary outcomes, such that the event A is the phenotype of some event $A' \subseteq S$ and $\overline{\overline{A}}' / \overline{\overline{S}}' \simeq \rho A$: such a hypothesis explains with a probability suitably close to 1, not the actual sequence of A, $\bar{A}$ observed, but the relative frequency of A in the sample. But there is a grave defect in this explanatory scheme. It is incoherent. For the notion of a space of irreducible outcomes is language-relative; to say that a case is ultimate is to say that it is, to use Carnap's terminology, a Q-predicate in the particular (monadic) language being considered. Now within this language one may evaluate, as far as is possible, a set of non-deductive inferences by Bernoulli's methods, and use his theorem to explore the limiting characteristics of inferences of the form

$$\left|\sum^{n} \frac{x_i}{n} - p\right| < \varepsilon \mid S^n \quad .$$

But it is a conceptual confusion to regard the description of the space of outcomes as refuted (even in a weak probabilistic sense) by the observation of a sample mean falling outside a suitably small interval around the term p and to use this 'refutation' as a stimulus to discover a 'real' underlying space of elementary events. The 'predictions' given by Bernoulli's theorem are relative to a particular characterisation of the possibility space, and that characterisation is only weakly constrained by the 'real' features of the experiment. For example, one may consider the space of outcomes of a die-cast to be the set {even, odd}, or {one, not-one}: the richness of one's language determines the fineness of the characterisation. Relative to the characterisation chosen, inferences to relative frequencies are then evaluated. Observations discrepant with that class of inferences whose probability is highest show not that the characterisation adopted is false, or even probably false, but that Bernoulli's theorem considered as a rule of inference is unsound.

Bernoulli's enterprising attempt to demonstrate the redundancy of a theory of physical probabilities in the explanation of statistical phenomena fails, and with it the possibility of using this theory of logical probability as anything but a theory of a weak entailment relation. It is not my purpose to discuss other reasons for Bernoulli's failure to adopt a theory of physical probabilities; and, indeed, any such discussion is well in the realms of the speculative (though no doubt a belief in the completeness of Newtonian theory with respect to true descriptions of physical events played a part).

What, if any, then, are the practical uses of his theory? First of all, we know that it defines a system of coherent betting quotients, in the sense of de Finetti; but then so does every function satisfying the general axioms of probability, and it is difficult to see why the classical measure defines, where it itself is defined, a more rational system than any other measure. Second, it does not furnish a theory of inductive logic satisfying the sorts of requirements that Keynes and Carnap deemed necessary.

There are two reasons why this is the case. First, the class of distinct canonical models of most sets of sentences is infinite, and Bernoulli's theory is capable of evaluating only those inferences whose data include a finiteness postulate. Second, and more importantly, this theory of weak inference is essentially non-inductive. I shall explain what I mean by this statement with two examples.

Let us consider an urn containing n black and white balls in an unspecified proportion. A language to describe states of the urn might contain as its only extralogical constants n distinct individual terms a_i, $1 \leqslant i \leqslant n$, and one one-place predicate P. The set up is described by the sentence

$$(1) \qquad \bigwedge_{1 \leqslant i < j \leqslant n} (a_i \neq a_j) \wedge \forall x \Big(\bigvee_{1 \leqslant i \leqslant n} x = a_i \Big)$$

The number of distinct interpretations of the language satisfying both (1) and a sentence attributing P to r individuals and $\bar{P}$ to the remaining $n-r$ is exactly one. There are 2^n conjunctions of (1) and such a 'state description', and hence there are 2^n models of (1). Relative to the conjunction of (1) with any m-fold sample, $m < n$, the probability that a_{m+1} will have the property P is exactly $\frac{1}{2}$, independent of how many individuals in the sample had P. The reason for this is that the classical measure makes the random variables x_i, $i = 1, \ldots, n$, with n_i taking the value 1 on Pa_i and 0 on $\bar{P}a_i$, independent.

The second example concerns the case where our language lacks individual names. Here the set up is described by the sentence

$$(1') \qquad \exists x_1 \ldots x_n [\bigwedge_{i<j\leqslant n} (x_i \neq x_j) \wedge \forall x [\bigvee_{i\leqslant n} x = x_i]]$$

Now the conjunction of (1′) with the sentence S asserting that there are exactly $r<n$ instances of the predicate P has exactly one model up to isomorphism; hence there are $n+1$ distinct models of (1′). The 'sample sentence' asserting that r out of $m<n$ individuals have P has as many models as there are integer solutions of

$$x+y=n-m; \qquad x, y \geqslant 0$$

namely, $n-m+1$. Hence relative to (1′) and any m-fold sample asserting that there are just r out of m instances of P observed, the probability of at least one further instantiation of P is independent of r, and is equal to $(n-m)/n-m+1$, which tends to 1 as $n \to \infty$, for fixed m. The probability that all the remaining $n-m$ individuals will instantiate P is $1/(n-m+1)$ which tends to 0 as $n \to \infty$.

If, then, we take as our criterion of an 'inductive measure' one which is sensitive to the number of successes in the sample, the classical measure is not inductive.

4. Conclusion

The dream of Leibniz and Bernoulli, and of Keynes and Carnap, was to develop a theory of a weak consequence relation, which would simultaneously be a theory of rational credibility in the truth of conjectures, satisfying certain intuitively desirable criteria. Specifically, these celebrated figures were adherents to a programme of research intended to define a number-valued function, defined on a set of sentences closed under the usual logical operations, which would determine both a measure of the strength of a deductive relation between pairs of sentences and the odds, conditional upon the truth of one of the pair, appropriate to a wager on the truth of the other. To use Lakatos's terminology, the programme appears to have been characterised by a 'hard core' asserting the uniqueness (presumably up to the choice of a scaling coefficient) and identity of the measure of deductive strength and the credibility measure.

The subsequent history of the programme has certainly seen doubt cast upon the thesis that there is such a unique measure of credibility. Indeed, out of Ramsey's criticism of Keynes, and in particular out of his scepticism concerning the intuitability of the probability relations postulated by Keynes, grew, as he tells us, his own theory of partial belief. Ramsey, rather than taking as his starting point the thesis that there exists an everywhere defined, unique reasonable degree of belief, or credibility, which he, in my opinion correctly, regarded as untenable, commenced with the weaker thesis that individuals' expectations are constituted by a set of partial beliefs, differing in intensity. He suggested a method of measuring these intensities and demonstrated[11] that if one's *system* of partial beliefs so measured satisfied an intuitively desirable condition, then it could be characterised mathematically as a pair $(\mathfrak{A}, P)$ where $\mathfrak{A}$ is a subset of an algebra $\mathfrak{A}'$ of events, or statements describing them, and P is the restriction of a finitely additive probability measure P' on $\mathfrak{A}'$ to $\mathfrak{A}$. If one were forced to contemplate all the elements of $\mathfrak{A}'$, P would extend to some particular probability measure P''. de Finetti, working independently, obtained the same result.[12] It was subsequently shown that the class of all probability systems with finitely additive P-measure is exactly the class of admissible ('consistent', in Ramsey's terminology, 'coherent' in de Finetti's) systems of partial belief.

Ramsey's and de Finetti's work inaugurated a new programme from which have stemmed many interesting results purporting to demonstrate properties of these constrained systems of partial belief, as they adapt to the accumulation of new information. I say purporting, because a method is adopted of representing these changes of belief through time which is questionable, at the least. It consists in committing the individual to a set of conditional bets which involve conditions supposed to exhaust the range of his total possible experience. Just as the selection of a set of canonical absolute betting quotients is taken to represent his belief structure at a given time, so the conditional bets ostensibly represent the way his beliefs will be modified in the event of the conditions becoming actualised. Once this mode of representation is granted, it is not difficult to show that the same intuitively desirable property of his unconditional bets which determine their probabilistic structure, entails that his system of conditional bets will constitute a conditional probability system. An objection, which in my opinion is a considerable one, to this procedure of representing his changes of belief is that it involves, as I remarked, the specification within a fixed language of his total possible future experience, and it commits him for all subsequent times to the way at some initial time he considered this range of possibilities as bearing on the set of events upon whose occurrence he will bet. This seems to me, as it has done to others, unrealistic.[13] On the other hand, the theory of belief-statics, so to speak, i.e. the theory of an individual's partial beliefs at some fixed time, and its representation as a system of betting quotients, seems to me to be unexceptionable. The way this system becomes modified through time may then be obviously assessed by examining it at distinct times; no elegant mathematical theory is forthcoming, however, of belief-kinematics so construed.

The early attempt to define a measure of rational belief in conjectures has issued, then, in this theory which sets no constraints upon one's degree of belief in any particular proposition, only upon the global characteristics, as it were, of one's belief system. Whether this is the whole truth about 'rational credibility' or not, the old idea of a credibility measure supported by, or reflecting, a logical, deducibility relation between arbitrary pairs of sentences, which will be in any sense inductive, i.e. will yield some interesting ordering among conjectures corresponding to the strength of the logical relation holding between them and a set of observation reports, cannot, if we adopt the plausible theory of such a relation developed by Bernoulli, be upheld. For the measure of the strength of this relation is not, as we have seen, defined at all for most pairs of sentences, the cases where it is being of negligible significance in both practical affairs and science. Moreover, where it is defined, it is essentially non-inductive; in a sufficiently large universe, particular predictions are assigned a value practically independent of the constitution of the same evidence (in the case where our language contains names for all the individuals in the universe of discourse, the values are independent of all sample parameters) and the same holds for the general sentences also; as we have seen, the model for the probability assignments is that of random sampling with replacement.

We can conclude (I think) that the attempt to base the 'art of conjecture' upon a notion of weak deducibility has failed. It has failed not because there does not exist a well-defined and intuitively plausible notion of weak deducibility: I have attempted, in my discussion of Bernoulli's work, to argue that there does exist such a notion. The attempt failed because the theory of weak deducibility will simply not carry the burden assigned to it.

London School of Economics

NOTES

* Imre Lakatos was a friend, colleague and teacher: my debt to him, and, as far as this paper is concerned, to his (1968), is enormous.

1 Popper (1972), pp. 391–404.

2 The types of argument for the additivity and multiplication 'laws' exemplified in Cox (1962), Good (1948) Appendix 111, Lucas (1970) and Reichenbach (1949), pp. 65, 66, and those in de Finetti (1937) and Ramsey (1931), were unknown to Keynes, and apparently also to Carnap in 1950. Those of the first group indicate the very general conditions which determine the characteristic formulas of the probability calculus; de Finetti and Ramsey show that degrees of belief constrained by a fairly minimal rationality condition satisfy the same system.

3 Quoted by F. N. David in her (1962).

4 Quoted and translated from Galileo's fragment *Sopra le scoperte dei dadi* by F. N. David, *op. cit.*

5 Sigwart (1895), p. 219.

6 It is developed in Laplace's (1819).

7 I have argued elsewhere (Howson, 1975) that the contradictions issuing from the Principle of Indifference fall into two distinct classes: those due to the indeterminateness of the language relative to which the probability function is defined; and those ineradicable contradictions associated with non-linear transformations of variables defined on continuous possibility spaces (the so-called paradoxes of geometrical probability).

8 Laplace (1819), p. 15.

9 Poisson (1837).

10 *Op. cit.*, pp. 137, 138.

11 Ramsey (1931), 'Truth and Probability'.

12 de Finetti (1937), pp. 102–110.

13 Cf. Hacking (1967).

BIBLIOGRAPHY

Bernouilli, J., 1713, *Ars Conjectandi*, Part IV (transl. by Bing Sung), Harvard University Statistics Department, 1966.

Borel, E., 1965, *Elements of the Theory of Probability*. Translation of *Eléments de la théorie des probabilités*.

Carnap, R., 1950, *Logical Foundations of Probability*.

Cox, R. T., 1962, *The Algebra of Probable Inference*.

David, F. N., 1962, *Games, Gods and Gambling*.

de Finetti, B., 1937, *La Prévision: ses lois logiques, ses sources subjectives*. Translated into

English and reprinted in Kyburg and Smokler (eds.), *Studies in Subjective Probability*.
Good, I. J., 1949, *Probability and the Weighing of Evidence*.
Hacking, I., 1967, 'Slightly More Realistic Personal Probability', *Philosophy of Science* **34**, No. 4.
Howson, C., 1975, 'The Rule of Succession, Inductive Logic and Probability Logic', *The British Journal for the Philosophy of Science*, **26**.
Keynes, J. M., 1921, *A Treatise on Probability*.
Lakatos, I., 1968, 'Changes in the Problem of Inductive Logic' in Lakatos-Musgrave (eds.), *The Problem of Inductive Logic*.
Laplace, P. S. de, 1819, *Essai philosophique sur les probabilités*.
Lucas, J. R., 1970, *The Concept of Probability*.
Neyman, J., 1952, *Lectures and Conferences in Mathematical Statistics*.
Poisson, S. D., 1837, *Recherches sur la probabilité des jugements en matière civile et en matière criminelle*.
Reichenbach, H., 1949, *The Theory of Probability*.
Ramsey, F. P., 1931, *The Foundations of Mathematics*.
Sigwart, H. C. W., 1895, *Logic*.

KURT HÜBNER

DESCARTES' RULES OF IMPACT AND THEIR CRITICISM

An Example of the Structure of Processes in the History of Science

1. INTRODUCTION. THE DUBIOUS CRITICISM OF DESCARTES' RULES WITHIN THE FRAME OF CLASSICAL PHYSICS

Since Huygens it has been said that six of Descartes' rules of impact are wrong. This seems to be completely clear and we might close the files on that case. What happened here is obviously considered as the simple replacement of error by truth. Contrary to this unanimous opinion, however, we should regard this case as a particularly characteristic example of the intricate structure of processes in the history of science.

To show this let us first select two of the Cartesian rules of impact, the second and the fourth.

The second rule states: If two bodies, *A* and *B*, move towards each other with the same velocity and *A* is somewhat bigger than *B*, only *B* will recede after the collision and both bodies will move with the same velocity in the direction in which *A* was moving.[1]

The fourth rule of impact states: If *A* is bigger than *B* and does not move at all, *B* will never be able to put *A* into motion, no matter how high the velocity of *B* against *A* might be. On the contrary, *A* will repulse *B* in the opposite direction.[2]

Perhaps someone not trained in physics might still think the second rule of impact plausible. The fourth one, however, would certainly be rejected by everyone, since it contradicts the most simple experiences. Descartes is not bothered at all by this. In connection with the seventh rule of impact he says something which seems to be valid for him generally: "This does not need examination, since it is obvious by itself."[3] Boldly he plays off reason against experience, and this is done so provocatively that one should wonder how something like this could be possible at all. Indeed, it is surprising that nobody ever raised such an obvious question.

A modern physicist will of course not only refer to the games with billiard balls and marbles in order to disprove Descartes' rules. Already

R. S. Cohen et al. (eds.), Essays in Memory of Imre Lakatos, 299–310. *All Rights Reserved*
Copyright © 1976 by D. Reidel Publishing Company, Dordrecht–Holland

Huygens had to make use of all his sagacity and needed quite a big theoretic apparatus to falsify the Cartesian rules of impact. And in doing so he had to agree with Descartes that common experience is not such simple evidence as it may seem at first. So it may be doubtful indeed whether the rules of impact are manifest 'per se'. Nevertheless it is by no means obvious that they are already wrong 'per probationem'.

How would a modern physicist examine the two rules of impact mentioned?

Let us begin with the second rule. The first step is to transfer its premisses into the language of mathematics. '*A* bigger than *B*' is replaced by '$m_1 > m_2$', where m_i marks the inertial mass of one of the bodies. Furthermore '$u_2 = -u_1$' means that the velocities of both bodies are the same *before* the collision, but in opposite directions. If v_i are the velocities *after* the collision, then we can set up the following two axioms:

(1) $$m_1 u_1 + m_2 u_2 = m_1 v_1 + m_2 v_2$$

(2) $$u_1 + v_1 = u_2 + v_2 .$$

From this and from '$u_2 = -u_1$' we infer mathematically

(3) $$v_1 = \frac{(m_1 - 3m_2)}{(m_1 + m_2)} u_1$$

and

(4) $$v_2 = \frac{(3m_1 - m_2)}{(m_1 + m_2)} u_1 .$$

If we look at the numerators of the fractions, on the right-hand side of the Equations (3) and (4) we find three possibilities

(a) $m_1 > 3m_2$; (b) $m_1 = 3m_2$; (c) $m_1 < 3m_2$.

If we start from (a) and take u_1 as positive, v_1 must be positive as well in accordance with (3); m_1 therefore continues to move in the same direction after the collision. However, in accordance with (4) v_2 is also positive. m_2 is therefore repulsed in the direction in which m_1 moved. Both results fit in with Descartes' statements. But under condition (a) $v_2 > v_1$. For let it be supposed that $m_1 = 3m_2 + \delta$. If this equation is put into (3) we get

$$v_1 = \frac{3m_2 + \delta - 3m_2}{3m_2 + \delta + m_2} u_1 = \frac{\delta}{4m_2 + \delta} u_1$$

and if we put it into (4) we get

$$v_2 = \frac{9m_2 + 3\delta - 3m_2}{3m_2 + \delta + m_2} = \frac{6m_2 + 3\delta}{4m_2 + \delta} u_1 .$$

$v_1 > v_2$, however, is in contradiction to Descartes' second rule of impact, according to which both bodies should move with the same velocity after the collision. Analogously we could demonstrate that in the cases (b) and (c), too no correspondence with this rule is obtainable.

If we examine the premisses of the fourth rule of impact with the same axioms as those mentioned under (1) and (2), we will find out that contrary to Descartes' result the bigger motionless body will be pushed in the direction of the smaller moving one.

So according to what has been said already the physicist, in order to disprove Descartes, will not only point out the simple evidence of daily experiences, but will also make use of the axioms (1) and (2), which he believes to be the right ones. Everything else is just a simple logical conclusion from this and from the boundary conditions Descartes had set up (the premisses of his rules of impact). Descartes is thus falsified in a way that makes him look like a student, who fails to solve the examination question. He does not know to handle the rules of impact within the context of classical physics and therefore he does not obtain the results that were expected. To express it in another way: It seems as if Descartes could be beaten on his own territory. He should have known better! This opinion is reflected particularly clearly in the works of Huygens, who seems to have believed that he could refute Descartes with Descartes himself. We could also say: He believed he could explicate Descartes' system in the right way, while Descartes had failed to do this correctly.

2. Descartes' conception of force and velocity. Time and movement

It must first be pointed out that the critics of the Cartesian rules of impact define the premisses of these rules quite differently from Descartes himself. It is simply assumed that these premisses contain certain data about the momentum of bodies (i.e. about the product of their mass and velocity). Therefore the conservation law of momentum (axiom 1) seems to be applicable here. Descartes, however, never speaks of momentum. He speaks of something entirely different. His investigation of the rules of impact (*Princ.*, Pars Sec., XLII, 19) begins as follows: "Here we have to observe exactly what kind of force makes a body act or react to another one." And a little further on we read the following explanation: "On the one hand, this force has to be measured both by the size of the body in which it is and by the surface separating it from another one. On the other hand this force has to be measured by the velocity of the movement and also by the character and the contrary way the various bodies collide with each other." As we see Descartes does not use the term of the 'inertial mass'. Does he at least define velocity in the same way as classical physics does?

After having declared duration as an attribute *within things* (in rebus ipsis) he says: "Some of the attributes and modes are within things, the others, however, only exist in our thinking. If we separate time and duration for this reason and say that time is the number of movement, then this is only a certain mode of thinking."[4] Descartes gives the following reasons for this: "In order to measure the duration of everything we compare it with the duration of those largest and most regular movements which determine the course of years and days; and this duration we call 'time'. Generally speaking we can say that this way only a certain mode of thinking is added to duration."[5] Obviously duration is strictly distinguished from time measured. Duration is something 'within things'. Time only exists 'within our thinking'. Now what does Descartes' term 'velocity' refer to? Does it refer to 'duration' or 'time'? Does he mean something that is within things, i.e. a 'modus in rebus extensis'? Or is it something that only exists within our thinking, i.e. a 'modus cogitandi'?

Velocity, however, is not only connected with time or duration, it is also connected with movement. Let us see what he says about movement: "But if we consider what should be understood by movement, not according to the common use of the term but in accordance with the truth, in order to specify its essential nature, then we can state the following: *Movement is the transfer of one part of matter or of a body away from the vicinity of bodies that touch it directly but are considered motionless, towards the vicinity of other bodies.*"[6]

Furthermore Descartes writes: "Finally I have added that this cannot be a transfer away from the vicinity of any arbitrarily chosen kind of contiguous bodies. The transfer only takes place from the vicinity of *those bodies that are considered motionless*. The transfer itself, namely, is mutual. It is impossible to imagine that the body *AB* will be transferred away from the vicinity of the body *CD* without imagining that *CD* will simultaneously be transferred away from the body *AB*. And on both sides exactly the same force and action is required."[7]

Therefore movement is something relative for Descartes. Movement can be used only to refer to something that is *regarded* as motionless. However, we can also consider the moving body as motionless and, conversely, the motionless body as a moving one. Do we not have to conclude that Descartes thinks of this kind of 'movement' as a 'modus cogitandi'? Is it not necessary, therefore, to distinguish 'movement' as Descartes understands it here from another kind of movement, which is in *rebus ipsis*? A kind of movement which would not be determined by our more or less arbitrary measurement of time or our arbitrarily chosen systems of reference?

I think we have indeed to draw this conclusion. At least we have to do so, if part XXXVI of the second part of the *Principia* which comes directly after the theory of movement, is not to be entirely incomprehensible, as apparently all interpreters have thought till now. There we find a central part of Descartes' metaphysics. Since matter and extension are the same thing for him, they can only be put into motion with the help of God. And since God is perfect and will never change, he will keep the total amount of movement constant within the universe. How this happens was intended to be shown by the rules of impact. However, this constancy of movement guaranteed by God would become meaningless, if movement could be determined only relatively. In that case this constancy can never be achieved. To put it in modern terms: Descartes' rules of conservation cannot be true, if, for example, we start out from a rotating system of reference. However, if God is the origin of movement, then for him movement can never be relative. It will always be 'in rebus' for him. Only for us will it be a 'modus cogitandi'.

In this context I want to mention the following interesting passage in the *Principia*: "We recognize the perfection of God not only because he is unchangeable in himself, but also, because he only acts in the most constant and the most unchangeable way. This is done in such a determined way that we should never think of any change in his works except for those changes that are certainly observed by clear experience and divine revelation and which we understand or believe to happen without any change in the creator himself. Therefore we must not find any instability in God himself. So in absolute accordance with reason we have to conclude that God, having moved the parts of the matter in various ways, in the process of creation he will also conserve all matter in the same way and in the same relation as it was in the beginning. That means, he will also keep the same amount of movement within it."[8]

Descartes makes a difference here between changes that are shown by obvious experience (evidens experientia) and changes that are observed by divine revelation (divine revelatio). Being aware of his low opinion of experience, an opinion which is expressly confirmed in the context of his rules of impact, we cannot have any doubts as to what kind of changes he believes to be caused by God himself: Only the true changes take place under his influence, not the apparent ones. The apparent changes are presented by our senses and in any arbitrarily selected relativity as a 'modus cogitandi'; the true changes, however (in rebus), are caused by God himself and therefore presented by divine revelation.

3. Descartes' distinction between time and movement 'in rebus ipsis' and as 'modus cogitandi'

To sum up: The force which according to Descartes is effective in an impact has nothing to do with momentum as we understand it. This force does not refer to the inertial mass. Nor does it refer to a velocity, which depends on human time measurement and on a possible observation of bodies that are only moved relatively. *On the contrary, Descartes' rules of impact describe fundamental processes within nature as God sees them.* They refer to a duration and to a movement 'in rebus' or 'sub specie aeternitatis' and not as a human 'modus cogitandi'.

Consequently the contradiction which according to Koyrè and Mouy exists between Descartes' theory of the relativity of movement on the one hand and his divine conservation rules on the other hand ceases to exist. We can also understand now that Descartes did not commit an error which could be disproved by simple experience and which he certainly could not have overlooked. Furthermore we no longer have to consider his reference to the relativity of movement as cunning tactics, as Koyrè had assumed; tactics by means of which Descartes allegedly intended to conciliate the church with the astronomy of Copernicus and with the movement of the earth. Tactics like that would indeed have made Cartesian mechanics look contradictory and obscure. All these contradictions, difficulties, obscurities and far-fetched hypotheses, however, dissappear, if we accept the following interpretation so near at hand and presumably overlooked just for that very reason: Descartes' rules of impact do *not* refer to relative movement as a 'modus cogitandi', by which common experience is necessarily guided.

Correspondingly Descartes separates that part of his *Principia* which deals with the visible world (de mundo adspectabili) from the other part, which deals with the principles of substances (de principiis rerum materialium). The part on the visible world begins as follows: "We have found some principles of substances which are unquestionably true, since we did not take them from the prejudices of our senses but from the light of reason. Now we will have to examine whether we can explain all processes in nature by these principles alone."[9] Reason free of any doubts recognizes the invisible on which the visible is based. The invisible has to be the starting point for all our interpretations of the visible. Reason knows about the true foundations of our senses and knows that it is one with the light of divine revelation. We now understand why Descartes was not interested even in the clear observation of facts as he expresses it most provocatively in his fourth rule of impact.

4. The correct criticism of Descartes' rules concerns his judicial and normative principles

Thus criticism of Descartes must not consist of the assertion that his rules of impact are *wrong* insofar as we mean such rules which can be useful for human experience and which are based on the term 'momentum' = m.v. For Descartes never speaks of this. Therefore all we can accuse Descartes of is just this: that he never speaks of it; that he is only concerned with 'heavenly' instead of 'earthly' matters. We can rightly say that there is no legitimation for his exuberant and excessive rationalism which he gives himself up to. It is impossible to understand how the rules of impact can be 'clare et distincte' for our reasoning. As Huygens has already pointed out, these rules even contradict each other. Above all, however, we can refer to the following: We find two kinds of rationalism in the Cartesian system. The one kind claims to be effective technologically and practically, it even pretends to have built up the conditions for such an effectiveness. The other kind of rationalism is based on an extreme apotheosis of reason. Divorced from earthly things it seeks for pure theory and pure cognition of God. But both are firmly separated in Descartes' work by an unbridgeable gap. This creates a disharmony in his system which is confusing and unsatisfying.

The rationalism guaranteed by God and therefore directed towards divine truth becomes most evident in the rules of impact. The other kind of rationalism, however, directed towards practical use is expressed, for example, in the following quotation (*Discours de la Méthode*, part six), to which I confine myself here for the sake of brevity. "As soon as I had discovered some general basic terms of physics,... I thought that I was not allowed to conceal them without seriously violating the rule of furthering everything that contributes to the best advancement of mankind. My discoveries have told me that it is possible to acquire knowledge, which can be of great usefullness for our lives. Instead of that speculative philosophy which is thought in the schools, a practical philosophy could possibly be found which would give us exact knowledge about the force and the effect of fire, water, air, the stars, the matter of the universe and all other bodies everywhere in the same way as we know about the various techniques of our craftsmen. And therefore we could also apply this philosophy in the same way to all aims for which they are suitable and become the rulers and owners of nature."

This and other parts of his work express his almost emphatic demand for a science that mainly serves the technological and practical solution of problems. So we find an unbridgeable gap between this demand and Descartes' vision of divine mechanics which he develops – divorced from experience – in the chapter on the principles of substances.

At the beginning I said that the Cartesian rules of impact and the subsequent criticism of them offer an interesting example of the intricate structure of processes in the history of science. Now finally I want to explain this.

First we can say: The generally shared opinion of Huygens that he had disproved Descartes empirically is part of the old cliché of progress in science. It is normally assumed that a theory which replaces another one must be closer to the truth. However, historical reality differs from such an idea. In this context it is not important that the Cartesian rules of impact have certain immanent contradictions, which could also be removed. For the point *here* is that in any case it cannot be said in accordance with the cliché just mentioned that the progress achieved by Huygens was based on an *empirical falsification* of Descartes or on *the detection of new facts*. As it has been shown already, Descartes' rules defy all empirical falsification and they refer to something different from those of Huygens. Furthermore it can be shown in the following way that no detection of new facts was implicated in the progress made by Huygens.

(1) Huygens and his successors did not follow Descartes any longer, mainly because they rejected his decision to accept only those sentences as scientific that were "clare et distincte" in the light of reason. Thus Huygens expressly declared he did not agree with this "κριτήριον veri" of Descartes. On the contrary he maintains again and again that his own rules of impact "are in complete accordance with experience", while the rules of Descartes are "contrary to all experience."[10] Therefore the first change that takes place refers to a *judicial decision* as I call it: a decision about the criteria according to which theoretical statements should be accepted or refused.

(2) The same is true of decisions about the desired aims of science. These I want to call *normative decisions* for the sciences (which of course are closely connected with the judicial ones). As I have already shown, Descartes had two such scientific aims. On the one hand he wanted to discover the divine principles of construction, which he believed could be revealed by reason. On the other hand he wanted to contribute to the technological and practical use of science. He did not succeed in connecting both aims harmoniously. Now Huygens wants to construct his rules of impact in such a way that they can be realized empirically by experiment. His intention becomes evident when he determines first of all the relative system of reference, for which his rules are to be valid. And this intention is so natural for him that in his opinion Descartes seems to be almost incomprehensible, since he never thought of such a possibility. He no longer sees that Descartes could have wanted to achieve something entirely different. Descartes still follows the traditional aim of science to find the divine basis of the universe, only in a new way which is characterised by his mechanics. In Huygens work and later this aim does not vanish altogether. But it is practically hidden by the other aim mentioned,

or at least both are closely connected and not clearly separated from each other any more.

(3) Only the changed judicial and normative conditions lead to the setting up of new axioms about the conservation rules from which finally the special rules of impact can be deduced. In *this frame* they are correct and can be applied even today.

As we clearly see, Huygens did not reject Descartes because he had discovered new facts. The discovery of new facts followed the change of principles and not vice versa. These principles opened up a new way of interpretation, of asking questions and receiving answers. In short they opened up a new kind of knowledge.

Now we have to face the question as to why this mutation of principles happened at all. In my opinion the answer is given by the disharmony of the Cartesian system mentioned above. There the expectation was stimulated – in a very fascinating and suggestive way – that it would be possible to handle nature practically and empirically. As it turned out, however, this expectation could not be fulfilled within this system. The reason is that Descartes was still too much involved in the scholastic theological way of thinking (shown particularly clearly in his *Meditationes*). So on the one hand he set up rules of impact in order to form a basis for the science of physics. This new kind of physics, it was claimed, would be of the highest practical usefullness. On the other hand, Descartes deprived his rules of all empirical realisation and use, because he constructed them from a quasi divine point of view. Human time measurement and human determination of movement were not to be applicable to them. Of course this could not be satisfying. Theology and combinations of reason, divine revelation and mechanics look like an erratic within the Cartesian system. Pascal was quite right when he said that Descartes just wanted God to push the world and afterwards he was of no use to him any more.

So the progress achieved by Huygens consists of his ability to wipe out the fundamental inner disharmonies of the Cartesian system and *not* of falsifying it empirically or improving it on the basis of new empirical discoveries. The price paid for this progress was a system mutation. On the other hand this mutation was accomplished with the help of all those elements of the Cartesian philosophy which could still be used within the new frame. There they are interpreted in a different way and appear in a new light. One of those elements is what is referred to as the basic thought of mechanicism: The reduction of all material processes to pressure and impact, the principle of inertia first developed by Descartes, the Euclidean structure of space etc. The new system is built up with the estate left the bankruptcy of the Cartesian system, so to speak. The problems set up by Descartes, the disharmonies that could not be removed by him, all this is solved by using only what is left from what he himself had offered.

5. Conclusion. The Paradigmatic Structure of the Change from Descartes to Huygens

Elsewhere I have called the development of historical processes in science a self-movement of system-sets.[11] By this term I wanted to express that neither absolute statements on facts (that means statement which are not given in the frame of a system) nor allegedly absolute valid principles cause the development of science. This development is caused rather by the attempt to remove disharmonies and instabilities within historically given systems. On the other hand substantial parts of these systems will be conserved during this process.

The change from Descartes to Huygens is a good example. We see that new theories are not caused by new experience. On the contrary, new experiences are caused by new theories, which are developed by using parts of the old ones. They give us a new horizon of experience, in other words, they give new conditions for making experiences possible. The example of Descartes and Huygens also supports another thesis of mine expounded in the same article: The attempt to remove disharmonies and instabilities within given systems (self-movement of a system-set) takes place through the adjustment of one part of the whole context of systems to another part that does not fit in with it. And wherever this harmonisation happens we can observe progress by mutation in the judicial, normative and axiomatic principles of a system. As far as Descartes is concerned, this would mean that his rules of impact become coordinated with the aim of his normative principles, which demand the empirical and practical usefulness of science.

Philosophisches Seminar
der Universität Kiel

NOTES

[1] Descartes, *Principia Philosophiae*, Pars secunda, XLVII, *Oeuvres*, Bd. VIII.
[2] *op. cit.*, XLIX.
[3] *op. cit.*, LII. There we read: Nec ista egent probatione, quia per se sunt manifesta.
[4] *op. cit.*, Pars prima, LVII.
[5] *op. cit.*, LVII.
[6] *op. cit.*, Pars secunda, XXV.
[7] *op. cit.*, XXIX.
[8] *op. cit.*, XXXVI.
[9] *op. cit.*, Pars tertia, I.
[10] quoted from Mouy, *Le Dévelopment de la Physique Cartésienne*, Paris 1934, p. 193.
[11] Hübner, K., 'On the Question of Relativism and Progress in Science', *Man and World* **7** (1974), No. 4, Martinus Nijhoff, The Hague. Concerning the terms 'judicial and normative principles'. Cf. also my article: 'Duhems historistische Wissenschaftstheorie und ihre Weiterentwicklung', in 9. Deutscher Kongress für Philosophie, Meisenheim 1969, and in *Philosophia Naturalis* **13** (1971).

I. C. JARVIE

TOULMIN AND THE RATIONALITY OF SCIENCE

The last paper I heard Imre Lakatos deliver[1] was a scathing critique of Stephen Toulmin's 1972 book *Human Understanding*. Shrewdly spotting the Wittgensteinianism of its emphasis on context, practice, and the proliferated polymorphism of its account of rationality, he predicted that the massive three-volume project would never be – could not in the nature of the case – be finished. He suggested that since the history and practice of science are, according to Toulmin, infinitely various and diverse, cases can be endlessly explored, examples must endlessly proliferate, and all descriptions would have to end with, 'and so on', 'etc.', or '...'. For Toulmin, Imre suggested, a full stop had become a category mistake.

As always, when he was on target, Imre could sum it up brilliantly and throw in a joke for good measure. Attempting, as I then was, to fathom the mystery of Toulmin's ideas, and whether they ought to be taken seriously, his example was stimulating. I hope he would have liked the resulting paper. I dedicate it to his memory.

1. Introduction to the Problems

The rationality of science is a vague idea, which fact may explain why it is so much discussed. If to be rational means no more than trying to apply reason to things, then a wide range of human action can lay claim to rationality. Above all, perhaps, intellectual pursuits, where minds reason together. Are all claimants equally deserving, or are some more rational than others? Traditionally, two main candidates have found themselves nominated for the position of the acme of all enquiry and hence of rationality: mathematics and science.[2] If these claimants are deserving, on what does their case rest? Alas, there are competing characterizations of what makes each of these subjects rational: e.g. Socratic dialectic; a priori axioms developed deductively; unprejudiced intuition; induction from experience. It was the brilliant achievement of Imre Lakatos to argue plausibly that the rationality of mathematics exemplified

R. S. Cohen et al. (eds.), Essays in Memory of Imre Lakatos, 311–334. *All Rights Reserved*
Copyright © 1976 by D. Reidel Publishing Company, Dordrecht-Holland

Socratic dialectic.[3] This fitted together beautifully with the theory of his colleague, Sir Karl Popper, that the fundamental rationality of science also was Socratic dialectic. If they are both right,[4] then mathematics and science need not be rival models of rationality, and other attempts to characterize them would have a stiff challenge to meet.

Why does the tradition pick out mathematics and science to exemplify rationality – rather than, say, music, literary criticism, history, philosophy, bureaucracy, law or what not? To a certain extent this is due to other traditions: the Greek (and possibly immemorial) distinction between the reasonable and passionate sides of human nature, the intellect and the emotions, knowledge and art. Knowledge was to be gained by the use of reason (i.e. thinking), and the laws of reason were embodied in deductive logic.[5] Even if we have revised our views about the connection between logic and thought,[6] obedience to the basic rules of logic – such as that contradictions are to be abjured – has continued to be regarded as a necessary condition of rational discourse. Observance of the rules of logic, and of associated ideals such as simplicity and rigour, is thought to be most highly developed in mathematics and science.

An obvious objection to any exclusive laying claim to rationality on behalf of mathematics and science is that if rationality is the application of reason to a task, and a rational attitude is an open-minded one, then such things as the creation of art could be seen as exemplifying these qualities. Artists may not follow logic, but that does not empty their actions of rationality. To this, I think it would be said that art does not yield progressively more power over nature. In vulgar terms, the *success*, both cognitive and practical, of mathematics and science, sets them in some people's eyes apart from and above other intellectual endeavours. Whether their obedience to logic is what renders them progressive and powerful is moot.

As though perversely to add to our confusion, Stephen Toulmin argued some time ago in his book *The Uses of Argument* (1958) that jurisprudence deserved to be a candidate exemplar of rationality and that its rationality had nothing to do with obedience to formal logic. Jurisprudence is progressive and powerful, so the example was a striking one. To capture and characterize the rationality embodied in jurisprudence, Toulmin argued, required a new informal logic, flexible, able to incorporate the diverse modes of argument found in a living discipline. Just as it would be difficult to give any neat, rounded-off, systematic general characterization of the rationality of jurisprudence using formal logic, he went on, perhaps we needed to re-think the whole association of logic and rationality in general. Rational behaviour is not only goal-directed, but also rule-following; but the rules are many, diverse, at different levels of generality and applicability and discourse, and do not contain any specifiable, universally essential, timeless, ingredients. Indeed, as disciplines innovate, they will employ new ways of arguing; hence logic will have to innovate too.

This, at any rate, is what I made of his curiously blurred and unspecific theme. Rational assessment is possible, but only within the fields where the arguments do their work. A general logic is useless. We can isolate and describe arguments in each field, and examine how they do their job, how they function, but we cannot explain their working by appealing to principles of argument in general. Fields advance by thinking up new and better ways of arguing, methods

> not further justifiable – at any rate not by argument: the fact that they have established themselves in practice may have to be enough for us. (In these cases the propriety of our intellectual methods will be what the late R. G. Collingwood called an "absolute presupposition".) (pp. 257–8)

Surprisingly, Toulmin's first major foray into the issue of rationality caused nary a flutter: I find it received only four reviews in major philosophical journals, and three were severely negative.[7] The author was unfazed. In the 1963 preface to the paperback edition he reports that "students of jurisprudence, the physical sciences and psychology, among others", whose interest in reasoning has some practical starting point, have warmly welcomed the book. Less than ten years later we find that in *Human Understanding* he is prepared to carry his programme further and argue that logic does not even capture the necessary conditions[8] for the rationality of science. His explanation of why logic has for so long been thought to be central to rationality is that philosophers were over-impressed by geometry and geometry as it so happens could be captured in formally set-out arguments.

My aim in criticising Toulmin's views is partly to correct his errors and partly to get clear my own submissions on the topics of rationality in general and the rationality of science in particular. Toulmin, I fear, in elbowing aside logic, succeeds in undermining both the rationality of science and the possibility of discriminating rationality in general. He makes it possible for astrologers and witch-hunters to claim rational equality with science.[9] This is not what he intends. He is grappling with two problems which are indeed daunting and seem to require drastic measures: scepticism and relativism. To defeat them Toulmin wants to find a way to justify our claims to knowledge and our value judgements. In my opinion justificationism is a strategy for defeat.[10] The interest of Toulmin's progress from *The Uses of Argument* to *Human Understanding* is that it constitutes a *reductio ad absurdum* of the programme of justificationist rationality, as I shall try to show in what follows. My plan is to consider the arguments of both books separately, (Sections 2 and 3), before going on to criticise them and articulate my own views (Sections 4, 5 and 6).

2. Toulmin on Logic and Rationality in General

At the heart of the notion of rationality is the idea of a rational argument, usually characterized as one which obeys the rules of deductive logic. Some philosophers maintain that inductive arguments are constantly and successfully used in everyday life, indeed they are *the* way we learn from experience. The fact that the concept of validity has to be stretched[11] if it is to incorporate these important everyday arguments suggests to those philosophers its inadequacy for appraising arguments in everyday life. Hence standard logic, which centres on the notion of validity, is declared to be inadequate to the tasks of everyday life. It is too strong, too austere, rules out too much.

A conclusion of this sort underlies *The Uses of Argument.* Toulmin offers two criticisms of the standard logical concept of validity. One is 'analyticity': if the premisses of an argument validly entail the conclusion (i.e. no counter-example exists), then the inference can be rewritten as a tautological (analytically true) conditional statement. The other is universality: for an argument form to be valid in *any* field it must be valid in *every* field. Thus valid argument forms have to be both very trivial and very general. These conditions of analyticity and universality Toulmin finds too strong to be met by perfectly good everyday arguments deployed in law, science, morals, etc. To evaluate these arguments, criteria are required which are substantial rather than analytic, field-dependent rather than universal (or field-invariant).[12] The problem is that standard validity excludes many intuitively acceptable inferences for good reasons, and these Toulmin nowhere discusses. If validity is to be stretched to cover hitherto prohibited 'inferences', we are entitled to enquire as to the disadvantages as well as the advantages. Let us look at Toulmin's reasoning.

After asking what logic is, Toulmin asserts that:

> Logic is concerned with the soundness of the claims we make – with the solidity of the grounds we produce to support them – or, to change the metaphor, with the sort of *case* we present in defence of our claims. (p. 7)

This is a clear enough statement of what, following Bartley[13] I would call logic as justification. The task of logic, the point of drawing inferences, is to justify claims. Towards the end of his book (pp. 219–223) Toulmin suggests that logic, because it is universal and analytic (that is, insists that premisses entail conclusions), sets impossibly high standards for justification. He argues that, e.g. astronomical predictions do not pass the test of 'idealised' logic because their theoretical premisses can only be based on evidence drawn from the past and present, while the predictions refer to the future. Historical claims will not pass muster because they involve inference from presently accessible evidence to inaccessible past events. General claims, psychological claims, moral claims founder in similar ways. In the end, even the evidence of our senses can be doubted, because the logician can always deny the inference from the way things look and sound, to the way things actually *are*. Such is Toulmin's *reductio ad absurdum* argument on logic's claims to be a standard of justification. The demand of logic that premisses entail conclusion is too strong to be met outside of mathematics. If we adopt logic as our standard of justification, we cannot escape scepticism. Thus it transpires that Toulmin's 'working logic' is ultimately part of a programme to solve the important traditional problem of defeating the sceptic. The sceptic challenges us by asking of any claim such questions, as "how do you know?", "what are you basing your claim upon?" We can only silence him it seems, if we succeed in *justifying* some of our claims. But what are our standards of justificatory argument: when *is* a claim justified? The standards of argument philosophers usually invoke are the rules of logic – logical standards. Toulmin thinks he has shown that if we use the standards of logic none of our claims to knowledge can meet them. From the facts and theories of science down to claims about direct experience, all is uncertain:

If we are going to hold out for analyticity, therefore, we shall find a general problem arising over all fields of argument other than analytic ones. Claims to knowledge, however well-founded they may appear in practice, are never going to come up to the philosopher's ideal standard. Once we have accepted this ideal, there seems no hope of salvaging our everyday claims to knowledge – pure mathematics apart – without resorting to philosophical rescue-work of a drastic kind. (p. 223)

Toulmin goes on to show that various attempts at such philosophical rescue work cannot succeed. He then suggests his own solution, which is that if logical standards are knocked off their 'field-invariant' pedestal (abandoning analyticity and universality), and 'field-dependent' standards are substituted, the problem does not arise. Instead of standards being the idealised abstractions of the logician (analytic), they stem from the actual practices of experts in the field (substantial), be it law, biology, or demography.[14]

3. The Problem of the Rationality of Science in Particular

Volume one of *Human Understanding* is a "critique of collective reason", part of a "comprehensive reappraisal of our working ideas about rationality" made necessary by philosophy's unbalanced concentration on logical form rather than rational function or intellectual adaptation. It is not clear whether Toulmin wants to break up philosophy's love affair with geometry and physics but certainly,

in science and philosophy alike, an exclusive preoccupation with logical systematicity has been destructive of both historical understanding and rational criticism. Men demonstrate their rationality, not by ordering their concepts and beliefs in tidy formal structures, but by their preparedness to respond to novel situations with open minds – acknowledging the shortcomings of their former procedures and moving beyond them. Here again, the key notions are 'adaptation' and 'demand', rather than 'form' or 'validity'. (vii–viii)

Openly declaring his aim to be, to be more historical, more empirical and more pragmatic than Plato, Toulmin sets out on a five-hundred-page-plus journey, in the course of which we learn about the intellectual ecology of man. Science, for Toulmin, is a body of collectively evolved and sustained concepts which are there to further our inquiries into the way the world is and how we can manipulate it.

The discussion proper begins with a consideration of the problem created by the variety of human knowledge claims, and the lurking danger of relativism. Toulmin is not prepared to see science as simply the knowledge claims of a few European countries and their heirs, with no more claim on anyone's attention than the witchcraft, flat-earthism, or reincarnation theories of other cultures. Science is not just the rational intellectual enterprise par excellence of our society, it is also the pre-eminent one on a cross-cultural basis. But how can this position be defended from sceptical attack? He looks at the arguments of Collingwood about absolute presuppositions and of Kuhn about paradigms, and finds they are both helpless before relativism. Neither can convincingly account for fundamental changes in the rational enterprise of science. It is, Toulmin suggests, in change that we show our rationality; and hence a theory of the rationality of science will have to be a theory of scientific change.

To help him get at such a theory Toulmin introduces a distinction (pp. 123–4) between theoretical principles (e.g. universal gravitation); and disciplinary principles, which define the basic intellectual goals of a science (e.g. explain all physiological functions in chemical terms). As long as these defining goals are agreed, Toulmin argues, actual theories can be very far apart and yet still part of a continuous subject matter, namely, theories that address themselves to the same disciplinary tasks (or problems?). There will not, however, be a neat system of propositions which can organize the many current concepts of a discipline into a systematic whole. Fragments of a discipline, or particular theoretical sections may be so systematisable, but not the whole thing. Rather as in society a particular institution may have a quite systematic organization, but different institutions may have no clear connections, so in a discipline.

> If the component elements of an entire society – or an entire science – are... tightly interrelated... there will then be no way of modifying them piecemeal or one at a time: the only chance of radical change will lie in rejecting the entire 'system' as a whole, and starting afresh (p. 129)

In place of such over-systematic, discontinuous, revolutionary procedures Toulmin wants to substitute evolutionary ones which recognize that concepts (like institutions) are introduced not together and for one purpose, but from time to time and for differing purposes. Moreover, if they survive today it is because they still serve their original purpose or because they fulfill a new one. Either way, the presumption is in their favour until 'legitimate scientific occasions' lead us to think otherwise.

Much of the rest of the book is spent trying to show that evolution is not a metaphor, but a process actually taking place in the phenomena of rational endeavour. Man, as it were, confronts nature, and sets himself certain tasks. Sometimes he finds his understanding insufficient to carry out these tasks and this stimulates him to reconsider, modify or replace them. The multiplicity of men and the multiplicity of the tasks they set themselves, guarantees that the concepts and theories they develop will be diverse. A discipline comes into being when some very general disciplinary ideal is set up. The problems then generated are a measure of the gap between the ideals and current capabilities:

'Problems = Explanatory Ideals – Current Capabilities'.

Actual disciplines develop partly under pressure from internal, intrinsic or intellectual factors, and partly from external, extrinsic or social factors. Toulmin contrasts the development of science in China and Ancient Greece. China has no physics because it never had a Galileo; it never had a Galileo because it never had a Euclid. It never had a Euclid because the arrangement of social institutions was inadequate to the development of an autonomous discipline of natural philosophy and an autonomous class of natural philosophers. The classical Greeks, however, had the 'intellectual self-confidence' to tolerate these developments. Hence social factors affect the occasions and incentives for intellectual innovation; the scientist's own judgement selects the 'ripe' field for it. They will then offer conceptual variants which can only be judged informally against the disciplinary ideals as to whether they "do a better explanatory job" (p. 226).

4. Criticism to Toulmin on Logic and Rationality in General

Toulmin holds that the category of rational argument is broader than the category of logical argument. No doubt it is. That there are arguments in use which logic has not captured no one denies. What special ones does Toulmin have in mind? About the only examples of argument that Toulmin bothers to give are these: from the fact that Petersen is a Swede and the further fact that most Swedes are not Roman Catholics, one can conclude that almost certainly Petersen is not a Roman Catholic; and, that because Harry was born in Bermuda, and because, with certain exceptions, everyone born in Bermuda is a British subject, Harry is a British subject.[15] In the third chapter of *The Uses of Argument*, Toulmin re-analyzes such arguments to distinguish a *claim* (C) from the *data* (D) used to back or justify or establish the claim. The move from data to claim will be allowed by the warrant (W).

Toulmin's notion of warrants is possibly the most contentious logical idea in the book, and one which lies at the heart of several other mistaken views. As I understand it, Lewis Carroll, in his Tortoise and Achilles paper,[16] showed that if someone fails to see how a conclusion follows from some premisses, he cannot be convinced by the addition of further super-major premisses warranting the inferential move. Toulmin thinks that the story shows the doubting Tortoise fails to grasp the *backing* of the warrant; whereas, in fact, the quarrel is a meta-linguistic one about inference-rules.[17] The problem is that a general rule of inference cannot be discussed within a particular argument which employs that rule. The Tortoise needs to be shown that one can't accept the premisses and the rule of inference and deny the conclusion *consistently*. In the inference 'Harry's hair is red, therefore Harry's hair is not black', Toulmin identifies the warrant authorising the step of this inference as:

(W) If anything is red it is not black.

Carroll's point is trivialised if it is taken that the doubting Tortoise does not grasp the backing of (W). Carroll even allows him to say he agrees to any such premiss. It is the *move*, the step, the rule of inference the Tortoise fails to grasp. (W) is not, of course, a rule of inference, but a suppressed premiss. Even with this premiss, the inference:

If anything is red it is not black
Harry's hair is red
———————————————
Harry's hair is not black

does not state the rule of inference which authorises the move or step. This rule is the rule of *modus ponendo ponens*. I conclude that, like the

Tortoise, Toulmin does not grasp the consequences of abandoning the rule, and hence the reasons we adopt it in the first place.

The confusion of rules of inference with hypothetical major premisses is a common error and a disaster.[18] There are several reasons why a supermajor premiss should not be confused with a rule of inference. A trivially straightforward reason would be that rules are in the meta-language and the premisses of the inference are in the object-language. Toulmin seems unaware of this distinction. But there are less artificial reasons: logical, methodological and philosophical. A logical reason is the stipulation that all rules of inference must be demonstrable (provable tautologies) and while that could arguably be true of, 'If anything is red, it will not also be black', it can hardly be true of, 'A man born in Bermuda will be a British subject'. Toulmin says, so much the worse for logic. Another logical reason is this: if what warrants an inference is a conditional with the premiss of the desired inference as antecedent and the conclusion as consequent, then all inferences can be rendered formally valid (if I read him correctly, Toulmin himself draws this conclusion in the middle of p. 119 as an argument against standard validity). If all inferences can be made formally valid, validity is emptied of its content and the distinction between good and bad arguments disappears.

The methodological reason is that making theories, for example, inference tickets, renders them irrefutable (they warrant inferences, they do not *say* anything), mere calculating devices, and this is instrumentalism. There are standard objections to instrumentalism, but Toulmin embraces it nonetheless.

The philosophical reason turns on the simple question of what a warrant is, raised earlier. Any conditional statement whatsoever can function as a warrant on Toulmin's definition, as a statement legitimising, as he says, a step from (D) to (C). This is absurd; all putative inferences can be validated in this way. Warrants can even, Toulmin says, have different force depending on their backing. Conclusions necessarily follow, or are merely possible, or impossible. This breathtaking claim ignores the fact that, even in inductive logic, no-one has succeeded in producing a coherent programme along these lines.

All this has further significance in Toulmin's scheme. His argument about Petersen being almost certainly a Roman Catholic does not obey any rule of inference at all, nor could it, since in this argument true premisses can go together with a false conclusion so truth is not transmitted. No rule of inference exists which makes this a valid inference or a good or sound pattern of argument in any field. The conclusion in no sense 'follows' from the premisses. All Toulmin's warrants and backing ('A Swede can be taken to be almost certainly not a Roman Catholic', 'The proportion of Roman Catholic Swedes is less than 2%'), are simply additional data marshalled in varying ways to convince the sceptic. But additional data rarely convince anyone: doubts are not drowned by data. The real difference between warrant-data, and backing-data, is simply that one lot is supposed to answer scepticism about the inference to the claim, the other, scepticism about the premisses of the inference. Warrants, Toulmin says, are implicit, general and sometimes qualified. Clearly they differ from rules of inference as I have outlined them, since these are explicit, universal and cannot be qualified: *there are no degrees of validity*. Inference rules *must* be a field-invariant, otherwise their claim to transmit truth and retransmit falsity under all substitutions of their terms would not be upheld. Toulmin's argument against this notion (pp. 119–20) is as follows: there are arguments we should all accept which do not involve reshuffling of the information in their premisses and moreover, all arguments can be rendered formally valid by formulating them with their warrants built-in as premisses, therefore validity cannot be a matter of formal properties!

That what Toulmin calls 'warrants' are a confusion of object-linguistic premisses and meta-linguistic rules of inference is here revealed most sharply.

Toulmin also wants a distinction between warrant-using and warrant-establishing arguments. Science, he explains, is concerned with the latter. Warrants now get confused with scientific theories. A challenge to a warrant is seen as an attack on the data used to establish it. This view goes back to Mill, who put forward the idea of scientific theories as means of inferring from particulars to particulars. Ryle took up the view, as have many conscious instrumentalists. But when I deny some scientific warrant, to use Toulmin's terminology for a moment, like 'friction generates heat', I am not at all attacking the *data* backing this warrant. Indeed, I may not doubt for a moment that those data are sound. What I am doing is producing a new piece of data: a counter-example which shows the warrant or theory to be *false*. This, of course, leads into the basic criticism of instrumentalism which is that many theories known to be false do the job of 'warranting inferences' or producing calculations as well as and sometimes more easily than other, rival theories not yet shown to be false. What rationale is there then for preferring theories not yet shown to be false to those known to be false? The instrumentalist has no answer. So change and progress have to go with *success*.

One could attack all this in another way. If Toulmin is going to loosen validity, to extend that honorific to arguments logicians would reject out of hand, he must at least give us criteria to decide when an argument is not valid.[19] A theory of inference must at least distinguish good inferences from bad. Perhaps this is the purpose of his discussion aimed at loosening the notion of analytic argument, the analytic syllogism. But the new criterion of analyticity turns out only to be a classification (p. 133), which does not provide a means of deciding when an argument is valid and why. Toulmin nowhere gives us contrasting examples from science or law of *real* (as opposed to ridiculous) arguments he considers valid and invalid and then explain to us what the difference is. He seems prepared to leave it to the users. If the scientists use it, then it must be OK; if not, not. If lawyers make those inferences, that is good enough for him. If it is good enough for common sense, it should be good enough for us. The real is rational, perhaps.

A reason that Toulmin neglects real cases of argument in particular fields may well be that on closer examination it turns out that there *are no* alternative logical standards. What evidence is there that lawyers or scientists use patterns of argument that do not pass logical muster? None that I can find, but evidence that they employ logic abounds. When a lawyer cross-examines a witness, a major aim is to reveal *inconsistencies* in his testimony; hearsay is inadmissible because facts at one remove can get distorted, i.e. *inconsistent* accounts of what happened may proliferate; in summing-up the judge seeks to explain the meaning of the law and how the case comes *under* it, how the law *applies*; in briefs, judgements, opinions, and so on, chains of reasoning are used to ensure *consistency* with precedent, etc. Scientists too are much concerned with *deducing* applications from a new theory to see if they are *consistent* with one another, with the corollaries of other theories and with statements of what is currently thought to be fact. Refutation of theories and crucial experiments proceed entirely in accord with standard logic. These sorts of examples flood to mind. Toulmin needs to make a case that there is sound reasoning going on in these fields that logic cannot account for. The possibility cannot be ruled out; but it hardly goes without saying.

Underlying Toulmin's book is the conflict between logic and the justification of claims. Logicians often say that nothing but the tautologies of logic and mathematics can be proved. They agree with Toulmin that the only claims which meet logical standards of proof are vacuous tautological statements that are true solely because of their logical structure. No substantive argument can ever have a proved conclusion. Toulmin's book shows that if the words 'conclusively verified' or 'justified' are substituted for 'proved' the result is the same. Here Toulmin has exposed a genuine dilemma of justificationism. Between the position of scepticism, which he recognises and rejects, and the position of justificationism, which he holds but does not characterise, there exists at least one other possibility, and that is non-justificationism; or, in other words, an epistemology which treats all our knowledge claims (even those of logic and mathematics) as tentative, hypothetical, or conjectural.[20] Before I expand on that, I want to criticise Toulmin's views on the rationality of science.

5. Criticism of Toulmin on the Rationality of Science

Despite its display of learning and its fluency, *Human Understanding* volume one is open to shoals of criticisms[21] and objections to everything from its broadest claims, down to the minutiae of its formulations and index.[22] Out of these I shall focus on two points only – Toulmin's attempt to view science as a social institution, and the claim that logic is not at the heart of scientific rationality and hence of rationality in general. Each fault carries over from *The Uses of Argument*. There the 'jurisprudential analogy' is regularly hinted at but never explored and developed. Here again, in *Human Understanding*, the analogy between the concepts of a discipline and the institutions of a society is made, but never carried further. Toulmin concedes that a "concept is an intellectual micro-institution" (p. 166), but seems unaware that, e.g. Gellner has written extensively[23] on this in a highly pertinent way. Gellner discusses both the functional theory of society and concepts, and the explanation of social and conceptual change. He argues that to say of institutions or concepts that they have survived and thus must perform a function, is either truistic or absurd; for conceptual incoherence may be fostered and exploited by groups involved in rapidly changing situations. Not only would Gellner's work instruct Toulmin about the real concerns of the social sciences, about which he is so condescending (382–86); not only would it indicate the superfluity of his admonitions about the application of Darwinism to society (348–353); it would also show that he could make his comparison of China and Ancient Greece much sharper and more plausible. To a social scientist, the contrast between the two societies is that between social stability, especially of the status system, and social change. Popper remarks that classical Greek thought may owe a lot to commerce, for commerce brings cultures into contact, opens possibilities, creates ambiguous or interstitial statuses, creates a base of wealth and power independent of land or patronage, and generally can threaten an established (intellectual or political) system.[24] Athens was above all a trading centre.

China, by contrast, was a culture so strong that even invaders joined it rather than change it. China was very large, satisfyingly diverse, and consciously self-sufficient. When Lord Macartney journeyed to Peking in 1793 to request permanent diplomatic and easier trade relations, the emperor replied:

> As to what you have requested in your message, O King, namely to be allowed to send one of your subjects to reside in the Celestial Empire to look after your country's trade, this does not conform to the Celestial Empire's ceremonial system, and definitely cannot be done. Hitherto,... men... allowed... to come to the capital... were obliged to adopt the costume of the Celestial Empire,... were confined within the Halls, and were never allowed to return home. These are the fixed regulations of the Celestial Empire... There never has been any precedent for allowing them to suit their own convenience. If... you... desire... to study our culture... it would be of no use since your country has its own customs and regulations, and would certainly not copy Chinese ones.[25] The Celestial Empire, ruling all within the four seas, does not value rare and precious things... We have never valued ingenious articles, nor do we have the slightest need of your country's manufactures.

There can, I think, be little doubt that China's poor intellectual showing compared to Ancient Greece has much to do with this social stability and intellectual complacency. The progress of science obviously owes much to social change, as seventeenth century England, eighteenth century France, and the last one hundred years attest. One wonders what is supposed to be explained by calling this progress 'evolutionary'. Notoriously, in the social sciences evolutionism is rejected because it lacks explanatory power; lacking an equivalent of microbiology it explains nothing; yet it imposes a spurious and irrelevant order on events *(post hoc ergo propter hoc)*; and is a poor guide to action. It fosters a quietistic if not conservative outlook: what we have has evolved to do its job, if it does not, it will mutate. How much more convincing is Popper's view that science is revolution in permanence, a means of change and improvement in cognition in which ideas die in the stead of people.

A revolutionary view of science allows science rationality and, moreover, also allows rationality to the individual acting to forward science. Whereas, if science is seen as an evolved society of concepts (institutions), and concepts, like institutions, are instruments for achieving men's purposes, what can be the aim of the society of science that it rationally pursues? Can evolution have an aim? Is evolution rational? In fact, Toulmin's instrumentalism slips from time to time, when he allows that science explores and explains (e.g. p. 150). Disciplinary ideals like, "all physiological functions are to be explained in chemical terms" (p. 124), or, explain the physical properties of matter in terms of 'atomic substructure' (p. 152), are hardly instrumentalistic. And yet, Toulmin will not allow that either concepts or theories are ever true or false, except indirectly (whatever that means, pp. 168–9). If we ask how, then, the functional adequacy of concepts can be appraised, Toulmin becomes positively Polanyian.[26] Disciplinary ideals, absorbed in apprenticeship, will be used to weigh and debate new ideas. This is vague, so Toulmin cites the physicist Chew and the biologist Delbrück as innovators taken seriously. When we look closely at what he says about them, we find that Chew argued that elementary particle theory is flawed because there seems to be an infinite regress; Delbrück says that biology as he found it was descriptive and not explanatory. These are cases, Toulmin says, of the individual creating a turning point in science, not evoked by the profession or by external social forces. They stepped outside current theory, but their acceptance of disciplinary ideals ensured they were taken seriously. A simple-minded reader might think that what Delbrück and Chew had offered was quite straightforward: devastating general or transcendental criticisms of current theories in their fields and suggestions for novel attacks on the problems of the science, namely those of "our ideas about the world at variance either with Nature or one another" (p. 150).

By this time, the reader cannot but be confused about where Toulmin stands. Toulmin himself is. How can we "improve our ideas about the natural world step by step" (p. 150), in a "dialectical succession of 'conjectures' and 'refutations'" (p. 140), when "questions about the empirical 'truth' or 'falsity' of theoretical principles in science – as such – do not arise" (p. 169)? We remember that Toulmin's ostensible concern is to avoid relativism, and to give a new account of the rationality of science; so far, we have learned that science is ordered into professional disciplines (in *The Uses* it was 'fields'), and that what demarcates a discipline are its disciplinary ideals; disciplinary ideals develop in certain (unspecified) social situations; and new disciplines or radical changes in disciplines come about when individuals challenge not current theories but the ideals themselves. How all this differentiates science from witchcraft and religion it is difficult to see. They are often highly organized professional disciplines, with clear disciplinary ideals (explain all events as the work of the gods, or the forces of witchcraft); there are elaborate apprenticeship initiations; and there are often widely differing, even competing, theories within the discipline.

Much the same could be said on behalf of those 'would-be' disciplines the social sciences. They even evince signs of progress and (evolutionary?) continuity. By what standard could we judge them insufficiently evolved? Toulmin is floundering back into relativism because he is trying to justify the claim of science to 'intellectual authority'. But justification of science because it is evolved is no stronger than justification of other social institutions because they have evolved; and that is no justification at all.[27] Rather, the search for their function and the structure into which they fit has only then begun. Just as evolution does not legitimate social states, it also does not legitimate cognitive states.

Not only does Toulmin vacillate over the truth *versus* usefulness issue; he cannot make up his mind whether scientists are concerned with theories, theoretical ideas, explanations, or 'concepts'. He sometimes uses them interchangeably. Admittedly, a concept is a rather nebulous concept; but not completely so. Being a Wittgensteinian, perhaps Toulmin should attend not to the concept but to the use, the context, that is to say theories in which the concept is embedded. Plucked from such contexts, how can concepts be compared? Within such contexts, it is theories, not concepts, which compare.

In his frantic attempts to escape relativism (pp. 317 and 492), Toulmin argues that where different cultures have parallel rational enterprises their concepts, judgements, and ideas *can* be compared:

> we can establish close similarities of intellectual function between the concept of 'impetus' in fourteenth-century scholastic discussions of change, the concept of 'quantity of motion' in Newton's dynamics, and the concept of 'momentum', as used in Einstein's twentieth-century physics. This done, it is quite clearly legitimate – having regard to those functions – to ask how far, and in what respects, the relativistic concept of 'momentum' represents an intellectual advance on Newton's *quantitus motus* and Buridan's 'impetus'. (pp. 492–3)

How can a *concept* constitute an 'intellectual advance'? A concept can hardly answer a question or solve a problem. A concept like 'impetus' does not spring into the pages of intellectual history from nowhere. It comes as part of a system of theoretical ideas. Toulmin cites Collingwood, and Collingwood said every answer presupposes a question. A context. The study of concepts and their history is intellectually vacuous; it ignores the real history, the context of stormy critical debate over theories proposed as solutions to problems.

This brings me to my second main line of criticism, the role of logic in science. Toulmin writes from time to time of debate, criticism, explanatory power – apparently not noticing the dependence of such notions on logic for their rationale. Yet Toulmin repeatedly repudiates logic (p. 44). In its place he wants to put something called "informal" argument. It is not clear why he is so down on logic. When he speaks of logic he seems either to mean medieval syllogistic, or formal axiomatic systems, both small branches of a large subject. No doubt lawyers and judges, scientists, philosophers and ordinary people eschew syllogisms (and formalization). Who can blame them? So far as I know, neither Aristotle nor the logical positivists commended them for general use. But what of the broad claim that logic is something like the hidden structure of rational debate: that which explains what is going on, why certain rules are observed, such as avoiding vagueness, incoherence, inconsistency, taking criticism seriously, assessing the force of an objection, etc.? (Each of these rules is violated in Toulmin's work.) True, Toulmin's logical straw man formalization was sometimes claimed to be the desirable form for rational science. But it would make *Human Understanding* pathetically *déjà vu* to read it as a criticism of an extreme branch of the dead tree of logical positivism, mistakenly identified as 'logic'.

In place of a hidden structure which explains events by rationally reconstructing them, Toulmin proposes lengthy and detailed investigations into the case by case history of science, where we will seek out the origins of concepts, their development, study the informal discussions about them and illustrate again and again how science rationally opts for the most functionally advantageous. What can conceivably be the point of such explorations of the endlessly proliferating polymorphism of scientific practice? To verify that it has evolved? History of science would become even more long-winded and pointlessly detailed than it already is. What explanatory light would be thrown on the rationality of science? Philosophy of science would cease to exist, since no *general* logical and methodological structures are allowed to be discovered. The public will be left with the view that the rationality of science is on a par with the rationality of nature, and that this is the best of all possible worlds, up to this point in the evolution of human cognition.

All this strikes me as ludicrous. And it all stems out of and hence reflects adversely back on the assumption that the rationality, if any, of science inheres in its 'intellectual authority', its grounds, its backing, its justification. For this we have to become hedging relativists, equivocators, fundamental intellectual conservatives. If this is where justificationism leads us, Toulmin's book is an effective *reductio ad absurdum* on itself. And so it is. For if no general logic underlies science, then its practice can only be justified by itself; but then practice as it were can only be compared to others if they have closely similar aims, and 'similarity' here is circular too; and if practice justifies itself there can hardly exist a separate measuring stick by which it can be judged, hence logic is ruled out. The circle of ideas has many roots and many confusions. Central of them all is the attempt to justify; to say science is rational because it has authority – whether that authority is traced to the indubitability of its empirical or intellectual base, or is traced to the authority of its having evolved. All such justifications, and every pattern of justificatory argument is defenceless against sceptical/regressive/relativist arguments.

6. Logic as the Hidden Structure of Rationality

In a way, scepticism and justificationism are not separate positions, but two versions of the same justificationist epistemology.[28] The sceptic and the justificationist would *agree* that we do not have knowledge unless we can justify it. They *differ* simply in that the sceptic thinks the task is hopeless and hence declares that we have no knowledge, all is uncertain; while the justificationist, faced with the infinite regress argument that every attempt to justify any claim simply leads to the further necessity of justifying the justification, and faced with the problem of induction, nevertheless struggles gamely.

Non-justificationism is thus a very radical position. It denies the assumption shared by the sceptic and the justificationist that we do not have knowledge unless we can justify it. It asserts that we cannot justify anything substantial; and yet we have knowledge. This is done by rejecting the traditional identification of knowledge with what *I* know, or believe, or am certain of, and instead identifying knowledge with the most severely tested of surviving theories, regardless of whether anyone believes them or not. The non-justificationist wants everything tentative: he declares the sceptic wrong to say we have *no* knowledge; the justificationist wrong to say we have any *certain* knowledge, or intellectual authority. We have a great deal of uncertain, tentative, hypothetical knowledge, some of it reliable, some not, almost all of it certainly in the long run likely to prove false. Science does not lay claim to authority; hence its rationality can inhere only in its change.

So the way out of Toulmin's dilemma – the rational authority of science versus scepticism and relativism – is the existence of a third possibility: instead, reject justificationism. Then it is possible to grasp the crucial role of logic in the rationality of science, and rationality in general.[29] Two things: logic is both a theory of proof or justification *and* a theory of criticism. Traditionally, the concentration has been on it as a theory of proof. Aristotle saw logic as among other things a basic part of the method of getting at *certain* knowledge, of penetrating to the *essence* of things. In this, logic has proved a disappointment. Interesting things like science, ethics, common sense cannot, it turns out, meet *logical* standards of proof. Does this mean they cannot be proved *tout court*? Many philosophers obviously thought so or they would not have strived so valiantly to reconcile logic *and* inductive 'proof'; they would simply have fallen back on other standards of proof like 'beyond a reasonable doubt'. Why was logic a disappointment, then? Because it turned out to be a theory of valid inference; and inference, or derivability, or deducibility is not to be confused with proof, as Aristotle knew. A *proved* statement is true because it is validly derived from true premisses. But the theory of valid inference (logic) does not usually provide the proof that the premisses are true. It turns out, in fact, that only the tautologous premisses of logical and mathematical systems can be proved to be true. Logic which started out to be a theory of proof ends up by showing that almost nothing can be proved; that only statements which say nothing about the world are provable! So far from being a *route* to certain knowledge, logic shows that in an important sense we have *no* certain knowledge of the world. This is a very important result for non-justificationism.

But besides being a theory of proof, besides transmitting truth from the premisses to the conclusion, the theory of valid inference also includes the rule of transmission of falsity from the conclusion to the premisses, and this can be regarded as (an important element in) a theory of criticism.[30] For the non-justificationist this theory of criticism is part of the explanation of how we can learn from experience, if at all, and of how science has progressed.

To the non-justificationist the fundamental epistemological situation is this. We can never prove that any non-tautological statement is true. We can, however, prove that two contradictory statements cannot be true together. Hence, if we have a way of discovering inconsistencies between the various statements containing knowledge claims, we have a means of weeding out some of the false ones. Of course, all the weeding out is hypothetical. Having shown that two statements are inconsistent, it may be difficult to find out which of them is false, but that one of them must be, is incontestable. So our task is to make hypotheses: hypotheses about the world, and logical hypotheses about which of any in a set of inconsistent hypotheses is false – not useless; false.

Logic, then, is part of the theory of criticism. By means of the *modus tollens* it is able to transmit falsity from the conclusion of an inference to its premisses. It shows that it is inconsistent to assert that the premisses of an inference are true and the validly derived conclusion is false. Of course, that the conclusion is false is also a hypothesis. In choosing to explore it, again the question arises of its consistency or otherwise with other statements. If it appears perfectly consistent with all unrefuted statements under serious discussion, and there are non-justificationary arguments in its favour (empirical content, simplicity, aesthetic appeal, etc.), then it may be taken seriously, i.e. be a fit subject for further criticism – *not* for acceptance or 'authority'.

Logic as a theory of criticism cannot be regarded as in any sense a theory of how in practice we do, or ought, to *think* or to *reason*. Logic is there to help us unpack the consequences of statements we are considering, with a view to discovering inconsistencies between those statements and others, and then transmitting the falsity thus disclosed back through the system. To show all this involves much formalization and refinement of intuitive notions like inference, and consequence. Much of the time in practice we may never refer to these refinements, or proceed in accordance with them. Nevertheless, our rationality can be checked against the hidden structure. That there is a hidden structure to our intellectual, problem-solving endeavours, is of course, a hypothesis. That our best intellectual achievements hide a structure that obeys the rules of logic, is another hypothesis. Hypotheses are fair game for criticism. I submit that Toulmin has made no criticism of the first hypothesis at all, and no effective criticism of the second. Instead he has tried to suggest an alternative hypothesis that is simply a non-starter; he has refuted it himself. As a hidden structure logic is normative: an available standard, but not one to be implemented all the time in practice. In much argument we proceed intuitively; our intuitions hopefully refined by the findings of logic; only where intellectual or other crisis makes scrutiny important do we need to expose and discuss the hidden structure.

York University, Toronto

NOTES

[1] At an L.S.E. seminar presided over by John Watkins, in the autumn of 1972.

[2] Plato, Descartes, Leibniz, Kant, Frege and Russell favouring mathematics; Aristotle, Bacon, Locke, Hume, Kant, Russell and many modern philosophers of science opting for science.

[3] Imre Lakatos, 'Proofs and Refutations', *British Journal for the Philosophy of Science* **14** (1963–4), 1–25, 120–39, 121–45, 296–342.

[4] As I believe they are.

[5] Mill John Stuart, *System of Logic*, Introduction.

[6] The best discussion of this I know is in Alan Musgrave's Ph.D. thesis, 'Impersonal Knowledge: A Criticism of Subjectivism in Epistemology', University of London, 1969. A section has appeared as 'George Boole and Psychologism', in *Scientia* (July-August 1972), 1–16.

[7] By S. Korner in *Mind* **68** (1959), 425–7; by D. J. O'Connor in *Philosophy*, **34** (1959), 244–5; by J. C. Cooley in *Journal of Philosophy* **56** (1959), 297–319. The favourable review is by Will, F. L., in *Philosophical Review* **69** (1969), 399–403. *The Australasian Journal of Philosophy*, the *British Journal for the Philosophy of Science*, the *Journal of Symbolic Logic, Philosophical Quaterly* and *Philosophy of Science* appear to have ignored it.

[8] At no point does Toulmin observe the necessary/sufficient distinction on this point. Is it seriously held anywhere that logic is sufficient for rationality or identical with rationality?

[9] See Korner, *op. cit.*

[10] As pointed out by Wisdom, J. O., in 'The Sceptic at Bay', *British Journal for the Philosophy of Science* **9** (1958), 159–63; and Bartley III, W. W., 'Achilles, the Tortoise and Explanation in Science and History', *British Journal for the Philosophy of Science* **15** (1962), 15–33.

[11] For the notion of concept-stretching see Lakatos, *op. cit.*, especially p. 314.

[12] In suggesting that a universal and analytic logic sets impossibly high or idealised standards for justification, Toulmin is obviously making an attack on the concept of validity. I have heard him deny this, but the evidence is in his text: "Once we have been struck by the apparently superior cogency of analytic arguments and tempted to demand analyticity as a condition of either 'deductiveness' or 'validity', we cannot consistently allow substantial syllogisms to pass without criticism. A valid analytic syllogism *cannot* in its conclusion tell us anything not already included in the data and warrant-backing, so a syllogism which involves a genuinely substantial set can – from our present point of view – be justified only by begging somewhere in the data and backing the very conclusion we are intending to establish." (p. 150).

[13] Bartley III, W. W., *The Retreat to Commitment*, New York 1962; 'Rationality *versus* The Theory of Rationality', in M. Bunge (ed.), *The Critical Approach to Science and Philosophy*, New York 1964, pp. 3–31.

[14] "What has to be recognised first is that validity is an intra-field, not an inter-field notion. Arguments within any field can be judged by standards appropriate within that field, and some will fall short; but it must be expected that the standards will be field-dependent, and that the merits to be demanded of an argument in one field will be found to be absent (in the nature of things) from entirely meritorious arguments in another" (p. 255).

[15] Neither of these arguments strike this reader at least as serious paradigms of reasoning in *any* field. Moreover, they are of greatly unequal merit – certainly many sociologists and philosophers of science would baulk at a logic which made the Petersen argument valid. Cooley remarks in a footnote to his review (op. cit.) that Toulmin's book is in danger of making all arguers like Molière's hero: they turn out to have been talking logic all along.

[16] Carroll, Lewis, 'What the Tortoise said to Achilles', *Mind*, N.S. **4**, 278.

[17] See Bartley, 'Achilles, etc.', *Op. cit.* See also Russell, Bertrand, *Principles of Mathematics*, London 1903, p. 35.

[18] Why it is common, and how the two came to be confused, is explained by Popper (*Conjectures and Refutations*, London 1903, p. 203) in the following passage:

> The fact that, to every well-known rule of inference there corresponds a logically true hypothetical or conditional formula of some well-known calculus... has led to confusion between rules of *inference* and the corresponding conditional formulae. But there are important differences. (1) Rules of inference are always *statements about statements*, or about classes of statements (they are 'meta-linguistic'); but the formulae of the calculi are not. (2) The rules of inference are unconditional statements about deducibility; but the corresponding formulae of the calculi are conditional or hypothetical 'If... then' statements, which do not mention deducibility or inference, or premisses

or conclusions. (3) A rule of inference, after substitution of constants for the variables, asserts something *about* a certain argument... namely, that this argument is valid; but the corresponding formula, after substitution, yields a *logical truism*; i.e. a statement such as 'All tables are tables,' although in hypothetical form, as for example, 'If it is a table, then it is a table' or 'if all men are mortal, and all Greeks are men, then all Greeks are mortal.' (4) The rules of inference are *never* used as premisses in those arguments which are formulated in accordance with them; but the corresponding formulae *are* used in this way. In fact, one of the main motives in constructing logical calculi is this: by using... those hypothetical truisms which correspond to a certain rule of inference... *as a premiss* we can dispense with the corresponding rule of inference. By this method we can eliminate all the different rules of inference – except one, the [*modus ponendo ponens*] (or two, if we make use of the 'principle of substitution,' which, however, can be avoided).

[19] Or better, criteria of what counts as an *argument*.

[20] The alternative or third view is due to Popper.

[21] See especially those by Cohen, L. Joathan, in *British Journal for the Philosophy of Science* **24** (1973), 41–61; and Larry Briskman, *The Philosophical Quarterly* **24** (1974), 160–9.

[22] A reviewer has virtually to construct his own index: Key terms like 'disciplinary ideals', institutions, concepts as institutions and 'genealogy of problems' have no entry.

[23] Gellner, E., 'Concepts and Society', *Transactions of the Fifth World Congress of Sociology* **1** (1962), 153–83; 'Sociology and Social Anthropology', *Transactions of the Sixth World Congress of Sociology* **2** (1967), 49–83; *Thought and Change*, London, 1965.

[24] Popper, K. R., *The Open Society and Its Enemies*, London, 1945 *et seq*., Vol. I, pp. 176 *et seq*., (1962 ed.).

[25] Cranmer-Byng, J. L., 'Lord Macartney's Embassy to Peking in 1793 from Official Chinese Documents', *Journal of Oriental Studies* **4** (1957–8), 117–187.

[26] Polanyi is mentioned once in the book, but not in connection with his theory of Science as a profession, which anticipates much of Toulmin. Similarly, Popper and Agassi, who have discussed science as a social institution, and suggested that rationality has to do with critical open-mindedness, with explicit willingness to change one's views, are referred to *for other things*.

[27] "The antecedent state does not produce the subsequent one, but the relation between them is exclusively chronological." Durkheim, Emile, *The Rules of Sociological Method*, Chicago 1938 (original 1895, paperback 1964), p. 118.

[28] This is clear from the work of Bartley III, W. W., 'Rationality *versus* The Theory of Rationality; *op. cit*.

[29] Cf. Martin Hollis, 'Reason and Ritual', *Philosophy* **43** (1967), 231–47.

[30] A theory that is still developing, see Settle, Tom, Agassi, Joseph, and Jarvie, I. C., 'Towards a Theory of Openness to Criticism', *Philosophy of the Social Sciences* **4** (1974), 83–90, and the literature cited there.

EUGENE KAMENKA AND ALICE ERH-SOON TAY

PARTICIPATION, 'AUTHENTICITY' AND THE CONTEMPORARY VISION OF MAN, LAW AND SOCIETY

I

Part of the contemporary crisis in law, justice and morals in Western post-industrial societies is the appearance of an unusually strong and wide-spread revulsion from the style, attitudes, presuppositions and arrangements implied in the ideal of a society governed by laws and not by men. The fear is the fear of dehumanization in a mass collectivity in which structures have become so vast, the ramifications of technological change so pervasive and irreversible, problems so complex and knowledge so specialized that the individual feels threatened in his ability to cope with, or even to understand, the things that go on around him. His life is changed so immediately and directly, his expectations are transformed at such speed and with such relentlessness that history, technology and 'progress' now appear as man's enemies rather than his friends.

There has consequently been a certain retreat from the life and values of sustained enterprise, objective understanding, technical rationality and personal responsibility, with their implied recognition of and respect for external and internal social and physical necessities as frameworks for action. There has been growing disenchantment with those 'rational-legal' modes of life, administration and goal-seeking which Max Weber, and many others, saw as the specific Western contribution to history and civilization. Objectivity has now become, in many quarters, a dirty word, a synonym for the unfeeling and inhuman. The emphasis is on personalization – of administration (evinced by the longing for 'charismatic' leaders), of news reporting, of both popular and intellectual discussion with their gross over-valuing of 'authenticity' and 'commitment', of education (as 'self-expression') and of law. The extreme form of that personalization is the belief that only the sufferer can prescribe the remedy, that only the worm can know the heel, or even more metaphysically, that all knowledge is shot through with subjectivity and that only the irrational, inexplicable action is a truly free action. Compare

R. S. Cohen et al. (eds.), Essays in Memory of Imre Lakatos, 335–358. *All Rights Reserved*
Copyright © 1976 by D. Reidel Publishing Company, Dordrecht–Holland

this with Hegel's insistence – against Jacobi and Schleiermacher – that 'feeling' is a disorderly storehouse into which the subject heaps, *pell mell*, all its perceptions and experiences, 'unmediated', unexamined, as it accumulates them in its contacts with the objective world, or Marx's practical and theoretical hammering of the point that the subjective is distinct from and ancillary to the objective, that the class must be one *in itself* before it can be one *for itself*, and that the politics of the proletariat will be determined by its *objective* conditions.

Much of the revulsion from the individualistic, liberal democratic concept of the rule of law presents itself as a conscious demand for a return to the face-to-face society, the organic community of living social bonds and commonly-shared ideologies and interests. It presents itself, in short, as an elevation of the non-commercial, pre-industrial *Gemeinschaft*. Its conceptual opposite, the pluralistic, commercial-individualistic society based on contract, personal responsibility and the rule of law, which Tönnies called the *Gesellschaft*, is no doubt, as Marxists like to say, a bourgeois phenomenon linked with the ascendancy of cities, trade and manufacture, of internal and external markets, and with the decrease in importance of the agricultural household, of kinship, locality and common customs and traditions. But the *Gesellschaft* and its ideology were also a revolt against the bondage of family, status and religion, against the linked personalization and communalization of life, politics and government in the *Gemeinschaft* of feudalism and of traditional pre-industrial societies generally. The divorce of law from morality, and of religion from both, the sharp separation of the private and the public, of the legislative, the judicial and the executive, the doctrine of representative Government and the abstraction and de-personalization of law and legal administration were charters of liberty without which the character, the achievements and the expectations of modern man would be totally unthinkable. These charters made the American Revolution reverberate throughout Europe, helped to bring down the *ancien régime* and made Feuerbach genuinely consider and Marx probably insincerely claim to be considering emigration to America. We cannot understand what the United States meant to 19th-century Europe unless we recognize that the movement from status to contract did and does represent a genuine liberation of personality, capacity, thought and opportunity. It is just such a movement and not any flight back into the *Gemeinschaft* which an increasing number of women are demanding today. If we want to know why, we have only to compare the life, opportunities and expectations of Bella Abzug with those of Hester Prynne.

Nevertheless, the liberties of the *Gesellschaft* which have made possible the demands and expectations of modern men and women were gained at a price. That price has been discussed by Karl Marx and by many more recent thinkers, sometimes thoughtfully, sometimes less so, under the heading 'the alienation of man in modern bourgeois (or industrial) society'. The political liberation of man as a citizen was accompanied (at least in principle) by the liberation of civil society, of the world of industry and trade, from the restraints of religion, morality and politics. Formal political and legal equality concealed and even facilitated real economic and social inequality. Behind the Republic of the Market lay the Despotism of the Factory; behind freedom of trade lay the doctrine of the intrinsic social, political, economic and cultural inferiority of colonies and dependencies.

It may be true that the ideals and the view of man which increasingly make gross social and economic inequality, cultural and physical deprivation, despotism and domination morally and politically unacceptable are the ideals and the view of man brought into history by the *Gesellschaft*, by its conception of man as a free moral agent capable of liberty and equality, bound, in his social life, only by laws that he is also capable, as a rational being, of framing, understanding and obeying. It may also be that these are not ideals to be given up lightly – even if, like the concept of the juridical person as a free and autonomous agent, they are in actuality fictions. Nevertheless, men notice the shoe only where it pinches. *Gesellschaft* law is seen by many in Western societies today as impersonal, inhuman, abstract, itself a form of alienation that tears man out of his living context, fails to see him as a man but recognizes him only as a debtor, a criminal, a rate-payer, a contracting party, in short, as the holder or ower of specific and limited legal rights or duties vis-à-vis another, rights or duties that are external to his personality. As we have argued elsewhere,[1] and as we shall also suggest below, there are important reasons connected with the scale of modern economic activity, with the growing power and expanding rôle of the State and with the attendant bureaucratization of advanced industrial society which make the pure *Gesellschaft* paradigm no longer an effective basis for the legitimation of all fundamental legal and political attitudes in our society or an effective source for the organizing concepts on which the whole of our legal and administrative system can be based. But Karl Renner, Max Weber and Josef Schumpeter, who saw these developments clearly, are not the heroes of modern popular and radical ideology, which is much happier talking about 'community' than about the State, about 'participation' than about administration and planning, about 'self-expression' than about emulation. There *is* that remarkable longing for the personalization of law and legal proceedings, for the restoration of man to a place in the organic community that recognizes him, judges him, and cares for him, as a total person, and that makes justice, at least in principle, the work of the whole community, and not a specialized branch of learning and experience. It is here that the admittedly limited appeal of a sentimentalized picture of the People's Republic of China lies. Just as barefoot doctors seem more human because they are simpler as people, more like ourselves, more accessible than the high priests of a complex and difficult science, so people's courts and people's judges seem more human than the bewigged and gowned representatives of a complex and learned art which still believes that men must be judged by universal principles grounded in and shaped by long-mulled-over and carefully recorded experience and that hard cases make bad law. The sentimentality and superficiality of this picture is reflected in its comparatively fleeting impact; it is always being undermined by concrete, *real* developments. It is thus that the romantic radical passes, in quick succession, from the achievements and alleged brotherhood of the Soviet Komsomol to Yugoslavia, Cuba, the Congo, China, Chile, and even Bangladesh, just as the critics of the medically trained pass from one 'healer' to another.

One basic *motif* in current attitudes, then, is simply the flight, in fantasy, from New York to Marlboro Country, from the computers that have replaced the dark Satanic mills to social units where persons still count as persons, with their personal hopes, fears and scars, where they can understand their environment and act upon it, knowing what the effects will be. It is a feeling that has run through the whole history of industrial and post-industrial society, whose dreams and fantasies have always been of a non-industrial, non-specialized and non-fragmented world. It is a feeling which lies, as we shall see, at the very heart of socialism as a historical movement: it has also lain at the very heart of Tory conservatism and of the ideologically conservative but practically modernizing Fascist right. It is there in the best-selling fiction of the last 150 years; it accounts for the popularity of the crime novel, and especially of the American crime novel, with its Philip Marlowes and Perry Masons – men who counterpose personalized, 'bespoke', *ad hoc* justice to abstract, impersonal, large-scale and therefore potentially impotent or corrupt law, action to a world of rules. These men still live in a world where their actions count; they are contrasted with men, no matter how important or powerful, who are to be seen and pitied as men driven by something outside themselves, who are part of a system that makes it impossible for them to do what the private detective can do – act as a person. The *Gemeinschaft* theme is in many respects becoming stronger and more self-conscious – such novels as *The Godfather* knowingly and deliberately take as their theme the *Gemeinschaft/Gesellschaft* dichotomy.[2] But while the fantasies intensify, those who live in actual *Gemeinschaften* continue to vote with their feet in the opposite direction.

II

The crisis of liberalism and individualism, of *Gesellschaft* law and of the *Gesellschaft* concept of liberty under law, we have been suggesting, is the product of great and important changes in modern social and economic life. These point not backward to the *Gemeinschaft*, but to a new and much more complex paradigm. Ideologists, however, and popular ideologists above all, see the present in terms of the myths of the past – they go back to old clichés when confronted by new, disturbing and complex developments and thus often help to revitalize them, but at the *expense* of genuine understanding and thought. It is thus that new developments can be made part of an old, unfulfilled hope and that the first, and most superficial, reaction to the current crisis and to demands for participation is to treat them as part of the final, glorious coming of socialism, as the enthronement of justice, equality, public interest, the common good and the general will and the banishment of injustice, inhumanity and self-seeking individualism. It is all this which is being suggested, but not spelt out, by shifting discussion from the problem of the relationship of the individual and the State (in contractual terms, concerned with rights and duties) to the new language of 'participatory democracy', 'community needs', 'community development' etc. etc. This question cannot be looked at seriously without first examining the history of socialism.

The history of socialism can be written as the story of egalitarian strivings, as ancient, perhaps, as society itself: it can be made into the record of the age-old protest against oppression and injustice. As such, it will always be with us; it will have no end. The history of socialism is also part of the history of protest, uprising and revolt, which have their own logic in which popular participation looms large, but only in short-lived transitional periods. In our view the history of socialism is best written and understood not as the history of populism, but as the record of a specific historical development in European society which begins when men bring into relation the political achievements of the French Revolution with the social consequences of the birth of industrial society and the industrial proletariat.[3] The Great French Revolution that swept aside the government and law associated with a system of privileges and estates had proclaimed – in one 'world-historical' act of liberation – the slogan 'Liberty, Equality, Fraternity'; it had fixed in an egalitarian, universal political Constitution the Rights of Man and of the Citizen. This political liberation, however, as we have said, resulted in a far-reaching economic 'liberation': 'Liberty, Equality, Fraternity' could and did lead to a social order that was profoundly individualistic, weighted in favour of a minority of property owners. As Lichtheim puts it:

> For the bulk of society, then made up of peasants and artisans, economic freedom – in the sense of an uncontrolled market in commodity values, operating in accordance with its own impersonal 'laws' – held danger as much as promise. To the proletariat, already in existence on the eve of the industrial revolution in the shape of a mass of paupers deprived of 'active' citizenship, this kind of freedom signified virtually nothing beyond the bare right to sell one's labor to the highest bidder. Economic liberalism thus conflicted with social democracy, unless it could be shown that all members of society stood an equal chance of attaining to ownership of property. Such an assertion was more plausible in the America of Jefferson and Jackson than in the France of the July Monarchy, or the England of the 1832 Reform Bill, whence the decisive impact of socialist doctrines in Western Europe and their relative failure to attract attention in the United States.[4]

From the disenchantment of the French *sansculottes* and the radical demands of those who identified with them, from the English social and economic critics of the miseries of the new industrial revolution, from the Saint-Simonian cult of the *industriels*, grew a movement that saw itself as the class ideology of the industrial proletariat. The term 'socialism' made its first appearance in the early 1830's, among radical sects in Western Europe. It defined itself through its criticism of bourgeois society, of bourgeois rights and of the economic liberalism associated with these. It brought together, as aspects of one movement, the weavers' revolt in Germany, the abortive 1830 revolution in France and the Chartist agitations in Britain. It fused, or held in loose relation, the Ricardian labour theory of value, the doctrines of Fourier or Saint-Simon and the radical egalitarianism of the extreme democrats of the French revolution. The link between socialism proper as the critique of bourgeois society in the light of the industrial revolution and of its own roots in the Enlightenment and the prophets, utopians, radical reformers and rebels of earlier days lay in their common hostility to private property, to the inequality of economic position and social opportunity that flowed from the uneven distribution of property and to the power this gave some men over others.

Until the rise of the industrial system, the protest against existing society tended to be a protest against the two most important features of traditional agrarian society – its coercive hierarchy of authority and its distribution (or non-distribution) of property in land. Both these features appeared to be based upon and safeguarded by law. The heroes of socialist pre-history were thus led into inevitable and quite fundamental hostility to law, at least to law as mankind had hitherto known it. The distinguishing mark of the 'progressive' thinker, of the utopian socialist or enlightened democrat, came to be his belief in the intrinsic goodness and sociability of man, and/or in the view that property was the true source of all significant evils. Not infrequently he saw law, together with private property, as either the cause or the symbol of human misery. For Rousseau, natural man, following his spontaneous inclinations, was sociable and benevolent; civilization introduced evil. For Thomas Paine,

> Society is produced by our wants, and government by our wickedness; the former promotes our happiness positively by uniting our affections, the latter negatively by restraining our vices ... Government, like dress, is the badge of lost innocence; the palaces of kings are built on the ruins of the bowers of paradise.[5]

Earlier, in 1755, in the Code of Nature at first falsely attributed to Diderot, the utopian 'socialist' or pre-socialist Morelly had written:

...The only vice that I perceive in the universe is *Avarice;* all the others, whatever name they be known by, are only variations, degrees, of this one; it is the Proteus, the Mercury, the basis, the vehicle, of all the vices. Analyze vanity, fatuousness, pride, ambition, duplicity, hypocrisy, dishonesty; break down most of our sophistic virtues into their component parts, and they all resolve themselves into the subtle and pernicious element, *the desire to have* ....

Now, would this universal plague, this slow fever, *private interest*, ever have been able to take hold if it had found no sustenance, nor even the slightest dangerous ferment?

I believe that no one will contest the justness of this proposition: that *where no property exists, none of its pernicious consequences could exist* ....

... if you were to take away property, and the blind and pitiless self-interest that accompanies it, you would cause all the prejudices and errors that they sustain to collapse. There would be no more resistance, either offensive or defensive, among men; there would be no more furious passions, ferocious actions, notions nor ideas of *moral evil.* If any were to remain, or if some vestige of it were to re-emerge, this would only be the result of the merest accident, one of the smallest consequence. Minor oppositions of will, obfuscating ever so slightly the light of reason among the opponents, would, far from weakening the domination of natural beneficence, only cause men to have a greater sense of its importance ....[6]

With the rise of the industrial system, the attack on property gained a new impetus. It was helped by the concentration of the propertyless in the new industrial towns and barracks, but it rested, above all, on the perception that ownership, now of capital and of machines, had, in these new conditions, vastly increased social ramifications. These ramifications stood in direct contradiction to the egalitarian hopes and pretensions of the French Revolution and to the scientific and economic optimism surrounding the industrial revolution. The machines that were to make man rich and free were in fact making or keeping thousands poor and dependent. By the late 1820's, some of the followers of Saint-Simon had developed his unflattering contrast between the 'bourgeois' and the *'industriels'* into a far-reaching attack on the role of bourgeois private property in industry. To make the ownership of industrial capital and of the means of production a private concern was *immoral* because it enabled one man to exploit the labour of others; it was *uneconomic* because it failed to provide for the proper planning of industry, for the optimal allocation of resources throughout that vast industrial workshop into which each nation was being turned. Bourgeois society treated property as a private function, when the industrial system was converting it into a social function. The Saint-Simonian Bazard, in the lectures later assembled in the *Exposition* of Saint-Simonian doctrine, was the first to use the phrase 'the exploitation of man by man'; he combined the moral and the economic attacks on private property in the industrial system into a broad indictment:

If, as we proclaim, mankind is moving toward a state in which all individuals will be classed according to their capacities and remunerated according to their work, it is evident that the right of property, as it exists, must be abolished, because, by giving to a certain class of men the chance to live on the labour of others and in complete idleness, it preserves the exploitation of one part of the population, the most useful one, that which works and produces, in favour of those who only destroy.

...

[Then] a *social* institution is charged with these functions which today are so badly performed; it is the *depository* of all the instruments of production; it presides over the exploitation of all the material resources; from its vantage point it has a comprehensive view of the whole which enables it to perceive at one and the same time all parts of the industrial workshop.

...

The social institution of the future will direct all industries in the interest of the whole society, and especially of the peaceful labourers. We call this institution provisionally the general banking system, while entering all reservations against the too narrow interpretation which one might give to this term.

The system will include in the first instance a central bank which constitutes the government in the material sphere; this bank will become the depository of all wealth, of the entire productive fund, of all the instruments of production, in short of everything that today makes up the mass of private property.[7]

The vital ideological force of Marx's thought lay in his apparent reconciliation of these two strands – of the vision of the undifferentiated, unstructured community and of the acceptance and promotion of a planned and industrialized world.

III

The negation or non-recognition of law, N. S. Timasheff has reminded us,

may be in the concrete or in the abstract. In the first case the content of law is submitted to criticism. The existing legal order, it is felt, should be destroyed and replaced by another. This is the usual form of revolutionary propaganda. It may be successful, but even success would not destroy law; only its content would be changed.

In the second case, law as an institution is criticized. Not only should the existing legal order be destroyed, but it should be replaced by another nonlegal order. The triumph of such propaganda would mean the end of law as such, and since law is a cultural, historical phenomenon, the possibility of its abolition cannot be denied *a priori*.

The abolition of law is the focal point of anarchistic doctrine (in the broad sense of the term).[8]

The history of socialism is a complex story. The doctrines that have been and can be incorporated within the term 'socialism' are not mutually consistent and not always internally coherent. They were shaped by men of differing circumstances and temperament, in specific countries and specific historical situations. In France, the extreme democratic wing that emerged from the upheavals of the 1790's stimulated, through Babeuf's follower Buonarroti, the conspiratorial revolutionary socialism of a Blanqui, with its 'dictatorship of true Republicans', to be followed by the anarchist and egalitarian communist society. The Paris Commune appealed to the 'participatory' tradition of the Great Paris Commune of 1793, the Commune of Hébert and Chaumette, but it also established a Committee of Public Safety, indicted its own general, and suppressed one newspaper after another, as the Bolsheviks later suppressed the 'maximalists' of Kronstadt and their demand for genuine, elected soviets *after* as well as *before* the assumption of power. In the same France, Saint-Simon and some of his disciples had first looked to their friends in French banking circles to bring about the planning and rationalization of industry and bequeathed to socialism the reformist, State-centred attitudes of a Louis Blanc, attitudes that anticipated the evolutionary democratic socialism that took shape in the 1860's. Fourier's *phalanstères* and Proudhon's rejection of the State stood in direct contradiction to the technocratic ideals of the socialist Saint-Simonians. In England, Robert Owen approached the conclusion of his Address to the inhabitants of New Lanark with the words:

Continue to obey the laws under which you live; and although many of them are founded on principles of the grossest ignorance and folly, yet obey them, – until the government of the country (which I have reason to believe is in the hands of men well disposed to adopt a system of general improvement) shall find it practicable to withdraw those laws which are productive of evil, and introduce others of an opposite tendency.[9]

Only fourteen years later, however, in conjunction with the agitation for the Great Charter, the 'hewers of wood and the drawers of water' in England were demanding to know in what way the Reform Bill could benefit them and were expressing their lack of faith in the country's laws and institutions in a wave of violent protest. As Dicey put it in 1905:

> The fifteen years from 1830 to 1845, which may well be termed the era of the Reform Act, were among the most critical in the history of England. The time was out of joint. The misery and discontent of city artisans and village labourers were past dispute. No Act of Parliament could remove at a stroke the wretchedness and pauperism created by the old poor law. The true cure contained in the new poor law of 1834, with its drastic severity, its curtailment of outdoor relief, and its detested Bastilles, increased for the moment the sufferings of the poorest amongst the poor, and excited intense popular resentment. The wages earned by the labourers in the country were miserably low. The horrors connected with factory life were patent. Widespread was the discontent of the whole body of wage-earners. It is recorded in a series of state trials for sedition, for conspiracy, or for treason, extending from 1832 to 1843. There was rick-burning by labourers in the country, there were acts of violence by trade unionists in the towns. The demand for the People's Charter was the sign of a social condition which portended revolution. To us who know that several points of the People's Charter have passed into law without causing social or political disturbance, the thought may occur that Chartism loomed too large in the eyes of contemporaries. But the men of 1832 understood the time in which they lived. The cry for the Charter told of bitter class hatreds and of widespread dissatisfaction with the whole constitution of society. Men who have known England only during the years of prosperity and of general goodwill which have followed the repeal of the corn laws, can hardly realize the urgency with which the 'state of England question' thrust itself upon the attention of the public between 1832 and 1840. It was a terrible question enough; it was nothing else than the inquiry, how, if at all, was it possible to alleviate the miseries and remove the discontent of the working classes? [10]

Socialism may thus be revolutionary or evolutionary, anarchist or étatist. Whether its criticism of law is – in Timasheff's terms – 'concrete' or 'abstract', a criticism of a particular legal system or of law in general, is not always clear. Certainly, as a movement of (or on behalf of) the deprived and the oppressed, socialism of all types reflected the latter's intense antagonism for the existing State and the existing system, and for those who seemed to know how to manipulate the rules of the system to their own advantage. The Great French Revolution had swept away the centuries-old legal traditions of the *ancien régime*, with their estates, privileges and feudal dues, and had enacted a series of constitutions followed by a code that was to be the glory of modern Europe. Socialists, however, stood in a relation of radical criticism to the rights of the citizen and the civil and political arrangements embodied in these constitutions and this Code. They rejected not only feudal law, but bourgeois law as well. Was their rejection of bourgeois law, then, 'concrete' or 'abstract'?

The Rousseauan-anarchist tradition in socialism gave a clear answer to this question. It believed in the fundamental and natural co-operativeness of men, in the triumph of reason and of the general will. In a truly human society, there would be no need for a coercive State and a system of law. In a speech to the German Workers' Educational Association in London, in June 1845, the quasi-religious German communist Wilhelm Weitling declared: "In my opinion, everyone is ripe for communism, even the criminals. Criminals are a product of the present order of society, and under communism they would cease to be criminals. Humanity is of necessity always ripe for revolution, or it never will be."[11] At that time, neither Weitling nor the London German Workers' Educational Association had yet made close acquaintance with Karl Marx, but that anarcho-communism of which Weitling was a simple-minded exponent was also to find its place in the infinitely more sophisticated Marxian system.

If the anarchists and anarcho-communists clearly rejected law in the abstract, the connection between this rejection and socialism as the critique of bourgeois society remained unclear. Proudhon's famous slogan 'Property is Theft' implied – as Marx was quick to note – that very concept of property which Proudhon was claiming to expose, for theft is the taking of something that 'belongs' to another.[12] Was the abolition of property to be followed by legislation, creating a new legal system for a socialist society? Morelly's attack on property and the desire for property was followed by an elaborate model code – the Code of Nature – with its sacred and fundamental laws, distributive or economic laws, agrarian laws, edile laws, police laws, sumptuary laws, etc., etc.[13] The étatist socialists clearly thought in terms of State regulation that would direct and control the material foundations and social life of a socialist society. Yet socialism, as Dicey's discussion of 'collectivism' notes, did not come upon the world with a socialist theory of law and in England (and elsewhere) it has never attained to one. The early socialists, no doubt, had their attention engaged by other matters, economics and the problem of smashing or gaining State power. As revolutionaries they naturally provided no blue-print for the future, as evolutionary reformers they worked within an existing system. There was, however, a fundamental ambiguity in the pre-Marxian socialist relation to law that cuts deeper than this. Socialists saw the existing legal system, and all major legal systems before it, as involving, in a fundamental and pervasive way, the concept of property and of individual rights as opposed to social responsibilities. When the concept of property and of the independent self-contained individual was excised from law, as from social life, what would remain? Some veered to the view that the new order would be based on new social customs and habits rather than law, on moral conviction and social pressure rather than on formal indictments and the possibility of formal litigation. Courts would give way to Assemblies. Others, like Morelly, saw the new society as based on a complex and pervasive set of *regulations*, administrative arrangements that defined duties, allocated resources and conferred or denied material and political rights, but which converted all legal problems and disputes into matters for administrative and political decision. Anyone who studies the history of socialism in the twentieth century knows which view has triumphed, and which has remained in permanent opposition.

IV

The central problems of socialism, then, have changed remarkably little, even if the world has changed far more. The Marxian synthesis, a tremendous act of force and faith, has once again come apart to reveal its disparate components. The talk is of private interest giving way to 'the social interest', of the values and ideals enshrined in Roman and Common Law giving way to the values and ideals of 'society', especially of 'contemporary society'. But the central contradiction of socialism or, at least, the central tension within it, lies precisely in its attempt to give any content to the notion of 'society' and its central weakness in its theoretical fixation on 19th-century problems, on property rather than administration. Socialists vacillate, and have always vacillated, between a backward-looking elevation of the pre-industrial *Gemeinschaft*, transmuted in fantasy to an organic but spontaneous and ahistorical community without hierarchy, authority or oppressive ideology, and the elevation of a society organized as a rationally planned workshop on bureaucratic-administrative lines, with the State as the centre and foundation of all values and all activities. This is the contradiction noted but hardly solved by Mao Tse-tung when he says:

Within the ranks of the people, democracy is correlative with centralism, and freedom with discipline. They are the two opposites of a single entity, contradictory as well as united, and we should not one-sidedly emphasise the one to the denial of the other. Within the ranks of the people, we cannot do without freedom, nor can we do without discipline; we cannot do without democracy, nor can we do without centralism. This unity of democracy and centralism, of freedom and discipline, constitutes our democratic centralism.

The overwhelmingly bureaucratic-administrative character of the Soviet regime, and its increasingly frank elevation of bureaucratic-administrative values, have starkly dramatised one side of socialism and at the same time lost the Soviet Union the admiration of most radical critics of Western society. The language of Chinese propaganda and of the new 1975 Chinese State Constitution, on the other hand, has about it a homely simplicity and betrays a seeming concern with the people, and especially with the problem of popular participation in the work of a mass society, which understandably strikes responsive chords among those who fear dehumanization and the loss of control over themselves and their environment in the highly fragmented and specialized society of the West. But the language of the Chinese Constitution is also the language of the seventeenth-century Puritan Divines which led to and made possible the Salem witch trials. Listen to the Reverend Thomas Hooker,[14] a man sincerely beloved by his colleagues and a saint according to his own lights:

Christ has appointed church-censures as good physic to purge out what is evil. All men are made watchmen over the welfare of their brethren, and by virtue of their consociation and combination have power over each other and a judical way of process against each other in case of any sinful aberration.

In its complete politicization of all aspects of human life, in its elevation of the Party and its ideology over the rest of society, in the pervasiveness and the ruthlessness of its pressures on every human individual in a nation or group of nations of more than 800 million persons, the People's Republic of China comes closer to Orwell's *1984* than either Nazi Germany or the U.S.S.R. It is morally superior in its much more sparing use of murder as a weapon of control. But the number of suicides in China both before and after the Revolution does help to bring out how intolerable life in a *Gemeinschaft* can be, how inescapable its pressures, how hopeless the outlook for those who seek either to protest or to withdraw. It does matter *what* one participates or is forced to participate *in*. 'One's own shirt', says a Russian proverb, 'is nearest one's back' – we resent and fear our own discomforts and lack imagination for the agony of others, especially, as Hume remarked, if they live in a time or a place or in conditions not at all like our own. We yearn for the *Gemeinschaft* if and to the extent that we have never known it; we thrill to millions of Chinese marching in uniform because they are sufficiently *unlike* us to preclude total identification.

Those who long for community, of course, wish to dissociate it from the structured, hierarchical, *Gemeinschaft* of feudalism and they do not often mention the comparatively unhierarchical but even more intolerant *Gemeinschaft* of Puritan New England. The history of the Soviet Union and of the People's Republic of China help to remind us that the concept and the reality of *Gemeinschaft* have a logic of their own. So, of course, have the concept and the reality of revolution. Revolutions are about power, not community. It is not by accident that the Chinese theoretical journal *Hongqi* (The Red Flag) reminded its readers on 4 February, 1967 – in the very hour that the ultra-leftists of the Cultural Revolution were proclaiming the need for communes – that a real revolution is not the act of abolishing all authority, but that of replacing one authority by another:

> Some persons oppose all authority. This is an expression of the inherent bad characteristics of the petty bourgeoisie, an expression of anarchism ... Engels wrote, 'Have these gentlemen' – the anti-authoritarians – 'ever seen a revolution? A revolution is certainly the most authoritarian thing there is. It is an act whereby one part of the population imposes its will upon the other part by means of rifles, bayonets and cannon' ... [15]

The history of England, like the history of any country, contains enough examples of cruel and sustained coercion, of judicial injustice and even murder, of the suppression of speech, assembly and the right to sympathize with great causes. Nevertheless, the ideals of the Common Law have been otherwise and it is not preposterous to connect the formation and strength of those ideals with the comparative weakness of thoroughgoing revolutionary ideology, the comparative absence of genuine revolutionary breaks in England's history, and the supplementary, rather than central, role of *Gemeinschaft* in the legal and political traditions of English society. For even without revolution, *Gemeinschaft* as a total society is an ideal that necessarily degenerates or carries with it features that are not part of the ideal. The social cement of the *Gesellschaft* is the social and economic interdependence of a host of particular individuals and activities, having their own specific character, and the framework of laws within which these move. The social cement of the *Gemeinschaft* is not provided by some mystical, spontaneous cohesion that springs up in an undifferentiated mass, drawn together as it were by the flare of camp-fires or the rhythm of marching to song. The social cement of the *Gemeinschaft* is religion or ideology, status and hierarchy, expressed, in modern conditions, by the organization of a disciplined State and Party apparatus, the creation of a centrally supervised ideology and the placing of 'responsible persons' in all politically significant areas of social life. The Maoist *Gemeinschaft* does not abolish bureaucracy, hierarchy or status – it simply ensures that these, too, shall have no security and no independent power vis-à-vis the leader (or leadership) who represents the ultimate and fundamental unity in the collectivity, who – with their 'apparatus' of control – are the real and necessary cement that holds together masses not bound by law.

V

The *Gesellschaft* of abstract individualism, whose only social relationships are based on the strict and calculable *quid pro quo*, is, as Tönnies himself saw, an ideal type, a *Normalbegriff*. There has been no society, no matter how commercial, in which the cash nexus has driven out all other ties. The *Gesellschaft* was made bearable by the existence of innumerable *Gemeinschaften*, from the family, through the working men's club or small-scale trade union, to the University, the traditional firm and the 'neighbourhood'. The Victorian world that preened itself on its respect for freedom of trade, freedom and privity of contract, commercial expansion and the cash nexus was a society almost uniquely dominated by public reverence for the Holy Book and the Law, geared through its public schools to produce neither the merchant prince nor the efficiency expert, but the district officer. That society held up to itself, or at least to its middle and upper classes, no self-congratulatory image of a (current) Great Society or New Frontier, but a conception of that Other Society against which men were to judge themselves and their times, from which they were to fashion standards of conduct and criticism directed at themselves. That Other Society, the leading concepts and figures of which reverberated through school books from London to Sydney, from Toronto to Fiji, was Rome, the Rome of the Republic and the Augustans, of which R.H. Barrow [16] could write in 1949, knowing that most educated readers would still share his sympathy:

> The Romans were 'a law-inspired nation', but the law was of their making and they imposed it on themselves. And, as the fundamental ideas of that law are studied, they will be found to enshrine the ideals and qualities which the Romans of the earliest times set before themselves, now broadened and refined and made of universal application. Respect for eternal values, the will of the gods (*pietas*), and their expression as objective 'right' in the practical things of human life – respect for human personality and human relationships (*humanitas*), whether in the family or the state or the circle of friends, springing from a regard for the personality of each individual and issuing in the maintenance of his freedom (*libertas*) – respect for tradition (*mores*) that holds fast to what has been handed down because it contains accumulated wisdom which no one moment and no one man can supply – respect for authority (*auctoritas*), not as obedience to superior power, but as regard for the judgment of men whose experience and knowledge deserve respect – respect for the pledged word (*fides*) and the expressed intention, the faith of the Romans by which 'with their friends and such as relied on them they kept amity', and 'the most sacred thing in life'.
>
> Respect for these things presupposed training (*disciplina*), the training of the home, of public life, of life itself, and the training which comes from the self (*severitas*). And training of this kind creates a responsible cast of mind (*gravitas*) which assigns importance to important things, so that, when once the hand is placed to the plough, a man does not look back and falter, but keeps to his purpose (*constantia*). These are the qualities which make up the genius of the Roman people.

There are affinities here with the Protestant ethic, but there is more – the sense of history and of subtle balance, of the relevance of character and institutions, the voice in short of the *Gemeinschaft* as it is in fact and not in fantasy. That sense constitutes the formal difference between the Common Law and the Code Napoléon, both used as the foundation stone for a *Gesellschaft*, but pointing in very different directions, breathing a very different spirit. [17]

The reluctance of Western radicals to face up to the real logic of the *Gemeinschaft* is connected with a second *motif* in current attitudes, the elevation of 'authenticity', of 'self-determination', of the self as the measure of all things. The very same history and technology that have made modern man feel helpless and powerless have at the same time laid the foundations for an incredible extension of the concept of human potentialities and human self-determination: will, says the vulgar imagination, took us to the moon; the Presidential word unleashed an atom bomb. Modern technology has thus consummated and infinitely magnified the modern (Cartesian) divorce between user and used and encouraged, in some, the complete and utter elevation of man or consciousness or feeling as pure, undetermined or self-determined and omnipotent subject. Beneath the more extreme and strident forms of this lies something no less horrible than the outrages of recent modern history and modern technology. This is the utter emptiness of the new post-industrial Ego, with its extreme and consequently unstable and insecure individualism, with its mingling of the cult of self-expression with the demand for instant satisfaction through external stimulation ('capture my interest'), with its quite extraordinary incapacity for conscious self-criticism and self-evaluation. All this represents more than the triumph of Max Stirner over Karl Marx; it consummates and bears witness to the unmistakable ending of one of the greatest and most creative periods of European culture. The Stirnerite Ego has no conception of culture: it flees from its 'self-determined' and therefore characterless 'personality' to the ideal of an undifferentiated, spontaneous, womb-like 'community', just as it flees from objectivity and the de-personalized tradition of critical discussion and enquiry to physical and psychological 'encounters', to nature mysticism and Zen Buddhism, as it flees from the sense of complexity and inherent tragedy that stands at the centre of all great art and culture to 'participation' in a single but allegedly 'total' and all too often 'mind-bending' and 'mind-blowing' 'experience'. Even in religion, an index of rather than a contributor to human culture, we have the flight from historical institutions, with their accumulated experience of the fragility of human society and human relationships and of the complex intertwining of virtues and vices, ends and means, to the Pentacostal enthusiasm of a truly primitive Christianity, made even worse in its encouragement of self-satisfaction by the rejection of a mysterious transcendence, by the insistence that God is in me, with me and for me, that Jesus was precisely the sort of superstar that I in principle, today or tomorrow, might also be. Feuerbach, for all his anthropotheism, his reduction of religion, law and society to Man, had some sense of culture, rooted for him in love and language and expressing itself, in modern conditions, in science and politics. It is this sense of culture which constitutes the 'metaphysical essence' that Stirner accused Feuerbach of retaining – what Stirner sees as 'metaphysical' is that which cannot be spun out of his solitary Ego, that which transcends and shapes particular individuals and which has love, curiosity and communication, as concrete and complex activities, among its constituents. The search for personal identity and 'authenticity', in conditions of the loss of culture and social setting, and the fantasy-elevation of an unstructured, ahistorical, *imaginary* community, as empty of content as the selfdetermined ego, are thus intimately related.

What, then, are the alternatives? The price of liberty is, indeed, unceasing vigilance; it cannot be gained by surrendering it to others to exercise on our behalf. The cost of material progress and satisfaction is effort; a total shift of attention from the problems and the morality of production to the problems and morality of distribution builds no foundations for social progress. The substitution of a crude and half-baked, unhistorical, sociology for history and philosophy as the central language of our time has sharply reduced the level of perceptiveness and responsibility with which social problems are discussed and with which social values are formulated. The new emphasis on 'dramatising' social ills, on the demonstration, the highly emotional protest, in short, on getting coverage in the media, has been valuable in overcoming a lack of visibility as a problem of the poor, the black, etc. But it also reinforces this tendency to vulgarise and simplify all problems, or simply to shirk them. The feeling of power that it gives to the people involved is illusory: the 'demo' is aimed at 'the authorities', at getting *them* to do something: it belongs with the *jacquerie* and not with the revolution: it presupposes the omnipotence of the State and the administration. The prevalent 'word-magic' which approaches the problem of women by insisting that 'workforce' must replace 'manpower' in our language, or that the distinction between the married and the unmarried can be waved away or made irrelevant by adopting the deliberately meaningless pseudocontraction 'ms', has the same element of fantasy-gratification, of a flight from real and often intractable problems. The real changes in the human condition – and they have been incredibly pervasive and revolutionary – are to be understood through that very process of technology and of serious political administration from which the contemporary radical flees, or on which he seeks to make sporadic, flanking assaults. To understand it we have to substitute thought for ideology, sustained grappling for instant dramatisation, and appreciation of complexity for simple and comforting 'black-and-white' thinking. That is what culture means: And whatever its shortcomings, there has been no social institution more geared, as part of culture, to the appreciation and taking into account of the complexity of human affairs, and the intractability of conflict, than law, expecially perhaps the Common Law.

The central problem for socialism – and for man – in the second half of the 20th century, then, is not the problem of property, an area in which classical liberalism has collapsed and in which, as Karl Renner foreshadowed, the distinction between the private and the public is already breaking down. It is not only that the State in most modern industrialized societies is very much stronger than private property and has already curbed its power on every side. It is rather that all the evidence shows that this is an accelerating development. The scale of property has become so vast, the sources from which it draws its wealth so multifarious and pervasive, and its social effects and ramifications so great, that modern man is having increasing difficulty in thinking of property as private, as the concretization of an individual will reifying itself in land or objects, as a walled-in area into which others may not enter. There is, in other words, a shift of attention from the property whose paradigm is the household, the walled-in or marked-off piece of land, the specific bales that make up a cargo or consignment, to the corporation, the hospital, the defence establishment, the transport or power utility whose 'property' spreads throughout the society and whose existence is dependent upon subsidies, State protection, public provision of facilities, etc. In these circumstances, a view of society and a view of property as a collection of isolated and isolatable windowless monads that come into collision only externally and as a departure from the norm becomes untenable. Property becomes social in the sense that its base and its effects can no longer be contained within the framework of the traditional picture. The major sphere of social life passes from the private to the public, not merely in the sense that more and more activity is State activity, but in the sense that more and more 'private' activity becomes public in its scale and its effect, in the sense that the oil company is felt to be as 'public' as the State electricity utility, the private hospital and the private school, with their growing need for massive State subsidies, as public as the municipal hospital and the State school. This explains one of the most striking of modern phenomena – the decline in respect for private property, the popularity of the sit-in, of the demand for *access* as independent of ownership and as something that ought to be maintainable against it. It is also the most significant sense in which socialist thought and action are emerging triumphant.

Society, then, we are arguing, is being revolutionized just as and largely because technology is being revolutionized. There are no longer boundaries that one can draw around one's self and one's possession. It is the much more urgent sense of social interconnection, of the destruction of social and individual boundaries, which accounts for the declining confidence among political ideologists in gradual and limited social and legal reform, in Karl Popper's concept of piecemeal social engineering.

Among lawyers, the predilection for radical and sweeping reform, for wholesale transformation of the law, is no doubt very much less marked. Nevertheless, the Law Commissions in England and those reforming and revising the laws in other countries are going, under the pressure of social demands, very far beyond the mere tidying-up and rationalization of existing law, and even beyond cautious piecemeal social engineering. It is not only that more and more private areas become public areas and that more and more private law becomes public law. It is also that the whole range of legal reforms is fundamentally undermining what many have seen as the specifically legal tradition in society – the tradition connected with a conception of the *distinction* between law and morality, between law and administration and between law and politics, with the doctrine of the separation of powers, with the concepts of justifiability, competence and jurisdiction and the admittedly defensible presumption in favour of the rights of the citizen against those of the State and the rest of society. Paradoxically, it is in the United States, where radicals are constantly seeking to rely on the values and provisions of a Lockean Constitution, that the individualistic treatment of these matters and the attempt to base all rights on property by recognizing a 'new property' are still strong.

In reality, as liberalism as a socially formative and legitimating ideology collapses, no coherent social ideology has yet risen to take its place. Socialism fails because it has produced neither a theory of law nor a theory of culture: in both it is parasitic upon and ambivalent toward what it calls bourgeois civilization. The radicals who know that they are no longer socialists in any classical sense vacillate, as it suits them, between *Gemeinschaft*, *Gesellschaft* and bureaucratic-administrative values: while they speak of community, self-expression and authenticity, they are in fact helping to demolish the last of the *Gemeinschaft* structures in modern society – from the family to the University – and to strengthen the simplification of society into the State as a set of administrative structures on the one hand and the population as an unstructured mass of individuals on the other. It is there that the problems, for a contemporary social theory and for contemporary mankind, begin.

The Australian National University
The University of Sydney

NOTES

[1] In our paper 'Beyond Bourgeois Individualism: The Contemporary Crisis in Law and Legal Ideology', presented to the World Congress on Philosophy of Law and Social Philosophy held in Madrid in September 1973; reprinted in *Archiv für Rechts- und Sozialphilosophie*, Beiheft (Neue Folge) No. 8 and in Eugene Kamenka and R.S. Neale (eds.), *Feudalism, Capitalism and Beyond* (A.N.U. Press, Canberra, Edward Arnold, London, 1975), pp. 126–144. There we also set out more fully our conception of the distinction between *Gemeinschaft, Gesellschaft* and bureaucratic-administrative ideologies and arrangements.

[2] On this point see the interesting paper by R. F. Brissenden, 'From *Black Mask* to *The Godfather*: The Crime Novel in the United States', presented to the Seminar on the Transformation of Law, Justice and Morals held in Canberra on 2–4 August 1975 and to be published in R. R. Brown, Eugene Kamenka and Alice Erh-Soon Tay (eds.), *Freedom and Responsibility: Perceptions of Man, Law and Society Today* (Edward Arnold, London).

[3] This is the theme of much of the late George Lichtheim's writing. See esp. his *The Origins of Socialism*, Praeger, New York, 1969, at p. 3.

[4] Lichtheim, *op. cit.*, p. 4. America has finally caught up, but the 'demonstration' effect came from blacks and other minorities, backed by the highly ideologized products of the U.S. education system, not from the proletariat.

[5] Thomas Paine, *Common Sense and Other Political Writings*, ed. by N. F. Adkins, Liberal Arts Press, New York, 1953, pp. 54, 55. See also the brilliant discussion of the popular hostility to law in American rural and frontier society between 1800–1840 in Perry Miller, *The Life of the Mind in America*, Gollancz, London, 1966, pp. 99–116. Miller describes this hostility as something more than a memory of Tom Paine, as a mood more pervasive than

either political party could control, as 'something deep, atavistic, persistent in the community' (Miller, p. 103). In America, this feeling took special (highly individualistic) forms, and flourished in conditions where the distribution of property was not an issue. However, it drew strongly on the protest against external authority and against a social caste (the legal profession) which appeared to make justice a conspiracy of the learned and powerful against the helpless integrity of the common man. It was linked, as one might expect, with revivalism.

[6] Morelly, *Code of Nature*, as translated in Albert Fried and Ronald Sanders, *Socialist Thought: A Documentary History*, Edinburgh U.P., 1964, pp. 18, 19. The classical statement of the contrary view is Mandeville's *Fable of the Bees* (1714), with its insistence that social progress and other *public* benefits depend upon avarice and other *private* vices:

> ...fools only strive
> To make a great and honest hive.
> T'enjoy the world's conveniences,
> Be fam'd in war, yet live in ease,
> Without great vices, is a vain
> Eutopia seated in the brain.
> Fraud, luxury, and pride must live,
> While we the benefits receive:
> ...
> So vice is beneficial found,
> When it's by justice lopp'd and bound;
> Nay, where the people would be great,
> As necessary to the state,
> As hunger is to make 'em eat.

Bernard Mandeville, *The Fable of the Bees*, ed. and abridged by Irwin Primer, Capricorn Books, New York, 1962, p. 38 (the poem which forms the nucleus of the work was first published in 1705). 'What I have laboured to prove throughout the book', Mandeville wrote later in *A Letter to Dion* (1732) "[is] that luxury and the vices of man, under the regulations and restrictions laid down in the *Fable of the Bees*, are subservient to, and even inseparable from, the earthly felicity of the civil society": Mandeville, *op. cit.*, p. 6.

[7] Celestin Bouglé and Elie Halévy (eds.), *Exposition de la Doctrine de Saint Simon* (originally published in 1830 and 1831 by Saint-Simon's disciples), pp. 255, 261, 272–3, here cited from the translation in Lichtheim, *op. cit.*, pp. 52–3. Saint-Simon's own use of the term 'industriels' combined into a single social class or interest group the industrial entrepreneur, the scientist and manager and the industrial worker. His disciples in the 1820's were calling for the conversion of private property from an absolute right into a social function alterable at will; they were not, however, radical egalitarians in the tradition of Babeuf and his 'Conspiracy of Equals', calling for the total abolition of property. Their incipient socialist critique of bourgeois society flowered into socialism in the 1830's. The *Exposition* and Bazard's public lectures familiarized the French public with the central tenets of the new faith: public ownership and the abolition of social inequality; *Le Globe*, the journal of the Saint-Simonian Leroux, coined the term *socialisme* in February 1832. Marx thought that history had overtaken the Saint-Simonians, that they were but a prelude to his own truth. He was wrong.

[8] N. S. Timasheff, *An Introduction to the Sociology of Law*, Harvard University Committee on Research in the Social Sciences, Cambridge, Mass., 1939, p. 367.

[9] Fried and Sanders, *op. cit.*, p. 183.

[10] A. V. Dicey, *Lectures on the Relation between Law and Public Opinion in England during*

the Nineteenth Century, 2nd ed., Macmillan, London, 1948, pp. 211–3.
[11] Cited in Boris Nicolaevsky and Otto Maenchen-Helfen, *Karl Marx: Man and Fighter*, Methuen, London, [1936], pp. 113–4.
[12] Proudhon himself was not guilty of contradiction; he made it clear that the 'property' he was denouncing as theft was property in the means of production and the labour of others, not property in general. This is now the position of all communist parties and governments, though the Chinese avoid the commonly-used term 'personal' or 'individual' property.
[13] For the (abridged) text of these, in English, see Fried and Sanders, *op. cit.*, pp. 20–31.
[14] As cited in Ludwig Lewisohn, *The Story of American Literature*, Modern Library, New York, 1939, p. 9.
[15] See Eugene Kamenka, 'The Paris Commune and Revolution Today', in Kamenka (ed.), *Paradigm for Revolution? The Paris Commune 1871–1971*, A.N.U. Press, Canberra, 1972, pp. 98–99.
[16] *The Romans*, Penguin Books, London, 1949, at pp. 2, 8–9.
[17] The Moral Code of the Builder of Communism, much bandied about in the U.S.S.R., like genuinely socialist morality generally, still has much more in common, verbally at least, with the morality of Republican Rome than with the strident but empty emphasis on 'self-assertion' of the Free Speech Movement or of the New Educational 'Philosophy'.

NORETTA KOERTGE

RATIONAL RECONSTRUCTIONS

History of science without philosophy of science is blind; philosophy of science without history of science is empty.
– LAKATOS

1. INTRODUCTION

Lakatos' rather innocent sounding paraphrase of Kant (1971, p. 91) announced a bold theory about the dialectical interaction between history and philosophy. His theory, as put in final form in his (1971), contains two theses which I wish to examine:

THESIS 1. The historian's account of the past should be dominated by the best available *normative* theory of rationality. Since factual history is only a "caricature of its rational reconstructions" (1971, p. 122), the historian is advised to put what *should* have happened in the body of the text; what *really* happened is relegated to the footnotes.

THESIS 2. However, the philosopher's normative account of scientific method must be appraised according to its usefulness and fertility in reconstructing the *actual* development of science (1971, p. 118). Philosophy of science should not be *aprioristic*; rather it should be grounded on actual scientific practice (1971, p. 121).

Perhaps no other aspect of Lakatos' philosophy illustrates so well his love of dialectics and Hegelian syntheses. It is clear that Theses 1 and 2 must be taken together – they qualify and counter-balance each other. However, for purposes of analysis I am going to discuss them separately. I am mainly concerned with the question of the legitimacy of introducing rational reconstructions into the history of science. I will also comment briefly on Lakatos' criterion for appraising rival theories of scientific method.

2. CONSERVATIVE RECONSTRUCTIONS

Even the most epistemologically conservative would be prepared to ad-

R. S. Cohen et al. (eds.), Essays in Memory of Imre Lakatos, 359–369. *All Rights Reserved*
Copyright © 1976 by D. Reidel Publishing Company, Dordrecht-Holland

mit, I believe, that very often the historian is called upon to 'piece together' some object or episode from the past and that in doing so, he must rely on theories or generalizations which go beyond the available facts. Examples:

(a) The museum curator who glues together potshards is, of course, *assuming* that they all belong to the same pot and that there is only one correct way to juxtapose them. (Only a philosopher would be inclined to call such an assumption a theory.)

(b) If some pieces of the pot are missing, the archeological conservator will use his general knowledge of pots of the particular period to *infer* the shape and design of the missing fragments. He may also tacitly rely on more universal principles such as "Other things being equal, people don't design pots which leak badly." For the convenience of the museum visitor, he may *fabricate* replicas of the missing bits. However, the modern parts will be carefully labelled as such.

(c) The archeological systematizer may construct a typology of pots of a particular culture and illustrate his books and display cases with diagrams or models of various 'ideal types' of pot. In general, the most useful sort of ideal type (or paradigm pot) will *not* be an exact copy of any existing pot, although it should, of course, strongly resemble actual specimens.

Similar sorts of reconstructions take place in all branches of history. The archivist must decide which pages of manuscript go together (cf. a). The papyrologist must interpolate missing letters (cf. b). The editor of a critical edition will collate several imperfect versions of a work and try to arrive at a definitive edition – the ideal version which is presumed to capture the original author's intentions (cf. c). All three will automatically assume (unless there are strong, explicit grounds for doubting it) that the long-dead writer did not write total gibberish. Whether the reconstruction involves collation or interpolation or idealization, the historian relies on certain minimal attributions of rationality to the actors in the past – he assumes their actions were appropriate to their problem situations.

3. Radical reconstructions

But if rational reconstructions, i.e. interpretations of the past predicated on assumptions of rationality, are part of the standard stock-in-trade of the historian, why should historians react so violently to the historiographic theories of Popperians? (For a typical example, see Smeaton and Finlay's review (1962) of Agassi's *Towards an Historiography of Science* (1963) and Holton's discussion (1974) of Lakatos.) Historians are quite correct in thinking that the radical reconstructions which have been proposed first by Agassi and now by Lakatos, are very different indeed from the modest reconstruction techniques which they are accustomed to using. But have they been correct in their evaluation of the new techniques? I will distinguish various types of radical reconstructions and discuss their merits (and demerits) in turn.

3.1. *Bold Reconstructions*

A historical account may be consistent with all the available data and yet be very bold and fallible, either because it goes *far beyond* the evidence (e.g., a whole skeleton inferred from one bone) or because it is based on a very controversial theory (e.g., Manuel's psychoanalytic biography of Newton).

Bold reconstructions can be both provocative and productive. As Agassi pointed out in his (1963), they direct historical research into new areas and may result in the uncovering of new historical materials which will either confirm or refute the underlying speculations. Many historians are uncomfortable with bold history (Finlay's comparison of Agassi-type history to historical novels is intended to be pejorative) because they believe it will not 'last' or make a 'permanent contribution'. But as long as the underlying conjectural theories are made explicit and the extent to which the account is (or is not) supported by the evidence made clear, bold reconstructions need not mislead the reader in any way and may lead to very exciting new historical insights.

3.2. *Procrustean Reconstructions*

It is one thing to ride far beyond the facts on the back of a philosophical hobby-horse. It is quite another to stretch or mutilate or contradict the evidence in order to make it fit one's theory. Clio can certainly survive and may even thrive on wild imagination which is tempered by criticism. But the deliberate distortion of data to make them fit one's ideological system, whether it favors rationality *or* irrationality, would lead to the destruction of history.

Does Lakatos' theory of rational reconstruction advocate the use of Procrustean methods? If so, then historians would be right to view it with misgivings.

It must certainly be admitted that Lakatos' own historical efforts showed an ever decreasing emphasis on factual accuracy. (I am not here concerned with the question of whether the references he does cite are correct or not. My point is that there are simply fewer factual allusions in his later work.) His early 'Proofs and Refutations' (1963–64) was carefully designed so as to avoid confusion between the reconstruction and what really happened. It was written as a dialogue between fictional characters, Alpha, Beta, etc. Footnotes then documented similarities and differences between what was going on in the play and actual historical developments.

However, in his 'Falsification and the Methodology of Research Programmes' (1970) he introduced the very misleading convention of linking the names of historical figures, e.g., Prout, with views which the historical characters themselves explicitly denied. Thus on p. 138, the text says that Prout recognized certain anomalies, while in a footnote we read, "Alas, all this is rational reconstruction rather than actual history. Prout denied the existence of any anomalies. For instance, he claimed that the atomic weight of chlorine was exactly 36." I will not enter here into the question of the extent to which this footnote accurately reflects Prout's views. The general point is to note that this style of historical writing still permits the attentive reader to distinguish what is *intended* to be fact from what is deliberate fiction. But later in the essay, Lakatos even gives up the practice of providing factual footnotes for the reader. For example, he admits that his account of Bohr is a 'caricature' to be taken with "tons of salt" (p. 140), but fails to provide a single compensating grain of truth in the footnotes! The reader knows *that* what he has been told about Bohr is partly false, but unless he has independent historical knowledge of the episode he has no idea *which* parts of Lakatos' account contradict readily available evidence.

But why would a historian ever wish to write an account which he *knows* to be false and which he does not believe will be vindicated by further research? (Note that when Prout reported the atomic weight of chlorine as 36 even though the best experimental data available gave another value, it was because he *thought* the contemporary experimental techniques were inaccurate. What Lakatos advocates doing is quite different.)

Kuhn raises this question (1970, p. 256) and concludes that Lakatos is really a philosopher who uses quasi-historical examples to illustrate philosophical theses but is not interested in real history at all. This may have been true of the earlier Lakatos ('Proofs and Refutations' is primarily an essay in philosophy of mathematics and not a history of polyhedra), but I think that in his 'History and Its Rational Reconstructions' Lakatos was seriously proposing a revolutionary historiographic approach. It involves the use of what I will call Galilean reconstructions.

3.3. *Galilean Reconstructions*

Since the time of Galileo it has been standard practice in the physical sciences to use ideal laws. I think it is plausible to argue that Lakatos was trying to introduce this fruitful Galilean innovation into the discipline of history. For example, so this line of argument would go, Galileo's law of the pendulum ignores air resistance and friction although these 'external' factors are always operative. Likewise, Lakatos' theory of the appraisal of research programmes ignores psychological and sociological factors although scientists are always (more or less) human. One could extend the parallel by noting that part of the opposition to Galileo's methodology stemmed from an empiricism which was too narrowly and pedantically construed.

Let us explore this possibility in more detail. Although the method of idealizations in physics has been enormously successful, it must certainly be used with considerable tact or the result will be Procrustean, not Galilean. The major problem connected with idealizations was faced directly by Galileo himself. He wished to criticize Aristotle's law of falling bodies because it did not correctly describe what happened. Yet he knew that his own law would also be refuted because it neglected the resistance of the medium which was always present in actual cases of free fall. The problem of how to have an empirically sound science and yet use idealizations was not an easy one to solve. (For an account of Galileo's struggle with this question, see my 'Galileo and the Problem of Accidents'. Many problems still remain. For example, in what sense can one claim to be a realist if one's theories are about non-existent entities, such as ideal gases?)

If we are to transplant the method of idealization from physical science into history, we must be sure to carry along with it guidelines for using it responsibly. Here are a few caveats which are generally heeded by physical scientists. One should list the 'external' factors which are known to operate in experimental cases but which are ignored by the idealized theory. (Thus Galileo admitted that air resistance affected the rate of free fall.) If one can find instances in which the effect of the external factors is predicted to be negligible, one should use such cases to test the ideal law directly. (Galileo's intricate description of parchment-lined inclined planes is an attempt to minimize friction and other 'accidents', as he called them.) If no such possibility exists, one tests the ideal law indirectly. (For example, Galileo suggested that one test his law of falling bodies by comparing the velocities of dense and light bodies in increasingly rare media. He predicted that as the medium offered less and less resistance the behavior of 'heavy' and 'light' objects would become more and more similar. If this were so, he said, then one would be justified in assuming that, in a vacuum, they would undergo exactly the same acceleration.)

To put the main point briefly – physicists who use idealizations should be able to give at least rough estimates of the *magnitude* of the effects of factors which the ideal law ignores. This is generally easier to do when the 'internal' factors (which the law describes) and the 'external' factors (which are being neglected) act independently.

But it is just this quantitative description of the limits of accuracy of the idealization which is almost always missing in the historical case. How can one *measure* the extent to which Galileo's acceptance of the Copernican system was based on its scientific merits and how much it was motivated by broader ideological concerns? What fraction of the rate of increase of scientific knowledge in the 17th century was caused by the formation of scientific societies? How large was the influence of the Protestant ethic? What component of the growth vector was the result of the craft tradition? Just by trying to *state* these familiar historical controversies in mathematical terms, we immediately see how far we are at present from being able to answer them with any sort of quantitative precision. Such issues may be subjected to rational debate and one may even be able to draw some conclusions as to which are the *major* factors. But one might well be reluctant to introduce idealizations unless one has some rough estimate of the amount of distortion which results.

Without some check, how could one compare the legitimacy of competing idealizations?

Although there are certainly difficulties in practice (though not, I think, in principle) in introducing idealizations into history, I think much of the resistance within the trade union of historians to responsible reconstructions stems from another source – this has to do with their conceptions of the aim of historical inquiry. Again I will use an example to explain what I have in mind. If one is writing a vignette about a dearly beloved Volkswagen, one will probably play down the general characteristics which it shares with all VW's and focus instead on its peculiarities – the way it coughes on cold mornings, the time it climbed an icy hill when all the other cars had stalled, etc. The historian of science may well adopt the same attitude towards his or her subject – we may be more interested in learning about Newton's fascination with alchemy and his vacillating views towards the ether than in hearing a recital of the attributes which he shared with the standard scientific stereotype. But of course Lakatos and others are quite free to construct and articulate the stereotype too.

Granted that there is a place for Galilean reconstruction in history, we are still left with the problem of arbitrating between accounts which are based on different philosophical viewpoints.

4. Reconstructing Rationality

It is to solve the problem of appraising competing idealized accounts of the growth of science that Lakatos introduces his meta-criterion and produces his most dazzling display of dialectics (1971). Although one must use philosophical theory to write history, one must also use the past to judge the adequacy of philosophical theory (1971, p. 109). One prefers that idealization which leads to the "discovery of novel historical facts" and which successfully "reconstructs" a "growing bulk" of history (p. 118). As I have described his position so far, it would *appear* that what Lakatos is proposing is a free competition between historians of different persuasions – there should be a Darwinian struggle between Popperian history and psychoanalytic history, history by Marxists, and anarchists, and Lakatosians. Then we would see which historiographic programme is most successful in turning up interesting new results and in accounting for what happened.

However, this is *not* what Lakatos has in mind. The episodes in the past which are to 'test' the competing historiographic approaches are to be pre-selected *according to their rationality*! As Lakatos puts it, competing accounts of the growth of scientific knowledge are to be evaluated with the help of "normatively interpreted history" (p. 92). The aim is to reconstruct "value-impregnated history" (p. 118) and to systematize the "basic normative judgments" of the scientist (p. 118).

It is not at all clear *which* norms are to enter into the construction of the data base against which theories of science are to be tested. Should we use Lakatos' norms or Feyerabend's? Should we use the norms explicit in what Newton said or those implicit in what he did? I think Lakatos may have thought that it would be easy to find a non-controversial normative common denominator. Even if this were the case (which I doubt), what Lakatos *cannot* do on the basis of a history of normatively selected episodes is to claim superiority over a Kuhnian, or Marxist, or psychoanalytic account. These latter approaches hope to comprehend the *entire* development of science, not just tell the story of certain carefully selected episodes. One cannot claim that there are no wild flowers in England if one only visits greenhouses!

I have argued that Lakatos' challenge to externalists or descriptivists, as we might better call them, is unfair because the historical data base which is to decide between them is deliberately rigged so as only to include 'rational' episodes.

I believe his challenge to competing internalist accounts is also unfair and inappropriate. It is unfair because Popper and others have sought to give *independent* arguments for or against the rationality of certain methods. Lakatos has made the *a priori* assumption that the best example of present (and past) scientific method cannot be improved upon. Therefore, he rejects *out of hand* any theory of rationality which would reform orthodox scientific practice.

The strength of Lakatos' commitment to the claim that the growth of science has been a rational process is clearly illustrated in the following passage. (He has just argued that a Popperian account does not fit history of science very well.)

But if – as seems to be the case – the history of science does not bear out our theory of scientific rationality, we have two alternatives. One alternative is to abandon efforts to give a rational explanation of the success of science ... and try to explain *changes* in '*paradigms*' in terms of social psychology. The other alternative [is to give a] *sophisticated* version... of falsification and thereby rescue methodology and the idea of scientific *progress*. This is... [the way] I intend to follow (1970, pp. 115–16).

But the assumption that our present ideas about rationality and scientific method are already perfect is absurd. Where would logicians be if they had rejected any theory of inference which did not reconstruct all of the arguments of the great Euclid as valid? Or what if medical theory had restricted itself to the articulation of the methods used by the best surgeons alive in 1850? Yet these absurdities are strictly analogous to what Lakatos' metacriterion dictates to philosophers. It is also a strangely anti-historical demand, for the history of science provides many examples of increasing sophistication (and at times degeneration) in scientific methodology. A notable example is provided by the development of theories of experimental error. For instance, when Lavoisier describes the weights of products and reactants in a fermentation, he gives data which appear to agree perfectly to within six significant figures (1789, pp. 133–37). In a later aside (p. 140), he admits that this accuracy is spurious and results from a judicious application of the law of conservation of mass which he takes as an "incontestable axiom" (p. 130). This is one very simple example of a procedure practiced by a great scientist which was acceptable then, but would now be judged irresponsible.

The philosopher has much to learn from the history of science, but he must always maintain a critical distance. To paraphrase Marx, the duty of philosophy of science is to *improve* scientific practice, not describe it.

5. Epilogue

When Lord Acton was editor of the Cambridge Modern History series he sent each of the contributors a statement of policy:

Our scheme requires that nothing shall reveal the country, the religion or the party to which the writers belong (1898, p. 397).

In order to fulfill this laudable aim of objectivity, historians have often tried to purge their accounts of all theoretical superstructure. Yet as Butterfield emphasized, the business of an historian is to discover "the interconnectedness of events" (1969, p. 8). Unless he shows connections, his chronicle will only be a "rope of sand" (p. 14).

So historians *must* use theories, not only as a sieve to select facts, but also as the mortar which glues them together. The goal of objectivity is best served by welcoming theory in an open, explicit way, while at the same time keeping a critical eye on it.

Agassi suggested that historians emulate scientists by using *bold* theories. Lakatos suggested that they go even further and use *idealizations* – theories which deliberately neglected factors known to be operative. I think much writing in the history of science could be improved by a bit of bold conjecturing and radical reconstructing – but only if it is accompanied by the severe criticism and stringent testing which also goes on in the sciences.

Lakatos is perhaps best known for his bold conjectures, his provocative caricatures, and the zany 'potted histories' with which he opened his lectures. One was forced to learn any number of things in self-defense! But he was also a very conscientious editor and critic. He used to say that he wanted the book review section of the *BJPS* to become so authoritative that the reviews would be read long after the original books had been forgotten. We can only dimly guess at the wit and wisdom which his review of the present book of essays would have contained.

Indiana University

BIBLIOGRAPHY

Agassi, J., 1963, *Towards an Historiography of Science*, Vol. 2 of *History and Theory*, Mouton & Co., The Hague.

Butterfield, H., 1969, *Man on His Past*, Cambridge University Press, Cambridge.

Dalberg-Acton, J. E. E. (Baron Acton), 1898, 'Letter to Contributors to the Cambridge Modern History' in W. H. McNeill (ed.), *Lord Acton: Essays in the Liberal Interpretation of History*, University of Chicago Press, Chicago, 1967, pp. 396–99.

Holton, G., 1974, 'On Being Caught Between Dionysians and Apollonians', *Daedalus* **103**, 65–81.

Koertge, N., 'Galileo and the Problem of Accidents', forthcoming in *Journal of the History of Ideas*.

Kuhn, T. S., 1970, 'Reflections on My Critics' in Lakatos [1970], pp. 231–78.

Lakatos, I., 1963–64, 'Proofs and Refutations', *The British Journal for the Philosophy of Science* **14**, 1–25, 120–139, 221–245, 296–342.

Lakatos, I., 1970, 'Falsification and the Methodology of Scientific Research Programmes' in I. Lakatos and A. Musgrave (eds.), *Criticism and the Growth of Knowledge*, Cambridge University Press, Cambridge, pp. 91–196.

Lavoisier, A. L., 1789, *Elements of Chemistry* (English transl. from the French by R. Kerr), Dover Publications, New York, 1965.
Smeaton, W. A. and Finlay, S., 1962, 'Review of J. Agassi, *Towards an Historiography of Science*', *Annals of Science* **18**, 125–129.

ISAAC LEVI

A PARADOX FOR THE BIRDS

Does '*a* is a red herring' confirm 'All ravens are black'?

In some contexts, it does not.

Context 1. *X* has already settled as part of his corpus of knowledge that all ravens are black. He then makes an observation and adds to his corpus '*a* is a red herring'. Clearly, the observation report does not confirm 'All ravens are black'; for the generalization is already an established part of the corpus of knowledge and is as confirmed as it could be. Of course, an observation of a black raven contributes nothing to the confirmation of 'All ravens are black' either.

Context 2. *X* includes in his corpus of settled knowledge 'Some ravens are not black'. Once more, neither '*a* is a red herring' nor '*a* is a black raven' confirms 'All ravens are black'.

Clearly, if '*a* is a red herring' is to confirm 'All ravens are black', the truth value of the generalization cannot be settled as a matter of knowledge prior to the observation.

But even in situations where the truth value of 'All ravens are black' is not decided by deduction from the corpus of knowledge, we cannot conclude that red herrings, black ravens or anything else will confirm 'All ravens are black'. Data confirm hypotheses only in contexts where these hypotheses are taken to be potential answers to some question taken as seriously meriting investigation. To illustrate, consider the following context:

Context 3. *X* wants to find out whether some crows are black or not. On at least one plausible reading of *X*'s question, the only consistent potential answers to his question are representable by 'Some ravens are black' and 'No ravens are black'. There is another response which can, in an extended sense, be regarded as a potential answer. *X* can suspend judgement between 'Some ravens are black' and 'No ravens are black'. On some occasions, suspension of judgement may be warranted by the evidence rather than the two alternatives and, in this sense, should be considered a potential answer.

R. S. Cohen et al. (eds.), Essays in Memory of Imre Lakatos, 371–378. *All Rights Reserved*
Copyright © 1976 by D. Reidel Publishing Company, Dordrecht-Holland

In Context 3, neither '*a* is a red herring' nor '*a* is a black raven' confirms 'All ravens are black'. The reason is that 'All ravens are black' does not represent a potential answer to the question under investigation.

To be sure, data consisting of a sufficiently large number of red herrings will warrant the conclusion that no ravens are black. A single black raven will suffice to entail 'Some ravens are black'. But, given the demands of the question, neither sort of data provides confirming evidence for 'All ravens are black'.

Thus, we have the following necessary conditions for '*a* is a red herring' to be a confirming instance for 'All ravens are black':

(1) 'All ravens are black' should be consistent with the corpus of settled knowledge prior to observation and not entailed by that corpus.

(2) 'All ravens are black' should be a potential answer to the question under investigation.

These necessary conditions, however, are not jointly sufficient.

Context 4. *X* wants to know whether all ravens are black or none are. On many occasions, *X* would formulate the question in this way only if he already assumed as part of his corpus of knowledge that there are ravens so that the joint truth of these two hypotheses is ruled out prior to observation. However, there may be contexts where the matter is not settled so that 'All ravens are black' and 'No ravens are black' are *not* contraries.

In that kind of situation, *X* should regard 'There are no ravens' to be a potential answer to his question; for it is equivalent to concluding that 'All ravens are black' and 'No ravens are black' are both true. Moreover, the contradictory of a potential answer should also be a potential answer. Hence, 'There are ravens', 'Some ravens are black' and 'Some ravens are not black' should be potential answers.

To avoid confusion, it is important to remember that adopting a potential answer as a conclusion to be added to a corpus involves a commitment to more than adding a single sentence to the corpus. It involves adding also all those sentences which are not as yet in the corpus but which are deductive consequences of the corpus together with the specific sentence used to represent the conclusion. We can say that the sentence *h* is 'accepted as strongest' relative to the corpus *K*.

Thus, when h is accepted as strongest, hvg is added to the corpus K. But it is not accepted as strongest. To accept hvg as strongest is to add hvg to K without adding h and without adding g. It is to suspend judgement between h and g. Thus, when we represent potential answers by single sentences, we regard these sentences as potentially acceptable as strongest relative to the question under investigation.

Thus, in Context 4, accepting 'there are no ravens' as strongest is a potential answer. So is accepting 'There are ravens and they are all black' as strongest. So is suspending judgement between these two sentences – i.e., accepting their disjunction as strongest. Since the disjunction is equivalent to 'All ravens are black', accepting the disjunction as strongest is to adopt the same potential answer as accepting 'All ravens are black' as strongest.

To obtain a complete set of potential answers to a given question, we may begin by specifying a set of sentences (called 'the partition') h_1, $h_2, \ldots, h_n$ (I shall suppose that n is finite) such that the corpus K entails that at least and at most one member of the partition is true and such that each member of the partition is consistent with K. Any consistent potential answer generated by such a partition is tantamount to the set of sentences added to K by adding a disjunction of a nonempty subset of the partition as a strongest accepted sentence and forming the deductive closure.

In Context 4, I suppose the partition to consist of (i) 'There are no ravens', (ii) 'There are some ravens and they are all black', (iii) 'There are some ravens and none are black' and (iv) 'Some ravens are black and some are not'.

When we say that data confirm some sentence, we may mean one of two things: (a) We may mean that the data, in some sense, favor a potential answer whose adoption involves adding the sentence mentioned to the corpus. In this sense, if data d confirms h, d confirms all the deductive consequences of h as the special consequence condition requires. (b) We may mean that the data favors adoption of a potential answer whose adoption entails accepting as strongest the sentence the data are said to confirm. In this sense, if d confirms h, it does not confirm the deductive consequences of h unless they are equivalent given the corpus K to h.

I submit that talk of confirmation is of primary relevance when we are concerned to evaluate potential answers. If that is so, when we say that *d* confirms *h*, we have in mind primarily confirmation in sense (b). That is to say, we have in mind to report that data *d* favors the potential answer consisting of accepting *h* as strongest and all its deductive consequences. At any rate, I shall interpret talk of data confirming sentences in this way in the sequel.

Thus, if '*a* is a red herring' confirms (i), it does *not* confirm 'All ravens are black' even though accepting 'All ravens are black' as strongest is a potential answer and is deductively implied by (i). The point is that if '*a* is a red herring' favors accepting (i) as strongest over the other potential answers, it cannot favor accepting 'All ravens are black' as strongest.

These considerations imply that the Hempelian explication of 'confirms' is unsatisfactory as it stands. '*a* is a red herring' entails the development of (i) and, hence, confirms (i) according to Hempel. But it also confirms, in Hempel's sense, the disjunctions of (i) and (ii), of (i) and (iii), of (i) and (iv), of (i), (ii) and (iii), of (i), (ii) and (iv), of (i), (iii) and (iv) and of (i), (ii), (iii) and (iv). The datum sentence can confirm at most one of the potential answers represented by accepting one of these sentences as strongest.

Nonetheless, we may attempt to stay close to Hempel's approach and require entailment of the development of a hypothesis as a necessary condition for its confirmation by data.

Of all the sentences representing potential answers in Context 4 whose developments are entailed by '*a* is a red herring', (i) and sentences equivalent to it given the corpus *K* are logically the strongest. The potential answer involving accepting (i) as strongest is itself stronger than any potential answer involving accepting as strongest any other sentence not equivalent to (i) whose development is entailed by '*a* is a red herring'.

Assuming that the corpus *K* itself contains no information as to the existence or nonexistence of ravens, *K* together with '*a* is a red herring' will not furnish an adequate evidential basis for accepting (i) as strongest. At least, it will not do so under ordinary circumstances. However, if the data consisted of reports affirming the existence of a sufficiently large number of red herrings, accepting (i) as strongest will be warranted by the data. Actually, I am not prepared to endorse this claim without reservation but shall assume it for the sake of the argument. We may say, therefore, that '*a* is a red herring' confirms (i) in the sense that sufficient replication of this datum should ceteris paribus warrant accepting (i) as strongest. More generally, a sufficiently large number of replications of datum *d* should, on this view, warrant accepting the strongest potential answer such that the development of the sentence accepted as strongest is entailed by *d*.

Consider the datum '*a* is a black raven' in Context 4. By our criterion, this datum confirms (ii). Notice, however, that '*a* is a black raven' also entails the falsity of (i). This means that the initial corpus *K* plus '*a* is a black raven' entails the equivalence of (ii) and 'All ravens are black'. From the point of view of the investigator who has added '*a* is a black raven' to his corpus and, hence, has concluded via deduction that either (ii), (iii) or (iv) is true, the potential answers to his question which remain are equivalent to accepting as strongest the disjunction of some subset of these three sentences. The strongest of these potential answers whose development is entailed by '*a* is a black raven' is accepting (ii) as strongest or, equivalently, accepting 'All ravens are black' as strongest. Hence, we may conclude that '*a* is a black raven' confirms 'All ravens are black'.

Thus, '*a* is a black raven' confirms 'All ravens are black' in Context 4 even though '*a* is a red herring' does not. Moreover, since 'All ravens are black' is equivalent to 'All nonblack things are nonravens', '*a* is a black raven' confirms 'All nonblack things are nonravens' even though '*a* is a red herring' does not.

Four contexts have been identified where '*a* is a red herring' fails to confirm 'All ravens are black'. There are, of course, contexts where it does. I shall cite two of these.

Context 5. Suppose *X* is concerned to find out whether all nonblack objects are ravens or none are. The potential answers might plausibly be generated from the following partition: (v) 'Everything is black', (vi) 'There are nonblack things and they are all ravens', (vii) 'There are nonblack things and none are ravens', and (viii) 'Some nonblack things are ravens and some are not'.

'*a* is a red handkerchief' eliminates (v) by deduction. Relative to a corpus containing this information, the strongest sentence whose development is entailed by '*a* is a red handkerchief' is (vii) or its equivalent (given *X*'s knowledge) 'All nonblack things are nonravens' or its equivalent 'All ravens are black'. In Context 5, '*a* is a red handkerchief' confirms 'All ravens are black'.

On the other hand, '*a* is a black raven' confirms (v). (v) entails 'All ravens are black' but is not equivalent given the corpus cum data to it. Hence, '*a* is a black raven' fails to confirm 'All ravens are black'.

Context 6. *X* asks whether all ravens are black or not. Here the partition consists of 'All ravens are black' and 'Some ravens are not black'.

In this context, both '*a* is a red herring' and '*a* is a black raven' confirm 'All ravens are black'. Thus, we have hit upon a context which reproduces the discontents of the raven paradox. I contend that anyone who is discontent does not take the question seriously but is interested in other questions with other potential answers.

The potential answers to the question raised in Context 6 are, of course, potential answers to the questions raised in Contexts 4 and 5. An investigator who begins by asking the question posed in Context 6 might be induced to shift to one of the other questions on the grounds that an answer to one of these two questions will furnish an answer to his initial question as well.

Suppose, however, that *X* is interested in the question raised in Context 4 and *Y* in the question raised in Context 5. How can they modify their problems so as to pursue a joint inquiry in which the list of potential answers includes all the potential answers of Contexts 4 and 5? There are, of course, many ways to do this. The following is a simple and prima facie plausible suggestion.

Context 7. Consider the predicates Q_1 = 'is a black raven', Q_2 = 'is a black nonraven', Q_3 = 'is a nonblack raven' and Q_4 = 'is a nonblack nonraven'.

Let the partition from which potential answers are generated be sentences which specify for each Q-predicate whether its extension is empty or not. Since the extension of at least one Q-predicate must be nonempty, there will be 15 'constituents' (in Hintikka's terminology) belonging to the partition. Every potential answer will be a case of accepting as strongest a disjunction of a subset of this set of 15 constituents. Thus, accepting 'All ravens are black' as strongest is equivalent to suspending judgement between all constituents stipulating that Q_3 is empty.

In Context 7, '*a* is a red herring' confirms the constituent which specifies that Q_4 is the sole nonempty Q-predicate – i.e., the hypothesis that everything is a nonblack nonraven. '*a* is a black raven' confirms 'Everything is a black raven'. Neither datum confirms 'All ravens are black'.

On the other hand, '*a* is a black raven, *b* a red herring, and *c* a black limousine' confirms 'There are black ravens, nonblack, nonravens and black nonravens but no nonblack ravens'. Given the datum sentence, this is equivalent to 'All ravens are black'.

Throughout this discussion, I have assumed that whether a datum confirms a hypothesis depends on the question under investigation and the potential answers identified for that question. Relative to some questions, '*a* is a red herring' confirms 'All ravens are black'. Relative to others, it does not. When the relativity of confirmation to the question under consideration is ignored, it is not surprising that an air of paradox is generated.

It may be objected that an investigator might at one and the same time be considering two or more questions – e.g., the one raised in Context 4 and the one raised in Context 6. If he had sufficient observations of red herrings to warrant adding (i) to his corpus relative to Question 4, those observations would not warrant his adding (i) to his corpus relative to 6. What should he do?

On the view I favor, when data warrants a potential answer relative to one question regarded seriously and fails to warrant the same potential answer relative to another, the conclusions recommended relative to each question should be held in abeyance until the set of questions taken to be serious can be modified in some suitable way. In our example, perhaps, the investigator might shift to the question raised in Context 7.

Is the selection of questions to be regarded as serious a mere matter of whim? I think not. Just as a corpus of knowledge is subject to critical review and revision, so is an investigator's problem situation.

I do not have a clear account to offer of how problem situations are to be revised. I suspect that something resembling Imre Lakatos' conception of a research program will be central to an adequate analysis.

My failure to offer criteria for ascertaining what is a serious question will undoubtedly leave many dissatisfied with my 'solution' to the raven paradox. My aim, however, has not been to solve the paradox but to advocate a problem shift which is, I hope, nondegenerate. I contend, as Lakatos, I believe, would, that the major task of epistemology is to account for the revision and improvement of knowledge. How knowledge should be revised depends on what counts as a serious problem. Hence, furnishing criteria for revising problem situations is itself an important task for epistemology.

Thus, the dependence of my 'solution' of the raven paradox on the conception of a serious problem is no objection to it but furnishes some small support for the view that the concept of a serious problem is an important epistemological concept deserving closer scrutiny than it customarily receives. The main lesson we should learn from the paradox of the ravens is that we should drop it for more urgent philosophical concerns.

Columbia University

PEGGY MARCHI

MATHEMATICS AS A CRITICAL ENTERPRISE

How does mathematics grow? According to tradition, mathematics grows as truths accumulate. Imre Lakatos criticized this traditional answer in a bold and revolutionary way. He said that mathematics grows as conjectures are made and criticized. In Lakatos' picture of mathematics, truths never get a chance to accumulate because they are always in danger of being refuted by newly discovered counterexamples. In fact, those parts of mathematics which under the traditional view were regarded as truths, that is, theorems, proofs, and so on, are under Lakatos' view seen as completely conjectural. In fact, according to Lakatos, all mathematics is conjectural. None of it is truth accumulated.

Under the traditional view, the growth of mathematics is unproblematic: mathematicians discover truths and with each discovery another truth is added to the mathematical corpus. Under Lakatos' view, the growth of mathematics is problematic. If all mathematics is conjectural, in what does the growth of mathematics consist? If there are no truths to accumulate, how does mathematics grow? This is a problem which I shall discuss and answer in the second part of this paper. In the first part, I shall explain the importance and revolutionary nature of Lakatos' ideas about the growth of mathematics by comparing them with Popper's ideas about the growth of science. This comparison will form the background for my solution to the problem of how mathematics grows by trial and error.

1. Lakatos and Popper Compared

Lakatos' ideas on how mathematics grows are often interpreted as Popper's ideas of how science grows, transformed and modified to make them suitable to describe mathematics and its growth. There is a great deal of truth in this view, in my opinion. Both Popper and Lakatos start with the problem of how to refute the justificationists who dominate their respective fields. For Popper, the justificationists are the inductivists,

R. S. Cohen et al. (eds.), Essays in Memory of Imre Lakatos, 379–393. *All Rights Reserved*
Copyright © 1976 by Margaret L. Marchi

those who say that scientific knowledge comes from facts, and that we can know that such knowledge is true because it comes from facts. For Lakatos, the justificationists are the formalists, those who say that mathematical knowledge comes from consistent axioms, and that we can know that such knowledge is consistent precisely because it comes from the axioms. In Popper's case and in Lakatos', the predominance of justificationists in their respective fields is a twentieth century backlash, a backlash against the unsettling realization that neither physics nor mathematics could ever be as secure and reliable as each looked in the nineteenth century.

Popper's problem background starts with Einstein and his refutation of Newton's theory. The whole of the nineteenth century has been characterized as a century of refutations of Newton's theory. But despite these frequent refutations, Newton's theory managed to retain sufficient support among scientists that the basic method of science which it supported was not threatened, and neither was the security and reliability of science. With Einstein's correct prediction that light could be bent by a strong gravitational field, the consensus of the scientific world was that Newton had been refuted, and furthermore, this refutation brought home to many people the fact that we could no longer rely on our scientific theories, no matter how good they seemed. Newton's theory had stood for nearly two hundred years, and yet had been refuted. Einstein's theory might also be very good, yet we would never again rely upon it as we had once relied on Newton.

Popper comes to fill the methodological gap which the refutation of Einstein's theory leaves. The gap is this: if science can never be relied upon again, how can it grow? What is the method of science? After Hume, very few people thought that Baconian induction was the method of science. But almost everyone thought that science did find the truth, had found the truth, would continue to find the truth, and that this truth was somehow connected with and somehow came from the facts. The actual method was not known. That is, people did not know how the truth came from the facts, but they knew that it did. When Newton's theory was refuted, this methodological theory was seriously called into question. Newton's theory was supposed to have been arrived at in the prescribed manner. It was supposed to come, somehow, from the facts. It was supposed to be a model which all other theories were supposed to follow.

When it was refuted people changed their minds about whether science could find the truth, and about whether scientific knowledge was secure in any sense. They realized that at any moment, any scientific theory could be refuted. Some people did not like to see the foundations of science and scientific truth broken in this way and they tried to reinstate a method for science which would somehow produce the truth from the facts. It is against these people that Popper first directs his arguments in *Logic of Scientific Discovery*.

There was a similar revolution in mathematics to the one which I have described for science, but it was less dramatic, and perhaps less definite. It is harder to describe this revolution, if that is what it was, in definite terms, because mathematics does not have the same well articulated methodological tradition that science does. But the tradition is strong nevertheless. Before the nineteenth century, mathematics and its growth were not much discussed by philosophers and mathematicians, except to say that mathematics was the perfect science, the science to which all others aspired. Mathematics may not have been to every one's taste – Bacon, for example, was hostile to it – but everyone agreed that it was true, that it was perfect, and that of all the sciences it was the most successful. Growth for such a perfect and true science was simple: mathematics grew as mathematical truths accumulated.

Unfortunately for this lovely picture, but fortunately for mathematics, non-Euclidean geometry was discovered in the first part of the nineteenth century. Bertrand Russell says that its discoverers wanted specifically to refute Kant's metaphysics, but whether this is true or not, the discovery of non-Euclidean geometry and the discovery of its consistency, seemed to show that Euclidean and non-Euclidean geometry cannot both be true. Georg Cantor finished the job which Riemann had begun, namely, the job of demolishing the idea that mathematics is true. Cantor discovered that infinite sets seem to be an essential part of mathematics. In the nineteenth century many people were opposed idea to Cantor's discovery, considering that infinite sets were an impossibility and that nothing as solid as mathematics could be based on anything as flimsy and impossible as infinite sets. To settle the matter, mathematicians began to investigate the question, "What are the essential assumptions in mathematics?" We now call this subject the foundations of mathematics. Mathematicians discovered that the essential assumptions of mathematics are very strange indeed, so strange in fact that it is very difficult to say in good conscience that they are true. Yet all of mathematics is supposed to come from these essential assumptions, in fact, to come from them by a process of deduction. But if we don't know that the essential assumptions of mathematics are true, we cannot know that the mathematics which comes from them is true.

At the end of the nineteenth century, the theory that mathematics is true and that mathematics grows as truth accumulates had collapsed. To resurrect mathematics and restore it to its place as queen of the sciences, some mathematicians tried to replace truth in mathematics with some other kind of certainty or guarantee of reliability. Russell tried to show that mathematics was a part of logic. If logic could be shown to be reliable, then so could mathematics. Hilbert tried another tack. He said that what was important as a guarantee in mathematics was not truth but consistency. If a set of mathematical axioms could be shown to be consistent, then this was a sufficient guarantee that the mathematics derived from these axioms was reliable. Brouwer had yet another approach, that of basing mathematics on our perception of time and thus giving it a reliable foundation. Whatever the hopes that mathematicians had for these new methods for guaranteeing the reliability of mathematics, and whatever the benefits gained by mathematics by its introduction – and there were many – these new methods failed to provide the guarantees which they sought. In the rest of this paper, I shall refer to these methods under the collective name of axiomatization.

Lakatos comes to fill the methodological gap which the failure of axiomatization leaves. The gap is this: If mathematics does not grow by the process of accumulation of truths, and since the truth of mathematics cannot be guaranteed by axiomatization, how then does mathematics grow? With the old idea that mathematics can be guaranteed refuted, no ready explanation of the advances in our knowledge of mathematics is available. As in science, there was a backlash in mathematics. Some people just didn't like it that mathematics could not be guaranteed and they tried to reinstate the idea that mathematical knowledge could be guaranteed if only it were properly derived from the correct axioms of mathematics. It is against these opponents that Lakatos first directs his arguments in *Proofs and Refutations*.

To the questions How does science grow? and How does mathematics grow? Popper and Lakatos give similar solutions. Popper says that all our knowledge of science is conjectural and that our knowledge of science grows by a process of trial and error. That is, we try one theory, find out if it solves our problem and if there is any evidence to refute it. If there is, we declare error and try another theory. If we find no refuting evidence, we still have no guarantee that tomorrow no such evidence will be found. According to Popper, science progresses as old theories are refuted and new and better ones come to take their places. These new theories are in turn replaced by others and so the process of trial and detection of error continues.

Lakatos says that all our knowledge of mathematics is conjectural and that our knowledge of mathematics grows by a process of trial and error. That is, we try one mathematical theorem, find out if it solves our problem and if there are any counterexamples which refute it. If there are we declare error and try another mathematical theorem. If we find no counterexamples, we still have no guarantee that tomorrow no counterexamples will be found. According to Lakatos, mathematics progresses as old theorems are refuted and new and better ones come to take their places. These new theorems are in turn replaced by others and so the process of trial and detection of error continues.

If from the exact parallelism of the two preceeding paragraphs, the reader gets the idea that Lakatos has adapted Popper's ideas to suit mathematics, well and good. It is probably no accident that the work of Lakatos' in which these ideas are contained is called *Proofs and Refutations*, reminding us strongly of Popper's title, *Conjectures and Refutations*. I would however be doing Lakatos an injustice if I implied that somehow Lakatos' ideas are lifeless copies of Popper's or that somehow there is really nothing very original about Lakatos' ideas and work. Precisely the contrary is the case. *Proofs and Refutations* is a work of very great originality and Lakatos' ideas about the nature of mathematical growth are bold and revolutionary.

Perhaps the easiest place to see Lakatos' originality is when we consider how the history of mathematics is changed by the new perspective of Lakatos' ideas. Just as in the traditional view, mathematics grew as truths accumulated, so too the history of mathematics was the history of the accumulation of these truths. Although it is true that all histories of mathematics written according to this precept are boring, because they reduce history to mere lists of successful inspirations, that is not my contention here. Rather, I want to repeat Lakatos' point, namely, that such histories distort the truth. They do not tell us truly what actually happened, either in the cases of individual mathematicians or in the case of the growth of mathematical knowledge as a whole. In *Proofs and Refutations* Lakatos gives a beautiful example of such a distortion. According to traditional histories of mathematics, Euler's theorem was discovered by Euler and proved by Poincaré. But if you check the actual history of the theorem, as Lakatos did, you discover that Euler's theorem was proved many different times and in many different ways, and furthermore, the theorem was reformulated many times. Why? It was proved so many times because every proof until Poincaré's was found to be mistaken and it was reformulated so many times because every formulation until Poincaré's was refuted by counterexamples. The many different proofs and reformulations are either ignored by traditional historians of mathematics, if they are insensitive, or apologized for, if they are sensitive. But according to tradition, what counts are the truths which are discovered and properly, these are what should be reported in histories of mathematics.

Such a traditionalist view may be correct, if the epistemology underlying it is sound. But in the case of mathematics, we have no reason to think that mathematics is true or certain or reliable and every reason to think that it (or at least parts of it) are false or uncertain or unreliable. Mathematical knowledge is conjectural and at any moment someone may discover a counterexample which will refute even the safest, best confirmed theorem. This, as I have been saying, is Lakatos' view of the status of mathematical knowledge. From this new view, Lakatos says that the history of mathematics needs to be thoroughly revised. Lakatos would agree with the traditionalist that Poincaré's proof of Euler's theorem is an important contribution, even though he would not agree that it was the truth. But he would insist on including in the history, the various trials and errors of Poincaré's predecessors, the improvements which they made to the theorem and to its proof, the counterexamples which they discovered and the resulting restatements of the theorem. These steps, according to Lakatos are just as important as the result which we provisionally accept today.

In a sense, they are even more important, in that they remind us that the search for knowledge, mathematical or any other kind, is a very human process. Mathematicians often like to pretend that there is something which differentiates great mathematicians from ordinary mortals. That thing is mathematical insight. I would certainly grant that some mathematicians are hugely gifted and perhaps even this gift is so great as to be different in kind from the intelligence which most of us possess. Be that as it may, ordinary mortals sometimes want to understand how a certain great mathematician came to make a certain discovery. According to the insight theory, it is impossible to understand this, unless an ordinary mortal is satisfied with the explanation that a great mathematician had a good insight and thus was able to make that particular discovery. Lakatos' overview of mathematics as trial and error cannot explain the entire psychological process of discovery, but it can shed light on it, light which is not shed by the insight theory. Lakatos would say, if you want to know why a particular discovery was made, ask first, what problem the mathematician was trying to solve. Ask second what were the solutions of his predecessors, if he had any, or what were his own previous unsuccessful solutions. Ask third how these unsuccessful solutions failed. Ask fourth what qualities a successful solution must have. Of course, none of this information can tell you how the successful solution to the problem was discovered. But according to Lakatos, great discoveries do not come out of vacuums. They come from people trying to solve problems, problems that they and perhaps others have previously failed to solve. Simple as this observation is, it humanizes mathematics greatly. Considering mathematics' bad reputation on this score, Lakatos has done mathematics a great service.

There is another way in which Lakatos' overview of trial and error humanizes mathematics. When we see the growth of mathematics as trial and error we can see mathematicians as part of an intellectual community. Under the traditional view, it was possible for a mathematician to be completely isolated and to know and care nothing about the work of his colleagues. To make a contribution to mathematics all he needed to do was discover something true. Under Lakatos' view, our solitary mathematician is still possible. He can find his own problems, invent his own solutions, discover his own counterexamples, all by himself. But he is the exception. Most mathematicians know about at least some of the work of their colleagues, and spend at least some of their time trying to criticize and improve it. In addition mathematics as a whole is a community enterprise. Since it is difficult to criticize one's own work effectively, other mathematicians take over this task. In fact, it is part of the mathematical tradition to investigate and criticize the work of colleagues. And this sharing of work is essential to the growth of mathematics. Not so in the traditional view. In this view, the only thing which is essential is the discovery of the truth. What makes Lakatos' view the more human is that it is not nice to be isolated, and it's somehow more human to think of mathematicians cooperating even if this cooperation is friendly-hostile, rather than sitting alone and lonely in their ivory towers.

To conclude this section, I offer my own assessment of Lakatos' achievement in *Proofs and Refutations*. Lakatos' ideas that mathematical knowledge is conjectural and that mathematics grows by trial and error are perhaps the most important discoveries ever in the methodology of mathematics. Its consequences for mathematics and its history are huge, as I hope earlier parts of this section indicate. Yet important as Lakatos' ideas are, they remain sadly uninvestigated and uncriticized. Why? One reason, of course, is that it is difficult to criticize a good idea – precisely because it is a good idea. However, there is another reason I think. Seen from a distance, Lakatos' ideas on mathematical growth are clear and simple; but seen from close-up, they are a jumble. Their strength and brilliance are at once apparent. But still it is difficult to reconcile the clear and simple over-view which Lakatos gives of mathematical growth with the welter of detail and diverse cases that we see in close-up. The source of the difficulty in moving from a clear wide-view picture to a clear close-up picture is this: Lakatos identifies a number of different elements in the critical structure of mathematics, but he does not give a consistent explanation of how they interact within the trial and error structure. He describes situation in which an error in a theorem is detected by a counterexample, others in which theorems are refuted, not proved, by their proofs, and still others where counterexamples are refuted by theorems. As Lakatos quite rightly says all our knowledge of mathematics is conjectural – and this includes knowledge of theorems, proofs and counterexamples. But how does mathematics grow amid the welter of such confusion? Popper answers this question for science, but Lakatos does not answer it for mathematics. In the next section, I shall discuss this deficiency in Lakatos' theory and provide a possible solution.

2. THE GROWTH OF CONJECTURAL KNOWLEDGE OF MATHEMATICS

Returning to the comparison of Popper and Lakatos, we now come to a question which both Popper and Lakatos have some difficulty answering. The question arises like this – and for the moment I shall speak only of Popper. According to Popper, scientific knowledge grows as scientists make conjectures and have them refuted by contrary evidence. How is such a process growth? It is certainly not growth in the traditional sense. No truth has been discovered. In fact, and this seems to make the situation worse, not only have we not discovered anything true, since the conjecture which we hoped would be true turned out to be false, but we can't be sure that we have actually refuted our conjecture. That is, the refuting evidence might itself be false. So where are we? According to Popper, we should not expect to find the truth and so whatever our standard is for growth, it should not be the standard of discovery of truth. Instead he proposes another standard – that of approximation to the truth. He says that every new conjecture is progress if it is a better approximation to the truth than the conjectures which it comes to replace. If a conjecture is refuted, then we know that in one important way at least, it was not a good enough approximation to the truth. Our next conjecture will, we hope, be framed in such a way that it will not be refuted in the same way as the previous conjecture was. By improving conjectures in the light of refutations, we make them better and better approximations to the truth.

As for the objection that we can't be sure that the conjecture is really refuted, since the refuting evidence might be wrong instead, Popper says that this is, of course, always possible. But according to him, the way that we will make our conjectures the best possible approximations to the truth is for us to assume that the refuting evidence is right and that it is the conjecture which is wrong. If we fail to take the refutation of our conjecture seriously, by questioning the refuting evidence, then we are unlikely to try to improve our conjecture and thus unlikely to make progress.

There is much to debate in this explanation of Popper's of how conjectural knowledge can grow. But that is not my purpose here. Rather, let me point out that Popper does offer a clear and definite, even if controversial solution to what is really a very important and pressing problem. Furthermore, the solution which he offers makes at least intuitive good sense, for we can all appreciate that approximations can both improve and be of real value in helping us to solve our problems – even if approximations are not the final and complete truth. Lakatos does not offer any solution to this problem, even though his ideas about the growth of mathematics leave him in the same situation – that is, if mathematics grows by a process of conjecture and criticism, how is such a process growth, especially when we know in advance that no conjecture is true for sure and no criticism is final. This is, in my opinion, a very serious defect in his theory, and one which has made his theory very hard to apply to the actual history of mathematics. It is hard to apply because Lakatos gives no standards for what constitutes growth and thus it is very difficult to identify in a particular historical incident what according to Lakatos would count as growth.

I have made the claim that Lakatos does not give a theory of what counts as growth. Since I take this lack to be a serious defect (though remediable) in his theory, I should like to explain just what Lakatos does say about growth in *Proofs and Refutations*, to show first that what Lakatos has to say about growth is very good and yet that good as it is, it is not specific enough. It's worth considering what Lakatos says about growth because he does have implicit standards, and does try to capture the essense of his standards with rules.

In *Proofs and Refutations*, after he has explained that mathematical knowledge is always being revised in the light of new discoveries, Lakatos goes on to try to explain what kind of revisions are better. That is, if you are a mathematician with a refuted theorem on your hands, what sort of changes should you make to it to improve it. What sort of change will improve it the most? You can, says Lakatos, change the theorem so as to make the refuting instance that you have just discovered, an exception, i.e. all *x*'s are *y*'s except for *x*'s such as *z*. But, says Lakatos, though many mathematicians have done this (the exceptions are usually in the definitions of terms rather than in the statements of the theorems, but it amounts to the same thing) it's not a good policy. Such changes don't do anything to locate the reason why the theorem was vulnerable to refutation from such-and-such type of counterexample. A better change would be one which excluded the counterexamples in a deeper or more organic way.

An even better change would be one which would explain not only why the entities which satisfied the theorem satisfied it and why those which didn't, didn't. In other words, Lakatos says that a theorem which makes a deep or organic or structural or conceptual change is more of an improvement than a theorem which is changed by an *ad hoc* addition.

So far so good. As it stands this is a theory, albeit a little vague, of what constitutes growth. Unfortunately Lakatos then destroys this theory with his correct contention that everything is conjectural and that there is no recipe for growth. He points out that while we may prefer the explanatory theorem, the *ad hoc* theorem may be true, that our counterexamples may be mistaken and not our theorems, that by the nature of mathematical entities, we simply may not be able to explain why some entities satisfy our theorem and others don't. The precept that we improve our theorems when we make changes to their deep structure thus fails before the onslaught of Lakatos' thorough-going scepticism.

Can we not say the same thing about Popper? That is, if Popper is the thorough-going sceptic that he claims to be, how can he be so definite about his definition of what counts as progress in science? Popper replies that science searches for the truth, and even if it cannot ever find it, the fact that truth is the goal of the enterprise, determines what rules scientists should follow to make progress towards the goal. This answer is controversial, but it is clear. One might think that Lakatos would make a similar reply. He might say, following Popper, that mathematics searches for the truth and so on. But Lakatos does not wish to say this, even though he often seems to come close. The reason is that truth as an aim of mathematics was given up in the nineteenth century, and furthermore, the metaphysical position that seems to accompany a mathematician who searches for the truth is Platonism, a metaphysical position which is now passé. Since Lakatos does not, cannot explicitely accept truth as a goal of mathematics, he cannot follow Popper's definition of what counts as progress. Yet paradoxically, critics of Lakatos accuse him of being a Platonist and dismiss his methodology on that account. In my opinion, this charge is made with a certain justice: the flavor and savour of Lakatos' recommendations are those of a truth-seeker. But I doubt that this means he is a Platonist and that this is fatal to his method.

Now I shall offer a small modification of Lakatos' ideas which will, I believe, solve the problem which I have been discussing, namely, how does mathematics grow when all knowledge is conjectural? I take the germ of my solution from Popper. According to Popper, science tries to find true explanations of facts. For this purpose, scientists invent theories and these theories are refuted by tests, i.e. by experiments which bring to light still other facts. Given this structure for science, Popper can define his idea of progress, i.e. better and better approximations to the truth.

Lakatos has no similar idea of what constitutes progress in mathematics – for the reason that, unlike Popper, Lakatos does not fix any one part of mathematics as the "to be explained." Consequently, no sort of better and better approximations, to the truth or to anything else can be Lakatos' definition of progress in mathematics. Because he never fixes any part of mathematics, even provisionally, he cannot apply something similar to Popper's answer to the question of how conjectural knowledge can grow. We know what counts as tests for Lakatos, that is, thought experiments. Thought experiments are usually counterexamples and sometimes proofs and very occasionally theorems. We also know what counts as an explanation for Lakatos. Theorems explain, in the sense of predict, why a certain entity satisfies them. They may also explain in a deeper way. Proofs explain theorems in the sense that a theorem is deducible from it – just as an instance of a theorem is deducible from the theorem. What we do not know is what counts as the things to be explained. What are the explicanda of mathematics?

What I suggest is the following: that Lakatos continue to take thought experiments as tests, proofs and certain high level theorems as explanations. For the things to be explained, I suggest that Lakatos take theorems. That is, theorems in mathematics are under my suggestion analogous to facts in science. If we make Popper's and Lakatos' systems analogous in this way, then we can apply Popper's ideas about what counts as growth in mathematics. These ideas may or may not be appropriate for mathematics. The question has yet to be investigated. But it is a promising one, and one that I hope to investigate myself – not in this essay however.

Instead I would like to investigate a prior question, namely the question of whether making theorems in mathematics and facts in science analogous is a good step. In what sense is a mathematical theorem a fact of mathematics? Intuitively at least, a theorem seems to be very different from a fact. But on closer investigation, I have found that in three important senses, at least, theorems in mathematics are very like facts in science. These three senses are: (1) a theorem is something universally recognized and accepted and in this sense a fact; (2) a theorem is a task and in this sense a fact; (3) a theorem's acceptance rests on a provisional compromise between dogmatism, infinite regress and psychologism and in this sense is a fact.

(1) A theorem is something universally recognized and accepted and thus a fact. This sense of fact is perhaps the most common. It is, for example, a fact that the shortest distance between two points is a straight line, that $2+2=4$, that the derivative of x^2 is $2x$. Of course, none of these facts is true without qualification. Yet they are all facts because they are universally recognized and accepted, with appropriate qualifications, of course. In science similarly accepted and recognized facts, such as that the atomic weight of water is 18 and that a water molecule is composed of one oxygen atom and two hydrogen atoms, are the facts which scientists attempt to explain with their theories. When in *Logic of Scientific Discovery* Popper speaks of basic statements, this is one of the senses of fact he has in mind.

Are mathematical theorems facts in this sense? I think the answer is clearly yes. The fundamental theorem of algebra, for example, was universally recognized and accepted, so much so that as proof after proof of it failed, new proofs were still brought forward. On might think that repeated failure (until Gauss) to prove the fundamental theorem might be taken by mathematicians as evidence that the theorem was false. But not so. Apparently the fundamental theorem was so widely recognized and accepted that mathematicians continually sought new proofs for it. Mathematicians seem to have reacted in much the same way with Euler's polyhedron theorem, judging from the numerous attempts to prove it and mathematicians' apparent disinclination to consider it false and disregard it. The results of the infinitesimal calculus are another example of mathematical theorems which were universally recognized and accepted as facts, even though their basis, the idea of infinitesimals was known to be inconsistent. Mathematicians were so convinced that the results of the infinitesimal calculus were true that they searched until they found a consistent basis for them. Which brings me to the next sense of fact.

(2) A theorem is a task and thus a fact. This sense of fact is explained in J. Agassi's important paper, 'Sensationalism.' Task-facts are facts

which are thought to be true but which are still being tested. Goldbach's conjecture, that every even number is the sum of two primes, is a fact in this sense, as is Fermat's theorem that $x=2$ is the only integer solution to $a^x+b^x=c^x$. The four color problem is another such example. We know of no exception, no map which needs more than four colors to color it in such a way that no two countries that touch are the same color. Yet we cannot prove that we need only four colors. To mathematicians who were opposed to complex numbers, the fundamental theorem of algebra was a fact, i.e. recognized and accepted, which they wished to disprove. They hoped, I think, to be able to show that the regularity stated by the fundamental theorem was more apparent than real, or in other words, that there are other interpretations which would fit the facts just as well but which would not involve complex numbers. A famous task-fact in science concerns the atomic weight of chlorine, found in the last century to be 35.5, instead of some integer value as theory predicted. No mistake in experimental procedure was found, yet scientists hoped to disprove the result. Today we know the result is correct, and we know the explanation, namely, that chlorine is of two types, one of weight 35 and one of weight 37.

(3) A theorem is a fact because its acceptance rests on a provisional compromise between dogmatism, infinite regress and psychologism. This sense of fact is explained by Popper in *Logic of Scientific Discovery*. The problem which this sense of fact comes to solve is the problem of the basis of experience, as proposed by Fries and Popper. The problem amounts to this: What is the proper basis for accepting statements of scientific fact? One basis for acceptance is dogmatism – we accept statements of scientific fact without question and without justification. But Fries is opposed to this alternative. He says that statements of scientific fact should be justified in some way. If we try to justify statements of scientific fact by reasoned argument, however, this lead to an infinite regress of justifications and thus to no justification. The reason that justification by reasoned argument is impossible is that to justify a statement of fact in this way required the use of other statements. These statements too must be justified. And so must the statements which justify the statements which justify the original statement. And so on. For justification to be effective, we must be able to finish the process, which we cannot do if we seek to justify by reasoned argument. Given that dogmatism is

unacceptable and justification by reasoned argument is impossible, Fries turned to psychologism as the proper basis for accepting statements of scientific fact. Psychologism is the doctrine that statements can be justified by perceptual experience. Statements of scientific fact are justified, and properly so in Fries' opinion, because they are statements about our experience.

Popper does not accept psychologism as a solution and thus, the problem of the proper basis for accepting statements of scientific fact is again raised. Popper's solution is neither psychologism, nor a resort to reasoned argument alone, nor dogmatism, but a way of considering the acceptance of scientific facts which takes something from each of these three ideas – hopefully the best parts. Popper says that scientific facts are an expression of our experience, but we only accept a scientific fact until we have some reason to question it. Thus we accept our experience but not dogmatically. If we question a fact, we make some kind of empirical test. This could lead to an infinite regress if we insisted on testing the tests which test the test and so on. But at any point we can stop testing because we don't seek a final justification but only a state of affairs where we have answered all the questions we think are important pertaining to the reliability of scientific facts.

The basis on which we accept mathematical theorems is, I think, very similar to what Popper describes as the basis for accepting scientific facts. Thus theorems are facts in this sense. Mathematical theorems are expressions of our experience, our mental experience in this case rather than physical experience. We do not accept our mental experiences dogmatically. We question them when we have cause to. But we don't feel compelled to question our experiences until we reach a final justification. That is, in accepting mathematical theorems, we rely neither on psychologism, nor on the justification of reasoned arguments nor on dogmatism. Thus we accept theorems as facts in mathematics on the same basis as we do facts in science.

The identification of mathematical theorems as the things to be explained in mathematics is a bold step and one which will drastically change our view of the history of mathematics. But I think the change will be a good one, no least because it finally makes Lakatos theory a viable historical model.

New School of Liberal Arts
Brooklyn College, CUNY

ERNAN MCMULLIN

THE FERTILITY OF THEORY AND THE UNIT FOR APPRAISAL IN SCIENCE*

In both rationalism and empiricism, the two 'classical' forms of philosophy of science originating in the seventeenth century, it was assumed that appraisal in science had to be a matter of applying logical rules. To justify a scientific statement was to infer it from the first principles (rationalist) or the evidence (empiricist). To appraise was to certify the starting point of the inference as true principle or reliable evidence, and then evaluate the validity of the inference (deductive or inductive) drawn from it. There are two major assumptions here: first the foundationalist one that science must rest on foundational statements, themselves unchallengeable, and second, the logical one, that inferability is the proper test of a scientific claim.[1] If science is to be the definitive sort of knowing that nearly everyone from Aristotle to Kant thought it could and should be, these assumptions are difficult to avoid.

1. Appraisal by a logic of confirmation

Three different sorts of proposition have been proposed as the object of appraisal in logicist theories of science. The rationalist would want the statement to be a *theorem*, deductively derivable from the intuitively-based first principles of his science. The empiricist might (like Bacon or Mill) stress inductive inference, and then his candidate as the major subject for scientific appraisal would be an *empirical generalization*. Or else (like Carnap, or for that matter, Descartes in some parts of his work) retroduction might be the preferred method, where an hypothesis is validated by way of the truth of the inferences drawn from it. This would make an *hypothesis* or *theory* the object of appraisal.

In all three cases, the unit for appraisal is a proposition or set of propositions capable of explicit statement. And appraisal takes the attractively straightforward form of testing the alleged relation of inference, or in the more sophisticated system devised by Carnap, the relation of confirmation. Hempel puts the matter with his accustomed clarity:

R. S. Cohen et al. (eds.), Essays in Memory of Imre Lakatos, 395–432. *All Rights Reserved*
Copyright © 1976 by D. Reidel Publishing Company, Dordrecht-Holland

We shall adopt this conception of confirmation as a relation between sentences here for the following reasons. First, the evidence adduced in support or criticism of a scientific hypothesis is always expressed in sentences, which frequently have the character of observation reports; and second, it will prove very fruitful to pursue the parallel ... between the concepts of confirmation and of logical consequence.[2]

He then goes on to "restrict the evidence-sentences which form the domain of the relation of confirmation to sentences of the character of observation reports". The 'test' of an hypothesis thus falls into three phases: the making of suitable observations, confronting the hypothesis with the observation reports, the decision to accept or reject the hypothesis in the light of the reports, or to suspend judgement temporarily. Hempel is mainly concerned with the second phase which:

is of a purely logical character; the standards of evaluation here invoked – namely the criteria of confirmation, disconfirmation and neutrality – can be completely formulated in terms of concepts belonging to pure logic.[3]

The third phase, a decision, can be construed as pragmatic, but he thinks it worthwhile to attempt to reconstruct it in 'purely logical terms'. This would require the formulation and establishment of some general 'rules of acceptance' which would provide criteria for the acceptance or rejection of an hypothesis by reference to the kind and amount of confirming or disconfirming evidence contained in the sum of the observation reports.

The criteria need not all (he adds) be concerned with the deductive relations between hypothesis and evidence. They could also:

refer to such additional factors as the simplicity of the hypothesis in question, the manner in which it fits into the system of previously accepted theories, etc. It is at present an open question to what extent a satisfactory system of such rules can be formulated in purely logical terms.[4]

Hempel was more cautious than other members of the Vienna Circle would have been in leaving this an open question. Even though a logic of confirmation had not as yet been adequately formulated, it seemed clear to Carnap that for scientific rationality to have attained the results it had, such a logic had to underlie it and ought therefore, in principle, be capable of discovery.[5] But there is another feature also of this passage that deserves attention. Hempel shows himself aware that the confirmation relations are *not* simply analogous with those of deducibility between hypothesis and evidence. There are other aspects of hypotheses: their coherence and their simplicity, to which scientists frequently appeal in citing reasons for their acceptance. And these aspects cannot be reduced to relations of inference. Still, both coherence and simplicity have a formal sort of ring to them; it sounds vaguely plausible to hope (as a good logicist should) that they too could be explicated in formal terms (though admittedly not in terms of an inferential logic alone).

The crucial word in the passage, however, is an easily-overlooked 'etc.'. Are there, perhaps, other criteria which would be even more difficult to reduce to 'purely logical terms'? One immediately suggests itself, a criterion to which working scientists were beginning to attach great weight, as far back as the seventeenth century. Thus, by the time Herschel and Whewell came to set down what they regarded as the basic principles of scientific methodology, it had come to seem the most important single criterion of the truth of an hypothesis:

> The surest and best characteristic of a well-founded and extensive induction ... is when verification of it springs up, as it were, spontaneously, into notice, from quarters where they might be least expected ... Evidence of this kind is irresistible and compels assent with a weight which scarcely any other possesses.[6]

And again:

> The evidence in favor of our induction is of a much higher and more forcible character when it enables us to explain and determine cases of a *kind different* from those which were contemplated in the formation of the hypothesis. The instances in which this has occurred indeed impress us with the conviction that the truth of our hypothesis is certain. No accident could give rise to such an extraordinary coincidence. ... The cases in which inductions from classes of facts altogether different have thus jumped together belong only to the best established theories which science contains. And as I shall have occasion to refer to this peculiar feature in their evidence, I... will term it the *consilience of inductions*. ... No example can be pointed out, in the whole history of science, so far as I am aware, in which this consilience of inductions has given testimony in favor of an hypothesis afterwards discovered to be false.[7]

This emphasis has continued, among theorists of science sensitive to the realities of scientific practice, at least. Campbell writes:

> But actually there is, in addition, a third test of the truth of a theory, which is of great importance. A true theory will not only explain adequately the laws that it was introduced to explain; it will also predict and explain in advance laws which were unknown before. All the chief theories in science (or at least in physics) have satisfied this test.[8]

This is, he notes, how *ad hoc* theories are discovered and rejected. Yet the logical relation of implication does not admit any distinction between propositions known before and those known after the formulation of the hypothesis. H either implies E or it does not; whether E was thought of when H was being formulated is of no consequence to the relation of confirmation, if confirmation is bounded by the simple deducibility of E from H. Yet it *should* affect confirmation, it would seem, if what one wishes to do is to explicate the notion of confirmation implicit in scientific practice.[9] There can be no doubt that the discovery that an hypothesis predicts some new unthought-of domain carries more weight than if that same domain had been part of the evidence, to account for which the hypothesis was originally proposed.

2. THE FERTILITY OF THEORY

But what if the persistent logicist decides to regard this criterion as a type of logical rule? One could divide the observation-reports into two classes, those for which the hypothesis was originally meant to account, and the others which have come to notice only subsequently (or perhaps, if a stronger condition be preferred: which have been discovered in consequence of the hypothesis). Then one could say that $C(H,E)$ is stronger than $C'(H,E)$, where C is the sort of confirmation afforded by the original evidence, and C' is that given by 'new' data. It would not be amiss to regard such a rule as 'logical' in the same sense in which Chisholm, for instance, speaks of the rules uncovered by his explication of belief-statements as an epistemic 'logic'. As it stands, the rule is a very weak one, of course; it is not clear how one would proceed in order to get a more precise delineation of the relation between C and C'.

But the logicist has a still more serious issue to face. Campbell continues:

> As a matter of fact, it is very seldom that a theory, *exactly* in its original form, predicts any laws except those that it was proposed to explain; but a very small and extremely natural development of it may make it predict new laws.[10]

Campbell's reassuring phrase, 'very small and extremely natural', cannot conceal the fact that the development of the original theory is not just an exploration of its logical implications. Rather, it is a matter of analogy, of tentative suggestion, of plausible modification in the light of a particular opportunity for expansion or in the face of a troublesome anomaly. The development is 'natural' in the sense that it is shaped by the structures of the original theory; but it is not so natural as to be rule-governed. There is need for a creative act of imagination, a 'what if' in which the resources of the theory are exploited to construct a new possibility, which on testing proves to be a fruitful insight.

Campbell's example is drawn from gas theory, where the suggestion that the size of the gas molecules ought to make a difference at certain ranges of the variables and thus should be specified (instead of being left unspecified or idealized as mathematically pointlike, as in the original theory) proved enormously fertile. Campbell rightly stresses that this is the *normal* way in which theoretical science develops, at once much more than simple logical implication and much less than total replacement of one theory by another.

It is striking that this feature of scientific theory, its *fertility* (to give it its commonest name), is hardly ever mentioned in the positivist discussions of confirmation, even though it had been a commonplace in earlier analyses of science, particularly those coming from scientists. Reichenbach and Carnap argued that the *only* criterion of theory-confirmation ought to be agreement with observed facts; the theory would thus be the 'most probable' one, in a sense to be defined within a formal theory of inductive probability. Von Mises and Bronowski rejected this criterion as inadequate; such a 'theory' might be no more than a record of observations made, without any sort of theoretical power. Bronowski proposed the criterion of formal simplicity, since:

> it is the internal unity and coherence of science which gives it truth, and which makes it a better system of prediction than any less orderly language ... Science is a process of creating new concepts which unify our understanding of the world.[11]

The unit for appraisal here would still be the theory, considered as a set of propositions, but appraisal instead of being a matter of logic would involve the application of some sort of (unspecified) aesthetic or intuitive criterion of internal coherence.

Frank construed 'simplicity' in a rather different way:

> If we investigate which theories have actually been preferred because of their simplicity, we find that the decisive reason for acceptance has been neither economic nor aesthetic, but rather what has often been called 'dynamic'. This means that the theory was preferred that proved to make science more 'dynamic', i.e., more fit to expand into unknown territory.[12]

To us it would seem important to distinguish between this 'dynamic' criterion and that of simplicity. Frank reduces them to a single criterion, just as Whewell earlier did. "What makes a theory is the quality of being simpler and shorter than the record of observations". Different reasons can be given for such a criterion: aesthetic, economic, dynamic; he makes the last-named the basic one. This allows him to conclude that:

The acceptance of a theory is always the result of a compromise between the requirement of 'agreement with facts' and of 'simplicity'.[13]

This is misleading, since the consideration affecting the scientist's decision (i.e. the criterion proper) is the 'dynamic' one of incorporating new territory. It is true that 'simplification' (in a far-from-simple sense) would ordinarily be a consequence of this extension. But it would *not* be correct to infer that 'simplicity' was thus the criterion governing appraisal of the theory. Frank's construal of the 'dynamic' criterion as no more than another version of the familiar and comfortable criterion of simplicity is easy to understand. The assumption underlying the classical views of theory-appraisal (that the unit for appraisal is a well-defined set of propositions to which the chosen criterion can be immediately applied) was a natural and convenient one; to have admitted the fertility-aspect of theory to be something other than a systemic criterion akin to that of simplicity might have forced him to face a question that the logicist tradition had never really taken seriously.

What is called the 'fertility' of a theory in this context is not something that can be ascertained by inspecting the theory at a single moment in its development or by assuming it to be constituted by a definite set of propositions. There is an ambiguity here that has proved very troublesome. 'Fertility' sounds as though it might designate the *potential* of a theory for further (as yet untested) development. If this were what it were, it would be discovered by examining the theory here and now, and estimating its imaginative resources (its 'neutral analogy' in Mary Hesse's phrase) for future extension and modification. It would *not* require one to trace the career of the theory over a period of time. The estimation of fertility would be a tentative, future-oriented affair. But this is *not* what is meant when fertility is spoken of in the context of confirmation. There it means performance, not potential; it is estimated by the *actual* success of the theory in opening up new areas, in meeting anomalies, and so forth. Thus it is *past*, not future, oriented; it is *proven* 'fertility' (*P*-fertility) that confirms the truth-value of a theory not its as-yet-untested promise (*U*-fertility). To estimate the *P*-fertility of a theory, one has to retrace its career and see how successful it has been in suggesting the right modification at the right time and in allowing incorporation of new areas not originally foreseen.

The unit for appraisal here is thus the theory, taken not as a timeless ensemble of propositions, but as a set of propositions and an accompanying model which are developed and modified over a period of time, with successes and failures of various sorts to take into account. When scientists speak of a 'good' theory it is this sort of assessment above all that they have in mind: a good theory is one which successfully guided research over an extended period of time, not only in the sense of providing correct predictions (an *ad hoc* theory could do this) but also in the sense of imaginatively suggesting plausible modifications in the original theory/model, modifications that allowed predictions to be made which would *not* have been directly deducible from the original hypothesis.

The criterion of *P*-fertility, if taken seriously, must lead one to question the adequacy of the logicist approach to confirmation. It forces one to abandon the assumption so central to logicism that the unit for appraisal is a well-defined set of propositions whose implicational structures are entirely open to scrutiny. Furthermore, it suggests that quantitative assessment of degree of confirmation and the search for formal axioms that this demands must be set aside as delusive. What carries most weight is the potential the theory has shown to allow creative scientists to go *beyond* the original statement. It is precisely *not* the predictions that were implicitly contained in that statement that most impress as a testimony to its truth-value; the crucial evidence is that borne by the predictions which were *not* logically contained in the original but which were made possible by imaginative (rather than logically rule-guided) and plausible modifications of that original.

The resources of a theory are evidently of two sorts: the implications derivable from it by means of formal rules, and the resources it displays as *metaphor* i.e. as a tentative conceptual juxtaposition of elements capable of suggesting to the creative mind a whole range of possible further developments of the theory itself. In evaluating a theory, both sorts of resource must be tested against the 'evidence'. But the manner of test is not the same in each case. In the first it can be simply put as a logical relation between H and E: E confirms H, if H implies E. But in the second, it is no such simple affair: E confirms H, if H 'suggested' a

H' such that H' implies E. Now while the inveterate logicist could write the key premiss here as: $H\,s\,H'$, it is obvious that s is not a purely logical relation. To evaluate a sequence where 'suggestion' of this sort is supposed to have occurred, it is necessary to examine the historical record. 'Suggestion' is not a univocal affair; there are endless ways in which it may happen and these may carry quite different weight when a question is asked about the degree to which a particular (successful) 'suggestion' confirms the original H. Reference to the historical record cannot, therefore, be avoided; one must know the exact sequence of events. What is important is the precise quality of the guidance the metaphor has given: the blind alleys, the degree to which the successful modifications went beyond the original proposal, the appropriateness of these modifications considered as 'plausible' extensions (in virtue of hindsight) of that proposal, and so on. The experienced scientist will have a certain 'feel' for considerations such as these, and they will influence his judgement, especially when he is asked to choose between two theories, each of which claims a high degree of predictive success.

Why exactly *should* he rest so much weight on this claim? Someone who defends a generally instrumentalist view of physical theory would find this question very hard to answer. Is not the value of a theory ultimately decided by the extent of the correct predictions it can presently enable us to make? Why should the past history of the theory, the precise sequence by which it came to be what it is, have any relevance to the assessment of present value? What is at issue here is, of course, the more basic question of what is meant by the 'value' (the 'truth', the 'adequacy', etc.) of a theory. We shall return to this later, but enough has already been said to make it clear that instrumentalists or positivists are unlikely to emphasize the practical importance of the fertility criterion in scientific appraisal. Realists on the other hand will find this criterion congenial, indeed will find implicit in it perhaps the best (some would say the only cogent) argument for their theory of science.[14]

3. The historicist turn

As one would expect, then, the fertility criterion is stressed by those whose general orientation to science is realist; this would, of course, include most scientists. Support for it has fluctuated among philosophers, depending on developments elsewhere in theory of knowledge and also on the degree of dependence on paradigms drawn from the history of mechanics rather than from the structural sciences. After Campbell came the long positivist interlude, 1930–1960. The historicist period which followed (and in which we still are) has not been much more favorable to a realist interpretation of science than its predecessor. Kuhn and Feyerabend stress the incommensurability of competing theories, and deny that conceptual change in science constitutes progress, in the sense of a convergence on truth. The historicist is hardly any better able than the positivist was to see a special significance in the ability of a good theory to predict novel and unexpected results nor to help us understand what it is that makes it possible to separate *ad hoc* theories from good theories.

In *The Structure of Scientific Revolutions*, Kuhn construes science as a problem-solving activity rather than as an effort to understand physical structures. Its cumulative aspect resides only in "the list of problems solved by science and the precision of the individual problem-solutions", both of which continue to grow.[15] Because "the unit of scientific achievement is the unsolved problem", paradigm-changes are governed by:

> two all-important conditions. First, the new candidate must seem to resolve some outstanding and generally recognized problem that can be met in no other way. Second, the new paradigm must promise to preserve a relatively large part of the problem-solving ability that has accrued to science through its predecessors.[16]

Handling of anomaly, conservation of previous problem-solutions: no hint of anything like a fertility-criterion here. In his concern with the sociological conditions of theory-acceptance and theory-change, Kuhn is led to underestimate the important epistemic distinction between two types of problem-solutions: those for which the theory (or theoretical modification) was originally designed, and those which have led to the discovery of new and unforeseen problems and their solutions.

Toulmin is even less anxious to commit himself on the criteria of appraisal (or selection, in his metaphor) than is Kuhn. Like the latter he sees science as a collective human activity; to discover in what its rationality consists, one must therefore begin (as one would in the case of other group-activities) with the group itself: how it is formed, how its discipline is enforced, how its standards alter, and so forth. By contrast with the logicist accounts of science:

> our own more historical analysis treats scientific disciplines as comprising informal populations of logically independent concepts, and leads us to ask, rather, in what manner a 'fruitful' discipline must develop.[17]

But 'fruitful' here has nothing to do with the fertility-criterion discussed above. Though some sort of 'selection-pressure' has to operate in order that science should evolve, it is not necessary to suppose that this pressure can be specified in terms of a set of epistemological criteria by means of which the best-'adapted' conceptual variants are singled out for survival. It is sufficient that selection *does* somehow operate; what we can then expect is an evolving intellectual discipline, in which both rational and causal factors play complex and continuously changing roles.

If one pushes the question of selection further, Toulmin argues, one encounters two very different sorts of problem-context. First there are the 'straightforward' cases:

> in which conceptual changes are made as the direct outcome of communal problem-solving activities, in a science whose disciplinary goals are sufficiently agreed. In these clear cases, the selective perpetuation of certain conceptual variants can be explained in 'rational' terms, by pointing out how the successful innovation helped the scientists concerned to achieve their collective goals.[18]

Unfortunately, however, these goals are not fixed, and they are frequently in conflict with one another. One may, if one wishes, look to the "various different kinds of merit" that scientists seek for in their theories: "predictiveness, coherence, scope, precision, intelligibility and the rest":

> But whatever criterion we pick for analysis, there will always be other cases to which it does not apply; and, more typically, considerations of several incommensurable kinds will be relevant to any particular choice.[19]

And these are the 'straightforward' cases; in the others (where there are failures of rationality, or where the criteria of rationality themselves are at issue), one has even less chance of isolating criteria of selection, particularly not criteria which would lay claim to an intrinsic plausibility on their own account. In short, it is not merely the fertility criterion which is scouted here, but 'logical' criteria generally, i.e. criteria in whose support something more could be said than that they represented one facet of practice at a particular epoch. There is no discussion of *why* the criteria we find embodied in scientific practice might be appropriate there; so strong in Toulmin's repugnance towards the logicist approach that he prefers to stress instead the many difficulties in the way of formulating or applying criteria.

The historicist turn taken by philosophy of science in the past decade has led philosophers to be much more sensitive to the complexities of historical scientific practice than they were, especially when they come to discuss the nature of scientific theory-appraisal. Most have found the KFT critiques of traditional views on appraisal too radical. In all the excitement, a more modest historical thesis has been all but lost from sight, one which has to do with the work of the scientist rather than that of the philosopher. To say that the unit for appraisal on the part of the working scientist is not a theory considered as a timeless set of propositions but rather the theory taken over its entire career to date is to impose upon the *scientist* the task of historian (if 'historian' is not too solemn a word when the 'career' in question is usually fairly short). This is a quite different (and logically independent) thesis from the KFT one that the *philosopher* must take into account the socio-historical nature of science in his attempt to formulate a theory of science. Toulmin tends to conflate the two theses, but it is important that they be kept separate. The grounds for the two are quite different; the first, besides, suggests that appraisal is to be taken seriously. And if our argument in Section 2 is correct, the main reason why a historical unit is needed is in order to allow the use in logical appraisal of one particular criterion, a criterion moreover for whose appropriateness independent philosophic reasons can be given. It is easy to see why an all-encompassing historicism would not find the stress on an historical unit of appraisal particularly congenial, despite its directing attention to the historical dimension of science. It does so in the interest of specifying an important and complex criterion of positive appraisal. And such an aim the thorough-going historicist is likely to distrust.

4. Popper: Corroboration and Fertility

A thorough-going falsificationist would presumably be even less happy with the fertility criterion, since it purports to give positive warrant for theory-acceptance. And he would be the least likely of all logicists to require (or even tolerate) an historical unit of appraisal. Once anomaly is perceived, *modus tollens* does the rest. On the face of it, then, we might not expect to find in Popper's work much support for the theses being argued here. But Popper (it perhaps does not need to be said) was never quite thorough-going in his falsificationism. In *The Logic of Scientific Discovery*, two entirely different outcomes of appraisal are described. The primary one is, of course, falsification. A theory is said to be falsified if a reproducible effect is discovered which contradicts it.[20] The unit of appraisal is the theory at a moment in time, and falsification is a straightforward logical relation between theory and evidence.

But Popper also allows a theory to be positively 'corroborated' in proportion to the 'severity' of the tests it has successfully survived. "Severe tests" involve "sincere attempts to refute the theory"; their severity depends upon "the degree of testability and thus upon the simplicity" of the theory.[21] Predictions in "a new field of application" would make the theory more 'universal', thus 'simpler', thus more 'testable'; if successful, they would thus corroborate much more strongly than would predictions in the original field.

There are some troublesome ambiguities about the concepts employed in this argument. Yet there is clearly a hint of the fertility-criterion here, though it is obscured by the falsificationist language Popper employs. A novel prediction is more improbable initially than one that conforms to the data on which the hypothesis was originally based. Because it is more improbable, it renders the hypothesis more liable to falsification. If it proves out, however, the fact of its having been more daring in the first place must be held to count somehow in its favor. Without some such concept as corroboration, Popper would have been unable to make sense of the fact that scientists are not *just* interested in being daring, i.e. in inventing hypotheses that lend themselves to quick overthrow. They want theories which will last; when Popper stresses that they try "to prove that their anticipations were false",[22] he does not mean that this is the outcome they *desire*. What they desire is that the anticipations prove *correct*. But the corroboration these latter give will be proportionate to the effort expended in testing, i.e. trying to refute, them and to the 'risk' the original anticipations involved, which is what makes that effort and that risk worthwhile.

What Popper was struggling to enunciate by means of the rather vague concept of corroboration was a *positive* note of appraisal for theory which would be much weaker than the notes of appraisal he associates with the classical verificationist theories of science, i.e. truth or, failing that, precise probability-measure. The novelty and variety of the predictions a theory generates are significant first because they subject the theory to severer test, an important desideratum if corroboration be proportionate to severity of test. Thus, even antecedent to any actual test, the more venturesome theory will be the 'better' theory. But if the predictions prove correct, the theory can be called a 'better' one in the stronger sense of 'actually corroborated'. The emphasis here is on survival in the face of test rather than on intrinsic fertility, reflecting the difference between a falsificationist and a verificationist perspective. Survival of the novel hypothesis counts positively because the hypothesis was antecedently improbable rather than because it testifies to the internal resources of the theory out of which this novelty came. Popper's use of the novelty criterion derives from his view of the relation between testing and antecedent probability, not from a realistic epistemology which would suggest that fertility is a property to be expected in a theory with realistic import.

The main difference between truth and corroboration, Popper says, is that corroboration varies over time, whereas truth is a timeless property:

> If yesterday we appraised a statement as true which today we appraise as false, then we implicitly assert today that *we were mistaken yesterday*; that the statement was false yesterday – timelessly false – but that we erroneously 'took it for true' ... (Whereas) we can never simply say of a statement that it is as such, or in itself, 'corroborated' ... We can only say that it is *corroborated with respect to some system of basic statements* – a system accepted up to a particular point in time ... Thus we must attach a subscript, as it were, to every appraisal of corroboration – a subscript characterizing the system of basic statements to which the corroboration relates (for example, by the date of its acceptance).[23]

How far has Popper moved here towards a historical notion of theory-appraisal? The unit of appraisal is itself still the theory or hypothesis, considered as a single well-defined set of propositions. And the estimate of degree of corroboration is still taken to be:

> a logical appraisal and therefore also timeless; for it asserts that a certain logical relation holds between the theoretical system and some system of accepted basic statements.[24]

But this latter system is itself a changing one; it grows with time and thus degree of corroboration will depend on when the appraisal is made. More important, the 'logical relation' is construed broadly enough (it would appear) to allow historical sequence to count. Popper rejects Keynes' view that it is irrelevant to the issue of hypothesis-appraisal whether a supporting instance is known before or after the formulation of the hypothesis.[25] Unfortunately, he does not develop this point, but it is consistent with his claim that severity of test (and therefore degree of corroboration) depends on initial improbability. This latter clearly depends on whether or not the supporting instance was or was not a part of the original *explanandum*. The 'historicism' here is obviously a very modest one. Appraisal is held to be both timeless and time-bound: timeless in that it involves timeless logical principles in regard to corroborative efficacy, time-bound in that it has reference to a set of basic statements whose extent depends on the moment chosen, and the date of formulation of each of which relative to the date of formulation of the hypothesis is significant for the purposes of appraisal.

In the essays collected in *Conjectures and Refutations*, Popper is struggling to retain the falsificationist thrust of his earlier writings, but is more than ever concerned that this not commit him to an instrumentalist view of scientific theory. In 'Three Views Concerning Human Knowledge' (1956), he tries to find a middle ground between essentialism and instrumentalism, both of which he rejects:

> This third view ... preserves the Galilean doctrine that the scientist aims at a true description of the world or of some aspects, and at a true explanation of observable facts; and it combines this doctrine with the non-Galilean view that though this remains the aim of the science, he can never know for certain whether his findings are true, although he may sometimes establish with reasonable certainty that a theory is false.[26]

This sounds like some form of critical realism. And Popper wishes it to be so understood. But it is of a rather special kind because the connection with reality is negative rather than positive: only in *falsification* do we discover "the points where we have touched reality". A good theory asserts something about reality only in the sense that it tells us that "events of a certain kind cannot happen". It is when our theories *clash* with reality that we know that there is a reality "and this is why the realist is right".[27] It is ironic that even though he has rejected verificationism, he retains its most dubious feature: that assertion must be 'assertion-with-certainty', so that only falsification allows us to make (negative) 'assertions' about reality.

Instrumentalism, he notes, is inadequate to account for the fact that theories often predict 'new kinds' of events. Such predictions must be regarded as 'discoveries'; lest this be taken positively he goes on to explicate it as "creating new situations for new kinds of tests", not a positive insight, then, but a means of continuing the testing process.[28] In a later paper, 'Truth, Rationality and the Growth of Knowledge' (1960), he continues to explore the topic of novelty, now for the first time making explicit a distinction like that between proven fertility and untested fertility discussed in Section 2 above. These are, in fact, now presented as two of the three requirements for the growth of theoretical knowledge. *U*-novelty is construed as strong testability, as in LSD. But what of *P*-novelty? Why should a falsificationist make it a requirement for a good theory that it should have passed 'new and severe tests'?[29] In a defensive footnote, Popper concedes that there is a 'whiff of verification here", but insists that "we have to put up with it, if we do not want a whiff of some form of instrumentalism".[30]

P-novelty is requisite for scientific progress. Science would stagnate and become pointless if it could not extend by means of new kinds of prediction. The emphasis on falsification ought not (he says) lead us to think that scientists produce theories "so that they can be superseded". For the aim of science is to "discover the truth", and theories are serious attempts to do just this. For them to be genuinely explanatory, they have to be "genuine guesses about the structure of the world".[31] And *P*-novelty serves as the crucial criterion, partly because a theory which is not only testable but true *would* be fertile in this way (so that it is a necessary, though not, of course, a sufficient condition of a good theory), and partly because *P*-novelty increases the truth-content of our theories, i.e. brings them in some sense "nearer to the truth".[32]

In an inductivist theory of science, Popper argues, there is no way of justifying the special value of new predictions (in the sense of "facts derived from the theory but previously not known"), because "if the value of a theory would lie merely in its relation to its evidential basis" – and he assumes the 'inductivist' to be committed to this – "then it would be logically irrelevant whether the supporting evidence precedes or follows in time the invention of the theory".[33] The HD-method, likewise, he urges, stresses only that the theory should "explain *known* experience". In short:

Successful *new* prediction – of new effects – seems to be a late idea, for obvious reasons; perhaps it was first mentioned by some pragmatist, although the distinction between the prediction of known effects and the prediction of new effects was hardly ever made explicitly. But it seems to be quite indispensable as a part of an epistemology which views science as progressing to better and better explanatory theories, that is, not merely to instruments of exploration, but to genuine explanations.[34]

This is something of a *tour de force*. The inductivist or the HD theorist would assuredly respond that it is *his* epistemology that supposes science to be "progressing to better and better explanatory theories", and that Popper has had to modify his original falsificationism quite drastically to allow himself to be associated with such a verificationist sentiment. Furthermore, there is no reason why inductivism or HD-theory should not sponsor the fertility criterion (we have seen above that it long antedated pragmatism). Popper's argument would hold against logicism only (i.e. the view that appraisal should be in terms of the logic of inference only). But inductivism and HD-theory are no more committed – indeed, rather less committed – to this than is falsificationism. It is true that most of the classical verificationists did *not* stress the fertility criterion. But this was assuredly not because they were verificationist (as Popper assumes), but because they were also, as it happens logicist. There is no necessary tie between the two.[35]

If stress on the fertility criterion is to be characteristic of an epistemology of progressive explanation, then one would expect to find it among verificationists first. Popper is so used to attacking verificationism that he attempts – unwisely – to utilize against it his new-found *P*-fertility criterion, apparently not realizing that if this criterion is to tell either way in the ancient verificationist/falsificationist dispute, it is more likely to tell against him than for him. It is *logicism*, not verificationism, that is his target now. As critics of logicism and defenders of *P*-fertility, verificationists have, however, at least as much credibility as a falsificationist, albeit reformed.

One can see this shift of focus in Popper's closing observation, an observation one could associate more easily with verificationism (either inductive or HD) than with his own earlier view:

It is my thesis that it is the *growth* of our knowledge, our way of choosing between theories, in a certain problem-situation, which makes science rational. Now both the idea of growth of knowledge, and that of a problem-situation are, at least partly historical ideas. This explains why another *partly historical* idea – that of a genuine prediction of evidence (it may be about past facts) not known when the theory was first proposed – may play an important role here, and why the apparently irrelevant time element may become relevant.

The focus here is on historicism, not falsificationism; the enemy is logicism, not verificationism. Though he does not admit it, alliances are changing. To reach even the very limited historicism of this passage, he has had to take a much more positive view of the aim of science than he did, and to modify the goal alluded to in the title of his first book of a 'logic' of appraisal in science. But there is a troubling tension here: in the last analysis, is *P*-fertility important because of its indicating the successful overcoming of potential refutation, or because of its demonstrating the internal resources – and hence the realistic import – of the theory? The significance of the historicist turn depends sensitively on the epistemological/ontological commitment prompting it. Despite what Popper says in the passage just quoted, the historical character of the notions of knowledge-growth and problem-situation hardly suffices of itself to explain why novelty in prediction – and thus the time-order of hypothesis and verified consequence – is so important for the scientist. Furthermore, it is not enough to historicize the order of hypothesis and verification. The hypothesis itself must be treated as a historical entity, whose changes over a period are neither deductively predetermined on the one hand nor arbitrary on the other. What is needed is a shift of emphasis from 'novelty' to 'fertility', a further move away from the over-simple view that the varied kinds of heuristic power a good theory displays over the course of time can be understood entirely in the testability framework of falsificationism.

5. The Methodology of Scientific Research Programs

This is accomplished, in part at least, by Imre Lakatos who has insisted more than has any other philosopher on making the unit for appraisal in science a historical one, a 'research program' or series of connected theories, rather than a single theory. Indeed, he has even made this the defining feature of his entire methodology. Lakatos argues that Popper's theory of science is inapplicable to actual scientific practice, mainly because he overestimates the falsifying force of single instances. One can, however, easily amend Popper's definition of science so that it no longer rules out essential gambits of actual science. I tried to bring about such an amendment, primarily by shifting the problem of appraising theories to the problem of appraising historical series of theories, or rather, of 'research programmes', and by changing Popper's rules of theory-rejection.[38]

It is the historicizing of the unit of appraisal that he perceives as his most significant move beyond Popper. In his original presentation of the MSRP, he describes it as a form of 'sophisticated falsificationism', a 'new rationale of falsificationism', which can rescue the idea of scientific progress, where naive falsificationism failed. This (he says) is "Popper's way, and one I intend to follow".[39] The new falsificationism differs from the old, both in its rules of acceptance of theory and in its rules of falsification. For a theory to be acceptable, it must not only be falsifiable, but must also be 'corroborated', in two senses: it must have excess empirical content (make new predictions), and some at least of these predictions must have been verified. We have already seen that Popper draws this distinction in his discussion of corroboration in CR; Lakatos is more emphatic than Popper in making corroboration a *necessary* condition for a good theory.

'Sophisticated falsificationism' differs much more, however, in its rules of falsification. Indeed, one wonders why the term 'falsification' is used other than as a gesture of piety to the tradition. A theory is 'falsified' only if there is a rival theory which can make all the correct predictions that the first can, and is, in addition, corroborated (in both senses) over the first theory. 'Falsification' is now reduced to a judgement of lesser degree of corroboration, a far cry from its original neat logicist sense. In this way, Lakatos hopes, one can attain a workable compromise between falsificationism (which *did* allow scientific theory to progress, but at the cost of an unreal simplification of the process of theory-rejection) and conventionalism (which correctly pointed to the difficulties involved in theory-falsification as well as to the conventional elements in the accepted practice of retaining theories in the face of anomalies, but at the cost of calling into question the whole notion of a 'real' progress in science).

The conventionalist elements in the new view are quite striking: not only are the chosen observational 'basic statements' assumed provisionally to be beyond dispute (as in Popper's LSD), but the theory is maintained indefinitely in the face of anomalies until a 'better' theory is available. And even when there are rival theories in competition (Lakatos, like Popper, assumes this to be the normal case), each is maintained for a sufficient time to allow a judgement in regard to corroboration to be made. Complete conventionalism is avoided by requiring that theoretical modifications in the face of anomaly be 'progressive', not *ad hoc*, i.e. that they lead to corroboration both 'theoretical' (increased content or prediction-extent) and 'empirical' (verification of some of the novel predictions). If they do not satisfy both these conditions, the research program is 'degenerating', though it will not be abandoned until a better one is available.

Clearly, then, to appraise a theory means looking at its past record. One must ask by "what sort of change it was brought about", and if there are rivals, how the records compare.[40] This means that appraisal *must* be of a historical sequence; it is not just a matter of determining temporal priorities of hypothesis and verification (Popper in CR), but the unit of appraisal itself *is* the sequence, not the theory taken as a logical entity at a moment in time. This shift, Lakatos suggests, was implicit in Popper's original notion of corroboration, though never made explicit by him. It arises from taking *novelty*, both untested and proven, as the critical test of theory and even as the core of 'falsification' (now described rather as 'rejection') itself. Progress in science is to be measured "by the degree to which the series of theories leads us to the discovery of novel facts".[41] Genuine explanation is entirely contingent upon novelty: "a given fact is explained scientifically only if a new fact is also explained with it".[42] Empiricism itself has been given a new focus:

> The time-honoured empirical criterion for a satisfactory theory was agreement with the observed facts. Our empirical criterion for a series of theories is that it should produce new facts.[43]

The reasons for the shift are, in sum, first that a good theory is not always in agreement with all the observed facts, and second, that agreement with the facts is insufficient as a criterion either to exclude *ad hoc* hypotheses or to decide between two theories both of which are in roughly the same degree of agreement with the facts. So a different criterion is needed, and that is the positive one of relative degree of corroboration. This is a much weaker criterion than the original falsificationist one, which did not allow time for a theory to "outgrow infantile diseases", such as anomalies, inconsistencies, *ad hoc* moves.[44] It takes time for a theory to display its resources in the face of challenge. This of itself would not, however, suffice to show that a historical unit of appraisal is needed: it might seem no more than the advice to "give it time before a standard falsificationist-logicist critique be applied to it". This would be no more than a cautious falsificationism; the properly *historicist* thesis is not just 'give it time', but 'see what it does during the time and just how and in what order it does it'.

But this is still somewhat vague, and a logicist might well want to know something more about the significance of following the temporal sequence. After all, he is perfectly prepared to allow that if T' accounts for everything that T does and in addition *correctly* predicts something that T doesn't, T' is to be preferred over T. If this is all that 'novelty' amounts to, there is no need to move to a historical unit of appraisal. He may even be willing to concede that the temporal order of hypothesis-formulation and of verifying observation ought to be taken into account, by the sort of extension of the original logicist program discussed in Section 2 above. But Lakatos evidently wants to make a stronger point than this; to estimate the corroboration a research-program has received, one has to follow the stages its 'positive heuristic' has gone through, and see what sort of guidance it has given. But how is one to do this?

Here Lakatos has to thread his way between logicism and conventionalism. If the 'positive heuristic' is entirely contained within the original hypothesis, a deductive analysis of the consequences of the hypothesis and a systematic observational test of them is all that is needed. There is no need to trace historical development, since it could be no more than an unrolling of consequences implicitly present from the beginning. On the other hand, if the positive heuristic is *not* fully derivable from the original hypothesis, how can one exclude the possibility that the path it follows is not (to the extent that it differs from what was implicit in the original) a conventional one? In other words, 'novelty' is systematically ambiguous: it may mean no more than 'new predictions' ('new' that is, relatively to the earlier theory to which it is being compared), in which case the logicist can still handle it, or it may mean "predictions not contained at all within the original hypothesis" in which case, we could be back either in falsificationism or conventionalism, since the modification of theory leading to the novel predictions appears to be an arbitrary one. Its success, if it does succeed, would therefore testify, not to the resources of the original theory but only to the (possibly *ad hoc*) value of this particular modification.

Lakatos cannot quite make up his mind how to handle this difficulty. He speaks both of 'predetermination' of the positive heuristic, and of the possibility of 'creative shifts' in it.[45]

> The basic unit of appraisal must be not an isolated theory or conjunction of theories but rather a *research programme*, with a conventionally-accepted (and thus by provisional decision 'irrefutable') *hard core* and with a *positive heuristic* which defines problems, outlines the construction of a belt of auxiliary hypotheses, foresees anomalies, and turns them victoriously into examples, all according to a preconceived plan.[46]

It sounds here as though the modifications from one 'theory' to the next within a 'research program' are somehow precontained; elsewhere he speaks of "the instructions which are laid down" to enable the scientist to construct "a chain of ever more complicated models" in response to the anomalies that arise.[47] True, the scientist cannot tell in advance which anomalies *will* arise. But in a good research program, Lakatos seems to imply, he will have a blueprint telling him what his response should be to any particular anomaly or other failure of theory.

The 'predetermination' thesis is convenient if one wishes to explain how the sequence of modifications corroborates the 'research program', i.e. the series of 'theories', or more exactly, the 'preconceived plan' that gives unity and continuity to the research program. But it faces two grave difficulties. First, if 'predetermination' were to be taken literally, we would be back in the old logicist position, as already noted. If it means that ready-made alternatives are to be found within the program which will enable one (if the program is a 'good' one) to turn anomalies into victorious supporting evidence, once again the historical sequence will not matter, except in one respect: the relative order of hypothesis-proposal (the 'preconceived plan') and verification of consequence in each case. Second, such 'predetermination' seems to run counter to all that we know of the history of successful scientific theories. It is by no means the case that when anomalies arose in these, the scientist had available to him within the research-program itself a well-defined schedule of possible theoretical moves.

Lakatos himself appears to realize this, for in other passages he describes the positive heuristic as being far less definitive than this:

The positive heuristic consists of a partially-articulated set of suggestions or hints on how to change, develop the 'refutable' variants of the research-programme, how to modify, sophisticate, the 'refutable' protective belt.[48]

This seems much closer to the experience of science than does the 'predetermination' thesis. But it raises a question about how corroboration is to work. How does the imaginative modification of a theory faced by a new result corroborate the *original* hypothesis in which this modification was no more than an unarticulated hint? Or how does it corroborate the research program as a whole, other than in the banal sense of a correct prediction finally attained to? Lakatos has no real help to offer on this crucial question. Nor should this surprise us, for he has retained just enough of the paraphernalia of falsificationism to make an effective answer very difficult for him to formulate.

6. Inadequacy of the 'Research-Program' Label

Before we come to develop this point further, it would be well to underline the inadequacy of the label, 'research-program', which Lakatos has chosen to designate the unit of appraisal in science. He chose it because of its historical overtone; it conveys the impression of being the product of an activity spread out in time and shared in by a well-defined group of people with similar training and similar goals. Its affinity with Kuhn's notion of paradigm is obvious. Lakatos refuses to use the word 'theory' for the unit of appraisal; indeed, he considers this a 'category-mistake',[50] since appraisal of relative corroboration can only be directed to a 'series of theories'. He accepts his definition of 'theory' from the logical positivists: a well-defined set of explanatory propositions, considered independently of any historical or developmental aspect. And he rightly refuses to admit that 'theory' in this sense can be the ultimate object of scientific appraisal.

But why *should* 'theory' be used in this sense? It is certainly *not* the sense in which it is used by scientists, who will speak of the 'wave-theory of light' for example, as something which developed over several centuries. No scientist would say that the successively modified forms of the Bohr theory of the H-atom should be called different *theories*. Even less plausible would it seem to say (as Lakatos does) that each 'falsifies' the previous one on the grounds of its having a higher degree of corroboration.[51] Here the falsificationist origins of the MSRP are at their most evident. Ordinary usage would, in fact, sanction the very sense that Lakatos needs for his unit of appraisal, namely, a historical explanatory system which may undergo numerous modifications over the years, and yet be presumed still the 'same' theory, on the basis of some usually not-very-clearly-defined continuities of basic structure.

'Theory' is thus a perfectly suitable term for what Lakatos has in mind, as long as one emphasizes that the term is used in the scientist's, not the logician's, sense. But whether or not 'theory' is the term one chooses, the label 'research program' just will not do. Or at the very least, it will require one to stipulate a sense which is remote from its ordinary usage. And if this is to be permitted, why not stay with 'theory'? A research program is *not* the unit of appraisal we seek. Many research programs (e.g. in cancer research) are almost purely fact-finding, and have no unified explanatory theory guiding them or in the process of being tested by them. Other research programs involve a mix of loosely-related theories bound together by the practicalities of the research situation involved. Often, the term will be applied to the team-effort, e.g. in high-energy physics, of a laboratory group engaged in a single, perhaps quite narrow, problem. Another competing laboratory, working on the same problem with different equipment and utilizing somewhat different strategies, would be said to have a 'different' research program. The recent simultaneous discovery of a new 'particle' at Stanford and at Brookhaven[52] could in this way be described as the product of two different research programs. Everyone who submits a proposal to a funding agency requesting support for a piece of scientific research presents a 'research program'; no matter how detailed, how general, how short-term, how untheoretical, the proposal is, standard usage would sanction its being thus described.

In short, the term is a pragmatic one, applicable in different ways in different contexts. To appraise a research program could involve factors like proper planning, co-ordinated team-work, cost-effectiveness and the like. Progressive problem-shifts would be only one among many possible desiderata. There is, of course, no reason why Lakatos should not appropriate the term to his own purpose, stipulating his own sense for it. But one must insist, in this event, on noting that the sense *is* a stipulated one, and that there might well be other terms more appropriate for the task.

The two main objections to identifying the normal unit of appraisal in science with a 'research program' affect, however, the substance of the MSRP and not just its terminology. First, a research program ordinarily terminates when the problematic is exhausted, when interest switches elsewhere (for whatever reason). Little work is being done today in some of the most active problem-areas of experimental physics and chemistry of a century ago; the assumption in such cases is that we have a good theory which accords well with the data available. This does not preclude the possibility that research might push to a further level of experimental sophistication or theoretical depth in these areas. But the theory in possession is assumed to be a 'good' one at its level of structural specification. The appraisal of *theory* and of *research-program* would quite evidently be different in such areas. To appraise a research-program is to ask about the *potential* a program has for further development in the light of its past history and its present resources. To appraise a theory ordinarily means to ask about its truth-value. How confident can we be not just about the accuracy of the predictions it makes but about the insights it purports to provide into the structures of the world?

The 'research-program' metaphor has unmistakeably falsificationist origins. Appraisal is assumed to be future-oriented, to have to do with potential falsifiability. We are back in the world of LSD, where the 'best' theory was the one that 'stuck its neck out' most, where it was assumed that there ought always be rival theories in contention, so that appraisal had to do with choosing between them and corroboration was a means to that end. But in *real* science, this is the exception rather than the rule. Take any textbook of general chemistry or biology and see in how few of the chapters are rival theories noted as being in active present contention. It is of course, the areas of contention that are of most research interest. But is appraisal directed to research interest or to truth-value? We shall return to this question in the next section.

There is a second major objection to the 'research-program' label. What precisely *is* the research program? Lakatos identifies it more than once as a series of theories. But how is a *series* to be appraised? Each 'theory' is after all replaced by the next, according to Lakatos. Ought the *last* 'theory', then, bear the main weight of appraisal? Not if we are to take seriously what he says about the historical character of the unit of appraisal. Clearly, appraisal is to be directed to some single entity which manifests itself (so to speak) over an extended period of time, as the original theory comes to be modified in the face of new situations. What matters most, then, are the conceptual resources of the original hypothesis. To appraise positively is to say that the theory has over the course of its career exhibited the ability to extend, modify, meet challenges, to a degree that warrants confidence in its corroboration. What exactly *is* corroborated? In one way, it is the theory as it *now* exists, which presumably comes closest to giving a good account of the real. In another way, it is the *original* theory which proved to have the creative resources needed, thus testifying to its non-*ad-hoc* character. In yet another way, it is the entire career, considered not as a sequence of competing theories but as an example of continuous *growth*.

It is evidently some sort of *continuity* that allows one to speak of its being the 'same' theory (or 'same' research program) throughout, and thus that makes it possible to appraise the temporal sequence as though it were a unity. It is crucial, then, to specify where the continuity should be sought. Lakatos is not of much help: "This continuity evolves from a genuine research programme adumbrated at the start".[53] Adumbrated? We will need some account of 'adumbration' if we are to understand what is going on in theory-appraisal. The falsificationist framework that Lakatos never quite escapes makes it seem as though each separate member of the 'series of theories' must be evaluated separately; it is hard to work up from this sort of atomism to the holistic concept he is bidding us seek.

He is convinced of the importance of the continuity theme:

> The members of such series of theories are usually connected by a remarkable *continuity* which welds them into research programmes. This continuity – reminiscent of Kuhnian 'normal science' – plays a vital role in the history of science.[54]

Indeed, yes. But more must be said about where it resides and how it is to be identified, since it is finally the *continuous* entity that is being appraised. And the term, 'research program', is to say the least unhelpful in this regard. The 'continuities' of a research program are of many kinds, most of them irrelevant to the matter of scientific appraisal of epistemic value. To appraise a research program is *far* more complicated than to appraise the theory or theories involved in the program (though theory-appraisal will, of course, be *part* of what is involved in appraising many kinds of research programs).

One further objection may be mentioned, though it is not as substantive as the two preceding. The notion that the scientist is dealing with a series of theories, each of which has to be appraised relatively to its 'predecessor' introduces an unnecessary ambiguity into the already complex problem of appraisal. For now, 'appraisal' is exercised in two quite different ways: there are rival 'theories' (i.e. a theory and its own tentative modification to meet a new situation) and rival 'research programs'. But this is a misleading way of describing the scientist's situation. When he suggests a possible extension of his theory, he does *not* regard this extension as a rival theory; its success would *not* refute, or falsify, his original view (as Lakatos' falsificationist bias impels him to suggest). In fact, quite the opposite: its success will testify to the positive value of the theory that suggested it and which is now not so much superseded as enlarged. This is the whole point, in fact, of taking an extended unit of appraisal.

To put this in another way. If the *continuities* in our unit of appraisal are of importance (as Lakatos maintains), we cannot also speak of 'rivalry' between the successive 'theories' (i.e. between the successive stages of the same theory, to use the more customary terminology). Or at least, if we do, we must mean something altogether different from the sort of rivalry that exists between two genuinely different theories proposed to account for the same data. Not only does it run counter to usage to describe the career of, say, the Bohr theory from 1911–1926 as a 'series of theories', but it positively obscures the very point we want to emphasize in the context of theory-appraisal, namely, that what the scientist engaged in appraisal is most interested in is the manner in which the resources of the original theory have lent themselves to continuous and fruitful exploration.

'Research program' is thus a bad choice of label. 'Theory' is still a much better one. It is the scientist's own choice, for one thing. Of course, it has disadvantages, notably its appropriation by logicians and formalistically-inclined philosophers of science to designate a timeless set of propositions entirely open to logical scrutiny. And it suggests a certain conceptual unity. In any actual appraisal, one will be concerned not only with the 'foreground' theory, but also with all the auxiliary hypotheses that underlie one's attempt to test the theory in this particular context.

Lakatos' rather too neat dichotomy between 'hard core' and 'protective belt' was intended to explain the fact that in testing, different parts of our system are accorded different degrees of vulnerability (a *spectrum* would be a rather better metaphor for this than the core/belt one). The 'hard core' is not being tested; we have decided in advance not to allow *modus tollens* to touch it (negative heuristic). But in this event (Popper would surely want to ask), in what sense is the *hard core* corroborated? The shifts that occur and that constitute the history of the 'research program' are all (Lakatos is forced to say) in the 'auxiliary hypotheses', the 'refutable variants'. The 'positive heuristic' has nothing to do with the hard core (he reminds us); it concerns only the protective belt.[55] Is it the research program or its protective belt only that is corroborated then? To be corroborated, "the programme as a whole should display an intermittently progressive empirical shift", as well as a consistently "progressive theoretical problem-shift".[56] But the hard core, by definition, cannot display *any* shift. In the falsificationist perspective of Lakatos' presentation, the hard core is simply *not* corroborated, since no risk was taken with it.

What is needed here, once again, is more attention to the sorts of modification that mark the career of theory, as well as to their sources. Lakatos describes the typical theory-modification (the sort of one on whose success corroboration depends) as "adding a novel auxiliary hypothesis".[57] But this is altogether too external a way to describe what usually happens. It sounds as though a discrete testable item has been added, to which the credit can be given if it works; if it doesn't, it is just dropped and no harm done. But, of course, this is *not* what Lakatos wants to say; the whole point of his suggestion about a temporal unit of appraisal is that modifications are *not* discrete hypotheses tacked on, nor are they 'auxiliary', if by that one implies that their genesis is independent of the 'hard core' of the program.

One ought, therefore, to avoid dividing the theory into 'hard core' and 'auxiliary hypotheses'. But there is still an important sense in which test is carried out not on isolated theories, but on a complex of theory and auxiliary hypotheses, the latter now being understood not in Lakatos' sense as the more vulnerable (or less conventionally protected) parts of the theory, but rather as the set of theoretical assumptions, over and above those that form an explicit part of the theory itself, which are being implicitly tested whenever the theory is tested. *These* 'auxiliary hypotheses', unlike those of Lakatos, are *less* likely to be modified than is the theory itself when anomalies arise. They are closer to being the 'hard core', at least in the sense that the aim of the research program is to develop the *theory*; it is true that the auxiliary hypotheses *may* have to be modified, but to the extent that this modification (e.g. of the theories of measurement used) is not in any way inspired by the theory, the outcome does not help in the appraisal of the theory.

Ought the principal unit for appraisal be the theory and auxiliary hypotheses, taken over time, or the theory alone? Remembering Duhem and Quine, we might be led to choose the former, since what is affirmed at any moment in the development is the theory *plus* whatever other assumptions the theory or its application require. (The truth of the predictions tested depends just as much on these latter as on the theory itself). But to the extent that we move to a historical unit of appraisal, it is the theory *alone* which is being appraised, since it is to *its* resources only that we are looking in evaluating the theoretical modifications that have occurred. In that sense, the auxiliary assumptions are regarded as a sort of invariant background, even though under other circumstances it might be one or more of these on whose historical development (and consequent degree of corroboration) we might wish to focus.

7. Two types of theory-appraisal

It is important, as we have already seen, to distinguish between two quite different features one might look for in a theory and thus two different senses of 'appraisal' in this context. One might be interested in the *truth-value* of the theory. To what extent has the theory been corroborated? Does it conform reasonably closely to the structure of the real? Is one warranted in accepting the existence of the theoretical entities it postulates? Though verification in the strong sense of definitive proof is beyond the reach of scientific method, it is still possible to ask questions like these, and expect helpful answers, admittedly in such qualitative terms as: 'highly corroborated', "there is good reason to suppose that electrons (i.e. entities partially modelled by our current theories) do exist", and so on. To say that we are interested in the truth-value of a particular theory does not mean that we want to know whether it is true or false. These terms are obviously inapplicable to theory. A much more flexible set of responses than the simple two-valued one of classical logic (or classical epistemologies) is needed. This is, however, not our topic here. It is sufficient to note that such questions *can* legitimately be asked and answered, and that (as Popper realized) the applicability of notions like progress or growth to scientific change depends on that legitimacy.

An appraisal which answers to a question like one of these we can call an *epistemic* appraisal. Admittedly, an *epistēmē* of the physical world is not within our reach, as Aristotle believed it to be. But there is a scaled-down sort of *epistēmē* which *is* available to us. It is provisional but reliable, secure but open to steady improvement in depth and comprehensiveness. When we look at the past record of a theory and estimate its *P*-fertility, it is with a view to epistemic appraisal. If the resources of the theory have shown that it was able to cope with the unexpected, the argument is that this gives us reason to believe that the structures postulated in the theory correspond reasonably well to the structures of the real. How strong a reason will depend upon just what sort of guidance the theory *did* offer, how far beyond the evidence on which it was originally based it has brought us, and so on.

'Established' theories are those appraised highly in this sense. The kinetic theory of gases, the double-refraction theory of the rainbow, all the 'safe' theories with which our elementary textbooks abound, would furnish an abundance of examples. The restricted theory of relativity, the gene theory of heredity ... the list is long. How about Newtonian mechanics? In the falsificationist tradition reflected in Lakatos and Feyerabend for instance, it is said to be 'refuted'. But surely this is rather too harsh an assessment for the most highly corroborated theory in the entire history of science? Is not this a remnant of the logicism which both Lakatos and Feyerabend have rejected? Since our purpose here is only to illustrate what is meant by 'epistemic appraisal', however, this tempting byway must alas! be left unexplored.

There is a second, and very different, sort of demand one can make of theory. One can ask: what is its research-potential for the future? how likely is it to give rise to interesting extensions? Does it show promise of being able to handle the outstanding problems (inconsistencies, anomalies, etc.) in the field? Is it likely to unify hitherto diverse areas, or perhaps open up entirely new territory? These questions have to do with *potential*, with what we called above *U*-fertility. They are much harder to answer than the first kind, and answers of the very nature of the case will have to be extremely tentative. Yet questions like these are of enormous importance in the planning of scientific work. This sort of appraisal we shall call *heuristic appraisal.* It is directed to the as-yet unexplored heuristic possibilities of the theory.

The two sorts of appraisal are obviously not unrelated. Part of an epistemic appraisal is concerned with *P*-fertility, i.e. with the way in which the original *U*-fertility of the theory proves out. But they are quite distinct nonetheless. To say that a theory has a high degree of heuristic potential tells us nothing of its epistemic status. Indeed, the greater its heuristic potential, the more problematic its present epistemic status may well be, since it contains so much in the way of untested suggestion. Likewise, to speak of a theory as 'established' tells us nothing of its heuristic potential for the future. Some of the best-established theories have very little research-potential; of course, some new perspective may open up a way to reconsider and perhaps reconstruct them, and then the heuristic appraisal will change.

This distinction is perhaps obvious, but falsificationism tended of its very nature to obscure it, and we are still trying to break clear from the resultant confusions. The criterion of falsifiability led quite naturally to a stress on heuristic potential. If a theory is no longer producing new insights, it is not lending itself to possible falsification, and so the associated 'research program' can be said to be 'degenerating'. This is a matter of heuristic appraisal; to say that a research program is degenerating in this sense, has nothing whatever to do with *epistemic* appraisal, of course.

Lakatos uses the term, 'degenerating', in two rather different ways. A research program may be degenerating when it "ceases to anticipate novel facts";[58] This may lead us to abandon it (since our decision to maintain the 'hard core' against rejection was from the beginning a 'conventional' one, in his sense of the term), but only if another theory is available. When an anomaly is handled by a 'problem-shift' that decreases content instead of increasing it (i.e. is *ad hoc*), this is said to be a 'degenerating problem-shift'.[59] Since 'problem-shift' and 'research program' are supposed to be synonymous,[60] this is confusing, because an established theory not presently heuristically lively would be said to be 'degenerating' (it would seem) in the first sense, but not in the second. Or is the term only to be used when two rival research-programs are in contention, and it is a matter of deciding which one is providing more novelty (more 'content')?

Lakatos does insist that if a research program is not 'progressive', it must be degenerating,[61] but since he appears to assume that appraisal takes place only when a choice must be made between rival theories, it is *possible* that he would not consider a relatively static established theory, without serious rivals, to be 'degenerating'. But his reader would be more likely to make the opposite assumption. The mere fact that he takes research-programs to be, by definition, 'problem-shifts' means that if the problem is *not* shifting, there isn't a research program, and thus the matter of appraisal – heuristic appraisal, obviously – does not arise.[62]

If, on the other hand, one begins not from the criterion of falsifiability but from the requirement of corroboration, the emphasis will be on epistemic, rather than heuristic, appraisal. 'Empirically progressive' problem-shifts are ones which have proved to be successful *in retrospect*; it is not a matter of estimating potential, then, but of assessing achievement. An established but presently static, theory would certainly be able to call in its support an 'empirically progressive' problem-shift. How would it be appraised? Heuristically low, epistemically high, presumably. But Lakatos, by committing himself to a single univocal notion of appraisal, has deprived himself of the possibility of a consistent answer here. The tension which he himself underlines between 'P_1' and 'P_2', between the falsificationist and the corroborationist sides of Popper's work, surfaces equally destructively here in the MSRP itself.

8. Novelty or fertility?

A similar tension is evinced in the notion of novelty, which is so crucial to the MSRP. From the falsificationist standpoint, novelty is advantageous simply because it increases falsifiability; novelty is taken to show itself through predictions derivable from the theory being appraised, so that the theory itself is falsified if the 'novelty' is not verified. The more daring the hypothesis, the greater the 'content-increasing' possibility it possesses, and the more 'theoretically progressive' it is. The appraisal of novelty depends only on its 'content', on how much extra it commits the theory to: the more states of affairs it excludes, the higher the appraisal. [63] But this is far too mechanical, too logicist, a notion of novelty to do justice to the realities of appraisal in science.

First, it supposes that the 'novel prediction' is logically contained in the original theory, so that it is refuted if the prediction is. But in practice, 'novelties' are much more tentative than this; they are *suggested* by the hypothesis when faced by anomaly or other problems, but not in such a way that their failure automatically involves the hypothesis as well. Second, there is no hint as to how the 'novelty' relates to what went before; in fact, it would not seem to matter: the wilder the conjecture, the better, just so long as it is 'content-increasing'. The requirement of continuity is altogether absent from this crude notion of novelty as no more than 'something new', something not predicted by the rival hypothesis. Even if one requires that some of the novel predictions must be verified for a positive appraisal to be given,[64] it still does not seem that the notion has the depth required for the assessment of a research program over a period of time. By reducing a research program to a series of theories, Lakatos is able to reduce appraisal to: (a) checking that each theory has more 'content' (predicts more) than its predecessor, and (b) showing that some at least of these novel predictions prove correct in each case.[65] Taken this way, it would not be a matter of looking at the whole, then; it would be sufficient to test the parts.

But elsewhere, as we have seen, Lakatos will stress the importance of the continuity of the research program as a whole, and will suggest that appraisal may depend on how well it all has been 'adumbrated' from the beginning. But what sort of adumbration, short of actual predetermination, would be sufficient to provide unity to the whole and thus allow the whole to be corroborated by the predictive success of the parts? The answer may lie in the notion of *fertility*. What the scientist begins with is a 'fertile' theory, one capable of giving rise to the desired growth, given the right conditions. The modifications are not 'precontained'. But the potentialities to produce them are there, and this is why the corroboration of the theory is, in part, a corroboration of the original metaphor from which the career of the theory (or of this episode in the career of the theory) began.

The term, 'metaphor', reminds us of what it is in the theory that permits the subsequent developments to be appraised as an organic unity. A good metaphor allows the imagination to work, and guides it in certain directions. It is not a merely vague or indefinite object of thought. It directs the mind, not coercively but tentatively, by analogies and hints. The mind's response to it will be highly individual, in consequence. It will be a search, not a resting in any one place. A powerful poetic metaphor may be developed by the poet himself; the poem in that case – take *Paradise Lost* or the *Divine Comedy*, for example – stands or falls with the resources of that original metaphor and how successfully it has been exploited. The poet may develop the metaphor explicitly, take the opportunities it offers. The reader cannot do this in quite the same way; he cannot extend or reconstruct the poem. But his imagination is challenged to explore, to transfer knowledge-elements from one part of his experience to another.

A good scientific theory has this same quality. It is to be found primarily in the model associated with the theory, a postulated explanatory structure whose elements are capable of further imaginative development. And the possible lines of development are suggested by the model itself, though the outcome will depend on the creative talent of the scientist and on the sorts of challenge he faces from the 'facts'. The important thing to notice for our purposes is that the epistemic appraisal of the model will have to take into account how successfully the model has guided research so far, what sorts of extensions and modifications have taken place, and how much they owe to the imaginative resources of the undeveloped original. The unity of this process is not logical; it is closer to being organic although the creative reworkings of the scientist are so essential to it that *organic* is not quite the right description either. The bond between the original and what comes of this original in consequence of an active period of development is such that an appraisal of one is an appraisal of the other: a good model, after a fruitful career, has developed into the basis for a highly corroborated theory. If one is appraising a theory that has already had a chance to develop, one of the criteria (not, of course, the only one) is the success of the original model in inspiring this development.

But if our object is epistemic appraisal of the theory here and now, why should it matter what the sequence of changes has been that led up to it? Why should one be concerned with the *original* model and what its resources were? After all, this is *not* the model one now has to work with. Why should not one simply evaluate the predictive success, the relation of evidence to hypothesis, of the theory as it now stands? This is where the intuitions of working scientists have diverged so much in the past from the views of logicist philosophers of science:

From a logical point of view, the strength of the support that a hypothesis receives from a given body of data should depend only on what the hypothesis asserts and what the data are: the question of whether the hypothesis or the data were presented first, being a purely historical matter, should not count as affecting the confirmation of the hypothesis.[66]

Maybe it 'should not count' but the fact is that it *does*, which indicates the inadequacy of the 'logical point of view' dictating this sort of analysis of confirmation. What counts (as we have seen) is not only relative order of observation and hypothesis (the falsificationist notion of 'novelty' can allow for this), but the details of the entire historical development linking the original model with the theory as it now stands.

The reason for this has to be sought in a realistic account of models and their function in science. A 'good' model in this view is one which has a 'good fit' with the structure of the real. How would such a model show itself to us as a 'good' one? By displaying the *U*-fertility and later the *P*-fertility one would expect that a model having a satisfactory correspondence with the object to be explained *would* display. Note that a 'good' model is not simply one with a high degree of heuristic promise. *U*-fertility is not enough, although it may be enough to persuade us to invest our effort in it. What is needed is *P*-fertility, a proven record. And the testimony to this record is *not* just the predictive ability (the 'content') of the theory at the time of appraisal, nor any sequence of increases in content, of the kind the falsificationist relies on. Rather, it is the imaginative guidance the model has given the researcher in meeting and overcoming conceptual and empirical challenges, that testifies to its 'truth', and therefore to the truth of its associated theory.

9. Conclusion

Much more would need to be said about this epistemic function of model.[67] But perhaps enough has been said here to show the interconnection between the three themes that have formed the topic of this essay: the fertility of theory, the historical unit of theory-appraisal, and the realism of theoretical models. No account of one of these can be complete without the others. There is no way to justify to the logicist the adoption of a historical unit of theory-appraisal without taking into account not just the role of novelty in science (this is enough only to warrant the thesis that the relative order of hypothesis and observation must be taken into account in science), but the importance of *fertility*. *P*-fertility can be estimated only by following the entire career of the theory (specifically, the model) to date, 'following', that is, as the scientist might, who would know just how well his model has worked to direct research and overcome anomalies. But *P*-fertility becomes an index of truth only if one adopts a realistic view of science. The logical structures of falsificationism or of statistical testing are not enough of themselves to show why long-term *P*-fertility should be taken so seriously. Finally, the success of this mode of appraisal, in which one of the criteria is model-fertility tested over the appropriate time-span, is the best argument for taking the theoretical models of the scientist realistically. Not, of course, as description, but as metaphor, which is the only way that complex things forever somewhat out of reach *can* be known.

Notre Dame University

NOTES

* I would like to express my appreciation to the National Science Foundation for support of a period of research at Cambridge University under grant GS-41320, during which time the first draft of this essay was completed.

[1] McMullin, 'Empiricism at sea?', *Boston Studies in the Philosophy of Science*, Vol. 14, 1974, pp. 21–32.

[2] Studies in the Logic of Confirmation' (1945), reprinted in *Aspects of Scientific Explanation*, New York, 1965, p. 22.

[3] *Ibid.*, p. 41.

[4] *Ibid.*, p. 42.

[5] This is the recurrent theme of such works as *Logical Foundations of Probability* (1950). He writes in his Autobiography: "I believed that the logical concept of probability should supply an exact quantitative explication of a concept which is basic in the methodology of empirical science, i.e. the concept of the confirmation of an hypothesis with respect to a given body of evidence". *The Philosophy of Rudolf Carnap*, ed. by P. Schilpp, LaSalle (Ill.), 1963, p. 72.

[6] John Herschel, *Preliminary Discourse on the Study of Natural Philosophy*, London, 1842, Sec. 180; quoted in *Theories of Scientific Method*, E. Madden (ed.), Seattle, 1960, p. 177. Herschel uses the term 'felicity' for this quality of theory, and notes that the history of the wave theory of light provided a succession of such felicities.
[7] William Whewell, *The Philosophy of the Inductive Sciences Founded Upon Their History*, London, 1847, Vol. 2, pp. 65–67. Whewell associated the consilience of inductions also with the criterion of *simplicity*: in a true theory, "all the additional suppositions tend to simplicity and harmony", *op. cit.*, p. 68.
[8] *What is Science?* (1921), reprinted New York, 1952, p. 87.
[9] A. Musgrave, 'Logical Versus Historical Theories of Confirmation', *British Journal for the Philosophy of Science* **25** (1974), 1–23; E. McMullin, 'History and Philosophy of Science: A Marriage of Convenience?', *Boston Studies*, Vol. 32, 1975, pp. 515–531.
[10] *Ibid.*
[11] *The Common Sense of Science*, New York, 1951, pp. 130, 133.
[12] *Philosophy of Science*, Englewood Cliffs, N. J., 1957, p. 352.
[13] *Op. cit.*, p. 353. He admits that two further criteria: "agreement with common sense experience", and "fitness to support desirable moral and political doctrines" have deeply influenced scientific development, and discusses their limitations as well as arguing for their legitimacy in certain contexts.
[14] This connection between the fertility criterion, a historical unit of appraisal, and a realist theory of science, was argued in my 'What Do Physical Models Tell Us?', *Logic, Methodology and Philosophy of Science*, ed. by B. van Rootselaar and W. Staal, Amsterdam 1968, 389–396, and later in 'The History and Philosophy of Science: A Taxonomy of Their Relations', *Minnesota Studies in the Philosophy of Science*, R. Stuewer ed., Vol. 4, 1970, pp. 12–67.
[15] Chicago, 1962, p. 169.
[16] *Op. cit.*, p. 168.
[17] *Human Understanding*, Princeton, 1972, p. 229.
[18] *Op. cit.*, p. 231.
[19] *Op. cit.*, p. 229.
[20] London, 1959, p. 86.
[21] *Op. cit.*, p. 267.
[22] *Op. cit.*, p. 279.
[23] *Op. cit.*, p. 275; his italics.
[24] *Loc. cit.*
[25] *Op. cit.*, p. 272.
[26] London, 1962, pp. 114–115.
[27] *Op. cit.*, pp. 116–117.
[28] *Op. cit.*, p. 118.
[29] *Op. cit.*, p. 242.
[30] *Op. cit.*, p. 248.
[31] *Op. cit.*, p. 245.
[32] *Op. cit.*, p. 246.
[33] *Op. cit.*, p. 247.
[34] *Loc. cit.*
[35] Unless one identifies – as Popper inclines to do – verificationism with the view that scientific verification, properly carried out, should be definitive, i.e. lead to provably true assertions. But relatively few inductivists or HD-theorists have maintained this; their notion of 'verification' has more often amounted to some sort of intuitive 'probabilification', or progressively improvable approximation.

[36] *Op. cit.*, p. 248.
[37] In a series of five overlapping articles: 'Changes in the Problem of Inductive Logic' (abbrev. CP), *The Problem of Inductive Logic*, ed. by I. Lakatos, 1968, pp. 315–417; 'Criticism and the Methodology of Scientific Research Programmes' (abbr. CM), *Proc. Arist. Soc.* 69 (1968), 149–186; 'Falsification and the Methodology of Scientific Research Programmes' (abbrev. FM), *Criticism and the Growth of Knowledge*, ed. by I. Lakatos and A. Musgrave, London, 1970, pp. 91–195; 'History of Science and Its Rational Reconstructions' (abbrev. HS), *Boston Studies in the Philosophy of Science*, Vol. 8, 1971, pp. 91–136; 'Popper on Demarcation and Induction' (abbr. PD), *The Philosophy of Karl Popper*, ed. by P. Schilpp, La Salle (Ill.) 1974, pp. 241–273.
[38] PD, p. 248.
[39] FM, p. 116.
[40] FM, p. 118.
[41] *Loc. cit.*
[42] *Op. cit.*, p. 119.
[43] *Loc. cit.*
[44] PD, p. 249.
[45] FM, p. 137.
[46] HS, p. 99. The same passage, with the omission of the phrase about 'auxiliary hypotheses', occurs in PD on p. 248.
[47] FM, p. 135.
[48] *Loc. cit.*
[49] It is clear from the exposition above that the MSRP is a development of ideas latent in the notion of corroboration in Popper's LSD, and more specifically in his discussion in CR of progress in science. Lakatos first presents it as Popper's, calling it 'sophisticated falsificationism'. But it is clearly at variance with other strands in Popper's doctrine. Lakatos makes an unhappy effort in CM to distinguish between three 'Poppers' in recognition of the evident tension in Popper's work between the falsificationist, the conventionalist and the corroborationist themes. The MSRP supposedly represents the views of 'P_2', the corroborationist 'Popper', with a few changes (CM, pp. 165–67), notably, as we have seen, the shift to an explicitly historical unit of appraisal. But Lakatos drops this attempt in HS, and in PD underscores rather the differences between Popper's 'naive' falsificationism and the MSRP (see especially the Note on p. 244). It is not worth pushing the question of historical antecedents any further at this point: it is evident that Lakatos drew on Popper for all of the characteristic themes of his MSRP, but that he carries several of them much further than Popper did, and that their final systematization in the MSRP is sharply at odds with the mainline falsificationism of LSD.
[50] CM, p. 179; FM, p. 119.
[51] CM, p. 164. In a 'progressive problem-shift', each theory "produces facts not entailed by its predecessor", and in this way "falsifies" (i.e. leads us to reject) the predecessor as an adequate account of the total evidence available: "Any 'verification' of the $(n + 1)$th version of the programme is a refutation of the nth version", FM, p. 137.
[52] *Science* 186 (1974), 909–911.
[53] CM, p. 167.
[54] FM, p. 132. See also p. 183.
[55] FM, p. 135.
[56] FM, p. 134.
[57] *Loc. cit.*
[58] FM, p. 134.

[59] CM, p. 164.
[60] FM, p. 118.
[61] *Loc. cit.*, also p. 133; CM, p. 164.
[62] It is worth noting that the unit of appraisal for a *heuristic* appraisal would not be (or at least would not necessarily need to be) a historical one, to return to what is supposed to be the principal distinguishing mark of the MSRP.
[63] CM, p. 163.
[64] What Lakatos calls 'acceptance$_2$', *loc. cit.*
[65] CM, p. 164.
[66] C. G. Hempel, *Philosophy of Natural Science*, Englewood Cliffs (N.J.), p. 38. It is significant that the example Hempel takes to support his case is the Balmer formula: a non-theoretical numerical hypothesis. Would it have mattered to the confirmation of the formula whether a particular spectral line subsumed under the formula was known before or after Balmer first formulated his 'law'? But the answer to this is irrelevant to the confirmation of *theory* (suppose one took the Bohr theory instead of the Balmer formula?). The confirmation of a numerically-expressed law like those of Bode or Balmer is a very special sort of case, the only one perhaps for which the criterion of formal simplicity on which Hempel relies so heavily might suffice. But even there, most scientists would still want to say that the historical order mattered.
[67] We have touched on it already in the essays mentioned in Note 14, and in 'Models and Analogies in Science', *New Catholic Encyclopedia* (Suppl. Vol.), 1974, pp. 295–299. A more detailed account is in preparation.

ROBERT K. MERTON†

THE AMBIVALENCE OF SCIENTISTS*,†

It is a high privilege to be invited to give the lecture that annually commemorates the first great president of this first authentic university in the United States. This is especially so for someone from a comparatively new academic discipline, a discipline which was installed in this university a short time ago with the appointment of Professor James S. Coleman (who, it is agreeable to note, is himself outstanding among the graduates of our department at Columbia). I like to think that the double recognition thus accorded sociology by the Johns Hopkins is only another bit of testimony to the far-seeing vision of Daniel Coit Gilman. We should remember, for one thing, that he had prophetically described the "study of man in his relations to society" as one of "the modern humanities"; that later, only three years after he became president of this new university be agreed to become president of the American Social Science Association; and finally, that, in his famous address on 'Research' almost sixty years ago, he conceived the expanding domain of science to include sociology which (in his words) was beginning to "employ the scientific method, with increasing success, and [to] demand recognition in the surrogate's court, as the next of kin." But since I have not been asked here to present the sociologist's credentials as kin to the older sciences represented at the School of Medicine, I must proceed to my subject.

1. Ambivalences in the Social Institution of Science

Science, and the men who create science, can be examined from different though connected points of view. There is the standpoint of the body of scientific knowledge itself. This requires little interest in men of science, centering only on the fruits of their work. What has been found out? how sound is the evidence? what are the implications for new knowledge? Beyond this is the perspective of the philosopher and logician of science who concerns himself with the assumptions underlying scientific ideas and the logic of inquiry. Still another angle of vision is that of the psy-

R. S. Cohen et al. (eds.), Essays in Memory of Imre Lakatos, 433–455. *All Rights Reserved*
Copyright © 1976 *by D. Reidel Publishing Company, Dordrecht-Holland*

chology of the scientist, focused on the processes of learning and of that special kind of learning that is creative investigation. These and other perspectives on science contribute each to our understanding of the life and work of scientists, but I shall say nothing about them there.

Instead, I shall stick to my last and examine a restricted set of questions in the sociology of science. From this perspective, science appears as one of the great social institutions, coordinate with the other major institutions of society: the economy, education and religion, the family and the polity. Like other institutions, science has its corpus of shared and transmitted ideas, values and standards designed to govern the behavior of men attached to the institution. The standards define the technically and morally allowable patterns of behavior, indicating what is prescribed, preferred, permitted or proscribed. The culture of science refers, then, to more than habitual behavior; its norms codify the values judged appropriate for scientists.

Now, one characteristic of social institutions is that they tend to be patterned in terms of potentially conflicting pairs of norms. This sets a task for those governed by the institution to blend these imposed inconsistencies into reasonably consistent action. This is what I mean by saying that sociological ambivalence is imbedded in social institutions generally and, in its distinctive fashion, in the institution of science as well. Consider only, in swift review, some institutionally defined pairs of norms and note the tension that can be generated by potential inconsistency within each pair.

(1) The scientist must be ready to make his newfound knowledge available to his peers as soon as possible, BUT he must avoid an undue tendency to rush into print. (Compare Faraday's motto: "Work, Finish, Publish" with President Gilman's anticipation of Ehrlich's motto: "Viel arbeiten, wenig publizieren.")

(2) The scientist should not allow himself to be victimized by intellectual fads, those modish ideas that rise for a time and are doomed to disappear, BUT he must remain flexible, receptive to the promising new idea and avoid becoming ossified under the guise of responsibly maintaining intellectual traditions.

(3) New scientific knowledge should be greatly esteemed, BUT the scientist should work without regard for the esteem of others.

(4) The scientist must not advance claims to new knowledge until they

are beyond reasonable dispute, BUT he should defend his new ideas and findings, no matter how great the opposition. (Compare C. N. Yang on the scientific credo: "it will not do to jump to hasty conclusions" and Pasteur: "do not fear to defend new ideas, even the most revolutionary.")

(5) The scientist should make every effort to know the work of predecessors and contemporaries in his field, BUT too much reading and erudition will only stultify creative work. (Compare Cayley's omnivorous reading of other mathematicians, – that same Cayley who with Euler and Cauchy, is among the three most prolific mathematicians of all time – with the irritation of his collaborator, Sylvester, at being expected to master what others had done. And, of course, there is always Schopenhauer's sin against the Holy Ghost: "to put away one's original thoughts in order to take up a book.")

(6) The scientist should pay scrupulous attention to detail, BUT he must avoid the excessive accuracy of the pedant, fastidious only when it comes to inconsequentials.

(7) Scientific knowledge is universal, belonging to no nation, BUT each scientific discovery does honor to the nation that fostered it.

(8) The scientist should recognize his prime obligation to train up new generations of scientists, BUT he must not allow teaching to preempt his energies at the expense of advancing knowledge. Of course, this reads just as persuasively in reverse. (Remember the complaints about Faraday that he had never trained a successor as Davy had trained him and consider the frequent criticism of scientists who give up research for teaching.)

(9) Young scientists can have no happier condition than being apprenticed to a master of the scientific art, BUT they must become their own men, questing for autonomy and not content to remain in the shadow of great men. (Remember Kepler's ambivalence toward Tycho Brahe; Sir Ronald Ross's ambivalence toward Manson, his devotion to the master pushing him to extravagant praise, his need for autonomy pushing him to excessive criticism; in psychiatry, Bouchard's ambivalence toward Charcot and the secessionists Jung and Adler displaying theirs toward Freud; Sir Everard Home's mixed feelings about John Hunter, and the rest in the long list of ambivalent apprentices in science.)

And so, on and on, with norms garnered from the literature of science which can be paired into actual contradictions, potential contradictions or near-contradictions. Before turning to one conspicuous kind of ambivalence experienced by scientists, I should first say a word about the spirit in which this inquiry is conducted. Embodying as they do some of the prime values of world civilization, scientists have been placed on pedestals where they have no wish to be perched – not, at least, the greatest among them. This is not the result of a conspiracy, not even a conspiracy of good will. It is only that men of science have been pictured, through collective acts of piety, as though they were more than human, being like gods in their creativity, and also as less than human, being deprived of the passions, attitudes and social ties given to ordinary men. As a result, scientists have been dehumanized by being idealized and, on occasion, idolized. Contributing greatly to this centuries-long process of distortion are the pious biographers who convert indubitably great men of science into what Augustus de Morgan once described as "monsters of perfection." Yet an honest appreciation would see them as people, not gods, and so subject to the pressures, passions and social relations in which people inevitably find themselves. Rather than deny their human qualities, we must examine them. As the historian of science, A. C. Crombie, observed for one of the most notable cases in point:

> We must completely misunderstand Newton the man and we run the risk of missing the essential processes of a mind so profoundly original and individual as his, if we exclude from our field of historical investigation all those influences and interests that may be distasteful to us, or seem to us odd in a scientist.

Did my time and your patience allow, it would be possible to consider, first, how potentially contradictory norms develop in every social institution; next, how in the institution of science, conflicting norms generate marked ambivalence in the lives of scientists; and finally, how this ambivalence affects the actual, as distinct from the supposed, relations between men of science. Instead, I shall consider the socially patterned sources of only one kind of ambivalence – toward the claiming of priorities in cases of multiple scientific discovery. This one type is not necessarily typical of the rest. But by examining it in the needed detail, we can perhaps catch a glimpse of the ways in which ambivalence is socially generated by each of the other potentially opposed pairs of norms in the institution of science: erudition *vs.* originality, apprentice-emulation *vs.* personal autonomy, and so on.

2. Ambivalence toward priorities in scientific discovery

As is well known, men of science typically experience multiple independent discoveries as one of their occupational hazards. These create occasions for acute stress. Few scientists indeed react with equanimity when they learn that one of their own best contributions to science – what they *know* to be the result of long hard work – is "only" (as the telling phrase has it) a rediscovery of what was found some time before or "just" another discovery of what others have found at about the same time. No one who systematically examines the disputes over priority can ever again accept as veridical the picture of the scientist as one who is exempt from affective involvement with *his* ideas and *his* discoveries of once unknown fact. The value of observing the behavior of men under stress in order to understand them better in all manner of other situations need not be emphasized here. By observing the behavior of scientists under what they experience as the stress of being forestalled in a discovery, we gain clues to ways in which the social institution of science shapes the motives, social relations and affect of men of science. I have tried to show elsewhere how the values and reward-system of science, with their pathogenic emphasis upon originality, help account for certain deviant behaviors of scientists: secretiveness during the early stages of inquiry lest they be forestalled, violent conflicts over priority, a flow of premature publications designed to establish later claims to having been first. These, I suggest, are normal responses to a badly integrated institution of science, such that we can better understand the report of a sample of American "starred men of science" that, next to "personal curiosity," "rivalry" is most often the spur to their work.

In saying that the social institution of science is malintegrated, I mean that it incorporates potentially incompatible values: among them, the value set upon originality, which leads scientists to want their priority to be recognized, and the value set upon humility, which leads them to insist on how little they have in fact been able to accomplish. These values are not real contradictories, of course – 'tis a poor thing, but my own – but they do call for opposed kinds of behavior. To blend these potential incompatibles into a single orientation and to reconcile them in practice is no easy matter. Rather, the tension between these kindred values creates an inner conflict among men of science who have internalized both of them. Among other things, the tension generates a distinct resistance to the systematic study of multiple discoveries and associated conflicts over priority.

This resistance is expressed in various ways: by seeking to trivialize the subject, by regarding the conflicts over priority as rare or aberrant, by motivated misperceptions of the facts of the case or by an hiatus in recall and reporting. Such resistance often leads to those wish-fulfilling beliefs and false memories that we describe as illusions. And of such behavior the annals that treat of multiple discoveries and priorities are uncommonly full. So much so, that I have arrived at a rule-of-thumb that seems to work fairly well. The rule is this: whenever the biography or autobiography of a scientist announces that he had little or no concern with priority, there is a reasonably good chance that, not many pages later in the book, we shall find him deeply embroiled in one or another battle over priority. A few cases here must stand for many.

The authoritative biography of that great psychiatrist of the Salpêtrière, Charcot, states that, despite his many discoveries, he "never thought for a moment to claim priority or reward." Our rule of thumb leads us to expect what we find: some 30 pages later, there is a detailed account of Charcot insisting on having been first in recognizing exophthalmic goiter and a little later, emphatically affirming that he "would like to claim priority" [the language is his] for the idea of isolating patients suffering from hysteria.

Or again, Harvey Cushing writes of the brilliant Halsted that he was "overmodest about his work, indifferent to matters of priority..." Alerted by our rule-of-thumb, we find some 20 pages later in the book where this is cited, a letter by Halsted about his work on cocaine: "I anticipated all of Schleich's work by about six years (or five)... [In Vienna], I showed Wölfler how to use cocaine. He had declared that it was useless in surgery. But before I left Vienna he published an enthusiastic article in one of the daily papers on the subject. It did not, however, occur to him to mention my name."

But perhaps the most apt case of such denial of an accessible reality is provided by Ernest Jones, writing in his comprehensive biography that "Although Freud was never interested in questions of priority, which he found merely boring, he was fond of exploring the source of what appeared to be original ideas, particularly his own..." This is an extraordinarily illuminating statement. For, of course, no one could have known better than Jones – "known" in the narrowly cognitive sense – how very often Freud turned to matters of priority: in his own work, in the work of his colleagues (both friends and enemies) and in the history of psychology altogether. In point of fact, Freud expressed an interest in priority on more than 150 occasions. With characteristic self-awareness, he reports that he even dreamt about priority and the due allocation of credit for accomplishments in science. He oscillates between the poles of his ambivalence toward priority: occasionally taking multiple discoveries to be practically inescapable, as when he reports a fantasy in which "science would ignore me entirely during my lifetime; some decades later, someone else would infallibly come upon the same things – for which the time was not now ripe –, and would achieve recognition for them and bring me honour as a forerunner whose failure had been inevitable." At other times, he reluctantly or insistently acknowledges anticipations of his own ideas or reports his own anticipations of others; he "implores" his disciple Lou Andreas-Salomé to finish an essay in order "not to give me precedence in time"; he admonishes Adler for what he describes as his "uncontrolled craving for priority" just as he admonishes Georg Groddeck for being unable to conquer "that banal ambition which hankers after originality and priority"; over a span of 40 years, he repeatedly reassesses the distinctive roles of Breuer and himself in establishing psychoanalysis; he returns time and again to his priority-conflict with Janet, reporting that he had brought the recalcitrant Breuer to publish their joint monograph early because "in the meantime, Janet's work had anticipated some of his [Breuer's] results"; he writes nostalgically about the days of "my splendid isolation" when "there was nothing to hustle me... My publications, which I was able to place with a little trouble, could always lag far behind my knowledge and could be postponed as long as I pleased, since there was no doubtful 'priority' to be defended"; again and again, he allocates priorities among others (Le Bon, Ferenczi, Bleuler, Stekel being only a few among the many); he even credits Adler with priority for an error; and, to prolong the occasions no further, he repeatedly intervenes in priority-battles among his disciples and colleagues (for example, between Abraham and Jung), saying that he could not "stifle the disputes about priority for which there were so many opportunities under these conditions of work in common."

In view of even this small sampling of cases in point, it may not be audacious to interpret as a sign of resistance, Jones's remarkable statement that "Freud was never interested in questions of priority, which he found merely boring..." That Freud was ambivalent toward priority, true; that he was pained by conflicts over priority, indisputable; that he was concerned to establish the priority of others as of himself, beyond doubt; but to describe him as "never interested" in the question and as "bored" by it requires the extraordinary feat of denying, as though they had never occurred, scores of occasions on which Freud exhibited profound interest in the question, many of these being occasions which Jones himself has detailed with the loving care of a genuine scholar. It is true that Freud appears to have been no more concerned with this matter than were Newton or Galileo, Laplace or Darwin, or any of the other giants of science about whom biographers have announced an entire lack of interest in priority just before, as careful scholars, they inundate us with a flood of evidence to the contrary. This denial of the realities they report and segregate seems to be an instance of that keeping of intellect and perception in abeyance which so typically reflects deep-seated resistance.

Such resistance has obvious parallels with other occasions in the history of thought, not least with psycho-analysis itself, when amply available facts with far-reaching theoretical implications were regarded as unedifying or unsavory, ignoble or trivial and so were conscientiously ignored. It is a little like the psychologists who once ignored sexuality because it was a subject not fit for polite society or regarded dreams, incomplete actions and slips of the tongue as manifestly trivial and so undeserving of thorough inquiry.

Complicating the problem in the case of multiple discoveries and priority-conflicts is the fact that investigation requires the detached examination of the behavior of scientists by other scientists. Even to assemble the facts of the case is to be charged with blemishing the record of undeniably great men of science, as though one were a raker of muck that a gentleman would pass by in silence. Even more, to investigate the subject systematically is to be regarded not merely as a muck-raker, but as a muck-maker.

The behavior of fellow-scientists involved in priority-disputes tends to be condemned or applauded, rather than analyzed. It is morally judged, not systematically investigated. The disputes are described as "unfortunate" with the moral judgment being substituted for the effort to understand what the disputes imply for the psychology of scientists and the sociology of science as an institution. At least since Goethe, we find references to "all those foolish quarrels about earlier and later discovery, plagiary, and quasi-purloinings." We are free, of course, to find this behavior unfortunate or foolish or comic or sad. But these affective responses to the behavior of our ancestors- or siblings-in-science seem to have usurped the place that might be given to analysis of this behavior and its implications for the ways in which science develops. It is as though the physician were to respond only evaluatively to illness, describe it as unfortunate or painful, and consider his task done or as though the psychiatrist were to describe the behavior of schizophrenics as absurd and to substitute this sentiment for the effort to discover what brings that behavior about. The history of the sciences shows that the provisional emancipation from sentiment in order to investigate phenomena methodically has been a difficult task, that it has been achieved at different times in the various sciences and at different times for selected problems in each of the sciences. I suggest that only now are we beginning to emancipate the study of the actual behavior of scientists from the human tendency to respond to that behavior in terms of the sentiments and values which we have made our own rather than to examine it in reasonably detached fashion.

Contributing to the substitution of sentiment for analysis is the often painful contrast between the actual behavior of scientists and the behavior ideally prescribed for them. When confronted with the fact that their discovery is a rediscovery or, much worse, when confronted with the suggestion that it is a plagiary, their behavior scarcely matches the image of the dispassionate man of science, exclusively absorbed by his scientific work. It is often seen as ugly, harsh and greedy for fame. And in the bitter social conflict that ensues, the standards governing behavior deteriorate. One or another of the discoverers caught up in a multiple – or often a colleague or fellow-national – suggests that he rather than his rival was really first and that the independence of the rival is at least unproved. Grouping their forces, the other side counters with the opinion that plagiary had indeed occurred, that let him whom the shoe fits wear it and furthermore, to make matters quite clear, the shoe is on the other foot. Reinforced by group loyalties and sometimes by ethnocentrism, the controversy gains force, mutual charges of plagiary abound, and there develops an atmosphere of thoroughgoing hostility and mutual distrust.

This is not exactly in accord with the ideal image of scientists and particularly, of the greatest among them. When we identify ourselves with the role-models provided by great scientists of the past and by lesser as well as outstanding ones of the present, we find it painful to observe their behavior in these situations of conflict. Regarded in terms of sentiment rather than of understanding, it may seem a bit sordid for a Galileo to engage in seemingly egotistic attacks on one Grassi who tried "to diminish whatever praise there may be in this [invention of the telescope] which belongs to me"; or to go on to assail another who "attempted to rob me of that glory which was mine, pretending not to have seen my writings and trying to represent themselves as the original discoverer of these marvels"; or finally, to say of a third that he "had the gall to claim that he had observed the Medicean planets ... before I had [and used] a sly way of attempting to establish his priority."

For all of us who harbor the ideal image of the scientist it can be only disconcerting to have Edmond Halley forthrightly described by the first Astronomer Royal, John Flamsteed, as a "lazy and malicious thief" who manages to be just as "lazy and slothful as he is corrupt." Or to have Flamsteed assert that he found Newton "always insidious, ambitious, and excessively covetous of praise."

Almost all those firmly placed in the pantheon of science – a Newton, Descartes, Pascal, Leibniz or Huyghens; a Lister, Faraday, Laplace or Davy – have at one time or another been caught up in these fierce disputes. As we approach our own day, we hear an echo of these angry and agitated words reverberating through the corridors of the peaceful temple of science. Since these episodes involve our contemporaries and often our associates, they become, we must suppose, even more painful to observe and more difficult to analyze with detachment. Even the social scientists who may not be directly involved, at least for the moment, feel acutely uncomfortable. Uneasy and distressed, they can hardly bring themselves to study this behavior. For when sociological analysis is stripped bare of sentiment, it often leaves the sociologist shivering in the cold. And to respond with detachment to these hot conflicts of their associates becomes all the more difficult. The study of multiple discoveries and priorities accordingly remains undeveloped.

The disputants themselves manifest ambivalence toward their own behavior. Even while he is assembling documents to prove his priority, for example, Darwin registers his mixed feelings, writing Lyell: "My good dear friend, forgive me. This is a trumpery letter, influenced by trumpery feelings." In a postscript he assures Lyell that "I will never trouble you or Hooker on the subject again." The next day, he writes: "It seems hard on me that I should lose my priority of many years' standing." Then, a few days later, he writes again to say: "Do not waste much time [on this matter]. It is miserable in me to care at all about priority."

Freud recognizes his own ambivalence when he writes of his work on the Moses of Michelangelo that, having come upon a little book published in 1863 by an Englishman, Watkiss Lloyd, he read it

> with mixed feelings. I once more had occasion to experience in myself what unworthy and puerile motives enter into our thoughts and acts even in a serious cause. My first feeling was of regret that the author should have anticipated so much of my thought, which seemed precious to me because it was the result of my own efforts; and it was only in the second instance that I was able to get pleasure from its unexpected confirmation of my opinion. Our views, however, diverge on one very important point.

This degree of self-awareness is a far cry from the ambivalence of a Descartes who manages to write that he "does not boast of being the first discoverer" and then proceeds to insist on his priority over Pascal or to beg his friend Mersenne "to tell him [Hobbes] as little as possible about ... my unpublished opinions, for if I'm not greatly mistaken, he is a man who is seeking to acquire a reputation at my expense and through shady practices."

The ambivalence toward claims of priority means that scientists are contemptuous of the very attitudes they have acquired from the institution to which they subscribe. The sentiments they have derived from the institution of science, with its great premium upon originality, makes it difficult to give up a claim to a new idea or a new finding. Yet the same institution emphasizes the selfless dedication to the advancement of knowledge for its own sake. Concern with priority and ambivalence toward that concern together register in the individual what is generated by the value system of science.

The self-contempt often expressed by scientists as they observe with dismay their own concern with having their originality recognized is evidently based upon the widespread though uncritical assumption that behavior is actuated by a single motive, which can then be appraised as 'good' or 'bad,' as noble or ignoble. They assume that the truly dedicated scientist must be concerned only with advancing knowledge. As a result, their deep interest in having their priority recognized by peers is seen as marring their nobility of purpose as men of science (although it might be remembered that 'noble' initially meant the widely-known). This assumption has a germ of psychological truth: any reward – fame, money, position – is morally ambiguous and potentially subversive of culturally esteemed motives. For as rewards are meted out – fame, for example – the motive of seeking the reward can displace the original motive, concern with recognition can displace concern with advancing knowledge. But this is only a possibility, not an inevitability. When the institution of science works effectively, and like other social institutions it does not always do so, recognition and esteem accrue to those scientists who have best fulfilled their roles, to those who have made important contributions to the common stock of knowledge. Then are found those happy circumstances in which moral obligation and self-interest coincide and fuse. The observed ambivalence of scientists toward their own interest in having their priority recognized – an ambivalence we have seen registered even by that most astute of psychologists, Freud – shows them to assume that such an ancillary motive somehow tarnishes the 'purity' of their interest in scientific inquiry. Yet it need not be that scientists seek only to win the applause of their peers but, rather, that they are comforted and gratified by it, when it does ring out.

Occasionally, a scientist senses all this and vigorously challenges the assumption underlying the shame over interest in recognition; for example, a Hans Selye who asks his peers:

Why is everybody so anxious to deny that he works for recognition? In my walk of life, I have met a great many scientists, among them some of the most prominent scholars of our century; but I doubt if any one of them would have thought that public recognition of his achievements – by a title, a medal, a prize, or an honorary degree – played a decisive role in motivating [or one might add, sustaining] his enthusiasm for research. When a prize brings both honor and cash, many scientists would even be more inclined to admit being pleased about the money ('one must live') than about the public recognition ('I am not sensitive to flattery'). Why do even the greatest minds stoop to such falsehoods? For, without being conscious lies, these ratiocinations are undoubtedly false. Many of the really talented scientists are not at all money-minded; nor do they condone greed for wealth either in themselves or others. On the other hand, all the scientists I know sufficiently well to judge (and I include myself in this group) are extremely anxious to have their work recognized and approved by others. Is it not below the dignity of an objective scientific mind to permit such a distortion of his true motives? Besides, what is there to be ashamed of?

Dr. Selye's final question need not remain a rhetorical one. Shame is experienced when one's identity and self-image is suddenly violated by one's actual behavior – as with the shame we have seen expressed by Darwin when his own behavior forced him to realize that recognition of his priority meant more to him than he had ever been willing to suppose. To admit a deepseated wish for recognition may seem to prefer recognition to the joy of discovery as an end in itself, activating the further awareness that the pleasure of recognition for accomplishment could, and perhaps momentarily did, replace the pleasures of scientific work for its own sake.

On the surface, this hunger for recognition appears as mere personal vanity, generated from within and craving satisfaction from without. But when we reach deeper, into the institutional complex that gives point to this hunger, it turns out to be anything but personal and individual, repeated as it is with slight variation by one scientist after another. Vanity, so-called, is then seen as the outer face of the inner need for assurance that one's work really matters, that one has measured up to the hard standards maintained by the community of scientists. It becomes clear that the institution of science reinforces, when it does not create, this deepseated need for validation of effort. Sometimes, of course, this need is stepped up until it gets out of hand; the desire for recognition becomes a driving lust for acclaim, megalomania replaces the comfort of reassurance. But the extreme case need not be mistaken for the modal one. In general, the need to have accomplishment recognized, which for the scientist means that his knowing peers judge his work worth the while, is the result of deep devotion to the advancement of knowledge as an ultimate value. Rather than necessarily being at odds with dedication to science, the concern with recognition is usually a direct expression of it.

In this company, I cannot keep from citing evidence of the composite devotion to science *and* to the recognition of priority-rights that was forthrightly supplied by the first professor of chemistry and the second president of the Johns Hopkins. Just before the close of the first academic year, Ira Remsen writes a letter to President Gilman that begins: "I beg leave through you to make the following request of the Trustees of the University" and then goes on to note of the work in the Chemical Laboratory:

> At the present juncture it is desirable to publish preliminary announcements describing what we have thus far done and what we intend to do. It is desirable mainly for two reasons; 1st, that we may be recognized as soon as possible as belonging to the working chemists of the country; 2nd, that the results of our labors may be insured to us, or, in other words, to establish our priority.

It may not be too much to suppose that we see here the origins of the influential *American Chemical Journal*, inaugurated by Remsen two years later.

President Gilman not only provided new outlets for scientific and scholarly publication but institutionalized the incentives of both faculty and graduate students to publish the results of their research. He established the practice of printing a *Bibliographia Hopkinsiensis* in the annual Register, which recorded the "books and articles published by members of the Johns Hopkins University, written during the connection of the author with the university, or based on work carried on while here ... it is obvious that a man's reputation was, in large part, based on his bibliography." Once again, we can note a fusion of the interest in advancing knowledge and the interest in recognizing the contributions of individual scientists and the collective contributions of the University. The fusion of these interests can of course give way to fission. This only requires an institutionalized reward-system that aggravates the itch to publish[1] by assigning merit to the mere length of a bibliography.

3. The Eureka Syndrome

All this can be seen in a somewhat different context: the deep concern with establishing priority or at least independence of discovery is only the other side of the coin of the socially reinforced elation that comes with having arrived at a new and true scientific idea or result. And the deeper the commitment to a discovery, the greater, presumably, the reaction to the threat of having its novelty denied. Concern with priority is often only the counterpart to elation in discovery – the Eureka syndrome. We have only to remember what is perhaps the most ecstatic expression of joy in discovery in the annals of science; here is Kepler on his discovery of the third planetary law:

What I prophesied 22 years ago as soon as I found the heavenly orbits were of the same number as the five (regular) solids, what I fully believed long before I had seen Ptolemy's Harmonics, what I promised my friends in the name of this book, which I christened before I was 16 years old, what I urged as an end to be sought, that for which I joined Tycho Brahe, for which I settled in Prague, for which I spent most of my life at astronomical calculations – at last I have brought to light and seen to be true beyond my fondest hopes. It is not 18 months since I saw the first ray of light, three months since the unclouded sun-glorious sight burst upon me! I will triumph over mankind by the honest confession that I have stolen the golden vases of the Egyptians to raise a tabernacle for my God far away from the lands of Egypt. If you forgive me, I rejoice; if you are angry, I cannot help it. The book is written, the die is cast. Let it be read now or by posterity, I care not which. It may well wait a century for a reader, as God has waited 6000 years for an observer.

We can only surmise how deep would have been Kepler's anguish had another claimed that he had long before come upon the third law. So, too, with a Gay-Lussac, seizing upon the person nearest him for a victory waltz so that he could "express his ecstasy on the occasion of a new discovery by the poetry of motion." Or, to come closer home, William James "all aflame" with his idea of pragmatism and hardly able to contain his exhilaration over it. Or, in more restrained exuberance, Joseph Henry, once he had hit upon a new way of constructing electro-magnets, reporting that "when this conception came into my brain, I was so pleased with it that I could not help rising to my feet and giving it my hearty approbation."

In short, when a scientist has made a genuine discovery, he is as happy as a scientist can be. But the peak of exhilaration may only deepen the plunge into despair should the discovery be taken from him. If the loss is occasioned only by finding that it was, in truth, not a first but a later independent discovery, the blow may be severe enough, though mitigated by the sad consolation that at least the discovery has been confirmed by another. But this is as nothing, of course, when compared with the traumatizing charge that not only was the discovery later than another of like kind but that it was really borrowed or even stolen. Rather than being mutually exclusive, joy in discovery and eagerness for recognition by scientific peers are stamped out of the same psychological coin. They both express a basic commitment to the value of advancing knowledge.

Perhaps scientists are learning to live with the stresses of multiple discoveries. This is suggested, at least, by a preliminary result of a methodical study of the subject. From among the multitude of multiples, Dr.

Elinor Barber and I have undertaken to examine 264 intensively. Of the 36 multiples before 1700 in this list, 92% were the object of strenuous conflicts over priority; this figure drops to 72% in the Eighteenth century; remains at about the same level (74%) in the first half of the Nineteenth century and declines to 59% in the latter half, reaching a low of 33% in the first half of this century. It may be that scientists are becoming more fully aware that with vastly enlarged numbers at work in each field of science, a discovery is apt to be made by others as well as by themselves.

4. Cryptomnesia ('unconscious plagiary')

Further complicating the already complex emotions that attend multiple discoveries is the phenomenon of so-called 'unconscious plagiary'. The potpourri term itself testifies to the admixture of moralizing and analysis that commonly enters into discussions of the subject. It is compounded of a loosely-conceived psychological component ("unconscious") and a legal-moralistic component ("plagiary," with all its connotations of violating a code and attendant guilt). As a concept, "unconscious plagiary" is just as obsolete in psychosocial studies as is the concept of "insanity," which has been relegated to the sphere of law, where it continues to lead a harrowing existence. The neutral and analytical term, cryptomnesia, serves us better, referring as it does to seemingly creative thought in which ideas based upon unrecalled past experience are taken to be new.

The fact that cryptomnesia can occur at all subjects the scientist to the ever-present possibility that his most cherished original idea may actually be the forgotten residue of what he had once seen or read or heard elsewhere. At times, this fear of cryptomnesia may lead a scientist to doubt his own powers of recall and his own originality.

Among the many cases in point, consider only these few. William Rowan Hamilton, the mathematical genius who had invented the quaternions (in part, independently invented by Grassmann), had had the experience at age 19 of learning that his theory of optical rays was a rediscovery. He developed a lifelong preoccupation with the twin fear of unwittingly plagiarizing others and of being plagiarized. As he put it on one of the many occasions on which he turned to this subject in his correspondence with de Morgan: "As to myself, I am *sure* that I *must* have often reproduced things which I had read long before, without being able to identify them as belonging to other persons." Or again: "...am I to quarrel with Dickens, or figure in one of his publications of a later date? Where is the priority business to end? I am sick of it as you can be; but still, in anything important as regards science, I should take it as a favour to be *warned*, if I were inadvertently exposing myself to the charge of plagiarizing." This, from the creator of the theory of vectors.

Turning from mathematics to psychology, we find Freud examining his own experience, recalling that he had been given Börne's works when he was 14 and still had the book 50 years later, so that although he "could not remember the essay in question," which dealt with free association as a procedure for creative writing, "it does not seem impossible to us that this hint may perhaps have uncovered that piece of cryptomnesia which, in so many cases, may be suspected behind an apparent originality." Freud was profoundly aware of the basic uncertainties about originality generated by the ubiquitous possibility of cryptomnesia. Elsewhere he writes: "My delight was proportionally great when I recently discovered that the theory [of the 'death instinct'] was held by one of the great thinkers of ancient Greece. For the sake of this confirmation I am happy to sacrifice the prestige of originality, especially as I read so widely in earlier years that I can never be quite certain that what I thought was a creation of my own mind may not really have been an outcome of cryptomnesia." It was this sort of thing, no doubt, that prompted the irrepressible Mark Twain to declare: "What a good thing Adam had – when he said a thing he knew nobody had said it before."

Still another recurrent phenomenon contributes to uncertainty about the extent of one's originality. The scientist or scholar may unwittingly borrow ideas from himself. Many have found, to their combined chagrin and disbelief, that an idea which seemed to have come to them out of the blue had actually been formulated by them years before, and then forgotten. An old notebook, a resurrected paper, a colleague cursed with total recall, a former student – any of these can make it plain that what was thought to be a new departure was actually a repetition of the scientist's own earlier innovation. Of many such instances, I note only a few, some of a century or more ago, others of contemporary vintage:

Joseph Priestley records with chagrin that "I have so completely forgotten what I have myself published, that in reading my own writings, what I find in them often appears perfectly new to me, and I have more than once made experiments, the results of which have been published by me."

The ingenious and jovial mathematician, Augustus de Morgan, has his own lively version of this pattern of experience: "I have read a Paper (but not on mathematics) before now, have said to myself, I perfectly agree with this man, he is a very sensible fellow, and have found out at last that it was an old Paper of my own I was reading, and very much flattered I was with my own unbiased testimony to my own merits."

Or let us come to a student of de Morgan's, the 'mathematical Adam' who minted countless new terms of mathematics, who, after he was forced to retire from Woolwich as 'superannuated' at the age of 56, languished for six years, and then took up the invitation of a new president of a new University to establish graduate work in mathematics, thus reinvigorating the subject in the United States once and for all. It has been told of James Joseph Sylvester that he "had difficulty in remembering his own inventions and once even disputed that a certain theorem of his own could possibly be true."

Or consider a brace of cryptomnesic borrowings from self in our own day:

The Nobel laureate, Otto Loewi, reports having waked in the middle of the night, jotting down some notes on what he sensed to be a momentous discovery, going back to sleep, awaking to find that he could not possibly decipher his scrawl, spending the day in a miserable and unavailing effort to remember what he had had in mind, being again aroused from his slumber at three the next morning, racing to the laboratory, making an experiment and two hours later conclusively proving the chemical transmission of nervous impulse. So far, so good; another case, evidently, of the pattern of subconscious creativity unforgettably described by Poincaré. But some years later, when Loewi, upon request, reported all this to the International Physiological Congress, he was reminded by a former student that, eighteen years before that nocturnal discovery, he had fully reported his basic idea. "This," says Loewi, "I had entirely forgotten."

And to advert to Freud, as I have so often done if only because his intellectual experience is uncommonly documented, Jones reports several instances of his "obtaining a clear insight which he subsequently forgot, and then later suddenly coming across it again as a new revelation." As Freud observed in another connection [in a fashion reminiscent of a comparable remark by Marx], "it is familiar ground that a sense of conviction of the accuracy of one's memory has no objective value..."

If cryptomnesia is possible in regard to one's own earlier work, then it is surely possible in regard to the work of others. And this can undermine the calm assurance that one has, in truth, worked out a new idea for oneself when confronted with another version of the same idea worked out by someone else.

Various contexts may affect the probability of cryptomnesia in relation to one's own work. It may be more probable, the more a scientist has worked in a variety of problem-areas rather than narrowly restricting his research focus to problems having marked continuity. Looking at this hypothesis, not in terms of the individual scientist but in terms of the relative frequency of self-cryptomnesia in different sciences, we should expect it to be more frequent in the newer sciences, with their large variety of prime and largely untapped problem-areas. In these sciences, investigators can move from one to another area with substantial gains in knowledge, contrasting with the older, better established sciences where continuous digging is more often the practice. To the extent that these patterned differences in choice of research problems occur, we should expect more cryptomnesia in relation to one's own work in the social sciences.

The frequency of such cryptomnesia should also be affected by the social organization of scientific work, which seems to affect every aspect of multiple discoveries in science. When research is organized in teams, it will be less likely, we must suppose, that earlier ideas and findings are altogether forgotten. For if some members of the team forget them, others will not. Moreover, repeated interaction between collaborators will tend to fix these ideas and findings in memory.

The conspicuous changes in the social organization of scientific research should have a marked effect on ambivalence toward priorities in science. The trend toward collaboration in research is reflected in patterns of publication, with more and more research papers having several authors rather than only one. The extent of this change differs among the various disciplines. The sciences which have developed cogent theory, complex and often costly instrumentation and rigorous experiments or sets of observations have experienced this change earlier than the sciences less developed in these respects.

By way of illustration, consider the patterns of publication in a few of the sciences and other disciplines, based upon tabulation of the number of authors of papers in leading journals. The results, brought together by Harriet Zuckerman in a still unpublished paper, are in brief, these: in physics, of the papers published during the decade of the 1920's, 75% were by single authors; in the next decade, 56%; then, in the 40's, 50% and finally, in the 1950's, single-authored papers declined to 39%. A similar pattern in biology begins later and develops at a slower rate: with 90% single-authored papers in the 20's declining to 73% in the 50's. Even mathematics witnesses the growth of collaboration, with 95% of papers in the 1920's by one author declining to 82% in the last decade.

The social and behavioral sciences exhibit two distinct patterns: economics, anthropology and political science manifest only negligible change, almost all papers being by single authors in the first period, with 90% or more in the most recent period. Psychology and sociology, in contrast, have marked tendencies toward collaborative work: in the 1920's, 98% of the papers in sociology were by single authors, this declining by decades to 92%, 89% and finally to 72%. The trend in psychology for the same period is even more marked: from 84% to 55% most recently. All this is, of course, in decided contrast to such a subject as history where collaborative research-papers (as distinct from textbooks) account for no more than 1 to 3% of the total.

Although the facts are far from conclusive, this continuing change in the social structure of scientific research, as registered by publications, seems to make for a greater concern among scientists with the question of "how will my contribution be identified" in collaborative work than with the historically dominant pattern of wanting to ensure recognition of priority over others in the field. Not that the latter has been wholly displaced, as we have seen. But it may be that institutionally induced concern with priority is being overshadowed by structurally induced concern with the allocation of credit among collaborators. A study of a team of 30 economists and behavioral scientists found, for example, that "the behavioral scientists were apt to be less concerned about 'piracy' and 'credit' than economists. This difference may be due to the greater emphasis on joint authorship in the behavioral sciences than in economics."

For our purposes, the import of these changes in collaboration is, first, that the degree of concern with priority in science is probably not historically constant; second, that it varies with the changing organization of scientific work; and third, that these changes may eventually and indirectly lessen the ambivalence of scientists toward obtaining recognition of their originality of contributions.

Nevertheless, though scientists *know* that genuinely independent discoveries occur, many of them, as we have seen, fail to draw the implications of this for their own work. For reasons I have tried to intimate, they find it difficult, and sometimes impossible, to accept the fact that they have been anticipated or that a contemporary has come to the same result just at the time they did, or that the others were truly independent of them. As we have also seen, the values incorporated in the social institution of science and the penumbra of uncertainty that surrounds the independence of thought combine to prevent the ready acceptance of events that undercut one's assurance of unique originality, an assurance born of the hard labor required to produce the new idea or new result.

The reasonably detached study of multiples and priorities may do a little to counter these tendencies toward dismay, self-contempt or suspicion. For, as we have seen in the case of Freud trying to rouse himself from his ambivalence toward having been anticipated by Watkiss Lloyd, independent discoveries do seem to lend confirmation to an idea or finding. Even W. R. Hamilton, tormented his life long by the fear that he was being plagiarized or by the anxiety that he himself might be an 'innocent plagiarist,' managed on at least one occasion to note the secondary benefits of multiple discovery when, in an effort to dissolve his ambivalence, he wrote Herschel:

> I persuade myself that, if those results had been anticipated, the learning it would have given me no pain; for it was, so far as I could analyze my sensations, without any feeling of vexation that I learned that the result respecting the relation of the lines of curvature to the circular sections was known before. The field of pure, not to say of mixed, mathematics is far too large and rich to leave one excusable for sitting down to complain, when he finds that this or that spot which he was beginning to cultivate as his own has been already appropriated. [And now comes his hard-won and, sad to tell, temporary insight:] There is even a stronger feeling inspired of the presence of that Truth to which we all profess to minister, when we find our own discoveries, such as they are, coincide independently with the discoveries of other men. The voice which is heard by two at once appears to be more real and external – one is more *sure* that it is no personal and private fancy, no idiosyncratic peculiarity, no ringing in sick ears, no flashes seen by rubbing our own eyes.

And then, unable to contain himself, Hamilton goes on to announce in the same letter that he had anticipated the work on ellipsoids by Joachimstal in "a long extinct periodical of whose *existence* he probably never heard, with a date which happened to be a *precise decennium* earlier..."

If the fluctuating ailment of that genius Hamilton proves that the awareness of multiple discoveries is no panacea for ambivalence toward priority, his moment of insight suggests that it might be some small help. The mathematician, R. L. Wilder, is, to my knowledge, the only one who has seen this clearly and has, to my mingled pleasure and discomfiture, anticipated me in suggesting that the study of multiples may have a therapeutic function for the community of scientists. Since he has anticipated my observation, let me then borrow his words:

I wish to inquire, above the individual level, into the manner in which mathematical concepts originate, and to study those factors that encourage their formation and influence their growth. I think that much benefit might be derived from such an inquiry. For example, if the individual working mathematician understands that when a concept is about to make its appearance, it is most likely to do so through the medium of more than one creative mathematician; and if, furthermore, he knows the reasons for this phenomenon, then we can expect less indulgence in bad feelings and suspicion of plagiarism to ensue than we find in notable past instances. Mathematical history contains numerous cases of arguments over priority, with nothing settled after the smoke of battle has cleared away except that when you come right down to it practically the same thing was thought of by someone else several years previously, only he didn't quite realize the full significance of what he had, or did not have the good luck to possess the tools wherewith to exploit it... [Yet] it is exactly what one should expect if he is acquainted with the manner in which concepts evolve.

All this only touches upon one type of ambivalence exhibited in the feelings and behavior of scientists. For to lay siege to the problem of ambivalence need not mean to conquer it. But, just as the hour allotted me inevitably draws to a close, I can report that inquiry into other expressions of ambivalence finds the same pattern of institutionally induced cross-currents of sentiment. Perhaps enough has been said to warrant the belief that for an understanding of how scientific knowledge develops, we need an intensive and methodical study of multiple discoveries and attendant conflicts over priority, rather than to neglect this study altogether or to come to it only when we plunge, as emotionally involved participants, into conflicts over rights to intellectual property. After all, one of the roles assigned the sociologist is to investigate the behavior of all manner of men, including men of science, without giving way to the entirely human tendency to substitute for that investigation a clucking of tongues and a condemning of that which is and ought not to be.

Columbia University

NOTES

* Since Imre Lakatos expressed interest in this lecture shortly before his lamented death, it is reprinted in his memory. The third Daniel Coit Gilman Lecture delivered at The Johns Hopkins University School of Medicine on September 25, 1962.

† University Professor, Columbia University, New York, N.Y.

† The investigation providing the basis for this lecture has been greatly aided by a fellowship from the John Simon Guggenheim Memorial Foundation.

[1] Physicians of the soul will see beneath this plain English phrase and recognize the

malignant disease, known, since the days of Juvenal, as the *insanabile scribendi cacoëthes*. Its etiology is obscure but epidemiological evidence affords some clues. There are indications that its frequency increases steadily in institutions that lavish rewards upon the prolific author of scientific or scholarly papers. Age seems to be an important predisposing factor as the result of a basic social process: with the passing of years, scientists who have published significant work are actively solicited for still more publications. Nevertheless, the general liability to the disease seems less widespread than the nothing-to-report syndrome (although, on occasion, the two have a way of coinciding). Sucklings are rarely attacked. A few scientists escape in their early professional years; others escape until full maturity; a good many never take it. But with the vast growth in the number of scientific periodicals, the disease threatens to become endemic. Attacks are recurrent, never conferring immunity. Susceptibility may be determined by intrapsychic injection of the toxin – what might perhaps be most appropriately called the *Merton test*. With a positive reaction, signs appear in ten minutes (or less) after seeing one's name in print, reaching toward an asymptotic maximum with each successive injection. The local reaction subsides temporarily but swiftly returns. The source of infection is often undiscovered in given cases, particularly when insufficient attention is devoted to the social ecology of the patient. *Carriers* are important, especially those who have been abundantly rewarded for effusions of print. Onset is as a rule sudden, preceded by a slight, scarcely noticeable, publication. The fever to publish is intense; rising rapidly, it may within a few years reach the degree of 15 or 20 publications annually. The articles are unusually dry and to the reader's eye may give a sensation of acute boredom. Complications and sequelae are too numerous to be examined here.

ALAN MUSGRAVE

METHOD OR MADNESS?

*Can the methodology of research programmes be rescued from epistemological anarchism?**

Imre Lakatos encouraged his students (and I like to count myself one of them) to criticise his ideas. And while criticism was not always received with *complete* calm, one invariably learned something from the exercise. I well remember how, shortly after I had obtained my first degree, he deemed me totally illiterate and set about trying to remedy the situation. He bought me some mathematics books (I still have them), and sat me down in his office to solve problems. At the time he was writing 'Proofs and Refutations': I would pass my latest effort across to him, to be mercilessly exposed, and he would pass his latest page of manuscript across to me, expecting it to receive the same treatment. He led me to believe that my exercises were just as important as his manuscript, and my modest improvements to his English just as important as his radical improvements to my mathematics. Both beliefs, were, of course, mistaken. But I shall always be grateful to him for having made me hold them.

I have decided that my paper should continue in the same vein, despite the fact that I can no longer hear his reactions. Specifically, I shall present three inter-connected criticisms of his influential 'methodology of scientific research programmes'. The first criticism concerns Lakatos's claim that scientists make the 'hard cores' of their research programmes 'irrefutable by methodological fiat'. I shall argue that if this is meant as a *historical* claim, it is refuted by what Lakatos himself calls "the classical example of a successful research programme"; while if it is meant as a piece of *advice* to scientists, it has obvious defects.

The second criticism is connected with the first, and concerns Lakatos's notion of the positive heuristic of a research programme. According to Lakatos, the positive heuristic of a research programme can *anticipate* empirical refutations and can give directions, in advance, about how they are to be handled. I shall argue that Lakatos has over-estimated what positive heuristics can achieve in this regard, and has consequently over-estimated the degree to which theoretical science can acquire *autonomy* from the verdict of experiment. Nevertheless, his idea of 'heuristic power'

R. S. Cohen et al. (eds.), Essays in Memory of Imre Lakatos, 457–491. *All Rights Reserved*
Copyright © 1976 by D. Reidel Publishing Company, Dordrecht–Holland

is an important one: I shall suggest that where such power exists it stems from a special source.

My third criticism concerns the question of whether Lakatos's methodology is in fact a methodology in the old-fashioned sense: whether, that is, it issues in advice to scientists. I shall argue that Lakatos once had sound views on this matter, but was led, mistakenly in my opinion, to renounce them. In renouncing them, he has gone a long way towards epistemological anarchism; I hope to reverse this tendency.

1. NEGATIVE HEURISTICS: ARE 'HARD CORES' RENDERED 'IRREFUTABLE BY FIAT'?

According to Lakatos, scientists make the 'hard cores' of their research programmes 'irrefutable by fiat'. The negative heuristic of the programme forbids them to direct the arrow of *modus tollens* towards the hard core. Instead, anomalies must be dealt with by making modifications in the protective belt, such modifications, ideally, being constructed in the light of the positive heuristic of the programme. Thus, in particular, Newton's theory (his three laws of motion and law of gravitation) was made irrefutable "by the methodological decision of its protagonists: [they decided that] anomalies must lead to changes only in the 'protective' belt of auxiliary hypotheses and initial conditions".[1]

There are parallels between this view and Thomas Kuhn's earlier one. Indeed, Kuhn has claimed that Lakatos's views closely resemble his own, once we replace 'hard core of a research programme' by his 'paradigm' (in one of the many senses of that term) and 'work in the protective belt' by his 'normal scientific puzzle-solving'.[2] The claim has substance. The only difference on this point is that Kuhn, following Polanyi, uses psychological terminology, and speaks of the normal scientific community being 'committed' to its paradigm. Lakatos, on the other hand, speaks not of 'commitments' but of methodological decisions.[3] The difference is not *merely* terminological: a scientist could make a decision not to modify some part of a theoretical system without 'committing himself' to the truth of it.[4] Yet despite this difference, the practical outcome is the same: the persistance of the hard core (paradigm) through the various versions of the programme (normal science) is explained by a dogmatic decision not to modify or renounce it.

But is this explanation correct? Is it the case, to use Lakatos's own example, that Newton's theory survived for over 200 years because Newtonians took a decision not to modify or renounce it in the face of empirical difficulties? In fact, this is not the case. Throughout the history of the Newtonian research programme there were suggestions that Newton's law of gravity required modification.

In 1747 Clairault found that the amount of precession of the moon's perihelion was twice the predicted value, and suggested that Newton's law required another term involving the fourth power of the distance. But in 1749 he recalculated, and showed the modification to be unnecessary (Euler confirmed this in 1750).[5] Then both Euler and Lagrange declared that anomalies in the secular acceleration of the moon might necessitate some modification of Newton's law. Laplace showed this to be unnecessary in 1787. Grant remarks that the publication of Laplace's *Mécanique Céleste* "forms an important landmark in the history of Physical Astronomy. The Theory of Gravitation, after being subjected to a succession of severe ordeals, from each of which it emerged in triumph, finally assumes an attitude of imposing majesty, which *repels all further question concerning the validity of its principles*".[6] But Grant is wrong: despite its great success, the theory of gravitation had not acquired the status of a 'hard core' even by 1800.

The next serious difficulty was the anomalous motion of Uranus. Several solutions to the problem were proposed. Bouvard suggested that the early, 'pre-discovery observations' of Uranus, which he used to calculate the anomalous orbit, might be erroneous. But this assumed that when early astronomers observed Uranus they made errors significantly greater than their margins of error in other cases; and the anomalous motion soon reasserted itself when only modern observations were used. For the same reason, the idea that Uranus might have collided with a comet shortly before its discovery was discounted. Other suggestions were that Uranus's motion might be affected by a Cartesian cosmic fluid far out in the solar system, or by a large moon; but neither of these could account for the observed irregularities. The idea which ultimately succeeded was, of course, the postulation of an extra planet beyond Uranus. And we all know how Adams and Leverrier adopted this solution, solved the inverse perturbation problem, and successfully predicted the optical discovery of Neptune. Another apparent defeat for Newtonian astronomy had been turned into its most famous victory.

So far, this story fits well into Lakatos's pattern. The blame for the anomalous motion of Uranus was shifted, not on to Newton's theory, but on to the 'protective belt' of auxiliary hypotheses and initial conditions. And we can even find Leverrier, after surveying the various solutions to the problem, declaring in 1846 that to modify the law of gravitation would be "a last resort to which I would not turn until all other potential causes for the discrepancies had been investigated and rejected".[7] But although the law of gravitation might have been something like a 'hard core' for Leverrier, it was not so for others. In 1846 the Astronomer Royal, Airy, suggested that the anomaly might be accounted for by postulating that "the law of force differed slightly from that of the inverse-square of the distance".[8] Bessel also favoured this solution at one stage.[9] Not every Newtonian astronomer treated Newton's law as his 'hard core'.

It might be objected that Clairault, Euler, Lagrange, Airy and Bessel, in contemplating modifications to Newton's law of gravity, merely showed themselves not to be true Newtonians like Laplace, Adams, or Leverrier. But this would turn Lakatos's historical claim "Newtonians rendered Newton's theory unfalsifiable by fiat" into a tautology: "Newtonians (that is, those who refused to amend Newton's laws) refused to amend Newton's laws". Definitions are not important, but it is surely better to define a Newtonian as one who contributed to the Newtonian research programme (as Clairault, Euler, Lagrange, *et al.* surely did).[10] It might be objected that Clairault, Euler, *et al.*, merely acted *irrationally*. But what if their modifications to the law of gravity had been successful? It might be objected, finally, that these modifications were only minor ones. This is true; but they remain exactly the sort of modification which, according to Lakatos, occurs only in the 'protective belt'. It is true also that these modifications turned out to be unnecessary, in that the various anomalies were more satisfactorily explained by making modifications elsewhere. Historians addicted to success seldom record such false trails; but Clairault, Euler, Lagrange, and Airy pursued them nonetheless. Newtonians did not turn defeats into victories before some had regarded them as defeats.

The discovery of Neptune was a staggering success, and it was hailed as such: it was "the most outstanding conceivable proof of the validity of universal gravitation", "the most noble triumph of theory which I know of", and "one of the greatest triumphs of theoretical astronomy".[11] Nichol was so impressed with it that he formulated a methodological rule which is worth quoting in full:

> ... there is a rule in philosophy... that admits of no dispute... we are never entitled to challenge the universality of laws that, within our experience, have nowhere failed – *until every other mode of overcoming the difficulty has proved of no avail*. If the law of gravity must be challenged, then, the time for that is not at the beginning of our consideration of this difficulty but *after* we have tried every circumstance, which – the law remaining entire – might affect the manner of its working, and so have demonstrated that what is now an apparent contradiction, may not be only one of its more recondite and least obvious results.[12]

By 1850, then, Newton's law of gravity had become, for Nichol and for others, something like a 'hard core'.

Nichol's rule was followed in dealing with the next great difficulty which faced Newtonian astronomy, the anomalous precession of Mercury's perihelion.[13] Leverrier tried all kinds of explanations.[14] Perhaps the data were wrong: but recalculation using only the most precise transit observations failed to remove the anomaly. Perhaps the value for the mass of Venus was wrong: but increasing it by a tenth only removed this anomaly by introducing others. Perhaps the gravitational effects of an intra-Mercurial planet were responsible: but the search for the missing planet, which was christened 'Vulcan', eventually proved unsuccessful.[15] Perhaps the anomalous precession was produced, not by a single planet, but by several of them hitherto mistaken for sun-spots: but again, observation failed to reveal these bodies. Leverrier died in 1877, leaving the problem unsolved.

Others continued in the same vein. Perhaps the discrepancy was produced by "multitudinous small bodies individually invisible ... mere interplanetary dust": but such a belt of dust, besides representing an *ad hoc* hypothesis, would have to be tilted at a mechanically impossible angle to do the required job. Perhaps the sun's mass is not distributed spherically (a conjecture recently revived by Dicke): but this contradicted the observed shape of the sun. Again, all these attempts fit well into Lakatos's account: they all involve modifications in the 'protective belt' of auxiliary hypotheses and initial conditions. But again, these were not the *only* responses to the problem.

Some supposed that the anomaly might be explained if the gravitational force between two bodies was not merely a function of their masses and the distance between them, as Newton had said, but also of their relative velocities. Asaph Hall suggested that the exponent in Newton's law must be slightly increased. Simon Newcomb favoured this solution, and wrote, in the 1910 edition of the *Encyclopaedia Brittanica*:

> ... the enigmatical motion of the perihelion of Mercury has not yet found any plausible explanation except on the hypothesis that the gravitation of the sun diminishes at a rate slightly greater than that of the inverse square – the most simple modification being to suppose that instead of the exponent of the distance being exactly −2. it is −2.000 000 161 2.[16]

Newcomb admitted that the effect of Hall's adjustment would only be detectable in the case of Mercury, so that this solution was dangerously *ad hoc*. But an *ad hoc* adjustment is still an adjustment: Newcomb had not made Newton's law of gravity 'unfalsifiable by fiat' – on the contrary, he thought that the Mercury perihelion had falsified it.

It did, of course, take a long time to reach this point. Newcomb's verdict was based upon the failure of all existing proposals for accommodating the anomaly without modifying Newton's law. And it takes time to work out, examine and reject such proposals. Thus we can agree with Lakatos when he argues that whether or not an experiment or anomaly qualifies as a *crucial* one only emerges with hindsight.[17] Adherents of a research programme may at first regard any anomaly as a minor one, which will easily be accommodated by some minor adjustment in the 'protective belt'. It is only when attempts to do this have failed, when more and better scientists have been attracted to the problem and have also failed, that the anomaly will come to be regarded with hindsight as a crucial one.

But here a question arises. Can we ever be in a position to say that *all* ways of accommodating some anomaly within a research programme have failed? All *extant* attempts may have failed, but may there not always be others yet to be tried? In the 1840's Leverrier and Nichol had said that to tamper with Newton's law of gravity in order to deal with the Mercury anomaly was "a last resort" to which they would not turn "until all other potential causes for the discrepancies had been investigated and rejected". But can *all* other potential causes for such discrepancies *ever* be investigated? By the 1890's Newcomb thought that they had, was reluctantly forced into the 'last resort', and chose an *ad hoc* modification of Newton's law of gravity. But perhaps he was not persistent or imaginative enough: perhaps there is no assignable limit to the ways in which any anomaly can be accommodated within a research programme. If so, then no anomaly is *really* crucial. Verdicts like Newcomb's are symptoms of a lack of persistence, rather than reflections of the objective state of affairs (the state of the critical discussion).

This seems to be the view Lakatos has taken of the matter. He says that no real distinction can be drawn between insignificant or small anomalies, and striking or major or crucial ones.[18] He admits, of course, that adherents of a victorious programme will *call* the experiments which they can explain and their rivals cannot 'crucial' ones against their rivals. But this will just be propaganda: such an experiment is merely an anomaly for their rivals, no more serious than any other anomaly:

> *There are no such things as crucial experiments*... when one research programme suffers defeat and is superseded by another one, we may – *with long hindsight* – call an experiment crucial if it turns out to have provided a spectacular corroborating instance for the victorious programme and a failure for the defeated one (in the sense that it was never 'explained progressively'... within the defeated programme)... But if a scientist in the 'defeated' camp puts forward a few years later a scientific explanation of the allegedly 'crucial experiment' within (or consistent with) the allegedly defeated programme, the honorific title may be withdrawn and the 'crucial experiment' may turn from a defeat into a new victory for the programme.[19]

Elsewhere he suggests that the success of a research programme depends mainly on the imagination, and the amount of research money, possessed by its adherents:

> If two teams, pursuing rival research programmes, compete, the one with more creative talent is likely to succeed... The direction of science is determined primarily by human creative imagination and not by the universe of facts which surrounds us. Creative imagination is likely to find novel corroborating evidence even for the most 'absurd' programme, if the search has sufficient drive... A brilliant school of scholars (backed by a rich society to finance a few well-planned tests) might succeed in pushing any fantastic programme ahead, or, alternatively, if so inclined, in overthrowing any arbitrarily chosen pillar of 'established knowledge'.[20]

These are drastic conclusions. They suggest that the *main* trouble with Ptolemaic astronomy was that its practitioners were not quite brilliant or determined enough. They suggest that phlogistic chemistry could well be revived, and a few pillars of established knowledge overthrown, if enough brilliant scholars with big enough research grants could be set to work on it. I doubt it. But how can these drastic conclusions be avoided? Only by rehabilitating the influence exerted on the direction of science by "the universe of facts which surrounds us": in other words, by rehabilitating 'crucial experiments' and 'experimental refutations' of a research programme. And this can be done if we take seriously some of Lakatos's own ideas. Consider again the anomaly with Mercury's perihelion, and the Newtonian research programme. Using Newton's laws and a finite number of auxiliary assumptions, a value for the precession of Mercury's perihelion could be deduced. The predicted value turned out to be wrong. Those who wanted to retain Newton's laws had a finite number of ways to deal with the anomaly, each of them involving a change in one or more of the finite number of auxiliary assumptions. More important, the positive heuristic of their programme told them that there was, basically, only one kind of strategy to adopt: if the anomaly is to be accommodated within the Newtonian programme, then it must be attributed to the mechanical or gravitational effects of some mass or masses not hitherto taken into account.[21] Fifty years of research produced no satisfactory explanation along these lines: the masses of known bodies could be adjusted to explain this anomaly, but only at the cost of invalidating other Newtonian explanations[22]; and postulations of new masses were *ad hoc*, found to be inconsistent with Newtonian principles, or refuted. The anomaly became more and more pressing. The Newtonian research programme had reached a 'natural saturation point'[23]. Finally, at least one Newtonian concluded that there was *no* way in which Mercury's perihelion could be accounted for without modifying Newton's law of gravity. For him, then, Newton's law of gravity had been refuted by the anomalous motion of Mercury's perihelion.

It goes almost without saying that to judge that a research programme (or rather, its 'hard core') has been refuted by a 'crucial experiment' is to make a highly fallible judgement. Perhaps one of the alternative solutions has been wrongly rejected, because of some computational or experimental error. Or perhaps some alternative solution which is in accordance with the positive heuristic of the programme has not even been tried. Such a judgement can, therefore, always be criticised by rehabilitating the wrongly rejected solution, or by developing the untried one. But the fallibility of such judgements does not mean that they cannot be made.

The Mercury perihelion was recognised as a crucial anomaly, meaning an especially *important* one, *before* relativity theory provided an explanation of it. Of course, if 'crucial' means 'helping us to *decide between rival programmes*', then it could not recognised as crucial until after the relativity programme had explained it. Lakatos is right that crucial experiments in *this* sense can only be recognised by comparing rival programmes. Yet the crucial importance of some anomalies, and the insignificance of others, can be recognised in the absence of any rival programme.[24] An anomaly is important if the more obvious, and less revolutionary, ways to accommodate it within a programme fail. And to judge that an anomaly is serious is to give a perfectly objective, rationally arguable, assessment of the state of the critical discussion.

Lakatos's main contention is that no matter *how* serious the anomalies besetting a programme, they will lead no one to reject or eliminate (cease to work in) that programme. A programme will only be eliminated if there is a rival programme which promises to do better. Now this may be correct so far as the 'Newtonian programme' and the Mercury anomaly is concerned[25]. But a scientist *may* try to initiate a new programme because of severe anomalies in the old: the anomalous weight-increase during combustion played a decisive role in the inauguration, by Lavoisier, of the oxygen programme. Severe anomalies *may* lead a scientist to eliminate (cease working in) a research programme. And it will not do for Lakatos to pooh-pooh this by saying that anomalies provide mere psychological stimulus "like an odd-shaped tree at the seaside, or a serpent seen in a dream".[26] Unlike odd-shaped trees, and dreamed-up serpents, anomalies in the old programme represent objective *problems* which a new programme must solve if it is to represent progress.[27]

I conclude that in the Newtonian research programme scientists did not follow Lakatos's methodology, and render Newton's laws unfalsifiable by fiat. Moreover, this methodology is clearly not one which scientists *ought* to follow. Lakatos seems to be recommending that scientists select certain of their hypotheses, christen them a 'hard core', and decide *in advance* not to modify or renounce them in the face of difficulties. He tells us little about how these hypotheses are to be chosen[28]. And therefore his methodological rule, stated generally gives *carte blanche* to any group who want to erect some pet notion into a dogma. But most pet notions are mistaken, and may quickly be found to be mistaken. To decide in advance that the blame for any mistake must be shifted elsewhere, is to open the door to obscurantism.

I feel sure, however, that these remarks fail to catch the spirit of Lakatos's enterprise. His 'hard cores' are not meant to be any odd hypothesis at all. They are meant to be deep and fertile hypotheses, which have the ability to stimulate important mathematical research and to emerge victorious from empirical trials[29]. But such ability can only be recognised with hindsight; which means that a 'hard core' worthy of the name can be detected only *after* it has proved its worth. As we have seen, by 1850 Newton's laws were being treated by some as a 'hard core', simply because of their previous success. But if Newton's laws only emerged as a 'hard core' in 1850, then the preceding centuries of research did not abide by the canons of Lakatos's methodology of research programmes.

Perhaps the crucial distinction between 'hard core' and 'protective belt' is intended not as a methodological device (a source of advice to scientists) but as a historiographical one (a source of advice to historians). Perhaps we are being told that when we come to write the history of research programmes like Newton's, we will find, as a matter of fact, that 'hard cores' of hypotheses were present in all the successful explanations which were produced. We surely do find this: some hypotheses have persisted over long periods despite many apparent refutations of them[30]. But has Lakatos satisfactorily explained this feature of the history of science (which he calls *continuity*)? His explanation is that 'hard cores' persisted because scientists took methodological decisions to retain them. And one can certainly 'rationally reconstruct' history in the light of this explanation.

But to reconstruct the history of Newton's theory in this way would be to falsify it. Some Newtonians at some points did take a methodological decision to retain Newton's laws unchanged in the face of an anomaly – and some did not. The former were usually successful, and the latter were not. But it is only by ignoring half of the actual history that it can be made to fit Lakatos's methodology.

Besides, there is an alternative, and very simple, explanation of the 'continuity' of Newton's theory: it survived simply because it was a very good theory; it was not made unfalsifiable by fiat, but was difficult to falsify in fact. This explanation, though rather trite, at least does not make a methodology out of our good fortune in having had a Newton. In adopting this explanation we do not run the risk of allowing theories to survive by methodological fiat which do not deserve to survive.

To sum up: the rule "Do not blame refutations on the hard core of your research programme" was not actually followed in the "classical example of a successful research programme", is not a rule which ought to be followed, and is not necessary to explain continuity in the history of science.

2. POSITIVE HEURISTICS: CAN THEY 'PRODUCE AND DIGEST' EMPIRICAL REFUTATIONS?

So far I have dealt with what Lakatos calls the 'negative heuristic' of a research programme, the rule which forbids us to direct the *modus tollens* at the 'hard core'. But the more interesting and original feature of his methodology is his idea that research programmes have a *positive heuristic*. This is how he describes it:

> Few theoretical scientists engaged in a research programme pay undue attention to 'refutations'. They have a long-term research policy which anticipates these refutations. This research policy, or order of research, is set out – in more or less detail – in the *positive heuristic* of the research programme... the positive heuristic consists of a partially articulated set of suggestions or hints on how to change, develop the 'refutable variants' of the research programme, how to modify, sophisticate, the 'refutable' protective belt.
>
> The positive heuristic of the programme saves the scientist from becoming confused by the ocean of anomalies. The positive heuristic sets out a programme which lists a chain of ever more complicated *models* simulating reality: the scientist's attention is riveted on building his models following instructions which are laid down in the positive part of his programme. He ignores the *actual* counterexamples, the available '*data*'... A '*model*' is a set of initial conditions (possibly together with some of the observational theories) which one knows is *bound* to be replaced during the further development of the programme, and one even knows, more or less, how. This shows once more how irrelevant 'refutations' of any specific variant are in a research programme: their existence is fully expected, the positive heuristic is there as the strategy both for predicting (producing) and digesting them. Indeed, if the positive heuristic is clearly spelt out, the difficulties of the programme are mathematical rather than empirical...
>
> Thus the methodology of scientific research programmes accounts for the *relative autonomy of theoretical science:* a historical fact whose rationality cannot be explained by the earlier falsificationists. Which problems scientists working in powerful research programmes rationally choose, is determined by the positive heuristic of the programme rather than by psychologically worrying (or technologically urgent) anomalies. The anomalies are listed but shoved aside in the hope that they will turn, in due course, into corroborations of the programme. Only those scientists have to rivet their attention on anomalies who are either engaged in trial and error exercises or who work in a degenerating phase of a research programme when the positive heuristic ran out of steam.[31]

Now what I want to concentrate on here is the claim that the positive heuristic of a research programme can 'anticipate' empirical refutations, as well as the way to deal with them. If correct, this claim means that (at least in ideal cases) the experimenter plays no role in the development of a research programme, so that theoretical science is autonomous in a very strong sense indeed! I think that the claim is incorrect.

It is worth noting here that Lakatos's first example of a 'positive heuristic' forging ahead without any regard for empirical facts occurs in his 'Proofs and Refutations'. Of course, it comes as no surprise to the reader of that brilliant case-study of informal, quasi-empirical mathematics to discover, quite late in the piece, that the mathematical properties of polyhedra could have been worked out without actually inspecting a physical polyhedron.[32] Indeed, a philosophically-literate reader is likely to have been wondering what observations of actual cubes, picture-frames, and the rest were doing in mathematics anyway. But now Lakatos tells us that research programmes in empirical science can be as independent of empirical facts as those in mathematics. And this should give us pause.

Let me explain what I think has gone wrong with reference to Lakatos's own examples. The first example is the Newtonian research programme, about which he writes:

> Newton first worked out his programme for a planetary system with a fixed point-like sun and one single point-like planet. It was in this model that he derived his inverse square law for Kepler's ellipse. But this model was forbidden by Newton's own third law of dynamics, therefore the model had to be replaced by one in which both sun and planet revolved round their common centre of gravity. This change was not motivated by any observation... but by a theoretical difficulty... Then he worked out the programme for more planets as if there were only heliocentric but no interplanetary forces. Then he worked out the case where the sun and planets were not mass-points but mass-*balls*. Again, for this change he did not *need* the observation of an anomaly... This change involved considerable mathematical difficulties... Having solved this 'puzzle', he started work on *spinning balls* and their wobbles. Then he admitted interplanetary forces and started work on perturbations. At this point he started to look more anxiously at the facts. Many of them were beautifully explained (qualitatively) by this model, many were not. It was then that he started to work on *bulging* planets, rather than round planets, etc.
>
> Most, if not all, Newtonian 'puzzles', leading to a series of new variants superseding each other, were forseeable at the time of Newton's first naive model and no doubt Newton and his colleagues did foresee them: Newton must have been fully aware of the blatant falsity of his first variants.... This shows once more how irrelevant 'refutations' of any specific variant are in a research programme: their existence is fully expected, the positive heuristic is there as the strategy both for predicting (producing) and digesting them.[33]

Now Lakatos is clearly right to say that Newton did not need an observer to tell him that planets do not move precisely in elliptical orbits. For his theory told him this: elliptical orbits could be obtained from it only by making obviously false, indeed incoherent, assumptions (that there is only one planet of zero mass). It would be to misunderstand Newton's theory to take that first 'model' as yielding a genuine Newtonian prediction which must be compared with the evidence.[34] What Newton's theory, combined with obvious facts about the solar system, does predict actually contradicts Kepler's law of elliptical orbits.

What then is the point of that first 'model'? Clearly, it is the first step in the long and difficult mathematical task of finding out what Newton's theory does predict about the solar system. The other models mentioned by Lakatos are later steps in the same process. Simplifying assumptions were made which were known to be false, and Kepler's law derived as a first approximation.[35] Then the simplifying assumptions were removed one by one, until finally a Newtonian prediction was reached.

I suggest, therefore, that the Newtonian 'positive heuristic' described by Lakatos is not a strategy for producing and digesting empirical refutations; rather, it is a strategy for solving, by a method of successive approximation, the difficult mathematical problem of calculating what Newton's theory asserts about planetary motions. Lakatos is quite right that 'falsifications' play no role in this process; for the various 'models' produced do not represent Newtonian predictions to be compared with the evidence, but are rather steps on the way to such predictions.[36] The 'autonomy of theoretical science' exists all right – but no anti-empiricist or anti-falsificationist lessons can be drawn from it.

Lakatos does, of course, concede that there came a point when Newtonians did actually pay attention to empirical evidence: after perturbations were taken into account they "started to look more anxiously at the facts".[37] He suggests, however, that this was somehow an accident: the 'positive heuristic' happened to run out of steam at that point, so that the programme entered its 'degenerating phase'.[38] But could the heuristic have *failed* to run out of steam at that point, as Lakatos seems to think?[39] I think that Newtonians *had* to look anxiously at the facts at that point simply because they had then managed to deduce predictions from their theory. The heuristic did not just happen to run out of steam; rather, the logico-mathematical problem of deriving empirically testable predictions had been solved.

It is just not true that refutations of any specific variant of a research programme can be produced and digested by a clearly spelt out heuristic. The anomalies in the orbit of Uranus could not have been foreseen by any Newtonian heuristic, however well articulated. And without those observed anomalies, no Newtonian would have dreamt of postulating an extra planet beyond Uranus. The discovery of Neptune was a theoretical triumph all right; but it was only made possible by a stubborn fact. Lakatos's claim that "if the positive heuristic is clearly spelt out, the difficulties of the programme are mathematical rather than empirical"[40] is a mistake. He is to be congratulated, however, for emphasising how much of the activity in a programme like Newton's is purely mathematical, and how difficult it is to find out what a precise theory like Newton's predicts.[41] This much autonomy can be granted to theoretical science.

Let us now look, briefly, at some other research programmes and at their positive heuristics, to see if the points made so far apply universally. Lakatos's second example, Prout's programme, immediately poses a problem. We have seen that Newton's positive heuristic as described by Lakatos was mainly a plan for solving the mathematical problem of finding out what the theory predicts. But Prout's hypothesis poses no analogous mathematical problem, and therefore contains no analogous detailed heuristic[42]. In fact, the positive heuristic of Prout's programme is a rather puny affair, consisting merely of a hint about how apparent refutations of Prout's hypothesis might be handled. It could be articulated as follows: "Blame apparent refutations upon impure samples, and hence upon inefficient purifying techniques".[43]

I think that heuristic hints like this can be found, to a greater or lesser degree, in all research programmes, and that their importance cannot be over-estimated. It is such hints, rather than tactics for solving logico-mathematical problems, that Lakatos really means by 'positive heuristics'. The analogous hint for the Newtonian research programme would be something like: "Blame anomalies upon the disturbing effects, either mechanical or gravitational, of masses not hitherto taken into account". Analogous hints from Lavoisier's oxygen research programme might be: "Whenever fixed air (carbon dioxide) is obtained from the reduction by heat alone of a metallic calx (oxide), blame this upon carbon impurities in the sample", or "Whenever inflammable gas (hydrogen) is obtained from a combustible substance, blame this upon the decomposition of water present in the substance".[44] Hints like this can often be laid down in advance of empirical refutations. But they 'anticipate' or 'produce' those refutations only in the Pickwickian sense that they allow that refutations might be forthcoming. And they 'digest' refutations only in the sense that, by following them, a scientist may be able to account for an anomaly in a non *ad hoc* way, predicting a novel fact which subsequently gets confirmed (as in the Neptune case).

Although such hints can often be laid down in advance of the apparent refutations to which they will be applied, I do not think they can always be laid down at the outset of the research programme. Sometimes a heuristic hint will depend upon a discovery made *after* the programme has got underway (like that of the decomposition of water, which figures in my second example from the oxygen research programme). The 'positive heuristic' of a research programme, its collection of hints about how apparent refutations might be handled, will typically develop and be articulated along with the programme itself. The exceptions to this, the cases where the positive heuristic can be spelled out in great detail at the outset, stem, I believe, from a special source.

This brings me to Lakatos's third example, Bohr's research programme. Lakatos describes its positive heuristic as follows:

> Bohr's plan was to work out first the theory of the hydrogen atom. His first model was to be based on a fixed proton-nucleus with an electron in a circular orbit; in his second model he wanted to calculate an elliptical orbit in a fixed plane; then he intended to remove the clearly artificial restrictions of the fixed nucleus and fixed plane; after this he thought of taking the possible spin of the electron into account, and then he hoped to extend his programme to the structure of complicated atoms and molecules and to the effect of electro-magnetic fields on them, etc., etc. *All this was planned right at the start: the idea that atoms are analogous to planetary systems* adumbrated a long, difficult but optimistic programme and clearly indicated the policy of research.[45]

It is clear from this that the strength of Bohr's positive heuristic, as opposed to Prout's or even Newton's, stems from the *analogy* between atoms and planetary systems which lay at the basis of Bohr's programme. It is because Bohr knew, from an already worked-out planetary theory, about circular and elliptical orbits, their motions through a plane, the rotation of planets on their axes, and so on, that he could envisage that analogous elements might play a role within the atom, and provide the key to the atomic spectra.

Mind you, we must not over-estimate the power even of a well-articulated heuristic like Bohr's. For the question with any analogy is: "How far does it obtain?".[46] Bohr had no idea, at the outset, in what respects his orbiting electrons resembled planets. His quantum postulates do not imply, for example, that electrons move in non-circular orbits (contrast

Newton's laws which, together with accepted initial conditions, do imply that planets cannot move in circles). And hence Bohr's first model (unlike Newton's first model) was not a mathematical approximation known in advance to be erroneous. Bohr needed the experimenter to tell him that his first model was inadequate; his positive heuristic could not predict this result (though it did, perhaps, render it a not very surprising one). Similarly, Bohr could not know in advance that electrons, like planets, had spin; and in fact, he seems to have opposed Sommerfeld's electron spin model.[47] Analogical arguments are notoriously weak ones: they can properly issue only in hints about what might have to be taken into account, not in predictions about what will have to be taken into account.

I suspect that in other cases in which an elaborate positive heuristic can be laid down at the outset of a research programme we will find that an *analogy* is responsible. Peter Clark has reconstructed the positive heuristic of the 'atomic kinetic research programme', and there the analogy employed is between the molecules of a gas and a large collection of billiard-balls interacting in accordance with the laws of Newtonian mechanics[48]. John Worrall has contrasted wave optics and corpuscular optics in the early nineteenth century, the former exploiting an analogy between rays of light and waves in an elastic medium, and the latter an analogy between rays of light and streams of small particles (obeying known laws of particle mechanics). Worrall says that the corpuscular programme was heuristically superior to the wave programme:

> Because of the comparatively less developed state of the mechanics of elastic media relative to the mechanics of rigid particles the heuristic of the wave programme was... less definite than that of the corpuscular programme.[49]

Here we see clearly that the heuristic strength of a programme often depends upon how far it can exploit an analogy with entities *for which we already have a worked-out theory*. But heuristic strength is clearly a matter of degree, and I do not think we should *deny* the status of 'research programme' to cases where such an analogy cannot be exploited, and where the positive heuristic has to develop piecemeal as the programme develops.[50] (Denying them this status would have the odd result that before any *genuine* research-programmes can get under way, a lot of science must have already gone on to provide the basis, *via* analogies, of their necessary heuristic strength.)

To sum up. Lakatos, in an excess of anti-falsificationist zeal, has overestimated what positive heuristics can achieve in the development of research programmes. Newton's heuristic, as he describes it, was not so much a strategy for 'producing and digesting refutations' as a strategy for deducing predictions. While a positive heuristic *can* give hints about how future anomalies might be handled, the theoretician still needs the experimenter to *produce* those anomalies. Where the positive heuristic can be spelled out in some detail from the start, this results from the exploitation of an analogy: but even here the theoretician needs the experimenter to tell him *in what respects* his analogy holds. Theoretical science is not so 'autonomous' as Lakatos occasionally suggests, and the development of a research programme remains a subtle dialogue between the theoretician (armed with his heuristic) and the experimenter (armed with the 'facts'). This, after all, is what gives science its empirical character.

3. Appraisals and Advice

So far I have discussed Lakatos's views about developments within a *single* research programme, and about the role played in those developments by his positive and negative heuristics. But of course, his central idea is that the history of science is the history of *competing* research programmes, and his central problem is how one research programme is *eliminated* or *rejected* in favour of another one[51]. He makes it clear that "the pragmatic meaning of 'rejection' [of a programme]" is *"the decision to cease working on it"*.[52]

Like Hanson and Kuhn before him, Lakatos claims that a programme will only be rejected or eliminated if there is a rival programme to take its place – and that this holds no matter how many unresolved anomalies beset the first programme.[53] A programme which reigns in splendid isolation reigns supreme (and Kuhnian 'normal science' is "nothing but a research programme that has achieved monopoly"[54]). But unlike Hanson and Kuhn, Lakatos is anxious to lay down objective criteria for comparing competing research programmes, and for saying that one of them is better than another one.

Lakatos's criteria for appraising research programmes are rather complex, and I will give only a *very* swift summary of them. A transition from one theory or version of a programme to the next (or a 'problem shift') is *theoretically progressive* if the second theory predicts a novel fact or facts, and *empirically progressive* if at least one of these novel predictions gets confirmed (otherwise it is theoretically and/or empirically *degenerating*). A research programme itself is (consistently or intermittently) theoretically and/or empirically progressive up to a certain time if its problem shifts up to that time have been progressive (otherwise it is, consistently or intermittently, theoretically and/or empirically degenerating). One research programme is better than a second at a certain time if the first has progressed up to that time and the second has degenerated, or if the first has progressed 'faster' up to that time than the second.

Armed with these criteria, it is possible for a scientist to be able to judge, in certain cases, that one research programme is better than another (at a certain time). It goes without saying that his judgement will be a *fallible* one: he may, for example, be wrong in saying that some particular problem-shift was theoretically degenerating, for matters like these are not easy to determine. But it is not the difficulty in applying Lakatos's criteria that concerns me, but rather what follows when they *have* been applied. So let us suppose that all scientists are agreed (perhaps wrongly) that at a certain time research programme A is better than research programme B. The question is: what follows from this for their practice?

The answer might seem a straightforward one: if scientists are agreed that research programme A is better than research programme B, then they should eliminate B in favour of A, that is, they should cease working on B and switch their energies to developing A. It might seem, in short, that there is a straightforward connection between the *appraisal* of research programmes in the light of Lakatos's criteria and *advice* to scientists about which programme to pursue.

But matters are not so straightforward as this. Lakatos has gradually reduced the extent to which his methodology issues in advice to scientists, until finally he denies that he gives any advice worth speaking of at all. Thus he writes:

my 'methodology', older connotations of the term notwithstanding, only *appraises* fully articulated theories (or research programmes) but it presumes to give advice to the scientist neither about how to *arrive* at good theories nor even about which of two rival programmes he should work on. My 'methodological rules' explain the rationale of the acceptance of Einstein's theory over Newton's, but they neither command nor advise the scientist to work in the Einsteinian and not in the Newtonian research programme.... when it turns out that, on my criteria, one research programme is 'progressing' and its rival is 'degenerating', this tells us only that the two programmes possess certain objective features but does not tell us that scientists must work only in the progressive one....

I, of course, do not prescribe to the individual scientist what to try to do in a situation characterised by two rival progressive research programmes [or, we might add, by two rival research programmes one of which is progressive and the other degenerating]: whether to try to elaborate one or the other or whether to withdraw from both and try to supersede them with a Great Dialectical Leap Forward. Whatever they *have* done, I can judge: I can say whether they have made progress or not. But I cannot advise them – and do not wish to advise them – about exactly what to worry about and in which direction they should seek progress.[55]

Lakatos goes on to explain that the only piece of advice he has to offer scientists is that they should be *honest* about the merits and demerits of their research programme: "a public record should be kept ... of known anomalies and inconsistencies".[56] But so long as the defects of a research programme are not hushed up, continued work on it is perfectly rational no matter how poorly the programme performs. An honest policy in science is a rational policy.[57]

I find this latest position of Lakatos an extremely odd one.[58] He develops an elaborate account of what is good science and what bad, but refuses (apart from 'Thou shalt not lie') to give any advice to scientists. It is as if the author of a cookbook were to conclude his graphic account of how to make a good steak-and-kidney pie by saying "Of course, I am not telling you what to do (but whatever you do, keep a record of it)". There was a time when Lakatos would have found such a position odd too, for he once wrote that:

... the appraisal of any finished product is bound to have decisive pragmatic consequences for the method of its production. Moral standards, by which one judges people, have grave pragmatic implications for education, that is, for the method of their production. Similarly, scientific standards by which one judges theories, have grave pragmatic implications for scientific method, the method of their production.[59]

Now I do not want to claim (nor do I think Lakatos was claiming in this passage) that an appraisal *logically* implies a piece of advice. Lakatos is right to insist that "*methodology* is separated from *heuristics*, rather as value judgements are from ought statements".[60] Lakatos's recent position is, in other words, a *logically consistent* one – but it is an odd one all the same.

It is especially odd that he, of all people, should adopt it. His original problem, remember, was to say when a research programme ought to be eliminated (where to eliminate a research programme means to cease working on it). He expressed the hope that his methodology "might help us in devising laws for stemming ... intellectual pollution", and emphasised how important it was to "retain the *determination* to eliminate, under certain objectively defined conditions, some research programmes".[61] And even in his 1971 paper, he says:

> ... the methodology of scientific research programmes has more teeth than Duhem's conventionalism: instead of leaving it to Duhem's inarticulated common sense to judge when a 'framework' is *to be abandoned*, I inject some hard Popperian standards into the *appraisal* of whether a programme progresses or degenerates or of whether one is overtaking another. That is, I give criteria of progress and stagnation within a programme *and also rules for the 'elimination' of whole research programmes*... If a research programme progressively explains more than a rival, it 'supercedes' it, and the rival can be eliminated...[62]

But if we only appraise programmes, without giving any advice about which should be pursued, then this is all hot air and Lakatos simply has nothing to say in answer to his original problem. He still claims that his rules of appraisal "explain the rationale of the acceptance of Einstein's theory over Newton's".[63] But his rules of appraisal would equally well have explained (or rather, have failed to explain) the rationale of the acceptance of Newton's theory over Einstein's. If what I say gives a man no indication about how to act, then whatever he does is consistent with my (non-existent) advice and I can hardly claim that what I told him explains his behaviour. In 1968 Lakatos complained about Carnap's inductive logic in the following terms:

> The abdication of the inductive judge is complete. He promised to hand down judgement on the rationality of beliefs; now he is ending up by trying to supply a calculus of coherent beliefs on whose rationality he cannot pronounce. The inductive judge cannot claim any longer to be a guide of scientific life in any relevant sense.[65]

I am tempted to use Lakatos-1968 against Lakatos-1971, along the following lines:

> The abdication of the Lakatosian methodologist is complete. He promised to hand down judgement on the rationality or otherwise of pursuing some research programme; now he is ending up by trying to supply rules to appraise research programmes on the rationality of pursuing which he will not pronounce. The Lakatosian methodologist cannot claim any longer to be a guide of scientific life in any relevant sense.[65]

Some, of course, welcome 'the abdication of the Lakatosian methodologist'. It is with glee that Feyerabend exclaims "scientific method, as softened up by Lakatos, is but an ornament which makes us forget that a position of 'anything goes' has in fact been adopted".[66] But I do not welcome it, and I regard it as an aberration given Lakatos's general philosophical position. How can his methodology help us to stem intellectual pollution, if it has no repercussions for intellectual practice? We must try to extricate Lakatos from anarchism, by taking a critical look at the arguments which have led him, unwittingly, towards it.

Notice that these arguments all concern the difficulty of advising scientists which programme they should *try to develop*. Feyerabend is right that Lakatos came to adopt an anarchistic position on *this* issue. But Lakatos does not take an anarchistic view of *existing* theories or research programmes. An anarchist would, presumably, maintain that any existing theory or programme is as good (or as bad) as any other. Lakatos, on the other hand, provides criteria in the light of which one theory or programme is to be judged superior to another. He merely doubts that these appraisals can yield any advice about future research (which is why Feyerabend deems them 'verbal ornaments'). Let me now review the arguments which led Lakatos to this conclusion.

The first argument need not detain us long. It is that a comparison of the merits of two competing research programmes will often give no clear-cut result. Even where it does, the verdict will be difficult to achieve and *fallible*. The judgement that a particular version of a programme was *ad hoc*, or that although not *ad hoc* it was refuted, may always be challenged by showing that some novel consequence has been overlooked, or an experimental result wrongly accepted. But the fallibility of Lakatosian appraisals is no reason for refusing to base advice upon them. A scientist who disagrees with a particular appraisal will be unlikely to *take* the advice to which it leads; but it is then up to him to show that the appraisal was, indeed, mistaken.

The second argument to the effect that heuristic advice cannot stem from Lakatos's appraisals bases itself, not on the fallibility of those appraisals, but on their backward-looking character.[67] We can judge (fallibly) that one programme has progressed up to a certain time, while its rival has degenerated. But it is possible that if both programmes are pursued a little further, our judgement will have to be reversed: the progressive programme will start to degenerate, while the degenerating one will stage a comeback. In this situation, is not *any* advice totally arbitrary?

Feyerabend states the point with characteristic verve:

Now it is easy to see that standards of this kind have practical force only if they are combined with a *time limit* (what looks like a degenerating problem shift may be the beginning of a much longer period of advance). But introduce the time limit and the argument against naive falsificationism reappears with only a minor modification (if you are permitted to wait, why not wait a little longer?). Thus the standards which Lakatos wants to defend are either *vacuous* – one does not know when to apply them – or they can be *criticized* on grounds very similar to those which led to them in the first place.

In these circumstances one can do one of the following two things. One can *stop* appealing to... standards... Or one can retain such standards as a *verbal ornament*, as a memorial to happier times when it was still thought possible to run a complex and often catastrophic business like science by following a few simple and 'rational' rules. It seems that Lakatos wants to choose the second alternative.[68]

As we have seen, in response to this criticism Lakatos deprived his standards of practical force, and adopted a position of 'anything goes'.

Feyerabend is clearly right that to specify a time limit is unsatisfactory: a rule like "Eliminate a programme which has degenerated up until now if it does not pick up within the next *n* years (or, within its next *n* versions)" is bound to be arbitrary. As well as being arbitrary, they are unnecessary: if we are to give practical force to any appraisals, we might as well give it to the purely backward-looking ones. But if a programme *might* always be revived, however badly it has fared in the past, can we give any 'practical force' to an appraisal of its past performance?

Of course we can. The fact that our judgements of past performance may be reversed in the light of future developments is no reason not to base advice upon them. Consider the following parallel. We stand at a cross-road, and ask a bystander which of two roads to take to get to our destination. He gives us a lengthy account of the merits of one road and the demerits of another, based on his present knowledge of them both. We would find it pretty odd to be told, after all this, that no advice was being given because it was just possible that, within the next five minutes, the 'good road' would be blocked by an avalanche. And if, after taking the good road, we find that an avalanche has occurred, we will not *blame* the bystander. He gave us the *right* advice given his knowledge of the situation at the time. Avalanches are unpredictable; that does not mean that people are never *well*-advised to take one route rather than another. A research programme in the doldrums *may* stage a comeback, and its rival *may* start to degenerate. But this does not mean that any programme is as promising, at any time, as any other.

But what are we, as methodologists, to say to the individual scientist who does not want to give up a degenerating programme? He may accept that, if we appraise their performance up till now, a rival programme has done much better than his own. And yet, without disputing the soundness of this verdict, he hopes that his work will reverse it. Do we advise him to relinquish this hope, and deem him irrational if he does not accept our advice? Priestley adhered to phlogistonism after every other reputable chemist had given it up in favour of Lavoisier's programme: are we to condemn him as irrational? Many philosophers and historians of science have reached this verdict.[69] But Lakatos is reluctant to do so, and for a very good reason:

> ... there is something to be said for at least *some* people sticking to a research programme until it reaches its 'saturation' point; a new programme is then challenged to account for the full success of the old.[70]

This brings me to the third argument for 'the abdication of the Lakatosian methodologist', which we might state as follows: we should not, on the basis of our appraisals, condemn the stubborn individual (or individuals) who wish to persist with a degenerating programme; therefore, our appraisals cannot issue in any heuristic advice at all.

I think the argument is invalid: its premise is true, but its conclusion is not. The invalidity of the argument can be shown if we distinguish between heuristic advice addressed to *individual scientists*, and heuristic advice addressed to *science* or to *the community of scientists as a whole*. Kuhn has been emphasising for years that science is a social or community enterprise. This being so, there is nothing to prevent different scientists (or even individual ones) from simultaneously pursuing different lines of research. This diversity will ensure (in so far as such things *can* be ensured) that no promising line of research gets neglected, and thus serve as the "community's way of distributing risk":

> If a decision must be made under circumstances in which even the most deliberate and considered judgement may be wrong, it may be vitally important that different individuals decide in different ways. How else could the group as a whole hedge its bets?[71]

Kuhn applies this general point specifically to the question of eliminating one theory (research programme) in favour of another because of anomalies or apparent refutations of the former:

> If everyone agreed in such judgements, no one would be left to show how existing theory could account for the apparent anomaly as it usually does. If, on the other hand, no one were willing to take the risk and seek an alternate theory, there would be none of the revolutionary transformations on which scientific development depends.[72]

It seems to me that if the methodology of research programmes (or indeed, any other methodology) is to provide advice or directives, then these must be addressed to science, or if you prefer, to the community of scientists, *as a whole.* Such community-directed advice would not forbid an individual (or even a group of them) from persisting with a degenerating programme when a progressive rival is available, or from working within a budding programme which at the time is inferior to an established rival. It *would* forbid wholesale persistance with a degenerating programme, or premature mass-conversion to a budding one (both of which would have to be explained on 'externalist' grounds). We cannot condemn Priestley for his die-hard adherence to phlogistonism; but we could condemn the community of late eighteenth century chemists had they all done the same. As Grünbaum puts it:

Since scientific inquiry is conducted by a *community* of scientists, ... practices that would indeed be irrational if adopted by that community as a whole or by a majority of it need not necessarily be irrational when only a certain gifted minority engages in them.[73]

I conclude, then, that an appraisal like "This programme has degenerated up till now, while its rival has progressed" can and should lead to heuristic advice of the following kind: "Science should devote most of its resources to the progressive programme".[74]

At this point those with a keen nose for such things will smell an inductivist rat. Surely, they will say, if we advise science to devote more energy to a progressing programme than to a degenerating one, we must be assuming that the progressing programme will continue to progress and the degenerating one to degenerate. But assuming that the future will be like the past is crude inductivism. Herbert Feigl puts it so:

Professor Lakatos is – and, I think – cannot help being, a second-level inductivist... There can be little doubt that he wishes to serve (at least) in a critical and/or advisory capacity to scientists.... But if he is to fulfill the critical and/or advisory functions, what else can he do but watch the course of the 'shifts' and *extrapolate*?[75]

Now crude inductivism is sinful enough on *any* level. But on Feigl's 'second level', where it concerns the essentially unpredictable growth of science, it is a mortal sin of the first order. Have we rescued Lakatos's methodology from the frying-pan of anarchism, only to see it cast into the hell-fire of inductivism? I think not.

Occasionally, in megalomaniac moments, I try to put myself in the shoes of a late eighteenth-century chemist, faced (say in 1785) with the problem of advising a bright young man which of the two extant research programmes to pursue, the progressing oxygen programme or the degenerating phlogistonist one. When I consider the oxygen programme, I can make quite a list of *open problems*, of varying degrees of difficulty, which the young man might tackle. And for many of these, heuristic hints can be given about how the investigation might proceed. Here are some of them:

(a) the oxygen theory of acidity predicted that the oxidation of organic compounds should produce acids – does it? (The oxygen theory of acids is false. But by 1790 Lavoisier could claim that it had opened up a new field of inquiry, and that nearly thirty new acids had been discovered.)

(b) the oxygen theory of acidity predicted that muriatic or marine acid contained oxygen, yet despite repeated efforts no one had succeeded in extracting it, and Berthollet had just presented a rather unsatisfactory paper on the matter – *can* oxygen be extracted from this acid, and if not, why not? (Muriatic acid is hydrochloric acid, which contains no oxygen.)

(c) Priestley claims to have extracted 'inflammable air' from charcoal – can this be done, and if so, does the inflammable air come from water contained in the charcoal or is it something other than the 'water-former' (hydrogen)? (Carbon monoxide was discovered in 1800, after Lavoisier had left a vacant place for it in his list of chemical elements.)

(d) Metallic oxides, upon reduction, occasionally yield 'fixed air' (carbon dioxide) instead of oxygen – can you show that this is because they contain carbon impurities?

(e) Is heat a *substance* (caloric), as Lavoisier said, or a 'mode of motion', as Laplace said? If the latter, how are we to explain changes of stage, or the evolution of heat in combustion?

(f) The number of chemical elements (defined by Lavoisier as the last products of chemical analysis) is growing apace – can and should any simplification be achieved, and if so, how?[76]

When, on the other hand, I consider the phlogiston programme I am at a loss. The latest versions of that programme make no novel predictions, but merely accommodate discoveries made by their rivals in an *ad hoc* way. There are no novel predictions to be tested. Of course, one could repeat old experiments which had been successes for the rival programme, in the hope that you might find something wrong. And if the rival programme makes further new predictions which get confirmed, these facts will have to be accommodated somehow. But if the programme *has* degenerated, if its heuristic *has* run out of steam, then its 'problems' are posed by its rival and the programme gives no hints for their solution. I need hardly say what my advice would be.[77]

Now assume that the situation I find myself in my imaginings is a typical one. Suppose that a progressive research programme is throwing up a rich store of interesting unsolved problems, and hints about how to solve them, while a degenerating programme is not. Take seriously, in other words, Lakatos's idea of *heuristic power*.[78] Then, I think, we can provide a purely *deductive* argument for the sort of heuristic advice which I have been defending. The argument runs as follows:

Premise 1 (Supreme ought-statement): Science ought to devote energy to investigating unsolved but solvable scientific problems.

Premise 2 (Heuristic power): A progressing research programme throws up more unsolved but solvable problems than a degenerating one.

Premise 3 (Appraisal): Research programme A is progressing and research programme B degenerating.

Conclusion (Heuristic advice): Science ought to devote more energy to research programme A than to research programme B.

Of course, there is no guarantee that following such advice will lead to success. For all we know, science may peter out into a series of arid specialisms, or, worse still, become so unfashionable that all intellectual energy is devoted to spreading intellectual pollution. Imre Lakatos fought hard against both of these things throughout his life. But such is the world we live in that not even his tenacious defence of intellectual standards is sure to succeed.

University of Otago

NOTES

* I have benefitted from helpful comments on previous versions of this paper by Peter Clark, Martin Frické, John Watkins and John Worrall.

[1] Lakatos (1970), p. 133.

[2] See Kuhn (1970b), p. 256, and (1971), p. 137. The differences between Kuhn and Lakatos, apart from terminological ones, boil down, it seems to me, to two:

(a) *Proliferation:* Lakatos urges, following Popper and Feyerabend, the necessity of *competing* research programmes, and claims that what Kuhn calls 'normal science' is "nothing but a research-programme that has achieved monopoly" (cf. Lakatos, 1970, p. 155).

(b) *Objectivity:* Lakatos provides objective criteria for appraising competing research programmes (whether contemporaneous or successive). Kuhn gives no such criteria, and once suggested, *via* the 'incommensurability thesis', that none could be given.

[3] Lakatos's position here is, in fact, identical to that elaborated by Popper for the special case of the 'rationality principle' in social science. Popper recommends that this principle be rendered 'irrefutable by fiat', so that when a theory containing it runs into difficulties the blame is shifted on to the other parts of the system (the 'situational analysis'). Popper has, of course, special reasons for making what is, after all, an exception to his general position: see his (1967). For an interesting discussion of the research programme in economics that resulted from taking the rationality principle as a 'hard core', see Latsis (1972).

[4] And this is not just an idle possibility: see below, note 10.

[5] On Clairault's proposed modification, see Grosser (1962), pp. 47–8, Hanson (1962), p. 361, footnote 2, Kuhn (1962), p. 81, or Thackray (1970), pp. 157–8. Clairault hoped that his amended law might have excess content, and be able to explain phenomena of chemical attraction.

[6] Grant (1852), p. vii (my italics).

[7] Cited by Grosser (1962), p. 100.

[8] Cf. Grosser (1962), p. 48, or Hanson (1962), p. 361, footnote 2. That Airy had proposed to modify the inverse square law partly explains his rather offhand treatment of young Adams, who had used the unmodified law to predict the existence of a new planet.

[9] Cf. Nichol (1848), pp. 62–3.

[10] As John Watkins pointed out to me, Lakatos's views risk falling into the following circularity: the 'hard core' of a research programme consists of those propositions held irrefutable by its adherents, and an adherent of a research programme is one who holds its hard core irrefutable. The circle is broken by the definition of 'Newtonian' suggested in the text, and by the definition of 'hard core' suggested in note 30 below (both of which require historical hindsight for their application). Many scientists who contributed to the Newtonian research programme (and who were therefore 'Newtonians' in my sense) had deep-seated metaphysical objections to Newton's law of gravity, and to the action-at-a-distance which it seemed to involve. Such people can hardly be said to have committed themselves to Newton's law. But one can contribute to a research programme without committing oneself to its hard core (which shows the historical and methodological irrelevance of 'commitments'). Metaphysical objections to Newton's law were one reason why some were quite happy to tinker with it should the need arise; for them it could only be 'accidentally true' anyway. And in fact, some proposed modifications (for example, the introduction of a finite velocity of propagation for gravitational force) were prompted not by empirical difficulties, but by the desire to make the law metaphysically more acceptable (on this, see North (1965), Chapter 3).

[11] For these contemporary reactions, see Grosser (1962), pp. 119–21.

[12] Cf. Nichol (1848), pp. 63–4. Clerke says of the discovery of Neptune "By it the last lingering doubts as to the universal validity of the Newtonian law were dissipated" (Clerke, 1885, p. 107).

[13] There were other, minor anomalies also. Encke found that the orbit of the first short-period comet to be discovered did not conform to the predicted one: "After every possible allowance had been made for the pulls, now backward, now forward, exerted upon it by the several planets, there was still a surplus of acceleration left unaccounted for. Each return to perihelion took place about two and a half hours sooner than received theories warranted." (Clerk, 1885, p. 122).

[14] For details, cf. Hanson (1962).

[15] But not before Lescarbault, on Leverrier's recommendation, had been presented with

a gold medal by the French Academy for 'observing' it! The search for Vulcan was a protracted one: cf. Fontenrose (1973). Some said that Vulcan could not be seen because it always hid behind the sun; but this contradicted Kepler's third law (since Vulcan would then have the same period as the earth), and it contradicted Newton's theory (as was proved by Liouville in 1842).

[16] Cf. Newcomb (1910a), p. 384; also his (1910b), pp. 154–5.

[17] For Lakatos's views on 'crucial experiments', see his (1970), especially pp. 154–177, and his (1974).

[18] Cf. his (1975), pp. 2–8.

[19] Lakatos (1970), p. 173. (Incidentally, an explanation *within* a programme is not the same as an explanation *consistent with* that programme – the latter is much weaker, and I propose to ignore it as a slip.) Elsewhere we have: "There is no predictable or ascertainable limitation on human imagination in inventing new, content-increasing theories [within a given programme] or on the 'cunning of reason'... in rewarding them with some empirical success..." (Lakatos, 1970, p. 158, footnote 2).

[20] Lakatos (1970), pp. 187–8.

[21] Lakatos does a disservice to logic, not to mention history, by suggesting that the positive heuristic of Newton's programme contained the idea that *electro-magnetic forces* might influence planetary motions. Could it be counted a success for Newtonian *gravitational* theory that *electro-magnetism* had to be invoked to explain some anomaly? Positive heuristics may be pretty flexible, but we must not make them *too* flexible or they will have no role to play at all. In this connection, Lakatos's talk about 'creative shifts' in positive heuristics represents, in my opinion, a degenerating shift in the methodology of research programmes. 'Creative shifts' are better regarded as changes to a new research programme altogether (though one may always, unhistorically, detect similarities between the old and the new). Actually, Lakatos was led by Noretta Koertge to admit that 'creative shifts' were an anomaly for him, since any creative shift is *ad hoc*$_3$ in his sense (see Lakatos, 1971b, pp. 176–7). He hoped to accommodate this meta-anomaly in a non-*ad hoc* fashion. I doubt that it can be accommodated, and think that 'creative shifts' should be given up. The point is not unimportant because whenever Lakatos talks about the actual or possible rehabilitation of a superseded programme, he usually admits that a creative shift is necessary to affect the rehabilitation (see, for example, Lakatos, 1970, pp. 137, footnote 1, or p. 164). One wonders what 'creative shifts' might rehabilitate phlogistonism, or Galenic physiology, or black magic?

[22] It was different with the Ptolemaic programme: there the theory of each planet was worked out *separately*, and there was no danger of adjustments to the theory of one planet upsetting that of another. (This given an instrumentalist view of Ptolemaic astronomy; a realist view of the epicycles, combined with a 'principle of plenitude', did introduce some constraints.) The more coherent or unified the heuristic of a programme, the less 'play' there is in it, and the sooner a 'natural saturation point' can be reached.

[23] The term is Lakatos's (see his (1963–4), Part IV, pp. 327–8). But in 1970, after emphasising (see Lakatos, 1970, p. 155, footnote 2) the desirability of pursuing a research programme until its natural saturation point is reached (until, I would say, its 'hard core' can be refuted), he says (see Lakatos, 1970, p. 158, footnote 2) that there is no such thing as a 'natural saturation point' and that he uses the expression ironically. I do believe in them, and I think they provide the objective basis for what Kuhn calls a 'crisis period' and what Lakatos calls 'the degenerating phase of a research programme'.

[24] Kuhn describes the emergence of the 'crisis state of a paradigm' as follows: "In the area for which it is advanced the paradigm is known to be in trouble. That trouble has

repeatedly been explored, and attempts to remove it have again and again proved vain. 'Crucial experiments' – those able to discriminate particularly sharply between the two paradigms – have been recognised and attested *before the new paradigm was even invented.*" (Kuhn, 1962, p. 153, my italics).

[25] It is pretty well established that Einstein, who devised the new programme, did not do so to deal with the Mercury anomaly. The problems which led Einstein to devise relativity theory were of a more abstract kind, 'theoretical anomalies' and not empirical ones (cf. Zahar, 1973, pp. 223–7).

[26] Lakatos (1971b), p. 174.

[27] For more on this, see my (1973), Section 5.

[28] All he says is: "The actual hard core of a programme does not actually emerge fully armed like Athene from the head of Zeus. It develops slowly, by a long, preliminary process of trial and error. In this paper this process is not discussed" (Lakatos, 1970, p. 133, footnote 4).

[29] I shall neglect the fact that one of his own examples, Prout's hypothesis, does not fulfil these conditions.

[30] We can therefore definite the 'hard core' of a research programme, not by reference to methodological decisions made at the outset, but rather as those hypotheses which, as a matter of fact, have persisted through several successful explanations. This means, of course, that we can only recognise a research programme (or locate its adherents) *with hindsight*.

[31] Lakatos (1970), pp. 135–7. Later he says: "in the positive heuristic of a powerful programme there is, right at the start, a general outline of how to build the protective belts: this heuristic power generates *the autonomy of theoretical science*" (Lakatos, 1970, p. 175).

[32] See Lakatos (1963–4), Part IV, Sections (c) and (d).

[33] Lakatos (1970), pp. 135–6. I shall not dispute the historical accuracy of this account, but merely the anti-falsificationist and anti-empiricist lessons Lakatos draws from it.

[34] Newton and his followers aided and abetted the misunderstanding. Kepler's law appears as a *theorem* of the *Principia* (Book III, Proposition XIII), so his readers might be forgiven for thinking that he was asserting that it held true of the solar system. Halley was the first to be misled, and claimed in his review of the *Principia* that in it "the verity of the Hypothesis of Kepler is demonstrated" (cf. Cohen, 1958, p. 410). Yet in the text to his Proposition XIII, Newton drew attention to its falsehood.

[35] These simplifying assumptions do not form a 'protective belt' in Lakatos's sense: for what do they protect from what? Such assumptions must be distinguished, however, from simplifying assumptions of a different kind, which involve neglecting some factor which, it is claimed, will have no appreciable effect on the predicted result. These assumptions *are* involved in the making of predictions, and we might call them 'empirical simplifying assumptions'. Confusion results because both kinds are presented in the same way, and because often a simplifying assumption begins as an empirical one, gets refuted, and lives on as a mathematical one.

[36] From a purely logical point of view, this needs qualification. The first Newtonian model allows us to predict that planets will actually move in some sort of *approximation* to ellipses. And there are possible states of affairs which would contradict even this very vague prediction: a planet reversing the direction of its motion, or moving in a square orbit. Had such a state of affairs, *per impossibile*, been observed, then Newtonians would have had to take it seriously right from the start. But of course, Newton knew in advance *from observation* that planets do not behave like this.

[37] Lakatos (1970), p. 136.

[38] Lakatos (1970), p. 137.
[39] He says that most, *if not all*, Newtonian puzzles were foreseeable by the heuristic, and that refutations of *any* specific variant of a programme can be produced and digested by a "clearly spelt out" heuristic (Lakatos, 1970, p. 136).
[40] Lakatos (1970), p. 136.
[41] A point emphasised also by Truesdell in his (1960) and his (1967). Incidentally, it is instructive that Lakatos uses the Kuhnian term 'puzzle' to refer to the mathematical problems which arise in articulating a theory like Newton's. For such problems *do* satisfy the main conditions for Kuhnian 'puzzles': Newton's theory *does* guarantee that the problem of computing, say, the Newtonian orbit of the moon has a solution; hence failure to solve this problem reflects, not upon Newton's theory, but on the mathematical competence of the 'practitioner' involved (it is only the failure of the calculated orbit to conform with the real one that can reflect upon Newton's theory). Of course, these 'puzzles' are not trivial or easy to solve (Kuhn stresses, in his (1970b), p. 246, that he never made ease of solution a defining feature of his 'puzzles'): on the contrary, they may be deep mathematical problems whose solution requires great originality.
[42] A point made already by Noretta Koertge in her (1971), p. 164.
[43] See the discussion of Prout's programme in Lakatos (1970), pp. 138–40.
[44] For further details of these, and other, hints, and of the role they played in the battle between the phlogiston and oxygen research programmes, see my (1976).
[45] Lakatos (1970), p. 146 (my italics).
[46] Using Mary Hesse's terminology (see Hesse, 1966, p. 8), there is always a *negative analogy* (the respects in which atoms are not like planetary systems) as well as a *positive analogy* (the respects in which they are); merely to draw an analogy does not tell us in *which* respects it holds. We may say that it is the (conjectured) positive analogy which provides Bohr with the hints contained in his positive heuristic. Hesse's central thesis is that properties "about which we do not yet know whether they are positive or negative analogies... are the interesting properties, because... they allow us to make new predictions" (Hesse, 1966, p. 8). I take this to mean that the conjecture that an analogy is positive in some specified respect provides a heuristic hint for the construction of a theory, which may then make a new prediction.
[47] Lakatos is admittedly guarded on this point (Bohr "*thought of* taking the *possible* spin of electrons into account"), and he tells us in a footnote that in fact Bohr accepted the idea only in 1926 (Lakatos, 1970, p. 146, footnote 2; see also Kuhn, 1971, p. 143).
[48] For details, see Clark (1976). Clark claims that the rival programme, thermodynamics, had a much *weaker* heuristic because it did not exploit any analogy.
[49] See Worrall (1976), Section 2(d). The whole sentence is italicized in the original.
[50] Besides, heuristics are many-splendoured things: the heuristic of Einstein's research programme, as reconstructed by Zahar (Zahar, 1973, pp. 223–7) bears little resemblance to any of the sorts of things we have discussed so far (see my (1974), p. 14, footnote 1).
[51] See Lakatos (1970), pp. 154–5.
[52] Lakatos (1970), p. 157, footnote 1.
[53] I have discussed this claim in my (1973), Section 5, and will say no more about it here.
[54] Lakatos (1970), p. 155.
[55] Lakatos (1971b), pp. 174 and 178.
[56] Lakatos (1971b), p. 174. When he wrote this Lakatos must have forgotten that he *also* said that journals should refuse to publish papers emanating from a degenerating research programme, and that foundations should refuse money to its practitioners. I shall forget this too – for the moment.

[57] Lakatos does not distinguish between honesty and rationality, or between dishonesty and irrationality (see his (1971a), p. 105). I agree with Richard Hall that one can be honest but irrational, and dishonest but rational (see Hall, 1971, p. 152).

[58] I sympathise with Philip Quinn when he says that "it is hard to see what point methodological appraisals of scientific theories and research programmes could possibly have if such appraisals were completely decoupled from heuristic advice, if they had no practical force". (Quinn, 1972, p. 143).

[59] Lakatos (1968), p. 343.

[60] Lakatos (1971a), p. 123, footnote 2. Lakatos uses the terms 'methodology' and 'logic of discovery' in idiosyncratic senses: they do not mean 'theory of method', but rather rules for the appraisal of finished products (see his (1971a), p. 92). He points out that the original hope of 'methodology' was to provide "a mechanical book of rules for solving problems", and that this hope has (rightly) been given up. But does this mean (as Lakatos suggests) that we must no longer hope to give any methodological advice (or in his terminology, heuristic advice) *at all*? To give advice like "Work on this problem (or this research programme) rather than that one" is not to dream of mechanical rules for solving problems.

[61] Lakatos (1970), p. 176, footnote 1, and p. 177 (the italics are mine). The 'intellectual pollution' Lakatos is referring to is contemporary social psychology, which, he claims, is a degenerating (because '*ad hoc*$_3$') research programme.

[62] Lakatos (1971a), p. 100. In the footnote to this passage Lakatos says that *during the protracted rivalry* between two research programmes "it is rational to work in either (*or, if one can, in both*)". Quite so. But the text to which this is appended only makes sense if the protracted rivalry can come to an end, so that one programme can be eliminated (no longer worked in). In case any doubt remains about Lakatos's original intention to give advice to scientists, recall that he also said that "rules governing the (scientific) *acceptance* and *rejection* of theories or research programmes... function as a *code of scientific honesty* whose violation is intolerable" (Lakatos, 1971a, p. 92).

[63] Lakatos (1971b), p. 174.

[64] Lakatos (1968), p. 372.

[65] Of course, Lakatos only abdicated the methodologist's throne to sit upon a new one, that of the historiographer of science. His growing reluctance to tell scientists what to do was matched by his increasing enthusiasm for telling historians of science what to do. I think his ideas are powerful enough for him to occupy both thrones. (Seriously, I shall not discuss in this paper Lakatos's subtle views on the interaction between the philosophy of science and the writing of its history, which form the main topic of his (1971a).)

[66] Feyerabend (1970), p. 229.

[67] Kuhn misses the backward-looking character of Lakatos's appraisals, and says that he provided no method of making such appraisals at all: "Lakatos has yet to... specify criteria which can be used at the time to distinguish a degenerative from a progressive research programme... Otherwise, he has told us nothing at all." (Kuhn, 1970, p. 239). Lakatos replies, quite rightly: "Actually, I *do* specify such criteria" (Lakatos, 1971a, p. 104). The question remains, however, whether the application of these criteria has any repercussions at all for scientific practice.

[68] Feyerabend (1970), p. 215.

[69] The usual view is that Priestley was too old, too dogmatic, blinded by prejudice, a view which is belied by the perfectly rational, balanced and good-humoured way in which he continued to defend phlogistonism. I cannot understand Kuhn's attitude to "elderly hold-outs" like Priestley: though they are "unreasonable to resist for as long as they did", there is no "point at which resistance becomes illogical"; though "the man who continues

to resist after his whole profession has been converted has *ipso facto* ceased to be a scientist", there is no "point at which resistance becomes... unscientific" either (Kuhn, 1962, p. 159).

[70] Lakatos (1970), p. 155, footnote 2. See also Lakatos's insistence, against Richard Hall, that "one of my crucial – and repeatedly made – points is exactly the negation of such a *universal* rule" as 'scientists should drop a research programme if things go badly enough for it" (Lakatos, 1971b, p. 180). In a similar vein, Lakatos argues that a budding research programme should be pursued, and given a chance to develop, despite the fact that if it is appraised with respect to an established rival it comes out badly (Lakatos, 1970, pp. 156–7).

[71] See the 'Postscript-1969' to the second edition of Kuhn's (1962), p. 186.

[72] Kuhn (1970), p. 241. These recent statements of Kuhn's may seem to conflict with his early account of how 'anomalies' are dealt with in periods of 'normal science'. For more on this exegetical puzzle, see my (1971).

[73] Grünbaum (1975), p. 89. Lakatos, commenting on Grünbaum's view that it is wise for the *scientific community* not to "put all its eggs into one basket" but rather to "distribute the risk", reveals that he has not really grasped the distinction between community-directed advice and advice to individuals. Lakatos takes Grunbaum to be giving the latter:

> "... *nobody has yet offered a theory of rationality for the individual scientist telling him which, among competing research programmes, he should choose to work on, or whether and when he should try to start one on his own.* All that Grunbaum says... is that it 'would be unwise to put all one's research eggs into one basket'; but... this trivial statement does not help the individual scientist in his decision *in the least*. For the individual, it is usually *wise* to put all one's research eggs into *one* basket, in the sense that it is unwise to give up a programme too easily; to acquire competence in the techniques of a serious research programme takes many of one's best years." (Lakatos (1975), p. 10; the last sentence quoted occurs in the original as a footnote.) This merely repeats Kuhn: "There can be no set of rules of choice adequate to dictate desired *individual* behaviour in the concrete cases that scientists will meet in the course of their careers" (Kuhn, 1970b, p. 238).

Incidentally, while "Do not put all your research eggs into one basket" may be pretty trivial if construed as a piece of advice to an individual, it is not trivial construed as a piece of community advice: thus construed, it is the heuristic equivalent of the methodological principle of *proliferation*, whose importance Lakatos has often stressed.

As Lakatos says, no methodologist has yet offered "a theory of rationality for the individual scientist". And when you think about it, there is something rather odd about a a *general* methodological position issuing in advice about what *particular* scientists should do. For what it is rational for an individual scientist to do will depend upon a vast number of idiosyncratic factors: his training, his ambition, his (and other's) estimates of his ability, his colleagues, his equipment, the availability of funds, etc. etc. All that a general methodological position like Lakatos's can say to an individual is: "Whatever you do, be honest, and try to live up to my standards of good science".

[74] I do not think Lakatos should quarrel with 'community-directed advice' such as this, for he gives some of it himself. He wants to make life quite difficult for die-hard adherents of degenerating programmes: "Editors of scientific journals should refuse to publish their papers which will, in general, contain [I would prefer: if, as is likely, they contain] either solemn reassertions of their position or absorption of counterevidence (or even of rival programmes) by *ad hoc*, linguistic adjustments. Research foundations, too, should refuse money [I would prefer to add: for projects of this kind]" (Lakatos, 1971a,

p. 105). These directives are addressed to individuals, but to those individuals who man the *social institutions* of science (journals, research foundations, etc.). Like talk of 'keeping public records', it is community-directed advice (a point made already by Kuhn in his (1971), p. 145).

[75] Feigl (1971), p. 146. Professors Grünbaum and Hempel, who also have keen noses for inductive arguments (and seem to *like* the smell), raised a similar objection in discussions at the Nafplion Colloquium.

[76] Details of these various problems, and of some others, can be found in my (1976).

[77] I hope I need hardly add that if someone does not take my advice, and tries instead to rescue phlogistonism from the doldrums, I shall merely wish him well (while doubting that he will succeed). Priestley was such a person. Yet it is not an accident that after 1785 Priestley produced nothing that was either theoretically or experimentally new (and admitted as much), Nor need we invoke his premature senility to explain this: the fact was that he adhered to the wrong theory, the degenerating research programme.

[78] An idea which enables us to give an epistemological rationale to Kuhn's view that the issue in 'paradigm debates' is:

> "... which paradigm should in the future guide research on problems many of which neither competitor can yet claim to resolve completely. A decision between alternate ways of practicing science is called for, and in the circumstances that decision must be based less on past achievement than on future promise. The man who embraces a new paradigm at an early stage must often do so in defiance of the evidence provided by problem-solving. He must, that is, have faith that the new paradigm will succeed with the many large problems that confront it, knowing only that the older paradigm has failed with a few. A decision of that kind can only be made on faith." (Kuhn, 1962, pp. 157–8).

BIBLIOGRAPHY

Buck, R. and Cohen, R. S. (eds.), 1971, *Boston Studies in the Philosophy of Science*, Vol. 8, D. Reidel Publishing Company, Dordrecht, Holland.

Clark, P., 1976, 'Atomism versus Thermodynamics'. In Howson (ed.), (1976).

Clerke, A. M., 1885, *A Popular History of Astronomy during the Nineteenth Century*, Adam and Charles Black, Edinburgh.

Cohen, I. B. (ed.), 1958, *Isaac Newton's Papers and Letters on Natural Philosophy*, Cambridge University Press, London.

Feigl, H., 1971, 'Research Programmes and Induction'. In Buck and Cohen (eds.) (1971), pp. 147–150.

Feyerabend, P. K., 1970, 'Consolations for the Specialist'. In Lakatos and Musgrave (eds.) (1970), pp. 197–230.

Fontenrose, R., 1973, 'In Search of Vulcan', *Journal for the History of Astronomy* **4**, 145–158.

Grant, R., 1852, *History of Physical Astronomy*, Henry G. Bone, London.

Grosser, M., 1962, *The Discovery of Neptune*, Harvard University Press, Cambridge, Massachusetts.

Grünbaum, A., 1975, 'Falsifiability and Rationality', mimeographed.

Hall, R. J., 1971, 'Can We Use the History of Science to Decide between Competing Methodologies?'. In Buck and Cohen (eds.) (1971), pp. 151–159.

Hanson, N. R., 1962, 'Leverrier: The Zenith and Nadir of Newtonian Mechanics', *Isis* **53**, 359–378.

Hesse, M., 1966, *Models and Analogies in Science*, University of Notre Dame Press, Notre Dame, Indiana.

Howson, C., 1976, *Method and Appraisal in the Physical Sciences: The Critical Background to Modern Physical Science* (Proceedings of the Nafplion Colloquium on Research Programmes in Physics and Economics, Vol. 1), Cambridge University Press, London.

Koertge, N., 1971, 'Inter-Theoretic Criticism and the Growth of Science'. In Buck and Cohen (eds.) (1971), pp. 160–173.

Kuhn, T. S., 1962, *The Structure of Scientific Revolutions* (International Encyclopaedia of Unified Science, Volume 2, number 2), Chicago University Press, Chicago (Second Edition, enlarged, 1970).

Kuhn, T. S., 1970a, 'Logic of Discovery of Psychology of Research'. In Lakatos and Musgrave (eds.) (1970), pp. 1–21.

Kuhn, T. S., 1970b, 'Reflections on my Critics'. In Lakatos and Musgrave (eds.) (1970), pp. 231–278.

Kuhn, T. S., 1971: 'Notes on Lakatos'. In Buck and Cohen (eds.) (1971), pp. 137–146.

Lakatos, I., 1963–4, 'Proofs and Refutations Parts I–IV', *British Journal for the Philosophy of Science* **14**, 1–25, 120–139, 221–245, 296–342.

Lakatos, I., 1968, 'Changes in the Problem of Inductive Logic'. In Lakatos, I. (ed.) (1968), pp. 315–417.

Lakatos, I., 1970, 'Falsification and the Methodology of Scientific Research Programmes'. In Lakatos and Musgrave (eds.) (1970), pp. 91–195.

Lakatos, I., 1971a, 'History of Science and Its Rational Reconstructions'. In Buck and Cohen (eds.) (1971), pp. 91–136.

Lakatos, I., 1971b, 'Replies to Critics'. In Buck and Cohen (eds.) (1971, pp. 174–182.

Lakatos, I., 1974, 'The Role of Crucial Experiments in Science', *Studies in History and Philosophy of Science* **4**, 309–325.

Lakatos, I., 1975, 'Anomalies versus "crucial experiments". (A rejoinder to Professor Grünbaum)'. Mimeographed.

Lakatos, I., (ed.), 1968, *The Problem of Inductive Logic* (Proceedings of the International Colloquium in the Philosophy of Science, London, 1965, Vol. 2), North-Holland Publishing Company, Amsterdam.

Lakatos, I. and Musgrave, A. (eds.), 1970, *Criticism and the Growth of Knowledge* (Proceedings of the International Colloquium in the Philosophy of Science, London, 1965, Vol. 4), Cambridge University Press, London.

Latsis, S. J., 1972, 'Situational Determinism in Economics', *British Journal for the Philosophy of Science* **23**, 207–245.

Musgrave, A. E., 1971, 'Kuhn's Second Thoughts', *British Journal for the Philosophy of Science* **22**, 287–297.

Musgrave, A. E., 1973, 'Falsification and Its Critics'. In Suppes, Henkin, Moisil and Joja (eds.) (1973), pp. 393–406.

Musgrave, A. E., 1974, 'Logical versus Historical Theories of Confirmation', *British Journal for the Philosophy of Science* **25**, 1–23.

Musgrave, A. E., 1976, 'Why did Oxygen supplant Phlogiston? (Research Programmes in the Chemical Revolution)'. In Howson (ed.) (1976).

Newcomb, S., 1910a, 'Gravitation'. In *Encyclopaedia Brittanica*, eleventh edition, Vol. XII, pp. 384–5, Cambridge University Press, London.

Newcomb, S., 1910b, 'Mercury'. In *Encyclopaedia Brittanica*, eleventh edition, Vol. XVIII. pp. 154–5, Cambridge University Press, London.

Nichol, J. P., 1848, *The Planet Neptune: an Exposition and History*, John Johnstone, Edinburgh.

North, J. D., 1965, *The Measure of the Universe: A History of Modern Cosmology*, Clarendon Press, Oxford.

Popper, K. R., 1967, 'La Rationalité et le Statut du Principe de Rationalité'. In *Les Fondements Philosophiques des Systèmes Economiques*, Bibliothèque Economique et Politique, Payot, Paris.

Quinn, P., 1972, 'Methodological Appraisal and Heuristic Advice: Problems in the Methodology of Scientific Research Programmes', *Studies in History and Philosophy of Science* **3**, 135–149.

Suppes, P., Henkin, L., Moisil, Gr. C., and Joja, A. (eds.) 1973, *Logic, Methodology and Philosophy of Science IV* (Proceedings of the Fourth International Congress for Logic, Methodology and Philosophy of Science, Bucharest, 1971), North-Holland Publishing Company, Amsterdam.

Thackray, A., 1970, *Atoms and Powers: An Essay on Newtonian Matter-Theory and the Development of Chemistry*, Harvard University Press, Cambridge, Massachusetts.

Truesdell, C., 1960, 'The Program toward Rediscovering the Rational Mechanics in the Age of Reason', *Archive of the History of the Exact Sciences* **1**, 3–36.

Truesdell, C., 1967, 'Reactions of Late Baroque Mechanics to Success, Conjecture, Error, and Failure in Newton's *Principia*', *Texas Quarterly* **10**, 238–258.

Worrall, J., 1976, 'Thomas Young and the "Refutation" of Newtonian Optics: A Case Study in How Philosophy of Science can solve Historical Problems'. In Howson (ed.) (1976).

Zahar, E. G., 1973, 'Why did Einstein's Programme supersede Lorentz's?', *British Journal for the Philosophy of Science* **24**, 95–123 and 223–262.

H. R. POST

NOVEL PREDICTIONS AS A CRITERION OF MERIT

Any contribution to a volume in honour of Imre Lakatos should, I suppose, be polemical. I have selected the question of novel predictions as a criterion of merit of theories because it brings out a difference of views which is, in my opinion, more important than the shift from 'theory' to 'research program'. The difference I am referring to is between what might be called 'objectivist' philosophy of science and what might be called the 'sociological' or 'externalist' approach. The former view is 'internalist' in that its criteria of evaluation are internal (i.e., refer to the theory itself), except that it also uses the criterion of 'agreement with experience'. The latter view introduces psychological, sociological or historical criteria (possibly in conjunction with criteria acceptable to an objectivist).

The issue may be put in the language of Lakatos's school in the following way: "What constitutes degeneracy in a research program?" Imre's position was not always clear-cut, but I detected a drift towards what I call the sociological side on the occasions on which he visited Chelsea. Is the mere *historical fact* of lack of progress (the failure to produce new results) in a research program over a certain period tantamount to degeneracy? Apart from the question of technological constraints, an affirmative answer to this question would make degeneracy a sociological phenomenon; the mere fact that no, or few, or only relatively uninspired or incompetent scientists actively pursued a certain program during a certain period would automatically render it degenerate. Of course, this sociological phenomenon may be rationally grounded. It may be, but it *need* not be.

In my view, philosophy of science must be able to judge scientific activity critically. If we judge a program by its sociological success we abrogate our position and render the judgement circular: that theory is to be accepted, that is in fact accepted. This 'posivistic' attitude is not made more acceptable by hedging references to an 'elite' of scientists. Such reference is irrelevant even when we are considering the validity of a *criterion* as such. Defining the elite as the most influential scientists

R. S. Cohen et al. (eds.), Essays in Memory of Imre Lakatos, 493–495. *All Rights Reserved*
Copyright © 1976 by D. Reidel Publishing Company, Dordrecht-Holland

again renders the judgement circular. On the other hand, references to objective though sociological criteria such as volume of publications, or possession of a Nobel Prize, does now allow an independent objective basis for judgement either. It is only on the basis of defining criteria internal to science, independent of sociology, that we may hope to proceed to the central problem of philosophy of science: "Why does science succeed (in progressively making more correct predictions)?"

My naive objectivist approach is in accord with one way of reading Popper, viz as a search for *internal* criteria to define and evaluate science. Popper's reference to "design"[1] in his second requirement (of novelty) need not be interpreted psychologically. Nor need we insist, as Popper does, on "the prediction of phenomena which have not so far been observed". We are not concerned with chronology or even the epistemic position of the author of the theory. The much-abused "ad-hoc" stratagem (depending on the prior knowledge of certain data by the author of the theory) cannot be excluded by any formal criterion. But this type of psychological "ad-hoc-ness" is quite irrelevant to objective appraisal and not necessarily bad when used heuristically. In comparing the merit of two theories we should use internal criteria: we must consider the merit of increased coverage of data, *and*, for a given coverage, the merit of having a logically weaker set of hypotheses.

Some part of Popper's theory of theories may form a basis for objective appraisal. Imre rightly considered strict falsificationism unrealistic and gave a more realistic account of actual history of science in terms of research programs. The *historiography* of science is indeed improved by considering research programs rather than single theories. But research programs, for better or worse, are more open to 'reconstruction' than theories. I do not suppose that the author of a theory would necessarily agree with Imre's reconstruction of the 'core' of 'his program'; but even an author's say-so outside the theory itself would be a dubious as well as a subjective psychological standard. In common with all good history, history of science should not exclude even psychological factors relating to individual scientists, or groups of scientists. However, such considerations must not form part of the objective appraisal, which is a function of *philosophy* of science. Of course, history of science may well illuminate philosophy of science and vice-versa; but we must be clear in any one instance which of the two approaches we are using.

It is, I think, no accident that Imre moved more and more towards the sociological side. He was brilliant in understanding the motives and ideological adherence of people and in labelling schools. His introduction of the notion of research program was a descriptive contribution of real value. But the task of descriptive historiography must be sharply distinguished from the normative task of philosophy of science.[2]

Chelsea College

NOTES

[1] 'Conjectures and Refutations', p. 241.

[2] Imre certainly intended his methodology of research programs to be normative as well as descriptive. But Imre's[3] mere declaration of faith in Popper's 'Third World' does not render his methodology objective. Zahar's[4] 're-definition' of novelty reiterates a reference to 'design' in a context that makes his criterion historical and not internal.

[3] *Methodology of Scientific Research Programmes*, 'Criticism and the Growth of Knowledge', Cambridge University Press, p. 179.

[4] *B.J.P.S.* **24** (1973), 103.

WILLARD QUINE

WHITHER PHYSICAL OBJECTS?

What counts as a physical object? We think first of bodies, but the notion of a body is both too vague and too narrow. It is too vague in that we are not told how separate and cohesive and well rounded a thing has to be in order to qualify as a body. And it is too narrow, since for ontological purposes any consideration of separateness and cohesiveness and well-roundedness is beside the point. Rather let us understand a physical object, for a while, simply as the aggregate material content of any portion of space-time, however ragged and discontinuous.

This liberal notion of physical object spares us the pointless task of demarcating bodies. Also it brings further benefits. It neatly accommodates mass terms such as 'sugar' and 'air' and 'water'. Such a term cannot be said to name a body, but it can quite well be construed as naming a physical object. We can identify sugar with a single large and spatio-temporally scattered physical object, consisting of all the sugar anywhere, ever. Correspondingly for air and water.

With a little stretch of the imagination, this notion of physical object can even be made to accommodate physical processes or events, on a par with bodies: simply as more or less scattered spatio-temporal manifolds. A ball game, for instance, might be identified with the scattered sum of the appropriate temporal segments of the players, taking each player for just the duration of his play.

Such an account of events has the possible drawback that it does not distinguish between different events if they happen to take up just the same portion of space-time. If a man whistled some song all the while he was walking to the bus stop, and not a moment longer, then presumably the event of his whistling the melody and the event of his walking to the bus would both be identified with the same temporal segment of the man. This outcome will be unwelcome if one feels that the event of the whistling and the event of the walking should be distinguished. However, it is not clear to me that they need to be. It is perhaps unnatural to identify them, but it is not clear to me that we lose anything. We still

R. S. Cohen et al. (eds.), Essays in Memory of Imre Lakatos, 497–504. *All Rights Reserved*
Copyright © 1976 by D. Reidel Publishing Company, Dordrecht–Holland

have the *general* distinction between whistling the song and walking to the bus, because sometimes people do walk to the bus without whistling the song, and even vice versa. Thus we are still conveying information about the man when we say he whistled the song all the way to the bus. We do not trivialize this statement by identifying the events.

The notion of physical object that I have proposed raises, however, certain questions. I described a physical object as the material content of any portion of space-time. What then is the ontological status of these portions of space-time themselves? They would be an unappealing sort of object, surely; too intangible, too abstract, too much a figment of the mind. We would like to exclude them from our ontology, and at this stage surely we can. When I described a physical object as the material content of any spatiotemporal region, this was just my way of showing how broad a scope I intend for my notion of physical object. We may take the material content and let the regions go.

Another problem arises, however, over the very notion of material content. Even in that quarter nowadays all is not well. The notion of matter wavers when we get down to the level of electrons. It does not always make sense to identify an electron over time. That is, it is sometimes wholly arbitrary to say whether two point events *a* and *b* are moments in the career of one continuing electron or moments rather in the careers of two different electrons. The point event *a* is followed by *b*, and meanwhile another nearby point event *c* is followed by *d*; and it can be a wholly arbitrary matter whether to assign *a* and *b* to one electron and *c* and *d* to another, or to assign them crosswise: *a* and *d* to one electron, *c* and *b* to another. What are objective are, at best, the point events. Their integration into world lines, or enduring particles, is only a matter of convenience, and arbitrary in varying degrees.

The naive conception of matter thus tends to dissolve in modern physics. In fact its dissolution has gone yet farther, to judge from an account by Heinz Post.[1] Not only can it be arbitrary to identify an electron over time; there seems even to be something not quite right about distinguishing between coexisting electrons in space. Thus consider two boxes, which I shall call the east box and the west box, and two electrons, *x* and *y*. Common sense recognizes four ways in which *x* and *y* could occupy the boxes: they could both be in the east box, or both be in the west box, or *x* could be in the east box and *y* in the west, or vice versa.

But statistical findings show, according to Post, that these last two apparent possibilities have to be counted as just one: the fact of x being in the east and y in the west must be identified with the seemingly opposite fact of x being in the west and y in the east. The moral is that we should think not of individual electrons x and y at all, but of states of the boxes: states of being singly or doubly occupied, or, to put it more innocently still, states of being singly or doubly affected. The east box may be doubly affected, or the west box may be doubly affected, or both boxes may be singly affected, and there the matter ends. There is no question of which electron, hence no question of electrons; only a question of singly or doubly affected boxes.

All this is said to apply to other elementary particles as well as electrons. Matter evidently goes by the board. We are left rather with a field theory, a theory of the distribution of states over space-time.

Physicists regale us with yet further testimony ill fitting the notion of a particle. There is talk of cyclic resolution of particles, as if particles could be components of components of themselves. Furthermore there is the well-established convertibility of matter and energy. Manifestly the physicists' age-old attachment to matter has relaxed. Matter is quitting the field, and field theory is the order of the day.

What then of our physical objects? They were to have been the material contents of portions of space-time. Must we now take as our objects simply those portions of space-time themselves, subject to whatever distributions of local states they happen to be subject to? A little while ago it seemed that space-time regions were pretty flimsy affairs, and that our ontology would be better without them. But now our physical objects have themselves gone so tenuous that we find ourselves turning to the space-time regions for something to cleave to.

There are some curious considerations to sort out at this point. Reification of space-time regions is one thing; reifying spatial regions would have been quite another thing, and quite unacceptable. There can certainly be no such things as regions of space. If there were, we could make sense of absolute rest and absolute motion. A body would be absolutely at rest as long as it continued to fill the same region. Since bodies are really at rest or in motion only relative to one another, and not absolutely, the notion of a real spatial region must be untenable.

Space-time regions, on the other hand, in four dimensions, present no such difficulty. The career of some body can be identified with a fixed portion of space-time, specifiable in terms of any of various systems of space-time coordinates. Some choices of coordinates will represent the body as having moved, others not.

It is interesting to reflect that the four-dimensional view is thus forced on us already by the classical or Leibnizian relativity of motion, not just by Einstein's relativity. Einstein's theory of course demands it doubly.

Here, then, is one way to ontologize physicalism. The objects are the space-time regions themselves. It is over these regions that the various physical states are distributed. And now what about these states: do they have to be reckoned to our ontology as objects of a further abstract sort?

At this point we must distinguish two possibilities. One possibility is that there are just a limited number of states to reckon with. In this event our language will include a simple predicate representing each state, and there will be no need to recognize the states themselves as objects. The mere use of a word, after all, carries no commitment to a corresponding object. It is only when we have to quantify over a multitude of objects, such as physical objects or space-time regions, that we are committed to the objects themselves as values of our variables.

Perhaps, along with the limited number of simple states of space-time regions, we shall need to reckon also with a limited number of relations of regions. This will not alter matters. Each relation can be accommodated linguistically by a polyadic predicate, innocent still of any presumption of a corresponding object.

But another possibility is that there are infinite ranges of states to reckon with; namely, variations in the intensity of various measurable properties. In this event we shall still not need to extend our ontology to include such things as states or properties or intensities, but we shall need to include the *numbers* that measure the intensities.

The admission of numbers and other abstract mathematical objects is an eventuality that has to be faced, melancholy though it be. There is no clear way to make natural science work without mathematics, nor to make mathematics work without its objects.

Set theory is a familiar way of integrating the whole universe of mathematical objects – numbers, functions, and the rest. Now once we have reluctantly admitted all the ontology of set theory, we may get some consolation from a curious bonus that comes through: we can thereupon dispense with the other part of our ontology, the space-time regions. For, now that we have the full mathematical apparatus, we can invoke Cartesian coordinates and identify each space-time point with a mere quadruple of real numbers. Predicates that formerly attributed states to points or regions will now apply rather to quadruples of numbers, or to sets of quadruples.

Thus consider, for a homely example, the ascription of temperature. What admits of a temperature is not a point, strictly speaking, but a small spatiotemporal region. A dyadic predicate is called for, predicable of a number and a set of quadruples of numbers: thus '$Fx\alpha$', meaning that the mean temperature in degrees Kelvin of the region whose coordinates comprise the set α is x. The notion of temperature and its scale of measurement are compacted in the monolithic predicate 'F'. The business of objective reference devolves now upon the mathematical variables 'x' and 'α', whose values are pure numbers and sets of quadruples of numbers. Such now are the denizens of our universe.

Such a system seems arbitrary in two respects: in its units of measure and in its choice of a fixed frame of reference. However, the arbitrariness of units can presumably be remedied by settling on cosmically significant ones: units that make for the simplest theory. As for the fixed frame of reference, it would simply not obtrude where it is unwelcome, namely in the most theoretical levels of physics. Laws at that level would quantify generally over quadruples of real numbers, picking out none specifically. The specificity of the coordinates would make itself known only when one descends to coarser matters of astronomy, geography, geology, and history, and here it is perhaps appropriate.

There remains the question of *ground elements*. Take the members of my sets; then take the members of those members, if such there be, and so on down, until you get to rock bottom: to non-sets, to individuals in some sense. These are the ground-elements; and what are they to be? Not physical objects; they gave way to space-time regions. But space-time regions gave way in turn to sets of quadruples of numbers; so nothing offers. However, this is all right. Since Fraenkel and von Neumann, a set theory without ground elements has even been pretty much in vogue. There is the empty set, there is the unit set of the empty set, there is the set of these two sets, and so on. We get infinitely many finite sets in this way. Then we take all the finite and infinite sets having these as members.

Continuing thus, we suffer no shortages. This is known as pure set theory, and I seem to have ended up with this as my ontology: pure sets.

Our physical objects have evaporated into mere sets of numerical coordinates. This was an outcome, we saw, of physics itself. But it was always a live option, even apart from any theoretical difficulty over physical objects. When the physical objects were still there, their coordinates uniquely determined them and so could be used in systematically naming them. Given this much, the ontological shift is an easy further step. Formally viewed, it is just a verbal regrouping of a complex term. Instead of analyzing the term 'temperature in degrees centigrade of the object whose coordinates are α' into a long but logically simple functor 'temperature in degrees centigrade of' and a complex term 'the object whose coordinates are α', we analyze it into a still longer but logically simple functor 'temperature in degrees centigrade of the object whose coordinates are' and a simple term 'α'.

Carnap was propounding such a *Koordinatensprache* already in 1934,[2] and not because of constraints on the notion of physical object from the side of physics; for the scheme has also a certain intrinsic appeal. Numbers and other mathematical objects are wanted in physics anyway, so one may as well enjoy their convenience as coordinates for physical objects; and then, having come thus far, one can economize a little by dispensing with the physical objects.

However, this attraction has usually given way before our robust sense of the reality of physical objects. As physicalists we have welcomed bodies with open arms. Ungrudgingly we opened the way to other physical objects too, however ill shaped and loosely knit, for at worst they were kinfolk. On the other hand the mathematical objects attained the ontological scene only begrudgedly for services rendered. A way of dispensing with them and making do with a strictly physical ontology would have been exceedingly welcome, whereas the opposite reduction did not appeal to us.

It is ironical, then, that we at length find ourselves constrained to this anti-physical sort of reduction from the side of physics itself. It is this I have wanted to bring out. Bodies were best, but they needed to be generalized to physical objects for reasons that rested on physical concerns: we wanted to provide designata for mass terms, and we wanted to accommodate physical processes or events. Physical objects, next, evaporated into space-time regions; but this was the outcome of physics itself. Finally the regions went over into pure sets; still, the set theory itself was there for no other reason than the need for mathematics as an adjunct to physical theory. The bias is physical first and last, despite the airiness of the ontology.

It is an ontology of abstract entities, but not of mental ones. If our sets are ideas, they are ideas in a Platonic rather than a mentalistic sense. Our ontological débacle, if débacle it be, is a triumph not of subjective idealism but of hyper-Pythagoreanism. Sets are numbers and other things of rather the same sort as numbers.

We must note further that this triumph of hyper-Pythagoreanism has to do with the values of the variables of quantification, and not with what we say about them. It has to do with ontology and not with ideology. The things that a theory deems there to be are the values of the theory's variables, and it is these that have been resolving themselves into numbers and kindred objects – ultimately into pure sets. The ontology of our system of the world reduces thus to the ontology of set theory, but our system of the world does not reduce to set theory; for our lexicon of predicates and functors still stands stubbornly apart.

The polysyllabic functor last cited is characteristic: 'temperature in degrees centigrade of the region whose coordinates are'. Let me abbreviate it as 'f'. It relates pure sets; we have '$f\alpha = x$' where α is a set of quadruples of numbers and x is a number. But there is no hope nor fear nor thought of translating this 'f' itself into the notation purely of set theory: into the 'ε' of membership and the symbols of logic. The relation of the coordinates of a place to the centigrade temperature of the place is indeed a relation of numbers, a relation of sets, but it is presumably not an arithmetical relation, not a set-theoretic relation.

By an extraordinary coincidence and a stretch of the imagination it could be, indeed: there may be some complex formula in the notation of pure arithmetic or set theory that just happens to pair the coordinates of all places correctly with their temperatures, throughout space-time. But we have no inkling of it – no inkling of a mathematical formula fixing the temperatures everywhere through all eternity. Our physical ideology remains unreduced to pure mathematics, for all the ontological debacle. We might most naturally react to this state of affairs by attaching less importance to mere ontological considerations than we used to do.

We might come to look to pure mathematics as the locus of ontology as a matter of course, and consider rather that the lexicon of natural science, not the ontology, is where the metaphysical action is.[3]

Harvard University

NOTES

[1] 'Individuality and Physics', *The Listener*, October 10, 1963. It is the Einstein-Bose effect.

[2] *Logische Syntax der Sprache*, pp. 11, 40.

[3] Acknowledgment is due Burton Dreben for stimulating the last portions of this paper. And a concluding word seems appropriate, still, on substitutional quantification; but only at footnote level, for readers conversant with pp. 104–114 of my *Roots of Reference* (Open Court, 1974). The conflict there noted, between objectual quantification over individuals and substitutional quantification over sets, subsides now that we have only the sets. Moreover the objection to substitutional quantification on the score of indenumerability seems spurious, as there explained. And surely substitutional quantification as applied to sets is more congenial than objectual quantification to the nominalist temper, even though not innocent of ontic commitment. The only remaining cause for hesitation over the substitutional version is impredicativity.

GERARD RADNITZKY

POPPERIAN PHILOSOPHY OF SCIENCE AS AN ANTIDOTE AGAINST RELATIVISM*

0. INTRODUCTION

0.1. Thanks largely to the pioneering work of Frege and Husserl, psychologism in *logic and meta-mathematics* is largely a thing of the past: the attempt to reduce the norms of logic to laws of thought is now merely a historical curiosity. In the *philosophy of the humanities and historiography*, "psychologism" in the form of historical relativism, has been subjected to a decisive critique by Heidegger, Gadamer and others (although, in our opinion, a residual of historical relativism has remained). It has been shown that understanding a text is a matter of understanding the "things themselves", and that the reconstruction of the author's thought is an entirely different matter. Moreover, a clear distinction is made between what a text means *(Textsinn)*, and what the author meant or thought *(Autorensinn)*. In a rational discussion we argue *with* a dialogue partner, we do not argue *about* him.

In *contemporary philosophy of science*, on the other hand, "psychologism" or "sociologism" has retained or more correctly *regained* a dominant position. The use of the history of science to challenge the philosophy of science has recently received a major impetus through the work of Polanyi and, above all, Kuhn, Feyerabend and Toulmin – to name only a few major figures. Even Imre Lakatos himself eventually arrived at a position according to which a methodology can be "falsified" (much like a scientific theory) by testing the rational reconstruction of the history of science to which it leads (historia rerum gestarum), by its "fit" to history (in the sense of res gestae) so to speak. And there is a whole style of thought, often with roots in Marxism, which propounds the view that the development of scientific theories itself is decisively influenced and even shaped by socio-political factors, such that methodological issues cannot be dealt with adequately without focussing on 'external' factors – thus rejecting the very distinction between internal and external. This is an "anti-objectivist" style of thought, which Imre Lakatos hated and which

R. S. Cohen et al. (eds.), Essays in Memory of Imre Lakatos, 505–546. *All Rights Reserved*
Copyright © 1976 by D. Reidel Publishing Company, Dordrecht–Holland

he wished to combat by means of an objective criterion of demarcation or of scientific quality.

It is against this background that Popperian methodology can be seen to be more timely than ever: as an antidote to the contemporary tendency to take a healthy reaction against the excesses of a purely "logicistic" methodology to the unhealthy extreme of not only overstressing the importance of non-logical aspects of theory-appraisal, but of even totalizing these aspects into the very principles of a methodology of science. Most of the criticisms of Popper's position rely heavily on the history of science and descriptions of actual research processes as a critical tool, and the popularity of this form of criticism, which, in my opinion often shows a striking resemblance to what G. E. Moore called the "naturalistic fallacy", can be explained only on the assumption that the character of Popper's methodology has been misunderstood. It is in these terms that *Popperian methodology and philosophy remain an urgent topic of discussion* more than forty years after the appearance of his *Logik der Forschung*.

0.2. In his early work Popper avoids both the concept of truth and the problem of ontology.[1] He learned of Tarski's semantic definition of truth from Tarski himself in 1935, and he lectured on it at Bedford College, London, the same year. It plays a prominent role in *The Open Society*. In 1963 he published his theory of verisimilitude, and in the late 1960's his ontology. Is this development to be understood as a reversal of his earlier position? The *thesis* to be developed here is that there is *no reversal* in Popper's thought, but rather *a systematic enlargement of the arsenal of weapons available in the fight against intellectual relativism*[2], which was and remains Popper's overarching concern. This development is to be understood as *a deepening of Popper's problems*; it does not constitute a degenerating research program, as Feyerabend would (like to) have it in his review-article of *Objective Knowledge*[3].

In Part 1 of this paper I wish *to criticize the tendency to carry the criticism of "logicism" far beyond its legitimate limits* and *to demonstrate the unity of Popper's philosophy*. In Part 2 I wish to give a rough *sketch of "Popperian" methodology in order to substantiate the thesis that it is pluralistic rather than a totalization of logical moves*. Should this be regarded as 'neo-Popperian' rather than orthodox – while Imre Lakatos *MSRP* would be "revisionist" – I do not mind, nor would Imre – who loved to play with such lables.

1. THE DISTINCTION BETWEEN METHODOLOGY AND THE DESCRIPTION OF ACTUAL RESEARCH; BETWEEN EXPLICATION AND APPLICATION, AND BETWEEN A DEFINITION OF TRUTH AND A CRITERION OF TRUTH; BETWEEN THE JUSTIFICATION OF A HYPOTHESIS AND THE JUSTIFICATION OF THE ACTION OF ACCEPTING A HYPOTHESIS

1.0. Popper's *Logic of Scientific Discovery* offers a methodology, that is, norms and rules as to how research *ought*[4] to be conducted in order to reach a *presupposed* goal. Methodology in this sense must be clearly distinguished from a *description* of actual research processes – past or on-going – and the methods actually used in such processes. Attempts to answer the questions concerning the appraisal of the results of research and the *justification* of such appraisals *(Geltungsfragen)*, by *describing* producers and production processes or by providing explanations of their behavior in terms of psychology or sociology are analogous to the attempt to justify or derive ethical sentences in terms of non-ethical, descriptive sentences: the famous "naturalistic fallacy". Popper diagnoses this fallacy in the philosophy of science as early as 1934.[5] This fallacy is related to psychologism/relativism in the sense that psychologism, understanding knowledge as the justified true belief, not only – to use Popper's terminology – confuses knowledge in the objective sense with knowledge in the subjective sense, i.e. reduces meanings to mental entities, but in so doing reduces questions of validity to questions about real producers and production processes, and thereby relativizes them: evaluation is propped up by description.

The presupposed *goal* of scientific research is stated by Popper variously as "theories with ever greater explanatory power"[6], "discovery of truth"[7], "Verisimilitude"[8], and the "deepening of the problems"[9]. Research is concerned with improving knowledge, problems, instruments (software and hardware), plans for implementation of various moments of research, etc., etc. But the *improvement of knowledge* holds a key position since only improved knowledge makes a deepening of the problems possible, and the same is by and large true of instruments, plans etc.

1.1. *How Can Methodological Rules Be Justified?*

1.10. One currently finds more and more arguments which attempt to justify such rules by referring to, broadly speaking, the history of science. However insofar as methodology gives advice there is only one type of adequate answer to the question *"Why should I follow these rules?"* The answer is: "*If* you follow these rules, then you can have a *better* chance of reaching your goal – the growth of knowledge – than if you follow other methodological rules." In other words, opting for a specific set of rules is justified by means of logical analysis and by reference to the fruitfulness of that procedure for the growth of knowledge. The only relevant form of critique is to argue that the claim of "better chances" of success is false. Questions of such validity must be clearly separated from questions which refer to producers or to the genesis and impact of the results of research. Similarly questions of the legitimation of the goal must be clearly separated from questions about the instrumental value of a procedure for reaching that goal. So far so good, but what is a methodology really? Giving advice is not its only function – and some would deny that it can do so –, and insofar as it does, the above hints have barely scratched the surface of the matter.

Levels of appraisal: theory-appraisal and methodology-appraisal. We all make as a matter of course appraisals such as 'Newton's theory is better than Galileo's', 'Einstein's better than Newton's', etc. We feel so justified in making such evaluations that these judgements are considered to be paradigmatic examples. That means we have an intelligible concept of Scientific Progress at least at the level of the explanandum. But how to justify, or better, how to give good reasons for such appraisals? In order to do this we have to articulate the methodology underlying any such appraisal. What if different methodologies give different answers? Then we will be faced with the problem of appraising competing methodologies. How to appraise a methodological theory-appraisal and how to appraise methodologies? We need a concept of Scientific Progress and as well a concept of Progress in methodology.

Before one can hope to give good reasons for his preference of one methodology over another, he must make clear what he means by "methodology", clarify his concept of Methodology. To this end he must state what he expects of methodology: what should a methodology do for him?

1.11. *Explication of the concept of Methodology*

We submit that a methodology has two main functions: appraisal and advice.

(1) *Appraisal function.* We concentrate here on theory-appraisal (which is basic to appraisal of instruments (software and hardware), of processes etc.). What do we *mean* when we evaluate scientific development as progress? The methodology should give us an *explicatum* (in the Carnapian sense) of the concept of Scientific Progress, the comparative concept of Scientific Quality. (The so-called demarcation problem is just quality control in black and white.) Such an explicatum – usually introduced by means of a definition of Progress – clarifies what we mean by 'progress' and above all proposes that the intuitive idea of progress be replaced, in the context of methodological appraisals, by an improved concept, a concept, which is a better tool for such appraisals. But we not only want to have a clear idea of progress, one that improves upon the intuitive notion, we also want to *recognize* progress. For this we would need criteria; and, if criteria cannot be had, i.e. criteria in the strict sense of at least in principle infallible decision-procedures for the application of the explicatum-term, then we must at least get *indicators*, fallible signs for the fact that a pair of theories exemplifies the relation predicate '... having a higher degree of scientific quality than ...'.

(2) *Advice-giving function.* We not only wish to know precisely what we mean by progress and to be able to *recognize* it (be the recognition oh-so fallible), we also expect that a methodology offer at least some broad procedural rules, strategic advice (or at least global warning against certain procedures detrimental to growth) which, if followed, will facilitate progress: facilitate bringing about a result that is evaluated as progress in the sense of the explicatum supplied by the appraisal function. This evaluation can be carried out only by means of the *indicators* supplied by the methodology. In this sense the appraisal function is primary. Since rules with a universal applicability are not to be expected, we also need a classification of typical research situations. This advice-giving has obvious similarities with technological prescription. Hence – since we also expect differences – we christen methodological prescriptions "quasi-technological" rules.

1.12. *On the special problems of methodology-appraisal*

(1) What can a methodological study of the phenomenon of explication in *empirical science* teach us for the explication of Scientific Progress? Explication of an "empirical" concept is but a limiting case of the replacement of a theory by an improved version of that theory, focussing on the processing of a single concept and thus of most value in the initial stages of inquiry.

What *criteria* for evaluating the success of an explication appear to be reasonable? (Of course this is itself a question of methodological appraisal, a limiting case of theory-appraisal as it were.) Yet a simple answer appears uncontroversial:

(a) The explicatum should be, in the context specified, *more fruitful* than the explicandum as a conceptual instrument. In the context of empirical inquiry this means that it should make possible the formulation of more law hypotheses than would have been possible by means of the explicatum: make possible production of novel knowledge, content increase.

(b) An increase in *exactness* and *precision* over the explicandum concept will count as an improvement only to the extent that this increase is of instrumental value for making the explicatum more fruitful. (Here we depart drastically from the meta-criteria of explicata proposed by Carnap.)

(c) There must be a *sufficient degree of similarity* between explicandum and explicatum: the new concept must be an improved version of the old concept and not of some other concept. Thus, we must clearly specify the concept we are trying to explicate, and this presupposes that the explanandum-concept has a sufficiently definite content, a sufficiently clear core meaning: otherwise there is no explicandum.

Following Carnap, the first step in the process of explication concerns the *clarification of the explicandum*. This is done by giving *paradigmatic* examples and by informal explanation commenting on the examples. The indicator that the meta-criterion of similarity has been fulfilled consists in the fact that the *paradigmatic* positive and negative cases of the application of the explicandum term *reappear* as paradigmatic cases of the application of the explicatum term. To require more would be to miss the very point of an explication: a maximum of similarity would merely indicate that the explicatum is identical to the explicandum, or at least almost qualitatively identical with respect to vagueness, etc. rather than being an improvement of it. Hence (c) is a *necessary* but not sufficient condition for judging the success of an explanation. It does *not* enable us to say that one explicatum is better than other explicata which also satisfy the similarity criterion.

Let us apply these meta-criteria to the explicatum of the comparative concept of Scientific Quality. For meeting the similarity criterion nothing more is required than the "matching" of the *paradigmatic* cases of judgements made in terms of the explicatum with the judgements of those cases made in terms of the explicatum. To propose a similarity criterion to the effect that one methodology is better than another if it offers "a coherent account of *more* old, isolated basic value judgements as rational" or that if it "enabled the historian to interpret more of the *actual* basic value judgements in the history of science as rational: it constituted progress" (Lakatos in Schilpp vol. I:251, italics are his) is self-stultifying: *it goes against the very point of the idea of explication.* It forgets that the wide fringe of vagueness of the explicandum term was one of the reasons why one wished to replace the explicandum by an improved concept.

Since the explicandum is a *normative* concept, description of the *actual* basic value judgements of the scientific élite is irrelevant for the task of clarifying the explicandum. We must make the value judgement ourselves, or endorse it. In doing so we shall invite others to join us, and to this end we have to outline *our good reasons* for accepting it. (This is the counterpart of the clarification of the explicandum by informal explanation in the case of the explication of an empirical concept.)

Since the "similarity"-requirement thus plays a very limited role, and since increase in exactness and precision are valued only to the extent in which they have instrumental value for "fruitfulness", *everything hinges* on the explication of the meta-criterion of *"fruitfulness", and this is an open problem*! This is a negative aspect of the analogy with explication in empirical science where there is a straightforward answer to the problem of fruitfulness. We can perhaps find a point of attack by considering the *possible contexts in which we need the explicatum of "Scientific Progress/Quality"*. Who are the interessees of such an explicatum? In what contexts will it be utilized? Here are just a few examples: (1) In producing a *prescriptive* or an *idealized* picture of science. This is the task of the *philosopher* for whom the *image of science* is an ingredient of man's self-conception, which the philosopher wishes to improve. (2) The explicatum is indispensable for those who focus on science as it actually is: who wish to produce a *descriptive* picture of science: The *historian of science* as well as the *sociologist of science* cannot even get started without some notion of what constitutes (indicators of) progress. (3) The active researcher may use the idealized/prescriptive image of science (as well as the descriptive image) to improve his understanding of his own activity as a scientist and as a methodologist on his own (for he cannot avoid appraising explanations, theories etc. as a matter of daily routine); and he may be particularly interested in trying out the *"quasi-technological" rules*, perhaps in terms of thought-experiments. (4) Anyone actively concerned with research policy making will be interested in these rules in order to learn about the 'inner logic' of research, with a view toward avoiding any interference which impedes progress, and he too will have to make appraisals.

(2) What can we learn from the positive and *negative aspects of the analogy* between *methodological* prescriptions and *technological* prescriptions? To forestall the Feyerabendian retort: It is of course the case that universalization will turn any rule into a hinderer of progress; there can be no rules which can regulate the application of rules; and rules in this context can never mean algorisms: any methodology which offers such rules is heavily overselling. On the other hand, to renounce advice-giving altogether is carrying modesty a bit too far.

The concept of Rationality pertaining to purposive-rational action (in Max Weber's sense) is based on the efficient use of means to reach a given goal. A technological prescription has to be justified in two respects: (a) with respect to the scientific knowledge on which the prescription is based – this knowledge, a law hypothesis, must not only be reliable, it must also be explained by the pertinent theory. (If it cannot be explained, then the fact that it "has worked up to now" at best supports the correlation upon which the prescription is based; but so long as we cannot explain it, we are not sure that the correlation represents a causal law and we cannot specify the realm of application in any detail.) (b) The prescription must be justified with respect to the given goal, i.e. the goal must be stated with sufficient clarity. On the other hand the legitimation of the goal must be clearly separated from the justification of the prescription, in order not to conflate the scientific problem (a) and the normative issue (b).

When it comes to the justification of a methodological rule this clear separation is no longer possible, for the explication, and even the task of legitimation of the goal is continuous with the methodology – of its appraisal function – and the recommendation of the explicatum of Scientific Quality makes evaluation of the ideal of science underlying it and hence an outline of its *philosophical grounding* unavoidable. Moreover, what sort of knowledge could be called on in the justification of a methodological rule? It must be knowledge about knowledge production: how scientific knowledge should be produced, i.e. precisely the prescriptive image of science which methodology is supposed to provide. The only relatively unproblematic step in the recommendation procedure appears to be that using a sort of 'praxio-logical' ("a-prioristic") style of argumentation: *Given* a specific ideal of science (e.g. "high content/scope/depth and truth"), a "quasi-technological" methodological rule can be critized only by questioning its instrumental value for the realization of that goal in the type of research situation at hand. The critic must make it plausible that the claim to instrumental value is mistaken. For this he can resort to thought experiments performed on examples culled from the history of science, but he must not directly appeal to that history as an arbiter (cf. below).

1.13. There is a point in making an evaluative comparison of scientific theories only if they compete, i.e. if their territories (fields of application) at least partially overlap. Analogously, a precondition for a meaningful comparison of methodologies is that they compete, i.e. that their goals overlap sufficiently, that the "ideals of science" underlying them at least partially coincide. That means that in appraising methodologies we cannot avoid facing issues in epistemology and in the philosophical groundplan underlying the methodology.

We submit the thesis that *in justifying a methodology*, in making it plausible that it fulfils the "meta-criteria" mentioned above (in the first place that special sort of "fruitfulness") to a sufficient degree, *the concept of truth is indispensable*.

According to Popper, the concept of truth is not required for methodology itself. (cf. note 1). He appears to mean this: The researcher decides whether a certain hypothesis has been falsified (pro tempore). Although he thereby implies that the hypothesis is "false" (that is his argued conjecture) in this context there is no need to specify the predicate "false" (and by implication the predicate "true") more closely. However, in order to legitimate a methodology – be it in its appraisal or its advice-giving function – the presupposed goal of science has to be stated. Progress in knowledge being considered to be of the essence of scientific research [10], one needs the notion that a specific hypothesis can be epistemically better, more correct, than another hypothesis. As his explication of the concept of Scientific Progress, Popper introduces the concept of Verisimilitude, [11] which welds together Empirical Content and Truth. It is to be used as a comparative [12] concept.

What role does the world-3 ontology play in connection with the specification of the goal, and hence in the justification of the methodology? The concept of propositional truth is essential for Verisimilitude. Tarski, according to Popper, was able to "rehabilitate" [13] this concept, which, although embodied in the semantic structure of any functioning language, nonetheless is philosophically very problematic, by means of his so-called semantic definition of truth.[14] Truth or Verisimilitude is predicated of a judgement, a hypothesis, a theory, etc. in its relation to the world to which it refers. This hypothesis etc. must, in this context, be understood as the meaning of a linguistic entity (sentence with *Darstellungsfunktion*), as "knowledge in the objective sense" (world-3). If it were taken to be a belief, a mental entity (world-2), one would be plunged back into all of the well-known difficulties of psychologistic epistemology.[15] Thus *the world-3 ontology performs an excellent service in grounding the legitimation of methodology*, in making a clear distinction between, on the one hand, the explication of the goal – the definition of Verisimilitude [16] –, and the explication of an indicator of verisimilitude (both world-3 entities) and, on the other hand, empirical consensus with respect to the value [17] of the indicator in a concrete case, and hence with respect to the supposed comparative degree of verisimilitude. With the help of the ontology one can see that conflating the two is a sort of category mistake, that it is essentially "the denial of world-3", as Popper eventually defines it.[18] Popper introduces the world-3 ontology primarily for that purpose rather than for the "Aufbau der Welt".[19]

Popper (along with Hempel and others) *differentiates* sharply between *explication* (in the course of which the explicatum may or may not be introduced by means of a definition) and *application* of the expression which designates the explicatum.[20] For example, we may have exactly specified the meaning of the expression 'true' by means of a *definition* (context of explication), and still lack a *criterion* for the correct application of the explicatum expression in a concrete situation (application). From the definition *alone* it is not possible to derive criteria. We may know exactly what we mean by an expression such as 'true' without being able to say with any certainty that any ("empirical") statement is true (or false). There is simply no algorism for applying the predicate 'true' in the sense of correspondence or correctness of a judgement: if criterion is taken in the strict sence of algorismic decision procedure, then *there is no criterion of truth and thus no method of justifying or grounding the truth of hypotheses or theories.*[21] Verificationism and its sister, "probabilism", are logically untenable positions – as Popper has demonstrated by his criticism of empiristic fundamentalism: there is no secure "base", no warranted basis sentences, and the transfer of warrant (truth) from a finite set of conclusions to a general statement functioning as key premiss in the corresponding deductions can cover only part of its information content, hence even if the deductive arguments are valid, the move cannot be conclusive, i.e. there can be no criterion of truth in the strict sense. So what are we to do? *We have to act*, to decide to accept this hypothesis and reject that one: we have to act, and wish to do so "reasonably".

We are thus led to look for *criteria* for the preference of or *acceptance* (rejection) of hypotheses and theories, i.e. for a method of justifying not theories but *actions* (world-2 entities). The fallibilist must sharply distinguish between the justification of a theory establishing it as true or "probable" and the justification of, giving good reasons for, the acceptance of a hypothesis as a "reasonable" course of action. The acceptance of an unfalsified (or not yet falsified) hypothesis (and sticking with it until something better comes along) in no way implicates that it is true. The *degree of corroboration*, the balance sheet of attempted falsifications if they were all negative, thus, a report *ex post* which gives an account of how things stand at a specific time, serves as a (fallible) *indicator* for the *degree of verisimilitude*[22]. The degree of corroboration defined in this way as a (fallible) indicator of the degree of verisimilitude is itself a world-3 entity, and has nothing to do with beliefs.

It may in no sense be confused with *consensus* within a research community, i.e. with world-2 entities. It goes without saying that in a *concrete* situation it is a group of researchers who *decide* that the balancing of accounts (world-3) is correct. Why is, sometimes or often, such a consensus reached? Popper (but not Kuhn) would say that such a consensus develops when there is no generally accepted argument against the arguments in favour of the interim acceptance of the hypothesis or theory in question. This factual or historical consensus is, of course, never definitive, and the decision may be revoked should the balance sheet itself be seen to have been defective. (Moreover, the new balance sheet at a later point in time will differ from the old one.)

According to Lakatos, these decision processes in actual research praxis bring an element of conventionalism into Popper's methodology. But as we have seen, it is a question of producing a consensus which is informed by the methodology, and which according to the methodology is always revokeable in light of the results of further tests. The concept of *degree of corroboration* is relativized with respect to a given point in time, to a specific *state of the science.* It is not relativized with respect to a particular factual consensus of a historical research group (world-2 entities). However, the *value attributed* to the degree of corroboration in a *concrete* research situation is relativized to that factual consensus. It is obvious that the concepts of truth and verisimilitude are not relativized.

1.14. *On the use of the history of science as an arbiter in the appraisal of methodologies*

If *'logicism'* is taken as a label for the view that in theory-appraisal it is *only* the *logical* aspect which matter, the opposite pole of *totalization* could be labelled *'historicism'*: The view that a purely descriptive approach can account for what scientists have done, and that the *past successful praxis of scientists*, history of science as a success story, *can serve as arbiter in appraising competing methodologies.* The relationship between methodology (prescriptive image of science) and history of science (descriptive image of science) is manyfold and complicated. We propose to begin by considering a typical "argument from the history of science", and then briefly look at the relationship in more general terms.

Such an argument generally takes the following form: (1) description of a piece of research, stating that the research situation was of such-and-such a type and that such-and-such results were achieved; (2) appraisal of these results as scientific progress; (3) the hypothesis that in this research process a certain method M was used and was responsible for the success. The difficulties with such premisses are well-known. For example: in order to do justice to (1), one must describe the research situation from the point of view of the researcher himself, since he did not have the benefits of hindsight available to the historian. A reconstruction of this situation as well as of the development of the research itself, however, is not possible without an at least implicitly presupposed methodological standpoint.

The historian wants to describe important developments. But deciding which ones are significant already involves an appraisal of scientific quality. While in (1) the methodology underlying such implicit appraisals might itself remain latent – as the historian's secret methodology!, in (2) its influence cannot remain hidden, because here a definition of Progress must be brought into play, a definition which is possible only as part of a methodology. (3) is problematic in terms of the well documented discrepancy between what scientists say that they do, what they think they have done, and what they in fact have done. Determining whether or not the scientist has proceeded in accordance with certain rules or has proceeded rather with the blind certainty of a sleepwalker, is a very difficult and sometimes impossible task. However for the argument to make its point, it is sufficient that the *virtual* developments of a reconstructed research situation are described, and that it can be shown that a certain development would have been facilitated (precluded) *if* a certain methodological rule had been followed. Leaving aside all these problems for the moment, assume that in a particular case (1) and (3) are correct, and that we have good reasons for accepting the evaluation in (2). Then what follows from the premises for the justification of M?

The situation appears to be analogous to the justification of a technology which is based merely on observed correlations for which no causal explanation has been given, and with reference to which it is thus not known whether or not the variables are causally connected. (This can still be the case in certain medical therapies, but not in advanced technologies.) In such a situation the technological *prescription* itself (which is analogous to a methodological rule) is not justified, but the *decision* to follow the prescription may well be (e.g. in a situation where one has nothing to loose). To justify M, to give good reasons why M leads to success, the methodologist has to embark on "praxio-logical" argumentation. To lay out what scientists *have* done in the past involves an indispensable moment of interpretation. Unless the methodology used in such interpretation is made explicit and laid open to criticism, interpretation runs the danger of becoming a process in which the historian – now acting as his own methodologist – reads his own point of view into the historical development. Even the selection of case histories may be and often is subterraneously influenced by the historian's philosophy of science. He will view them in the light of his preferred methodological position and perhaps assert that, contrary to appearances, all successful moves are really reducible to *X* strategies and hence, that, this part of history may be taken to support *X* point of view (be it falsificationist or *MSRP*-ist or one of the many sorts of inductivist methodology).

More important still, the attempt to justify *M* by means of an argument from the history of science has merely deferred the problem of methodological appraisal. The problem will rear its ugly head once more when premiss (2) has to be legitimated.

1.15. Having seen this, one also sees that the problem of the relations between philosophy of science and history of science is but a facet of the problematic of the relationship of practical philosophy (ethics etc.) and history (description of past actions and of on-going praxis). Philosophers are professionally concerned with the normative. Problems of judgement and evaluation require practical arguments, arguments which cannot provide an "ultimate foundation". Rational discussion about norms is possible, although the norms are not definitively, once and for all, determinable: decisions concerning evaluation, norms and ethics cannot be derived from covering laws. We can at best develop "well-founded" options which are supported by good arguments.

At this point philosophy of science can learn from practical philosophy. Methodologists are professionally concerned with that which is *normative* for science, with explicating Scientific Progress, etc. Where philosophy makes the finitude of man and the boundaries of happiness clear, methodology makes the fallibility of scientific knowledge and the limitations of the realm of application of methodological rules clear.

Just as practical philosophy shows that moral decisions are not theoretically decidable, methodology shows that there is no algorism for making decisions in scientific research. The demand for such "meta-rules" is just as unreasonable as the demand for certain knowledge, for true hypotheses. Thus, Kant notes that the "Anwendung von Regeln überhaupt verlangt Urteilskraft, die nicht durch Regeln gesichert werden kann".[23] In addition, the rules offered by methodology (see Figures 1 and 2, which present them as "models") must be schematic, since they hold for *all* research (which is not to say that the various moments and aspects of the "model" cannot be specified in greater detail for specific disciplines). Such methodological rules offer the scientist guiding principles, which not only leave him his full freedom of decision and responsibility, but even increase his freedom by allowing him to conceptualize the developments which are possible within certain kinds of research situations. Practical philosophy also shows that there is no non-formal moral principle which is applicable in every situation without regard to specific circumstances, or which has an unconditional priority over against all other principles. These two remarks about moral rules indicate that Feyerabend's thesis concerning the restricted validity of methodological rules is fully correct; but also that it is nothing new – and certainly no objection against Popper.

Of course any position in methodology is based upon a position in epistemology and ontology, whether the holder of the position is aware of it or not – or, less psychologistically, whether the structural relationships are latent or manifest. The position which is *antipodal* to the one argued here appears to hinge upon the attempt *to eschew the concept of propositional truth at all costs*. The rest of the syndrome seems to flow out of this basic gambit (basically an ontological gambit involving an anthropocentric world-view rather than a cosmo-centric one). In epistemology, a *consensus theory of truth* is offered (often combined with an "anti-metaphysical" stance: what is really real is "praxis"); and a *conventionalistic view of methodology*. In methodology itself: a "pragmatist" *view of scientific theory* (theory-appraisal in terms of "praxis" – a view which equates scientific merit with practical (technological) usefulness) leading to the *denial of the distinction between "internal" and "external"* criteria in the appraisal of science. (This latter view has consequences for research-policy making: the denial of the distinction between pure and applied research, a distinction which, indeed, pertains to science policy and not to methodology, leads to the recommendation that all science policy decisions be made solely on grounds of practical "usefulness" to "society", since *there is no pure science*.)[23a]

1.16. Against this background one can see clearly *where the conflation* of (a) truth – the problem of justification of a *hypothesis* in the sense of *justifying its truth-claim* –, and (b) acceptability – the problem of justifying, *giving good reasons for, the acceptance* of a hypothesis i.e. justification or legitimation of an *action, inevitably leads*: If justification of a hypothesis is required, i.e. guarantees for truth, then sooner or later comes the sobering insight that on these terms knowledge is simply impossible. The history of philosophy is full of illustration of the pendulum swing between dogmatic demands and radical scepticism. The upshot of these developments in epistemology would appear to be this: the problem of establishing the truth of a statement which turns out to be the problem of transfer of warrant from original (warranted) statements to other statements, leads to the trilemma (H. Albert) of vicious circle, infinite regress or (arbitrary) ascription of secure epistemic status to some special kind of statements (fundamentalism). The latter "solution" was adopted in classical intellectualism ("rationalism") and classical empiricism, both of which share the conviction that there is such a basis and that it is constituted by mental entities (*Evidenzerlebnisse*) (world-2). Now the sceptic can rightly point out that there is no empirical consensus with respect to these "basis" experiences. The way out appeared to be either postulating that such a consensus *would* be reached if some "therapy" were successfully deployed (Descartes, e.g.) or the socialization and *idealization* of these "Evidenzerlebnisse" (Apel), or even the attempt to replace the sentences describing such experiences by actions (Kambartel, Mittelstrass). (The latter approach jibes with the idea that the "knowledge-constituting interest" underlying the natural sciences is primarily or solely the so-called technical interest (Habermas).) But let us glance at the roots of the motives underlying such gambits.

Insight into the unrealistic nature of demands for guarantees for truth leads to insight into the fact that no criterion can be derived from a definition of truth. To philosophers who cling to such impossible epistemological requirements, this leads to the (false) impression that there is no relation at all between a definition of truth and criteria for the acceptance (rejection) of hypotheses. Retaining their excessive demands, they attempt to produce a new kind of definition of truth which will meet the following meta-criterion: a definition of truth is acceptable ("reasonable") if and only if it is applicable as a criterion for the acceptability of hypotheses.

The notion of consensus offers itself as an operational criterion. Thus, one might attempt to *define* truth in terms of consensus. A factual consensus, here and now, cannot do the job, for error is always possible. As a way out of what appears to them as the dilemma of either giving up the idea of an operational criterion of truth or relativizing the concept of truth, they turn to the idea of "consensus in the long run". Thus many philosophers think that relativism can be overcome only be transcending empirical consensus, by extrapolating or idealizing to the concept of an unlimited communication community. But doesn't this attempt to overcome historical relativism (historism) land us in historicism *if* it is taken to be relevant to methodology of science, in a metaphysics of history in the form of the (mere) *postulate* that the history of science is governed by a principle of general progress? Lakatos, in his (still unpublished) critique of Toulmin's *Human Understanding*, thought this to be the case, and he speaks of the "ad hoc stratagem of historicism: the Long Run": truth being reached at (and being defined as that which one has at) the end of the history of science. Be that as it may, one might also wonder whether such a formal principle is really enough to free one from historical relativism.[24]

Truth is defined by "pragmaticist" philosophers from C. S. Peirce to K.-O. Apel counterfactually as the character of those judgements with respect to which the ideal community of investigators *would* reach consensus *in the long run*. (E.g. in Apel's "transcendental pragmatics of language".) Such a concept of truth avoids – insofar as the "ideal communication community" is introduced as a regulative idea – the reproach of being psychologistic or an expression of historical relativism, but it provides no operational criterion of truth, and, much more importantly, in this form is totally unable to aid in justifying methodological rules for the acceptance of hypotheses. Moreover it does not either succeed in sidestepping the problem of explicating Truth in the ordinary sense, in the sense of *Urteilswahrheit*: for with respect to what should there be a consensus in the Ideal Communication Community if not with respect to truth-claims and validity-claims of judgements?

The historical relativism overcome on *one* level can sneak back in on another level: this approach skips over the entire methodological dimension. This is hardly surprising. The contemporary style of philosophy which transforms Kant's "transcendental unity of apperception" into the principles governing the "ideal communication community" (and extrapolates from the empirical subject to the "transcendental subject"[25]: principles which are prerequisites of any knowing), investigates the *transcendental* conditions of the possibility of knowing, argumentation, science, ethics, etc.) Yet, in spite of all differences and in spite of the fact that by and large these two philosophies do not even have a common subject matter, Apel's position is in certain important respects related to Popper's Critical Rationalism.[26]

What role does the concept of Consensus play within Popperian methodology? As we have seen, the only relevant concept of consensus is *empirical* consensus: the methodology provides the means for *legitimating* consensus, e.g. consensus about the comparative degree of verisimilitude of T_2 over T_1 in terms of their degrees of corroboration at the time of appraisal.

1.17. *The significance of methodology for social philosophy and the fight against moral relativism, and the relation of Critical Rationalism to the idea of the Ideal Communication Community*

The normative principle of Popperian methodology is "Critical Rationalism": the Socratic ideal of rational discussion, the ideal of enlightenment. It does not thematize the psychic attitudes of individuals or groups (world-2), but rather a *system of norms*: "the general idea of intersubjective criticism, or in other words, of the idea of mutual rational control by critical discussion"[27]. The rules for hypothesis-checking and theory appraisal[28] as well as the ban on immunization strategies[29] (in Figure 1 labelled "master-rule": "No rules must be used that would prevent falsifications") are derived from this principle.

All this is developed in terms of world-3. Realizing that as empirical subjects we always represent a real communication community, we see that as "critical rationalists" or rational discussion partners we are always involved in an idealization of a counter-factual anticipation: We assume that this real communication community at least to some degree approximates an ideal, which we might call the "Ideal Communication

Community" (Apel). When it comes to acting in terms of principles, we hope that we can by and large act as if we already represented such a communication community, something which is factually impossible for any number of (world-2) reasons. This figure of thought finds expression in Popper's philosophy in that he generalizes from the "scientific community" as that real communication community which most closely approximates the ideal, developing his philosophy of the open society of this basis.[30]

Popper's methodology and philosophy, oriented to permanent critique as enlightenment, aims at those real communication communities which stand under the regulative idea of Reason in the sense of the philosophy of the Open Society – and not in the sense of the utopia of a society of fully "emancipated" citizens ("herrschaftsfreie" Gesellschaft). Permanent enlightenment, however, means that no "contents", no substantive positions are privileged: that only the approach itself is privileged. The philosophy of the Open Society (J. S. Mill, Popper) intends to produce good reasons for institutionalizing the permanent enlightenment in the real communication community.

It is often held that the chief difference between Popperians and "transcendental pragmatists" (like Apel) lies in their different views with respect to the ultimate foundation problematic: whether or not the "Criticist Frame" can be "founded". Popper rejects such a "*Letztbegründung*". He is certainly correct in arguing that the realization of Reason rests on a *decision*: whether or not one acts ethically is, according to Kant, a question of the Good Will which cannot be taken care of by an "ultimate foundation". Apel emphasizes that the choice of the "Criticist Frame" is not an irrational act as Popper claims.[31] For it is a matter of *self-consciously affirming* that which is already presupposed in every argument and even in every meaningful action. The only real self-consistent alternative would be silence, a withdrawal from meaningful life and action as a sort of "quasi-argument' against all argumentation. Man possesses this very special, ultimate freedom of total negation. Popper agrees: in acting meaningfully we have by implication already shown our "faith in reason" (loc. cit.). This *faith* we may self-consciously affirm, but we are not obliged to do so – man possesses this freedom too; but we cannot "found" it or defend it by means of arguments since it is presupposed in all argumentation, even in all speaking and thinking. This what is impossible is consistently to reject it by means of sentences, by speech acts or thought acts: in any attempt to do so it would already have been presupposed. Hence what by many is regarded as a great divide boils down to a *verbal* difference. Popper's concept of "founding" (begründen) is narrower than Apel's: demonstrating that *Y* is a condition of the possibility of *X* would for Popper not suffice as an "ultimate founding" of *Y* not even if *X* is speaking meaningfully, so that the question of "founding" *Y* cannot be raised without presupposing *Y*, and (perhaps with an intention to provoke – provoke thinking) Popper uses 'irrational' in a wider sense than Apel and ordinary usage.

Apel tries to overcome *moral* relativism by developing a transcendental underpinning for the philosophy of the Open Society. Popper, on the other hand, argues that this *presupposes* the critique of *intellectual* relativism (psychologism), since the former is to a large extent based upon the latter. In other words, that *respect for the concept of truth* in the sense of propositional truth (*Urteilswahrheit*) *is a presupposition for an effective critique of moral relativism*, and this seems to us to be an important insight. He would hold that methodology – being normative – can and must learn from ethics, but also remind us that the philosophy of the Open Society was generalized out of methodology (this indeed is Popper's road, but it is perhaps not the only one).

2. A BRIEF SYNOPSIS OF POPPERIAN METHODOLOGY

2.0. The chief aim of this part of the essay is to make sure that the reader and I mean the same when speaking of "Popperian methodology".[32] Such a *co-understanding* cannot be presupposed in view of the different Popper interpretations in the literature; but it is a precondition for a relevant criticism of part I. I also wish to make good the claim that *Popperian methodology implies pluralism rather than pure falsificationism ("logicism")*. Two figures will be used as a means of exposition. Their function is two-fold: to epitomize the basic features of Popperian methodology and to identify problems for the methodologist. (Space does not permit here to discuss possible difficulties of the methodology.)

2.1. *Skeleton Outline of the Methodology (Figure 1)*

2.10. According to Popper, a research enterprise *starts from a problem*

(P_i in Figure 1). *Where does the problem come from*? Popper focusses on the type of research situation in which progress of a dramatic order may occur rather than on the "piece-meal" mode production of new knowledge characteristic of research enterprises approximating the Kuhnian ideal type of Normal Science. In the *typical situation of this type a*

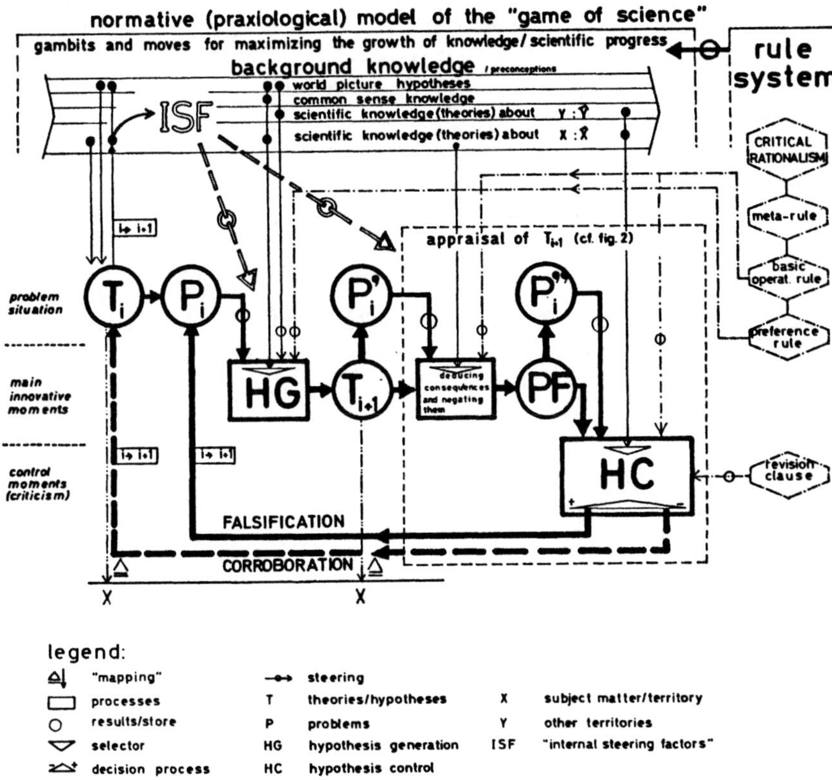

Fig. 1

"veteran" theory has run into difficulty, "clashed" with some experimental results. The "clash" may be more or less manifest, and more or less serious, with logical contradiction as a limit case. It may be latent, in particular in a type of research situation where pieces of knowledge produced in an "empirical" manner cannot be explained by the pertinent veteran theory. The residual may grow but resist explanation, as

e.g., spectroscopic knowledge did before 1913. Such a residual-knowledge which cannot be explained by the veteran theory is practically always in evidence; however, when it accumulates or manifestly conflicts with the veteran theory, it constitutes a challenge which can no longer be ignored. The afore-mentioned type of situation of manifest conflict typically arises when attempts are being made to apply a theory to a new territory. Indeed it *must* occur, sooner or later, since the kernel of the theory is a conceptual model of a real (physical, biological,...) system, and because it is in the nature of a model that it also has negative analogy.[33]

Thus a theory, once created, will *sooner or later* give rise to *an objective problem situation of that sort:* a problem which confronts us or which we literally have to dis-cover. In particular, in case of a "clash", it is not just a question, but a problem in the etymological sense of the word 'problema': situation in which one cannot remain. The case under discussion is just one example of the phenomenon that objective problems arise whether the researcher wishes it or not, which phenomenon serves as part of Popper's "argument from independence" for giving ontological status to World-3.[34]

2.11. *Where does the theory come from?*

Figure 1 sketches only two possibilities: the theory being one of the theories which, at the time, are accepted in the discipline concerned; and the situation where an *Ansatz* towards a theory has just been made – on the basis of the background knowledge, in particular on the basis of preconceptions about the general nature of the new "territory" (new sort of real system) which are derived from "world-picture" hypotheses. In the creation of theories about a hitherto unexplored realm "metaphysics" is *structurally indispensible*, and not merely indispensable as a part of psychological heuristics (world-2).[35]

Preconceptions about the general nature of the subject matter at the level of world-picture hypotheses ("cosmological" hypotheses) also play an important role in the first-mentioned type of case, even if they remain implicit. They constitute the kernel of a system of factors – preconceptions, evaluations,... – which may be called "Internal Steering Factors" (*ISF* in Figure 1).[36] Partially based upon the preconceptions is a programmatic conception about the discipline to which the theory belongs: evaluative statements about what the discipline (the "*X*-ology" in question, $\hat{X}$ in Figures 1 and 2) should be like, which may include an "ideal of science".

According to the methodology under discussion a research enterprise that is properly conducted is governed by the rules[37] derived from Critical Rationalism (right column of Figure 1); and thus the models presented by Figures 1 and 2 consist of the gambits and moves resulting from the application of these general rules to the various moments of research. The rules also specify the concept of "scientific progress" or the comparative concept of scientific quality, and the methodology of appraisal is spelled out in Figure 2 as the operational characteristics of the "Hypotheses-Checking/Control Station" *HC*. These methodological gambits and moves apply to *all* research enterprises worthy of that name. In "Normal Science" a *certain* substantive system of *ISF*, in particular a certain set of preconceptions (cosmological hypotheses), steers the concrete procedures of a certain ensemble of concrete research enterprises in more detail; it provides substantive criteria for the appraisal of products and gives these research enterprises a certain unity and direction: gathers them together in a research "direction", "tradition", "school", "style",....[38]

2.12. The proposed solution to the problem (P_i in Figure 1) is a tentative theory (T_{i+1} in Figure 1). *Where does T_{i+1} come from?* In Figure 1 the box "*H*ypothesis *G*eneration", *HG*, marks the "station" producing the tentative theory. Following Popper's usage, in Figure 1 'theory' and 'hypothesis' are used interchangeably. Among the general rules (applying to all research) the one particularly important for the operation of *HG* is that labelled *"preference rule": "Prefer theories (hypotheses) with high information content* (and hence more "falsifie-able" i.e. involving greater risks of being wrong, allowing more *severe* tests, theories whose absolute logical probability is as low as possible) *to hypotheses with a lesser amount of information!*" (Roughly: Nothing risked, nothing gained: the more "daring" the theory, the greater its potential for corroboration by tests.) (This rule of course follows from striving for severe tests.) However this rule gives only a very broad "policy" direction. In a concrete research situation the operation characteristics of *HG* are governed, in more detail, by the *ISF*, especially the (substantive) preconceptions ("cosmological" hypotheses). The *ISF* also steer the operation of the selector (marked in the way customary in cybernetic charts, by a "standing" triangle), i.e.

they guide the decision concerning which part of the background knowledge available at t is made use of. Of course all the input into *HG* are world-3 entities.

Within the scope of the operation characteristics of this type of "Hypotheses-Generator", a logical move may be accommodated which, if used as a move of appraisal, would be highly objectionable (because then appraisal and generation, *HG* and *HC*, would have been conflated): a logical move which is often called "inductive" and which might more properly be labelled "observationist-inductivist": from a set of data sentences to a generalization of the same kind. If the data are taken as basic only in the sense of not being problematized in the particular research context at hand, then, although the logical move is not governed by precise formal rules, the procedure may nonetheless lead to a probabilistic *conjecture*. This move is certainly made use of in experimental work. But it must not be forgotten that not even this procedure is purely "empirical" (as one objectionable branch of "inductivism" would have it) since it is guided by the *ISF* all the way, even if this guidance may not be prima facie recognizable. In the type of research situation Popper focusses on (situation with chances for a "big step forward", other than "normal science"), in situations where the essential output of the *HG* is a theory rather than a law hypothesis, the essential input will come from the preconceptions, from the "cosmological" hypotheses. For instance, a proposed analogy between a yet unexplored territory and some territory about which successful theories already exist may provide guidance: it will suggest a program of exploring positive and negative aspects of the proposed analogy. (This appears to be one aspect of the "positive heuristics" of the *MSRP*.) In general, for the operation of the *HG* practically everything is allowed – the checking and control, the appraisal comes afterwards.

2.13. The output of *HG*, the tentative theory T_{i+1} again raises, automatically, a new objective problem: *to appraise the quality of* T_{i+1}: What is the *scientific merit* of T_{i+1}? Is the development from T_i to T_{i+1} *progress?* If so, in what sense? Among the general rules, in particular the *"basic operational rule", the "falsification principle"*, or better "principle of severe testing", governs the appraisal. Theories as conjectures are testable against observation reports functioning as "evidence basis", but according to Popper, *not* justifiable in the sense of quasi-verificationistic or probabilistic theories of confirmation (a point which we shall question in connection with Figure 2). Figure 1 provides the skeleton outline of research, while Figure 2 gives a blow-up of this theory-appraisal, which constitutes the center of concern of Popper's methodology.

For the moment we shall consider the skeleton outline. The output of the first station within theory-assessment, a *potential falsifier* of T_{i+1}, again creates a new objective problem (P_i'' in both figures): to investigate whether or not the potential falsifier (*PF*) can be processed into an actual falsifier, and eventually to decide on what has been the outcome of the falsification attempt. The terminal station "*E*mpirical *H*ypothesis *C*hecking/*C*ontrol", *HC*, has as a final step a decision-taking procedure. There are a good many such decision-points. (In both figures some of them are indicated by means of the "decision triangles" customary in system-theoretical charts). Such moments play an indispensable role in the Popperian scheme: if only for this reason, the charge of "logicism" is unfounded. But so is the opposite charge of "conventionalism". We shall return to this when commenting on the appraisal moment to blow up of Figure 2.

In Figure 1 only the two clear-cut cases are presented: (1) Falsification, which automatically gives a new objective problem: P_i has been processed into P_{i+1}. With P_{i+1} begins a *new turn* in the research enterprise: an iteration of the feed-back loop just described. But on a higher level – if only because one of the inputs into *HG* is now the new knowledge gained by the preceding investigation, including the falsification itself. (2) If the falsification attempt has a negative result, the rule is to retain T_{i+1} until further notice, i.e. to let it replace T_i.

The rules also define in what sense T_{i+1} is "*better*" than its precursor T_i: very roughly, it "explains" as much as does T_i with at least the same precision; it does not encounter the same difficulties (which gave rise to P_i) – preferably it has also succeeded in incorporating the residual pieces of knowledge, which "clashes" (more or less manifestly) with T_i; a desideratum is also that it makes deduction of new pieces of knowledge possible ("additional information content").[39]

2.14. Figure 1 indicates also that the replacement of T_i by T_{i+1} may have a feedback, from T_{i+1}, to the background knowledge. In those cases where T_{i+1} turns out to be in serious conflict with the veteran theory, the successor theory could be produced inly in a research process in which a part of the *ISF* was disregarded.[40] In such cases the change over from the veteran to the successor theory *may* have *dramatic repercussions on the preconceptions* about the general nature of the territory, which in turn may lead to substantial changes in the *ISF* and eventually to *changes in the background knowledge* in general. (These are very rare developments sometimes labelled 'paradigm shifts' or 'scientific revolutions'.) If one wishes to use a term from the philosophic tradition, one might say that this process – be it piece-meal-development or dramatic shift – illustrates an aspect of the "hermeneutic circle" or "spiral"[41] – a (non-vicious) "circle" in the sense that the existence of background knowledge in its original, less adequate state was a prerequisite for the development of the successor theory and therefore for the improvement of the preconceptions, the "cosmological" hypotheses (in particular by elimination of false items from the preconceptions). Hence, even the preconceptions – the "metaphysics", the world-picture hypotheses – are "empirically" criticizable, although only in a *very* indirect way. The rule labelled "*revision clause*" in Figure 1 applies also to them even if they are more resistant to change than the other components of the "body of knowledge" at a given point of time – if this were not the case, one would abandon an *ISF* long before it has given a substantial part of the dividends in knowledge which could be derived from it. The revision clause – "for fallibilism" – can be supported by arguments from Critical Rationalism (but there are good reasons for its introduction also on other grounds; falsification, e.g., cannot be conclusive because one cannot conclusively establish the premisses of the falsifying argument, if only because of the "theory dependence" of the data sentences). It stipulates among other things that the verdict that is the output of the *HC*-station in Figure 1: "*T* is corroborated" or "*T* is falsified" –, must not be regarded as final either. Falsification is no exception.[42] On the contrary the researcher is under the obligation, in principle, to re-test, i.e. to attempt to falsify his falsifications (cf. Section 2.22 s.f. below). To think otherwise would be a denial of fallibilism.[43] Thus 'discorroboration' would be better than 'falsification'.

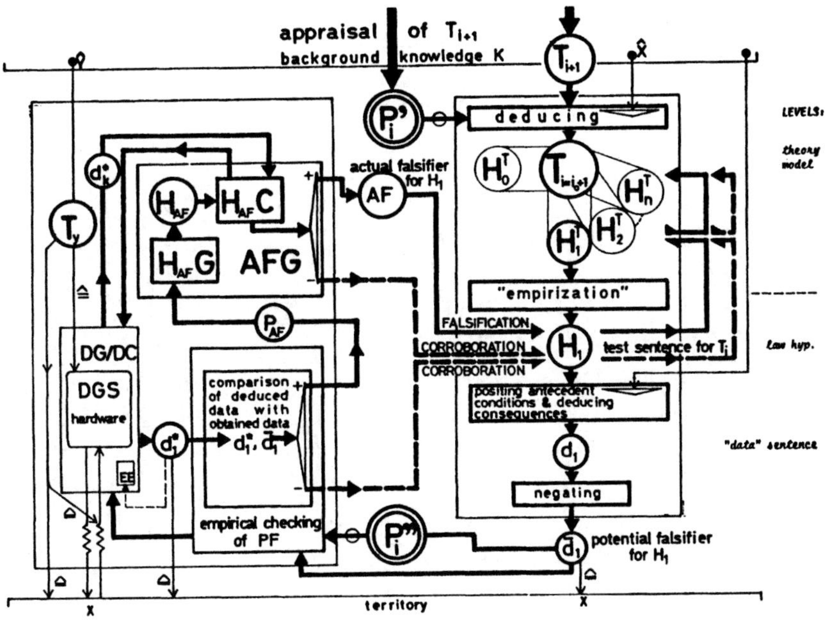

legend (cf. also legend of fig. 1)

H_i	hypotheses deduced from theory T (in the presence of K)	T_y	theories underlying the data-generating system
H_0	hypothesis to be explained	—⋀⋁→	"signals"
$H_1,...,H_n$	"predicted" hypotheses	EE	error elimination
d	"data" sentence, deduced	d^*	"data" sentence: output of DG/DC
DG	data generation	H_{AF}	hypothesis: candidate for actual falsifier
DC	data checking/control	$H_{AF}G$	actual falsifier generation
DGS	data-generating system	$H_{AF}C$	actual falsifier checking/control

Fig. 2

2.2. *Appraisal of the Tentative Theory*

2.20. Figure 2 presents a blow-up of the moment "appraisal of the tentative theory". The difference between theory (model) and law hypothesis is emphasized in a way not found in Popper, but fully compatible with his methodology. All theory-appraisal is governed, in the first place, by the "basic operational rule" (Figure 1), the *falsification principle:* "Try to falsify your tentative theories!" or better "Test your theories by severe tests?" The "theoretical appraisal" of the theory (right box of Figure 2) is governed by *its first aspect*: 'Prefer a more *severe test* to a less severe one!" – roughly, the more the information carried by the hypothesis *H* deduced from the theory *T*, in the presence of "background knowledge" K or, strictly speaking, of the selection from K used in deducing H from KT *goes beyond* the information contained in that selection from K, the more "severe" is the test suggested by that hypothesis. (Selection is guided by the *ISF* which supplies substantive criteria (cf. above)). That means that the sense of *"background knowledge"* relevant here is not the knowledge available at the time t when the deduction is being made, i.e. the set of hypothesis and theories not problematized at t, but the *premisses added* to T.

Within these additional premisses A a clear distinction must be made between *auxiliary theories* and hypotheses (general sentences) and *initial conditions* (singular sentences) because the criteria for appraisal of each will be different. (E.g. the famous explanation of the anamalous motion of Uranus by means of postulation of an extra planet – addition of one masspoint to the model of the planetary system – and its subsequent optical discovery concerns merely one of the *initial* conditions – hence it produced only a *minor* piece of novel knowledge.)

2.21. *"Theoretical" appraisal of T: logical or information-theoretical aspects of theory-appraisal*

2.211. The first station in the "theoretical" appraisal of T_{i+1} is *deducing* consequences of T_{i+1}, the presence of K. In Figure 2 the *output* of this station is presented as *a (growing) system of deductive patterns* such that (1) the premisses of each pattern contain T_{i+1}, (2) one of the patterns constitutes an explanation of the residual piece of knowledge, i.e. of that piece of knowledge, e.g. an experimental result, which could not be deduced from ("explained by") the predecessor of T or which manifestly "clashes" with the predecessor theory; and (3) the conclusions of the other patterns make predictions, i.e. are statements about the real system investigated which have not yet been tested, containing, hopefully, "virtual" new knowledge.

However to say so is but a very rough approximation. More exactly, each of the hypotheses deduced (H_i^T in Figure 2) of course still talks about the *model* (e.g., to use the simplest example, about a masspoint model of the planetary system) and not about the real system, certain aspects of which are to be mapped by the model. To put it at its crudest, the simplest logical form of such a sentence could be stated as: $(y)_{y \in M}\ S_y$, where '*S*' stands for a predicate introduced in a theoretical way (not "operationally defined") and '*M*' refers to the conceptual model in question (world-3). Before they can be empirically tested, H_i^T hypotheses must be *given empirical import*, transformed into law hypotheses proper – *"empiricized"* if that be the word. Again the simplest possible logical form of H_i – the "empirized version" of H_i^T – could be stated as: $(x)_{x \in A}\, Bx$, where '*B*' stands for an expression designating an observable, preferably, a measurable character (world-1) of a real physical system (as in the example alluded to above, the planetary system or a system consisting of earth-and-satellites, or sun-and-planets (natural and artificial) etc.), and '*A*' refers to the realm of application of the theory or the model, i.e. to real systems of a specified sort – (world-1). (E.g. in the above-mentioned example to all real systems (world-1) consisting of a "central" body whose mass is considerably greater than that of all the other bodies in the system).

Let *E* be a law hypothesis ('*E*' for 'explanandum') which at the time *t* when the deduction $TA \rightarrow H_0$ is made, is already "accepted". ('Accepted' meaning that it has a certain degree of corroboration meeting our level of acceptance, independent of *T* or, if you please, a certain degree of evidential support of its own, independent of *T*.) Moreover, let *E* be such that an "identification" is possible between H_0 and *E*, i.e. H_0 (the empirized version of H_0^T) is found, with respect to its form and the values of the magnitudes to be sufficiently like *E*. If this is the case then the pattern $TA \rightarrow H_0$ is regarded as an "explanation" of *E* by *T*, in the presence of *A*. A simple example will concretize what is meant: *E* could stand for the original, "empirically" discovered Galilean law of free fall (or the original Keplerian laws of planetary motion); H_0 stands for the law of free fall (or the laws of planetary motion) derived from Newtonian theory by deducing H_0^T and empirizing it to H_0. H_0, the Newtonian version of the law of free fall, although it logically contradicts *E*, is considered an improved successor of *E*, or *E* is considered an approximation of H_0. ('Considered' is to indicate that a moment of decision is involved; but one for which good reasons can be given.) Popper sees this phenomena of *improvement of the law hypothesis deduced or "explained"* (incorporated into the theory *T*) *achieved in the process of explanation* very clearly,[44] although he does not stress the difference between "theory" and "law hypothesis" as much as we wish to do here and think it advisable to do.[45]

The empirized versions of the hypotheses $H_1^T, \ldots, H_n^T$ differ from H_0 only in the respect that at time t', after the deductions have been made, they do not yet possess any degree of corroboration: nothing is known about their epistemic status. If A is not problematized, then H_1 functions as a *test sentence* for T. But how? At any rate we cannot directly "confront" H_1 with "reality" or with "experience of reality" (as the picturesque manner of speaking puts it).

2.212. In order to check H_1, from H_1 and certain initial conditions (singular sentences) consequences are *deduced* such that the conclusion of the argument is a *"data" sentence* ("basis" statement – pertaining to the "evidence basis" of T), i.e. a singular sentence the simplest logical form of which could be stated as 'Bx_1'. A simple example: by means of the so-called gas law an instance of the "isothermic compression" is derived, viz. it is predicted that in a certain container at a certain time the volume and the pressure will have certain specified values (the temperature being assumed constant): a sentence of the form, roughly: $(x_1, t_1, (v_1, p_1))$.) The *negation* of this deduced data sentence functions as a *potential falsifier* for H_1 because, if one of the sorts of states or processes *"outlawed"* by the law hypothesis H_1 is exemplified by only a single individual belonging to the realm of application, then the general hypothesis is falsified. ($-Bx_1$ being equivalent with $(\exists x)-Bx$.) Of course, strictly speaking, the deductive argument is valid (this is assumed) but, since the premisses cannot be established conclusively, the falsifying argument cannot be conclusive either, and may be challenged any time.

Hitherto the appraisal was concerned *only* with the *logical or information-theoretical* aspect of T – the appraisal of the epistemic aspect, the "empirical" appraisal (left side of Figure 2) has been postponed. The *falsificationist move* (negative hypothetico-deductive) has dominated the scene. Yet we feel that it ought to be *complemented* by a *positive hypothetico-deductive move*,[46] and that a temporal element may be allowed to enter, a temporal element in the degree of corroboration of T. When (let us assume) the "predicted" hypotheses have been corroborated, they function not only as corroborators but also as evidential support for T. And when the degree of corroboration of T has reached a certain "level of acceptance" (moment of decision!), then the pattern $TA \rightarrow H_0$ has changed from a virtual explanation of E into an explanation proper (approximative explanation). Then H_0 too (which was both corroborated and had an evidential support of its own, independent from T, *before* it was deduced) becomes an instance of evidential support for T – although *never a corroborator*. This seems desirable for the following reason: suppose that H_0 is the set of spectroscopic knowledge before 1913 and T the first *Ansatz* of Bohr's theory, then T should certainly get credit for having achieved an explanation of this hitherto residual knowledge, for the feat of succeeding in incorporating the residual into the pertinent system of knowledge. It "ought" to get credit because there are *good reasons* for this, and *not* because such an appraisal coincides with the unanimous judgement of the scientific élite. Thus if two competing theories have the same degree of corroboration but one of them a higher degree of evidential support, this latter would be "better" so far as logico-epistemic aspects are concerned.

However, falsificationist moves not only need to be complemented by positive hypothetico-deductive moves: in addition to explication of logical moves, a methodology must *recognize and account for* moments of *decision*. This becomes clear in playing through the falsificationist move: It consists basically in deducing e from TA, if $\bar{e}$, then TA is falsified, i.e. if $\bar{e}$ and A, then T is falsified. *This is all logic can tell*! Whether or not to accept $\bar{e}$ as authentic, whether or not to problematize A (the additional premisses), and eventually whether or not to "reject" the falsified T in the sense of ceasing to work with it or on its improvement, all this is a matter of *decision*. (In Figure 2 only a few decision points have been indicated in order not to complicate the diagram.) Decision-making in the face of incomplete information and uncertainty, hence risky – and fallible (that is one of the reasons for the "revision clause"). But there is no reason to call these moments 'conventionalistic', for the decision need not be based on conventions nor give rise to conventions. Nor is it justified to call these moments 'non-rational'. The rationality relevant to the type of problem situation and to the type of action under consideration is that of deliberation, of giving legitimating arguments for decisions arrived at, i.e. that of *giving good reasons*. For instance one may decide to stop re-checking a certain initial condition simply because there are at the moment no reasons for supposing it incorrect. This may turn out to be wrong – and get corrected. But a convention has been established only in a Pickwickean sense. Although moments of decision occur at many junctures in research, they are corrigible and they are controllable and often get corrected by "logic and experiment", by contact with the territory, with the subject matter so to speak. If these moments of decision were considered to be all that matters in research then this epistemological view – methodology itself would be a matter of conventions – could be labelled "conventionalism". It is as much a *totalization* as is its counter-pole: "logicism" (only logical aspects matter in theory-appraisal). *In the Popperian scheme the moment of decision is kept in balance by the theory-realism* underlying the logical moves (and here too the concept of truth is essential). A full-blooded conventionalism would imply an instrumentalist view of theory: then the aim of science would not involve truth, but consistency, coherence etc. and perhaps technological efficacy. The criticism of the decisions and of the good reason assays, as an appraisal of processes or actions, is of course distinct from theory-appraisal (in terms of logical, epistemic and other aspects).[47]

Figure 2 is to model the theory T over its history, *T with a career dimension: T is being developed in the course of testing it, in the course of producing new knowledge by means of it.* Only one type of development – needless to say that there are others – has been used for purposes of illustration, viz. one starting from a research situation in which E (in Figure 2 only indicated by H_0) constitutes the challenge, as residual or as clashing with the veteran theory. Then T in the pattern $TA \rightarrow H_0$ is really T_0 – typically an *Ansatz* towards a new theory, e.g. one central component of it. T_k could stand for a full-fledged theory which has grown out of T_0; and T_n may well stand for a so-called incorrigible theory, i.e. a theory which cannot be substantially improved but only superseded by a better one (in the sense that Newton's theory was "finished" quite early in its history).

The degree of corroboration of T – T over its history – changes: T has to be evaluated in terms of career. Imre Lakatos has placed this in the center (with Ernan McMullin I think that a Research Program in his sense is a theory over its history rather than a series of theories), but already in *Logik der Forschung* with the appraisal of a theory in terms of past performance and the emphasis on improvement through testing, this career dimension is clearly there.

2.22. *Empirical appraisal*
2.220. The *PF*, the negation of the deduced data-sentence poses an objective problem (P_i'' in Figures 1 and 2): the "empirical" appraisal of the *PF*, and hence of H_1, and thus, indirectly, of T_{i+1}. It is governed primarily by the *second aspect* of the "*Falsification Principle*". *Actual as well as potential falsifiers must be subjected to severe (empirical) falsification attempts.*

2.221. To begin with one has to ask: *Can the PF* ($\sim d_1$ in Figure 2) *be accepted*? Only if it has been corroborated; hence we must first attempt *to falsify* it! For this purpose we have to produce a data sentence which is the result of actual experiment and observation, and with which the negation of the deduced data sentence can be compared. For this we need *hardware* instruments, *technical data-generating systems*. (*DGS* in Figure 2). Such a data-generating system is based upon some physical theory (T_y in Figure 2). Hence, in this obvious sense all data are "theory-laden" to some extent,[48] or at least cannot be theory-neutral; but the data is dependent on T_y – on the various theories *used* in the appraisal of the theory under development, theories underlying the construction and manipulation of the hardware instruments as well as theories used as software instruments e.g., logico-mathematical theories and techniques used mainly in the "theoretical" part of theory-appraisal or statistical theories used in connection with data-generation – and *not* on the *T* being tested or processed. (These "auxiliary" theories, or even the hardware instruments themselves, can take the lead, and research then becomes "steered" by the instruments rather than being problem-oriented. Such a situation either has the character of an interlude (new hardware makes the production of new knowledge in an empirical manner possible) or turns into a sort of pathology of the discipline concerned.) Data sentences are in principle fallible, if only because the theory underlying the data-generating system needed for their production must be fallible. That theory may be problematized at any time, although in the context of assessing T_{i+1} it has to be presupposed: it would be stultifying to problematize several items simultaneously. The outcome of the comparison of the negation of the deduced data with the experimentally obtained data (singular observation report) is the decision whether or not the negation of the deduced data sentence ($\sim d_1$ functioning as *PF*) has withstood attempted falsification. (It is usually a matter of routine to decide whether or not the functor-values of the theoretically derived sentences "match" those of the data sentences obtained in an experiment). If $\sim d_1$ has not withstood the falsification attempt, it is to be rejected, and H_1 has been corroborated to that extent. If the negation of the deduced data sentence has withstood the empirical test, the researcher immediately gets a new objective problem, this time a *"practical" problem: How to get this result accepted by the scientific community*? ('P_{AF}' in Figure 2). From the point of view of logic a single singular sentence like $\sim d_1$, falsifies H_1 (assuming of course that the initial conditions are not (now) problematized). But in research praxis the singular sentence must first be "accepted" by the scientific community, to secure intersubjectivity.

2.222. For this purpose the *assumption* is made that the phenomenon described by $\sim d_1$ is a reproducible effect. From a sentence of the form 'Bx' as a clue, we conjecture a general statement whose simplest form is $(x)_{x \in A'} - Bx$, where $A' \in A$ or possibly, even $A' = A$. If corroborated, the latter type of statement would constitute an actual falsifier of H_1. How to get, if possible, this hypothesis corroborated? Attempt to *falsify* it! If the falsification attempt gives positive results, i.e. if other experimentors are unable to produce the effect, the scientific community will not accept the candidate for the role of the actual falsifier: the AF-candidate is rejected, and H_1 has been corroborated to a corresponding extent. If the attempt gives negative result, we have produced an authentic falsifier of H_1: unless another part of the premisses is problematized, H_1 has been falsified. (Here we see the *"master-rule"* (of Figure 1) in action, the rule which forbids immunization tactics: it forbids that H_1 is *not* considered falsified *without* some elements of the set of statements used in the argument – as auxiliary theory or as initial condition – being simultaneously problematized.) If $A' = A$ the AF is the contrary of H_1 – the clash is even stronger than a contradiction of H_1. If $A' \neq A$ then H_1 has been falsified at least in its present form. In accepting an AF we take a calculated risk because a falsifying hypothesis may turn out to be false: the action of accepting (rejecting) is, again, decision-making with all of the attendant risks, but supported by good reason assays.

Much like the attempt to explain E gave rise to an improvement of E or to an improved successor of E, viz. to H_0, *the falsification may give rise to an improvement of the derived hypothesis that has been falsified:* to the production of an improved successor to H_1, because it gives rise to an objective problem: to investigate its "realm of application or validity", or strictly, to increase its "precision" (*LScD* expression for degree of determinateness, information content of the predicates in the law hypothesis) and/or to specify more accurately the realm of application (*LScD* "level of generality") of the law hypothesis in question. For instance, to continue our simple example, where H_1 was a rudimentary version of the so-called gas-law (pv=constant), we may now get an improved successor H_1', stating, roughly, that for gases 'pv=constant' holds only for temperatures that keep within certain specified limit values. Attempts to falsify H_1' then in turn give rise to further specification, increase in information content.

If H_1 is considered falsified (never conclusively, for even if all the deductions involved are valid, there are (at least) the two decision-points just mentioned), then we are back to the "theoretical" part of theory-appraisal. This ground has already been covered in Section 2.21. Logic cannot tell us which part of the premisses to blame for the falsification. Popper focusses on the type of research situation where the testing of *T*, the processing of *T*, is at the center of concern and hence the additional premisses are not problematized without there being special reasons. This is the type of situation in which, following a crisis situation, a new theory has emerged which promises to be a major innovation, progress of an extra-ordinary magnitude: sometimes dramatized into a "scientific revolution". Kuhn focusses on the type of situation where a theory, or its "puzzle-solving" potential, is being exploited in the context of solving puzzles of a specified kind. But this type of situation cannot be perpetuated, and on this view the change-over to a new theory becomes *just* a Gestalt-Shift. Lakatos sees the Kuhnian puzzle-solving as work in the protective belt of an *RP* and accounts for the persistence of the hard core as the result of a methodological decision not to modify – and this is indeed a convention! In his rational reconstruction of the history he thus accounts for the continuity which Kuhn explains (psychologistically) in terms of the scientist's commitment to his paradigm. We would think that the continuity is the result of exploiting the *ISF*, especially the preconceptions, the "cosmological" hypotheses – without such a policy (world-3) there would remain only Penelope strategies[49] –, but that the theory over its history or, if your prefer, the series of theories produced under the guidance of the *ISF* has only limited possibilities of being processed and generalized, if only because its basic model must also have some negative aspects: when one tries to apply the theory to new realms of application, then these negative aspects eventually become central and the theory is bound to get into serious troubles.[50]

The recognition that the above-mentioned continuity is beneficial, and even indispensable, has given rise to the so-called Principle of Tenacity (Feyerabend, Lakatos): some researcher should try to make the most of their pet theory even if its recent performance has been deplorable. This principle can be found in Popper too.[51] At first sight it may look as if it conflicted with the so-called Principle of Proliferation (Pluralism): some should seek new alternatives even if the performance of the veteran theory has been fine. But both principles are necessary in order that the scientific community may achieve a desirable measure of diversification, distribution of risk. And even with a highly diversified portefeuilles future gains remain uncertain – to the despair of the science policy maker concerned with setting priorities for projects within a certain discipline.

What the theory of science can do is, in terms of the two illustrative figures, to make each of the "stations" in either of the figures subject to "blow-up" accompanied by case studies from the history of science. Yet the nature of the "stations" and of the methodology as a whole remains *prescriptive* rather than descriptive – as Popper has always emphasized;[52] and the case studies culled from the history (descriptive in intent even if an element of interpretation in unavoidable) cannot contradict let alone "falsify" a prescriptive image of science. They serve for purposes of illustrations and are indispensable for getting clues for the possible improvement of methodological gambits and moves, as sources of conjectures which have to be subjected to critical scrutiny. This once again brings home the point that the appraisal of methodologies is very different from the methodological appraisal of physical theories.

Ruhr-Universität Bochum

NOTES

* I wish to thank fil. lic. Gunnar Andersson for valuable discussions and constructive criticism of an earlier version of this paper.

[1] *LScD*, p. 273.

[2] The basic thrust at the bottom of Popper's entire philosophy is the fight against relativism (cf. *OS*, Vol. II, pp. 369ff. and *Sch*, Vol. II, p. 1080): against intellectual relativism in *Logic of Scientific Discovery, Conjectures and Refutations* and *Objective Knowledge*, and against "moral" and "practical" relativism – which is to a large extent based upon intellectual relativism and which, by relativizing the concept of truth, prepares the way for the historicistic myth of history – in *Poverty of Historicism* and *The Open Society and its Enemies*.

[3] In *Inquiry* **17** (1974), 500.

[4] *LScD*, pp. 50, 278f.

[5] *LScD*, p. 53.

[6] *LScD*, p. 61.

[7] *CR*, p. 229.

[8] *OK*, pp. 57, 59.

[9] *OK*, p. 258; *LScD*, p. 281; *CR*, p. 222.

[10] *CR*, p. 215.

[11] *OK*, p. 103, 143; *LScD*, p. 268, n*2; *CR*, p. 234.

[12] Verisimilitude alone, of course, is not sufficient for a full account of the goal. In the first place, knowledge that "approaches" truth, or an "adequate" theory, must be *relevant to the problems* to be solved in the research in question. Further, one requires an *evaluation of the problem itself*, e.g. in terms of the significance it possesses for the development of a particular scientific discipline. Finally, one must ask what kind of *contribution* a particular theory or discipline can make to the *improvement of our "world picture"* (in order to gain a better understanding of man and world – the most important task of science according to Popper, cf. *OK*, p. 153, the task of "getting to know the beautiful world and ourselves", *OS*, Add. 382, or what kinds of *application* this knowledge might have in *other disciplines or technologies*, which are valued in their own right. In short, one must explicate the ways in which a specific research project is embedded in "science as a whole", and the relevance of science itself for the life of man. This problematic leads us far beyond the problems of methodology and its justification (and Popper is thus justified in not treating them). (cf. Radnitzky, 1976)

[13] *LScD*, p. 274; *CR*, p. 223f.; *OK*, p. 328.

[14] The following is a somewhat pedantic note, which is included here merely to preclude certain types of objections. Tarski's concept of truth can be applied only to *type* sentences in *formalized* languages (not to instances of these forms, to token sentences in natural languages, where truth-value is context-dependent). Tarski defined a specific concept of truth for a specific formalized language, and he also offered a method for defining the concept for a specific class of formalized languages. Popper, of course, does not use Tarski's definition itself; he doesn't even state it. That is not only unnecessary for his purpose, it would also be totally inappropriate (Tarski needed 145 pages to present his definition). What he in fact *uses* is a kind of meta-criterion, which we might designate, following Carnap, as the *requirement of a sufficient similarity between explicandum and explicatum*: a "similarity criterion". Roughly: an explicatum of the concept of propositional truth is adequate if and only if we can derive, by means of the definition which introduces the explicatum expression into the meta language, every sentence of the object language which has the following form: the sentence '(write any sentence here, i.e. mention it)' is true if and

only if (write exactly the same sentence here, i.e. use it). Thus, (1) the assertion that Tarski's theory is a rehabilitation of the concept of propositional truth (cf., e.g. *OK*, pp. 44, 60, 308, 314, 323, 328) *could* easily lead to misunderstandings, while (2) the assertion that the application of Tarski's definition of natural languages is *trivial* (an assertion often used in the criticism of Popper's appeal to Tarski) *is* a misunderstanding. As to (1): the adequacy criterion mentioned above presupposes the concept of propositional truth, and implicitly recognizes that it is embodied in the semantical structure of every natural language (which could not function without it). As an adequacy criterion it cannot "add" anything to the explicandum. And this adequancy criterion is the only thing Popper de facto uses. If (2) is taken literally, it is simply meaningless, for it is impossible to apply the definition formulated in the formalized language for a formalized language to a natural language. If, on the other hand, the similarity criterion is meant, then it is (2) which is trivial, for it is clear that it is impossible to get more information concerning the explicandum out of the meta-criterion than was put into it in the first place. Tarski's definition serves Popper as a *paragon* (and the formal considerations serve for the "logical underpinning" of the theory of verisimilitide). In such a context it seems rather pointless to emphasize the well-known differences between formalized and natural languages.

[15] For a brief overview see, e.g. Musgrave (1974).

[16] The logical underpinning of the intuitive concepts Content and Verisimilitude has its well-known difficulties and dangers. As regards Empirical Content of Information, there is no measure of content since there is no unit for measuring amount of information (we have only examples which clarify the intuitive notion). Attempts to define Content in terms of Probability are likewise notoriously problematic. "*Absolute* logical probability" is intelligible only for *events* (such as dice throwing). (The label 'relative logical probability' appears to be a misnomer for "degree of confirmation on the basis of evidence".) To speak of the absolute logical probability of a *theory* has absolutely no cash value. When '*p*' occurs in such formulae as, e.g., Popper's $Ct(a) = 1 - p(a)$, the grammar of (p) is said to be given by the probability calculus; but the *semantics* of '*p*' remains an open question: in interpretation the *OL*-concept is simply not available – in everyday language it makes no sense to talk about the absolute logical probability of a theory! As regards *Verisimilitude* even if it should turn out that the *formal* underpinning of the concept of Verisimilitude (in particular the quantitative concept) cannot be developed to handle objections of the type brought forward by David Miller, Pavel Tichý and others, this would in no way be a catastrophe for Popperian methodology, or more strictly, for its *justification*, because in that context one can manage quite well with the intuitive concept of Verisimilitude. (In the methodology itself at any rate only comparative concepts are required, cf. also note 17.)

[17] It is a question of *comparative* judgements, not of numerical values. In the same vain in which Popper refuses to bring numerical values into play for the concept of absolute logical probability (as a concept "complementary" to the concept of Empirical Content) (*LScD*, pp. 119, 270; *OK*, p. 59), he does not envisage numerical values for the degree of corroboration (*LScD*, p. 268; *OK*, p. 59). As mentioned in note 14 the formal studies serve as the logical underpinning – with their help one is supposed to get a better grip on concepts like Verisimilitude. For the development of the methodology these studies are dispensible – hence they rightly are placed as Appendices.

[18] *OK*, p. 162.

[19] See, e.g. Radnitzky (1974c), esp. pp. 150–156.

[20] *OS*, App.; *OK*, Chapters 1 and 2.

[21] E.g. *CR*, p. 226; *OK*, pp. 46, 317.

[22] *OK*, p. 103; *CR*, p. 235; Lakatos in *Sch*, Vol., p. 258.

[23] Kant, *Urteilskraft*, 1795, VII.
[23a] A critique of this position is given in § 3 of (Radnitzky, 1976).
[24] Kuhn's position is a typical example of this form of historical relativism. We commit ourselves by a sort of leap of faith to an élite-constituted paradigm. According to Kuhn's paradigm-monopoly thesis there prevails consensus about the paradigm during periods of "Normal Science", a situation which can change during periods of "scientific revolution". (See the critique of Watkins (1970), pp. 35ff.) Toulmin seems to treat the concept of truth in terms of empirical consensus; while Kuhn (esp. in the last chapters of *The Structures of Scientific Revolutions*), being concerned with "development", not with development towards a special goal, loses the problematic of theory-appraisal over paradigm-shifts altogether. Likewise Toulmin (1972), e.g., p. 231, on the concept of sufficient degree of consensus, or p. 259 "What is 'justifiable' is what is found 'justifiable'" brings into play empirical consensus. The influence of Wittgenstein II is noticeable both in Kuhn and in Toulmin.
[25] From this point of view Popper is charged with losing the transcendental subject. This criticism, which may be justified, misses the point of Popper's ontology: the first question concerning Popper's ontology is whether or not it fulfills the role it was meant to perform, namely that of connecting the methodology (in particular criteria for the acceptance of hypotheses and theories) and a definition of truth (the concept of truth as correspondence), which definition is needed to specify the *goal* of the research praxis which is to be guided by that methodology. As already mentioned, the ontology was not designed to be used in the "Aufbau der Welt" – to say this does, however, *not* imply the denial of the thesis that the entities of Popper's World-3 have to be given *some* ontological status in any adequate ontological groundplan.
[26] It goes without saying that the correspondence notion of truth, without which it is in principle impossible to overcome historical relativism, is "relative to", *presupposes an already developed language*. (That Tarski's definition is relative to a (fixed) *formal interpretation*, which alone makes the syntactic system into a so-called formalized "language".) But it is timeless in the sense in which a proof, if correct, remains so completely independent from the existence or non-existence of a consensus about its validity, and remains so even if the human species should cease to exist. [Of course if other beings with minds sufficiently similar to the human intellect should later on come across the written proof and decipher it, it *would* be true for them too. To accept this is very different from insisting that the concept of truth is "meaningful" only as "true *for* a knowing subject" (no truth without a knowing subject, human subject that is – as the secularization of "knowledge existing in the mind of God"), which would oblige us to hold that the truth of the proof ceased to exist with the extinction of mankind, and came into existence again when the proof was made the intention of some other "minds" – which brings us close to the absurd ontological position of subjective idealism. This itself is an argument for ontologizing entities of Popper's World-3.] This is in no way to deny that the proof or theory is a human product, or that language, the existence of a communication community, etc. are conditions of the possibility of its being constituted as meaningful. The philosopher of science can appreciate the investigations of conditions of possibility of methodology itself as an important *philosophical* task, in spite of the fact that they are irrelevant for the development and justification of *methodology*. The "transcendental pragmatics of language" is interested in something completely different: in founding a minimal ethic with the aid of the concept of the ideal communication community, and thus in the *rehabilitation of practical philosophy carried out with a "practical" intent*.
[27] *LScD*, p. 44.
[28] *LScD*, p. 44.

[29] *LScD*, p. 54.
[30] The affinity between Popper's philosophy of Open Society and the Philosophy of the Ideal Communication Community of Apel and Habermas (contrary to the impression one gets from the debate in Germany between Albert and Habermas) is lad out in Vol. II of (Radnitzky, 1968, 1970).
[31] *OS* II, p. 231.
[32] I deliberately speak of a "Popperian" methodology (a methodology in a Popperian spirit) not of Popper's methodology to indicate that in this context what matters is not an exegesis of certain texts but the problems themselves.
[33] Cf., e.g. Radnitzky (1974b), pp. 44f.; Radnitzky (1974a), pp. 80, 86.
[34] E.g. *OK*, pp. 160, 118f., 147.
[35] Cf. *LScD*, pp. 38, 277f.
[36] Preconceptions at the level of world-picture hypotheses do not belong to the theories themselves. E.g. among the preconceptions characteristic for the *ISF* of "Newtonian" type of theory would be the ontological posit that the world (that world with which physics is concerned) consists of matter spread out in empty space; in such a world, for two bodies not in contact with each other (not even through intermediation of other bodies travelling from one to the other) to interact causally, the simplest thing to assume is "simultaneous action at a distance" as a causal connector, (cf., e.g., Radnitzky (1974b), p. 33) etc. For a discussion of the concept of "preconceptions" cf., e.g., Radnitzky (1974a), 212, Radnitzky (1972), pp. 141–147; on the role of *preconceptions in research* cf. Radnitzky (1974b) with that title, esp. pp. 27ff. and for examples from the "distilled history of science", pp. 32–43. Imre Lakatos confirmed to me in personal conversation that the "hard core" of an *RP* basically consists of such preconceptions; but there are passages in his writing which suggest that the hard core is made up mainly of the central components of the theory, which, over its history, constitutes the *RP*. Of course, the concept of the *ISF* is close to one of the many meanings of the Kuhnian "paradigm".
[37] Such as the "master-rule" mentioned in Section 1.7 a.i., note 29.
[38] Cf., e.g., Radnitzky (1974b), pp. 7f. The claim that the rules apply to all research should not be taken to imply any form of "unity of science" thesis. We only wish to imply that the *sciences humaines* – a very heterogenous group – too contain certain moments (which Apel calls 'quasi-naturalistic' or 'explanatory' moments) to which the model of Figures 1 and 2 may be adapted. Again, to claim this is not to deny that the typical *sciences humaines*, the *Geisteswissenschaften* have *additional methodological problems of their own*, which this model does not touch upon. Some of them very likely cannot be tackled without taking into consideration "knowledge-constitutive interests": there is a great difference between the task of "giving an explanation of understanding" (*Erklären des Verstehens*) and the task of "understanding explanation" (*Verstehen des Erklärens*).
[39] Cf. *OK*, p. 14; Lakatos and Musgrave (1970), p. 220; Radnitzky (1974a), p. 88.
[40] Cf., e.g., Radnitzky (1974b), pp. 45f.; Radnitzky (1974a), esp. p. 86.
[41] *OK*, p. 259: 'growth of our knowledge... as consisting throughout of corrections and modifications of previous knowledge", also *OK*, p. 71. "Erst ein Wissen bringt ein Nicht-Wissen zu Bewußtsein" (Popper, 1962); "we never start from scratch" – and moreover (the growth of knowledge), "is largely dominated by a tendency towards increasing integration towards unified theories" (*OK*, p. 262), likewise *LScD*, p. 276).
[42] *LScD*, p. 50, fn.*1.
[43] Fallibilism as the epistemological idea of the fallibility of human Reason again exhibits affinity with the position of the hermeneutics: the idea of human finitude in Heidegger, the recognition of the various limitations of man, of., e.g. (Gadamer, 1965) esp. pp. 340f.

The affinity is there – even if drawing attention to it may not be considered good taste among Popper aficionados.

[44] Cf., e.g., *OK*, pp. 200f.

[45] E.g. Radnitzky (1974a), 2212 and esp. fn. 42.

[46] Apart from the difficulties with the interpretation of those concepts other than relative frequency which are also called 'probability' merely because their grammar can be given by means of the probability calculus, but whose interpretation is highly problematic (cf. note 16), *theories of confirmation* appear to face a dilemma. The more of the information content of a hypothesis is "covered by" the information carried by the evidence, the better supported is the hypothesis, given that evidence. A maximum of support would obtain if (per impossibile) all of its information were "covered" in this manner. But then the hypothesis would not contain anything new, no information going beyond that of the evidence. However, we wish to have theories which condense a wealth of information – go far *beyond* the information of any finite set of data sentences – so that by means of them much new knowledge can be generated. It almost looks as if, speaking metaphorically, the champions of a theory of confirmation would like to play the game of science safe – but that would not be the game of science any more.

[47] In this context only the logical (information-theoretical) and epistemic aspects of "theory"-appraisal have been dealt with. When the problematic of appraisal is spelled out in more detail, it becomes necessary to propose (and try to legitimate) criteria not only for "knowledge" claims, but also for other sorts of results of research such as instruments (software and hardware), problems, strategies and actions such as decisions etc. We already hinted at that the knowledge component will be central (Section 1.0 s.f.). However, also *within* the component of the results "knowledge", one will have to distinguish various *levels* or types of knowledge: e.g., data as raw material (not yet candidates of incorporation into the "body of scientific knowledge"), law hypotheses, explanatory patterns, systems of explanatory patterns (such as the one declared to be the output of the first station of "theory-appraisal" in Figure 2 (cf. Section 2.211 a.i.)), theories (such as, e.g., Newtonian theory of gravitation), and systems of theories (such as, e.g., classical mechanics). The criteria for appraising scientific quality will differ somewhat according to level. In Radnitzky (1974a), §231 (pp. 88–92) the dimensions of appraisal have been sketched for the simplest case, for law hypotheses: beside criteria for logical and epistemic aspects, other criteria are required such as, e.g., ones applicable only in the context of research approximating Normal Science: "coherence of law hypothesis and veteran theory (hence with *ISF* underlying the veteran theory)", i.e. a substantive criterion, or the "relevance of *T* for the original problem", etc. It was pointed out there that for every legitimated criterion there is some type of research situation where it has to be kept in abeyance or even waived. (cf. Radnitzky, 1974a, §232, pp. 91ff.)

[48] *LScD*, p. 107 fn.*2; Radnitzky (1974b), p. 24; (1974a), §2212, p. 80, and fn. 75.

[49] Cf. §2.14 s.f. above; the *ISF* is resistant to change if only because the "root metaphors" are few, cf. Radnitzky (1974b), pp. 36, 31.

[50] Cf. note 33.

[51] *CR*, p. 312, fn. 1, 49; *OK*, p. 30.

[52] *LScD*, pp. 50, 278f.

BIBLIOGRAPHY

In the notes the following abbreviations are used of books by Karl Popper:

LScD: The Logic of Scientific Discovery.

OS: The Open Society and Its Enemies.

CR: Conjectures and Refutations.
OK: Objective Knowledge.
further: *Sch:* Schilpp, P. (ed.), *The Philosophy of Karl Popper.*

Apel, K.-O., *Transformation der Philosophie. Sprachanalytik, Semiotik, Hermeneutik.* 2 vols., Suhrkamp, Frankfurt, 1973. (Engl. transl. in prep. at Routledge & Kegan, London.)

Feyerabend, P., 'Popper's Objective Knowledge', *Inquiry* **17** (1974), 475–508.

Gadamer, H.-G., *Wahrheit und Methode*, Mohr, Tübingen, 1960, 2nd ed., 1965, 3rd ed. 1972. (Engl. transl. forthcoming from Sheed & Ward, London.)

Lakatos, I., 'Falsification and the Methodology of Scientific Research Programmes'. In Lakatos and Musgrave (eds.) (1970), pp. 91–195.

Lakatos, I., 'Popper on Demarcation and Induction', in Schilpp (1974), Vol. I, pp. 241–273.

Lakatos, I., Review article on Toulmin, (1972), Unpublished Manuscript, to be published in *Collected Works* (ed. by John W. N. Watkins).

Lakatos, I. and Musgrave, A. (eds.), *Criticism and the Growth of Knowledge*, Cambridge Univ. Press, London, 1970.

Musgrave, A., 'George Boole and Psychologism', *Scientia* (July–August 1972).

Musgrave, A., 'The Objectivism of Popper's Epistemology'. In Schilpp (ed.) (1974), Vol. I, pp. 560–596.

Popper, K., *The Logic of Scientific Discovery*, Hutchinson, London, 1959. (German original *Logik der Forschung* in 1934).

Popper, K., *The Open Society and Its Enemies*, 2 vols., Routledge & Kegan Paul, London, 1945ff.

Popper, K., *Conjectures and Refutations, The Growth of Scientific Knowledge*, Routledge & Kegan Paul, London, 1963.

Popper, K., *Objective Knowledge, An Evolutionary Approach*, Oxford University Press, London, 1972.

Radnitzky, G., 'Towards a "Praxiological" Theory of Research', *Systematics* **10** (1972), 129–185; reprinted in the paper edition of

Radnitzky, G., *Contemporary Schools of Metascience*, Regnery, Chicago, 1973 (original edition, Humantities, New York, 1968, 1970).

Radnitzky, G., 'From Logic of Science to Theory of Research', *Communication and Cognition* **7** (1964), 61–124. (1974a)

Radnitzky, G., *Preconceptions in Research*, Literary Services & Production, London, 1974. (1974b)

Radnitzky, G. 'Vom möglichen Nutzen der Forschungstheorie', *Neue Hefte für Philosophie* **6/7** (1974), 129–168. (1974c)

Radnitzky, G., 'The Intellectual Environment and Dialogue Partners of the Normative Theory of Science', forthcoming in *Organon*, No. 11, (1975).

Radnitzky, G., 'Prinzipielle Problemstellungen der Forschungspolitik', forthcoming in *Zeitschrift für Allgemeine Wissenschaftstheorie*. (1976)

Russo, F., 'Typologie du progrès des connaissances scientifiques', *Revue des Questions Scientifiques* **145** (1974), 345–363, 479–502.

Schilpp, P. (ed.), *The Philosophy of Karl Popper*. (The Library of Living Philosophers), 2 vols., Open Court, LaSalle, Ill., 1974.

Toulmin, S., *Human Understanding*, Vol. I: *General Introduction and Part 1*, Princeton University Press, Princeton.

Watkins, J., 'Against "Normal Science"'. In Lakatos and Musgrave (eds.) (1970), pp. 25–38.

ERHARD SCHEIBE

CONDITIONS OF PROGRESS AND THE COMPARABILITY OF THEORIES

1. I believe that there is progress and continuity in the development of physics. This is perhaps the deepest prejudice standing behind the considerations in this paper. As is well known, it has become necessary to defend this view if one belongs among its advocates because for some time past it is exposed to heavy fire from the guns of Thomas Kuhn and Feyerabend, – to mention only the two most effective champions of the 'new movement' in the philosophy of science. But I believe in meaning change, too. Therefore my chief problem is: How will a successful defense be possible? For several reasons I am unsatisfied with most of the reactions coming from the logical empiricist camp, the main reason being perhaps that their argumentation is too remote from the concrete difficulties presenting themselves, for instance, in trying to compare quantum mechanics or the theory of relativity with their classical predecessors. Such difficulties were pointed out by Feyerabend, but I cannot find any direct answers to them in the aforementioned reactions. The matter is different with Lakatos in so far as his methodology of scientific research programmes is a *positive attempt* to find a criterion for scientific progress. But a thoroughgoing elaboration of this methodology would bring to view that it can not simply outflank the difficulties of Feyerabend's incommensurability type. For one thing, they would appear in the comparison of the successive versions of one and the same research programme. And even if one would succeed in solving the problems to be encountered in this domain, the question would remain, how to compare the hard cores of different research programmes. At a seminar held together with Lakatos in Alpbach in 1973, I had the opportunity to discuss these questions with him, and he encouraged me to undertake further research work related to them. I hope that the following considerations, insufficient as they are, would have found the approval of this admirable man.

Let me begin by giving a somewhat modified formulation of two conditions (henceforth called the *P-conditions*) usually believed to be *necessary*

R. S. Cohen et al. (eds.), Essays in Memory of Imre Lakatos, 547–568. *All Rights Reserved Copyright © 1976 by D. Reidel Publishing Company, Dordrecht-Holland*

conditions for one (physical) theory *B* to constitute progress with respect to another (physical) theory *A* in a common domain of application *Γ*: With respect to a certain correspondence relation between the empirical successes and failures of *A* and *B* in *Γ*

(P_1) every success of *A* in *Γ* corresponds to a success of *B* in *Γ*,

but

(P_2) there are failures of *A* in *Γ* which correspond to successes of *B* in *Γ*.

The first condition expresses the conservation of successes of *A* when transition is made from *A* to *B*. It is therefore a condition of continuity: Would there be any successes of *A* reappearing as failures of *B*, we could hardly speak of *B* as constituting a progress with respect to *A*. The second condition expresses the transformations of at least some failures of *A* into successes of *B*. *Provided that* the first condition is fulfilled the second says that *B* is really an improvement upon *A*.

The foregoing formulation of the *P*-conditions differs from formulations to be found in other publications in that (1) I introduced the domain *Γ*, not meaning thereby the domain of (approximate) validity of *A* relative to *B*, but the domain with respect to which the whole comparison of *A* and *B* is to be performed, and (2) I do not speak (in (P_1)) of successes of *A being* also successes of *B* and (in (P_2)) of failures of *A being* successes of *B*, but rather of successes or failures of *A corresponding to* successes of *B*. By modifying the *P*-conditions in this way I already take measures against objections brought forward by Feyerabend.

A first argument is that "the growth of knowledge or, more specifically, the replacement of one comprehensive theory by another involves losses as well as gains" ([7], p. 219). One of Feyerabend's favourite examples within modern physics is the case of quantum mechanics vs. classical mechanics where it has indeed remained doubtful up to this very day whether the former theory can reproduce every success of the latter in the macroscopic domain ([5], p. 296ff). It may therefore be necessary to relativize the *P*-conditions by explicitly introducing a domain of application for which progress can be claimed. Since I do not want to enter this matter, let me only add one other reason for introducing *Γ*, relevant to the main parts of my paper. There a theory is not viewed as a statement, but rather as something which becomes a statement only when applied to this or that object. Adopting this procedure obviously necessitates the introduction of Γ.

As for a second objection of Feyerabend, I may quote from his [1] where on p. 28f he says with respect "to such comprehensive structures of thought as... Newton's celestial mechanics, Maxwell's electrodynamics, the theory of relativity, and the quantum theory":

What happens *here* when transition is made from a theory T' [A] to a wider theory T [B]... is something much more radical than incorporation of the *unchanged* theory T' (unchanged, that is, with respect to the meanings of its main descriptive terms as well as to the meanings of the terms of its observation language) into the context of T. What does happen is, rather, a *complete replacement* of the ontology (and perhaps even the formalism) of T' by the ontology (and the formalism) of T and a corresponding change of the meanings of the descriptive elements of the formalism of T' (provided these elements and this formalism are still used).

In its context this objection is directed against the traditional empiricist schemes of explanation and reduction in order to show that these schemes can not be used to yield conditions of progress of the form: B explains the successes (and even the failures) of A and it makes additional and successful predictions in cases where A failed. I have avoided using the terminology of explanations in formulating (P_1) and (P_2) because this would only multiply the difficulties. But it is quite clear from all the relevant papers written by Feyerabend that the theory expressed in the text quoted can be viewed as directly turning upon the P-conditions in their unspecified form, viz. with 'to be' instead of 'to correspond to'.

As a matter of course, for the time being my suggestion to choose the correspondence formulation is nothing but a *façon de parler*. Even at the end of the paper, the reader will be provided only with some hints as to which direction the solution of the problem may be found. In Section 2, I shall give a first 'to be'-interpretation of the P-conditions. In this trivial interpretation the P-conditions turn out to be equivalent to Popper's definition of (qualitative) verisimilitude. On account of recent results obtained by other authors they are thus at variance with the principle of fallibilism and therefore untenable. The investigation is nothing but a nice logical exercise. However, from the viewpoint of logical analysis which I shall adopt, it represents an ideal case of what can be achieved in this field as regards logical rigour. We are provided with a method showing what it would mean to dismiss the P-conditions also in the non-trivial cases: Produce verifiable conditions contradicting the *P*-conditions. Now, what Feyerabend has produced in this respect is his concept of *incommensurable theories*. According to Feyerabend, theories T' and T as they are described in the above quotation are incommensurable. At other places he characterises theories as incommensurable "when the meanings of their main descriptive terms depend on mutually inconsistent principles" ([4], p. 227, n. 19, also [8], p. 98ff) or when one of them "does not even permit to formulate statements expressing... states of affairs" described by the other theory and when therefore the two theories "cannot share a single statement" thus being "completely disjointed" ([6], p. 82). These indications taken together with the illustrations given by Feyerabend certainly have an intuitive plausibility. But they are not precise definitions. Therefore it remains somewhat questionable how the incommensurability of theories is to be used in order to discuss conditions of progress. It is not even clear whether that relation holding between two theories does not permit us to *meaningfully formulate* *P*-conditions for these theories, or whether, although this being possible, the *P*-conditions simply *are not satisfied* in that case.

In order to clear up the matter, and by now not being able to give a precise definition of the notion of incommensurability, I asked myself: Which cases are clear cases of *commonsurable* theories? Perhaps no two theories are commensurable and the whole thing is a matter of degree. Nevertheless, in Section 3 I begin an analysis of the concept of a physical theory, and I arrive at (if only) a general description of what I (and probably also Feyerabend) would call clear cases of commensurable theories. Accordingly a second 'to be'-interpretation of the *P*-conditions is possible, this time one which is compatible with fallibilism. The main idea is to analyse physical theories as consisting of two parts, one of which presupposes the other. Theories coinciding in the basic part will be commensurable, and the *P*-conditions can be realized by introducing explicitly the limitations set to the accuracy of measurements. The considerations in this section perhaps show some similarity to the only two relevant papers by Feyerabend in which he is more specific on a general level of thought ([2] and [3]).

The last section will be devoted to an outlook concerning the serious cases in which incommensurability comes into play. In order to enforce the possibility of progress and continuity also in these cases I suggest that we make use of a conceptual correspondence as it can be found even in pairs of theories like classical mechanics and quantum mechanics, or classical field theory and quantum field theory, etc. Although the basic parts of such theories (in the sense to be developed) are different and even somehow incompatible, this does not exclude the possibility that we may *analyse* such differences. And such an analysis eventually will enable us to relate the possible empirical successes and failures of the theories in such a way that the *P*-conditions in their 'correspondence'-form, as given above, will hold. Accordingly it will enable us to understand the successes as well as the failures of a theory which is superseded by another. As a matter of course, such an understanding will only be possible in the light of the new theory, and it may even take some time until it is accomplished. Moreover, at the time of the advent of a new theory a total embarrassment may occur in the minds of all adherents of the old theory. But this must not, and will not, by itself prevent progress in the sense that eventually everybody employed in a certain field of research will have understood what has been going on according to criteria which are more stable than the theories to be appraised by them.

In the following I shall assume the reader to be familiar with the elements of mathematical logic and, chiefly, model theory. As regards the foundations of logic, I shall adopt the semantical approach with the concept of a logical system as basic. A functionally complete set of logical constants will be assumed to be fixed in number and interpretation. Under this assumption the logical systems can be characterized by their underlying languages. The languages and, consequently, the theories and structures over them may be of any complexity as regards the sorts and types of their variables and descriptive constants. In particular, for the logical reconstruction of physical theories it will be adventageous to have many-sorted theories at hand. One thing, however, must be guaranteed at all events: it must be possible to define the fundamental relation of the validity of a formula in a structure. How this can be done for a very extensive class of logical systems is shown in [10]. For first order logic the reader may be referred to [17]. Many-sorted logical systems are treated in [11].

2. For the first interpretation to be given of the *P*-conditions, let us think of our world as being a huge structure of unimaginable complexity, not

yet equipped with a language. Let us furthermore think of the scientists as trying to acquire knowledge about this structure. For this purpose they introduce languages and give them a well defined interpretation in this or that part of the world. They will also make theories about these parts telling us what is going on therein. I shall assume that their theories are consistent. Thus the elements of our knowledge will be triples (L, A, u), L being a language and A and u being a consistent theory and a structure over L respectively. Let S be the set of sentences of L. With respect to A the set S is a direct union

(1) $$S = Q_A \cup R_A \cup U_A$$

where Q_A is the set of theorems of A (within S), R_A the set of their negations, i.e. the set of sentences contradicting A, and U_A the rest, i.e. the set of sentences undecided by A. With respect to u the set S is a direct union

(2) $$S = T \cup F$$

with T and F as the set of true and false sentences in u respectively. Finally we have the intersections Q_A^T, Q_A^F, R_A^T,... etc. of Q_A, R_A and U_A on the one side and T and F on the other. Q_A^T and Q_A^F are usually called the truth content and the falsity content of A respectively.

Let us now ask for the successes and failures of an enterprise as characterized by an (L, A, u). It is in the spirit of model theory to take the sentences of L as the things of which we say that they are possible successes and failures (with respect to an (L, A, u), of course). But which of them are the successes and which the failures? I think we can classify without hesitation the sentences of Q_A^T and R_A^F among the successes and the sentences of Q_A^F and R_A^T among the failures. The classification of the sentences undecided by A is somewhat problematic. We could argue that we may look upon the $\alpha \in U_A^T$ or $\alpha \in U_A^F$ as being successes because in the first case it has been avoided that there are true sentences among the sentences contradicting A, whereas in the second case false sentences have been prevented from being theorems of A. But with the same right it may be argued that these sentences are failures because in the first case true sentences have not been recognized as theorems of A and in the second case we failed to recognize false sentences as being refutable by A. With the first argument we would emphasize the avoidance of a catastrophe.

With the second argument we would point to the (possible) incompleteness of A.

There is one argument in favour of the second option. In our P-conditions we do not distinguish between big and small successes and failures. On the other hand, whether we take the undecided sentences as successes or failures, compared with the sentences in $Q_A^T \cup R_A^F$ (sure successes) and $Q_A^F \cup R_A^T$ (sure failures) they would at any rate be small such things. Thus, if we would take the undecided sentences as successes it could very well happen that, according to our condition (P_1) (conservation of successes), a 'big' success of the theory $A(\alpha \in Q_A^T \cup R_A^F)$ will become a small success of the theory $B(\alpha \in U_B^T \cup U_B^F)$. Contrary to the intentions pursued with the P-conditions we would then have a change for the worse and not a progress. If, however, the undecided sentences are classified among the failures, according to (P_2) the situation could only change for the better. Therefore it will be advisable to exclude the case where all undecided sentences of A are among the successes and to restrict the following investigation to the two main cases

$$(3)\qquad \begin{cases} \text{successes: } Q_A^T,\ R_A^F, \text{ either } U_A^F \text{ or } U_A^T \\ \text{failures: } Q_A^F,\ R_A^T, \text{ either } U_A^T \text{ or } U_A^F \end{cases}$$

and

$$(4)\qquad \begin{cases} \text{successes: } Q_A^T,\ R_A^F \\ \text{failures: } Q_A^F,\ R_A^T, U_A^F,\ U_A^T. \end{cases}$$

Having fixed the situation for the case of one triple (L, A, u), we will now have to face the comparison of two of them, say (L, A, u) and (L_1, B, u_1). I shall restrict the consideration to the case where $L_1 = L$ and $u_1 = u$. We then have two theories A and B about the same structure u over the language L.

Beginning with (3), it turns out that in each of these two subcases the P-conditions are equivalent to

$$(\mathrm{PP})\qquad \begin{array}{l} Q_A^T \subset Q_B^T \quad \textit{and} \quad Q_B^F \subseteq Q_A^F \quad \text{or} \\ Q_A^T \subseteq Q_B^T \quad \textit{and} \quad Q_B^F \subset Q_A^F. \end{array}$$

The proof for the first of the subcases of (3) is as follows. In terms of (3) the condition (P_1) becomes

$$(\mathrm{P}_{13})\qquad Q_A^T \cup R_A^F \cup U_A^F \subseteq Q_B^T \cup R_B^F \cup U_B^F.$$

Obviously, this is equivalent to

$$Q_A^T \subseteq Q_B^T \quad and \quad R_A^F \cup U_A^F \subseteq R_B^F \cup U_B^F.$$

Taking the complement (in F) of both sides of the inclusion at the right and taking account of (1) and (2) this is readily seen to be equivalent to

$$(5) \qquad Q_A^T \subseteq Q_B^T \quad and \quad Q_B^F \subseteq Q_A^F.$$

Thus (P_{13}) is equivalent to (5). Turning to (P_2) within the context (3), this condition now reads

$$(P_{23}) \qquad (Q_A^F \cup R_A^T \cup U_A^T) \cap (Q_B^T \cup R_B^F \cup U_B^F) \neq \emptyset.$$

Evidently, this is equivalent to

$$(6) \qquad Q_A^F \cap (R_B^F \cup U_B^F) \neq \emptyset \quad or \quad Q_B^T \cap (R_A^T \cup U_A^T) \neq \emptyset.$$

Looking at the identities

$$(7) \qquad \begin{aligned} Q_A^F &= (Q_A^F \cap Q_B^F) \cup (Q_A^F \cap (R_B^F \cup U_B^F)) \\ Q_B^T &= (Q_B^T \cap Q_A^T) \cup (Q_B^T \cap (R_A^T \cup U_A^T)) \end{aligned}$$

and taking account of the equivalences of (P_{13}) with (5) and (P_{23}) with (6), we see at once the equivalence of (P_{13}) and (P_{23}) with (PP). The reasoning for the other subcase of (3) is similar.

The matter is different in our second main case (4). Here the P-conditions are equivalent to

$$(PP') \qquad Q_A^T \subset Q_B^T.$$

To see this we start with the reformulation

$$(P_{14}) \qquad Q_A^T \cup R_A^F \subseteq Q_B^T \cup R_B^F$$

of (P_1) in the light of (4). Obviously, (P_{14}) is equivalent to

$$(8) \qquad Q_A^T \subseteq Q_B^T.$$

(P_2) becomes

$$(P_{24}) \qquad (Q_A^F \cup U_A^F \cup R_A^T \cup U_A^T) \cap (Q_B^T \cup R_B^F) \neq \emptyset,$$

and this is equivalent to

$$(9) \qquad R_B^F \cap (Q_A^F \cup U_A^F) \neq \emptyset \quad or \quad Q_B^T \cap (R_A^T \cup U_A^T) \neq \emptyset.$$

Since $\alpha \in Q_C^F$ iff $\neg \alpha \in R_C^T$ and $\alpha \in U_C^F$ iff $\neg \alpha \in U_C^T$ for any theory C, the members of (9) are equivalent to each other. Together with (8) the second of the identities (7) immediately gives the announced result.

Looking back at our previous result, the first thing to be remarked is that (PP) is essentially Popper's definition of qualitative verisimilitude ([13], p. 233). What has been shown can therefore be expressed by saying: In the present context, especially with one of the subcases (3) as the classification of successes and failures, progress in the sense of the above P-conditions is the same as approximation to truth in the sense of Popper's definition of qualitative verisimilitude. Of course, apart from the technical details, this does not at all come as a surprise. The really important thing about our result is that recently Miller and Tichý have shown that (PP) in its turn is equivalent to

$$(10) \qquad Q_A \neq Q_B, \quad Q_A^T \subseteq Q_B^T \quad \text{and} \quad Q_B^F = \emptyset$$

([12], [18], and also [9]). From this result it can immediately be seen that (PP) is at variance with the principle of fallibilism so much emphasized by Popper. For this principle states that, although progress may be possible in science, there will always be some falsities in any given state of its development. In our present context this can only mean that $Q_A^F \neq \emptyset$ and $Q_B^F \neq \emptyset$ which clearly contradicts (10). Now, the principle of fallibilism could itself be false, in which case it could do no harm to the definition (PP). But if anything mirrors the real state of affairs in science, then it is this principle. Therefore Popper's definition of verisimilitude has to go and with it our P-conditions in the present interpretation (P_{13}) and (P_{23}).

What about our second version (P_{14}) and (P_{24}) of these conditions? It proved to be equivalent to (PP′). In [18], p. 157 and [9], p. 165 it is reported that Popper, after learning of the Miller-Tichý result, suggested (PP′) as a new definition of verisimilitude. Our second result shows that this comes down to a move from (3) to (4) as regards the undecided sentences of our theories. But we are not better off with this alternative: In the papers mentioned it is shown that, although two theories A and B satisfying the new definitions can both be false, B would be – in this case – an extension of $A: Q_A \subset Q_B$. But in science, improving a false theory can not simply mean to replace it by an extension. On the contrary: the historically important cases show that as a rule the new theory contradicts the old one. Thus (PP′), too, is untenable, and the same holds for our second version of the P-conditions.

3. There are several reasons why the approach followed in Section 2 is doomed to failure. In a certain sense the whole construction is misconceived. At any rate it is too remote from the actual methods used in physics. In particular, we can not have all the sentences of a physical language as possible successes and failues without having an inductive logic at our disposal. But it is not necessary to enter so controversial a matter as is inductive logic. One can avoid this by restricting the possible successes and failures to sentences which are experimentally accessible. But even this would not be sufficient for our purpose unless experimental accessibility would include the limitations set to the accuracy of measurement: Unless there would be such limitations, none of our physical theories could have any success at all. We would just have to hit the truth with them. Thus, bearing this idea in mind, let us make a fresh start in order to find a second and more satisfactory interpretation of the P-conditions, compatible with fallibilism. I start with an analysis of the concept of a physical theory, proceeding in three steps.

In the first step I shall argue that in analysing physical theories as they are to be found in the textbooks of theoretical physics we always meet with a pair (u_0, A_0), where u_0 is a structure and A_0 is a theory. The language $L(A_0)$ of A_0 is an extension of the language $L(u_0)$ of u_0, including the possibility of introducing new kinds of variables. Furthermore, A_0 is logically contingent with respect to the theory $\tau(u_0)$ of u_0. The theory

(11) $$A_0' = \tau(u_0) \cup A_0$$

does *not* already contain what is usually called the fundamental laws of a physical theory. Rather the structure u_0 and the theory A_0' or, equivalently, the pair (u_0, A_0) is appointed to yield the *presuppositions* necessary for formulating these laws in a well defined manner. It is, if not impossible, then at any rate very inexpedient to formulate, say, the gas laws, the Schrödinger equation, Maxwell's equations, Newton's gravitational equations, etc. in such a way that any conceivable entity could meaningfully be submitted to the question whether it obeys these laws or not. Instead of this, establishing a physical theory always means to make a preselection of a certain *kind* of objects which sensibly could or could not obey the laws of the theory. Thus, in our general context (u_0, A_0) defines the set Δ_0 of all structures u over the language $L(A_0)$ such that the restriction of u to $L(u_0)$ is (identical with) u_0 and u is a model of A_0:

$$(12) \qquad \Delta_0 = \{u \textit{ over } L(A_0): u \mid L(u_0) = u_0 \quad \textit{and} \quad \mathrm{mod}(A_0, u)\}.$$

In this way it is decided (1) to look upon physical objects as (being described by) structures at all, and (2) to look upon the possible objects of a physical theory with the *core*, as I shall call it, (u_0, A_0) as (described by) the structures belonging to Δ_0 only.

The following example will illustrate the general idea. Let us take Newton's gravitational theory in its point mass idealization and with a fixed number **n** of point masses. Since the kind of the objects to be described by this theory are systems of **n** bodies moving in space, the structure u_0 will have to contain the Galilean space-time structure. Moreover it will have to contain sets of functions with time points as arguments and space points as values. The same is the case with respect to velocities and accelerations. There will be mappings representing time derivation, and we shall have to take measure for expressing conditions of differentiability. As regards the masses there must be a continuous variable indicating possible mass values. An order relation for masses and the addition of masses must be introduced, etc. So much for the structure u_0 and its language $L(u_0)$ in this case. To give a precise formulation of u_0 and $L(u_0)$ would, of course, be a tedious thing to do. But mind that everything which would have been introduced so far would have received a fixed meaning (at least) in the following sense: In introducing, say, a time variable, a set for possible time points, a functional constant for the time metric and the latter itself, we always imply that it is *one* (concept of) time which we are introducing: In Newton's gravitational theory we do not provide for the possibility of two objects whose time behaviour is referred to two different time metrics. Instead, all possible motions are related to one and the same time metric, whatever the physical meaning of that metric may be.

The matter is different for the second part of the core of Newton's theory: the extension of $L(u_0)$ to $L(A_0)$ and the theory A_0. Although we have fixed the *concept* of possible motions and the *concept* of mass already in establishing u_0, we must not pin down Newton's theory to this or that *actual* system of masses and motions. For this is a matter of contingency. It is at this point where the multiplicity of possible objects of

Newton's theory comes in. In extending $L(u_0)$ to $L(A_0)$ we have to exhibit descriptive constants $m_1, \ldots, m_n$ as well as $\mathfrak{U}_1, \ldots, \mathfrak{U}_n$ for the masses and the motions respectively, and the theory A_0 has to restrict the otherwise arbitrary interpretations of these constants in such a way that their denotations fall under the corresponding concepts already introduced with u_0.

Coming back to the general argument we are now ready for the second step. Besides the core of a physical theory there will be a fundamental law restricting the set of 'a priori' possible objects in an essential way. This is done by a theory A over the language $L(A_0)$ which is logically contingent with respect to the theory A_0'. Thus we shall have the further extension.

(13) $\quad A' = A_0' \cup A$

of A_0' and the corresponding restriction

(14) $\quad \Delta = \{u \textit{ over } L(A_0)\colon u \in \Delta_0 \quad \textit{and} \quad \mathrm{mod}(A, u)\}$

of Δ_0. The structures belonging to Δ represent the physical objects which, beyond their being possible objects of our theory, obey its fundamental laws. In Newton's theory A would be given by the gravitational equations. As well illustrated by these equations, in writing down A we express an essential relation between the denotations of the constants in $L(A_0) - L(u_0)$, and in doing this we (necessarily) make use of the symbols and meanings already introduced with u_0.

Up to this point the concept of a physical theory has been analysed in so far as such theories appear in textbooks of *theoretical* physics. In these books questions of physical meaning usually are assumed to be settled, and we are chiefly taught about the mathematics of physical theories, couched in a physical vocabulary. For our present purpose, however, the addition of a setting representing the possibilities of acquiring *experimental* knowledge about real objects is indispensible. Since the details of this third step of our analysis are a bit complicated the following presentation of the matter will be still more sketchy than it has been hitherto.

In addition to the core (u_0, A_0) and independently of the laws A, a set E of *experimentally accessible sentences* with roughly the following properties is introduced:

(1) The sentences of E are instances of certain formulas of $L(A_0)$. To get the formulas a finite subset of experimentally accessible constants or

observables is chosen from the set of all constants in $L(A_0) - L(u_0)$. Then a finite set of formulas is chosen with only the observables (among all constants of $L(A_0) - L(u_0)$) occurring in them. These formulas already have the form of sentences expressing the possible results of measurements. Typical forms would be

$$c_1 < c < c_2$$

where c is an object constant, or

$$\textit{there is a } t \textit{ with } t_1 < t < t_2 \textit{ and } x_1 < f(t) < x_2$$

where f is a functional constant. All other letters representing variables we still have to make instantiation possible. This is done by extending the language in a peculiar way, well known form logic (cf. [17], p. 18), namely by adding names for the possible values of all observables and all sorts of variables being involved. By substituting these names for the free variables in our formulas we get the sentences of E expressing the possible results of measurements. It goes without saying that the physical basis of this logical procedure is the interpretation of the observables and the complex sorts of variables by measurable physical quantities.

(2) For any finite subset $C \subseteq E$ we introduce the set

$$(15) \qquad \Delta_C = \{u \textit{ over } L(A_0) \colon u \in \Delta_0 \quad \textit{and} \quad \mathrm{mod}(C, u)\},$$

i.e. the set of possible objects in our 'world' Δ_0 for which C is valid. Then a $\Delta_C \neq \emptyset$ never will consist of only one structure u. In physics Δ_0 will always have a natural topology and the $\Delta_C \subseteq \Delta_0$ will even be open subsets in this topology. This corresponds to part of the limitation set upon the accuracy of measurement: A finite set of measurements will never single out one object only. Moreover, although the accuracy of measurement might be increased indefinitely in principle, in any given situation it will have an upper limit. Accordingly, each Δ_α will contain a minimal Δ_β, i.e. a Δ_β with no $\Delta_\gamma \subset \Delta_\beta$ where $\alpha, \beta, \gamma \in E$.

(3) Given any real object Σ of which we can reasonably assume that it is a possible object of (u_0, A_0), the sentences of E are thrown into statements about Σ. As regards the question how we can justifiably arrive at these statements the assumption will be made that the theory A_0' as well as further background knowledge but *not* the theory A may be used thereby. Thus the possible statements about Σ may be heavily theory

laden, and the possible successes and failures of (u_0, A_0, A), immediately to be introduced, are what they are (1) only with respect to the laws A, and (2) only under the assumption that the theory A'_0 is true.

Having the components u_0, A_0, A and E of a physical theory at our disposal, we are now ready to ask for the possible successes and failures of such a theory. The main idea in answering this question will be to restrict the possible successes and failures to entities which can be constructed from E alone and to make use of the fact that A will introduce correlations between the sentences of E. Apart from differences which are only of academic interest, the matter can be settled in the following way. Imagine a real object Σ being a possible object of (u_0, A_0) and, consequently, representable by a structure $u \in \Delta_0$. Then u could be used to simulate Σ in the following considerations. But since we do not know which u will represent Σ, it will be nearer to actual practice to assume next that we have acquired some knowledge about Σ represented by two subsets E_a and E_r of E containing the sentences expressing the so far accepted and rejected empirical statements about Σ respectively. Clearly, the sentences of E_a and E_r will be true and false sentences in u respectively.

We now consider pairs (C, α) where $C \subseteq E_a$, $\alpha \in E_a \cup E_r$ and α is logically contingent with respect to C and Δ_0, i.e. neither $\Delta_C \subseteq \Delta_\alpha$ nor $\Delta_C \cap \Delta_\alpha = \emptyset$. Let S be the set of these pairs. They will be our possible successes and failures (with respect to Σ, of course). There are the two subdivisions of S

$$(16) \quad \begin{cases} S = T \cup F \\ T = \{(C, \alpha) \in S : \alpha \in E_a\} \\ F = \{(C, \alpha) \in S : \alpha \in E_r\} \end{cases}$$

and (with Δ belonging to A according to (14))

$$(17) \quad \begin{cases} S = Q_A \cup R_A \cup U_A \cup W_A \\ Q_A = \{(C, \alpha) \in S : \Delta \cap \Delta_C \neq \emptyset \quad \textit{and} \quad \Delta \cap \Delta_C \subseteq \Delta_\alpha\} \\ R_A = \{(C, \alpha) \in S : \Delta \cap \Delta_C \neq \emptyset \quad \textit{and} \quad \Delta \cap \Delta_C \cap \Delta_\alpha = \emptyset\} \\ U_A = \{(C, \alpha) \in S : \Delta \cap \Delta_C \neq \emptyset, \Delta \cap \Delta_C \nsubseteq \Delta_\alpha \quad \textit{and} \\ \qquad\qquad\qquad\qquad\qquad\qquad \Delta \cap \Delta_C \cap \Delta_\alpha \neq \emptyset\} \\ W_A = \{(C, \alpha) \in S : \Delta \cap \Delta_C = \emptyset\}. \end{cases}$$

Evidently, (16) and (17) correspond to (1) and (2) respectively. Forming the intersections $Q_A^T = Q_A \cap T$ etc., we can once more make up our mind which of these sets are to be grouped together as the sets of successes and failures. Again the sentences of Q_A^T and R_A^F certainly will be among the successes, and those of Q_A^F, R_A^T and, additionally, W_A among the failures. As regards the undecided sentences we meet with the same ambiguity as in Section 2, expressed by (3) and (4).

However one may decide upon this matter, one thing turns out to be quite different from the situation as it appeared in Section 2. Let us take two physical theories having a common core (u_0, A_0) and a common set E of experimentally accessible sentences, but possibly different laws A and B. Then, with respect to a certain domain of application Γ, these theories can very well satisfy the P-conditions (in the 'to be'-formulation) without violating the principle of fallibilism and even without being compatible. The reader may convince himself of all the relevant details by taking, for instance, as A and B the laws of Rayleigh-Jeans and Planck concerning black body radiation or Newton's law of gravitation and a certain variant of it having an additonal long range factor e^{ar} as it has actually been suggested for cosmological application. Without further ado these laws can be supplemented by their corresponding cores (u_0, A_0) and E-sets, and they will turn out to be the same in each case. In the first case we have a finite-dimensional Δ_0, the three occurring dimensions being equated with the energy density, the frequency and the temperature respectively. In the second case A_0 is infinite-dimensional: it consists of all spatio-temporal and mass descriptions of an n-body system satisfying only certain conditions of smoothness. In both cases A and B are incompatible, i.e. $\Delta_A \cap \Delta_B = \emptyset$ with Δ as defined by (14). But in a certain region $\Delta' \subseteq \Delta_0$ the laws A and B have the same asymptotic behaviour ($h\nu \ll kT$ in the first case, and $ar \ll 1$ in the second). Owing to this asymptotic behaviour, A and B will be experimentally indistinguishable in Δ' and, in particular, all successes of A will also be successes of B. With the help of the common E-sets having the above properties, a precise meaning can be given to these statements. On the other hand, in other regions there can be failures of A which at the same time are successes of B. Finally, there is plenty of room in Δ_0 for possible failures of both A and B. (From a historical viewpoint the two last statements are, of course, fictitious for the second case.) A' and B' may even be false with respect to a certain $u \in \Delta_0$. Thus, our scheme is broad enough to give shelter to nontrivial cases of commensurable theories as regards the P-conditions.

4. The breakdown of our first reconstruction of the P-conditions in Section 2 had nothing to do with any difficulties inherent in physics. We just took the wrong road from the very outset. On the other hand, the success of our second approach in Section 3 was simply due to the fact that the consideration was restricted to relatively unimportant cases, if any of them are to be met with in physics. Nevertheless their investigation is useful for at least the following reason: We have come to know at least one type of case in which the P-conditions can be reconstructed within physics. They are pieces of physical theories having the same core and the same set of experimentally accessible sentences. These, then, at the same time seem to be pairs of commensurable theories, – if there is anything like that at all. Accordingly we must expect that cases of incommensurability will be localized somewhere within the region where the conditions developed above are no longer satisfied. It would, however, be rash to make our result the starting point of a *definition* of commensurability and, consequently, of incommensurability. One reason for this is that there may be, and as we shall see presently there actually are, other cases of commensurable theories. Moreover, as has been expressed already in Section 1, it may be doubted whether there is any clear-cut dichotomy between commensurability and incommensurability of theories. Together with the present state of research in this field there are reasons enough to limit further considerations to case-studies gradually moving away from the ideal conditions of Section 3 in the direction of ever growing difficulties. In the rest of this paper I can only touch upon some problems on this thorny path.

To start once more from scratch and intersperse a brief methodological consideration let us take for the moment the word 'theory' in a very wide sense such that, as Feyerabend will have it, also myths, political ideas and religious systems are theories ([3], p. 252, n. 5). In this situation there is no reason at all to complain of the difficulty or even the impossibility in making any comparison between two arbitrarily given theories, let alone a comparison with regard to progress. One is simply and voluntarily resigned to comparing, say, Newton's and Einstein's theories of gravitation with the help of means different from those which are available for a comparison between an ancient myth and one of these two theories of modern physics. Saying this is perhaps admitting that there is no universal demarcationism. But for the moment my point is a technical one.

In the situation envisaged above we could, of course, say e.g. that he who is in the possession of the general theory of relativity simply knows more than the believer in an ancient myth about the origin of the universe. But saying this is not to compare the two 'theories' in any *specific* way, and as long as a theory can be almost anything it will even be impossible to make a specific comparison between them.

If we now look back at our considerations in Section 2, we at once see a host of relations presenting themselves for a specific comparison between theories. The reason is that there we relied upon real (as opposed to nominal) definitions of the concepts of a theory and its interpretations as they are given in mathematical logic. These restricted and well defined concepts enable us to define *specific* relations between theories or interpreted theories such as logical equivalence, consequence, incompatibility, inclusion of truth contents etc. These relations are specific in the sense that their meaning is derived from the very definition of the basic concepts. They hold or do not hold between theories *qua* theories. *Vice versa*, theories are the sort of things that can stand in those relations. In this regard the interpretations given to the *P*-conditions in Section 2 were quite in order. They led to relations permitting a specific comparison between theories interpreted in the same structure. Moreover, we could have given a definition of incommensurability by calling two theories A and B incommensurable if they are logically incompatible ($Q_{A \cup B} = S$) and their common part contains only logical truths ($Q_A \cap Q_B = T_{\log}$). Theories incommensurable in this sense can not share a single statement (=theorem, except logical truths), they are completely disjointed and, together with their incompatibility, would thus have at least some of the characteristics of incommensurable theories as demanded by Feyerabend ([6], p. 82). More than this, it is a child's play to prove that for incommensurable theories none of the *P*-conditions in Section 2 are satisfied. So it seems that we could have found what we were looking for all the time: a relation such that *if* it holds between two theories then neither of them can be progressive with respect to the other. Now, *within* the context of Section 2 this may indeed be the solution of our problem. But, as we have already seen, this context is entirely *inadequate*: it does not lead to a comparison of theories specific enough for the intended applications in physics.

Therefore, a further restriction of the concept of a theory was necessary:

We arrived at the distinction between two parts of a theory, one being more fundamental than the other. Enriched by the concept of an experimentally accessible sentence, the concept of a physical theory analysed so far was sufficiently specific to yield pairs of theories which satisfactorily could be viewed as commensurable theories. Let us now see that they do not exhaust the concept of commensurability. As a rule not all the descriptive constants in $L(A_0)-L(u_0)$ will be observables in the sense introduced above. Beyond the observables there will also be constants to be interpreted by *theoretical entities*. Whenever this is the case for at least one of two theories with the cores (u_1, A_1) and (u_2, A_2), then for these theories to be commensurable it is sufficient that (u_1, A_1) and (u_2, A_2) coincide in that common restriction (u_0, A_0) of (u_1, A_1) and (u_2, A_2) which contains the observables and the sorts of variables belonging to them. Then the comparison of two laws A and B belonging to (u_1, A_1) and (u_2, A_2) respectively can be made exclusively in the domain Δ_0 assigned to (u_0, A_0) by (12). A case in point is Newton's theory of gravitation vs. a suitable generalization of Kepler's theory of planetary motions. The cores of these theories coincide in the Galilean space-time structure together with the (observable) functional constants $\mathfrak{U}_1, \ldots, \mathfrak{U}_n$ describing the motion of an n-body system. They differ with respect to the theoretical entities, these being the masses in Newton's theory and the wellknown constant of Kepler's third law in Kepler's theory. The details are given in [14] and [15].

The extension of the class of pairs of commensurable theories just indicated could, however, raise serious doubts when it is realized that the common part of the theories with respect to which the comparison is made can become very 'small' compared with the part being eliminated. Consider, for instance, two (non-relativistic) quantum theories of an hydrogen-like atom, one treating the atom as a one-particle system moving in a Coulomb field, the other treating it as a two-particle system with Coulomb interaction. It is not difficult to reconstruct these theories according to the general scheme of Section 3. If we assume that only the energy differences (and possibly the charge and the mass) are observable then the theories would turn out to be commensurable, the two laws for which the comparison is made being given by the well-known formulas for the energy levels. It could even be argued that the two-particle treatment is an improvement over the one-particle treatment: In the former, but not in the latter, the Rydberg constant is dependent on the mass of the nucleus, and this dependence can be observed by taking hydrogen atoms, singly charged helium ions, etc. On the other hand, there is a remarkable unlikeness between the *eliminated* parts of the theories chiefly characterized by their respective ψ-functions. In the one-particle theory the nucleus either is completely absent or could be viewed as a classical particle at rest at the origin. On account of the uncertainty relations neither idea can have a conceptual counterpart in the two-particle theory. Accordingly, there is no natural identification of the Hilbert spaces. Again, the energy operator H_1 of the one-particle theory formally *has* the counterpart $1 \otimes H_1$ in the two-particle theory. But there this operator does *not* function as the operator yielding the correct energy eigenvalues: Although, being a selfadjoint operator, it may have some meaning also in the two-particle theory, its meaning is not the (internal) energy.

These 'incommensurabilities' would be increased and aggravated if we now turn to Newtonian mechanics and its special cases on the one side and quantum mechanics or the theory of relativity as well as their special cases on the other. Looking, for instance, at general quantum mechanics and general classical (Hamiltonian) mechanics, their reconstruction on the line of Section 3 clearly shows that they have completely different cores containing completely different concepts of quantity and state. Whereas for Newton's and Kepler's theory mentioned above the respective theories A_0' defined by (11) are, if no longer identical, then at least compatible, they *contradict* each other in the case now before us. Irrespective of this fact (cf. [16]) *no common restriction of the two cores* exists (although there are common theorems of the theories A_0'). In spite of repeated attempts, no generally acknowledged quantization rules relating the two theories have been found. There even are speculations that the logic of quantum mechanics is different from classical logic. Since these are differences appearing on a very abstract level, let us consider another example in which differences between the cores of two physical theories lead to differences in the meaning of observation statements. A case in point is the Schwarzschild solution of Einstein's field equations in comparison with Newton's gravitational theory for one body moving in a central field. Here the experimentally accessible sentences are directly linked with the respective theory of the core. Since we had to allow that the latter be used in arriving at the former and since the theories of the two space-time structures are (logically!) incompatible, the experimentally accessible sentences of the two theories have different meanings.

Thus, with respect to incommensurability one has to admit what Feyerabend emphasizes. On the other hand, there is something to be emphasized that *he admits*, – if only in footnotes. For the case at hand he admits that the transition from Newton's to Einstein's theory "will not lead to new methods for estimating the size of an egg at the grocery store". He further admits "that distances that are not too large will still obey the law of Pythagoras" ([2], p. 270). For special relativity vs. Newton's mechanics, he admits the possibility "to establish correlations between the *sentences* of the two theories" ([6], p. 125, n. 197; also [7], p. 222, n. 3 and [8], p. 102, n. 36). Likewise it is conceded "that the relativistic scheme very often gives us *numbers* which are practically identical with the numbers we get from CM [=Newton's mechanics] ([7], p. 221). Generalizing these concessions we may say that even in the cases of incommensurable theories T and T', the following can be observed: (1) there is a certain correspondence between the concepts (or the sorts of variables and the formulas) of T and T', and (2) this correspondence can be made the basis of an approximative comparison of T and T' or, putting the approximations in temporarily fixed limitations of the accuracy of measurement, of a comparison on the line of our P-conditions in their 'correspondence'-formulation. The concepts of classical physics were not altogether extinguished by quantum mechanics and the theories of relativity. Rather they themselves, and not only classical physics as a whole, had their successors in quantum and relativistic physics. Admittedly, the relativistic concept of mass is different from its Newtonian predecessor. But in his arguments for this and other relevant differences, Feyerabend himself makes use to this concept-by-concept-correspondence: He does not argue that relativistic time is different from prerelativistic mass. And when he argues ([7], p. 221) "that the relativistic scheme... gives us *numbers* which are practically identical with the numbers we get form CM" and then continues "but this does not make the *concepts* more similar" we may be permitted to request: What do we know about the similarity of concepts that would forbid us to say that here, if anywhere, we have a case of the similarity of concepts?

To conclude, besides the fact of incommensurabilities, there is the other fact of concept-by-concept-correspondence accompanied by the possibility of numerical approximations. I can think of no reason to minimize the importance of the latter fact, so much the less as we have to put up with the former. Continuing the work sketched in Section 3, conditions of progress in the correspondence-form can be maintained. Feyerabend puts the right question when asking: "Why should it be necessary to possess terminology that allows us to say that it is the *same* experiment which confirms one theory and refutes the other?" ([7], p. 226). My answer would be: It is not necessary. On the contrary: In a certain sense and in certain cases it is not even possible. For let us imagine that we are pressed to answer the question: What, *in the last analysis*, do we mean by stating the advance of the perihelion of Mercury? According to the present state of our knowledge, I think the answer would either be given in terms of the Schwarzschild metric or in terms of the old Galilean metric of space-time. In the first (second) case it would be a contradiction in terms to say that the phenomenon *thus stated* is a failure (success) of Newton's (Einstein's) theory. There is, therefore, not *one* phenomenon which is at the same time a failure of Newton's and a success of Einstein's theory. But we can very well say that what appeared as a failure of Newton's theory (second formulation) reappears as or corresponds to a success of Einstein's (first formulation). Following up this way by further developing the analysis of Section 3, we could eventually turn the tables and say: Incommensurability will not have destroyed the possibility of progress. Rather it will render *great* progress possible: *Provided* there is progress in the sense of the modified *P*-conditions, it will be an important one whenever incommensurability takes part in it. Two seemingly hostile aspects of the development of physics would be reconciled.

Georg-August Universität

BIBLIOGRAPHY

[1] Feyerabend, P., 'Explanation, Reduction, and Empiricism', in H. Feigl *et al.* (eds.), *Scientific Explanation, Space, and Time* (Minnesota Studies, Vol. III), University of Minnesota Press, Minneapolis, 1962, pp. 28–97.

[2] Feyerabend, P., 'On the "Meaning" of Scientific Terms', *The Journal of Philosophy* **62** (1965), 266–274.

[3] Feyerabend, P., 'Reply to Criticism', in R. S. Cohen *et al.* (eds.), *Boston Studies in the Philosophy of Science*, Vol. II, Humantities Press, New York, 1965, pp. 223–261.
[4] Feyerabend, P., 'Problems of Empiricism', in R. G. Colodny, (ed.), *Beyond the Edge of Certainty* (University of Pittsburgh Series, Vol. 2), Prentice-Hall, Englewoods-Cliffs, N.J., 1965, pp. 145–260.
[5] Feyerabend, P., 'Problems of Empiricism', Part II, in R. G. Colodny (ed.), *The Nature and Function of Scientific Theories* (University of Pittsburgh Series, Vol. 4), University of Pittsburgh Press, 1970, pp. 275–353.
[6] Feyerabend, P., 'Against Method', in M. Radner *et al.* (eds.), *Analyses of Theories and Methods of Physics and Psychology* (Minnesota Studies, Vol. IV), University of Minnesota Press, Minneapolis, 1970, pp. 17–130.
[7] Feyerabend, P., 'Consolations for the Specialist', in I. Lakatos *et al.* (eds.), *Criticism and the Growth of Knowledge*, Cambridge University Press, 1970, pp. 197–230.
[8] Feyerabend, P., 'The Philosophy of Science – A Hitherto Unknown kind of Lunacy?', in K. Hübner *et al.* (eds.), *Natur und Geschichte*, Meiner, Hamburg, 1973.
[9] Harris, J. H., 'Popper's Definitions of "Verisimilitude"', *The British Journal for the Philosophy of Science* **25** (1974), 160–166.
[10] Kemeny, J. G., 'Models of Logical Systems', *Journal of Symbolic Logic* **13** (1947), 16–30.
[11] Kreisel, G. and Krivine, J. H., *Elements of Mathematical Logic (Model Theory)*, North-Holland, Amsterdam, 1971.
[12] Miller, D., 'Popper's Qualitative Theory of Verisimilitude', *The British Journal for the Philosophy of Science* **25** (1970), 166–177.
[13] Popper, K. R., *Conjectures and Refutations*, 5th ed., Routledge & Kegan Paul, London, 1974.
[14] Scheibe, E., 'The Approximative Explanation and the Development of Physics', in P. Suppes *et al.* (eds.), *Logic, Methodology, and Philosophy of Science*, Vol. IV, North-Holland, Amsterdam, 1973, pp. 931–942.
[15] Scheibe, E., 'Die Erklärung der Keplerschen Gesetze durch Newtons Gravitationsgesetz', in E. Scheibe *et al.* (eds.), *Einheit und Vielheit: Festschrift für C. F. v. Weizsäcker*, Vandenhoeck & Rupprecht, Göttingen, 1973, pp. 98–118.
[16] Scheibe, E., 'Vergleichbarkeit, Widerspruch und Erklärung', in R. Haller *et al.* (eds.), *Philosophie und Physik*, Vieweg, Braunschweig, 1975, pp. 57–71.
[17] Shoenfield, J. R., *Mathematical Logic*, Addison-Wesley, Reading, Mass., 1967.
[18] Tichý, P., 'On Popper's Definition of Verisimilitude', *The British Journal for the Philosophy of Science* **25** (1974), 155–160.

ABNER SHIMONY

COMMENTS ON TWO EPISTEMOLOGICAL THESES OF THOMAS KUHN

1. *The Structure of Scientific Revolutions*[1] and Kuhn's afterthoughts (1970a, 1974) have been subjected to so much penetrating criticism that any further examination might be supposed redundant.[2] There are, however, two epistemological theses in his work which have not, in my opinion, received adequate analysis. They are not essentially dependent upon considerations of Gestalt switches, theory-ladenness of observation, incommensurability, or the ambiguity of the word 'paradigm', which have received much critical attention.

The first thesis is that *the progress of science ought not be construed as the approach to a fixed goal which is the truth about nature.*

> But need there be any such goal? Can we not account for both science's existence and its success in terms of evolution from the community's state of knowledge at any given time? Does it really help to imagine that there is some one full, objective, true account of nature and that the proper measure of scientific achievement is the extent to which it brings us closer to that ultimate goal? (1970, p. 171).

Kuhn maintained this thesis in his replies to critics and even strengthened it somewhat:

> If I am right, then 'truth' may, like 'proof', be a term with only intra-theoretic applications. (1970a, p. 266).

The second thesis is that *the procedures of scientific investigation can be shown to be rational, and the appropriate sense of "rationality" can be explicated, only by drawing upon the substantive achievements of science.* One formulation of the thesis can be found in Chapter I of his book (1970), where he rejects the dichotomy of "the context of discovery" and "the context of justification" and the dichotomy between normative and descriptive disciplines. A more explicit formulation occurs as part of his answer to the charge of being an irrationalist:

> I have not previously and do not now understand quite what my critics mean when they employ terms like 'irrational' and 'irrationality' to characterize my views. These labels seem to me mere shibboleths, barriers to a joint enterprise whether conversation or research. (1970a, p. 263).

R. S. Cohen et al. (eds.), Essays in Memory of Imre Lakatos, 569–588. *All Rights Reserved.*
Copyright © 1976 by D. Reidel Publishing Company, Dordrecht-Holland

... existing theories of rationality are not quite right and... we must readjust or change them to explain why science works as it does. To suppose, instead, that we possess criteria of rationality which are independent of our understanding of the essentials of the scientific process is to open the door to cloud-cuckoo land. (*ibid.*, p. 264).

Another statement, which occurred in a discussion rather than an essay, is very explicit and extends the thesis from scientific methodology to epistemology in general:

Although my professional identity is as an historian of science, what is on my mind when I get involved with the sort of thing I am doing here today is ultimately epistemology. I really want to know what sort of thing knowledge is, what it is all about, and why it is that it works the way it does. Now in order to do that, it seems to me the right move (I am glad somebody else said philosophy is an empirical enterprise) is to look around and try to see what is going on and what it is that people who have knowledge have got. If I then think that what I discover when I look gives me certain sorts of understanding of what goes on – makes it plausible that knowledge should be the sort of thing it is and should develop the way it does – then I can legitimately say that from the examination of scientific communities I am beginning to become a better epistemologist. (1974, pp. 512–3).

The two theses ought to be decoupled, since I shall argue that the second is partly correct while the first is thoroughly erroneous. If the first thesis were accepted, and truth were taken to be irrelevant to the progress of science, then the second thesis would have to be construed as saying that scientific achievements provide proximate justifications for methodological principles but it is vain to seek for ultimate justifications, for there are none to be found – a conclusion which might indeed deserve the charge of 'irrationalism'. On the other hand, if the first thesis is denied, and scientific progress is considered to be evidence that serious people immersed in research are collectively sensitive to the truth, then we can understand the second thesis in an entirely different way: we could acknowledge that sensitivity to the truth is very difficult to explain, and that the organons, systems of inductive logic, and theories of scientific inference proposed hitherto somehow all fall short of the mark, but we might hope that looking more closely than methodologists are accustomed to do at the substantive achievements of science will help to provide the desired explanation. In other words, we should turn to the history of science not for the purpose of seeking a surrogate for the approach to the truth, but rather for illumination on the question of how this approach is possible for human beings.[3]

The message that I am trying to convey in urging that Kuhn's two theses be decoupled has been said from time to time in different ways, for example in the following passages:

> For that which the author had at heart throughout his studies of the history of science was to gain an understanding of the whole logic of every pathway to the truth. (Peirce, 1958, p. 175).

> My suggestion is that Galileo's own view of his subtle, flexible, and discriminating method has in common with Plato two things: First, the reconciliation of a 'mathematical' and 'rational' conception of science with a full recognition of the difficulty of coming to know: a reconciliation that is only possible through a crucial emphasis upon the process of inquiry itself. Second, a conviction that inquiry as such is not a profitable subject for positive doctrinal exposition ('Discourse on Method', 'Rules for the Conduct of the Understanding'), but has to be learned and taught through its practice – a process which, in turn, may be foreshadowed in dramatic representation. In the *Two New Sciences*, on this reading, Galileo has offered us his view of *what science is*, in the form of a dramatised commentary upon the central work of his own scientific life. (Stein, 1974, p. 397).

I hope that there will be some value in reiterating and reformulating the message.

2. The first of Kuhn's arguments for the irrelevance of truth to scientific progress is essentially that for the purpose of explaining the phenomenon of progress we have no need of the hypothesis that science is drawing constantly nearer some fixed goal (1970, pp. 170–1). Scientific progress consists in unidirectionality with respect to certain desirable characteristics:

> I believe it would be easy to design a set of criteria – including maximum accuracy of predictions, degree of specialization, number (but not scope) of concrete problem solutions – which would enable any observer involved with neither theory to tell which was the older, which the descendant. (1970a, p. 264).

Unidirectionality of this kind no more depends upon the existence of a fixed goal than does the "steady emergence of more elaborate, further articulated, and vastly more specialized organisms," which, according to neo-Darwinian evolutionary theory, can be accounted for by natural selection without postulating orthogenesis (1970, pp. 172–3). A second argument, not included in the first edition of *The Structure of Scientific Revolutions*, is this:

> There is, I think, no theory-independent way to reconstruct phrases like 'really there': the notion of a match between the ontology of a theory and its 'real' counterpart in nature now seems to me illusive in principle. (1970, p. 206).

I shall postpone for a while a consideration of Kuhn's first argument and start with an examination of his second, which is recognizable as a special version of a general type of skeptical reasoning that has frequently occurred in the history of philosophy. I do not know a commonly used name for this type of reasoning, but perhaps it could be called "arguments for the impossibility of transcendence", or more mundanely "trapped-in-a-box arguments". The box in which the subject is trapped varies from one version to another of the argument: the subject's own consciousness, the historical epoch in which the subject is born, or the cognitive peculiarities of the human race. In any case, transcendence is impossible, because in principle the only information accessible to the subject is contained within the box. An exemplary argument of this kind occurs in Hume's *Enquiry Concerning Human Understanding*:

> It is a question of fact, whether the perceptions of the senses be produced by external objects, resembling them: how shall this question be determined? By experience surely: as all other questions of a like nature. But here experience is, and must be entirely silent. The mind has never anything present to it but the perceptions, and cannot possibly reach any experience of their connection with objects. The supposition of such a connection is, therefore, without any foundation in reasoning. (1955, Sect. XII, Part I).

The parallel between Kuhn's argument and that of Hume is evident. If, therefore, one accedes to the conclusion that "the notion of a match between the ontology of a theory and its 'real' counterpart in nature now seems to me illusive in principle," how can one consistently resist Hume's skepticism concerning knowledge of external objects? Conversely, and more constructively, Hume's skeptical argument may fail to be compelling because of a generic weakness in arguments for the impossibility of transcendence, and in this way an answer to Kuhn's version of skepticism can be discovered.

The response to Hume which I find most persuasive consists of two steps. The first, which is usually not made explicit, is to establish that indirect or hypothetico-deductive arguments are legitimate in epistemological investigations. To disallow such arguments would "block the road to inquiry" by ruling out of consideration *ab initio* a large spectrum of epistemological positions, whereas none are censored by allowing such arguments. Specifically, it is conceivable that phenomenalism would be found, in an investigation using indirect arguments, to have greater explanatory power than any alternative to it which is proposed; but if phenomenalism were accepted because indirect arguments were ruled inadmissible, then it would have been effectively insulated from the possibility of critical examination. The second step, which is developed in innumerable versions in the literature, consists in taking advantage of the legitimation of indirect arguments and in weighing the explanatory power of phenomenalism against that of critical realism, etc., the crucial evidence being the apparent persistence of physical processes when gaps occur in observation of them, the concurrence of observers with regard to events, the reliability and precision of certain laws formulated in terms of physical concepts, and the non-existence of laws of comparable reliability and precision formulated directly in phenomenalistic terms. If, in reply, a variant of phenomenalism is presented which is agnostic regarding the existence of external objects but admits that experience is organized as if such objects exist and bear causal relations to phenomena, then a critical realist could claim as decisive a settlement as one can reasonably expect of philosophical disputes: the variant is clearly parasitical upon critical realism, and the difference between them has been diminished to the point where the question of choice is a *Scheinproblem*.

The argument of the preceding paragraph misses many of the subtleties and important cognate issues that can be found in the literature on the existence of external objects. (At least I hope that this is so, for it would be depressing to think that such an immense corpus could be distilled without loss to so small a residue!) Nevertheless, I believe that the argument is fundamentally correct and that its solid structure will not be essentially changed by refinements. It is evidently applicable as a counter-argument to any kind of "argument for the impossibility of transcendence", but the outcome of any particular application cannot be foreseen without examining the details; indeed, it is precisely the virtue of the kind of argument that I have sketched in answer to Hume that it moves freely by reflection and critical examination of evidence, thereby keeping open the road to inquiry.

What then should be said about Kuhn's particular version of the argument for the impossibility of transcendence? He surely is correct that what is "really there" is not directly accessible to us; but if indirect arguments are admissible, then one cannot say initially (whatever one may say at the conclusion of the analysis) that it is illusive in principle to consider "the match between the ontology of a theory and its 'real' counterpart in nature." Consider the hypothesis, which I shall call "the

Hypothesis of Verisimilitude",[4] that *for well-established theories in a science in which high critical standards have been achieved such a match exists to some good degree of approximation.* Then if philosophical inquiry is to be open, the Hypothesis of Verisimilitude and its negation should initially both be on the same footing. An appeal to indirect argument consists in examining which of them has the greater explanatory power. It may be noted that if the Hypothesis of Verisimilitude turns out to be superior in explaining the relevant evidence, then an answer will also have been given to Kuhn's first argument, which had been temporarily set aside and which consisted essentially of the challenge, "Does it really help to imagine that there is some one full, objective, true account of nature and that the proper measure of scientific achievement is the extent to which it brings us closer to that ultimate goal?" (1970, p. 171).

There are two massive bodies of evidence which fall naturally into place upon the Hypothesis of Verisimilitude, and which are very mysterious indeed upon its negation. Kuhn neglects the first of these bodies of evidence almost entirely, but the other he recognizes and does try to fit into his scheme.

The first is the evidence that once a certain stage of maturity is reached in a science, a remarkable continuity of thought is maintained even across scientific revolutions, exemplary instances being the transition from classical mechanics to relativistic mechanics, from classical mechanics to quantum mechanics, from classical electro-magnetic theory to quantum electrodynamics, from special relativity to general relativity, from phenomenological thermodynamics to statistical thermodynamics, from phenomenological valence theory to quantum chemistry, from the Darwinian to the genetic theory of natural selection, and from the phenomenological theory of the gene to molecular biology. In each case the theory that was "overthrown" was in fact retired with honor: its domain of approximate validity was not only delimited but reaffirmed, and its continuity with the new theory, not just in approximate agreement with regard to a certain class of predictions but also with regard to conceptual structure, was recognized. In short, what Bohr called "the correspondence principle" holds, *mutatis mutandis*, between the new and the old theories in all these instances of scientific revolutions.[5] The phrase "once a certain stage of maturity is reached", which was used above, will evidently be problematic for historians of science, but the case which I am trying to make is unaffected by the problem of dating, for even the foregoing short list of exemplary fulfillments of the correspondence principle suffices to constitute an impressive body of evidence for weighing the epistemological hypotheses in question. Nevertheless, the notion of maturity is important for the present considerations, for it suggests that a distinction ought to be made (without precluding the possibility of overlap and of intermediate cases) between those scientific revolutions which consisted primarily of the establishment of decisively new standards of investigation, and those revolutions which consisted primarily of conceptual generalization or of deepening the level of description.

The second body of evidence is just that which Kuhn cites in support of the unidirectional character of the development of science: that more and more problems are solved in ways that are considered to be satisfactory by the community of scientists, and the precision of solutions in general increases (1970, p. 171). Kuhn is so strongly impressed by the evidence of progress in the natural sciences in these and related respects that he cautions against pushing too far the conflation of the history of science with that of art or politics, towards which his argument had been drifting (*ibid.*, pp. 208–9).

It remains now to judge the Hypothesis of Verisimilitude and its negation as fairly as possible on the basis of the two bodies of evidence just summarized.

The first body of evidence consists of instances in which the correspondence principle has held after various sciences have reached maturity. If the Hypothesis of Verisimilitude is true, and if maturity is identified with the achievement of high critical standards, then it would follow that any theory which has satisfied those standards sufficiently well to be accepted as 'well-established' is a good approximation to the truth. Characterizing the approximation as 'good' imputes to the theory some validity within an appropriate domain, which is expected to be discernible from the standpoint of a theory yet closer to the truth. But this is just what the correspondence principle asserts. Thus, the Hypothesis of Verisimilitude accommodates the first body of evidence in an entirely natural manner.

My argument, however, is loose in texture, so that objections can reasonably be raised against its decisiveness. The source of the trouble is the occurrence of three vague expressions in the formulation of the Hypothesis of Verisimilitude: "well-established theories", "high critical standards", and "to some good degree of approximation." The effect of these vague expressions, it might be contended, is to make the content of the Hypothesis of Verisimilitude so ill-defined that any body of evidence drawn from the history of science would be compatible with it. In reply I would say that the vagueness should appear troublesome only to some one who conceives of methodology as a discipline to be developed entirely in advance of the sciences. According to Kuhn's own second thesis (to be defended, with certain qualifications, in the next section), the phrases "well-established theories" and "high critical standards" could be explicated by reflection upon exemplary scientific achievements. Similarly, exemplary instances of the correspondence relation will help to explicate the vague expression "good degree of approximation," though it may be wise to recognize the necessity for some reliance upon analogy in using this expression, just because the relation between any actual theory (a human artifact, however inspired) and the ideal theory which exactly matches the truth is an extrapolation from the relation holding between two actual theories.

That the second body of evidence falls into place naturally from the standpoint of the Hypothesis of Verisimilitude is almost a corollary of what has already been said. The displacement of a theory by one which is a better approximation to the truth will mainly preserve – possibly with refinements – the predictions and explanations achieved in the domain of validity of the former. One would expect, therefore, that revolutionary as well as normal development of a science subsequent to its maturity will, for the most part, increase the number of solved problems and the precision of solutions. To prevent misunderstanding it should be pointed out that eddies and retrograde motions do in fact complicate the unidirectional development of science, and that the explanation of unidirectionality offered on the basis of the Hypothesis of Verisimilitude can take these complications into account. A considerable amount of mathematical analysis or attention to fine physical details is sometimes required in order to correct the retrogression. (For example, Drude's classical treatment of conduction electrons yielded an explanation of Ohm's law and some remarkably good evaluations of electrical resistances; all of which seemed to be undercut by the quantum mechanical theorem of Bloch that electrons propagating in a periodic lattice experience no resistance; to which the correction was to take account of impurities and thermal fluctuations of the lattice.) Despite such complications, the central point is that the history of various sciences, after the achievement of certain critical standards, has been astonishingly progressive, and the Hypothesis of Verisimilitude removes some of the mystery from the progress without dispelling the sense of astonishment.

It is also interesting to consider a kind of scientific progress not emphasized by Kuhn: the typical history of experiments. A pioneer experiment is often delicate and hard to duplicate, so that its decisiveness is doubtful. In time, however, reliability is often increased to the point that the experiment is incorporated into the repertory of student laboratory work. The transformation is due to such improvements as the control of perturbations, the elimination of systematic errors, the correction of instabilities in the equipment, and the increase in sensitivity of detectors. From the standpoint of the Hypothesis of Verisimilitude, this kind of progress can be understood quite naturally: there is such a thing as the 'reality' to which the theory of interest refers, but the causal connections between this reality and the appearances in the laboratory are complicated and often masked by various factors, and the improvement of the experiment consists in the identification and control of these factors.

Both the evidence concerning the correspondence principle and that concerning the progress of science could only be regarded as fantastic networks of coincidences if the Hypothesis of Verisimilitude were false. The reason that this is so is virtually revealed by Kuhn's own analysis. He recognizes that scientific investigation is at least to some extent open to nature. Even his doctrine of the theory-ladenness of observation does not permit him to say that the full content of the observational report is under the control of the experimenter. But if nature plays a role in the results of experimental tests, how can either of the two bodies of evidence be explained unless nature is somehow cooperative? In a brief passage Kuhn says precisely this:

> What must nature, including man, be like in order that science be possible at all? Why should scientific communities be able to reach a firm consensus unattainable in other fields?... It is not only the scientific community that must be special. The world of which that community is a part must also possess quite special characteristics. (1970, p. 173).

In the immediately following passage, however, Kuhn veers away from examining the epistemological consequences of the correct and important admission that he has just made:

> That problem – What must the world be like in order that man may know it? – was not, however, created by this essay. On the contrary, it is as old as science itself, and it remains unanswered. But it need not be answered in this place. Any conception of nature compatible with the evolutionary view of science developed here. (*ibid.*)

The final sentence is a sleight of hand. Presumably "the evolutionary view of science developed here" includes the proposition that there is no "permanent fixed scientific truth" which is the goal of scientific development (*ibid.*); but there is indeed a "conception of nature compatible with the growth of science" (*ibid.*) which contradicts this proposition, namely, the conception implicit in the Hypothesis of Verisimilitude. To readers interested in paradigms Kuhn has provided a splendid one: a paradigm of the abortion of a viable line of reasoning at exactly the moment that it becomes embarrassing to the author!

The case in favor of the Hypothesis of Verisimilitude and against Kuhn's dismissal of the role of truth can be summed up in the following way. Long ago Leibniz compared the attempt to discover the laws of nature from reflection upon phenomena to the attempt to discover the rules by which a message has been encoded in a text. Suppose that we have such a text, and that after numerous conjectures the tentative decipherment has become more and more coherent. The success may be no more than a series of coincidences, so that the tentative decipherment is completely on the wrong track. But somehow it is more plausible that a good approximation to the correct rules of encoding has been found than that the long run of successes is coincidental. Kuhn's thesis that the truth plays no role in the progress of science is analogous to maintaining that progressive coherent decipherment could occur even though no such things as the initial message and the rules of encoding exist. To undertake a quantitative probability calculation concerning a hypothesis of such extreme implausibility would be supererogatory.

3. Kuhn's argument for the second thesis is far from explicit, but I shall try to reconstruct it without misrepresentation from a number of passages in his work.

(i) He makes a historical claim that the genesis of methodological principles is inseparable from substantive scientific discoveries. He says, for example, that in learning a paradigm the scientist acquires theory, methods, and standards together, usually in an inextricable mixture. (1970, p. 109).

It is evident from the discussion in Chapters 9–11 of his book that this assertion is intended to apply to the initial discovery as well as to later imitation and indoctrination. I do not deny that Kuhn's historical assertion is largely correct, but I wish to register disappointment that he does not present and analyze detailed historical evidence for it, which would be of great intrinsic interest and might throw some light upon the complex relations among scientific thought, the thought of pre-scientific civilized peoples, the reasoning of children, and "the mind of primitive man".

(ii) Kuhn is dubious about the possibility of applying explicit methodological principles, because of

the insufficiency of methodological directives, by themselves, to dictate a unique substantive conclusion to many sorts of scientific questions. (1970, p. 3).

The most explicit justification that I have been able to find for step (ii) is on pp. 145–7 of Kuhn's book, where he elliptically discusses the possibility of judging a theory by either the methods of probabilistic verification or by falsification. The following is essentially all that he has to say about the former:

In their most usual forms, however, probabilistic verification theories all have recourse to one or another of the pure or neutral observation-languages discussed in Section X. One probabilistic theory asks that we compare the given scientific theory with all others that might be imagined to fit the same collection of observed data. Another demands the construction in imagination of all the tests that the given scientific theory might conceivably be asked to pass. Apparently some such construction is necessary for the computation of specific probabilities, absolute or relative, and it is hard to see how such a construction can possibly be achieved. If, as I have already urged, there can be no scientifically or empirically neutral system of language or concepts, then the proposed construction of alternate tests and theories must proceed within one or another paradigm-based tradition. Thus restricted it would have no access to all possible experiences or to all possible theories. As a result, probabilistic theories disguise the verification situation as much as they illuminate it. (1970, pp. 145–6).

The *demonstrandum* in this passage is the impossibility of computing specific probabilities in crucial situations of conflict between theories. Evidently, the details of the various probabilistic methods are irrelevant to Kuhn's demonstration, since not even an outline is presented of any one of them, and in fact the description of one of the methods ("Another demands etc.") is so perfunctory that I cannot decide which one he is referring to. Kuhn obviously considers the decisive point in the argument to be the theory-ladenness of observation, expressed in the sentence beginning "If, as I have already urged". The elliptical character of his argument then has some justification, for the details of the various probabilistic methods, and of any other method that might be considered, are irrelevant to this consideration.

It is worth pointing out, however, in anticipation of later analysis, that Kuhn's *demonstrandum* about the computation of probabilities could be reached in an entirely different way, by pointing out that numerical values are underdetermined in all extant formulations of probabilistic methods, whether because of the range of choice left open as a matter of principle in personalist theories, or because there are too many admissible *c*-functions, or because the principle of indifference is insufficiently precise, or because in frequency theories there are too many possible choices of reference classes. Here would be a reason for asserting "the insufficiency of methodological directives" which does take account of the detailed structure of extant methods. In the case of the method of falsification, a similar demonstration of 'insufficiency' could rest upon the difficulty of giving unambiguous criteria for "basic statements" (Popper, 1961, pp. 100–111). If it is then objected that no general proof has been provided for "the insufficiency of methodological directives" in all formulations of scientific method, including those which the ingenuity of future methodologists may devise, then an answer could be given which Kuhn's empiricist approach to epistemology should find congenial: the shortcomings of methods so far devised in this respect constitute inductive evidence that those of the future will suffer the same defect. The argument suggested in this paragraph has the evident advantage over the one extracted from Kuhn's text that it is unaffected by analyses which challenge the theory-ladenness of observation.[6] It also has the virtue of applying to 'normal science' as well as to 'scientific crises'. Nothing in Kuhn's own argument precludes the possibility of unequivocal calculations of the probability of a theory upon given evidence, so long as both the theory and the evidence are governed by a definite paradigm. But any one who has struggled with the problem of devising 'objective' procedures for specific probability calculations would be euphoric at achieving even this much!

(iii) Kuhn sometimes goes beyond step (ii) by asserting that the methodological content of a paradigm can never be completely abstracted and articulated. Scientists are able to

> agree in their *identification* of a paradigm without agreeing on, or even attempting to produce, a full *interpretation* or *rationalization* of it. Lack of a standard interpretation or of an agreed reduction to rules will not prevent a paradigm from guiding research. (1970, p. 44).

He evidently is maintaining that the unabstracted residue of a paradigm compensates for the insufficiency of methodological directives which he had pointed out in step (ii). In order to elucidate the *modus operandi* of a paradigm, Kuhn refers (*ibid.*) to Polanyi's ideas on 'tacit' knowing'; but in spite of admiration for Polanyi as a phenomenologist of scientific investigation, I do not think that these ideas will provide an answer to the criticisms which will be raised below.

The main criticism is that all passages which can be taken as formulations of Kuhn's second thesis are obscure on the following questions: Even if one grants that the genesis of methodological principles is inseparable from substantive scientific developments, might not some of them be rationally justified in abstraction from actual scientific discoveries (so that the context in which they originated would only be psychologically relevant, like the figures drawn in sand for the geometer)? If so, which ones can be justified in this way and how? The most informative quotation that I can find discusses "a preliminary codification of good reasons for theory choice" as follows:

> These are, furthermore, reasons of exactly the kind standard in philosophy of science: accuracy, scope, simplicity, fruitfulness, and the like. It is vitally important that scientists be taught to value these characteristics and that they be provided with examples that illustrate them in practice. If they did not hold values like these, their disciplines would develop very differently...
>
> What I am denying then is neither the existence of good reasons nor that these reasons are of the sort usually described. I am, however, insisting that such reasons constitute values to be used in making choices rather than rules of choice. (1970a, pp. 261–2).

Even from this passage I cannot determine whether the goodness of the reasons mentioned is supposed to consist only in their being abstracted from several successful paradigms, as suggested by the opinion cited earlier that "philosophy is an empirical enterprise", or whether there is some other basis for their rationality. There is no need, however, to worry the text for an answer to this question, since Kuhn could be asked directly to say which interpretation he intended. His work deserves censure on this point whatever the answer might turn out to be, just because it treats central problems of methodology elliptically, ambiguously, and without the attention to details that is essential for controlled analysis.

What, then, ought to be said about the rationality of methodological principles? Or, put another way, if Kuhn's second thesis is partly correct, how exactly should it be understood and qualified? The questions are evidently difficult, and I shall attempt only an outline in reply. Complete answers would require a fully satisfactory theory of scientific method, which we still have not achieved. One thing that does seem clear, however, is the indispensability of a variety of considerations which are organized so as to complement each other. Specifically, all of the following seem to me essential for a thorough understanding of the rationality of scientific procedures.

3.1. *Deductive logic.* The scientific method draws freely and sometimes crucially upon deductive logic. Whatever the basis of the principles of deductive logic may be, they are so reliable in comparison with other elements of the scientific method that in the present context there is no 'real and living doubt' about their rationality. Even the extreme empiricists, who maintain that all statements are ultimately to be judged according to their contribution to a coherent account of experience, do not usually propose to derive the principles of deductive logic from specific achievements of science.[7] Consequently, Kuhn's second thesis ought to be qualified at least to the extent of admitting that those principles of deductive logic which are appropriated by the scientific method stand in no need of justification from the history of science. The most that can be retained of Kuhn's second thesis at this juncture is the observation that substantive scientific achievements may indicate what parts of deductive logic are important for scientific method. Thus, Aristotle recognized the validity of inference by *modus tollens* but considered it to be inferior to affirmative demonstration (*Posterior Analytics* I, 25, 87a); the reversal of his judgment may plausibly be attributed to the historical occurrence of exemplary eliminations of theories on the basis of observation (together, of course, with the revelation of weaknesses in the affirmations of Aristotelian physics).

3.2. *Probability theory.* Corresponding to the different senses of 'probability' there are various justifications of the axioms of probability, using arithmetic, measure theory, decision theory, etc. In no case is the justification based upon substantive scientific achievements. There is nothing to be retained here from Kuhn's second thesis except that scientific achievements suggested (or formed part of the intellectual milieu which suggested) the significance of the notion of probability for scientific investigation (see Hacking, 1975). There are still many problems concerning the role of probability in scientific method which are not completely solved. For example, if 'probability' is understood in the personalist sense, then the Dutch book argument of Ramsey and DeFinetti supplies a very convincing justification of the axioms, provided that it makes sense to bet upon the propositions to which probabilities are assigned, but the typical universal generalizations which occur in the natural sciences do not seem to be 'bettable'. There happens to be a surrogate for the Dutch book argument, namely the theorem of Cox, Good, and Aczél, but it depends upon premisses which are difficult to justify definitively, even though they are highly plausible (Shimony, 1970). Such problems, however, do not seem to me to lend any support for Kuhn's thesis, but only indicate that much remains to be clarified in decision theory when a game is played for truth rather than for monetary stakes (see Levi, 1967).

3.3. *A posteriori considerations*. Here is where Kuhn's second thesis is correct and important. His assertion of "the insufficiency of methodological directives" is nowhere better illustrated than by the notorious inadequacy of the axioms of probability theory, even with such reasonable supplements as the principle of indifference, to yield non-arbitrary numerical evaluations. The axioms and supplements also fail to yield a non-arbitrary rule for the tentative acceptance of a hypothesis – a circumstance which may be a virtue rather than a failing in practical decision-making, when utilities can be compared (Jeffrey, 1956; Carnap, 1963, p. 972), but which appears to be a serious inadequacy in the context of theoretical investigations. There may be no way of compensating for these insufficiencies that is justifiable in all possible worlds. But the counsel of wisdom may be that nothing of value is thereby lost, for surely we should be content with methods that work in the actual world. A moment's reflection, however, diminishes the appeal of this solution: if we knew what the actual world is like, the objectives of scientific inquiry would already be accomplished and there would be no need for the methods of science. Nevertheless, something can be salvaged from the suggestion: *we may know enough about the actual world, or at least have tentative suppositions upon which we may reasonably enough rely, to add specificity to a priori methodological principles.* Since the great achievements of science can reasonably be claimed to provide good approximations to the truth about the actual world (as argued in Section 2), we are justified in attempting to derive from them some of the specificity which is lacking in a priori principles. Thus, it is a commonplace that the prior probability of a simple hypothesis should be greater than that of a less simple one, but many methodologists have despaired of the utility of this rule just because of the difficulties of finding non-arbitrary a priori criteria for simplicity. If, however, serious attention is paid to Weyl's admonition that "we must let nature train us to recognize the true inner simplicity" (1949, p. 155), then it is possible to use such criteria as the order of differential equations and tensorial rank in making judgments of relative simplicity. The 'training' by nature consists of exemplary scientific discoveries in which the roles of differential equations, tensors, etc. are exhibited. Here, then, is strong support for Kuhn's assertion that paradigms guide research, and for part of his thesis that the procedures of scientific investigation can be shown to be rational only by drawing upon substantive scientific achievements. It must be emphasized, however, that the reason which has just been offered for the partial correctness of Kuhn's thesis is very different from Kuhn's own explanation that a scientist may have tacit knowledge of the methods implicit in a paradigm without being able to articulate them. Whether the scientist's knowledge is tacit or explicit, the crucial point is that it is warranted only by evidence that the actual world is constituted in a way that makes certain procedures appropriate.

3.4. *Dialectic.* The proposal to justify even part of the scientific method by a posteriori considerations raises serious problems about the structure of scientific knowledge. Unless one proceeds with some finesse, one might conclude that there are no rational grounds for the acceptance of those scientific achievements from which methodological principles are abstracted. But this conclusion is absurd (and therefore I credit Kuhn's protests against critics who attribute the opinion to him) for the obvious reason that some criteria, however rough, must be operative in the judgment that a theory is successful. It may be suggested at this point that those methodological principles which are drawn from deductive logic and probability theory will suffice to establish the criteria of success, so that nothing but a priori principles and experience are needed to identify the exemplary achievements upon which the a posteriori methodological principles are based. I very much doubt that the structure of scientific knowledge is as neat as this. Rather, it seems that some complex considerations of the interplay between the a priori and the a posteriori – which may appropriately be called "dialectic" – are indispensable. Much remains to be clarified about this dialectic. How, for example, are a posteriori guidelines to be used in prior probability evaluations at the same time that open-mindedness is maintained towards revolutionary hypotheses? Is it possible to formulate inductive arguments in support of inductive principles without vicious circularity? What role is played by vindicatory arguments, i.e., arguments of the form that nothing will be lost, and possibly something of value will be gained, if certain procedures are followed? (For a debate on the second and third questions see Teller (1975) and Shimony (1975).) The work of Imre Lakatos on the methodology of research programs (1970) is one of the important contributions to the study of this dialectic, even though he uses very different language to characterize what he is attempting to do.

3.5. *Moral factors.* There are principles of scientific methodology that are manifestly not so much intellectual as moral, for example that a scientist should not suppress data which are unfavorable to his own proposals and that the state should abstain from scientific censorship. Such principles can perhaps be regarded as special instances of a posteriori considerations, for they derive from our awareness of some of the evil proclivities of human beings. (They would not be articulated in the scientific method of the Houynhmhms, though perhaps they would be "tacitly known".) In a sense, however, all methodological principles are subsumed under moral ones:

> The most vital factors in the method of modern science have not been the following of this or that logical prescription – although these have had their value too – but they have been the moral factors. First of these has been the genuine love of truth and conviction that nothing else could long endure. Given that men strive after the truth, and, in the nature of things, they will get it in a measure etc. (Peirce, 1958, p. 56).

Here, then, is an outline of the elements of a scientific methodology, drawing upon the contributions of many workers, which presumes to account for the rationality of scientific procedures. It makes no pretense to completeness or to freedom from conceptual difficulties, and indeed a number of difficulties were explicitly pointed out. Despite the unfinished state of methodology, however, enough has been accomplished, I think, to establish firmly the main contentions of this paper: (i) Kuhn is partly correct in his thesis that the rationality of scientific procedures derives from scientific achievements, but his thesis needs qualification and refinement; and (ii) the most important qualification is to acknowledge what his other thesis denies, that the community of serious investigators exhibits collectively a wonderful sensitivity to the truth.

Acknowledgments. I am very grateful to Prof. Martin Eger for his helpful comments, and to the National Science Foundation for support of research.

Boston University

NOTES

[1] All references will be to the second edition of 1970, which contains "Postscript – 1969".

[2] See, e.g., Shapere (1964) and (1971), the essays in Lakatos and Musgrave (1970), some of the essays in Suppe (1974), Scheffler (1967) and (1972), and Quay (1974). The criticisms of the last two authors in this list come closest of any I know to the comments of the present article.

[3] Although my primary concern is with epistemology, it should be evident that the point of view which I am recommending also has consequences for the discipline of history of science. Specifically, human sensitivity to the truth should be taken seriously both as an *explanandum* and as an *explanans* in studies of scientific development. There is no question worthier of investigation by historians of science than one raised by Cohen: "Can we establish, by historical inquiry, what is the most likely social ensemble of human need, desire, conflict, prejudice, and training to produce a search for, and attainment and recognition of, true knowledge?" (1970, p. 233).

[4] The terminology is borrowed from Popper (1963) but with some change of meaning.

[5] See Tisza (1963), Shimony (1970), Post (1971), Koertge (1973), and Boyd (unpublished).

[6] Cf. Shimony (forthcoming).

[7] An exception is the proposal of Finkelstein and Putnam to change the logic of discourse because of quantum mechanical evidence. They are wrong, in my opinion, but the issue is too complex to discuss here. If they are correct, however, then Kuhn's second thesis would have wider applicability than I have allowed.

BIBLIOGRAPHY

Boyd, Richard, *Realism and Scientific Epistemology* (unpublished).

Carnap, Rudolf, 1963, 'Replies and Systematic Expositions', in P. A. Schilpp (ed.), *The Philosophy of Rudolf Carnap*, Open Court, La Salle, Ill.

Cohen, Robert S., 1970, 'Causation in History', in W. Yourgrau and A. D. Breck (eds.), *Physics, Logic, and History*, Plenum, New York and London.

Finkelstein, David, 1969, 'Matter, Space and Logic', in *Boston Studies in the Philosophy of Science*, Vol. V, (ed. by R. S. Cohen and M. Wartofsky), D. Reidel, Dordrecht and Humanities Press, New York.

Hacking, Ian, 1975, *The Emergence of Probability*, Cambridge University Press, Cambridge.

Hume, David, 1955, *Inquiry Concerning Human Understanding* (ed. by C. Hendel), Liberal Arts Press, Indianapolis.

Jeffrey, Richard, 1956, 'Valuation and Acceptance of Scientific Hypotheses', *Philosophy of Science* **23**, 237–246.

Koertge, Noretta, 1973, 'Theory Change in Science', in G. Pearce and P. Maynard (eds.), *Conceptual Change*, D. Reidel, Dordrecht.

Kuhn, Thomas, 1970, *The Structure of Scientific Revolutions*, second edition, University Chicago Press, Chicago.

Kuhn, Thomas, 1970a, 'Logic of Discovery or Psychology of Research?' and 'Reflections on my Critics', in Lakatos and Musgrave (eds.) (1970).

Kuhn, Thomas, 1974, 'Second Thoughts on Paradigms' and contributions to discussions, in Suppe (1974).

Lakatos, Imre, 1970, 'Falsification and the Methodology of Scientific Research Programs', in Lakatos and Musgrave (eds.) (1970).

Lakatos, I. and Musgrave, A. (eds.), 1970, *Criticism and the Growth of Knowledge*, Cambridge University Press, Cambridge.

Levi, Isaac, 1967, *Gambling with Truth*, Knopf, New York.

Peirce, Charles S., 1958, *Collected Papers*, Vol. VII (ed. by A. W. Burks), Harvard University Press, Cambridge, Mass.

Popper, Karl R., 1961, *The Logic of Scientific Discovery*, Science Editions, New York.

Popper, Karl R., 1963, *Conjectures and Refutations*, Harper and Row, New York and Evanston.

Post, Heinz, 1971, 'Correspondence, Invariance, and Heuristics', *Studies in History and Philosophy of Science* **2**, 213–255.

Putnam, Hilary, 1969, 'Is Logic Empirical?', in *Boston Studies in the Philosophy of Science*, Vol. V (ed. by R. S. Cohen and M. W. Wartofsky), D. Reidel, Dordrecht, and Humanities Press, New York.

Quay, Paul, 1974, 'Progress as a Demarcation Criterion for the Sciences', *Philosophy of Science* **41**, 154–170.

Scheffler, Israel, 1967, *Science and Subjectivity*, Bobbs-Merrill, Indianapolis.

Scheffler, Israel, 1972, 'Vision and Revolution: A Postscript on Kuhn', *Philosophy of Science* **39**, 366–374.

Shapere, Dudley, 1964, 'The Structure of Scientific Revolutions', *Philosophical Review* **73**, 383–394.

Shapere, Dudley, 1971, 'The Paradigm Concept', *Science* **172**, 706–9.

Shimony, Abner, 1970, 'Scientific Inference', in R. Colodny (ed.), *The Nature and Function of Scientific Theories*, University of Pittsburgh Press, Pittsburgh.

Shimony, Abner, 1975, 'Vindication: A Reply to Paul Teller', in *Induction, Probability, and*

Confirmation, Minnesota Studies in the Philosophy of Science, Vol. VI (ed. by G. Maxwell and R. M. Anderson, Jr.), University of Minnesota Press, Minneapolis.

Shimony, Abner, 'Is Observation Theory-Laden? A Problem in Naturalistic Epistemology', in a forthcoming volume edited by R. Colodny, University of Pittsburgh Press, Pittsburgh.

Stein, Howard, 1974, 'Maurice Clavelin on Galileo's Natural Philosophy', *British Journal for the Philosophy of Science* **25**, 375–397.

Suppe, Frederick (ed.), 1974, *The Structure of Scientific Theories*, University of Illinois Press, Urbana.

Teller, Paul, 1975, 'Shimony's A Priori Arguments for Tempered Personalism', in *Induction, Probability, and Confirmation, Minnesota Studies in the Philosophy of Science*, Vol. VI (ed. by G. Maxwell and R. M. Anderson, Jr.), University of Minnesota Press, Minneapolis.

Tisza, Laszlo, 1963, 'The Conceptual Structure of Physics', *Reviews of Modern Physics* **35**, 151–185.

Weyl, Hermann, 1949, *Philosophy of Mathematics and Natural Science*, Princeton University, Princeton.

FRED SOMMERS

LEIBNIZ'S PROGRAM FOR THE DEVELOPMENT OF LOGIC*

INTRODUCTION

In the summer of 1969 Imre and I were critically discussing the program – instituted by Frege – that had given direction to the development of logic in the twentieth century. Imre suggested to me that I look back into the history of logic for alternative programs that have remained undeveloped. I remember we spoke of Leibniz but I knew very little about him then. Later Imre kept at me to write on the theme, and parts of this paper were prepared for the journal he was editing.

I

It is generally held that Frege achieved a revolution in logic that compares with Newton's revolution in natural philosophy. As for Leibniz, he is recognized as having anticipated Frege in his insistence on making proofs formally explicit, in his appreciation of logic as a calculus, and in his use of algebraic or mathematical algorithms for logistic purposes. Though noteworthy as a precursor of Frege, Leibniz's own contributions to logic were (so goes the historical verdict) inevitably limited by his adherence to the older logic of terms, by his analysis of propositions in terms of subject and predicate, and by his view that syllogistic inference was at the foundation of logic.

I believe that one may accept much of this judgment. Leibniz did concentrate on the logic of terms to the neglect of propositional logic. He adhered to the subject-predicate analysis of categorical propositions according to which 'some *S*' is the subject and 'is *P*' or 'isn't *P*' the predicate. And he never questioned the fundamental character of syllogistic. Frege is opposed to Leibniz on each of these points. For Frege propositional logic is primary logic and it comes before his analysis of categorical propositions. In the case of propositions of the sort entering

R. S. Cohen et al. (eds.), Essays in Memory of Imre Lakatos, 589–615. *All Rights Reserved*
Copyright © 1976 by D. Reidel Publishing Company, Dordrecht–Holland

into syllogistic reasoning, Frege opposes the subject-predicate analysis replacing it with an analysis of function and argument. Finally, Frege showed that syllogistic itself corresponded to a logically arbitrary part of a more comprehensive predicate logic.

We may add to this contrast (hardly favorable to Leibniz) the discovery by Frege that singular propositions could be viewed as atomic elements in the analysis of general propositions. Here the difference between Leibniz and Frege is not well understood because Leibniz's view of singular propositions has been virtually ignored. According to Leibniz a proposition like 'Socrates is mortal' is both universal and particular and we are free to parse it either as 'every Socrates is mortal' or as 'some Socrates is mortal'. The view that singular propositions have the logical form of unversal propositions is traditional. In adding that they may also be construed as particular in quantity Leibniz distinguished them from general propositions by the very fact of having an indifferent or "wild" quantity. Moreover, since 'Socrates is mortal' is indifferently universal or particular we can explain why neither quantity is actually specified. Although Leibniz's innovation is of considerable logical importance, it is made within the Aristotelian tradition which holds that standard singular propositions are to be classified with the standard general categoricals as propositions whose subjects are expressions of form 'every *S*', or 'some *S*'.

Frege's analytical logic went in the opposite direction. Taking the singular form as primitive he decomposed the general forms into singular elements. For example, 'every man is mortal' is analyzed into "atomic" elements '*x* is mortal', '*x* is a man' and it is reconstructed as a function of these elements. For Frege elementariness of atomicity resides in the singular proposition. The non-atomic proposition is constructed of atomic propositions by means of formative operators, the quantifiers and the truth functions. The idea that singular propositions are syntactically more primitive than general propositions is characteristically Fregean. Its acceptance has been universal and it is perhaps worth mentioning today that the idea was quite foreign to Leibniz.

With the benefit of some hindsight we can say that Frege's doctrine of the syntactic elementariness of singular propositions is the key to his revolution in logic. If singular propositions are atomic, then the general proposition that we find in the classical syllogism needs to be analyzed into its atomic parts. The analysis brings us immediately to the truth functions. For example, 'every *S* is *P*' is immediately seen to contain propositional connectives such as 'if ... then ...' or 'or'. The primacy of propositional logic to the logic of terms is thus one consequence of the doctrine of the elementariness of the singular propositions in the analysis of general propositions. Russell who was, well acquainted with the difference between Leibniz and Frege on singular propositions remarks on its historical significance:

The first serious advance in real logic since the time of the Greeks was made independently by Peano and Frege – both mathematicians. Traditional logic regarded the two propositions 'Socrates is mortal' and 'All men are mortal' as being of the same form; Peano and Frege showed that they are utterly different in form. The philosophical importance of logic may be illustrated by the fact that this confusion – which is still committed by most writers – obscured not only the whole study of the forms of judgment and inference, but also the relation of things to their qualities, of concrete existence to abstract concepts, and to the world of Platonic ideas.... Peano and Frege, who pointed out the error did so for technical reasons...but the philosophical importance of the advance which they made is impossible to exaggerate.

In this connection it is worth quoting the passage from Leibniz in which he argues that singular propositions have the logical form of general propositions, their only distinction being that they have "wild" quantity.

How is it that opposition is valid in the case of singular propositions – ... since elsewhere a universal affirmative and a particular negative are opposed. Should we say that a singular proposition is equivalent to a particular and to a universal proposition? Yes, we should. So also when it is objected that a singular proposition is equivalent to a particular proposition, since the conclusion in the third figure must be particular, and can nevertheless be singular; e.g., 'Every writer is a man, Some writer is the Apostle Peter, therefore the Apostle Peter is a man.' I reply that here also the conclusion is really particular and it is as if we had drawn the conclusion 'Some Apostle Peter is a man'. For 'Some Apostle Peter' and 'Every Apostle Peter' coincide, since the term is singular.[1]

If Leibniz is right, there are no atomic propositions: even a singular proposition asserts or denies something of the form 'some/every *x* is/isn't *y*'. Frege does not appear to have questioned the existence of atomic propositions. In any case he did not believe that their existence was a matter that was subject to proof. Discussing the difference between individual constants and predicate letters Frege says:

Take the proposition 'two is a prime number'. Linguistically we distinguish two parts: a subject 'two' and a predicate is a prime number. It is important to grasp the point that the difference [between the two parts of the proposition] really cuts deep and must not be blurred. I call the second part unsaturated and the first part saturated. The decomposition into saturated and unsaturated parts is a primitive feature of logical structure, which must simply be recognized and accepted.[2]

According to Frege then there is a class of propositions wholly devoid of logical signs like 'every' or 'not'. According to Leibniz, every proposition must contain *some* "syncategorematic" or logical signs. The issue is of crucial importance. For Leibniz, the difference between subject and predicate is syntactically unproblematic; it is the difference between a phrase that contains a sign of quantity and one that lacks it. For Frege, the difference is also unproblematic, the asymmetry of subject and predicate however is not a difference of logical signs but of material elements and this latter difference is characterized as a "logically primitive phenomenon". Finally, if Leibniz is right, the subject and predicate have *terms* as common syntactical elements that are interchangeable. It is therefore possible to predicate any term that appears in either position. But if Frege is right, subjects and predicates are logically simple and primitively different with nothing in common. And the predication of a subject is simply ill-formed.

The question whether or not to postulate a class of propositions wholly devoid of logical signs is of obvious moment. Despite Frege's contention that the phenomenon of atomicity is primitive, its universal acceptance could hardly have been a matter of logical intuition. Perhaps the main reason for the centrality of the atomic proposition in modern logic is the technical one that Russell alluded to: the power and scope of predicate logic in which the atomic proposition figures as the element of logical analysis. The rules of formation for the most powerful logics we know begin with atomic forms and proceed with the construction of non-atomic sentences by the addition of sentence forming operators (the quantifiers and the logical constants). It was felt that the traditional syntactical doctrine of the singular proposition must be wrong; had it been right it should have been possible to construct a logic as powerful as Frege's in which singular propositions and general propositions have the logical form of an assertion or denial of a universal or a particular proposition (i.e., a proposition of form 'every *S* is/isn't *P*' or 'some *S* is/isn't *P*').

It may be acknowledged that a Leibnizian syntactical theory has the virtue of giving a unified account of categorical propositions, one that does not assume an absolute distinction of logical form for 'Socrates is mortal' and 'every man is mortal'. But historically, the price seems to be a very weak logistical system, one that is especially, inadequate for inference involving relations.

The inference power of modern logic was certainly an important factor for the triumph of Frege's logical syntax. Today one would point as well the other virtues of the standard logical grammar. Thus Quine remarks:

The grammar that we logicians tendentiously call standard is a grammar designed with no other thought than to facilitate the tracing of truth conditions. And a very good thought this is.[3]

As an example of the advantage of standard grammar over the grammar of natural language consider the sentence 'an American has walked on the moon'. The sentence has no part that explicitly signifies the existence of Americans. The existence of an American is nonetheless a condition for the truth of the statement made with this sentence and its canonical translation says as much. Frege's conceptual script has the virtue of semantic explicitness, a desideratum that is deemed essential to the claim that it is a *logical* syntax. This requirement would spell doom to the Leibnizian ideal of a logical syntax of natural language. For it is a fact that the natural languages are not semantically explicit at the level of syntax. The matter at issue here is deserving of extended discussion. For reasons of space my remarks must be brief.

A sentence of form 'some x is y' or 'the x is y' is said to have existential import if the statement made by its use cannot be true unless some x (the x) exists. It is undoubtedly true that many such sentences have existential import. But if we parse all such sentences to make this condition explicit and then treat the parsing as a presentation of logical form we find ourselves forced to judge certain statements false that do not have the existence of (an) x as a condition of truth. We are, for example, constrained to judge 'some frictionless motion on an inclined surface is constant' to be false on the grounds that there is no such motion anywhere in nature even though this sentence can be used to assert a truth in physics. We are similarly constrained to consider 'the beast captured by Bellerophon is a flying horse' false although we demand the opposite judgment of a student taking a "true-false" examination in classics. It is a familiar fact that sentences in the vernacular of form 'some *x* is *y*' often lack existential import. Even so, no one would deny that all such sentences have the same logical form. The same is true of sentences of form 'the *x* is *y*' or '*a* is *y*' (where '*a*' is a proper name). For we should like to give the same syntactical analysis to 'the beast captured by Bellerophon is a flying horse' as to 'the 37th president of the United States is a Republican' while allowing for the truth of both. Sentences of form 'the/some *x* is *y*' have a variety of truth conditions and the use of truth conditions to determine logical syntax is a questionable practice even when accompanied by the disclaimer that the expression whose syntax is semantically explicit is not meant to be faithful to the natural syntax of the original. We have already observed that this practice introduces a pervasive and constricting ontological idiom into logical syntax. In natural languages there is a clear distinction between 'the/some *x* is *y*' and 'the/some *x* that is *y* exists'. The former may have existential import but only the latter is an existential sentence. A logical grammar that preserves this distinction will lack semantic explicitness and the traditional grammar of logic allows for classes of sentences (the *S* is *P*, some *S* is *P*) that have no existential conditions written into their syntax. In allowing this, the traditional doctrine of logical grammar is sensitive to the variety of truth conditions for sentences within a given class and insensitive to the modern requirement that a logical syntax, worthy of the name, is determined by truth conditions.

Quine is historically right in saying that Frege's conceptual script (Begriftschrift) was designed to achieve semantic explicitness. This motivation played no important part in Leibniz's program. Moreover, Frege himself was aware of this difference between himself and his predecessors. In a summary of his views to Darmstaedter (1919) Frege says:

My particular conception of logic is initially characterized by the fact that I put the content of the word 'true' at the beginning and let it immediately be followed by the thought with respect to which the question of truth arises. In other words, I do not begin with concepts out of which the judgment is composed but I get to the parts of thought through the splitting (Zerfallung) of the thought. In this respect my conceptual script is different from similar creations by Leibniz and his successors in spite of its possibly misleading name (Begriftschrift).[4]

Here Frege is observing that where Leibniz would arrive at the concept-term elements '*S*' and '*P*' as the units of analysis of a sentence 'some/every *S* is/isn't *P*', he (Frege) would decompose it into sentential elements '*Sx*', '*Px*', to whose instances we may assign truth values. In introducing this mode of analysis Frege took the decisive step toward a semantically perspicuous syntax. It is only recently that philosophers have begun seriously to question whether the perspicuity of truth conditions in logical syntax is the unmixed blessing that Russell and Quine have taken it to be.

In comparing the logical grammars of Leibniz and Frege we could set aside as moot the question of the desirability of having the truth conditions explicit in a logical syntax. Even so we must face the first difference: the logistic advantage that standard grammar enjoys over the traditional logical grammar of Leibniz and other Aristotelian logicians. The advantage appears to be decisive and for this reason alone a logical grammar along Leibnizian lines would seem to be no more than a historical relic.

In much of what follows I attempt to show that the accepted assumption that a logical syntax along Leibnizian lines must be weak in inference power is an unwarranted assumption. I argue that, on the contrary, there are significant technical reasons to view Leibnizian syntax with great sympathy on purely logical grounds and I point to this as a reason to reconsider the condescending historical judgment which views Leibniz as an ingenious but limited logical thinker whose best work has been superseded by Frege.

We have observed that Leibniz's logical syntax, and in particular his view of the logical subject as a complex phrase (as contrasted with Frege's simple name theory) allows for the predication of singular subject terms. Nothing in the logical syntax of Leibniz forces him to treat 'Tully is Cicero' as a binary sentence used for predicating 'is identical with' of 'Tully, Cicero' (in the order given). Instead it is open to Leibniz and to anyone who adopts the Aristotelian syntax of singular propositions to view 'Tully is Cicero' as a monadic sentence on all fours with 'Tully is an orator'. One may well imagine Russell reminding us that before Frege 'Tully is an orator' and 'Tully is Cicero' were mistakenly believed to have the same logical form and how Frege taught us to see the radical difference between these two sentences. Certainly, given Frege's theory of the atomic propositions, it is no longer possible to view 'Tully is Cicero' as a monadic sentence. If 'Tully' and 'Cicero' are the same kind of expression, then the syntactic gap between subject and predicate prohibits a monadic reading of 'Tully is Cicero' and the only way to save the sentence is to read it dyadically with the so-called 'is' of identity.

For Leibniz, then, an "identity" sentence is a monadic singular sentence that happens to have singular terms in both subject and predicate position. For Frege, identity sentences must be dyadic on pain of ill-formation. In his most famous essay Frege examines the consequences of his doctrine that identity sentences are dyadic. The paradoxical nature of the identity relation (a relation that Leibniz has no need to recognize) leads him to distinguish between the sense and the reference of signs that occupy subject positions. This is not the place to discuss the distinction between sense and reference. However, in the context of the present discussion it is appropriate to note that the theory of dual signification for individual constants is a consequence of Frege's view that the logical subject is syntactically simple. The semantic complexity of the logical subject is the price Frege must pay for its syntactical simplicity.

The price is higher still for we find that the kind of technical reasons Russell alludes to for defending a difference of logical form for singular and general sentences will actually favor the older theory of logical syntax according to which it is possible to view identity sentences as monadic subject-predicate propositions with singular terms in subject and predicate positions. To see how this is so we shall examine the consequences of reading 'Tully is Cicero' monadically and contrast them with the familiar consequences of reading it dyadically.

There is first the fact that should not be overlooked that Frege needs the sign of identity as a new primitive sign of logic. Leibniz has no need of such a sign. There is second the fact that special *laws of identity* must be introduced as new axioms governing the sign of identity in standard first-order logics. Leibniz needs no new axioms for the monadic counterparts of Frege's dyadic sentences; in Leibniz each of the so-called laws of identity can be derived within the traditional term logic. This comes about because identity sentences, like other singular sentences, have indifferent quantity. We are always free to consider '*a* is *b*' (where '*a*' and '*b*' are singular terms) as having the form 'every *a* is *b*' or else as having the form 'some *a* is *b*'. To show that '*a* is *c*' follows from '*a* is *b*' and '*b* is *c*' we simply treat all sentences as universal. The so-called law of transitivity of identity then is proved by the fact the 'every *a* is *b*' follows *syllogistically* from 'every *a* is *b* and every *b* is *c*'. The monadic counterpart of the law of symmetry is easily proved by the fact that 'some *b* is *a*' follows from 'some *a* is *b*'. As for the law of reflexivity, that is simply an instance of what traditional logicians called the law of identity: every *a* is *a*. In brief, Leibniz can do without the identity relation and without the laws of identity; he can achieve the same logical results by treating '*a* is *b*' as having the form of a monadic sentence with wild quantity. Since the results in the logic of identity are in every way more economical than Frege's we may count them in favor of Leibniz's view that singular and general propositions have the same logical form.

There are other familiar places where Leibniz's theory of logical syntax has a decided advantage over Frege's. Using standard modern logic one must appeal to a primitive, underivable principle – the principle of existential generalization – in accounting for the inference from 'Mars is a plant' to 'something is a planet'. But in Leibniz's logic the inference is valid on syllogistic grounds with the addition of the platitudinous premise 'Mars is a thing'. We then have the valid syllogism: 'every Mars is a planet, some Mars is a thing, so some thing is a planet'. The use of wild quantity also explains the inference found in Aristotle: 'Pittacus is a good man, Pittacus is a wise man, so some wise man is a good man'. This inference is a valid syllogism if we read the premises with opposing quantities ('every Pittacus is a good man, some Pittacus is a wise man'). Here too, the Fregean cannot account for the inference without the use of the principle of existential generalization. Leibniz's logical syntax and his use of wild quantity for singular propositions need no appeal to any principle of universal instantiation to account for the inference 'everything is created, so Mars is created'. Here again we assume that the inference is an enthymeme with a missing premise: 'everything is created, Mars is a thing, so Mars is created'. (The choice of quantity for the minor premise and the conclusion is open.) In general, then, Leibnizian syntax, in which singular propositions have the logical forms 'some/every *x* is *y*', permits the syllogistic derivations '*a* is *P* (*a* is a thing) so something is *P*' and 'everything is *P* (*a* is a thing), so *a* is *P*' where Frege requires a primitive principle of generalization or instantiation. Here the technical advantage lies with Leibniz.

It is often said that Frege's greatest technical contribution in his treatment of inferences involving relations. And it must be acknowledged that Leibniz, like Jungius, whom he admired and who was the first logician to seriously tackle relations, simply failed to give an adequate account of inferences like 'every circle is a figure so everyone who draws a circle draws a figure'. Historians ascribe the failure to Leibniz's insistence than this inference and others like it are syllogistic in nature. But here the criticism is wide of the mark. I shall show that relational inferences can without strain be understood as a kind of syllogistic reasoning and that Leibniz's failure to handle relational inference is due instead to his failure to achieve an adequate notation for term logic. If this is correct, the historical judgment that Leibniz is no more than a precursor to Frege needs to be reassessed. I think it will be clear that an adequate notation for term logic does what Leibniz set out to do for relational inference. Moreover, we shall see that the notation which proves adequate to the development of a comprehensive term logic is already suggested by Leibniz. Unfortunately, Leibniz did not himself develop his suggestion to the point where it could be seen to be effective for logical reckoning. A good notation is, as Russell has aptly said, "a live teacher" and Leibniz' failure here deprived him of the well deserved fruits of his logical insights.

Leibniz always strove to give algebraic expression to the logical form of propositions. For example, he devoted much effort to an algorithm that treated 'every *A* is *B*' as '$A = AB$'. We shall develop another line which Leibniz himself did not pursue. The following passage is from his early work:

> Thomas Hobbes, everywhere a profound examiner of principles, rightly stated that everything done by the mind is a *computation*, by which is understood either the addition of a sum or the subtraction of a difference (*De Corpore* I.i.2). So just as there are two primary signs of algebra and analytics, + and −, in the same way there are, as it were, two copulas 'is' and 'is not'. [5]

We shall take this suggestion of Leibniz quite seriously and see where it leads. We note first that 'isn't' is a contraction of 'is not'. Since 'is' is a plus sign, the suggestion that the negative copula is a minus sign amounts to the suggestion that 'not' is a minus sign. So just as '$a - b$' is a contraction of '$a + (-b)$' so is '*a* isn't *b*' a contraction of '*a* is (not *b*)'. We now have a plus/minus representation of two syncategorematic signs; the plus sign transcribes the positive copula 'is' and its tensed variants and the minus signs transcribes the negative particle 'not' and other negative particles such as 'un-', '-less', 'not' and so forth. Using this notation we should transcribe a sentence like 'not a creature was stirring' as '$-(C+S)$'. The contradictory affirmation would be transcribed as '$C+S$'.

We recall that Leibniz holds that every categorical proposition has a logical subject with explicit or implicit quantity. Thus 'a creature was stirring' could be explicitly read as 'some creature was stirring'. To represent this in the algebraic notation we shall transcribe 'some' as a plus sign. We now transcribe 'a creature was stirring' as '$+C+S$' – a trivial and insignificant algebraic modification but one that serves to remind us that the logical subject of the proposition is 'some creature'. The contradictory sentence now becomes '$-(+C+S)$'.

'Not a creature was stirring' and 'every creature was non-stirring' are logically equivalent propositions. This suggests that the word 'every' may be implicitly defined thus:

$$\text{every } S \text{ is } P =_{\text{df}} \quad \text{not: some } S \text{ is not } P.$$

Algebraically the definition reveals that 'every' is a minus sign:

$$-S+P=_{\text{df}}-(+S+(-P)).$$

Extending the Leibnizian notation to the representation of the signs of logical quantity has proved instructive; we have found that 'some' and 'every' are opposed in a plus-minus way. We shall presently justify this by showing that these assignments are logically efficient. The following transcriptions illustrate the use of the notation:

every A is B
$-A+B$

some A is R to every B
$+A+R-B$

Whatever is R to an A is F to every B
$-(R+A)+(F-B)$

no A is B
$-(A+B)$

not an A is a B
$-(+A+B)$

Note that relational propositions are easily and naturally represented in the plus/minus notation. In this respect the plus/minus notation is decisively superior to those that were developed and studied by Leibniz himself. Nevertheless we consider the plus/minus notation to be Leibnizian in character and conception.

Leibniz hoped to unify term logic and propositional logic and he aspired to a notation that would reveal their isomorphism:

If, as I hope, I can conceive all propositions as terms, and if I can treat all propositions universally, this promises a wonderful ease in my symbolism and analysis of concepts and will be a discovery of greatest importance.[6]

In this spirit we may extend the plus/minus notation to the representation of propositional signs like 'and' and 'if...then'. We actually use 'and' and 'plus' as synonyms so it is intuitively acceptable to begin by representing 'and' by a plus sign. The compound expression 'bald American' could then be transcribed as '$A + B$' (for 'American and bald'). Similarly if p = 'roses are red' and q = 'violets are blue', the conjunction 'roses are red and violets are blue' would be represented as '$p + q$' and, generally, any expression of form 'x & y' would be transcribed algebraically as '$x + y$'. The analogy of 'x and y' to 'an x is a y' is an important one. To keep it clearly in mind we shall transcribe 'x & y' with *two* plus signs: $+ x + y$. This shows that 'and' and 'some...is...' are isomorphic functors for terms and propositions since both are represented by the ordered pair of signs '+, +'.

We know how to represent 'and' and we know how to represent 'not'. It is clear that we can now define a plus-minus representation for the other propositional connectives. I shall do this for 'if ... then'. Analogous methods give us representations for 'or', 'if and only if' and so forth.

The compound form 'not both p and not q' transcribes as '$+(+p+(-q))$'. Since this form is equivalent to 'if p then q' we may use it to *define* the latter and to get an algebraic representation for 'if...then...'.

Our procedure here is strictly analogous to the one used for getting a way of transcribing 'every...is...' by equating it to the algebraic expression that transcribes 'not: some...is...'.

$$-p+q =_{\mathrm{df}} -(+p+(-q))$$
if p then q $=_{\mathrm{df}}$ not (P & (not q)).

The equivalence reveals that the sign 'if...then...' is an ordered pair '$-$, $+$'. By a similar procedure we can show that 'x or y' is represented as '$--x--y$'.

The following transcriptions illustrate the use of the extended plus-minus notation.

if every A is B then some C is D.	$-[-A+B]+[+C+D]$
p if and only if q	$+[-p+q]+[-q+p]$
some A is B or C	$+A+<--B--C>$
every A is B and C	$-A+<+B+C>$

The general form of propositions is

$$\pm(\pm(\pm x)\pm(\pm y))$$

In this formula 'x' and 'y' may represent propositions. For example the expression 'if p then q' would have the explicit form '$+(-(+p)+(+q))$'. It would naturally be abbreviated as '$-p+q$'. The same general form may have terms as its elements. In that case the formula would be interpreted thus:

yes/not (some/every x/non-x is/isn't y/non-y)

The terms of the proposition may be simple or compound. If they are compound they themselves have the form '$\pm(\pm(\pm x)\pm(\pm y))$'. They may be relational or non-relational. A relational term has the form $R\pm X_1 \ldots \pm X_n$, with 'some x' or 'every x' as objects of the relation, R.

It is commonly assumed that a logical syntax along traditional (Leibnizian) lines must fail to disambiguate a sentence like 'every boy loves some girl' which may on the one hand be interpreted to say of some girl that she is loved by every boy and, on the other hand, to say of every boy that he loves a girl (some girl or other). In the first case we give 'some girl' a specific reading (some girl a certain girl). In the second case we give 'some girl' a non-specific reading (some girl some girl or other). In the notation we have devised, the difference between the specific and nonspecific reading would be decided by the order in which the subjects 'every boy' and 'some girl' appear in the transcription. The transcription of 'every boy loves some girl or other' is '$-B+l_1+G$'. The transcription of 'some girl is loved by every boy' is '$+G+l_2-B$'.

In algebraic transcription every subject is either positive or negative, having the form 'x' or '$-x$'. A relational sentence has two or more subjects and a positive subject may be specific or non-specific. The rule for a non-specific reading is this:

A positive subject is non-specific if and only if it is preceded by but not followed by a negative subject.

This rule corresponds to the convention, in standard logical grammars, for reading the existential quantifier that is preceded by but not followed by a universal quantifier. Thus '$(x)\ \exists y\ (Lxy)$' is the translation of 'everyone loves someone or other' just as '$-P+l_1+P$' is the transcription of 'every person loves some person or other'. '$\exists x\ (y)\ (Lyx)$' is the translation of 'someone is loved by everyone'. Its algebraic transcription is '$+P+l_2-P$'. Note that the transcription makes use of the converse form 'is loved by' (l_2). In the standard logical language the explicit use of the converse (passive) form is avoided. The effect of the converse is achieved by the difference of the order of 'x' and 'y' as these appear in the quantifiers and in the matrix. Thus 'Lyx' can be read as 'x is loved by y' and when the order of letters in the quantifiers is 'x, y' we so read it.

In algebraic transcription, as in the vernacular, the converse form is explicitly used and we rely on the analytic relations between the forms 'loves' and 'is loved by'. In general, if R_1 is an n-place relation, it will belong to a set $R_1 \ldots R_{n!}$ any two of whose members are converse to each other. For example, there are 3! ways of ordering the subjects of '...gives...to...' and so there are six sentences beginning with 'some x gives some y to some z' ($+X+G_1+Y+Z$) and ending with 'some z gets some y from some x' ($+Z+G_6+Y+X$) any two of which are analytically equivalent to one another. Schemas of this kind are analytically equivalent whenever the quantity of the subjects is uniform. Thus 'every boy loves every girl' and 'every girl is loved by every boy' are analytically equivalent but 'every boy loves some girl' and 'some girl is loved by every boy' are not equivalent.

As another illustration of the ability of the algebraic notation to disambiguate the specific and non-specific senses of 'some *x*' consider the difference between a sailor who reads a certain poem to each and every girl and one who reads every girl some poem or other. We use (1) to describe the former and (1*) to describe the latter.

(1) $+S+r_1+P-G,$
(1*) $+S+r_2-G+P.$

II. SENTENCES CONTAINING PRONOUNS

In translation from the vernacular a pronoun that is not cross-referential to a proper name will appear as a bound variable. The translation is said to reveal the logical form of the translated sentence and it is widely assumed that the grammars of natural languages obscure the true nature of pronouns precisely because they lack a standardized apparatus corresponding to the apparatus of quantifiers and variables of the standard logical languages that serves to make the nature of pronominalization explicit. We have then the thesis that pronouns (at bottom and unsuperficially) *are* bound variables and that this is explicated in logical translation.

If this thesis were correct the prospects would be dim indeed for a logical syntax of natural language along Leibnizian lines, for we could not present pronouns in our logical grammar without some device comparable to the quantifier variable notation and this mode of representation is a radical departure from the grammars of natural language. Once again we seem to be faced with the choice between accepting an unnatural but highly perspicuous and logically effective way of representing the logical form of a significant kind of discourse and some more natural traditional way which is logically vague and semantically unperspicuous. Fortunately for the Leibnizian program, it is not hard to get a logically effective way of representing pronouns that is close to the natural grammars. As for the bound variable view, it suffers from more than artificiality. I shall briefly discuss one grave objection to it. There are others.

If pronouns are bound variables they must occur in the same sentence as the antecedent subject which binds them and which they pronominalize. For example, 'Some man is at the door. He is armed.' becomes 'some man *x* (*x* is at the door and *x* is armed)', where the pronoun 'he' refers back to 'some man' and is seen to be bound by it. It is clear that any such binding requires that the pronoun and its antecedent be in the same sentence. Unfortunately for the bound variable theory of pronouns, it is, as often as not, simply not possible to construe pronominalization as taking place within a single sentence. Thus consider

> Some man is at the door. Is he armed?
> Some man is at the door. Arm him!
> Some man is at the door. So he climbed all those stairs.

In such cases it is not open to us to form a syntactical union of the sentence containing the pronoun with the sentence containing the antecedent. Nor is it plausible to say that 'he' in 'So he climbed all those stairs' differs in logical form from 'he' in 'and he climbed all those stairs'. Similarly 'he' in 'is he armed?' cannot differ from 'he' in 'he is armed'. Since 'he' in 'is he armed?' is not a bound variable, it is not a bound variable in 'he is armed' and we ought to look for another way of understanding the pronouns of the natural languages. This is not to say that we cannot, for certain logical purposes, translate 'he' as a bound variable. But we should do so with the awareness that our "translation", however useful, cannot be faithful to the original. The fact that the translation cannot be carried through in innumerable cases where the pronoun has exactly the same meaning as it has in some case for which the translation can be made to work should give pause to anyone who holds to the idea that translation as a bound variable reveals the nature of the pronoun in the vernacular.

According to the traditional doctrine, every logical subject is of form 'every *x*' or 'some *x*'. We apply this doctrine to pronouns and distinguish between the logical sign of the pronoun and the term (or proterm) that follows the sign. Thus the word 'he' in 'he is armed' can be taken as a whole subject and its logical form will include a sign of quantity or it can be taken as the term that follows the (implicit) sign of quantity. The ambiguity of 'he' is similar to the ambiguity of 'Socrates' in subject position. For 'Socrates' can be taken as a whole subject with implicit quantity or it can be taken simply as a singular term. Because every logical subject consists of a sign of quantity followed by a term, the traditional theory distinguishes between the modes of signification of the whole subject and the term it contains. Let *reference* be the mode of signification of a subject and *denotation* be the mode of signification of a term. In 'some man is at the door' the subject 'some man' *refers* to some man in a non-identifying way; the term 'man' *denotes* all men. A term is said to be distributed in a subject when the reference of the subject coincides with the denotation of the term; otherwise the term is said to be undistributed. Consider again the pronoun 'he'. Taken as a whole subject 'he' refers to whatever the antecedent refers, i.e., it refers to "the man in question". Taken as the term of this subject (the proterm) 'he' also *denotes* what is referred to by the antecedent subject. This is generally true of the proterm: it is defined to denote what the antecedent subject *refers* to. Thus every proterm is distributed in its pronoun and every pronoun has universal quantity in addition to the quantity of the antecedent subject. In effect, the pronoun for a definite subject 'some *x*' will always be an expression with "wild" quantity. Pronouns that cross-refer to 'every *x*' are another matter. They of course will have universal quantity. But the assignment of particular quantity to a pronoun whose antecedent is universal is unjustified in those cases where this would confer existential import on the sentence containing the pronomial subject. In such cases the assignment of particular quantity would illegitimately give the pronoun a mode of signification that is stronger than that possessed by its antecedent subject.

For the algebraic transcriptions of pronouns we must have some way of indicating the cross reference of the pronoun to the antecedent. To this end we affix to the antecedent subject a letter as an index. The occurrence of the letter attached to a subject will anticipate its subsequent pronominalization. We then use the same letter for the proterm of the pronoun. For example, "Ed came. Is he drunk?" would be transcribed thus:

$$\pm \mathrm{E}_1 + C. ? \pm I + D?$$

Other examples of transcribe pronominalization are

Tom loves himself

$$\pm T_i + l_1 \pm I$$

every owner of a donkey beats it
$-(O+D_j)+b\pm J$

if some being is divine then no other being is
$-[+B_k+D]+[-((-K)+D]$

A boy who was fooling her kissed a girl who trusted him
$+<+B_c+(f\pm M)> +k+ <+G_m+(t\pm c)>$

The algebraic notation that we have developed adheres closely to the logical syntax of natural language as it is regimented in traditional logic. In its treatment of categorical propositions, including singular propositions and identities, it is faithful to the philosophy of logic that characterized the best pre-Fregean logicians, and among these Leibniz was paramount. The notation extends the Hobbes-Leibniz idea of a plus/minus representation for the positive and negative copulas to other syncategorematic signs, in particular to the signs for 'every' and 'some'. The idea receives a further extension when we learn how to represent the truth functions as ordered n-tuples of plus and minus signs. In revealing a formal isomorphism for 'every...is...' and 'if...then...' (both of which are transcribed as '$-\ldots+\ldots$') and a formal isomorphism for 'some...is...' and 'and' (both of which are transcribed as '$+\ldots+\ldots$'), the notation meets Leibniz's specification for an adequate symbolization of logical signs. It still remains to show that the notation is logically effective and not a mere stenographic device for representing propositions in algebraic form. To exhibit its potential for inference we state some rules of transformation and examine some examples of logical reckoning with propositions algebraically represented. In particular we shall be interested to see how algebraic representation can help us to understand the logical processes in inferences that involve relational propositions, identities, and other inferences not comprehended in traditional formal logic.

III. PRINCIPLES OF THE PLUS/MINUS LOGIC

(1) *Laws of Commutation*

$+x+y=+y+x$

$--x--y=--y--x$

(2) *Laws of Association*

$+x+(+y+z) = +(+x+y)+z$

$--x--(--y--z)=--(--x--y)--z$

(3) *Laws of External Not-distribution*

$-(\pm x \pm y) = \mp x \mp y$

(4) *Laws of Internal Not-distribution*

$\pm x-(\pm y) = \pm x+(\mp y)$

$\pm x-(R\ldots+y\ldots) = \pm x+(-R)\ldots\pm y\ldots$

(5) *Law of Double Negation*

$x=--X$

(6) *Laws of Iteration*

$x=+x+x$

$x=--x--x$

(7) *Laws of And-Or Distribution*

$+(-x-x)+(-y-z)=-(+x+y)-(+x+z)$

$--(+x+x)--(+y+z)=+(--x--y)+(--x--z)$

(8) *Laws of Tautology*

$-x+x$

$-x_1+(-x_2+x_2)$

$+[-x_1+(-x_2+x_2)]+\ldots[-x_n+(-x_{n+1}+x_{n+1})]$

(9) *The Law of Syllogistic*

$-x+y$

$\ldots x\ldots$

$\ldots y\ldots$

The first seven laws express equivalence transformation. Each equivalence has a dual interpretation depending on whether 'x' and 'y' represent terms or propositions. For example, according to the laws of external not-distribution $-(-x+y)=+x-y$. This has two interpretations:

(a) not every x is y = some x isn't y
(b) not: if x then y = x and not $-$ y.

The Law of Syllogistic is the basic rule of inference for arguments with more than one premise. In effect it is a principle of substitutivity that permits the algebraic elimination of 'x' from a context '...x...' and its replacement by 'y' whenever a premise of form '$-x+y$' is also given. The following two inferences are typical applications:

Tully is Cicero	$-T+C$
Tully is a senator	$+T+S$
∴ Cicero is a senator	$+C+S$

if there is smoke then there is fire	$-p\ +q$
there is smoke	$-p$
∴ There is fire	q

Only transformations that have a basis in the rules of transformation are legitimate. By the laws of not-distribution we may always distribute 'not' to effect a change of sign in any expression within its scope. While this is true of 'not' it is not true of 'every' or 'if' which are represented by the minus signs in '$-x+y$' or '$+x+R-y$'. The right hand law equates '$\pm x-(\pm y)$' to '$\pm x+(\mp y)$'. But there is no analogous left hand law that equates '$-\ (\pm x)\pm y$' and '$+(\mp x)\pm y$'. Similarly no rule of transformation sanctions the logical equality of '$\pm x+R-(\pm y)$' and '$\pm x+R+(\pm y)$'. That '$-(-x)+y$' is not logically equal to '$+x+y$' is an implicit constraint of the logical algebra that distinguishes it from the familiar everyday algebra of addition and subtraction. On the other hand, logical algebra has a certain license not found in the everyday algebra: the Law of Iteration permits the replacement of 'x' by 'x and x' $(+x+x)$ and by 'x or x' $(--x--x)$. But in ordinary algebra 'x' and '$x+x$' are not interchangeable. As a practical matter in the actual reckoning of logical arguments it is especially important to avoid distributing minus signs that do not represent truly negative particles. Once this precaution is observed the coincidence of logic and ordinary algebra is quite striking.

The two meta-theorems that follow are justifiable by the principles:

(I) If two propositions of the same logical quantity are algebraically equal, they are logically equivalent.

(II) A syllogism or sorites is valid if (a) the sum of its premises is equal to the conclusion, and (b) either all of its propositions are universal in quantity or else it has exactly two particular propositions one of which is the conclusion.

Some equivalences that illustrate the application of (I) are:

every A is not B = no A is B
$-A+(-B)=-(A+B)$

some A isn't B = some non B is A
$+A-B=(-B)+A$

Some A isn't R to any B = Some A is non R to every B
$+A-(R+B)=+A+(-R)-B$

The second metatheorem provides an easy way of testing for the validity of syllogistic arguments: (1) first check whether the syllogism or sorites satisfies the requirements of quantity for if it is valid it will either have no particular propositions or else exactly two particular propositions one of which is the conclusion; and (2) if the argument satisfies the condition of quantity we next check to see whether its premises add up to the conclusion; if they do the argument is valid, if not invalid.

As an example consider whether the following sorites is valid:

no x is non y	$-(x+(-y))$
every non z is u	$-(-z)+u$
some non y is non u	$+(-y)+(-u)$
∴ some z isn't x	$+z-x$

The argument has exactly two particular premises one being the conclusion. So it passes the quantity test. The sum of its premises is equal to the conclusion. So it satisfies the algebraic condition. It is therefore a valid argument.

We now illustrate the power of the plus/minus notation by reckoning a variety of arguments including some that are beyond the competence of traditional formal logic.[7]

every A is B	$-A+B$
some C is A and D	$+C+(+A+D)$
∴ some C is B and D	$+C+(+B+D$

This argument is a fairly simple application of the syllogistic principle.

Zeus is immortal	$-Z+(-M)$
Socrates is mortal	$-S+M$
∴ Socrates is not Zeus	$-S-Z$

The conclusion of this argument is that Socrates is not identical with Zeus.

every circle is a figure	$-c+F$
everyone who draws a circle draws a circle	$-(d+c)+(d+c)$
everyone who draws a circle draws a figure	$-(d+c)+(d+F)$

This argument was unsuccessfully treated by Jungius and Leibniz because they failed to make explicit the tautological second premise. Once this premise is added we can simply apply the syllogistic law to cancel 'circle' and to replace it with 'figure'.

The next argument tacitly exploits the transitivity of the relation 'is brighter than'.

every human is brighter than every chimpanzee
a certain chimpanzee is brighter than every dog
every human is brighter than every dog

To reckon this argument we need a third premise that explicitly recognizes the transitivity of 'is brighter than'. In general, if R is a transitive relation the following formulas are analytical:

if some/every A is R to some/every B
then everything R to every/some A is R to some/every B
$-[\pm A+R\pm B]+[-(R\pm A)+(R\pm B)]$

Applying this omnibus formula to the second premise we add the analytical premise: if some chimpanzee is brighter than every dog then everything brighter than every chimpanzee is brighter than every dog.

$$-H+b-C$$
$$+C+b-D$$
$$-[+C+b-D]+[-(b-c)+(b-D)]$$
$$\therefore -H+b-D$$

The next example answers the oft-repeated charge that a traditional logic must prove incapable of handling inferences like 'some girl is loved by every boy; so every boy loves some girl or other'. The proof exploits the wild quantity of pronominalizations of 'some girl'. The second, third, and fourth premises are truisms. The second premise reads 'she is a girl'. The consequent of the fourth premise reads 'every boy loves her'.

(1) $+G_i+l_2-B$
(2) $+I+G$
(3) $-[+G_i+l_2-B]+[-I+l_2-B]$
(4) $\underline{-[-I+l_2-B]+[-B+l_1-I]}$
$-B+l_1+G$

When an indexed nominal expression is replaced by cancellation, the pronoun that refers to it will now refer to the new nominal that replaces it. For example, from the premises 'every man loves himself, Cicero is a man' we can infer 'Cicero loves *him*self' and the pronoun in the conclusion now has cross-reference to 'Cicero' which has replaced 'every man' by cancellation:

$$-M_j+l\pm J$$
$$\underline{\pm C+M}$$
$$\pm C_j+1\pm J$$

In traditional logic an identity proposition is a monadic proposition with singular terms in both subject and predicated positions. The final example exploits this and shows how to get by without treating identity as a relation governed by special axioms. In the proof we have added a truistic third premise which reads: if Cicero is envious of everyone else (everyone not himself) then Cicero is envious of eveyone other than Cicero (everyone who is not Cicero):

(1) every senator is envious of everyone else
(2) Cicero is a senator
(3) Cicero is not envious of Tully
Tully is Cicero

(1) $-S_i+e-(-I)$
(2) $\pm C+S$
(3) $-[\pm C_i+e-(-I)]+[\pm C_i+e-(-C_i)]$
(4) $\underline{\pm C_i+(-e)\pm T}$
$\pm T+C_i$

Russell once compared a notation to a live teacher. The algebraic concept script teaches us how to comprehend under one rule such seemingly different kinds of inferences as

(1) $a = b$, b is p, hence a is p
(2) if p then q, p, hence q
(3) every S is P, whatever is Q and not S is R to some S, hence whatever is Q and not P is R to some P.

The "oppositional" (plus-minus) notation teaches that inference from more than a single premise is a matter of substitution *salva veritatae*: given a premise '...x...' we may (substitute 'y' for 'x' and) infer '...y...' in every case where '$-x+y$' is also given. The schema '$-x+y$' ...x..., hence ...y...' is simple and comprehensive and it forcefully illustrates the generic character of the plus-minus notation.

IV. CONCLUSION

Perhaps enough has been said to indicate that the logical syntax of Leibniz, when given the kind of algebraic expression that is characteristically Leibnizian and one that was suggested by Leibniz himself albeit an incipient form, has considerable technical virtues of a purely logical kind. In addition, the algebraic formulation has two special Leibnizian features:

(1) It is abstract enough to exhibit the isomorphism of term logic and propositional logic.

(2) It makes it possible to apply the fundamental syllogistic principle to arguments with relational propositions, a kind of argument that Leibniz and Jungius were especially concerned to formalize.

The Leibnizian analysis of singular propositions is expecially powerful since it gives to traditional logic the ability to handle identity propositions without the addition of axioms defining the properties of an dentity relation. None of these features is thought to belong to any pre-Fregean kind of logic. Yet all that has been added to Leibnizian logic is the algebraic notation which he already had in embryo.

Leibniz certainly did not subscribe to Kant's judgment that logic was a finished science. His doctrine of "wild quantity" for singular propositions is only one example of an important but neglected contribution. On the other hand too much has been made of Leibniz's principle of the identity of indiscernibles. For, strictly speaking, the principle is trivial since 'is *a*' is tautologically true of (every) a and by the hypothesis that *a* and *b* are indiscernible 'is *a*' will also be true of (every) *b*. The indiscernibility of identicals is, formally speaking, the principle of the substitutivity of identicals. This too is a logical truism since, as we saw, it is simply a variant of P9 applied to a universal proposition with singular terms in subject and predicate positions: $-a+b=-[\ldots a\ldots]+[\ldots b \ldots]$. The current emphasis on Leibnizian laws of identity owes everything to the fact that in Frege's logic identity theory is a special theory requiring its own special principles. Ironically, Frege needs "Leibniz's Law" but Leibniz himself does not. In Leibniz the laws of identity are merely special instances of common principles within the classical logic of terms.

It is unfortunate that Leibniz did not carry out much of his own logistical research programs. These included the development of a neutral notation, conceived along arithmetical lines, that would transcribe the logical signs of both term logic and propositional logic. It included the development of a calculus – arithmetical in nature – within which arguments of all kinds could be made formally explicit. Most important, these things were to be done not for some artificial language but for the natural languages logically regimented in the traditional manner of regimentation that one uses for syllogistic. In the extended arithmetical notation we represent 'some *A* gave some *B* to every *C*' as '$+A+g+B-C$'. This expression is not a radical departure from the original. It is not a translation but a transcription that preserves the order of the original. Contrast this with the Fregean translation '$\exists x\ (Ax\ \&\ \exists y\ (By\ \&\ (z)\ (Cz \rightarrow Gxyz)))$'. The Fregean expression introduces new syncategorematic elements ('and', 'if... then') not found in the original sentence and it changes the order of the categorematic elements. From a purely linguistic standpoint, the algebraic representation is to be preferred over the quantificational formula as staying closer to the actual syntax of natural language. In this connection we remark that a theory of deep structure that comprehends logical structure as a linguistic universal will do better to choose traditional logic over modern predicate logic. There is, of course, no question of putting back the clock or of giving back any of the fruits of Frege's revolution in logic. Frege has given us the best developed and most powerful system of logic we have ever had. It is also the best examined from a proof-theoretic standpoint. Nevertheless the problem of which structures underlie the surface is at least partly an empirical one and it is entirely possible that the linguists will find the logical syntax of Leibniz a more suitable candidate for a theory of deep structure than the logical syntax of Frege and his successors. It would, after all, be too good to be true if the canonical forms of modern predicate logic just turned out to be the structures that underlie the actual languages. To assume this prospect in advance as some fashionable linguists do is to assume that Frege's syntax for a logically powerful but admittedly artificial formal language is the only syntax worthy of the name "logical syntax". In particular it would be to assume that a class of atomic sentences – sentences devoid of syncategorematic elements – are at the base of logic and language. But questions of universal structure and base elements are not ones to be settled in advance by postulations. And the brute fact is that the syntax of traditional logic is closer in structure to the syntax of natural language than any variant of the syntax of modern predicate logic.

There is also the question of how we actually reason, of what takes place in the brain as we move from premises to conclusion. This question belongs to cognitive psychology and the ultimate choice of a logical syntax will be partly determined by the answer. Here too the suggestion that we reason by cancellation of elements that have opposed signs is a plausible candidate for a theoretical description of the deductive process. For a number of reasons then, quite outside of logic, the possibility of a move to some form of neo-Aristotleian Leibnizian theory of logic and language should not be ruled out. In any event I hope I have shown that Leibniz's research program for logical syntax is viable, creative, and radically different from Frege's. Since this is so, it is time to stop thinking of Leibniz as an incipient but unsuccessful Fregean.

Brandeis University

NOTES

* This paper was inspired by Imre Lakatos and has been adapted for this memorial volume. A version of it will appear in *Frege-Studien*, vol. III, ed. Mathias Schirn; Frommann-Holzboog, Stuttgart. See also by the same author 'On a Fregean Dogma', *Problems in the Philosophy of Mathematics*, North-Holland Publ. Co., Amsterdam, 1967, pp. 47–81; 'Do We Need Identity?', *Journal of Philosophy*, **66** (1969), 499–504; 'On Concepts of Truth in Natural Languages', *Review of Metaphysics*, **23** (1969), 259–286; 'The Logical and the Extra-Logical', *Boston Studies in the Philosophy of Science*, ed. by Cohen & Wartofsky, XIV, 1973, pp. 235–251; 'The Calculus of Terms', *Mind* **79** (1970), 1–40; 'Existence and Predication', *Logic and Ontology* ed. by M. Munitz, NYU Press, 1973, pp. 159–174; 'Distribution Matters', *Mind* **95** (1975).

1 G. H. Parkinson (ed. and trans.), *Leibniz Logical Papers*, Clarendon Press, Oxford, 1966, p. 115.
2 'Grundlegung der Geometrie', trans. by T. Szabo, p. 12, *Philosophical Review* **LXIX**, 389.
3 W. V. Quine, *The Philosophy of Logic*, Prentice Hall, 1970, p. 35.
4 *Nachgelassene Schriften*, G. Frege, Hans Hermes, Friedrich Kaulbach (eds.), Hamburg, Felix Mainer, 1969, pp. 288f.
5 *Logical Papers*, p. 3.
6 Leibniz, *Logical Papers*, p. 55.
7 For further discussion on the validation of arguments in algebraic form, see the author's 'Calculus of Terms', 'The Logical and the Extra-Logical', and 'Distribution Matters'.

ROGER H. STUEWER*

ON COMPTON'S RESEARCH PROGRAM

1. Introduction

When Imre Lakatos sent me a copy of his and Alan Musgrave's *Criticism and the Growth of Knowledge*,[1] he inscribed it with the words: "With hopes for future cooperation and best wishes." That future cooperation, to my very deep regret, never occurred owing to Imre's sudden and premature death in February 1974. I know, however, what he had in mind. He wanted me to join with him in analyzing certain case studies in the history of physics in light of his methodology of scientific research programs. Furthermore, I know that one of the case studies Imre wanted to treat was my own on the discovery of the Compton effect, since I can vividly recall our discussing an early draft of my book, *The Compton Effect: Turning Point in Physics*,[2] after dinner at our home in Minneapolis in the fall of 1970. Now, however, collaboration is impossible. I therefore propose to simply summarize some of the highlights of Compton's research program here, and to invite others to consider its possible relevancy to Imre's methodology, or to philosophical-methodological questions in general. It goes without saying that if such connections seem to exist, my detailed study must be consulted to substantiate their existence, since my present discussion constitutes only a very rough and incomplete sketch of certain sections of my book.

The first general feature of Compton's work between 1917, the year after he received his Ph.D. degree from Princeton University, and 1922, when he discovered the Compton effect at Washington University, is that it divides naturally into two quite distinct research programs which, however, he pursued partly simultaneously. The central assumptions of the first, which he pursued between 1917 and 1922, consisted of certain basic assumptions from Thomson's classical theory of scattering. The central assumptions of the second, which he pursued between 1919 and 1922, consisted of certain basic assumptions concerning the relativistic

R. S. Cohen et al. (eds.), Essays in Memory of Imre Lakatos, 617–633. *All Rights Reserved*
Copyright © 1976 by D. Reidel Publishing Company, Dordrecht–Holland

motion of a charged particle through matter. I will discuss each of these research programs in turn.

2. Compton's Large Electron Research Program

To understand Compton's research programs, it is necessary to understand just how thoroughly Compton was committed to classical electrodynamics when he commenced his research career in the year 1916. Owing I believe primarily to O. W. Richardson's lasting influence at Princeton, Compton was simply not prepared to think in any other than classical terms at that time. There was, of course, nothing unusual or extraordinary about this attitude – virtually all of Compton's contemporaries took the universal validity of classical electrodynamics for granted, and as a natural (but logically unjustifiable) consequence either rejected Einstein's light quantum hypothesis outright, or did not take it seriously, in spite of Einstein's own persistent urging to the contrary. This attitude was by no means without foundation: On the one hand, few could question the solid experimental base and great heuristic power of Maxwell's theory, and von Laue's 1912 discovery of the crystal diffraction of X-rays seemed to conclusively demonstrate their *wave* nature. On the other hand, Einstein's statistical mechanical arguments for quanta lacked conviction, and several of the most prominent physicists of the day undercut the photoelectric effect as proof of Einstein's hypothesis by developing essentially classical, non-Einsteinian interpretations of it. Even Millikan rejected Einstein's *hypothesis* after he had conclusively verified Einstein's *equation* in 1915. Compton's attitude, in sum, was fully in the spirit of the times.

Compton embarked directly upon his first research program in 1917, after leaving his first position at the University of Minnesota to go to Westinghouse in Pittsburgh as a research engineer. He was always definite as to what initially stimulated his thought: it was C. G. Barkla's 1917 observation that when 0.145 Å X-rays pass through aluminium, they have a mass absorption coefficient of 0.153 $cm^2\ gm^{-1}$. Why was this seemingly prosaic observation so puzzling? The reason is easy to understand. Compton knew that the mass absorption coefficient, μ/ρ, may be written as the sum of two terms, the mass fluorescent coefficient, τ/ρ, and the mass scattering coefficient, σ/ρ, the latter of which was presumably equal to the *Thomson* mass scattering coefficient, σ_0/ρ. Now, under the assumption that the atomic number of an element is numerically equal to half its atomic weight, it is easy to show that the Thomson coefficient σ_0/ρ is equal to a constant, 0.188 $cm^2\ gm^{-1}$. How was it possible, therefore, that the *total* mass absorption coefficient, μ/ρ, was observed to be *less* than one of its constituent parts, the Thomson mass scattering coefficient σ_0/ρ?

That was the puzzle, and Compton's response to it inaugurated his first research program. He concluded that Thomson's classical theory of scattering had to be modified to allow for the *diffraction* of the incident radiation as well as its scattering. Thomson scattering, of course, involves the picture of electromagnetic radiation incident upon N electrons per unit volume in a substance, and the theory itself is based upon the following four fundamental assumptions: (1) that classical electrodynamics is valid; (2) that the scattering electron is a point charge; (3) that the scattering electron moves without constraints; and (4) that all electrons scatter independently. It is seen, therefore, that Compton's conclusion that *diffraction* was taking place was equivalent to his accepting all of Thomson's basic assumptions *except* the second. For diffraction to occur, he had to assume (2′) that the electron is *not* a point charge. Rather, its diameter had to be on the order of the wavelength of the incident radiation. It had to be a *large* electron. Thus, it was Thomson's assumptions (1), (3), and (4), coupled with Compton's new assumption (2′), that constituted the central assumptions of his first – what we may term his large electron – research program. He did not abandon these four assumptions until 1922.

To make quantitative predictions, Compton had to make one further assumption or auxiliary hypothesis, namely, one on the exact shape and rigidity of his large electron, since electrons of different shapes and rigidities would scatter and diffract the incident radiation differently. In fact, Compton first considered one particular model, then another, and yet another, all the while keeping his central assumptions intact. Furthermore, he did not relinquish his central assumptions in spite of the fact that there existed data which conflicted with them. In particular, as I have already mentioned, for diffraction to occur he had to assume that the diameter of the electron was on the order of the wavelength of the incident radiation, *i.e.* X-rays or perhaps γ-rays. He had to assume therefore that its diameter was on the order of 0.1 Å or perhaps 0.01 Å. In other words, Compton's electron had to have a diameter several orders of magnitude larger than the classical radius of the electron. Furthermore, if we recall that the Bohr radius of the hydrogen atom is about 0.5 Å, we see that Compton's electron was almost as large as the entire *atom*. These, to my mind, are rather striking conflicts. Yet, they were of little or no concern to Compton himself.

The fundamental reason is very straightforward: Compton's large electron scattering theory was a very successful, suggestive, and *quantitative* theory. He first assumed that the electron was a large, *rigid shell* of electricity. Calculating how electromagnetic radiation is scattered from such an electron, he found that the mass scattering coefficient σ/ρ went to zero as the radius-to-wavelength ratio a/λ went to infinity, and that σ/ρ went over to the Thomson value σ_0/ρ as the ratio a/λ went to zero. In other words, σ/ρ went over to the Thomson value in the point-charge limit, and dropped below the Thomson value as the wavelength of the incident radiation decreased – precisely as Barkla's observation demanded.

In spite of its success, however, this model was inadequate. It could not account for the fact, well-known by 1917, that scattered (or secondary) X-rays are emitted predominantly in the forward direction. Compton therefore assumed next that the electron was a large, *flexible shell* of electricity. Calculating the so-called angular distribution, he found that this model indeed predicted a forward-backward asymmetry in the scattered radiation. It also predicted the same general behavior of the mass scattering coefficient σ/ρ as the former model.

Several additional points bear comment. Note, in the first place, that Compton's shift from his first to his second model (rigid to flexible shell) was a very natural move – indeed, it was suggested quite naturally by the first model itself. Note, secondly, that this shift only involved Compton's auxiliary hypothesis. Note, finally, that his second model predicted at least one more novel fact than his first. Indeed, what had been a counter-instance for his first model became a corroborating instance for his second. Actually, Compton's second model predicted yet another novel fact. It predicted that the soft components of an incident inhomogeneous beam of radiation should be backward-scattered to a greater degree than the hard components. This represented a 'selective scattering' effect which Compton took to explain 'beautifully' the earlier 1910 observation of D. C. H. Florance that the backward-scattered radiation is softer on the whole than the forward-scattered radiation.

Once again, however, in spite of its success, *this* model was inadequate. It could not account for the correct mass of the electron, assuming as Compton did that its mass was entirely electromagnetic in origin. Thus, it is easy to show that the self-energy of any spherically symmetric distribution of charge of the assumed radius (Compton here assumed $a = 0.023$ Å) corresponds to an electromagnetic mass several orders of magnitude smaller than the known rest mass of the electron. Once again, Compton turned this counter-instance for his second model into a corroborating instance for a third by replacing his auxiliary hypothesis. Taking over a model proposed earlier by the American chemist A. L. Parson, Compton assumed that the electron was a large, *flexible ring* of electricity. This spherically asymmetric model accounted for the electromagnetic mass of the electron provided, as a calculation of D. L. Webster had indicated, that the thickness-to-radius ratio t/a of the ring was on the order of e^{-2560}!

While this result strikes one as very unreasonable today, Compton himself clearly regarded his shift to his new model to be a significant advance in his research program, since he actually pursued his large ring electron scattering theory for several years. At first, it was only a qualitative theory, but he soon succeeded in solving the scattering problem in detail quantitatively. He found that his ring electron model accounted for everything his earlier models had accounted for, in addition to the electromagnetic mass of the electron, and in addition to the results of a certain absorption experiment of A. H. Forman. This success prompted Compton to further exploit and refine his ring electron model in various ways. (1) He applied it to a calculation of the mass fluorescent coefficient τ/ρ. (2) By a clever curve-fitting technique, he calculated from experimental data what he believed to be the *precise* radius of the ring, namely, 0.0185 ± 0.0004Å. (3) He defended his model against various criticisms of certain English physicists. (4) He proved that the experiments he had carried out earlier at the University of Minnesota, which had indicated that the electron was the "ultimate magnetic particle," were fully consistent with his ring electron model.

This represented the stage of development in Compton's large electron research program at the time he left Westinghouse for the Cavendish Laboratory as a National Research Council Fellow in October 1919. It was at the Cavendish that he initiated his second research program, which I shall discuss in a moment, but first I would like to finish the present line of development. Thus, Compton eventually concluded that certain γ-ray scattering experiments he had carried out at the Cavendish now actually conflicted with the earlier experiments of A. H. Forman which I have just mentioned. As a result, Compton for the first time began to seriously doubt the validity of his large ring electron scattering theory. However, and this is the significant point at the moment, this new knowledge did *not* conflict with any of the *central assumptions* in Compton's large electron research *program*. Once again, it conflicted only with his auxiliary hypothesis, and once again Compton replaced it. He now rejected his large ring electron model for a fourth model in which the electron was assumed to be a large *sphere* of electricity. Thus, Compton now rejected his earlier theoretical objection that all spherically symmetric configurations are not consistent with the electromagnetic mass requirement discussed earlier. To justify this move, Compton appealed to two plausibility arguments: first, he noted that a spherical configuration was at least in keeping with our usual conceptions; and second, he noted that relativistic mass increases implied that not all of the electron's mass was electromagnetic in origin. Clearly, both of these arguments were also valid earlier, even at the time he had embarked upon his large electron research program. He was then, however, either unfamiliar with them or he did not consider them significant.

Compton proved that his new large spherical electron model, in contrast to his ring and shell models, predicted an angular distribution for the scattered radiation in the short wavelength region which agreed quantitatively with the new γ-ray scattering data he had obtained at the Cavendish. It is somewhat difficult to tell, however, how seriously Compton on balance really took his new model, since on the one hand his shift to it represented a clear set-back for his ring model, which he had used successfully for several years, while on the other hand it proved to him, as he explicitly pointed out, that his most fundamental assumption of all, that the electron was a *large* electron, remained valid and untouched.

That is, the central assumptions of Compton's large electron research program remained intact. Only his auxiliary hypothesis regarding the specific shape and rigidity of the electron had been affected. We will see that Compton rejected this entire research program only after he realized that *no* large electron model of a *definite* radius could fit *all* of the reliable experimental data. This was in 1922, after a very different research program, which I shall now discuss, had assumed the more prominent position in his mind. Ultimately, of course, *both* of these research programs became, in Compton's term, 'superfluous.'

3. Compton's 'Fluorescent' Radiation Research Program

Compton's second research program, which he pursued between 1919 and 1922, in part simultaneously with his first, grew out of experiments he carried out at the Cavendish Laboratory. These experiments, which were related to earlier ones of D. C. H. Florance and J. A. Gray, proved to Compton that when γ-rays are scattered they become softer or less penetrating. Moreover, he found among other things that the softest components in the secondary radiation were *softer* than the softest components in the primary radiation.

This observation was inexplicable on Thomson's classical theory of scattering, because that theory requires each component in the primary radiation to be transformed into a component of identical wavelength or hardness in the secondary radiation. Even Compton's large electron modification of Thomson's theory was helpless here, because while that theory permitted a 'selective scattering' effect which allowed the secondary radiation to be softer *on the whole* than the primary radiation, it did *not* allow any secondary component to become softer than the softest component in the primary radiation. Evidently a very different kind of explanation was required.

Compton finally concluded that the softened secondary γ-radiation consisted of a new type of 'fluorescent' radiation which had been excited in the scatterer. Thus, essentially by definition, he drew a sharp distinction between secondary radiation which was of the same hardness as the primary, and that which was softer than the primary. The former to Compton was 'truly scattered' radiation, while the latter was 'fluorescent' radiation.

This 'fluorescent' radiation, as Compton found in subsequent γ-ray experiments, in addition to being softer than the primary rays, had some very unusual properties: (1) its intensity steadily diminished as the scattering angle increased from 30° to 150°, that is, there was a very pronounced forward-backward asymmetry in its angular distribution; (2) the extent to which it was softened was independent of the nature of the scatterer; and (3) the softering also increased markedly as the scattering angle increased from 45° to 135°. These three properties were very unusual because each one *directly conflicted* with the corresponding property of the usual *X-ray* fluorescent radiations. Fluorescent X-rays, of course, were known to be emitted symmetrically, with a hardness which is *characteristic* of the scatterer, and with one which does not vary from angle to angle. Their only point of similarity with Compton's newly discovered γ-rays, in fact, was that they too were softer than the primary radiation.

This meant, ultimately, that Compton was faced with a very difficult problem in interpretation: What mechanism could possibly be invoked to account for the origin of these very unusual 'fluorescent' γ-rays? His answer to that question led to what we may regard as the central assumptions of his second research program.

Compton concluded that his very unusual 'fluorescent' radiation was being emitted by secondary β-particles which behaved like tiny oscillators, and which had been liberated in the scatterer by the primary radiation, and were moving forward at high velocities. The forward-backward asymmetry in the angular distribution of the 'fluorescent' radiation then arose from the well-known forward peaking in the radiation emitted by an accelerated charged particle moving at relativistic velocities. Its softening, and the angular variation in its softening, was then accounted for by the relativistic Doppler shift. That the softening was independent of the nature of the scatterer followed from the fact that β-particles liberated in *different* scatterers had *identical* properties. In sum, the central assumptions of Compton's second research program consisted of those assumptions which underlie the description of motion of a relativistically moving charged particle, itself considered to be a kind of oscillator. Once again, it is evident that none of these assumptions conflict with classical electrodynamics.

What we might regard as an auxiliary hypothesis in Compton's second research program was well hidden at its inception. It only emerged, in fact, after Compton had left the Cavendish and had carried out further experiments at Washington University in 1920. Compton's move to Washington University was very significant for him: he sensed more and more by this time that he was on the trail of something very fundamental, and he consciously and explicitly chose to go to Washington University, instead of one of the larger and more established centers of research, because he *wanted* to isolate himself. He wanted to be able to pursue his research program on his own, as he said, in the way in which one experiment leads to another. To a considerable degree, in other words, the autonomy of Compton's research program after 1920 was self-imposed.

Furthermore, beginning at the Cavendish, but especially after 1920 at Washington University, Compton's research program consisted of an intimate blend of theoretical and experimental work, each stimulating the other. Thus, having discovered his new 'fluorescent' γ-radiation at the Cavendish, one of the first questions Compton attempted to answer after arriving at Washington University was an experimental question: could *X-rays* also excite the same type of 'fluorescent' radiation in a scatterer? He answered this question by using a crystal to produce a homogeneous beam of X-rays, which he then scattered from different substances. He found that X-rays could indeed excite an unusual 'fluorescent' X-radiation completely analogous to his unusual 'fluorescent' γ-radiation, although the relative softening at different scattering angles was much less for X-rays than for γ-rays.

The very fact that X-rays could also excite this unusual 'fluorescent' radiation in significant amounts, however, presented a serious difficulty for Compton's β-particle oscillator interpretation. If, he asked himself, these β-particles or electrons emitted X-radiation so copiously when moving in a scatterer, why, as was well-known, were presumably identical electrons so inefficient in producing X-rays when incident upon the metal target in an X-ray tube? In an attempt to gain some insight into this question, Compton, with the help of C. F. Hagenow, undertook still further experiments to determine whether or not this unusual 'fluorescent' X-radiation was *polarized*. They concluded, much to their surprise, that it was essentially *completely* polarized. Thus, rather than gaining insight into the 'efficiency problem' above, Compton and Hagenow uncovered yet another 'polarization problem,' for it was well known that ordinary fluorescent X-radiations were completely *un*polarized.

It was in Compton's response to these *twin* problems that he revealed and replaced what we might regard as a hidden auxiliary hypothesis in his second research program. He now assumed that the β-particle oscillators, instead of emitting their radiation *any time* after their liberation by the incident radiation (his first model), they emitted it *at the instant* of their liberation (his second model). This solved the 'efficiency problem,' because it implied that ideally none of the β-particle's energy would be dissipated through heat losses while moving through the scatterer before radiating. It solved the 'polarization problem,' because it implied that the trajectories of the β-particles could not become highly irregular, as was usually the case, through multiple scattering effects before radiating. Thus, two counter-instances to Compton's first model were turned into corroborating instances for his second. We might also note that his second model (as well as his first) existed in the presence of conflicting data – specifically, all of the evidence mentioned earlier on the behavior of the ordinary fluorescent X-radiations.

Quite likely the most striking instance of how invulnerable Compton regarded the central assumptions of his second research program occurred in October 1921, when he was forced to respond to a very serious challenge which had been posed by certain recent experiments of S. J. Plimpton. Earlier, J. A. Gray had argued that only if X-rays consist of electromagnetic *pulses* could a primary beam of X-rays be transformed into a softer secondary beam, as a result of two given pulses being scattered by two separate electrons and subsequently recombining out of phase to form a broader and hence softer secondary pulse. By contrast, long, homogeneous wavetrains, according to Gray, could *never* give rise to softer secondary radiation. Now, Compton at this time *agreed* with Gray that such homogeneous wavetrains could not be transformed into wavetrains of longer wavelength *in the scattering process*. At the same time, however, he definitely *disagreed* with Gray that such homogeneous wavetrains could not *give rise* to secondary wavetrains of longer wavelength: they could, he believed, excite his unusual 'fluorescent' radiation. This distinction, Compton argued, could serve as the basis for a 'crucial test' between his own 'fluorescent' interpretation and Gray's pulse interpretation. As it happened, however, S. J. Plimpton had *actually carried out* a scattering experiment using *homogeneous* X-rays, and had reported finding *no* softer secondary radiation whatsoever. This result, of course, was totally disastrous to Compton's interpretation.

The upshot was that Compton immediately reperformed what amounted to Plimpton's experiment, actually found the softer secondary radiation that Plimpton had missed, calculated its change in wavelength from the observed change in absorption coefficient, speculated on the sources of Plimpton's experimental errors – and concluded that his own 'fluorescent' radiation interpretation had been vindicated once again! The decision of the 'crucial test' had been appealed against – and reversed. The fallacy in Compton's conclusion, of course, was that at this time he simply did not realize that a *third* and fundamentally very different interpretation of this so-called 'crucial test' was possible. This criticism, needless to say, may be readily generalized to apply to all so-called crucial experiments.

4. Compton's creative leaps

In the months following Compton's 'crucial test,' especially between early 1921 and October 1922, the importance of his second, 'fluorescent' radiation research program steadily increased in his mind, while the importance of his first, large electron research program steadily diminished. Compton had concluded by then that some of the experimental data upon which his ring electron model had been based was unreliable, and ultimately, as we have seen, he had been forced to relinquish that model for a large spherical one. But most seriously of all, he now saw that for *any* large electron model to fit *all* of the reliable data, the radius of the electron would somehow have to increase as the wavelength of the incident radiation was increased – a highly implausible situation. This, in fact, represented the first serious conflict with the major central assumption in Compton's entire large electron research program.

Concurrently, Compton's second, 'fluorescent' radiation research program was undergoing a series of transformations which would ultimately culminate in his famous discovery. These transformations, which I shall now describe, may be characterized as 'creative leaps' – they involved more substantial and more unpredictable developments in Compton's research program than had occurred in the past. As in the past, however, they had both their theoretical and experimental bases.

The first key change Compton made in his experimental program was, as we have already noted, his shift from γ-ray to X-ray experiments at Washington University. The second was his decision to use his Bragg spectrometer not as a *wavelength selector*, that is, not to produce a beam of homogeneous X-rays, but rather *as a spectrometer*, that is, to compare the spectra of the primary and secondary radiations. This simple re-positioning of his spectrometer crystal had profound consequences.

Compton obtained his first spectra in late 1921. They consisted of primary and secondary spectra of MoKα X-rays which had been scattered through about 90° by pyrex and graphite. Now, close examination both of Compton's original laboratory notebooks, and of his published report, leads to the surprising conclusion that Compton actually did not observe the small and now well-known shift of a few percent in the position of the MoKα peak. Instead, Compton concluded that the secondary spectrum contained a new peak displaced by about 35% from the primary MoKα peak. We now realize that this new peak is simply the *second order* of the scattered MoKα peak, but Compton did not realize that. He believed that this new peak had resulted from the excitation of his recently discovered 'fluorescent' X-radiation. He believed, in other words, that the wavelength of the primary MoKα peak was 0.708 Å, and that the wavelength of the secondary 'fluorescent' radiation peak was about 35% greater, or about 0.95 Å.

Now, instead of completely derailing Compton's research program, as might be anticipated, this misread data actually inspired the first of Compton's creative leaps. For in attempting to interpret the origin of this spurious 'fluorescent' radiation peak, Compton, for the first time, made use of the 'quantum relation' $h\nu$. He assumed, as before, that this 'fluorescent' radiation was being emitted by β-particle oscillators which had been instantaneously liberated in the scatterer. However, he now also assumed that they had somehow acquired energy $h\nu$ from the incident X-rays. Note that Compton did *not* assume that a *single* β-particle had acquired energy $h\nu$ from a *single* X-ray – he did *not* conceive of the process as a particle-particle interaction.

Carrying out Compton's calculation in detail for his specific experimental configuration, the relativistic Doppler expression reduces fortuitously to the simple Doppler expression $\lambda/\lambda' = 1 - v/c$. The β-ray velocity (here small compared to c) may now be eliminated by using 'conservation of energy,' $(1/2)\,mv^2 = h\nu$, with the result that $\lambda/\lambda' = 1 - \sqrt{2h\nu/mc^2}$. Substituting numbers, one obtains $\lambda/\lambda' = 1 - \sqrt{2(0.017\ \text{Mev})/0.51\ \text{Mev}} = 1 - 0.24 = 0.76$. This was to be compared with the 'observed' ratio $\lambda/\lambda' = 0.71$ Å/0.95 Å$=0.75$. Who could ask for better agreement, considering that we now realize, as indicated, that this entire episode actually constituted an incorrect theoretical confirmation of misread experimental data?

Perhaps, as I mentioned earlier, the most surprising aspect of this episode is that it did not completely derail Compton's research program. This seems understandable, however, when one observes that Compton in effect had here merely refined his auxiliary hypothesis by specifying the *amount* of energy acquired by the β-particle oscillators at their instant of liberation, all the while leaving the central assumptions of his research program intact. Note, however, that since Compton had recourse here to the 'quantum relation' $h\nu$, it might be also argued that to some extent he had changed a basic central assumption in his research program, namely, that classical electrodynamics is universally valid. My reason for rejecting this argument is that Compton himself was still firmly committed to classical electrodynamics at this time – he himself scarcely recognized the creative break he had made with his past work. Rather, in essence, he had simply tried something, and it had worked.

It was only a few months later, in fact, in March 1922, that Compton's classical biases were once again strongly reinforced. Compton, for the very first time, and in very important experiments, proved that X-rays may be totally internally reflected from certain substances. What could be more wave-like than total internal reflection? Moreover, the Drude-Lorentz dispersion formula, which describes this process, contains the same factor N for the total number of electrons per unit volume that appears in Thomson's classical scattering formula. Did this not prove that in *both* cases the incident X-rays were interacting with *groups* of electrons, and *not* with *single* electrons?

Shortly after carrying out these now-classical experiments, Compton began to prepare a report on X-ray phenomena for the National Research Council which was eventually published in October 1922. It was during this period of time that Compton achieved what we may term his second creative leap. It was closely analogous to his first. He now realized, however, that earlier he had misread his spectroscopic data, and that MoKα X-rays, when scattered through about 90° by graphite, undergo a shift in wavelength of only a few percent, not 35%. He now concluded, explicitly, that the ratio $\lambda/\lambda' = 0.969$.

Once again, however, the fundamental question remained: how should this observation be interpreted? To answer this question, Compton assumed, as before, that this secondary longer wavelength radiation was being emitted by β-particle oscillators at the instant of their liberation by the incident X-rays. Now, however, by undoubtedly generalizing the classical relationship between energy and momentum, Compton also assumed that the β-particles had acquired not energy $h\nu$, but *momentum* $h\nu/c$ from the incident radiation. Hence, now by 'conservation of momentum,' $mv = h\nu/c$, and the Doppler expression becomes $\lambda/\lambda' = 1 - v/c = = 1 - (h\nu/mc^2)$. Substituting numbers, we now have that $\lambda/\lambda' = 1 - 0.017\ \text{MeV}/0.51\ \text{MeV} = 0.966$. Once again, who could ask for better agreement, considering that we now realize, as indicated, that this episode actually constituted an incorrect theoretical confirmation of correctly read experimental data? The reason that Compton's research program once again did not become derailed seems understandable along the lines of our earlier discussion. Once again, Compton in essence had tried something, and once again it had worked.

Compton's third and final creative leap occurred sometime in November 1922. Meanwhile, he had truly struggled with the profound question of whether X-rays consist of electromagnetic waves or of light quanta. Ultimately he concluded that it was unreasonable to imagine that β-particles which had been liberated by the incident radiation could move at velocities approaching the velocity of light in a scatterer at rest, as his 'fluorescent' radiation-Doppler shift interpretation required. He therefore finally decided to see, as he put it, what would happen if he assumed that a *single* X-ray quantum interacted with a *single* electron or β-particle in the scatterer. In other words, he now definitely replaced the central assumptions of his 'fluorescent' radiation research program. The result was, as everyone knows, that he set up the appropriate equations expressing *both* conservation of energy *and* conservation of momentum, solved them simultaneously, and derived his famous expression for the change in wavelength. The Compton effect had been discovered.

There are several observations which I would like to make at this point in some detail but can only treat in a cursory fashion owing to limitations of space. First and formost, we have seen how relatively autonomous Compton's route to his discovery actually was. There is no evidence, in particular, that he read or was in any serious way influenced by Einstein's 1905 paper. Nowhere does Compton cite Einstein's 1905 paper in his own, nor does Compton even mention Einstein's name. The most that can be said is that Compton at one point decided to see what would happen if he tried a relationship, the 'quantum relation' $h\nu$, that was somewhere in the back of most physicist's minds at the time. The impression that one often gains from textbook accounts, that Compton in 1922 somehow became acquainted with Einstein's light quantum hypothesis, applied it to the scattering problem, and then verified the predicted change in wavelength by means of a clever X-ray scattering experiment, is very definitely incorrect.

Second, Compton's new quantum theory of scattering, which was based on entirely new central assumptions, rendered both his old large electron theory of scattering, and his old 'fluorescent' radiation-Doppler shift interpretation of the longer wavelength secondary radiation, in Compton's term, 'superfluous.' There was no longer any need to appeal to a 'selective scattering effect' or to a 'fluorescent' radiation hypothesis to account for the existence of the softened secondary radiation. Moreover, Barkla's reduced absorption coefficient, the preponderance of the scattered radiation in the forward direction, the angular variation in its hardness, the fact that its hardness was independent of the nature of the scatterer, its relatively high efficiency of production, and its essentially complete polarization, all became understandable on the basis of Compton's new quantum theory of scattering. In addition, it explained why the relative change in wavelength $\Delta\lambda/\lambda$ is much greater for γ-rays than for X-rays, it offered a possible explanation for the so-called 'limb effect' at the surface of the sun, and it predicted the existence and behavior of the so-called 'recoil electrons.' And Compton's new theory explained all of these phenomena *quantitatively*, without, as he noted, introducing any adjustable parameters whatsoever. In all of these respects, Compton's two older theories were superseded by his new one.

Third, there is a point to be made about the actual neutrality of some of Compton's most important experimental data. For it turns out that the *spectroscopic* data which Compton cited in support of his famous discovery was identical to that which he had cited earlier in support of his October 1922 Doppler shift interpretation of his 'fluorescent' radiation. He had not as yet taken any spectroscopic data at scattering angles other than 90°. In other words, Compton's spectroscopic data had not changed at all, while his theory had been fundamentally altered. The reason this situation could occur was, as I have shown, that Compton's *new* expression for the change in wavelength is simply the zero-order term of his *old* expression, when the latter is expanded in powers of $\alpha = h\nu/mc^2$. And for X-rays, the difference was simply not measurable. The same criticism actually applies to *all* of Compton's X-ray data at the time, leading to the rather surprising conclusion that Compton, if he had been pressed, would have had to appeal to his old γ-ray *absorption* data to distinguish his new theory from his old.

Finally, in spite of the general persuasiveness and elegance of Compton's new quantum theory of scattering, and in spite of the decisive tone of Compton's classical paper, Compton himself by no means changed from a classical physicist to a quantum physicist overnight. No one more than he, in fact, recognized the profound dilemma he had precipitated into physics. That dilemma is strikingly symbolized by the fact that roughly midway between the time Compton delivered his quantum scattering paper at a meeting of the American Physical Society on December 1 or 2, 1922, and the time he sent it to *The Physical Review* for publication on December 13, 1922, he *also* sent his total internal reflection paper to *The Philosophical Magazine* for publication on December 6, 1922. In other words, within the space of about two weeks, Compton published convincing experimental evidence for *both* the quantum *and* the wave nature of X-rays.

The dilemma Compton faced over the ultimate nature of X-rays was reflected in a remarkable note which he published in *Science* at about this time, in which he concluded that radiation, in general, must be its own medium of propagation, just as water is when it issues from a hose whose nozzle is being shaken. So profound was Compton's uncertainty at this time. His uncertainty, however, was simply a harbinger of what was to occur among physicists generally. The results included a profound re-examination of the ultimate nature of radiation and of the conservation laws, as well as a proliferation of specific theories of radiation. These, however, are only some of the aspects of my study which I cannot develop here.

University of Minnesota

NOTES

* School of Physics and Astronomy and Minnesota Center for Philosophy of Science, University of Minnesota. I wish to thank the National Science Foundation for support during the research and writing of this paper.

[1] Cambridge University Press, Cambridge, 1970.

[2] Science History Publications, New York, 1975. The reader is referred to this book for detailed substantiation of the statements and analyses presented in this paper.

HÅKAN TÖRNEBOHM

INQUIRING SYSTEMS AND PARADIGMS

1. Inquiring systems and paradigms

Theme

In this essay, I will outline a view on science in which the notion of inquiring systems and paradigms plays a key role. Kuhn's term 'paradigm' is here used to refer to a complex of factors which direct and control the work done within a field of research.

A researcher engaged in an investigation of a selected part X of the real world belongs to what I will call an inquiring system $\hat{X}$. X will be called its territory. The work done by the researcher and his colleagues within $\hat{X}$ will also be regarded as constituents of $\hat{X}$.

2. On inquiring systems

2.1 *Relations between an Inquiring System $\hat{X}$ and its Territory*

There are three kinds or relations between a territory X and an inquiring system $\hat{X}$.

A. 'Primary information' flows from X into $\hat{X}$ through various channels. The arrow $\overset{\sim}{\rightarrow}$ in Figure 1 represents flows of primary information.

B. Hypotheses, intended to map features of X, are framed in $\hat{X}$. The arrow $\overset{\triangleq}{\rightarrow}$ in Figure 1 represents mapping relations.

C. The work done in $\hat{X}$ may affect its territory. This is the case if, for instance, X is a social organization served by assets produced in a scientific study over X. The arrow $\rightarrow\!\!\!\!\ominus\!\!\!\rightarrow$ in Figure 1 represents effects exerted on a territory.

There are different kinds of primary information:

(1) $\hat{X}$ receives information from X in the form of measurement results whereby the channels are manmade technical systems in which instruments of measurement are incorporated. Theories are required in order to design the instruments, as well as to interpret the measurement results, and to correct systematic errors and to estimate accidental errors. As a

R. S. Cohen et al. (eds.), Essays in Memory of Imre Lakatos, 635–654. *All Rights Reserved*
Copyright © 1976 by D. Reidel Publishing Company, Dordrecht-Holland

consequence pieces of primary information are strongly 'theory laden'.

(2) X belongs to the past of $\hat{X}$. The sources of primary information are then traces of the past, in the form of historical documents or archeological findings or in the form of fossils etc. Before becoming pieces of information, however, a flint axe or a fossil must first be interpreted and described by researchers before $\hat{X}$ receives primary information from X.

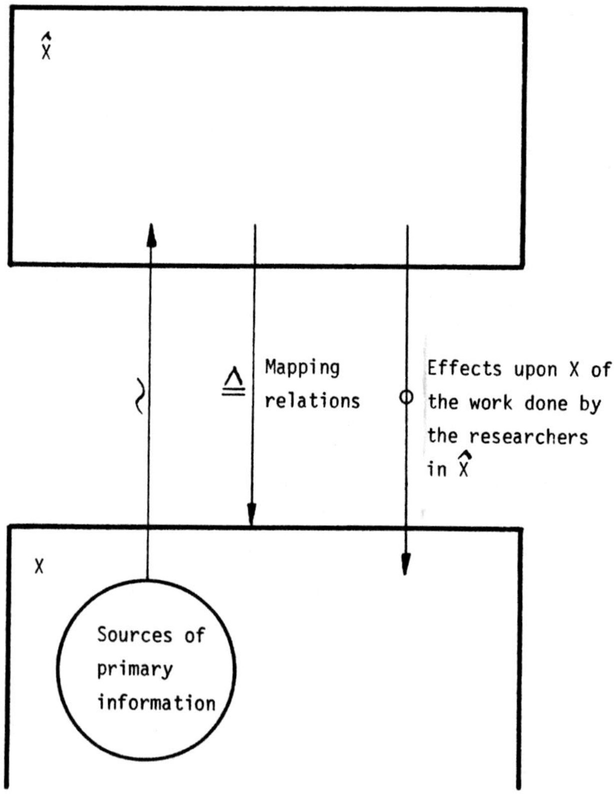

Fig. 1.

For subjects like archeology and paleontology reasonable assumptions about trace formation are thus required in order to produce primary information. Historians require knowledge about 'trace formation' in order to be able to assess the information value of the traces of the past. They need to know about the writers of the documents and how these are related to persons, events etc. dealt with in their writings. Primary pieces of information are thus 'theory laden' in historical fields of research also.

(3) The researchers in $\hat{X}$ carry out biological or geographical field studies. The primary information consists of observation reports on wild life or landscapes etc. made by explorers. In the early stages in the development of field studies, very little theoretical work is carried out. The pieces of primary information are therefore only to a minor extent affected by additional knowledge. Data are 'theory laden' only to a minor extent in case (3).

(4) X is a social organization; a family, an indigenous African village etc. The primary information consists partly of observation data of the kind mentioned under (3) and partly of reports from interviews with informants belonging to the territory X. These reports call for interpretations which require competence in the language natural to the informants as well as background knowledge about the social and cultural conditions of the informants and of their qualifications, such as familiarity with relevant features of X, and their veracity. The data are thus in this case also dependent on additional knowledge of a general nature.

We conclude that pieces of primary information are useful to researchers in an inquiring system only to the extent that they are supplemented by additional information. Data are 'theory laden' in the cases described above, with the possible exception of case (3) and then only in the early stages of research.

After having outlined the significance of the arrow $\overset{\simeq}{\rightarrow}$, I will deal very briefly with the two other arrows in Figure 1.

An arrow $\overset{\triangleq}{\rightarrow}$ goes from an hypothesis h framed within $\hat{X}$ and a feature of X. I can do no more in this short essay than merely remark that $\overset{\triangleq}{\rightarrow}$ represents epistemic mapping in cases where h is, and deserves to be, confirmed. I will regard epistemic mapping as being analogous to the relation that a correct map bears on a part of the surface of the earth. It is this view on knowledge that has suggested the term 'territory' employed throughout in this essay.

Not every inquiring system $\hat{X}$ affects its territory X, but if $\hat{X}$ belongs to the social sciences, and if we take $\hat{X}$ to encompass a major part of political economics, then it is evident that $\hat{X}$ does affect X in various ways. Knowledge about economic systems, as well as techniques of working within and upon them, produced within $\hat{X}$ exert major social effects. We may say, in general, that the social sciences furnish politicians and others with assets in the form of knowledge and know-how which do affect social development to a considerable extent.

2.2. *Informational Aspects of Inquiring Systems*

I distinguish, firstly, between primary and secondary information and, secondly, between internal and external information as can be seen from Figure 2.

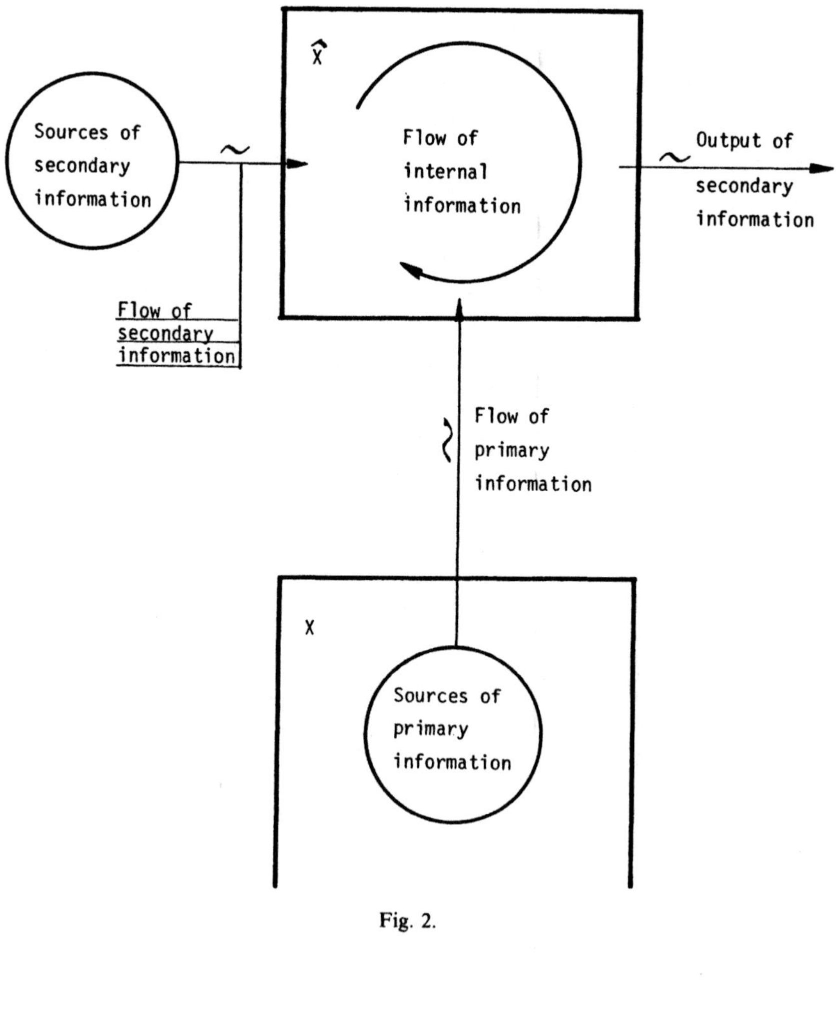

Fig. 2.

Imports from other inquiring systems, $\hat{Y}$ etc., into a system $\hat{X}$ will be called secondary information. This term will also be used for the content of research reports written inside $\hat{X}$ and intended to be used outside it. Information is termed external with respect to $\hat{X}$ if its sources or its intended users are to be found outside $\hat{X}$. If both sources and intended and actual users are inside $\hat{X}$, I will use the term internal information.

Figure 2 represents these various kinds of information used and produced by an inquiring system $\hat{X}$.

2.3. *Result Sequences*

Taking a look at the inside of an inquiring system $\hat{X}$, we find a collection H of hypotheses about X at a given time t. These hypotheses are surrounded by a field P of problems. The researchers are equipped with a set I of instruments at the time t. I contains 'software' instruments such as mathematical techniques, concepts etc. I may also contain 'hardware' instruments, such as instruments of measurement and other equipment, needed in order for the researchers to obtain primary information from and about their territory.

The hypotheses H, the instruments I and the problems P form a complex (H, I, P). The results at various instants of time of the work carried out within $\hat{X}$ consists of (H, I, P)-triads. These triads form a sequence. We observe various types of ingredients in (H, I, P)-sequences:

2.3.1. *Piecemeal knowledge-formation*

New pieces of knowledge are formed. This is the case if an hypothesis h which has not yet been tested in an earlier (H, I, P)-triad corresponds to an hypothesis h' (which may or may not have the same content as h) in a later (H, I, P)-triad in which h' is confirmed after testing.

Piecemeal knowledge-formation may briefly and schematically be described as follows:

(1) An hypothesis h is accompanied by a problem or question p at a time t: "Is h true or not?" or better "How true is h?"

(2) Problems of this kind prompt researchers in $\hat{X}$ to design testing plans.

(3) In the process of implementing a testing plan for an hypothesis h, the researchers bring together information from primary and secondary sources to be employed as evidence e pro or contra the hypothesis h.

(4) A verdict is passed on the relation between e and h; whether and to what extent e supports h, or whether and to what extent e undermines h.

(5) The verdict leads to a decision concerning h; either to accept h as true or to reject h, or to gather more evidence, or to modify h into an hypothesis h' such that favorable evidence for h is also favorable for h' and such that unfavorable evidence for h is either neutral or favorable for h'.

(6) If a researcher after testing an hypothesis h comes to believe that h is true, he procedes to argue on its behalf. If his colleagues, convinced by the arguments, also accept h, h is confirmed and a new piece of knowledge (in a descriptive sense of the term 'knowledge') has been produced.

2.3.2. *Theory formations*

Hypotheses which are not connected in an earlier (H, I, P)-triad correspond to hypotheses in a later one where they are ingredients in a deductive system or some other kind of synthesis. Theory formation is prompted by problems of a different kind from those mentioned under Section 2.3.1, i.e. explanatory problems. The instruments used for theory formation include techniques of deduction when syntheses have the logical form of deductive systems.

2.3.3. *Instrumentation*

Later (H, I, P)-triads include instruments which replace those found in earlier (H, I, P)-triads. Entirely new instruments may have been constructed. Changes in the I-parts of result sequences are usually prompted by the appearance of new problems.

2.3.4. *Problem evolutions*

It may happen that problems in an earlier triad have been replaced in a later triad, not by their solutions but by problems which are better articulated and better adapted to the resources available for solving them.

2.3.5. *Research planning*

How are (H, I, P)-triads transformed? This question concerns the arrows linking (H, I, P)-triads into a sequence. (Compare Figure 3.)

Researchers in $\hat{X}$ facing a (H, I, P)-triad at an instant t have to make up their minds about what to do next. They plan subsequent moves. A plan includes the following items:

(1) Problem selection, possibly accompanied by a refinement of the selected problem(s).

(2) Assembling instruments to be employed in solving the selected problem(s).

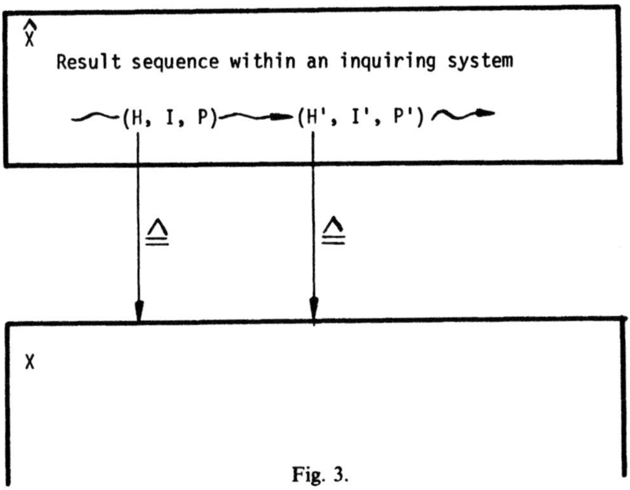

Fig. 3.

(3) Decisions concerning the ways and means of gathering pieces of information from secondary or primary sources for use as additional assets in solving the selected problem(s).

(4) Decisions concerning the procedures of using assets mentioned under (2) and (3) in order to arrive at solutions of the selected problem(s).

The various planning activities are thus adapted to (H, I, P)-triads, being the result of previous work. Planning of research is thus an opportunistic activity.

It would obviously be wrong to believe that researchers do their planning only once – at the initiation of a research enterprise. As a consequence an inquiring system does not have a definite goal which can be envisaged at its beginning.

2.3.6. *Running research programs*

A combination of a sequence of plans and their implementations with a sequence of results gives rise to a sequence of tetrads:

$$---\rightsquigarrow(H, I, P, P1) \rightarrow (H', I', P', P1')\rightsquigarrow---$$

It would seem plausible to characterize every tetrad in such a sequence as a research program and to refer to the sequence of tetrads itself $\rightsquigarrow(H, I, P, P1)\rightsquigarrow$ as a running research program. A running research program can thus be conceived of as a synthesis of research work consisting of planning activities accompanied by implementations of the plans on the one hand and result sequences on the other. We may picture such a synthesis as shown in Figure 4.

2.3.7. *Internal criticism*

Running research programs, as well as parts of them, are exposed to critical scrutiny within an inquiring system. The structure of the targets of the scrutiny is reflected in the set of criteria employed. The merits of a running research program are thus assessed by a set of criteria presupposing criteria on various parts of it. Critics within an inquiring system are likely to commend a running research program if they judge the sequence of results to be a satisfactory one.

The criteria employed in the assessment of sequences of results do presumably include the following ones:

A move from one (H, I, P)-triad to another one (H', I', P') is progressive to the extent that

(1) (H', I', P') contains more confirmed hypotheses than (H, I, P),
(2) (H', I', P') contains better instruments than (H, I, P) so that previously insoluble problems become tractable,
(3) the problem field P' contains items which are better articulated than corresponding items in P, and finally
(4) H' is better organized than H. H' may, for example, include

an acceptable explanatory pattern absent in H.

Criterion (1) moreover presupposes various criteria employed in the process of testing hypotheses, e.g. criteria regarding the information employed as evidence pro or contra the hypothesis being tested. Criterion (2) presupposes criteria regarding concepts, techniques and instruments of measurements etc.

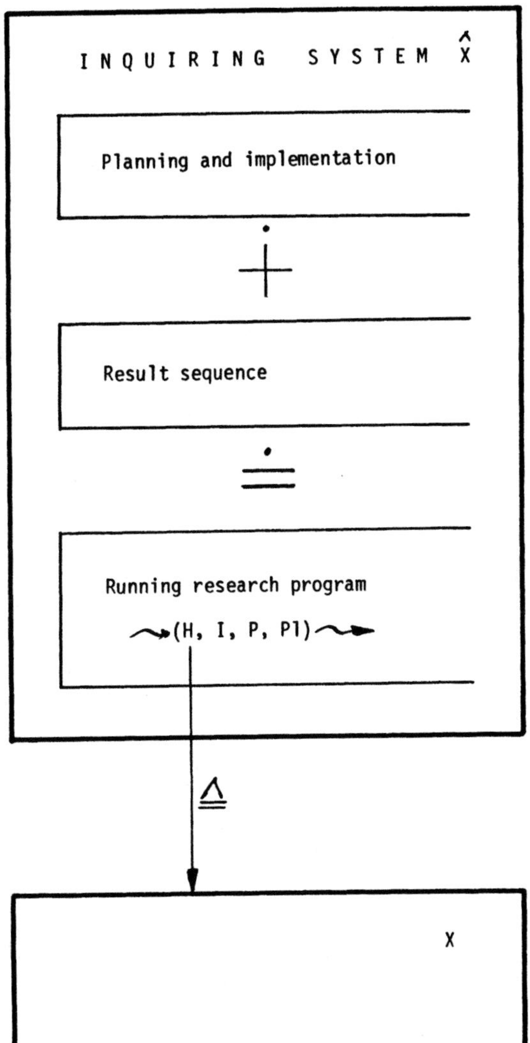

Fig. 4.

Criterion (4) presupposes criteria regarding explanatory patterns and theoretical structures.

Internal criticism within an inquiring system has two main functions:

(α) It serves to detect errors committed in running research programs.

It gives rise to problems of revision which enter into the problem fields and affect further moves in a subsequent running research program.

(β) It serves to promote some results to the status of 'public commodities'. If, for example, the 'umpires' accept an hypothesis h, it will be included in a stock of established knowledge.

Figure 5 pictures internal criticism within an inquiring system $\hat{X}$.

Not only running research programs are exposed to critical scrutiny. Input information from secondary or primary sources and output information from an inquiring system $\hat{X}$ are also subject to internal criticism.

The criteria employed in these cases, however, are distinct from those by which the merits of running research programs are assessed. In the case of input information, the researchers within $\hat{X}$ are mainly interested in the question as to whether it can be usefully inserted into their own running research program. In the case of outgoing information, they are concerned with the question as to what extent it is intelligible for properly qualified readers. Esthetic criteria may also play a role in assessing research reports in which terminal results of the work in $\hat{X}$ are presented to intended users outside $\hat{X}$.

2.3.8. *Initiation of an inquiring system*

How is the first research program (H_0, I_0, P_0, Pl_0) built? In H_0 is included general assumptions about the territory X which may be regarded as items of a world picture. In order to form a general view on X, to be included in H_0, the researchers look for positive and negative analogies between X on the one hand and other territories on the other. If Y has already been mapped by another inquiring system $\hat{Y}$ and if researchers, intent upon investigating the part X of the real world, come to believe that X is similar to Y in such and such respects, they are led to assume that such and such hypotheses, contained in reports from $\hat{Y}$, may also be applicable to X. Transfer of hypotheses from $\hat{Y}$ into H_0 is thus prompted by working hypotheses about positive analogies. Such transfers are not unrestricted since working hypotheses about negative analogies serve to set limits on the assimilation of information from $\hat{Y}$ into H_0. The factor I_0 in the first research program is composed of instruments already in the possession of the initiators of $\hat{X}$. Some instruments may also be imported from other inquiring systems just as parts of H_0 may come from external secondary sources.

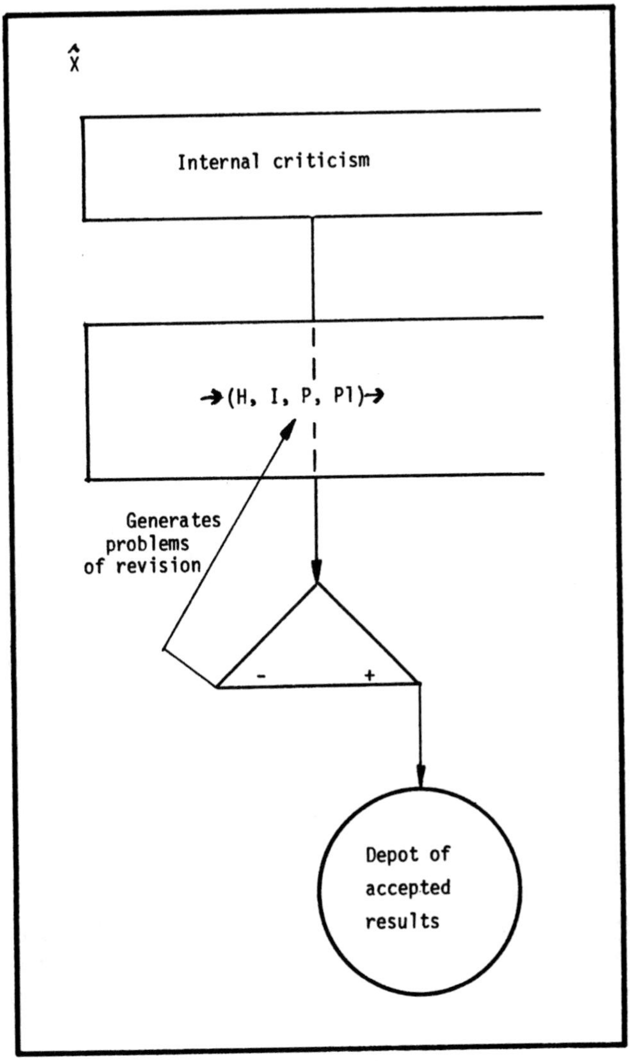
X̂
Internal criticism
→(H, I, P, Pl)→
Generates problems of revision
-
+
Depot of accepted results

Fig. 5.

The original problem field P_0 is in part determined by H_0, some items of which are problematical. We note that untested hypotheses and unassimilated confirmed hypotheses are accompanied by truth problems and explanatory problems respectively.

3. Paradigms

The work carried out within an inquiring system is directed and controlled by what I will refer to as a paradigm. A paradigm in my sense of the term is composed of various items.

3.1. *View on Research*

Researchers about to launch on an investigation, or who are working inside an inquiring system which has been going on for some time, profess views on their 'niche' of the world of science. Their conception of their particular fields of research has been acquired during a period of formal training and as a result of subsequent research experience. A view of research can thus be conceived of as being focussed on a region with which a researcher is familiar, a region surrounded by an environment known to him indirectly by way of readings and by way of conversations with fellow researchers.

A researcher's view of his place in the world of science is evaluative. He recognizes masters and masterpieces as well as items of research which he regards as warning examples. He sees a 'landscape', as it were, with mountains occupied by the greats and valleys inhabited by minor minds. Certain parts of the landscape are brightly illuminated. Others are clouded in mists. The region where he finds himself as an active researcher is most brilliantly lit. Its details are seen most clearly. Its surroundings are more hazy. A researcher's view on science affects his choice of tasks. He is presumably inclined to follow the examples of admired masters. His judgements on good or bad research will presumably be affected by comparisons with what he recognizes as masterpieces.

3.2. *Direction of Research*

A second factor of a paradigm, which in contrast to a view on science is voluntaristic rather than cognitive and evaluative, consists of a decision about which course to follow in an investigation.

A researcher may decide to make use of experimental methods in acquiring primary information from his chosen territory. He may decide to synthesize knowledge in the form of a deductive system. He may decide to employ a mathematical formalism in his attempts to organize a body of knowledge etc.

These are some examples of very general types of decisions concerning the direction of research to be followed by a researcher. Within a general framework of this kind, more specific and detailed decisions naturally have to be made before a researcher is in a position to launch on the investigation proper. He must make up his mind about the procedure of more specific kinds of experiments to be performed and a more specific formalism to be employed etc.

3.3. *World Picture*

A researcher intent on investigating a territory *X* assumes before his investigation that a specified pattern of features is characteristic of *X* and that this pattern may also be found in other territories *Y*, *Z* etc., some of which may already have been investigated. In other words, he holds the view that the territory *X* is linked to other terrirories by means of specific positive analogies.

A world picture is formed and as such plays an important role within the inquiry system *X*:

(1) Parts of the contents of the *H*-sequences in a running research program belong to a world picture.

(2) A world picture suggests to a researcher the kind of secondary sources from which he may obtain information to be assimilated and used in a running research program.

(3) A world picture supplies basic concepts which, in a refined form, will be employed in the process of forming theories.

(4) A world picture serves to mark out what researchers will regard as important tasks and results.

(5) General theories in a mathematical form may be traced back to world pictures. Indeed, a general theory may itself be regarded as a refined world picture. An historical route leads from qualitative mechanistic world pictures in early modern physics to fundamentals in classical mechanics and one broad road from Faraday's qualitative field conception to Maxwell's electromagnetic theory.

3.4. *Ethics of Research*

The three factors mentioned above, i.e. a view on research, a direction of research and a world picture are not the only ones which serve to govern and control research work within inquiring systems. It is evident that ethical and esthetic criteria are sometimes employed in critical assessments within inquiring systems. Researchers may be praised or blamed for possessing moral virtues or vices. Objectivity and impartiality for instance are classifiable as moral virtues and are not merely concerned with epistemology.

They have certain duties and responsibilities. They will be exposed to sanctions if they fail to conform to an established professional code. The recipients of secondary information put their trust in the competence and conscientiousness of its suppliers. If let down, they may run into various kinds of troubles in their own work. They may, for example, use as evidence confirmed hypotheses taken from outside their own inquiring system. If the suppliers of those hypotheses have failed to be sufficiently conscientious and careful, the recipients run a grave risk of passing wrong judgments and making incorrect decisions in the course of their own knowledge production. Confidence has to be matched by a feeling of responsibility with regard to others. The responsibilities of a critic vis-à-vis those persons whose work he is assessing are made up of his duty to assist in avoiding mistakes and to give them the go ahead to carry on with new tasks without undue delay.

We may distinguish between an internal and an external part of the ethics of research. An internal ethics is concerned with the maintaining of duties and responsibilities towards other researchers. An external ethics is concerned with duties and responsibilities towards people outside the community of researchers. If the work that is carried out within an inquiring system $\hat{X}$ affects the life and work of other persons than the researchers themselves, and if a person within $\hat{X}$ cares about those effects, in so far that his own work is affected thereby, then an external ethics can be said to be included in his paradigm. It is to be expected that service-oriented research is governed by paradigms which include an external ethics.

The primary concern of an external ethics has to do with possible unfavorable and/or favorable effects, in serving to direct research in such a way that harm is prevented and beneficial effects may be expected to occur. The requirements of an external ethics may tell researchers to use more than one language when they report their results; one language when they address themselves to their colleagues, and another when they write for interested outsiders.

3.5. *The Esthetics of Research*

There are several esthetic elements in the work done within an inquiring system. An experiment may be elegant. A theory may have beauty and a report may have a fine composition and an appealing style. The criteria employed in a critical assessment of the quality of running research programs and reports of results include esthetic ones. The esthetics of research may be characterized as being functionalistic. Techniques are regarded as elegant if they are effective and yet simple to use. A theory is regarded as beautiful if its formalism is elegant and its conceptual apparatus simple and yet powerful.

3.6. *Paradigms in Various Contexts in the World of Science*

Paradigms can be found in several contexts in the world of science.

(1) Paradigms are transmitted in the process of formal training and are affected by subsequent research experience.

(2) Paradigms govern and control running research programs in inquiring systems.

(3) Paradigms may give rise to the formation of various types of structures that are composed of several inquiring systems, for example research traditions and aggregated clusters of inquiring systems capable of producing families of theories, such as classical mechanics or quantum mechanics.

(4) Paradigms may be the subject of general deliberations, serving several functions: (a) to articulate them, (b) to submit them to a critical scrutiny, (c) to modify them, or (d) to build new ones, which then may be employed as means of steering and control within inquiring systems.

In the next section I will consider context (2).

3.7. *Paradigms within Inquiring Systems*

A paradigm within an inquiring system $\hat{X}$ may be represented as follows:

$$\text{Paradigm in } \hat{X} = (V\hat{X}, D\hat{X}, VX, E, Est)$$

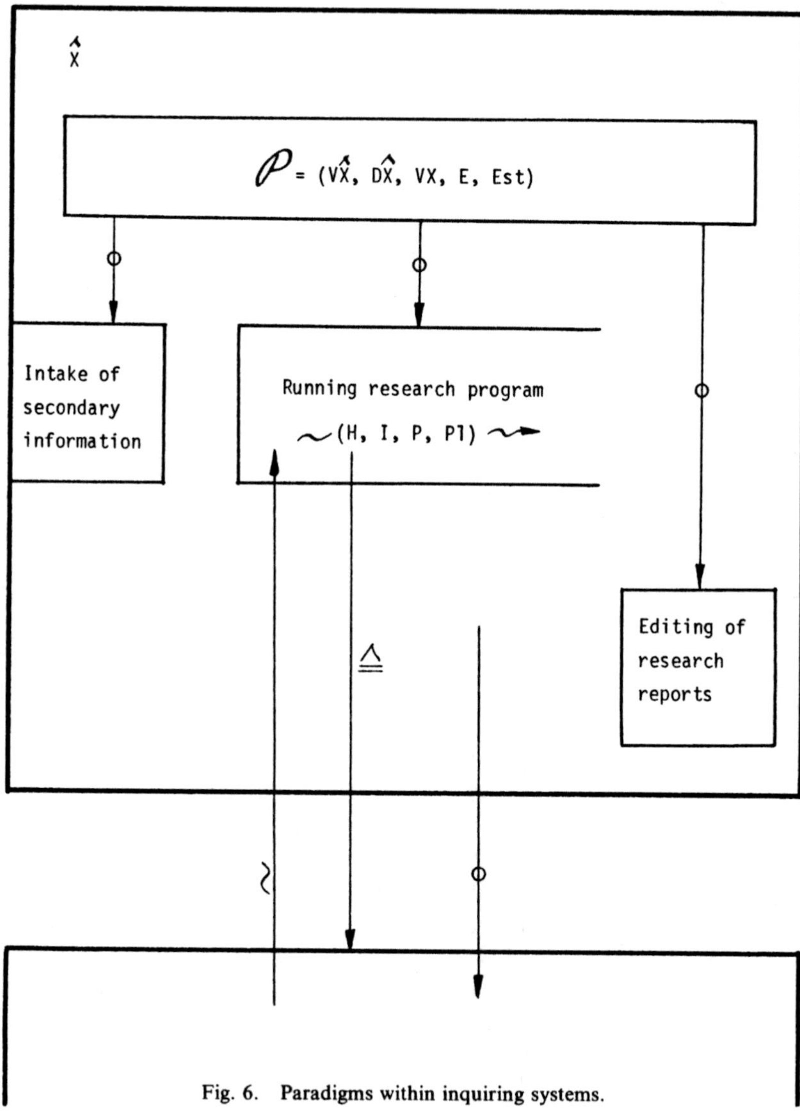

Fig. 6. Paradigms within inquiring systems.

where

$V\hat{X}$ is a *view* on $\hat{X}$ in an intellectual environment parts of which are made up of sources and 'sinks' of secondary information.

$D\hat{X}$ is the *d*irection of the research work within $\hat{X}$,

VX is a picture or *v*iew of the territory X, according to which X exhibits positive analogies with respect to territories which have been or are being mapped by inquiring systems which, accordingly, are regarded as potential sources of useful secondary information by the researchers within $\hat{X}$.

A paradigm affects the work done within $\hat{X}$ in various ways:

(1) Its first and third factors affect the gathering and utilization of secondary information to be assimilated into the running research program within $\hat{X}$.

(2) Its world picture supplies hypotheses to the H_0-factor of the first research program and also to the H-factors of its successors.

(3) Its world picture does also supply material from which fundamental concepts employed in the organization of subsequent H-factors in the running research program are fabricated.

(4) Its second factor determines criteria employed in the internal criticism of the running research program within $\hat{X}$. If a critic judges that forthcoming results deviate too much from the set course, i.e. $D\hat{X}$, he will require that earlier steps in a running research program are to be revised.

(5) The E-factor is likely to play a role in setting the course of the whole enterprise. In other words $D\hat{X}$ may have been formed in a process which includes ethical considerations: some directions of research are forbidden and others are approved of on account of deliberations which belong to the external part of E; some ways of obtaining primary information from X are outlawed for moral reasons, even if those avenues to X should be judged to be very useful in the research work within $\hat{X}$.

(6) The esthetic factor in a paradigm supplies some of the criteria employed in assessing the outgoing secondary information. It may also play a role in assessing running research programs. Clumsy procedures of solving problems may be castigated and elegant ones praised.

3.8. *The Role of Paradigms in the Formation of Compounds of Inquiring Systems*

The previous section has dealt with the effects of a paradigm within an inquiring system. In this section, I will consider very briefly how paradigms may serve to bring together several inquiring systems into a compound which I will call an aggregated inquiring system. Let $\hat{X}$, $\hat{Y}$, $\hat{Z}$,---

be inquiring systems mapping the territories X, Y, Z,--- respectively. Suppose that the world picture within one of the paradigms, e.g. VX, includes assumptions which are closely similar to assumptions contained in the world pictures VY, VZ--- etc. in the other paradigms. Under these conditions, the researchers in the various inquiring systems would be strongly inclined to exchange secondary information. One of the effects of such exchanges may be that the results of the work done within the different inquiring systems exhibit close similarities: families of theories are produced.

A theory may be characterized in this simple way:

A theory about X=(a language – formalism – L, a text T written in L about X) or for short

$$Th=(L, T).$$

If contents of the text parts of several theories, one T_X about X, one T_Y about Y etc. include a common part T_0 describing common features of the territories X, Y etc., the various theories may be regarded as members of a family linked together by their sharing a common world picture, i.e. T_0. A cluster of inquiring systems producing a family of theories may be regarded as forming an inquiring system of a higher order which I propose to call an aggregated inquiring system. The clusters which produced classical mechanics and electrodynamics respectively may serve as examples.

The ideal of science within the second part of a paradigm which prompts researchers to work towards the formation of aggregated inquiring systems may be called an integration ideal. Such an ideal is held not only in physics but also elsewhere in the world of science and is, I feel, commendable also in the field of meta-science.

3.9. *Changes in Paradigms within Inquiring Systems*

Paradigms do affect the work and its results within inquiring systems. Inquiring systems do in their turn exert effects upon paradigms within them. This is most evident in large scale aggregated inquiring systems:

Qualitative world pictures within early paradigms tend to be replaced by general theories expressed in a technical language, possibly a mathematical formalism. Such general theories serve as foundations of special theories and are indirectly tested when the latter are confronted with experience. These confrontations may result in clashes which call for revisions and reforms of general theories. Researchers do not seem to be keen, however, to modify general theories except in the early stages. They prefer to revise less general assumptions at occasions when theoretical anticipations are frustrated. An effect of this policy is that general theories tend to be longeaval, maybe lasting for generations. It is understandable why researchers should wish to protect general theories: if the content of a general theory *G* is revised, then a number of special theories, held together by *G* into a family, must also be revised in the interest of avoiding inconsistencies. As a family grows, so does of course the task of revising it.

The criteria employed by critics in their assessments of running research programs may be modified as an outcome of the development of new techniques which permit critics to assess results on the basis of more effective criteria than those supplied by the paradigm. A theory of differential equations is a warrant that a specified method produces a correct solution of a class of differential equations. A critic does not need a paradigm in those special cases where his criteria are backed by a theory.

3.10. *Special and General Methodological Discussions*

In the course of an investigation, researchers within an inquiring system do at times engage in discussions about the methods to be employed in solving selected problems. Such discussions may be regarded as media of tactical research planning. They contribute to the evolution of a running research program. Special methodological discussions presuppose a paradigm of research.

Another type of methodological discussions in the world of science takes place at occasions when researchers find themselves in dire difficulties calling for diagnosis and therapy. Researchers engage in general discussions also before committing themselves to engage in regular activities characteristic of an inquiring system. General methodological discussions serve to articulate and criticize paradigms and may also be breeding places for new paradigms. Researchers engage in strategical research planning when they build paradigms.

4. CONCLUDING REMARKS

I have outlined a view on the world of science. It may be summed up as follows: The world of science is composed of inquiring systems within which researchers investigate selected parts of the real world. Inquiring systems receive and give out information. The received information comes partly from a territory and partly from other inquiring systems. The output information is intended to be used by researchers and in some cases by other intressees. Inquiring systems contain running research programs directed and controlled by means of 'steering factors' here called paradigms. Paradigms may give rise to the formation of compounds of inquiring systems, capable of producing families of theories. Paradigms in their turn are modified by inquiring systems. They may be the topic of discussions which often have the character of strategical research planning preceding the establishment of new inquiring systems.

The view on science outlined in this essay may be regarded as the third factor within a prospective meta-scientific paradigm. It is my hope that meta-scientific studies, governed by a paradigm in which this view is included, will have beneficial effects.

University of Gothenburg

BIBLIOGRAPHY

Bunge, M., *Scientific Research*, Vols. I and II, New York 1970.
Churchman, C. W., *The Design of Inquiring Systems*, Cambridge, Mass. 1968.
Diederich, W. (ed.), *Theorien der Wissenschaftsgeschichte*, Frankfurt am Main 1974.
Händle, F. and Jensen, S. (eds.), *Systemtheorie und Systemtechnik*, München 1974.
Kuhn, T. S., *The Structure of Scientific Revolutions*, Chicago 1962.
Lakatos, I. and Musgrave, A. (eds.), *Criticism and the Growth of Knowledge*, Cambridge 1970.
Om komplexa system [*On Complex Systems*] (red. Göran Wallén), FOA P Rapport C 8280-10, Stockholm 1971.
Popper, K., *Objective Knowledge*, Oxford 1972.
Spinner, H., *Pluralismus als Erkenntnismodell*, Frankfurt am Main 1974.
Toulmin, S., *Human Understanding*, Vol. I, Oxford 1972.
Törnebohm, H., 'A Paradigm Shift in Physics', forthcoming in a *Festschrift for Gerard Frey*.
Törnebohm, H., 'An Essay on Knowledge Formation' in *Zeitschrift für allegemeine Wissenschaftstheorie*, (1975), No. 8.
Törnebohm, H., 'Inquiring Systems', in *Scientia* I-II-III-IV (1974).
Törnebohm, H., Paradigm i vetenskaperna värld och i vetenskapsteorin (Paradigms in Science and Meta-Science), No. 59 in Reports from the Department of Theory of Science.
Wartofsky, M. W., *Conceptual Foundations of Scientific Thought*, New York 1968.

STEPHEN TOULMIN

HISTORY, PRAXIS AND THE 'THIRD WORLD'

Ambiguities in Lakatos' Theory of Methodology

1. Personal introduction

This paper draws attention to a difficulty that arises in the interpretation of Imre Lakatos' writings about methodology and the philosophy of science, and suggests certain clues that may help us past that difficulty. The problem in question has a special importance for me personally, since it underlay (I believe) some unexpectedly violent expressions of disagreement that took place between the two of us at a number of public meetings, right up to November, 1973. And it is one that I have a special motive for puzzling about since, in so many other ways, Imre and I approached the philosophy of science along generally parallel lines.

What was it, then, about the arguments of historically-minded philosophers of science like Michael Polanyi, Thomas Kuhn and myself (very diverse intellectual characters, in most respects) that converted us in Imre Lakatos' eyes, into a 'school' – not to say a heresy, or 'dangerous ideological tendency'? What, indeed, could have roused him to such wrath, particularly if we consider (1) how closely his 'methodology of research programmes' appeared, in the eyes of onlookers, to resemble my own discussion of 'intellectual strategies' in science; and (2) the crucial roles that he gave to historical change, and to the collective judgment of mathematicians, at the conclusion of his own fundamental monograph, *Proofs and Refutations*?

It should not be surprising that, outside the walls of the London School of Economics, Imre's ideas about 'research programmes' should have been so easily equated with my own views on 'intellectual strategies'. Both accounts were developed in order to deal with the same question:

> How are we to recognize what directions of theoretical innovation in science are more or less rational/productive/fruitful etc., in this or that natural science, at this or that moment in its history?

R. S. Cohen et al. (eds.), Essays in Memory of Imre Lakatos, 655–675. *All Rights Reserved*
Copyright © 1976 by D. Reidel Publishing Company, Dordrecht-Holland

Furthermore, both accounts required the philosopher of science to begin from an accurate descriptive characterization of the 'programme' or 'strategy' embodied in a particular phase of theoretical development: e.g., Newton's analysis of central forces, the nineteenth-century wave-theory of light, or Darwin's account of the origin of species. And, in addition, neither account conceded to any currently successful program/strategy/paradigm any *intrinsic* authority, simply in virtue of being established. On the contrary, both accounts sought to show how the currently established directions of theorizing could be put to a critical test, so as to indicate just *how* fruitful, successful and/or 'progressive' they really were.

The chief point of difference between us (or so it seemed to me) lay in our views about the source and character of those final, 'critical' standards of judgment. At one stage in the development of Imre's philosophy of science, he was clearly tempted by the notion that these standards might be timeless and a-historical: i.e. that we might frame universal canons for recognizing 'progressive' from 'reactionary' directions of scientific change, as counterparts of Karl Popper's 'demarcation criterion'. Yet by 1973 (as we shall see) he had largely abandoned this notion. Still: my own counter-suggestion, that we are – on this ultimate level – driven back on to finding out by experience, as we go along, just what can constitute 'fruitfulness' in (say) quantum mechanics or physical cosmology or cell physiology or oceanography, at this or that stage in the science's development, apparently roused Imre Lakatos to wrath. He used to denounce this view as intolerably elitist – as having implications similar to those of Stalinism (P.S.A., Lansing, 1972), or as comparable to the views of *Der Stürmer* (U.C.L.A. Copernicus Symposium, 1973), or as invoking the sanctions of 'the Wittgensteinian thought-police' (cf. his unpublished review of my *Human Understanding*, Vol. 1, 1974).

In the course of these exchanges I was never able, for the life of me, to fathom just what was leading Imre to such extremes; and I was particularly perplexed, when I recalled how far my own views about the judgment of conceptual change in natural science had been strengthened by reading Imre's similar account, in *Proofs and Refutations*, of the judgment of conceptual change in mathematics. At times, I could only conclude that his rejection of anything associated with Ludwig Wittgenstein had been picked up by contagion, through his close association with Karl

Popper, and so was scarcely more than an historical curiosity – a late, irrelevant echo of Old Vienna:

> ...forgotten, far-off things
> and battles long ago.

For myself, having learned important philosophical lessons from both Wittgenstein and Popper, as well as from R. G. Collingwood, I could not see the two Viennese philosophers as locked in an irresoluble conflict.

All the same, this diagnosis was not wholly satisfactory. Evidently – as Imre saw things – there were genuine matters of principle over which Polanyi, Kuhn and I were united in error. We were all three more or less explicitly committed to 'elitism', 'historicism', 'sociologism' and 'authoritarianism'; and we all three failed to distinguish the current facts about the ('first world') physical activities and ('second world') mental judgments of working scientists clearly enough from the propositional relations in the 'third world', in terms of which those activities and judgments must ultimately be assessed.

How, then, did Imre come (I shall be asking here) to interpret this opposition – between the *activities and beliefs of scientists* and the *propositional relations of science* – in the particular way he did? What was the source of this point of view in his own philosophical development? And just how can we reconcile the things he said in his classic papers about science with the markedly more 'historicist' and 'elitist' position over mathematics that he advanced in *Proofs and Refutations*? If I could answer that question in a convincing manner, I could also put to rest the personal perplexities about Imre's opposition to my own *Human Understanding* and other related papers.

2. Consistency and Change in Lakatos' Development

The crucial issues that I shall be discussing concern this last relationship: viz., that between *Proofs and Refutations*, which was Lakatos' long initial monograph on the philosophy of mathematics, and the views on the philosophy of natural science and scientific methodology that he was putting forward in the mid and late 1960s. As we shall see, there are genuine parallels between his views on these two subjects – in places, his most recent views on natural science read like a transcription of his earlier views on mathematics – but there are also notable divergencies between them, particularly over the central standards of judgment.

For the purposes of my discussion, I shall divide Lakatos' arguments about methodology in science and mathematics into three historical phases, hoping to indicate just how far consistent, continuous views can, and cannot, be found in his writings, all the way from *Proofs and Refutations* to the latest papers, e.g., the UCLA paper (Nov. 1973) on Copernicus. The first phase embraces

(1) *Proofs and Refutations* (1963–64), which largely covered the same ground as Imre's Cambridge Ph.D. thesis (1961), and also the paper he wrote for the joint session of the Aristotelian Society and the Mind Association in 1962, on 'Infinite Regress and Foundations of Mathematics'.

In these early papers, Lakatos focused his attention on the methodology of conceptual change in mathematics. The 'Euclidean', 'empiricist' and 'inductivist' research programs with which he concerned himself, at this stage, were programs for the intellectual advancement of mathematics, and the people whose work came under his scrutiny were Cantor, Couturat, Hilbert and Brouwer. Galileo and Newton also came in for passing references (it is true) but only as mathematical physicists: it was the contemporary disputes between Gödel and Tarski, Gentzen, Stegmüller and post-Hilbertians, that he was most interested in adjudicating.

From 1965, we find Imre playing a different role. Beginning with the conference at Bedford College, London, in the summer of that year, we have

(2) a series of papers on the philosophy of natural science, delivered from 1965 to 1970, in which the focus had shifted from mathematics to physics and astronomy.

What was the reason for this shift? My own guess (which I shall explain below) is that it was Imre's response to the public controversy generated by Thomas Kuhn's theory of 'scientific revolutions': notably, to the direct confrontation between Kuhn and Popper at the Bedford College meeting. From this time on, Lakatos' methodology of 'scientific research programmes' was rapidly taking shape, with special reference to theoretical development in the physical sciences. This phase culminated in the revised form of Imre's own Bedford College paper, published in *Criticism and the Growth of Knowledge* as 'Falsification and the Methodology of Scientific Research Programmes' (1970). In these middle period papers, the terms Imre used to classify *scientific* research programs remained, significantly, the same quasi-logical terms that he had devised to classify alternative programs of *mathematical* discovery earlier: viz., 'inductivist', 'empiricist', 'falsificationist' etc. Aside from this shift of attention from mathematics to physics, the other chief novelty in these papers was a fresh hostility to 'historicism' in all its variants, and an emphasis on the timeless critical functions of reason and the 'third world', in mathematics and science alike. (Both features possibly reflect Lakatos' support for Popper, in his opposition to Kuhn's theory of 'paradigms', and the historical relativism to which Kuhn's earlier views so easily led.)

Finally, we have

(3) The papers of Imre's last two years, notably the Jerusalem paper and the UCLA Copernicus paper.

In these, we shall recognize the beginnings[1] of a further shift of emphasis. The motives for this second shift came from paying closer and more detailed attention to the *actual* intellectual strategies embodied in the changing theoretical research programs of physics and astronomy over the last three hundred years. We cannot adequately distinguish between the varied intellectual goals and ideals that led such physicists as Galileo and Newton, Maxwell and Einstein, to pursue their own chosen lines of thought, if we employ a *quasi-logical* terminology alone. The differences in intellectual strategy between them were not merely *formal* – one man playing the 'inductivist', another the 'falsificationist', a third the 'Euclidean', and so on – but *substantive*. The differences between their ideas and strategies were a consequence of differing *empirical* ideals of 'explanatory adequacy' and 'theoretical exhaustiveness'. So in these final papers, particularly those prepared in collaboration with Elie Zahar, we see Imre loosening up and adopting a broader and more substantive conception of the significant differences between rival 'research programmes'. (It was at this point that I saw a real chance of his moving to a position at which different 'research programmes' would correspond exactly to different 'intellectual strategies', in my sense of the phrase.)

Despite these successive shifts in the focus of Imre Lakatos' attention, however, certain things about his views remain unchanged throughout. Suppose we look at the text of *Proofs and Refutations* and his later writings, side by side. Suppose we compare, for instance, the final version of his paper on 'History of Science and its Rational Reconstructions', as delivered at Jerusalem in January 1971 and revised for publication in 1973. The opening words of the published paper are:

"Philosophy of science without history of science is empty; history of science without philosophy of science is blind." Taking its cue from this paraphrase of Kant's famous dictum, this paper intends to explain *how* the historiography of science should learn from the philosophy of science and *vice versa*.

Turning back to the introduction of *Proofs and Refutations*, we find the same idea, applied this time to the philosophy of mathematics:

Under the present dominance of formalism, one is tempted to paraphrase Kant: the history of mathematics, lacking the guidance of philosophy, has become *blind*, while the philosophy of mathematics, turning its back on the most intriguing phenomena in the history of mathematics, has become *empty*.

Similarly, the closing words of Lakatos' 1973 paper on the philosophy of science are an explicit quotation from his 1962 paper on 'Infinite Regress' in the philosophy of mathematics:

Let me finally remind the reader of my favorite – and by now well-worn – joke that history of science [mathematics] is frequently a caricature of its rational reconstructions; that rational reconstructions are frequently caricatures of actual history; and that some histories of science [mathematics] are caricatures both of actual history and of its rational reconstructions. This paper, I think, enables me to add: *Quod erat demonstrandum*.

In short: all the intellectual tasks that Imre Lakatos set himself from 1965 on in the philosophy of *science*, together with much of the terminology that he used to characterize *scientific* methodologies, simply re-applied, to the procedures of natural science, ideas developed initially for the discussion of *mathematical* methodologies and the philosophy of *mathematics*.

It is particularly intriguing to consider Lakatos' developing attitude towards Popper's problem of the 'demarcation criterion' and the standards of scientific judgment. During the middle period of his development ('Lakatos$_2$') he flirted with Popper's idea that philosophers have an obligation to lay down firm, permanent criteria for distinguishing 'science' from 'non-science', or 'good science' from 'bad science', *from outside* the actual experience of natural science: specifically, to insist in a properly *critical* way that the scientist must conform to certain 'rational' standards in the arguments that are the end-product of his work. But, in the final papers, he makes concessions to philosophers like Polanyi that are not easily reconciled with his earlier denunciations. For instance, in the 1973 revision of his Jerusalem paper, he explicitly rejects Popper's conclusion, that

there must be the constitutional authority of an *immutable* statute law (laid down in his demarcation criterion) to distinguish between good and bad science,

as being illegitimately *aprioristic*. On the contrary, Polanyi's rival position, viz. that "there must be – and can be – no statute law at all: only case law" turns out after all to have "a great deal of truth in it".

Until now all the "laws" proposed by the *apriorist* philosophers of science have turned out to be wrong in the light of the verdicts of the best scientists. Up to the present day it has been the scientific standards, as applied "instinctively" by the scientific *élite* in *particular* cases, which have constituted the main – although not the exclusive – yardstick of the philosopher's *universal* laws. But if so, methodological progress, at least as far as the most advanced sciences are concerned, still lags behind common scientific wisdom. It is not then *hubris* to demand that if, say, Newtonian or Einsteinian science turns out to have violated Bacon's, Carnap's or Popper's *apriori* rules of the game, the business of science should be started anew? I think it is.

In this final phase ('Lakatos$_3$') Imre's approach to the methodology of research programs was, therefore, becoming quite as *historicised* as Polanyi's or my own. Why, then, the indignation about our scandalous elitism, authoritarianism and so on? That is just the problem...

Ironically, even these final concessions to the authority of scientists' 'case law' represented merely a return to Imre's original position in regard to mathematics. At the climax of the dialog which forms the bulk of *Proofs and Refutations*, he insisted that a similar 'case law' has generated radical changes of intellectual strategy in the history of mathematics, also:

THETA: Come back to the point. You are unhappy about 'open', radical concept-stretching?

BETA: Yes. Nobody will accept this last brand as genuine refutation! I quite see that the mild concept-stretching trend of heuristic criticism that Pi uncovered is a most important vehicle of mathematical growth. But mathematicians will never accept this last, wild form of refutation!

TEACHER: You are wrong, Beta. They *did* accept it, and their acceptance was a turning point

> in the history of mathematics. *This revolution in mathematical criticism changed the concept of mathematical truth, changed the standards of mathematical proof, changed the patterns of mathematical growth*!...

So, from the outset, Lakatos was open to the idea that the concept of truth, the standards of proof, and the patterns of growth *in mathematics* must be analyzed and applied in ways that take account of their historical development; and also to the idea that the historical acceptance of changes in the ideas of 'truth', 'proof' and 'growth' by *working mathematicians* is itself a matter of crucial relevance for the *philosophy of mathematics.* If these positions are not expressions of that very 'historicism' and 'elitism' which Imre later objected to in other philosophers of science – we may be tempted to ask – what are they?

3. THE CONTENTS OF THE 'THIRD WORLD'

In the remainder of this paper, I shall suggest two possible reasons why Lakatos felt bound to draw so harsh a distinction between his own final position, on the one hand, and those of Michael Polanyi and myself, on the other. In this section of the paper, I shall raise some questions about the parallels – and lack of parallels – between the philosophy of mathematics and the philosophy of natural science: in particular, I shall argue that, because his first-hand experience was limited to the field of *mathematics*, Imre was mislead to oversimplifying the contents of that 'third world' in terms of which (for all good Popperians) the intellectual content, methods and products of *any* rational discipline must be appraised and criticized. Then, in the final section, I shall go on to indicate how such an oversimplification apparently led him to believe that all those positions in philosophy of science which attach central importance to the *praxis* of scientists are open to the charge of 'historical relativism', in the same way as (say) the first edition of Thomas Kuhn's *The Structure of Scientific Revolutions.* In return, I shall argue that an account of scientific praxis can perfectly well be given, which includes all the safeguards for 'rationality' supporters of the 'third-world' view can legitimately demand; which, as a result, clearly avoids the stigma of relativism; and which need face no objections or difficulties that do not arise with equal force about force about Imre's own final position.

To start with the comparison between mathematics and natural science: philosophers of science who have actually earned a living by working as natural scientists frequently find themselves ending up at cross-purposes with those colleagues who have come to the subject, rather, from mathematics or symbolic logic. I have discussed this phenomenon elsewhere: arguing (e.g.) that the general *philosophical* program of 'clarity through axiomatization', popular among empiricist philosophers of science in the 1920s and 30s, derived its charm and plausibility from confusing two very different things: viz., Hilbert's pursuit of axiomatization as an intrinsic goal or 'end' of *mathematics*, and Hertz's more utilitarian resort to axiomatization as a 'means' for overcoming theoretical confusions in mechanics, regarded as a branch of *physics*. The example of Frege's *Foundations of Arithmetic* likewise (in my view) led philosophers during the inter-war years to demand greater idealization and 'timelessness', in their analysis of science, than the actual nature of the natural sciences permits. Nor, despite their public repudiation of positivism and all its works, has either Popper or Lakatos succeeded in breaking away entirely from this feature of the Viennese inheritance. In Imre's case particularly, his original experience as a mathematician seems, instead, to have distracted him from the necessity for such a break.

There are two aspects of pure mathematics, however, which it shares with any authentic *natural* science only to a very questionable extent.

(1) The intellectual content of a systematic theory in pure mathematics can be specified to a high degree of approximation (even perhaps, in some cases, exhaustively) by the system of propositions in which that content is expressed. From the standpoint of mathematics, that theoretical system simply *is* the system of propositions, together with their connections. Matters of praxis – e.g., the practical procedures by which we are to identify or generate specific physical exemplars of the entities referred to within the system, whether dimensionless points, equilateral triangles, uniform velocities, or whatever – are 'external to' the system. Such matters of praxis, that is to say, have no direct bearing on the validity of the mathematical system in question, considered simply as a part of 'mathematics'.

(2) In some branches of mathematics (if not all) a further idealization is also possible: viz., we can imagine situations in which we are entitled to think of the *current* form of a mathematical system as being also its *final and definitive* form. When Frege developed his 'logical' analysis of

arithmetic, for example, he claimed just such a finality for it. At long last (he claimed) it was possible for philosophers of mathematics to "strip away" the "accretions" that had for so long "veiled arithmatical concepts, in their pure form, from the eye of the mind". This Platonizing move had the effect of cutting arithmetic off from its history. There was no longer any room to consider arithmetical conceptions *à la* Frege as historical products, which might provisionally claim to be *better* than their rivals-for-the-time-being. The only question Frege allowed was, "Is this analysis right?". Either he had *correctly* described the 'pure form' of arithmetical concepts – considered as timeless inhabitants of the 'third world' – or he was simply *in error*. Rather than see his account labeled as a temporary improvement which would presumably be superseded by some further conceptual shift, as mathematics subsequently developed, he would play for the highest and most 'absolute' stakes only.

Philosophers accustomed to working in formal logic and pure mathematics may quite naturally suppose, as a result, that the entities and relationships which it is their business to judge in 'rational' terms – which form the population of Popper's (and Plato's?) non-empirical 'third world' – comprise propositions, their constituent terms, and logical relations holding between them.[2] Yet it remains as dubious as ever whether that supposition is well-founded. Even in those natural sciences whose theories can be cast in a mathematical form, the empirical content of the sciences in question extends far beyond the bounds of those mathematical theories. For instance, the manner in which actual empirical exemplars of the entities discussed in any such theory are to be identified or generated is now – in direct contrast to the situation in pure mathematics – a problem 'internal to' the corresponding science: in fact, a problem on which the validity and acceptability of the resulting scientific theory may intimately and directly depend. (The rational status of contemporary physics hangs on the demonstrable existence of actual 'electrons', in a way that the rational status of geometry does not hang on the empirical discoverability of 'real-life dimensionless points'.) In the case of any empirical natural science, furthermore, it is a much less plausible historical hypothesis to suppose that its *current* form is also, at the same time, its *final and definitive* form. For example: even in kinematics, formulae and relationships that appeared capable of being 'known *a priori*' throughout the seventeenth and eighteenth centuries have been modified as a result of the development of relativity theory. Similarly, the only way of turning 'rational mechanics' into a branch of pure mathematics has been to divorce it from all genuinely empirical connections.

These two differences between mathematics and natural science have serious implications for the character and contents of the so-called 'third world' which plays so important a part in the arguments of Karl Popper and Imre Lakatos. If the intellectual content of any genuine natural science embraces, after all, not only *propositions* but *praxis* – not only its theoretical statements, but also practical procedures for their empirical application – then neither scientists nor philosophers can afford to confine their 'rational' or 'critical' attention to a *formal idealization* of its theories, i.e. to a representation of those theories as comprising merely systems of propositions and inferences structured in logico-mathematical form. For some philosophers of science, this is a disconcerting thought. They are so used to treating 'rational criticism' as a matter of 'formal validity', 'logical rigor', and the like, that the introduction of an historically-changing body of praxis onto the stage strikes them as a dangerous concession to 'irrationalism'; and, when Michael Polanyi claims furthermore that a substantial part of this praxis is known 'tacitly' rather than explicitly, their suspicions are only strengthened.

Still: it is time to challenge those suspicions, and to argue in reply that they rest on misunderstandings. The entire body of what is 'known' in a natural science is not captured in its theoretical terms and propositions alone; the application procedures for giving those theoretical ideas an empirical relevance, for instance, represent an indispensable constituent of the science; and, however far these procedures may tend to be left 'tacit' in actual scientific practice, this fact does not exempt them from being brought into the open and exposed to rational criticism.

We can, indeed, go over to the counter-attack. Far from more historically-minded philosophers of science failing to appreciate the importance of rational criticism or delivering themselves tied and bound to the relativists, many of them are perfectly aware of that importance, and go to great lengths to respect it. What divides (say) Polanyi and myself from Popper and Lakatos is our insistence that 'rational criticism' must be applied, not just to the scientist's *words*, but also to his *works* – not just to his theoretical propositions, but also to his empirical praxis – and that the canons of rational criticism accordingly involve, not just propositional 'truth' and inferential 'validity', but scientific adequacies and inadequacies of other kinds also.

So, if we are to retain the Popperian image of the 'third world', we must find some way of enlarging its scope. Since the intellectual content of a natural science embraces *both* its (linguistic) terms and propositions and *also* the (non-linguistic) procedures by which those ideas are given an empirical relevance and application, the 'third-world' model of science must make room for the essential praxis of natural science alongside its propositions and inferences, terms and 'truths'.

Lakatos consistently refused to make this concession. Given his mathematical temperament, he dismissed all allusions to praxis as an irrational surrender to empirical sociology or psychology. In this way, he ended up by caricaturing his opponents' views, and missing the point of their main arguments. Michael Polanyi can defend himself well enough without my help, so I shall here speak simply for myself.

The entire account of 'conceptual change' in science in volume I of *Human Understanding* was built around a distinction having precisely the 'critical' implications that Popper insists on, when he separates the 'third world' of rational criticism from the first and second (physical and mental) worlds of empirical fact: namely, the distinction between 'disciplines' and 'professions'. Everything about the sciences, considered as 'disciplines', is *directly* open to rational criticism, including those parts of their intellectual content that show themselves more in praxis than in propositions. By contrast, the institutional transactions that embody the activities of the sciences, considered as 'professions', are open to rational criticism only *indirectly*, by enquiring how far they serve the intellectual needs of the discipline they are supposed to promote. In general terms, accordingly, there is no difficulty about distinguishing the *praxis* of a science from its *politics*. Questions of praxis remain intellectual, or disciplinary matters: questions of politics are always institutional, or professional matters.

Far from my argument misguidedly *equating* these two things, I took good care to underline the differences between them at every appropriate point. (The book even contains separate chapters, dealing separately with problems about 'disciplines' and 'professions' respectively.) Far from yielding intrinsic, unchallengable authority to any particular master-scientist or an institution, likewise, I took good care to demonstrate that the activities and judgments of human scientists, whether as individuals or in groups, are always open to rational review. So I felt some incredulity – not to say, vexation – when I found Imre Lakatos' unfinished review of *Human Understanding* ignoring that central distinction and dismissing my entire position as one more instance of elitist authoritarianism.[3]

Why was Imre Lakatos unable to recognize my discussion of the relationship between 'disciplines' (with their intellectual content) and 'professions' (with their institutional activities) for what it is: viz., the basis for a functional analysis of 'rational criticism' in science? According to my first hypothesis, anyone who enlarged the 'intellectual content' of a science to embrace praxis as well as propositions – and so enlarge the sphere of 'rational criticism' to embrace something more than the analysis of propositional relationships – was guilty in Imre's eyes of some vicious kind of psychologism or sociologism. Yet this was just a mathematician's misunderstanding. Any analysis of rational criticism in natural science which is to do justice to the new elements that become relevant, the moment we leave the philosophy of mathematics for the philosophy of natural science proper, must acknowledge the new elements of praxis and discuss the considerations affecting their rational assessment. In giving the claims of rational criticism the respect and attention they deserve, therefore, we must not restrict their scope and application to matters of propositional logic, but admit into the 'third world' *all* those elements that can be critically judged against rational standards. *If the result is to transform the 'third world' from being a formal world of Being, comprising only propositions and propositional relations, into a substantial World of Becoming, comprising both linguistic/symbolic and non-linguistic/practical elements, so be it!*

There is ample corroboration of this first hypothesis in Imre Lakatos' own writings. His central artillery barrage against *Human Understanding*, for instance, begins with a paragraph that states my position very nearly correctly – but with some significant distortions:

For Toulmin the major mistake of most philosophers of science has been to concentrate on (third-world) questions of 'logicality' about propositions and their provability confirmability, probability and falsifiability, rather than questions of 'rationality' concerning skills and social activities which Toulmin calls 'concepts', 'conceptual populations', 'disciplines', and their cash value in terms of profits and losses.

The small but tell-tale misunderstandings evident in this passage show, firstly, in Imre's use of the phrases, 'social activities' and 'cash value', in place of my own terms, 'procedures' and 'fruitfulness'; secondly, in his explicit (though parenthetical) equation of 'third-world questions' with 'questions about propositions and their provability...'. Having sharply divided 'propositions and their provability' from 'procedures and their fruitfulness' in this way, Imre simply took it for granted that procedures (even *rational* procedures) have no place in the 'third world'. So my own insistence on giving the non-linguistic praxis of science as much attention as its linguistically-formulated propositions seems to have represented, for him, an affront to the proper *logical* claims of rationality and the 'third world'.

With this misunderstanding under his belt, Imre did not hesitate to denounce me as an *anti-rationalist*: as being "for pragmatism, for elitism, for authoritarianism, for historicism and for psychologism and sociologism". Yet, in doing so, he was begging the crucial philosophical issue: viz., *whether procedures and their fruitfulness may not be entitled to a place within the sphere of rational criticism, quite as much as propositions and their provability*. Imre clearly thought that 'procedures' were *not* so entitled, whereas I am equally clear that they *are*. From my own point of view, indeed, it is just as much a species of 'rational criticism' for us to pay attention to (e.g.) the intellectual fruitfulness of explanatory procedures in science as it is for us to scrutinize the inferential steps in a formal scientific argument. Far from the concern with praxis implying a species of 'anti-rationalism' in the philosophy of science, it represents a necessary middle way, by which we can properly defend the claims of 'rationality' against *both* the narrowness of formal logicians and mathematicians, from which Lakatos was not finally exempt, *and* the exaggerations of relativist historians, such as the early Thomas Kuhn.

4. Two variants of historicism

I have a second, complementary hypothesis, also, about the sources of Imre Lakatos' hostility towards philosophers of science who took the history and practice of science 'too seriously'. This second guess connects with his complaint that we were all of us committed to a vicious form of historicism. As I shall show, ambiguities built into Imre's use of the term, 'historicism', concealed the crucial issues at this point. (Parallel arguments could be constructed to rebut his other complaints, of 'psychologism', 'sociologism' and the rest.) Instead of Lakatos presenting a single clear definition of 'historicism', to which Kuhn, Polanyi and Toulmin all unambiguously adhere, and from which he can equally unambiguously dissociate himself, we must at the very outset differentiate in his arguments at least *two* distinct 'historicist' positions, having quite different implications for the rational analysis of scientific methodology. Once this distinction is made, we can see

(1) that the position defended in the first edition of Kuhn's *Structure of Scientific Revolutions* is 'historicist', in a much stronger and more objectionable sense than anything which Michael Polanyi or I would ever seek to argue;

(2) that, furthermore, in the only relevant sense of the term, Imre Lakatos' final position is just as 'historicist' as Polanyi's or my own.

Having overlooked or ignored this distinction, Imre assumed that any valid argument against Kuhn could at once be pressed into service against Polanyi and Toulmin also. Why did he do this? The underlying reasons apparently lead us back once again to the previous point: viz. Imre's mathematical preoccupation with 'propositions and their provability', and his resultant refusal to admit 'procedures and their fruitfulness' into the sphere of rational criticism on equal terms.

The *strong* variant of historicism can be illustrated by certain familiar features of Kuhn's initial position. Notoriously, Kuhn originally argued that natural scientists who operate from within the framework of different 'paradigms' will have no common ground for comparing the rational and intellectual merits of their respective views. During its period of scientific dominance, any particular 'paradigm' imposes corresponding, though temporary, canons of rational judgment and criticism on those scientists who work under its authority: for those who work beyond its authority, by contrast, its canons have no particular relevance or cogency. It is open to question, of course, whether Kuhn ever really intended to take up the position that he actually expressed in the first edition of his book. As Lakatos himself concedes,

Kuhn seems to be in two minds about objective scientific progress. I have no doubt that, being a devoted scholar and scientist, he *personally* detests relativism. But his *theory* can either be interpreted as denying scientific progress and recognizing only scientific change; or as recognizing scientific progress but as 'progress' marked solely by the march of actual history.

It is this last view – viz. that "scientific 'progress' is marked solely by the march of actual history" – which Imre rightly condemns as *viciously* historicist; though (as he well knew) my own arguments about conceptual change begin by rejecting precisely this form of 'historical relativism'.

The central question of this present paper can therefore be restated. Knowing very well that I shared his opposition to the *historical relativism* of Kuhn's position, why did Imre wilfully lump Polanyi's and my positions with Kuhn's, and argue that we could not really escape from *historicism*, however hard we might try? Compared with this question, Lakatos' other accusations of 'elitism' and the rest are so much secondary rhetoric.

Anyone who accepts a *strong* historicist position will also, quite naturally, accept equally strong variants of the other positions. From this point of view, for instance, the particular scientists and institutions whose views are authoritative during the dominance of a particular 'paradigm' will acquire a correspondingly *absolute* authority in the judgment of scientific issues; and such a conclusion will indeed be open to criticism, as 'elitist', 'authoritarian', etc., etc. (So, likewise, for 'psychologism' and 'sociologism': the reader can easily re-apply the same argument to those other terms.) The alternative, *weaker* form of 'historicism', by contrast, involves no such surrender of authority to any single scientist, group or epoch. All that it emphasizes is the fact that, in natural science as elsewhere, the criteria of rational judgment have themselves undergone an historical development; so that comparative judgments of rational adequacy, within different sciences, at different stages in their evolution, will be appropriate and relevant only if they pay attention to this *criteriological* history.

So far as this issue goes, the only 'historicism' to be found in my *Human Understanding* is the same as that which Imre presented triumphantly, as the final, fundamental insight about mathematics in *Proofs and Refutations*: viz., the insight that "a turning point in the history of mathematics" consists primarily in a "revolution in mathematical criticism", which has the effect of changing "the concept of mathematical truth, the standards of mathematical proof, and the patterns of mathematical growth".

In this sense, 'Lakatos$_1$' *itself* presents an 'historicist' position in the philosophy of mathematics: in respect of mathematical methodology, that is to say, the views put forward in *Proofs and Refutations* about mathematical criticism/truth/proof/conceptual growth were tied to the historical development of mathematics quite as much as my own views about scientific criticism etc. are tied to the historical development of natural science.

If anything, indeed, the historicism of *Proofs and Refutations* is stronger than my own. The closing pages of Imre's argument could well be read as characterizing mathematical 'revolutions' in terms quite close to Kuhn's. Without reading between the lines of Imre's position, and teasing out its detailed implications, we might be tempted to accuse his philosophy of mathematics of precisely those heresies that he himself finds in Kuhn's philosophy of science. (Does he not say, "Mathematicians *did* accept this revolution in mathematical criticism, and their acceptance was a turning point in the history of mathematics"? Does this not suggest that their *acceptance* was all that mattered? And how much more elitist and authoritarian can one get?) Yet such an accusation would be unjust. A more careful reading of Imre's text makes it clear that even these "revolutions in mathematical criticism" leave open the possibility of rational judgment, depending on whether they involve rational or irrational 'concept-stretching'. Such mathematical 'revolutions' accordingly take place *for reasons of their own kind.* And the central question for the relevant part of *Human Understanding* is, precisely, the corresponding question about 'turning points' in scientific change: viz., *what kinds of reasons* are appropriate, when changes in intellectual strategy lead to changes in the criteria of scientific criticism, and so to consequential modifications in "the concept of scientific truth, the standards of scientific proof and the patterns of scientific growth".

During Lakatos' middle period – what I have called 'Lakatos$_2$' – he was slow to extend to natural science the fully historicist analysis he had already put forward in the case of mathematics. Why was this? Why did he hesitate to re-apply the conclusions of *Proofs and Refutations* to natural science in all their richness, and so justify a correspondingly historicist analysis of the changing criteria of rational criticism in science?[4] I can establish no conclusive answer to this question from the actual text of Imre's early papers on the philosophy of science, so I am driven back onto a more speculative hypothesis. This is: that the initial public reception and intellectual impact of *The Structure of Scientific Revolutions* – notably, of the implicitly 'irrationalist' variant of historicism put forward in that first edition – set Imre Lakatos back on his heels. For some years (I recall) Imre was very ambivalent about *Proofs and Refutations*, and came quite close to disowning it. Those of us who admired the work, and exhorted Imre to reprint the original series of articles as a separate monograph, were puzzled by his reluctance. And, if we put Lakatos' views down alongside Kuhn's initial theory and note their superficial similarities, we can see in retrospect why he might have been anxious. What if his own views about the effect of 'mathematical revolutions' on the critical concepts of truth, proof and validity were to be read as having the same irrationalist implications as Kuhn's views about the effects of 'scientific revolutions'? Faced with this risk, it is easy to see why he may have felt the need for a more defensible position, in which his theory of 'scientific rationality' would be clearly exempt from any possible charges of historicism or relativism. In this respect, Popper's views about the 'third world', and about the 'demarcation criterion' for distinguishing between good and bad science, evidently provided a more secure line of defense.

In course of time, however, Imre recovered his courage, and ventured out again in his original direction. So, in 'Lakatos$_3$', we find him rejecting Popper's *a priori* 'demarcation criterion' as over-rigid, and conceding to the methodology of natural science something of that historical *relativity* (as contrasted with *relativism*) that he had earlier acknowledged in the case of mathematical methodology. At this final stage, for instance, he conceded that Polanyi's thesis, about the importance of 'case law' in the exercise of scientific judgment, "has a great deal of truth in it". And, despite all his additional glosses and qualifications about the need to balance up "the wisdom of the scientific jury and its case law" against the analytical clarity of "the philosopher's statute law", he ended up by unambiguously rejecting the views of

> those philosophers of science who have taken it for granted that general scientific standards are immutable and reason can recognize them *a priori*.

In this respect, at least, Imre's "criteria of scientific judgment" were quite as open to historical modification and reconsideration, in the light of philosophical criticism and scientific experience, as Michael Polanyi or I could ever require. Was it Lakatos' professional association with Elie Zahar that finally helped him back to this position? Or did he get there entirely under his own steam? That is another question. Either way – as I said at the UCLA symposium – *it was a pleasure to welcome Imre back to the real problems.*

What did I mean by this? Let me explain the point very briefly. Once Imre was firmly committed to 'Lakatos$_3$' – once he had admitted an element of 'case law' and 'historical relativity' into the criteria of scientific judgment – all his glosses and qualifications could not postpone forever certain fundamental problems that arise for anyone who accepts such an historical relativity. What, for instance, about "the problem of the long run"? What should we say about the possibility that our own current scientific judgments, and even our own current *criteria* of judgment, will be reconsidered and modified, at some later time, for reasons arising from future intellectual strategies that we cannot at present foresee? I shall set aside Imre's heavy irony about my 'Hegelianism', and his appeal to Maynard Keynes' well-known comment, that "in the long run we are all dead": despite Imre's refusal to accept the 'long run' problem as legitimate, in his review of *Human Understanding*, the argument he puts forward does not get him off the hook. For we can ask him, too:

> How are we to view potential conflicts in rational criticism between the most carefully-considered scientific views and criteria, reflecting the best state of scientific judgment *at the present stage in science*, and the retrospective views of scientists in centuries to come, whose judgments draw on the practical experience and theoretical reconsideration of the intervening years?

In particular: if we face the need for a strategic reappraisal in our methodology, how can we rationally justify laying our bets as we do, or anticipate the judgments of future scientists on the comparative fruitfulness of the strategic alternatives (i.e. alternative research programs) facing us today? Imre may not have admitted the legitimacy of that question; yet it arises for 'Lakatos$_3$' just as much as it does for my own *Human Understanding*.

One last question: How could Imre Lakatos have overlooked this implication of his mature views about scientific methodology? At this point (I believe) we must revert to my earlier hypothesis: viz. that Lakatos – like Karl Popper – accepted only a restricted population to his 'third world'. Anyone who conceives of this 'third world' as comprising propositions and their formal interrelations alone can think of it as somehow *atemporal*: as exempt from historical change and outside the empirical flux. From this atemporal standpoint, *philosophical* criticism just is *logical* criticism, concerned with the "provability, confirmability, probability and/or falsifiability" of propositions, and with the 'validity' of the inferences connecting them. But, once procedures and other elements of praxis are allowed into the 'third world', its *temporal* or historical character can no longer be concealed. For the 'long run' problem really confronts those who would confine the sphere of 'third-world questions' to logical or propositional issues just as severely as it does those who acknowledge 'rational procedures' also as legitimate objects of scientific appraisal. Even if we consider only the propositional content of a current science, together with its internal criteria of validity, proof and relevance, the resulting account can give us only *a representation of 'the third world' as currently conceived.* Despite the logico-formal or mathematical character of its internal relationships, *the totality of this 'world' will quite evidently be an historical existent as of A.D. 1975*, or whenever it may be. However much the propositions and inferences involved appear well-founded and 'in good rational standing' today, they will be in many ways unlike those comprised in the 'third world' which future scientists in (say) A.D. 2175 will be in a position to define. So, once historical relativity and 'case law' have entered our account of scientific methodology, the problem of accounting for comparative historical judgments of *rationality* becomes an inescapable one; and the pretense that the 'third world' is a world of *logicality* alone merely postpones the moment at which we confront that fact.

Need I say what regret I feel, that Imre's premature departure from the scene should have deprived me of the chance to take up these points with him once again, in person? As a respectful and friendly adversary, I shall continue to miss in almost equal measure the seriousness of his mind, and the pleasure of his invective! And I hope he would not have found the present 'rational reconstruction' of the history of his philosophy of science too gross a 'caricature', either of his actual development, or of its underlying rational justification.

University of Chicago

NOTES

[1] I say, the *beginnings* of such a shift, because I am naturally tempted to overemphasize any move that seemingly brought Imre's final position close to my own. In commenting on the Copernicus paper at Los Angeles, I teased him by suggesting that, just as Imre had fathered on Karl Popper a position ('$Popper_3$') identical with his own middle-period position, ('$Lakatos_2$'), so also the new position towards which he was moving ('$Lakatos_3$') would eventually turn out to be the same as '$Toulmin_2$'. As we shall see, however, Imre himself would probably have had reason for resisting the final shift to '$Lakatos_3$' as Popper has resisted the corresponding move to '$Popper_3$'.

[2] Given this starting point, for instance, the intellectual dream of a 'unified science', in which all our natural knowledge would be expressed in a single, final axiomatic system, is understandable.

[3] So little did he grasp the rational significance of separating disciplines from professions that he even succeeded in standing on its head a passing allusion I made to 'the oldest profession':

> Toulmin's first imaginary example of a 'compact discipline' is a Royal College of Prostitution (p. 405). But this College narrowly escapes from being branded as 'scientific', since its ideal is not an explanatory one.

[4] Ironically, reading *Proofs and Refutations* helped give me the courage of my convictions, at the stage when I was myself elaborating the views later published in *Human Understanding*.

JOHN TUCKER

AGAINST SOME METHODS

I want to argue against some methods but not all and first give some reasons for not wanting to argue against all of them. To claim to know that there are no methods is surely to claim more than can be cogently claimed since it would have to be based upon some argument which would itself presuppose the acceptance of some method of arguing. There cannot be a cogent argument against all argument. The search for method is, after all, a search for generalisations about our procedures and there is already an abundant supply of such generalisations over restricted ranges of procedures. My problem here is that of explaining why in certain cases the attempt to generalise breaks down. The area that I have in mind in which unrestricted generalisations are tried out is conventionally called Philosophy and my question might be put as: Under what conditions do philosophical generalisations break down? I am looking for an explanation of these failures.

With this question in mind I discuss two methodological proposals neither of which can be incorporated into the relevant body of genuine enquiry with which they are associated. These are: Platonism-in-Mathematics and the Mind-brain Identity thesis. In both cases I show that although the proposal is put forward as an overall precept of method, in neither case can the precept actually be followed. Both masquerade as methods yet neither can be applied. Both are instances of the same sort of breakdown, which is why I have selected them for common treatment. Both are intended to apply to an entire field of enquiry and both turn out to be sterile. Both have plausible aspects which recommend them to practitioners since both seem to give a positive directive for research, a directive which, at one stroke, removes a host of difficulties. Yet all of this is illusory since both are merely verbal readjustments which give no more than an illusory satisfaction. Now one of the deceptive things about these incoherent proposals is that they are closely associated with perfectly genuine disciplines, and this is how they gain their plausibility. But if Platonism is an empty verbal move then clearly it cannot

R. S. Cohen et al. (eds.), Essays in Memory of Imre Lakatos, 677–689. *All Rights Reserved*
Copyright © 1976 by D. Reidel Publishing Company, Dordrecht-Holland

be genuinely incorporated into Mathematics, and likewise, if the Identity thesis is empty verbiage it too cannot be geared in an effective way to neurophysiology. What is to be shown, then, is that both are indeed vacuous in this way. I argue that both of these alleged methods are vacuous in the same way, since in both there simply is no identifiable procedure at just the crucial place where, according to the views themselves, there ought to be one. Thus the Platonist needs a procedure for coming to know when a Platonic entity has been identified, and yet, I argue, no such procedure can be provided. It is not just that by an oversight that it has not yet been provided, it is that it cannot possibly be provided given the Platonist's own view. Likewise, the Identity theorist needs a procedure, not just an arbitrary fiat, in order to establish the identity of those mental states and brain processes which he wishes to identify, and here again I argue that no such procedure is possible. What we are left with is the reiteration of the word 'identify', unbacked by any procedure for carrying out the identification. Both of these views, in fact, proceed by means of an appeal to an arbitrary fiat in lieu of a specific procedure. In neither case is any real work done.

1. Why Platonism Cannot Be Incorporated into Mathematics

I have selected Platonism-in-Mathematics as an example of a methodological proposal which cannot be carried out. First I explain why it is desirable to confine the discussion to Platonism-in-Mathematics rather than to take on the entire issue of Platonism in general. The advantage of restricting the argument in this way is that we avoid a lengthy historical excursus and have to deal only with a contemporary use of the term 'Platonism' as applied to Mathematics. It is a fact that eminent practitioners such as Cantor and Gödel have been both mathematically fertile and explicit Platonists and this looks at first sight like evidence in favour of Platonism-in-Mathematics. Again, it is Cantor's Platonism-in-Mathematics that Skolem conceived of himself as refuting by mathematical means, and so it is inevitably with Platonism in this specially restricted sense that I am concerned. Skolem explained that the 'philosophical Platonism' with which he disagreed was the view that mathematical objects exist independently of the activities of mathematicians and cited the Löwenheim-Skolem theorem as mathematical evidence against this view[1]. Now there are two features of this disagreement between Skolem and the Platonist which have to be accounted for:

(i) the apparent fertility of Platonism-in-Mathematics
(ii) Skolem's illusion that he had refuted Platonism by mathematical means.

Both seem to be incompatible with the view that Platonism is incoherent. In the first place, if it is incoherent it cannot also be genuinely fertile, and in fact can have no real effects within Mathematics at all. It simply relies parasitically on the fertility of the genuine Mathematics with which it is associated and this is how the illusion that it is fertile arises. Again, if it really is an incoherence which is external to Mathematics then it cannot be refuted by internal mathematical means. In fact it is quite characteristic of the Platonist's manoeuvres that they are mathematically impregnable, consisting as they do in the reiteration, always after the event of course, that there is a Platonic entity corresponding to whatever has been provably asserted. By always adding this vacuous assertion to whatever has been achieved mathematically, the Platonist provides his view with a self-guaranteeing immunity. Now since Platonism is not falsifiable by mathematical means, then a fortiori, it is not falsifiable by the Löwenheim-Skolem theorem. Morever, Skolem cannot really have learned that Cantor's Platonism is false in this way since what he really learned from the theorem was that indenumerability is relative to explicity adopted rules. If and only if the appropriate rules are adopted do we get indenumerability, which is just what one would expect on the view that Mathematics is an essentially intra-linguistic body of procedures. It is also what one would expect from Cantor's diagonal procedure, which, when stripped of its Platonist trappings, turns out to be an intra-linguistic device for excluding an element of an infinite set from the rest of the elements by means of a systematic rule of exclusion. Which element, if any, is excluded depends entirely upon whether we adopt a suitable rule of exclusion. Evidently then, this entire procedure, from which the notion of indenumerability arises, is relative to the way in which the rules are set up, a conclusion which follows from an examination of Cantor's diagonal procedure alone, without appeal to anything more complex.

Skolem rightly pointed out that Cantor needed an absolute indenumerability which the Löwenheim-Skolem result shows to be unavailable and from this he drew the conclusion that he had refuted Cantor's 'philosophical Platonism'. Since, however, he could not possibly have done so by mathematical means, the question arises of precisely what he had achieved. In order to give an account of what Skolem had in fact done, I distinguish between

(i) Cantor's Platonist assertions proper

and

(ii) Cantor's assertions within naive set theory.

Skolem had shown something which affects naive set theory as such, but which does not affect Platonism as such. Skolem in fact showed that Cantor's treatment of indenumerability in naive terms is inadequate and needs to be supplemented by formalisation in terms of explicit rules which make the rule-relative character of indenumerability plain. But none of this has anything to do with Platonic entities. Naive indenumerability is superseded by relativised indenumerability without any mention of such entities. We can talk in the idiom of naive set theory without mentioning Platonic objects, and when we encounter the Löwenheim-Skolem theorem we are forced by it to modify our naive talk, but not forced to deny that there are Platonic objects. We were not mentioning them before and so we are not mentioning them now. In doing naive set theory we neither affirmed nor denied anything about them, and since we were not talking about them in the first place our modified way of talking cannot include the retraction of assertions which we never made. It follows that the Löwenheim-Skolem theorem cannot refute Platonism.

1.1. *Why Platonism-in-Mathematics is not a Genuine Method at all*

I now argue that Platonism-in-Mathematics is an incoherent proposal which cannot in the nature of the case be incorporated into Mathematics. My aim is to prise off this incoherent excrescence leaving the coherent body of Mathematics intact. What has to be removed is the requirement that mathematical objects exist independently of the activities of mathematicians, since this and this alone is the source of incoherence. This requirement is incoherent because no instructions which lead to the conclusion that mathematical objects exist in the required sense are possible.

Or, what comes to the same thing, there is no procedure for identifying the Platonist's alleged objects. We cannot identify them solely by an appeal to mathematical proof, since this is to beg the question. We cannot deal with the Platonist mathematically because he insists on making moves which render his view impervious to counterexamples. We can attack him only by getting at the extra-mathematical sources of his view. The ultimate objection is that there simply is no way provided for carrying out the leap from what is proved to the Platonic entity. Merely to assert that Platonism is the case is not to provide that procedure, but only to promise it. The Platonist never fulfills this promise, since the only 'procedure' that he ever offers just is the repetition of his original assertion of his Platonism.

Now the source of the Platonist's view is not to be found within Mathematics, since it is essentially a view about meaning which comes from outside of Mathematics. This is why Platonism-in-Mathematic is not a genuine methodological proposal but rather a view about meaning which, as a separate transaction, is added to Mathematics by way of an explanation of how it is meaningful. But this is not in general how Platonists see it. They see it rather as an overall methodology which is in direct conflict with rival methodologies such as Constructivism. Nevertheless, I think, a case can be made for the view that everyone concerned with the so-called area of 'Foundations of Mathematics' is in one way or another concerned with a view about meaning which he brings to Mathematics from elsewhere. The different views about meaning held by Frege, Russell, Hilbert, Brouwer, determined the foundational research programme which they adopted. Whether this concern with meaning is misguided or not is not the issue here. All that I am pointing out for the moment is that there has been a concern with meaning in the Philosophy of Mathematics and the derived area of so called 'Foundations' from the very first. And the very first candidate was Platonism.

1.2. *Why the Platonist's Parallel between Mathematical and Physical Objects Cannot Hold*

Certainly one of the deep-seated causes of Platonism-in-Mathematics, in distinction to reasons for it, is the temptation to think of mathematical objects as a sort of object, and to speak of them picturesquely in language taken from ordinary talk about material objects. And while no doubt as picturesque talk it is psychologically gripping it can never be more than merely picturesque since when we come to think about what is to be said about mathematical objects literally, instead of picturesquely, we find that the parallel breaks down. Nevertheless the attempt to draw the parallel, not as a merely picturesque illustration, but as a literally intended assertion is to be found in Gödel's account of his own Platonism-in-Mathematics [2]. But any attempt to draw the parallel only pushes the difficulties of explaining just what Platonism involves into another area, that of the philosophy of perception, which is at least equally involved and obscure. We can hardly expect to gain a greater understanding of Platonism from a comparison with something else which is itself ill understood. This is borne out by the account which Gödel gives, since in his account he assumes a representative theory of perception and draws the parallel with Platonism on that basis. But the representative theory of perception is itself a prime example of an incoherent view, the form of which consists in sawing off the branch on which we sit. Naturally, to establish this would call for a whole work on perception, but without waiting for this full account, we can at once point out the lack of parallel between the two key expressions in the argument, namely 'sensory experience' and 'mathematical experience'. For these are dealt with as comparable and parallel expressions, which they are not. The expression 'sensory experience' refers to a range of experience in one sense only of the term 'experience'. This is a sense in which we might, for example, draw a contrast with cognitive experience, and here we are making use of the term 'experience' in what might be called its phenomenal sense. Now this phenomenal sense is readily distinguishable from the sense of 'experience' in such expressions as 'scientific experience' or 'angling experience' since in these and similar expressions no phenomenal range of experiences *sui generis* is intended. The scientist and the angler do not have a special sort of phenomenal experience, they have the usual sort organised in the way appropriate to science, angling or whatever activity is engaged in. It is in this sense of 'experience' and not in the phenomenal sense what we speak of 'mathematical experience'. No one would seriously suppose that in the context of mathematics we meant a special range of phenomenal experiences which only adepts ever encounter. What the expression 'mathematical experience' really amounts to is that it is a way of talking about the activities of mathematicians. It is not a phenomenal expression at all. For this reason, then, the parallel between 'sensory experience' and 'mathematical experience' fails to hold and with it the parallel drawn by Gödel between physical objects and Platonic objects.

1.3. *Platonism and Objectivity*

The point of Platonism-in-Mathematics is to account for the objectivity of Mathematics and this cannot be shown via the obscurities of an attempted comparison with perception. What is called for instead is an intelligible account which must, I think, proceed on the following lines. Mathematics is an intra-linguistic activity and as such consists essentially of systems of rules, which are exemplified in entirely public performances of suitable organisms which are themselves publicly identified. It is because it is an essentially public system of rule-generating and rule-following procedures that Mathematics is objective already and so there is no call for the invention of dubious extra-linguistic entities in order to make it so. Now a Formalist-in-Mathematics will point out that this is what he has been saying all along, but nevertheless the view that I am sketching is more than just a rehash of traditional Formalism since I deny that it follows from the exclusive concern with rules that Mathematics is meaningless, and it is this last requirement that has always been the distinctive characteristic of the Formalist. Nor am I entirely convinced that we need to appeal to the nebulous notion of meaning at all here, but if we are to try to locate it then we surely must do so within our rule-generating and rule-following activities. But this is too big an issue to be pursued now.

2. Why the Mind-Brain Identity Thesis Cannot Be Incorporated into Neurophysiology

2.1. *The Cartesian Source of the Incoherence of the Mind-Brain Identity Thesis*

The fallacy upon which the drive towards identity is based is the fear that if we settle for less than identity we are committed to the horror of entities which fall outside of the scope of scientific enquiry. But this fear is no more than a Cartesian legacy. Just as Platonism cannot be incorporated into Mathematics, so the Identity thesis cannot be incorporated into neurophysiology for the same reasons of incoherence. The source of the incoherence in this case is the Cartesian notion of the absolute or intrinsic privacy of experience.

The real target here is the notion of intrinsic privacy, since if experiences are in some way intrinsically private then they cannot be incorporated into the sciences, and it was this apparent impossibility of incorporation which gave rise to the Identity thesis in the first place. It arose out of the Identity theorists' inability to accept the evident absurdity of this exclusion. I will be arguing that they were right in rejecting the exclusion of private items from the scientific arena but wrong in their way of going about setting things right.

I propose to show that the requirement that the privacy of experience is absolute or intrinsic is an incoherent requirement, and I will also show that the 'identification' proposed by the Identity theorists is itself incoherent and therefore useless as a remedy for the prior incoherence of the appeal to an intrinsic privacy. The Identity thesis is an incoherent move against an incoherent move: the incoherence is compounded. The difficulties with which the Identity theorist is concerned arise solely from the prior insistence of Cartesians on intrinsic privacy. But in rejecting the privacy of experience as laid down by Cartesians we must take care about what it is that we are rejecting since, here as elsewhere, a blind evasion cannot take the place of an intelligible account. If we do reject private items in a wholesale fashion we generate the equally familiar array of unpalatable consequences known as Behaviourism, but only because we have made no more than an evasive move. A natural, though inadequate, reaction to the intractable difficulties generated by the Cartesian barrier of intrinsic privacy has been that of impatient rejection. The Behaviourist looked at the Cartesian mess and decided to have nothing more to do with it, but, in his impatience threw out entirely inoffensive private items. We can, however, reject intrinsic privacy without rejecting privacy, a point which Behaviourists have never seen.

Now in order to settle the difficulties of the privacy of experience we must go back to the point at which the Cartesian fiat was laid down, and show that it was simply laid down in what was no more than a gratuitous manner. But against this view of the matter, it will be urged that, at least at first sight, it seems that we *find* that experiences are private, that we do not lay it down by fiat that it is so. What we find, however, is privacy but not intrinsic privacy, contrary to the Cartesian view. The evidence that we find privacy but not Cartesian privacy consists in the fact that we have a range of investigations of sensations and other private items in which the Cartesian strictures on the 'availability' of these private items are ignored. We have scientific investigations which, according to the Cartesian, we could not have. But here it is essential to avoid a mere appeal to the authority of our standard procedures, since there is no such authority. Instead of an appeal to authority, what we appeal to here is a shifting of the onus of proof. The form of the argument is that since there is nothing in our standard procedures which forces us to accept the Cartesian barrier, it is the Cartesian who must give compulsive reasons for doing so, and in order to do this he is called upon to do more than reiterate his assertion that there is an absolute barrier. This however is all that he ever does and in this respect he behaves exactly as the Platonist-in-Mathematics does. Both views have this in common, that they lack even the possibility of independent support, and the 'evidence' that is offered in their support turns out to be no more than an repetition of the view itself.

2.2. *The Contrast between the Standard Notion of Privacy and the Notion of Intrinsic Privacy*

By the standard notion of privacy I mean the privacy involved in saying that an after-image is private to me. Now in saying this I seem to be mentioning a positive relation which holds between me and a coloured expanse, a positive relation which is indicated in such idioms as 'experiencing it'. But when the positive relation of 'experiencing it' is appealed to in this way it applies equally well to my experiences of public objects too, and so fails to pick out what it is that leads me to say that the after-image is private. Whatever it is, it cannot be this positive relation which holds between me and whatever I experience. In fact, when we say in this standard way that an after-image is private, we are saying something essentially negative. We are denying something. The privacy of the after-image consists in its unavailability to others, and not in its positive relation to me. 'Private' points out that something is lacking and what is lacking are the correlations between our experiences which are present in the case of public objects. So 'private' here is a negative and relative term. In complete contrast to this, the notion of intrinsic privacy was intended as a positive and absolute term and these intended requirements are incoherent, as I now show.

If privacy were a positive relation of some sort we would have the following dilemma:

Either (i) it is a relation between a publicly identified self and the experienced item, or (ii) it is a relation between a private self and the experienced item. But, if privacy is a positive relation between a publicly identified self and something else, then it serves to place that some thing else firmly within the public context of identified public objects and so can hardly be a relation which confers absolute and intrinsic privacy. Again, if privacy is a relation between the item and a self which must already be understood to be private in the approved Cartesian manner, then privacy is presupposed in introducing the notion of privacy, which will not do either. This is the dilemma, and I conclude that the notion of intrinsic privacy is without application. It does not correspond to any procedure which can be intelligibly specified.

2.3. *Why the 'Disappearance Version' of the Identity Thesis Cannot Serve as a Solution to the Problem of Privacy*

I now show that the question of our standard ways of talking about sensations being superseded by new scientific ways of talking about them is beside the point. It is beside the point because these standard idioms neither affirm nor deny that privacy is intrinsic, so that their removal cannot have any effect on an issue which was generated by the appeal to intrinsic privacy. The difficulties against which the Identity thesis is directed are such difficulties and these cannot be resolved by any argument for or against our standard idioms.

Feyerabend has rightly stressed, in this connection, that no one way of talking is sacrosanct and has emphasised that here as elsewhere our standard ways of talking are destined to be superseded with the advance of the sciences [3]. Yet he also rightly points out that however much our language changes in this way, the pains will remain the same [4]. But there is a difficulty in maintaining both of these things, which arises from the fact that 'the same' is one of our standard expressions which will supposedly have been superseded. If it does make sense to say that the pains remain the same while theory changes then there is an unchanged link between the old and new ways of talking about pains which on the 'disappearance view' surely ought not to be there. On a consistent and radical 'disappearance thesis', at any rate, the very possibility of saying that the pains remain the same should be excluded. But if it is not excluded then we have not changed the most important feature of our way of talking about the pains, namely, *our way of identifying them*. This is a fundamental difficulty for the disappearance view.

Moreover, the same difficulty can be raised in a way which is independent even of the standard ways of saying 'the same'. We can raise much the same difficulty in a non-linguistic, or extra-linguistic, way. It is evidently a non-linguistic matter that I scratch at a particular area on my skin and it is more than a mere coincidence that our standard idioms also direct attention to that same area. If and when these idioms have been superseded I will still scratch in the locatory way that I now do. My public procedures in this situation will be what they now are and it is this stability of procedure which is significant, not which idioms I happen to use. This stability of procedure will not disappear, come what may, and it is this procedure which leads to identification. My scratching at a particular place is the public basis of my procedure for identification and this is not destined to disappear.

2.4. *The Discrepancy between Scientific Practice and the Requirements of the Identity Thesis*

The Identity thesis, so far from being scientific, is at variance with scientific practice and it is easy to see why this must be so since the Identity theorist is concerned with a problem arising solely out of the notion of intrinsic privacy which, in practice, neurophysiologists ignore. There is a striking difference between what these scientists do and what, according to the Identity theorists, they ought to do. Now this may sound suspiciously like no more than an appeal to the authority of what scientists actually do and I hasten to say that I do not think that any such appeal carries any weight at all. What I am appealing to here is not the authority of actual practice but rather the requirement that the practitioners in a given field need to be given some cogent reason for changing their current practices. The Identity theorist cannot in the nature of the case give them any such reason.

Now according to the Identity theorist sensations and the like cannot be incorporated into the sciences unless they are identified with brain processes, yet strangely enough we find a rapidly expanding literature on the investigation of sensations which have been carried out without this so-called 'identification'. The scientists involved have apparently no need for 'identification' here, and yet their researches grow apace simply because these scientists are not latter day Cartesians and so have no inherited methodological qualms as to what they are about. They do not share the Identity theorist's problem of intrinsic privacy and are instead concerned with the characteristically scientific activity of the assessment of rival theories of sensation. Moreover, these rival theories presuppose as an unquestioned premise that sensations fall within the scope of scientific enquiry.

Now suppose that we had adopted the Identity theory and were engaged in the investigation of cutaneous sensations. What effect would our adoption of the Identity thesis have on our research? Apparently the only difference that it could have would be that instead of saying, as we used to, that we are investigating how sensations arise from neural impulses, we now say that we are investigating these two aspects of one and the same thing. Yet everything that we do remains the same, so it seems that adopting the Identity view has no effect at all. Merely pronouncing the word 'identification' does not of itself change any of our standard procedures, and whatever we now distinguish by means of procedures for identification will continue to be distinguished and nothing whatever is affected by our saying that these distinct procedures are investigations of aspects of one and the same thing. This alone is a sure indication of the sterility of the Identity thesis. We still continue to make use of the same distinct procedures for identifying neural impulses and sensations and it is no more than an idle ceremony to declare them, by fiat, to be one and the same thing. There is no way of carrying out the 'identification'.

Finally, a fundamental objection to the appeal to identity here is that in comparable contexts we do not make use of it. When we say that lightning is an electrical discharge we are not even trying to say that lightning is identical with an electrical discharge. Only if we are already Identity theorists do we think that that is what we ought to be trying to say. What we are really trying to do, surely, is to incorporate *already identified* lightning within the wider context of electrical theory. Instead of sterile identity we have a fertile non-identity. We are never, in making such assertions, trying to establish a sterile identity. The proof that the notion of identity is not genuinely involved in such cases lies in the fact that our standard procedures for identification are not changed. We continue to identify lightning as we used to before we placed it within the context of electrical theory. In the same way, the need to incorporate our *already identified* sensations within neurological theory does not, and cannot, lead us to modify our procedures for identification of sensations, since they are already presupposed in what we are trying to do. And in general the inclusion of *already identified* items within a theory cannot of itself lead us to make new acts of identification. But once it has been shown that the notion of identity is not involved here, the point of the Identity theorist's manoeuvre has vanished.

3. Conclusion

The moral I wish to draw from all this is that here at any rate are two clear cases of sterile methodological proposals which fail to provide a procedure at just the point at which it is essential to have one. That there are other cases and that they are widespread throughout the Philosophy of Mind and the Philosophy of Mathematics will be argued at length elsewhere. This, I hope, is one of those generalisations that do not break down.

Institute for Communication Studies,
School of Television and Cinema,
National Iranian Radio and Television,
Tehran, Iran

NOTES

[1] T. Skolem, *Abstract Set Theory*, Notre Dame Mathematical Lectures No. 8, 1962, pp. 1–2.
[2] Gödel, K., 'What is Cantor's Continuum Hypothesis?' in P. Benacerraf and H. Putnam (eds.), *Philosophy of Mathematics*, 1964, p. 220. See also Kreisel, G., 'Mathematical Logic', *Lectures on Modern Mathematics*, Vol. 3 (ed. by T. L. Saaty), p. 186, for the appeal to 'mathematical experience'.
[3] Feyerabend, P., 'Materialism and the Mind-Body Problem' in C. V. Borst (ed.), *The Mind/Brain Identity Thesis*, 1970, pp. 142–156.
[4] *op. cit.* 150, line 10, "A new theory of pains will not change the pains".

J. W. N. WATKINS

THE HUMAN CONDITION: TWO CRITICISMS OF HOBBES

This paper is not in a field to which Imre Lakatos made any published contribution (at least since he came to England). But he was intensely interested in matters political, and he often urged that we should revive our department's traditional concern with political philosophy. This is an essay in that direction. (I realise, of course, that my two criticisms of Hobbes are small beer in comparison with Popper's criticism of Plato and Marx. But not everyone in the commonwealth of learning must hope to be a master-demolisher. It is ambition enough to be employed as an underminer of one or two erroneous ideas.) My other justification for contributing this piece to the present volume is that my old friend seemed to enjoy an earlier version of it which I gave as an inaugural lecture. At least, he told me that it had *surprised* him.

1. Two theses of Hobbes

For Hobbes, there is no such thing as spontaneous social co-operation among men; men are originally and by nature essentially *a-social*. That is one thesis of his that I will criticise.

The other is his thesis that men are originally and by nature devoid of any moral sense. True, men in a Hobbesian state of nature may bandy moral words about, but they use these words only as puffed-up expressions of their personal desires and aversions and not according to any interpersonal standard or rule. For rules have to be *made* by some authority; and this authority also has to be made, since no one enjoys authority by nature.

To say that men in a Hobbesian state of nature do not co-operate with one another may seem an under-statement. But in one respect it is an over-statement. While they are 'jostling and hewing one another'[1] in the war of all against all, they are also able to communicate with one another in a common language. There is at least *linguistic* co-operation in the state of nature: otherwise men could never create the authority who

R. S. Cohen et al. (eds.), Essays in Memory of Imre Lakatos, 691–716. *All Rights Reserved*
Copyright © 1976 by D. Reidel Publishing Company, Dordrecht-Holland

will create the rules that will make possible further kinds of co-operation under a commonwealth.

Reflections on linguistic co-operation have led me to counter Hobbes's two theses concerning the a-social and a-moral nature of pre-political man with two contrary theses. First: although a language is a cultural phenomenon, the creation of language required something pre-cultural, namely, a certain mutual attunement or pre-established harmony between the innate dispositions of different individuals, a mutual attunement that makes spontaneous co-operation possible. Second: although an adult person's moral outlook may owe much to cultural influences, it cannot owe everything to culture; for he could never have formed a moral outlook unless he had had some rudimentary moral sense in the first place.

Hobbes's account of the formation of language is individualistic in the following sense. Suppose that a Recording Angel had kept a record of all human utterances from the very beginning. And suppose further, that whenever he met with an utterance involving some departure (whether in vocabulary or in grammatical structure) from earlier speech-habits he recorded it in red and added the date and the speaker's identity. Then on Hobbes's view it would in principle be possible for the Angel, starting with any contemporary form of speech, to identify each of the many individuals whose separate linguistic innovations at various earlier dates had collectively made it possible. For Hobbes, a language is rather like a vast mosaic pavement on which many hands have worked: but each little tessera in it was put there by some individual. Analogously, every part of a public language, as it exists at a given time, owes its existence, eventually, to what were, originally, private innovations made by single individuals. (Hobbes did not explain how they managed to get their innovations publicly accepted. Perhaps they enjoyed some local authority.)

Against this I shall claim that the formation of language must have been an essentially social process at all stages, even the earliest ones.[2]

Before proceeding to this I want to say a word about my main intellectual debts in this paper. Its central idea has a close affinity to Chomsky's thesis that language-acquisition presupposes an 'innate mental structure' common to all children capable of acquiring a first language.[3] As a matter of autobiographical fact, however, the main influence came, not from Chomsky, but from a published lecture by Hayek[4] which I read soon after finishing a book on Hobbes. (It was at a seminar organised by Hayek, in April 1966, to discuss the significance of 'meta-conscious' rules and allied topics that I first presented the gist of the present paper.)[5] Three other authors who have greatly influenced this paper are Roger Brown, Konrad Lorenz, and V. C. Wynne-Edwards.

2. Rules that are shared and meta-conscious

I now leave Hobbes for the time being and turn to M. Jourdain, the man who, as everybody knows, discovered that for years he had been speaking prose without knowing it. Put into English Molière's actual dialogue goes like this:

> M. Jourdain: What? When I say: 'Nicole, bring me my slippers and give me my night-cap', is that prose?
> Professor of Philosophy: Yes, Sir.
> M. Jourdain: Good Heavens! For more than forty years I've been speaking prose without knowing it.

This is usually taken as a witty expression of the fact that 'knowing how' may exist without any corresponding 'knowing that'.[6] M. Jourdain knew how to speak in conformity with the rules of his language without being able to spell them out.

But there is another aspect to this, a social aspect. Everyone has heard of Mr. Jourdain, but his servant Nicole tends to get overlooked. Hayek has made an important point which might be put this way: when Nicole listened to what M. Jourdain was saying to her, she *heard* prose without knowing it.[7] Once made, Hayek's point is obvious: Nicole needed essentially the *same* know-how, the same unconscious grasp of linguistic rules, to receive M. Jourdain's commands as M. Jourdain needed to give them. Hayek uses the term 'meta-conscious rule' to denote a rule which a person *is* following while thinking or acting in a certain way, and which indeed he *must* follow if he is to think or act in that way, but which is no part of what he is currently thinking about. Borrowing this term we can say that human communication is governed by rules that are not only meta-conscious but have a shared, *social* character.

3. Rules versus Uniformities

Now Hobbes allowed that an *individual* may use unarticulated know-how – for instance, in playing tennis.[8] And I think Hobbes could have allowed that some things are essentially social in character – money, for instance, which he called "the blood of a commonwealth", a phrase which suggests that money circulates *between* people and has an *inter*-individual status. But his philosophy could not, I believe, accommodate the idea of rules that are *both* unarticulated and social, of rules that are obeyed collectively by people who are not consciously aware of them.

I can think of three ways in which someone might try to come to Hobbes's rescue, here.

First (and least interestingly), it might be said that M. Jourdain and Nicole were able to follow the same rules without being conscious of them, in speaking and listening to one another, because they had been so drilled in French grammar when they were school-children that it became second-nature to them. But perhaps Nicole never went to school; and if she did she must have been able to understand and speak some French, more or less grammatically, before she could be taught any French grammar.[9] In any case, languages were being spoken before there was study and teaching of the grammar of those languages. And we may add that the syntax of some languages are of such enormous complexity that it is doubtful whether anyone possesses the *know-that* that would fully correspond to the intricate *know-how* unconsciously possessed by native speakers of these languages. Consider Nootka, a language spoken by some Indian tribes in Vancouver Island. Admittedly, Nootka has been studied – by Benjamin Lee Whorf, on whose account I am relying; but we can safely assume that his studies had little feed-back on native speakers of Nootka. According to Whorf, individual words in Nootka have vague and essentially plastic meanings; but when suitably combined into phrases and sentences they mutually colour each other and fuse into a compound unit of meaning that is far more than the sum of its parts[10] (rather as in our language the combination of the words 'jelly' and 'fish' has a unitary meaning). Obviously, the syntax of such a 'polysynthetic' language, as Whorf called Nootka, is likely to be far more complex than that of what might be called a 'bead-like' language, that is, a language whose elementary meaning-units (or 'morphemes') are invariant when strung together in different ways (rather as 'fish' and 'bone' are when combined into 'fishbone').

A second and somewhat more interesting way of defending Hobbes would be to say that I create needless mystery and perplexity by my talk of people observing linguistic *rules* of which they are unaware. I ought to have spoken simply of uniformities among people's speech-habits. Then there would be no mystery; for there is nothing mysterious about a uniformity that exists without people being aware of it.

Well, I do not claim that 'rules' is the ideal word for what we are dealing with here, but I do claim that it is more appropriate than 'uniformities'. For one thing, they are by no means invariably followed. I am not thinking merely of accidental lapses. It has often been pointed out that the very mistakes in a child's speech often indicate that he is seeking (unconsciously, of course) to get hold of a linguistic rule and is not merely reproducing snatches of adult talk.[11] The child who says 'I walk-homed' instead of 'I walked home' has nearly got it. And when he *does* get hold of it the rule will not allow him to lapse back into his earlier incorrect mode of talking. Most linguistic rules are highly *effective*, even though they do not secure a perfectly uniform compliance. Of course, we do not notice the pressure that the rules of our native language exert upon us. Nor do we notice the pressure of the atmosphere, but it is there nevertheless.

True, there are linguistic rules which exert only a weak pressure, easily resisted. These are, so to speak, rules of linguistic etiquette, where the etiquette is becoming obsolescent. But there may be a *linguistic* explanation for their weakening hold. For instance: most of us would admit that it is strictly correct to say '*Whom* did you see?' but we tend in practice to say '*Who* did you see?' Sapir suggested that this may be due to the countervailing pressure of other linguistic rules: the interrogative 'who' has closer affinities with the interrogatives 'which', 'what', 'where', 'when' and 'how' than it has with the pronouns 'he' and 'they'. So why should it behave like these pronouns, with their variants 'him' and 'them', instead of being invariant like the other interrogatives? Our tendency to eliminate the interrogative 'whom' in colloquial speech looks like a tendency to iron out a minor aberration in the syntax of the English language.[12]

Another consideration which tells decisively against treating linguistic rules as mere uniformities is this. It has been stressed, especially by

Chomsky (who has searched out various seventeenth century and later anticipations of his ideas), that possession of a limited number of rules of grammar – and this is true of 'surface' grammar as well as of what Chomsky calls 'deep' grammar – empowers one freely to generate an unlimited number of different grammatical sentences.[13] We can conceive of an ideal linguistic community in which each sentence spoken is perfectly grammatical and also freshly minted and different from every sentence that had previously been spoken. One could hardly say that this community's linguistic rules are no more than linguistic uniformities.

More interesting questions are raised by a third way in which one might try to rescue Hobbes's individualistic theory of language. There is a strong *as-if* tendency in much of Hobbes's philosophising.[14] For instance, he does not say that the state *was* created by men covenanting with one another to establish a sovereign over themselves, but that the state may be regarded *as if* it had been so created. And it might be contended that his theory of language says, not that each part of the vocabulary and syntax of a language *was* originally invented by some individual, but only that language may be regarded *as if* it had been so created. On this view, to refute Hobbes we would have to show that it is *impossible* that language should have been created in this way. I will now try to show just that.

4. First steps in the invention of language

If the development of a language consists of a series of additions and changes, each originally introduced by an individual who subsequently got it accepted by other individuals, then there must have been an individual who took the very first step in the series. I will call him Alpha. Hobbes, always respectful of the Scriptures, declared that it was Adam who first invented names, for the creatures which God presented to him. But Adam's role, here, raises some knotty problems. Adam gave names to the creatures *at God's behest.* In what language did God communicate this behest? In Hebrew, perhaps? But how did Adam come to know enough Hebrew to understand God? Perhaps God gave him a Hebrew lesson not recorded in the Book of Genesis? But in what language did the lesson begin?

Since I cannot answer these questions, and since we are now concerned with a purely hypothetical, 'as-if' account of the origin of language, I will leave Adam and stick to Alpha. For simplicity's sake I will assume that Alpha lived alone with his cave-mate Beta, and that there was no serpent to distract them.

For Hobbes, speech consists of propositions and a proposition consists of names coupled together. So Alpha's very first step will be to invent a *name*. According to Hobbes, a name begins as a mark. A mark is some object or noise or inscription that an individual has arbitrarily chosen to serve him as a reminder of some other object or objects. A mark is not yet a name because it has only a private significance for its maker.

We assume, then, that Alpha has invented some marks for himself; his task is now to begin turning these into names, which means, in his case, getting Beta to use them in the same way as he does, so that they acquire a common or public significance.

Again for simplicity's sake, I will assume that by a happy co-incidence the language that is going to develop between Alpha and Beta is remarkably similar, so far as it goes, to our English language today. For instance, Alpha uses the vocal sound 'dog' as his mark for the dog that shares their cave. And he now seeks to get Beta to use it in the same way. He cannot, of course, do this by giving a verbal explanation since no language exists as yet. Perhaps he will try some ostensive method, such as repeatedly pronouncing 'dog' while pointing to the dog.

Now if Beta is to catch on to what he is doing, she must be psychologically prepared to receive what he is trying to transmit. The dog itself, presumably, will not catch on to Alpha's intention. Dogs do not seem to understand our ostensive procedures: they tend to gaze at the pointing forefinger instead of at the thing at which the finger is pointing. Beta needs to be better equipped than that. She must have some inklings about the significance of *pointing*; and she must already have learnt to invent private marks for herself. In short: Beta must be just about as inventive, and in the same sort of ways, as Alpha; otherwise, names will never get established in their cave.

I will pass over the next step in this hypothetical history of language – namely, the invention of what Hobbes called common names as opposed to proper names – and will go straight to the next great step, the invention of ways of *putting names together* to make assertions.

5. From Names to Sentences

In the sentence 'Not all the swans in Australia are white', only three of the eight words are names (common or proper); the others are what mediaeval logicians called syncategorematic words, words which serve to link the names together. Hobbes's treatment of such words was pretty high-handed. With one big exception he squeezed them in under the category of *names* by declaring them to be *parts* of names[15]: he treated the words 'all' and 'some', as in 'all men' and 'some men', rather as one might treat '-er' and '-ing' in 'singer' and 'singing'; and he treated 'not', as in 'not man', rather as Newspeak treats the 'un-' in 'unperson'.

The big exception was the copula: he insisted that words like 'is' and 'are' are neither names nor parts of names. When a proposition is expressed in "its most simple and categorical form ... the copulative word *is* must be expressed by itself, and not mingled in any manner with the subject or predicate."[16]

Hobbes did not regard the *words* 'is', 'are', etc. as indispensable: he mentioned that in Hebrew there is no equivalent word,[17] and that in English we might have a convention whereby 'A tree, a plant' conveys what we now convey by 'A tree is a plant'. Such a language would, however, still have a copula-*rule* for coupling names into propositions though it would not do this by means of a copula-word.

With Hobbes, the whole elaborate syntax of a language is epitomised by the copula. Arbitrarily affixed names provide the beads and the copula provides the one bit of string for threading them together. It is a pretty simplified account of language, but that does not matter for my present purpose.

Back, now, to Alpha who is about to make another linguistic breakthrough. Returning from a hunting expedition at sunset, he finds Beta inside the cave and excitedly exclaims to her: 'Sky is red.' The first proposition has been made. A new chapter has opened in the story of mankind. The human larynx has been used to make something that is *true or false.*

Beta understands 'sky' and she understands 'red', but what will she make of 'is'? Perhaps she will vaguely suspect – without, of course, being able to verbalise her suspicion – that 'is' is a new name that Alpha has invented. What can Alpha do to help her? He cannot *explain* 'is' to her – one cannot explain what the copula is without using the copula.

In 1797 a wild boy, aged eleven or twelve, was found naked and speechless in woods near Aveyron.[18] A certain Dr. Itard undertook to try to civilise him which meant, in the first place, teaching him French. He first taught Victor (as he called him) to distinguish and recognise such inscriptions as 'livre', 'plume', etc. Then he introduced a set of objects consisting of a book, a pen, etc., and taught Victor to pair off the card with 'livre' inscribed on it with the book, 'plume' with the pen, and so on. At first, Victor treated the inscriptions as proper names: he would pair off 'livre' only with the book that Dr. Itard had first presented to him and not with other books substituted for the original one. In time, however, he learnt to use 'livre' as a common name for books of various sizes and colours. He also learnt some adjectives: confronted with a big book and a small book and two cards inscribed respectively 'grand livre' and 'petit livre', he could match them correctly, and likewise with other pairs of big and small objects. He even learnt to associate certain verbs with corresponding actions.

Dr. Itard reported, however, that Victor's learning seemed eventually to reach a limit. It seems that he could not make the leap from words to *sentences*. He never, so to speak, grasped the copula. 'Le livre *est* petit' would have baffled him, just as we are supposing 'Sky *is* red' might have baffled Beta.

I can see no escape from the conclusion that Beta will catch on to Alpha's invention of the copula only if *she invents it too*. They must invent it together. An adequate picture of linguistic innovation would have to be less individualistic and more social and equalitarian than the Hobbesian picture of a wilful and inventive individual imposing his novel conventions, Humpty-Dumpty-style, on other people whose role is passive and receptive. If we think of an important linguistic innovation as a kind of mutation, then we should think of it as a mutation that occurs, not in this individual or that, but collectively within a linguistic community.

6. Vertical innovations

The same point can be made by an argument which does not involve any fairy-tale about how language began. But before I can present this argument, I must first introduce two distinctions.

The first distinction is between a less interesting and a more interesting kind of linguistic innovation. A linguistic innovation may be called *horizontal* if it merely extends an existing practice, and *vertical* if it introduces a new principle. Our imaginary case of a sudden advance from the ability merely to name things to an ability to connect names into sentences would be a clear example of a vertical innovation. Other cases may be less clear. Suppose that a language contains a primitive counting system in which one can count only to five (for numbers greater than five it possesses only some word like 'many'). If this were to develop so that one could count up to ten, I would call that a horizontal innovation; but if it were to develop so that one could identify *any* number, I would be inclined to call this progression to an unlimited and potentially infinite number-system a vertical innovation. The existence of borderline cases does not matter for my argument, which requires only that there have been *some* vertical linguistic innovations, whether or not we can specify which they were. And I take it that no one would claim that it was by horizontal extensions alone that the grunts of our remote forbears developed into the language of Shakespeare or of Einstein.

The second distinction I need is the well-known philosophical distinction between an object-language and a meta-language.

Anyone unfamiliar with this distinction can grasp it by first supposing some specialised sub-language – say, the language of chemistry – separated from the rest of our language. In this amputated language, which I will call L_0, we can talk about such extra-linguistic things as, say, acids, alkalis, hydrogen atoms, etc., but not about linguistic things like noun-phrases, predicates, quantifiers, etc. So one thing we cannot discuss in L_0 is L_0 itself. If we wish to make L_0 an object of investigation we shall need another language, a meta-language L_M, in which to discuss this object-language.

The question arises whether our L_0 and our L_M could be amalgamated into a unified, one-level language in which could be discussed not only acids etc. but also the properties of this language itself. Some forty years ago Carnap, using Gödel's method of arithmetization, showed that it is actually possible to construct a meta-language for a language L within L itself without generating contradictions, *provided* that the meta-language is confined to the *syntax* of L.[19] This is a severe restriction. In such a meta-language we could relate linguistic expressions occurring in

L only to other linguistic expressions occurring in L. We could not say anything about the *meaning* or *truth* of these expressions. That would involve *semantic* concepts, that is, concepts that relate linguistic expressions to something other than linguistic expressions.

Tarski showed that a formal meta-language containing semantic concepts (e.g. satisfaction, truth) must be constructed outside or above its object-language if antinomies are not to be generated: there has to be a language-hierarchy. Moreover, if the meta-language meets certain adequacy-requirements, it will be *richer* than its object-language. For whatever can be said in the object-language can also be said in the meta-language (by forming a name of the object-language sentence and declaring it true); but there will be things which can be said in the meta-language which cannot be said in the object-language (e.g. "The 'is' in the L_0 sentence 'Ammonia is a gas' is the 'is' of class-inclusion and not the 'is' of class-membership").

Tarski was dealing with formalised languages, but a similar kind of hierarchical principle holds for natural languages. Consider that fragment of the English language used in playing contract bridge. Suppose that you wish to explain to someone ignorant of the language of bridge how all the expressions in that language are used. You would need a language in which you could mention any expression in the language of bridge, so your meta-language would have to be at least as rich as it; and you would need to go outside the language of bridge in order to explain its expressions. So your meta-language would have to be richer than its object-language.

Now I come to my argument.

Let L_1 be a rather primitive natural language – too primitive, at any rate, for it to be possible to discuss L_1 in L_1. Let L_2 be this language as enriched by some vertical innovation (say: a negation sign has been introduced into it). Then by the foregoing considerations, it will not be possible for users of L_2 to *explain* this innovation within L_2 itself. The explanation of the innovation will have to await the further enrichment of the language: the language will have to have advanced to a new stage, L_3, before its advance from L_1 to L_2 can be discussed within it.

I conclude that when a new linguistic practice was introduced, our ancestors must – at least if it was a vertical innovation – simply have started practising it collectively without, at first, being able to explain it to one another or even to themselves. They must have collaborated spontaneously at a new level, as if by a kind of pre-established harmony.

It is said that the game of rugby was invented on the playing fields of Rugby in 1823 by someone breaking the then rules of football by picking up the ball and running with it. But surely there was more to it than that: his strange behaviour must have inspired some sort of spontaneous co-operation from other players. (Perhaps he was brought down by a tackle and then a scrum formed.) If the other players had merely stared at him incredulously, he would have succeeded only in looking silly.

7. The cunning of childish reason

Descartes suggested that God's original creation of the world seems less singular when we remember that God has gone on re-creating the world moment by moment ever since. One might likewise suggest that the original invention of language seems less singular when we remember that the same miracle takes place on a small scale in every child who learns to speak.[20] The miracle is obscured, partly by the fact that almost every child pulls it off, and partly by the fact that grown-ups are around speaking the language and nudging the child along the road to linguistic maturity. (He says, "Dog bark" and his mother says, "Yes, a dog is barking".)[21]

But I find it more impressive that the small child's untutored, unself-conscious, groping penetration of his native language must be far in advance of anything that can be explicitly taught him about it. What I said earlier about the need for Beta to be prepared for Alpha's innovations – which means, in effect, that she had to go through the same inventive process herself – applies *pari passu* to all children learning a first language.

For my part, I believe that children strive to *say* things almost from the beginning. They are not much interested merely in name-dropping: they want to speak prose.

We should not be deceived, here, by the fact that a small child's speech may consist essentially of *names* which he does not connect together with the help of syncategorematic words. His utterances may nevertheless have a syntactical structure. (After all, some adult expressions consist of a string of names silently structured by invisible bracketings; for example, 'main burner oil feed adjustment cover retaining screw' or 'tourist holiday coach death crash enquiry verdict appeal decision sensation'.)[22] Even a child's one-word utterances may have an implicit syntax, conveyed by their tone. A hungry child screams "Milk!". His mother gives him some. Later, gurgling contentedly, he points to the milk-bottle and says "Milk". After which, pointing at a jug whose contents he cannot see he says, with a different tone and expression, "Milk?" He already has an instinctive grasp of the imperative, indicative, and interrogative moods of the language that he is in the process of acquiring.

If small children are to catch on to and exploit the grammatical resources of the language they hear being spoken around them, they must be endowed with some factor – call it the *X*-factor – that enables them to do so. Obviously, this *X*-factor cannot be instilled into them verbally: it must already exist before they can make anything of adult speech. Nor, it seems, could it be instilled by pre-verbal means. To the popular idea that children are originally inducted into their mother-tongue by ostensive definitions Collingwood rightly objected that not only is it falsified by what goes on in the nursery but that:

> the supposed gestures of pointing and so forth are themselves in the nature of a language. Either the child has first to be taught this language of gestures, in order to help it in learning English, or it must be supposed to 'tumble to' the gesture-language for itself. But if it can do that, we want to know ... why it cannot (as in fact it does) 'tumble' in the same way to English.[23]

In short, it seems that this indispensable *X*-factor is not instilled or acquired; that it is pre-cultural and innate. This, of course, is Chomsky's thesis, which met with some strong opposition from empiricist philosophers.[24] But among biologists it is "a matter of course", according to Lorenz, that learning presupposes innate learning mechanisms "built into the organic system in the course of its evolution."[25] Paraphrasing Kant, Lorenz added that the innate is "What must be in existence before all individual learning in order to make learning possible" (p. 44).

Locke saddled the doctrine of innate ideas with the ridiculous rider that, if innate, they should operate from birth. But many innate biological capacities do not operate at birth. Assume that dogs have an innate ability to swim. This does not mean that a day-old puppy will start swimming if it is thrown into a pond. An innate capacity may need time to mature.

Remembering Dr. Itard's failure to awaken any constructive grammatical ability in Victor (who was then about twelve), I want to say that, as well as needing time to mature, an innate capacity that is not exercised may atrophy and become extinct. Empiricist philosophers from Hume to Goodman have looked with suspicion on the idea of an unactualised disposition that does not manifest itself in observable behaviour: "the distinction ... betwixt a *power* and the *exercise* of it, is entirely frivolous," Hume declared; "neither man nor any other being ought ever to be thought possesst of any ability, unless it be exerted and put into action."[26] The idea of an unexercised dispositional system that goes through a cycle of development, maturing and atrophying, would presumably be doubly suspect. But is a dog that never got into deep water to be denied an ability to swim since such an ability was not "exerted and put into action"? Surely not. And is it not reasonable to suppose that such a dog's swimming potential is stronger when it is young and full-grown than when it is a small puppy or old and blind?

If a capacity for 'tumbling to' the syntactical structure of a language can atrophy through non-use, that would help to explain the low upper limit on Victor's ability to learn French. Indeed, it would seem that even when, as almost invariably happens, the *X*-factor is fully exercised, it usually withers away once its main task is done; with the unfortunate result that most of us cannot acquire foreign languages in the same instinctive way that we acquire our first language.

I turn now to the second thesis of Hobbes which I said I would criticise, his thesis that men are, by nature, essentially a-moral.

8. Can morality be taught?

Let me credit Alpha with one more linguistic break-through. (This will be the last.) Beta is humming to herself as she stirs a bubbling stew-pot, but Alpha frowns, shakes his head, and says: "Eating people is wrong." The first moral pronouncement has been made.

Now there are philosophers (Hobbes was one of them) who take a debunking view of so-called 'moral' utterances, viewing them as no more than puffed-up descriptions or expressions of their authors' feelings. For these philosophers, "Eating people is wrong" would mean something like "Eating people disagrees with me" or perhaps, "Eating people! Ugh!"

I fully admit, of course, that people often wrap up their selfish interests in moralised language. But I claim that a philosophical account of moral language must be false if it asserts that moral expressions just *are* puffed-up expressions of non-moral likes, dislikes, etc. For such an account is self-refuting. For suppose, for a moment, that the words 'good' and 'evil' were always used in conformity with Hobbes's famous dictum: "Whatsoever is the object of any man's appetite or desire, that is it which he for his part calleth *good*; and the object of his hate and aversion, *evil*."[27] Then calling something good, or evil, could no longer be a *puffed-up* expression of a feeling of desire, or aversion. If everyone always used these expressions in this way, '*x* is evil' and 'I hate *x*' would be interchangeable and, moreover, *obviously* interchangeable expressions. The former would no more be a moral-sounding version of the latter than 'oculist' is a moral-sounding version of 'eye-doctor'. If a distinctive and genuinely moral language did not exist, a moral *sounding* language would not exist either. The very fact that one may debunk this or that particular use of moral language as pretentious or hypocritical means that one cannot so debunk *all* moral language.

I conclude without more ado that Alpha made a new *kind* of utterance, achieved a vertical breakthrough, when he first declared a certain practice *wrong*.

The pattern of my argument here is the same as before. Alpha cannot *explain* to Beta the peculiar meaning of this new word. All the words they share so far are non-moral; and a moral term can no more be explained in non-moral terms than the copula can be explained without using the copula. If Beta is to catch on to his meaning there must be something in her which corresponds to the something in him which prompted him to start speaking in this strange new way.

Again, what I have said about Beta applies *pari passu* to children: in this case, to children who come to understand something of the language of morals.

This yields an answer to the question posed at the start of Plato's dialogue the *Meno*, namely: Can virtue be taught? The concept of virtue does not have quite the resonance for us that it had for the Greeks. This question might be re-phrased for English ears as: Can morality be taught? The answer suggested by the foregoing seems to be this. It is one thing to try to inculcate a particular moral rule or value into someone who already has some inkling of morality, but quite another to try to inculcate some inkling of morality into someone who has none. Obviously, there are no insuperable obstacles to attempts of the first kind, though they may run into practical difficulties: the recipient of the moral instruction may be pig-headed or counter-suggestible (I hope that he would be if the moral rule or value in question is a bad one). But attempts of the second kind face an insuperable difficulty. A person whose thinking entirely lacks any moral dimension cannot be taught morality. (You may put the fear of God into him, but that is another matter.) A moral sense has something in common with a sense of humour: if you have got one it can be enlarged, but if you have not got one you cannot be given one. I can see no escape from the conclusion that the original seed from which an adult person's moral outlook develops must be inborn (though how it develops may depend very much on cultural influences). Freud's account, in *Civilization and its Discontents*, of the genesis of conscience as the internalisation of an aggressive father-figure may be correct as far as it goes; but there must be something missing from this account. Merely replacing an external bully by an internal bully (Freud says that the super-ego is *feared* by the ego which it *ill-treats* and *torments*)[28] will not suffice to generate a moral sense. The case would be different if the figure that gets internalised were tacitly regarded, not merely as pre-eminently powerful, but as enjoying some kind of 'moral' ascendancy as well; but to amend Freud's account in this way would be to postulate an already existing sense, however rudimentary and inchoate, of the difference between power and authority.

9. Group Selection and Morality

To say that something is innate is, by implication, to say that it has been evolved in the process of selection, which is in turn to say that possession of it has survival-value. But can this be so in the case of a moral sense? If a moral sense makes any difference at all, will it not tend rather to inhibit the ruthlessness with which an animal strives to stay alive? Would not the survival-prospects of tigers diminish if they began feeling moral qualms about catching their dinner?

I agree that tigers do not seem to have much moral sense (though perhaps we should make some exception for tigresses guarding their cubs). But lions are in another category. I quote from Roger Brown:

> In the wild, lions live and hunt in packs..., whereas tigers are solitary except when mated. The two species have a strong antipathy and when they are combined in a circus act, murderous fights are possible. The tigers are likely to lose these battles... The lions charge in a gang to aid their comrade and proceed from killing the original antagonist to finishing off the rest of the tigers one by one. But while the current tiger battles all the lions, the other tigers wait their turns, gazing disinterestedly up in the air.[29]

In the case of animals that live together in packs or herds or societies, 'Put self first' is by no means a maxim that is unambiguously underwritten by natural selection, at least according to the modern theory of group-selection. This theory emphasises the importance of the evolutionary selection of traits which, though they may be of no immediate or direct advantage to an individual member of a species, may be of great indirect advantage, since they enhance the survival-capacity of the collectives to which the individuals belong. Wynne-Edwards has argued in his great book[30] that the survival needs of a *population* (or stock or group) may impose severe constraints upon its individual members.

One over-riding need is that the population should not use up its means of subsistence as a result of over-expansion. It is with this need and its implications that Wynne-Edwards was primarily concerned. I will present his main idea by way of a contrast with Darwin's view as expressed in the chapter 'Struggle for Existence' in *The Origin of Species.* There, Darwin stated (with acknowledgement to Malthus) that any living population is always tending to increase in a geometrical progression; but this explosive tendency is checked by the ravages of competition: each species "is constantly suffering enormous destruction". This view seems to imply that on a walk through a forest one should keep stumbling over the remains of animals that have either dropped from starvation or been killed by competitors; scarcely a blade of vegetation should remain; any surviving animals should be ravenous.

An interesting implication of the idea of a potential population explosion liable to become actual as soon as restraints are relaxed was drawn by Volterra in 1928.[31] Consider the following simple prey-predator situation. Species *A* gets its food exclusively by preying on species *B* which is preyed on only by species *A*. The two species are confined together within a certain area (an island, say). Suppose that at time t_0 species *B* happens to be plentiful relative to species *A*, so that the latter has an ample food supply. Then species *A* should multiply rapidly. By time t_1 species *A*, now greatly enlarged, will be making such heavy depredations on species *B* that the latter population will be shrinking rapidly. By time t_2 species *A* faces mass-starvation, having largely destroyed its food supply: now *its* population will shrink rapidly. By time t_3 it will have shrunk so far that species *B*, if it has not been eliminated, will be permitted to expand and again to become plentiful relative to species *A*. Then the cycle will begin again.

Now consider a prey-predator situation involving species *A′* and *B′* which are similar to species *A* and *B* except for this crucial difference: *A′* and *B′* have evolved certain mechanisms and conventions which tend to inhibit population-expansion beyond an optimal density.

The concept of optimum population densities was introduced in 1922, a few years before Volterra's oscillation theory, by Carr-Saunders,[32] to whom Wynne-Edwards makes generous acknowledgements. Carr-Saunders argued that our primitive ancestors had evolved ways of limiting their fertility roughly to a population-density at which economic returns to the individual members of a territorial group could be maximised – expansion beyond this density would bring dimishing returns. Wynne-Edwards extended this idea to all living species.

Thus our species *A′* will maintain approximately that population-level at which its members, although well-fed, prey upon species *B′* at a rate low enough not to endanger the latter's re-stocking and maintenance of its population. On the supposition that species *B′* has also evolved population-checking mechanisms, and so on down, then the whole ecological situation should be essentially stable (so long as the habitat remains stable). No species will be engaging in 'over-fishing' (to use one of Wynne-Edwards' favourite terms). Obviously, individual members of *A′* will, on average, have better survival prospects than those of *A*.

Under the head of population-controls evolved by group-selection Wynne-Edwards brings a wide variety of phenomena which were well known but which could not easily be accounted for in terms of *individual* selection and propagation: sterile castes in insect-societies; peck-orders (which effectively deprive lower-ranking members of their ability to reproduce); territories[33] (which disperse animals relatively sparsely over the available terrain; animals unable to secure a territory are, again, effectively deprived of their ability to reproduce); emigration and colonisation. Wynne-Edwards finds among all levels of vertebrates and even among invertebrates a huge variety of 'social conventions', evolved through group-selection to limit population-density, compliance with which "appears to be the exact counterpart of what we recognise in our own behaviour as moral conduct."[34]

Group surrival requires not only that the group should not expand excessively but also, of course, that it should not be ravaged by internal conflicts. As is well known, the natural condition of any gregarious species is very unlike a Hobbesian state of nature. Konrad Lorenz has examined a huge variety of 'social conventions', also evolved through group-selection, presumably, for ritualising intra-specific conflicts and rendering them more or less innocuous. He points out that the inhibitions have to be especially reliable among those predators, such as lions and wolves, that are both fierce and gregarious.[35]

Ardrey has drawn attention to the fact that Darwin himself came (under Wallace's influence) to recognise that certain moral characteristics have had group survival-value: "When two tribes of primeval man ... came into competition", Darwin wrote in 1871, "the tribe including the greater number of courageous, sympathetic and faithful members would succeed better..."[36]

I do not equate morality with serving tribal or group interests; and in any case the behaviour of an animal that has been genetically programmed to inhibit or modify certain anti-social tendencies could hardly be called *moral*: the animal is doing what it is natural for it to do when it, say, skirts round a neighbours's territory or finds itself unable to sink its fangs into a defeated rival's proferred neck.

Nevertheless, the theory we have been considering does suggest that there is no abrupt discontinuity between the *de facto* rules, encoded in an animal's biological make-up, which help to regulate intra-specific behaviour in the animal kingdom, and the legal and moral rules which help (somewhat fitfully) to regulate behaviour in a human society. Hayek has pointed out that it may be very difficult to decide at what point a *de facto* rule becomes a norm:

Is an innate inhibition which prevents a man or animal from taking a certain action, but of which he is wholly unaware, a 'norm'? Does it become a 'norm' when an observer can see how a desire and an inhibition are in conflict...? Or when it leads to a conscious conflict between a particular impulse and a feeling that 'one ought not to do it'? Or when this feeling is expressed in words...? Or when, although not yet articulated as a verbal rule, the feeling is shared by all members of the group...? Or only when it is... laid down in articulated form?[37]

Rather as we have been obliged by recent investigations to give up the idea that language is a unique possession of *homo sapiens*, so we are obliged to give up the idea that there is nothing at all like our human morality among animal societies. The theory of natural selection, as now augmented by the idea of group-selection, by no means rules out the possibility that both a linguistic capacity and a moral capacity are inborn in men.

10. Methodological Individualism Reconsidered

The foregoing ideas call for a revision of "methodological individualism" at least as it was being formulated by me in various papers in the 1950s.[38]

Let *S* be any (human) society, institution, or social process, and let an *I*-predicate be a one-term predicate that is predicable of individual people; that is to say, it always makes sense, though it may be false, to predicate any *I*-predicate of any individual. *I*-predicates can designate physical, psychological, and psycho-physical properties of individuals. Thus 'tall', 'ambitious' and 'weary' are all *I*-predicates. Let an *R*-predicate be a relational predicate which designates a relation either between individuals ('*a* is a cousin of *b*') or between an individual and a thing ('*a* owns this house') or between individuals and things. There is no restriction on the type of relation that an *R*-predicate can designate – it may be spatial, physical, biological or social – provided that it is a relation into which individual people can enter.

Then my old position can be restated thus:

(1) An adequate description of *S* will essentially involve predicates – say, *S*-predicates – that are neither *I*- nor *R*-predicates.

(2) However, the *explanans* of an adequate explanation of the formation of *S*, or of the subsequent functioning of *S* or of changes in *S*, will essentially involve only *I*- and *R*-predicates. If *S*-predicates still figure in our *explanans* we have an 'unfinished' or 'half-way' explanation:[39] we could proceed to a deeper *explanans* containing no *S*-predicates.

(3) Moreover, explanations of the formation of properties designated by the *I*-predicates in our *explanans* for *S* will in turn essentially involve *I*-predicates but not *S*-predicates.

A nice illustration of the above is the explanation of the origin of money put forward by Carl Menger, a great exponent and practitioner of what later came to be known as methodological individualism.[40] Menger emphasised that money is "a *social* phenomenon"[41] (in our terminology: 'money' is an *S*-concept). He rejected the hypothesis that money was originally created by some legislative act or collective decision.[42] Rather, he explained it as developing in a rationally reconstructible way out of barter. (Barter, I take it, is something that can be described just with *I*- and *R*-predicates: one individual is willing to give up some *X* to get some *Y*; he discovers another individual who is willing to give up some *Y* to get some *X*...) An individual regularly engaging in barter at some traditional meeting place might notice that there is a ware, say *A*, that exchanges more often and readily than others; so if he now wants to barter some *B* for some *C* but finds no one who will give him some *C* for some *B*, he may first exchange *B* for *A* and only later exchange *A* for *C*. As for the *I*-predicates in this explanation:

> The economic interest of the individuals, therefore, with increased knowledge of their *individual* interests,... without legislative compulsion, *even without any consideration of public interest*, leads them to turn over their wares for more marketable ones, even if they do not need the latter for their immediate consumer needs.[43]

The more marketable, transportable, durable and divisible a good is, the more likely it is to assume this intermediary role. If a particular good becomes pre-eminent in this respect, it will turn into something like money. Eventually, its status may be regularised by law.

This illustration suggests that at least the above schema is not hopelessly wrong. But it needs revision.

First, a minor revision is needed in (2) above. If the conclusion of an individualistic explanation contains *S*-predicates while the premisses do not, the deduction will not go through. There is a need for some kind of bridging or co-ordinating assumptions.

But (3) above calls for a larger revision. According to it, a 'rock-bottom' (as opposed to an 'unfinished' or 'half-way') explanation of *S* will ascribe to individuals certain properties (designated by the *I*-predicates in its premisses) which are taken as ultimate and, as it were, God-given. It may be that such individuals when placed in certain relations will engender something social (e.g. money) as an unintended result of their actions and interactions; but that is a happy co-incidence. There is no pre-attunement of the basic dispositions of individuals to collective needs.

This latter idea was given a dramatic expression in Hobbes's account of the state of nature. Now there is a very curious thing about that account: the reader is given the impression that everyone in the state of nature is a full-grown *male*, or at least that any women in it are like Amazons.[44] By contrast, Aristotle's account of the origin of the *polis* begins with the union of male and female and the development of families (which later coalesce into villages); and this is effected by natural instinct.

Obviously, Aristotle is right: there is a pre-attunement at least of sexual dispositions to collective needs. Men and women are made for each other (though not in the precisely programmed way in which male and female spiders are made for each other).[45]

The neo-Darwinian theory of group-selection allows us to go further than Aristotle. It tells us that the pre-attunement of an organism's inherited endowment to the collective needs of the group or population to which it belongs is by no means confined just to its sexual endowment. Indeed, as we saw, a main function of mechanisms evolved through group-selection is precisely to restrain the operation of sexual instincts so as to check the growth of a population above its optimum level.

I still hold that the right way to explain social phenomena is to exhibit them as the (largely unintended) resultants of the activities of interacting individuals. But I now hold that the evolutionary process has adapted human capacities to the needs of social existence (and left us singularly ill-equipped for solitary existence: it is a marvel that Victor kept himself alive in the forest). Aristotle's view that a human society 'exists by nature' is nearer the truth than Hobbes's idea that a commonwealth is a contra-natural artefact.

Do my two criticisms of Hobbes give grounds for cheerfulness? Or are they tainted by the kind of optimism that Schopenhauer condemned? (Schopenhauer declared that optimism, "when it is not merely the thoughtless talk of such as harbour nothing but words under their low foreheads," is "a really *wicked* way of thinking.")[46] Am I not guilty of optimism in talking of spontaneous co-operation between men and of an innate moral sense within men when we are bombarded with news of war and torture and enslavement and crime?

Well, I think that my two criticisms do somewhat mitigate Hobbes's grim picture of the natural condition of men. And I do not think that they are optimistic. For there is a certain ambivalence about the human capacities to which I have been drawing attention. When I spoke of spontaneous *co-operation* I used the word in a stretched sense. In this sense, two football teams 'co-operate' in playing against each other; and if Nicole had snapped back at M. Jourdain: "Fetch your slippers yourself, you lazy old man!" there would have been no impairment of their linguistic co-operation in accordance with shared, meta-conscious rules. It is this co-operative capacity that enables men to wage war, a social achievement of which other animals are not capable.

As to my idea of an innate moral sense: this is far less sanguine than that of, for instance, Francis Hutcheson, the leading representative of the 'moral sense' school in the eighteenth century. Hutcheson conceived a moral sense as a kind of cognitive faculty which enables us to perceive the objective moral qualities of actions. Except that it improves with exercise, this faculty was supposed to operate in an essentially uniform way in all of us. If two people morally appraise the same action differently, this must be due to some factual disagreement:

> 'Tis our reason which presents a false notion ... to the *moral faculty*. The fault or error is in the opinion or understanding, and not in the *moral sense*: what it approves is truly good; tho' the action may have no such quality ... Nor do ... mistakes about the moral qualities of actions prove either that we have no *moral sense*, or that it is not uniformly constituted.[47]

Just as it is one thing to say that children are able to catch on to the distinction between truth and falsity and a very different thing to say that they can tell true statements from false ones, so it is one thing to say that children are able to catch on to the distinction between right and wrong and a very different thing to say that they can tell what is right and what is wrong. It is easier to instil falsehoods into people who are eager for truths than into people who do not care much about truth and falsity; and it is easier to instil obnoxious morals into people with a moral sense than into people who do not care much about right and wrong. I do not know when it was first discovered that the best way to get men to do wicked things is to tell them that they have a higher duty to do them. In *La Trahison des Clercs* Julien Benda claimed that it was only quite recently that 'clercs' began to engage at all systematically in "the intellectual organization of political hatreds". Moral beliefs can be dangerous things, especially when ideologically organised. In this century terrible crimes against humanity have been inspired by moral idealism perverted by "a doctrine invented in Europe at the beginning of the nineteenth century",[48] namely nationalism.

In the last few decades there has been a new *trahision des clercs:* mentors of the young have engaged in a new kind of dialectical didactic which is really a sort of philosophical doublethink. Young people have been bombarded by dialectical equations ('tolerance = repression', 'democracy = totalitarianism', etc.) whose tendency must be to induce moral stupidity.[49]

For my part, I believe that philosophy can help a little here – not by preaching 'sound' moral doctrines, but by helping to create a critical and unsolemn intellectual climate in which neither the higher nonsense nor the lower nonsense has a good chance of survival.

London School of Economics

NOTES

[1] *Lev.* p. 167 & *EW* iii, p. 108. References to Hobbes's works follow the pattern used in my *Hobbes's System of Ideas*, Hutchinson, London, 1965; second edition 1973. One reviewer of that book complained, with justice, that my criticisms of Hobbes were 'rather desultory'. I hope to go a little deeper this time.

[2] This counter-claim certainly appears to have adverse implications for a position which I have often advocated in the past, namely methodological individualism. I will touch on this question in Section 10.

[3] See, for example, Chomsky, Noam, *Language and Mind*, Harcourt, Brace & World, New York, 1968, p. 69.

[4] 'Rules, Perception and Intelligibility', *Proceedings of the British Academy*, xlviii, London, 1963. Reprinted as Chapter three in Hayek, F. A., *Studies in Philosophy, Politics and Economics*, Routledge & Kegan Paul, London, 1967.

[5] Also at this seminar was Mr. Robert Ardrey, whose *Territorial Imperative* was then in the press. His fund of animal observations opened my eyes to the extent to which animal behaviour suggests the presence of something suspiciously like 'moral' instincts, a matter I will touch on in Section 9.

[6] See Ryle, Gilbert, *The Concept of Mind*, Hutchinson, London, 1949, Chapter II.

[7] "From these instances where action is guided by rules... which the acting person need not explicitly know... we must now turn to the corresponding and no less interesting instances where the organism is able to recognise actions conforming to such rules or patterns without being consciously aware of [them]" (Hayek, F. A., *Studies in Philosophy, Politics and Economics*, p. 45.)

[8] *Lev*, p. 107 & *EW* iii, pp. 195–6.

[9] The paper of Hayek's to which I have already referred opens with these words. "The

most striking instance of the phenomenon from which we shall start is the ability of small children to use language in accordance with the rules of grammar and idiom of which they are wholly unaware. 'Perhaps there is', Edward Sapir wrote thirty-five years ago, 'a far-reaching moral in the fact that even a child may speak the most difficult language with idiomatic ease but that it takes an unusually analytical type of mind to define the mere elements of that incredibly subtle linguistic mechanism which is but a plaything in the child's unconscious'." (*Ibid.*, p. 43)

[10] Whorf, Benjamin Lee, *Language, Thought and Reality*, (ed. by J. B. Carroll), M.I.T. Press, Cambridge, Mass., 1964, pp. 233f.

[11] See, e.g. Church, Joseph, *Language and the Discovery of Reality*, Random House, New York, 1961, pp. 64–5, and Brown, Roger and Bellugi, Ursula, 'Three Processes in the Child's Acquisition of Syntax', *Harvard Educational Review* **34** (1964), 144.

[12] Sapir, Edward, *Language*, London 1921, pp. 166f. In the Newspeak of George Orwell's *Nineteen Eighty-Four* "*whom* had been scrapped as unnecessary" (Penguin edition, p. 244).

[13] Chomsky, Noam, *Aspects of the Theory of Syntax*, M.I.T. Press, Cambridge Mass., 1965, pp, 6, 8; *Cartesian Linguistics*, Harper & Row, New York: 1966, pp. 3f.; *Language and Mind*, p. 10 and *passim*.

[14] As was duly noted by H. Vaihinger, in his *The Philosophy of 'As If'* (transl. by C. K. Ogden), Kegan Paul, London, 1924, pp. vii–viii.

[15] *De Corp* I, ii, 11 & *EW* i, p. 22.

[16] I, iii, 12 & i, pp. 39–40.

[17] *EW* vii, p. 81.

[18] What follows is based on the account in Brown, Roger, *Words and Things*, Free Press, New York 1958, pp. 3f.

[19] R. Carnap, *The Logical Syntax of Language*, Routledge & Kegan Paul, London, 1937, Part II.

[20] "The language is 're-invented' each time it is learned", Chomsky, Noam, *Language and Mind*, p. 75).

[21] See Brown, Roger, *Social Psychology*, Free Press, New York, 1965, pp. 286f. for a case study of this sort of child-mother co-operation.

[22] These examples are taken from Halliday, M. A. K., 'Grammar, Society and the Noun' (An Inaugural Lecture delivered at University College London, 24 November 1966), p. 8.

[23] Collingwood, R. G., *The Principles of Art*, Clarendon Press, Oxford, 1938, p. 227.

[24] See the symposium by Chomsky, Putnam and Goodnam on innate ideas in *Synthese* **17** (March 1967). Goodman called it "an intrinsically repugnant and incomprehensible theory" (p. 27).

[25] Lorenz, Konrad, *Evolution and Modification of Behaviour*, Chicago University Press, 1965, p. 13.

[26] Hume, *Treatise* Bk. II, Pt. I, Sect. X (Selby-Bigge edition p. 311). For Nelson Goodman, see his *Fact, Fiction and Forecast*, Athlone Press, London, 1954, pp. 44f.

[27] *Lev*, p. 24 & *EW* iii, p. 41.

[28] Standard Edition, Hogarth Press, London, Vol. XXI, pp. 123f.

[29] *Social Psychology*, p. 19.

[30] Wynne-Edwards, V. C., *Animal Dispersion in Relation to Social Behaviour*, Edinburgh, Oliver & Boyd, 1962.

[31] I am relying on Wynne-Edwards's account, *ibid.*, pp. 389–390.

[32] He later became Director of the London School of Economics.

[33] On this large subject see Robert Ardrey's readable and scholarly *The Territorial Imperative*, Atheneum, New York, 1966.

[34] *Ibid.*, p. 131.
[35] *On Aggression*, Methuen, London, 1966, pp. 110–111.
[36] Quoted in Robert Ardrey, *ibid.*, p. 284.
[37] Hayek, F. A., *Law, Legislation and Liberty*, Vol. 1: *Rules and Order*, Chicago University Press, 1973, p. 79.
[38] Most of these papers have been assembled in Krimerman, Leonard I. (ed.), *The Nature and Scope of Social Science: A Critical Anthology*, Appleton-Century-Crofts, New York, 1969, pp. 457–472, 603–611, 621–624; and also in O'Neill, John, (ed.), *Modes of Individualism and Collectivism*, Heinemann, London, 1973, pp. 143–184.
[39] As I called it (*Brit. Jour. Phil. Science* **8** (1957), 106).
[40] So far as I know, this term was invented by Ludwig von Mises. See his *Epistemological Problems of Economics* (transl. by George Reisman) D. Van Nostrand, New York, 1960, pp. 40f.
[41] Menger, Carl, *Problems of Economics and Sociology* (ed. by L. Schneider and transl. by F. J. Nock; University of Illinois Press, Urbana, 1963, p. 152; italics in the original.
[42] In other words he rejected what Popper calls a 'conspiracy-theory" for money. See Popper, K. R., *The Open Society and its Enemies*, Chapter 14.
[43] *Ibid.*, p. 154, italics in the original.
[44] So far as I know Hobbes only once mentions the existence of women in the state of nature (*De Cive* ix, 1–6 & *EW* ii, pp. 115–8). He makes several references to Amazons.
[45] "A young salticid spider which ... approaches a female must neither mistake another species for his own, nor must he perform the signalling of his specific courtship dance in any other way than the one to which his female responds; otherwise, he would be eaten by her immediately. He has no opportunity in his short life to gain any information about what a female of his species looks like, nor what movements he must perform to inhibit her feeding reactions and to stimulate her specific mating responses." (Konrad Lorenz, *Evolution and Modification of Behavior*, p. 25.)
[46] *The World as Will and Idea* (transl. Haldane & Kemp), Vol. I, p. 420.
[47] Hutcheson, Francis, *A System of Moral Philosophy*, London, 1755, Vol. I, p. 91.
[48] Kedourie, Elie, *Nationalism*, Hutchinson, London, 1960, p. 9.
[49] Thus Marcuse, who advocates liberation through "thought in contradiction"[a] and condemns logic as an instrument of domination,[b] constantly equates liberal democracy with totalitarian authoritarianism. He has also declared that "The democratic abolition of thought... is brought about in the 'higher learning'",[c] that "the prevailing mode of freedom is servitude",[d] that "the realm of the irrational becomes the home of the really rational",[e] etc. See Herbert Marcuse, *Negations*, Allen Lane, London, 1968, pp. xx and xiii for passages flagged, respectively, by *a* and *c*, and his *One Dimensional Man*, Sphere Books, London, 1968, pp. 105, 80, 194 for, respectively, passages flagged by *b*, *d*, and *e*.

MARX W. WARTOFSKY

THE RELATION BETWEEN PHILOSOPHY OF SCIENCE AND HISTORY OF SCIENCE*

In the heat of a debate, it is often hard to see how one-sided and sometimes foolish the opposed positions are. The debate on the relation of the history of science to the philosophy of science is so patently an unreasonable one that one has to step back to see it in its full absurdity. Further, in order to understand why reasonable and often intellectually superior people have been caught in this absurdity, it will help to examine the alternative positions and to suggest how and why they developed as they did, both conceptually and historically.

The problem posed by the debate is the following one: *Two disciplines – philosophy of science and history of science – are both concerned with what is ostensibly the same thing – namely, science. Are the two related?* The problem in this form is vacuous. It is resolved vacuously in the way the question is put. If the two disciplines are *about* the same thing, then that is already *a* relation. But obviously a vacuous one. The issue is: *How* are they non-vacuously related? Or: *Are* they in fact about the same thing?

In this paper, I will argue:

First: that the relations are crucial ones, for both disciplines; crucial in the sense that a non-philosophical history of science is a fatally flawed history of science; and that an ahistorical philosophy of science is, similarly, totally inadequate to its task.

Second: that *such* inadequate and flawed history and philosophy of science do exist, and that the debate arises out of just these crucial inadequacies; but that its present terms do not lead to a resolution of the problem, but only confound it, or mystify it.

Third: that the conception of science, and of the activity of science – if you like, the ontology of science, what science is – is what is at issue; and that this conception needs to be recast. In brief, I will argue that science is neither simply a body of knowledge, in some stable form, whose anatomy can be understood by dissection and reconstruction. I therefore reject the rational reconstruction view as it is represented in

R. S. Cohen et al. (eds.), Essays in Memory of Imre Lakatos, 717–737. *All Rights Reserved*
Copyright © 1976 by D. Reidel Publishing Company, Dordrecht-Holland

the dominant philosophy of science of the last half-century. Nor is it simply a sequence of practices, discoveries, beliefs or forms of institutionalization. Nor is history of science a chronicle of *facts* concerning all of these. But further: neither is science, or its history, simply an internal or rational dialectic of problems and solutions, conjectures and refutations, proposals and replacements. I therefore reject a series of views on the history of science, as either factual (external) chronicle, or internal dialectic, though I accept some of the fundamental premises of such historical views – namely, that science changes, grows, or develops; and that theoretical criticism and replacement of theories are crucial to this growth.

The alternative ontology of science which I will propose is that it is an historically evolved mode of human cognitive action or *praxis*, – and that as such, it is essentially a teleological activity. Its character or nature cannot be understood apart from its purposes *in this* context – i.e. its purposes or cognitive praxis in socio-historical context. Further, that such a teleological mode of action is essentially theoretical and that is what distinguishes it from other modes of action, including other modes of cognitive activity; and that the growth and change of scientific theory has both an internal theoretical dimension, and an external one, having to do with the genesis of theory, the uses or practices to which the theory is put, and the conditions under which theories change or are replaced. Such an ontology of science, therefore, construes science essentially as a mode of cognitive human action or praxis; as such, teleological; and distinctively theoretical.

Fourth: I will argue that such a view of science requires, as *its* adequate theory of science, a philosophical history of science, and an historical philosophy of science, though what these terms mean will have to be elaborated more fully later. Insofar as I will characterize science as *cognitive* human action or praxis, the emphasis is on science as a knowledge-acquiring activity, or more specifically as a truth-seeking activity. Philosophically, this establishes *epistemology* as the broader context for a characterization of this activity. But because this truth-seeking activity is to be understood in the context of its historical development, what is required is an *historical epistemology*. My own argument is therefore part of a broader program of what I will call *historical epistemology*, and which is to be distinguished from some of its look-alikes, now emerging in philosophy and history of science – namely 'naturalistic epistemology' (Dewey, Quine, Campbell) or 'evolutionary epistemology' (J. Piaget, J. Herrick, Toulmin, Shimony, H. Yilmaz and to some extent Popper and Feyerabend). Broadly speaking, but in senses as yet unspecified, historical epistemology is Marxist in its orientation, and needs to be further characterized, especially with respect to other contemporary views which seem similar to it – e.g. those of G. Bachelard, Althusser, Foucault, and D. Lecourt, who represent some of the present discussion on these issues in France. But this remains to be done.

So much for the program of my paper and its context.

I

In order to examine the present debate on the relations of the history and the philosophy of science, it might be useful to do a typology of the relations which could be conceived to exist between them. It will be easily seen what historical and contemporary views fit these types, and I will sketch some of these examples, and their history. These views will be seen to fall into five types, the last of which I offer as an appropriate resolution of the problem. These are (1) AER, (2) DER, (3) AIR_1, (4) DIR, (5) AIR_2, (to be spelled out in due course).

(1) *Agreeable External Relations:* Since philosophy and history of science are *about* the same thing – science – the division between them is merely a division of labor. Philosophy deals with the philosophical aspects of science – i.e. the logical, epistemological and metaphysical aspects, and perhaps even the value-theoretical aspects – and history deals with the historical aspects – 'who did what, when, and with whom'. On this view, history and philosophy of science are *externally* related to each other, in an *agreeable* division. That is, each recognizes the other's domain, and accepts, the territorial division, either explicity or tacitly. There is, however, no relation between history and philosophy of science of an internal sort. Not only do they not impinge on each other at all, but rather the concepts and modes of discourse of one do not effect the concepts or modes of discourse of the other. And of course there is no debate. This view represents a philosophical fiction, like the *state of nature* in political theory. It never existed but it is useful as a starting point, and makes the schema symmetrical.

(2) *Disagreeable External Relations:* Here, hegemony is claimed either by one or the other or both of the disciplines, to the exclusion of the other. In effect, the claim is that the other is fundamentally inadequate to the task, or construes the common subject – science – in a wrong way. What this means is that there are two competing *theories* of science, mutually incompatible; and therefore, the debate becomes open. It is not over the methodological relation of the history to the philosophy of science, but whether science itself is to be construed in one or another way. Yet the *outward* form of this debate may become one between alternative *philosophies* of science, under the impact of historical considerations; or between two theories of the *history* of science, under the impact of philosophical considerations – and that in fact is what has happened.

Stage one of the debate concerned the *rejection* of historical contexts (as well as sociological and psychological ones) by the logical-positivist, and logical empiricist schools of philosophy of science. It may be necessary to characterize these views more fully later on, but for present purposes it is enough to say that this school – let's call it *rational reconstruction with an empiricist basis* – founded its view of science on the division between *contexts of discovery* and *contexts of justification*, taking the latter as its proper domain. Though this sounds like *agreeable* external relations and division of labor, (assigning the contexts of discovery to the history, sociology and psychology of science) it was not, and I will say why later. Justification demanded (a) the rational reconstruction of the *achieved* body of knowledge of contemporary science, i.e. the recasting of this body of knowledge – construed as a set of statements – in the logical form of premises and conclusions in a system of valid inference (ideally in axiomatic form); (b) an empirical or observational base for the verification, or confirmation, or corroboration or acceptability of scientific statements. (The details are unimportant, and so are the alternative verificationist and falsificationist alternatives here.) Justification was therefore to derive from a universal and subject-neutral way of linking theory to observation and experiment; and because universal, it was conceived as *ahistorical.* However, there was tolerance for alternative theoretical frameworks, as long as each could be constructed rationally and consistently, and could be linked by coordinating or interpretive definitions to the empirical base. And there was also tolerance as to how the empirical base could be chosen – i.e. as to what should count as protocol-sentences or basic predicates. This tolerance, however, led to a relativization of the reduction-base which has made the original project of justification bankrupt, at least on the terms originally conceived. (Details will be given upon request.)

The alternative *philosophy of science* which developed on the grounds of this very liberalization of the observational base itself, had its sources in two different contexts: (1) in the older philosophical relativism, rooted in classical skepticism, for which the simple notion of a subject- or context-neutral 'observation-statement' was always suspect. This view construed all observation as a function of context: e.g. interests and purposes, or conceptual frameworks, or cultural set. Being ahistorical, at least philosophically, this alternative conceptual framework had to rediscover this observational relativism in its available contemporary form, in the linguistic mode – namely in the Wittgensteinian distinction between 'seeing' and 'seeing as', or in its juicier versions, in Wittgenstein's "forms of language are forms of life" dictum – an incipient but underdeveloped sociolinguistic thesis already popular in the works of B. L. Whorf and Edward Sapir, and old hat to an old metaphysician like Collingwood. (I won't go into the venerability of this tradition here, since it is embarrassing.) (2) On the other hand, this alternative philosophy of science drew its sustenance from historical considerations: The philosophers of science had *discovered* (!) that theories change, that science replaces old theories with new ones, and that there is at least a problem in understanding the *growth* of science. The older notion that science progressed by incremental and cumulative improvements within the one true theory began to crumble before the historian's account of radical changes and replacements in theory.

This historical influence was aided and abetted by what may be characterized as a naturalistic epistemology claimed that, as a cognitive practice, theoretical scientific activity grew out of and was involved in the practical, social and psychological contexts of success and failure; in the animal behavior that scientific inquiry had its basis of trial and error; and that science was also rooted in, and influenced by the social contexts of ideology, belief, and in the common forms of group and institutional behavior. Therefore what entered into the fabric of this theoretical mode of social cognition which is science, were purposes and interests not always conscious, not always rational and not always explicit. Such a naturalistic epistemology thus linked the analysis of scientific inquiry to the more general analysis of inquiry as a social and natural mode of human activity; and though such a naturalism was not yet historical; it took the arena of social and cultural history into account.

The further weakening of both the logicist and the empiricist dogmas, by internal criticism (e.g. by Quine and Hempel), by the dissolution of the older analytic-synthetic distinction- all contributed to the growth of an alternative philosophy of science, (whether as Hansonian, Polanyian, Kuhnian, Feyerabendian, Lakatosian, Agassian, or Quinean. I will not go into the independent bases for this alternative view here, since they are treated elsewhere in this volume and in the current literature at great, if not enormous, length.)

Thus, within the philosophy of science, *Disagreeable External Relations*, in stage one, generates a split between two philosophies of science – one of which (rational reconstruction) had become all but canonical, in its ahistorical mode; the other of which threatened to usurp this hegemony, with a peculiar link to history of science.

Stage two of D.E.R. concerns the history of science itself: its disagreeable external relations were with the logical-positivist, logical-empiricist philosophy of science, wich it took to be as irrelevant to science (and not just the history of science) as the philosophers took history of science to be irrelevant to justificationist philosophy of science. Now to be fair, rational reconstruction was not opposed to history of science, and regarded it as useful to know – if not as a history of past mistakes, then at least as a resource for examples, especially in classical physics. What rational reconstructionism had no use for was *philosophical* history of science, that is, history of science construed from the point of view of a philosophy of history, specifically in Hegelian or dialectical modes. Such 'history' was regarded as metaphysical, in the worst sense. But even apart from this the *rational reconstructionist* philosophy of science has always been uncomfortable with history generally. It is a messy domain, perhaps the most recalcitrant of the social sciences, in terms of reduction to the manageable frameworks of the hypothetico-deductive or nomological model of explanation. A word needs to be said here in defense of the richer sociological contributions of early logical positivism, e.g. in the works of Edgar Zilsel, of Otto Neurath, and even of Hahn and Reichenbach. But that involves an assessment of the *earlier* Vienna Circle in contrast to its later epigonic stage – the one we are mainly concerned with.

It is useful to stop here for a moment to review the alternative *ontologies* of science presupposed by these different views, especially since I will argue for a fundamental ontological replacement later.

By an ontology of science, I don't mean a scientific ontology – i.e. an ontology of the domain over which science ranges, but rather the ontology of the entity *science* itself.

The rational reconstructionist ontology may be sketched thus: Science is a *body of knowledge*; or at a given stage, the synchronous time-slice of this body of knowledge; it is constituted, for purposes of characterizing it, by a set of statements related to each other logically, i.e. by rules of inference. These are the acceptable statements: some of them may turn out to be true, others false. But all are entertained as truth-claims or as proferred truths, or as verifiable, or falsifiable – the verificationist, confirmationist, corroborationist, falsificationist issue doesn't concern us here – all are brothers under the skin). The constitution of this set of statements is by means of the agreement of the scientific community as to their acceptability. But the *genesis* of these statements, their history, and the *actual* conditions under which they come to be agreed upon (as against the formulation of ideal or formal rules or conditions under which they *ought* to be agreed on) – all of these lie *outside* and *beyond* the task, (and therefore outside the ontology) of the rational reconstructionist view of science. They are rather held to lie in the domain of the context of discovery. The only statements about which we can say that they are generated *within* the system, are those which are deductive consequences of others. And if they are reconstructed as logical consequences, such statements may be said to already be *contained* in the system, and merely actualized for practical purposes by valid inference. Otherwise they are synthetic statements which may be *added* to the system from the outside, as observation or protocol statements – i.e. those based on subject-neutral or intersubjectivily acceptable experience, according to the canons of the science. So-called statements of initial conditions, measurement statements, and singular statements are of this sort. Although such statements may be weakly construed as 'historical', in that they are not timeless deductive consequence in the system or set of statements, and insofar as they may be time-indexed, yet they are not historical in any other sense, and their historicity (and their time index) is indifferent to the logic of the system.

In short, within such a rational-reconstructionist ontology of science, the *context of justification-context of discovery* dichotomy leaves only the context of justification as the appropriate domain of the philosophy of science. The ontology of science which is determined thereby is that of an *a*historical, synchronous (and time-independent) body of *achieved* scientific knowledge. The task of philosophy of science is thus taken to be rational reconstruction of this achieved body of knowledge – (or of the set of statements which express it) – ideally, in the form of a logical system, and ultimately, in an axiomatic form.

The ontology of science of the historian of science is sharply different. Here, the *events* of science, in their sequence and interconnection, constitute a diachronic content to be reconstituted from the record (or the *archive*, to use Foucault's term), and includes *both* the theoretical statements or systems, and also the actions, institutions, beliefs, motives and interests of the community of scientists. This sequence of events reconstructed from the record may be construed as a history of scientific theories, and of internal practices (e.g. modes of experiment, instrumentation, measurement). Or it may be construed in a wider sense, to include, as part of the history of science, the economic, political, social, technological, biographical, i.e. 'external' contexts in which scientific theories and 'internal' scientific practises arise and fall.

In this historian's ontology of science, the *logic* of the historically reconstructed sequence may be one of causal relation, either as a simple linear succession of causes and effects, or as a cumulative development, e.g. as the unfolding of reason, or of form, or of God's purpose, or of truth. Or again, the *logic* of the history of science it may be conceived dialectically, as one of conflicts and resolutions; or as the progressive revelation and elaboration of an invariant theme, whether *Identity*, or *Economy of Thought*, or *utility*. Each of these, then, may be taken as characterizing an ontology of science, i.e. the kinds of entities, events, processes, causal relations, properties, which constitute the domain of science, and the course and means of its development. In short (in any of the above modes), science may be seen to have a *thematic* history, or one which has a logic of a sort; or it may be reconstructed as sheer chronicle, allegedly recording nothing more than the uninterpreted facts, or the sequence or temporal order of events. The ontology in any case, is one of events, or of actions, or reasons, or motives, or purposes, but all of these as a diachronic process. Occasionally, the synchronic 'time-slice' 'history' of a given period seems to violate this condition: But even here, the account is not time-indifferent: the temporality of the events, and their sequence, is crucial.

I will argue later that both diachronic and synchronic ontologies are, as such, inadequate; though aspects of both need to be retained in an adequate ontology of science.

To return to our typology:

(3) *Agreeable Internal Relations: Classical:* This type is qualitatively different from the others. Here, by internal relations, I mean the interaction of philosophy and history of science, in such a way that each, in part at least, constitutes the other, or gives it the characteristics it has. This is where we should end, optimally, for then, we would presumably have an historical philosophy of science and a philosophical history of science. But I speak here only of the *classical* stage, which proceeds the *modern* or contemporary one, i.e. of *disagreeable internal* relations. The classical stage includes the classical philosophical histories of science and the classical historical philosophies of science. In fact, the classical stage is classical precisely to the effect that historical and philosophical *divisions* do not yet appear in it. It is like Anaximander's *indeterminate* in which the different elements are contained, but not yet separated out.

In the past, in what might count as a Golden Age, the history of science was itself construed as philosophical history, and indeed as a normative history. Such a history could be constructed as embodying an *a priori* norm of rationality; or, if not, the norm could be discovered and reconstructed by attention to the historical evidence itself. Here, Hegel sets the pattern: The notion of a dialectical history of ideas was first discovered by Hegel as the *modus operandi* of the history of philosophy itself. Though the models are much older – (i.e. of Reason, or God, or the Idea elaborating itself in the vehicle of history, in the actions or thoughts of men, or in the City of Man) – Hegel distinctively constructed the full *logic* of this development (first recognized in the history of philosophy) in his *Phenomenology* (as a science of consciousness), in the History of Nations, and of Society and Law (in his *Philosophy of*

History, and his *Philosophy of Right*) and *also* in his work on the philosophy of nature and of science in which nature itself partakes in this history, and becomes self-conscious in the reflection of this process in theoretical consciousness (in his *Philosophy of Nature* in the Encyclopedia). This is not, properly speaking, history of science as yet, though Hegel studied the history of the sciences and of mathematics for the materials of his *Naturphilosophie*. His interest was systematic rather than simply historical or reconstructive, as such. One might call this sort of construction *tendentious* history of science, in the sense that the history of science was seen as innately philosophical, as the vehicle for the elaboration of an Idea. This *tendentious* history of science was philosophical in the sense that Aristotle in the *Poetics* demarcated *Poetry* as philosophical by contrast to History as mere chronicle (the Ideographic-nomothetic distinction). What was sought, in history, were the evidences of a universal (or if one prefers, of an invariance). History was in the service of a Universal Idea,

1. whether the Idea was that of Reason unfolding itself by critical negation (Hegelian 'falsificationism' – which would lead one to characterize Popper as a crypto-Hegelian in a mode of psychological abreaction; and Lakatos as a closet-Hegelian, because *he* knew he was one, but didn't want anyone else to. But more on this later);
2. or whether the Idea was that of eliminative positivism, in which the history of science was a history of the overcoming and elimination of dogma, superstition and metaphysics by 'positive fact';
3. or whether the Idea was 'Economy of Thought' as the adaptive, biological form of a principle of least action as a strategy of cognitive success (e.g. as proposed by Ernst Mach);
4. or whether this idea took the form of the reduction of all phenomena to the intelligible or rational form of causal law, or of some overarching Principle of Identity (which Myerson saw revealed in the history of physical science).

In all these cases, and in others besides (e.g. in Poincaré, with *invariance*; in Peirce, with asymptotic approximation to reality, and increasing determinateness), there is seen to be some general or universal heuristic or methodological principle by means of which the history of science is organized, and through which it can come to be understood in its development. Poincaré and Duhem had both insisted on the essential role of theoretical *pre*conceptions within science itself (without which, as Poincaré had said, all experiments would be sterile; und upon which, as Duhem had pointed out, experiments so much depended). This same approach to the history of science was characteristic of the most fruitful and epochal historical works of this period (e.g. Mach's *Principles of Mechanics* and *Root and History of the Principle of the Conservation of Energy*; Meyerson's *Cheminement de la Pensée*, and *Identité et Realité*; Duhem's *Aim and Structure of Physical Theory*). The methodological principle of starting with a priori assumptions within science itself had already been proposed by Kant in two works: The *Metaphysische Anfangsgrunde der Naturwissenschaft* and *Von Übergnag von den Metaphysischen Anfangsgrunde der Wissenschaft zur Physik*. Why not, then, take this same approach to the study of the *history* of science?

Thus, in fact, with such names as Whewell, Mach, Duhem, Poincaré, Meyerson, Lalande, Brunschvicq we pass over from what might be construed as purely *a priori* reconstructions of science and the history of science in the service of an Idea, in Kant and Hegel, to a modified philosophical history of science and historical philosophy of science as two interweaving aspects of a *single* discipline. The internal relations of history and philosophy of science were not only agreeable, in this classical age; they were essential since the single unified discipline was constituted by this interrelationship.

(4) *Disagreeable Internal Relations – The Modern Period:* The idyllic classic age or Golden Age of the philosophical historians of science was followed by an unravelling or separation of the elements already contained in their thought. Mach was the great inspiration of the logical positivist Vienna Circle (i.e. the *Mach Gesellschaft*), and his methodological dicta concerning reduction of all theoretical formulas to sensory elements became a systematic basis for the new logical construction or reconstruction of the sciences ('Logische Aufbau') undertaken by Carnap and his followers. Duhem's so-called conventionalism, first developed by him as a solution to the problem of theoretical change in the history of science, passed over into a pragmatic relativism with respect to linguistic (and therefore also ontological) frameworks which we find in Carnap and Quine. These systematic concerns with the logical and linguistic reconstruction of science constituted a distinct break with the historical concerns of the earlier generation.

Lenin was among the first to launch a full scale realist attack upon what he saw as the epistemological consequence – (i.e. phenomenalist idealism) – of the new empiricist positivism which derived from Mach (in Lenin's attack upon Bazarov, Bogdanov, Yushkevich and others of the Russian school of empirio-critics, and upon Poincaré, Duhem, Karl Pearson and the young Philip Frank, among others – in *Materialism and Empirio Criticism* (1908)). But the *modern* strife between history and philosophy of science did not reemerge until some philosophers, in collusion with historians of science, provided the historians themselves with a *philosophical* rationale for their attack on positivism.

A most important early work here was E. A. Burtt's *The Metaphysical Foundations of Modern Science*, (1924, revised 1932). In this work, that very metaphysics which the positivist program (in those *same* years) was trying to eradicate and eliminate from rationally respectable science as a hangover from its *pre*-scientific stage, was placed instead at the very foundations of modern science – i.e. at its *historical* origins, in Copernicus, Kepler, Galileo, Descartes, Gilbert, Boyle and Newton. Nor was it ascribed to the 'backwardness' of these minds, as a remnant of their prescientific training. It was seen as essential to the theory-formation of classical physics itself. Duhem tried to push metaphysics out the front door (in *The Aim and Structure of Physical Theory*) but immediately invited it in the back door, splitting himself into positivist-scientist on the one hand, and 'believer' on the other. Popper, in his ambivalent campaign for a realist epistemology, also permitted a pale and anemic ontologism in through the back door: After first demarcating science from metaphysics, he then proceeded to find a role for metaphysics in the *critical* dialectic of scientific thought. His students and followers – Feyerabend, Agassi, Lakatos, Watkins pushed the door open further. And finally, because of their interest and work in the history of science, their increasing rejection of the demarcation criterion between science and metaphysics, and their sharpening anti-positivism and anti-sensationalism (anti-inductivist-empiricist), the influence of these followers and students of Popper upon the historians grew. The Popperians, ex-Popperians and post-Popperians at least took history of science and historiography of science seriously. (The two crucial works here are Lakatos' *Proofs and Refutations*, dealing with a case study in the history of mathematics, and Agassi's *Toward a Historiography of Science*).

Popper's effects were enhanced by those of Polanyi – himself a scientist of note, turned philosopher-critic-historian and methodologist of science. His strong anti-positivism, and his insistence on the personal and tacit dimensions of scientific thought – in discovery, and in the understanding of science – also fed the fires of the anti-positivist, pro-historicist mood of some historians of science. Also, two philosophers – both in one or another fashion historically oriented, both rejecting the received wisdom of logical reconstruction, and both strongly influenced by Wittgenstein – added insult to injury by resuscitating the contexts of discovery and rejecting the straitjacket of the context of justification. Toulmin, in *Foresight and Understanding* and Hanson, in *Patterns of Discovery*, both recovered themes which had been earlier rejected as psychological, and therefore beyond the pale.

There is more history to this story, but its happy and serendipitous culmination came when a bona-fide historian, Thomas Kuhn, with an excellent book on Copernicus to his credit, and with the influences of his colleagues – Leonard Nash and Paul Feyerabend, and of Polanyi as well – came out with a *blockbuster* – *The Structure of Scientific Revolutions* – an irritating, naive, confused and provocative work, which nevertheless, brought it all together in one glorious explosion.

The sociology of the reception of *The Structure of Scientific Revolutions* is an interesting subject in its own right, and I won't try to get into it here. But the results are well known. Anti-positivist, anti-reconstructionist *history* of science had found a naive but provocative formulation which shook up the philosophers of science. Small wonder. They knew little history – especially history of philosophy – so it all looked new to them. They were incompetent sociologists, so the sociology of science alleged in Kuhn's work simply had to be accepted enthusiastically or rejected violently on a priori philosophical grounds. What the philosophers *could* do was talk about paradigms (which they did endlessly. But the effect, one way or another, was to turn many of them to the history of science, and to afford a new legitimation to the philosophical reflection on this history, whether in criticism of or in agreement with Kuhn.

Well, here we are, with a Disagreeable Internal Relation between philosophy of science and history of science – disagreeable because the old *paradigm* – (excuse me, I mean *model*) – of inherently rational science to be explicitly rationally reconstructed, *and* the old model of the subject-neutral and firm reduction base in observation had been eroded from *within* philosophy, and on the grounds of an *historical* reconstruction of the growth of science.

The typology thus far is missing one component: the *Modern Stage* of *Agreeable Internal Relations.* But this will require a formulation of the internal relations between history and philosophy of science which is more than simply an account of the relations between two disciplines – and also more than the next stage in this schematic dialectic of their relations. It will require a more systematic *ontological* reconstruction of the entity called *science*, i.e., a better philosophy of science; so that the relation between the study of its history, and the study of its epistemological and logical character will flow from that reconstructed ontology itself. All I can give here is a programmatic sketch of a larger enterprise – if you like, a research program – and one which I noted briefly at the outset in setting forth my own theses.

II

Perhaps I should begin by rejecting two close approximations to the program: Kuhn's and Lakatos'.

Kuhn characterizes science as a social enterprise, constituted by men with beliefs, needs, ambitions; but all institutionalized in their modes of action as scientists – i.e. constituting a scientific community. The shifts and changes in this enterprise are seen as changes of a human sort – not only rational, but social-psychological as well. All of this is set into the context of *stages* of scientific growth, each with its investment in a technology – i.e. not only in the usual sense, but also in the sense of an articulated cognitive and practical technology of rules, procedures, accepted truths, and modes of conceptualisation and of practice. Therefore each such stage, i.e. so-called normal science, exhibits a built-in *tenacity* to remain unchanged. *Proliferation* [I borrow the tenacity and proliferation vocabulary from Feyerabend, who borrows it, in part, from Peirce] comes from within the community, but for reasons not yet clear. In the original version in *The Structure of Scientific Revolutions*, Kuhn has the shift take place when anomalies are obvious in the old framework – but not because they are. In this sense, he rejects the dialectic of critical negation as too formalist and rationalist a model of what *actually* goes on, which he recognizes as much messier. But in place of this 'clean' rational dialectic, Kuhn resorts to the irrational and nonrational as the motive power of historical changes in theory. Thus, he speaks of 'leaps of faith', and quotes Planck's famous remark about a theory dying only when its proponents die. There is a teasing authenticity to this humanized and socialized account of science. It gives a description of the historical sequence, and even suggests that since that *is* how science *did* grow, that's how it *should* grow. *Is* converts to *ought*, by the *success*-transformation. So Kuhn can say (in his 'Reply to Critics', in *Criticism and the Growth of Knowledge*) that his is not just a value-free descriptivist sociologism, but is, 'in a sense', also prescriptive.

Lakatos, in his merry way, had in the meantime cooked up a methodology for science which is frankly normative, by (his) contrast to Kuhn's (which he criticized as non-normative). Rejecting Popper_1 .... Popper_n, Lakatos wants to find, in the rational dialectic of criticism within scientific theory, that invariance which demarcates good from bad science. His unit is not the theory, and his criterion is neither truth nor falsifiability. It is, rather, whatever advances scientific research, broadens it, solves problems better than what preceded it. The unit is not the theory, but the program of research (more complex, but no clearer than the paradigm. The phrase 'research-programmes' is Agassi's originally, in this context.) So some of these are *progressive*, some *degenerating*. Not everything goes. And since *all* theories are *born* false – (a straight out Hegelian view, since all theories are incomplete, and if thus inherently fallible then false – no theories are privileged in this way *except* insofar as they *increase* knowledge. Dialectic there is – but history as such provides no norms. Only the principle of selection, within the history of science, of those programs which are *progressive* or which shift the cognitive activity of science in a progressive direction by leading to *new* knowledge – only such a selection principle yields the *normative* dialectic of good science, and hence, of rationality itself, in the example.

I cannot pretend to cope with these two views here, except to caricature them both: With Kuhn, we get history without dialectic; with Lakatos we get dialectic without history (by the use of history as no more than the source of examples of good science, or of the norm of science). The mediation is obvious. And also obvious is the requirement that we transcend these two views, and not merely combine them. What we need therefore is a dialectical history of science. But having said this, it should be clear that the norms of such a science must be derived from the historical context itself and cannot be imposed upon it by philosophical fiat, or by some supra-historical *a priori* conception of rationality. This can't be done by finding an invariance in the history, by some abstractive or inductive means. It requires, I think, a characterization of the *historical enterprise* called science, in a new way, in order to see in it the sources of its *intrinsic* normativeness.

I have to be very brief and elementary here. I start with the characterization of science as a mode of cognitive human action, or praxis. Like all human action or praxis, this activity is teleological. By this, I mean not only that science serves some function – for so do the lungs, and so do leaves on a tree, and that's not what I mean by teleology – but that it is *intended* to serve some purpose. That is, it is undertaken by conscious agents in order to serve some purpose. It is, in short, intentional activity. Moreover, like every mode of human action, it is social. That is to say, its context is social human existence, and its teleology is a social teleology. The needs it serves, and its conscious purposes in serving these needs, are social needs, and socially conscious purposes. As *cognitive* praxis, it serves these needs by means of the activity of acquiring knowledge. Science, therefore, is a practice of acquiring knowledge, whose presupposition is that such an acquisition of knowledge itself serves a social human need or purpose.

Now there are other modes of cognitive praxis – in fact, no form of *human* praxis is non-cognitive. For example, sheer skill or craft, e.g. knowing how to proceed, is cognitive in this sense. So too is all ordinary human action. I would argue that all, i.e. all the forms of social human interaction, from interpersonal to social and political activity, to theoretical activity, are cognitive, involving either language, or some other symbolic mode of communication, and requiring the socialization and learning of the rules and concepts of one's peers, one's parents, one's social group, and one's profession.

What is distinctive of science, as a mode of cognitive praxis, however, is that it involves *theory*. That is to say, what science acquires as *knowledge* is neither skill nor craft nor social rules, though the imprint of all of these is upon the scientific mode of cognition. What it acquires as knowledge is *theoretical* knowledge. *Science is thus the theoretical mode of cognitive human praxis*. The history of science is therefore the history of this distinctive mode of cognitive activity and of the conditions of its genesis, growth and change.

But what is theory? Here I want to be perversely brief, and give you my thesis, but not yet its argument: Theory is the mode of knowledge whose object or content is truth (and *not* convenience, utility, pleasure, or rightness, etc.). The very formation of the concept of truth is the genesis of theoretical knowledge. To know a skill, or a craft, or to know how to get on with others, is *not* a matter of knowing what is true. Such cognitive activity is instrumental in its own way, at its own level. But social development of the species introduces a distinctive value in human life, not present in other species as far as I can understand, namely, that of reflectively formulating assertions whose value (in both senses: as *survival* value, and as propositional value) is truth; or to put it differently, assertions which may be true or false, i.e. proferred truths, or truth-claims.

Part of the program of an historical epistemology is to account for the genesis of this feature of human cognitive praxis, not simply as a biological feature of adaptation, but as the product of human social evolution – i.e. as a postbiological or *historical* achievement. The behaviorist reduction of truth to a matching-function between two states (one a propositional state or a verbal behavior and the other a state of the environment), simply won't do here, though the instinct for correspondence is sound. The relation is much more complex. But suffice it to say here, sketchily, that only when an acquired cognitive achievement – e.g. a skill, a craft, a mode of action – can be detached, so to speak, from its actual performance, and reconstructed or represented symbolically, in a language, can the cognition *begin* to approach the conditions of a truth-claim, i.e. the condition of theory.

In linking the genesis of theory to the formulation of truth-claims, i.e. of assertions. I also want to make an instrumentalist claim without proposing an instrumentalist theory of truth: Cognition is instrumental in serving human needs and purposes. Its warrant is its social utility. But cognition would have no utility, if *truth* had no utility, i.e. if theoretical knowledge had no utility. The condition for the genesis of theory as a mode of cognitive praxis is the practicality of theory in human life, both as an instrument for the satisfaction of the needs of human existence – i.e. for the mastery of nature as a resource for human life – and for the purposes of social organization of human activity for these purposes. An older formulation puts it thus: It is not that what is practical is true, but rather, that what is true is practical. But that is in fact too pat a credo. Rather, let me reformulate it thus: that theory, as the cognitive mode of the search for truth, is practical. Looking for the truth is what is practical and efficacious about science. Finding it is the hope of science. Science may then be characterized as rational hopefulness: Not empty wish, but deliberate activity or practice in the service of realizing such a hope.

What I am proposing here, only programatically and sketchily, is what I would characterize as an historical-materialist theory of the genesis of theory, or of theoretical cognitive praxis. The apparent instrumentalism of truth in such a view, as I set forth here, is then replaced by a realist emphasis: As a representation in some symbolic form of a mode of action or practice, the theoretical formation has its truth-value in the adequacy of the representation: i.e. in its practical conformity to a successful mode of activity, which it institutionalizes, so to speak, in the representation. It is *not*, therefore, an unmediated representation of some external state of affairs, but rather one which is mediated by the practice or mode of action which it represents. The sheer externality of a state of affairs becomes 'objective' for us, then, only as it is mediated by our practice. What we can know is therefore always conditioned by the way that we come to know it. In a sense, our knowledge of the 'external' world is a knowledge of what this externality is amenable to, in our incursions and interventions upon it.

An action itself succeeds or not – or more complexly, a process of action, or a rule of action or a strategy – succeeds or not, depending on how well it realizes the external (in this sense, objective) constraints of the world in which the operation or action proceeds. That world, in turn, is independent of the action. The action doesn't construct it; it reconstructs or transforms. It is in the activity of transforming an external world to human use, by means of action or praxis, that practical knowledge of that world is acquired. It is by means of the *representation* of successful action that theoretical knowledge of that world is acquired.

Theory, then, is that form of action which is mediated by representation in a language, i.e. symbolically. Or, to put it in a much more traditional form, theory is reflective practice – or the very practice of reflecting on practice. In full dialectical jargon, one would say that theory is practice become self-reflective. All human practice or action is already cognitive, though not yet theoretical. It is a means of acquiring knowledge of the world by transforming it to meet needs, purposes, interests. This transformation, in turn, is a necessary life-activity – it is the life activity of the species, its mode of reproducing its existence. But such cognitive activity achieves a distinctive human form when it becomes reflective: When it can be embodied in a representation, communicated, transmitted as acquired knowledge from one generation to the next – in short, when it becomes theoretical. In this sense, humans are distinctively theoretical animals.

To begin in this way with the genesis of science is ambitious, if not arrogant. But if one is to build into the conception of science both its teleology and its normativeness, on the one hand, and also its historicity on the other, these features must be built into the genesis of science itself. Please forgive, therefore, the briefness and perhaps the density of this sketch. It is at best, programmatic here. Still, certain conclusions follow from it, in this genetic ontology of science:

First, that theorizing is a natural requirement of human action, and not an extraneous one.

Second, that it proceeds from and depends on cognitive praxis which is not yet theoretical – but then, it comes to inform this praxis itself, as conscious expectation in the imagination becomes a distinctive feature of human practice, and the intentionality of human action is achieved in this reflective and teleological mode.

Third, that *all* modes of such theoretical praxis are already included in what I would call science and that science, in this broad sense, is of much greater antiquity than standard histories or philosophies of science allow (unless for purpose of historical demarcation, one wants to call such early, yet *already* theoretical modes of cognition *proto*-scientific).

Fourth, that the normativeness of such theoretical practice is built into its teleology, It is intrinsically teleological; its norm, and its practical utility, its *truth* (thus far inadequately specified here);

Fifth, and most crucial for our purposes: such theoretical cognitive activity or *praxis* changes and develops with the changes in the needs and purposes it serves; and may be cumulative, only insofar as there are perduring historical needs and pruposes which have to be met at *all* times. Its dynamic character – its proliferation, to use Feyerabend's term – is a function of the historical evolution of human needs, as societies change and grow; its stable character – its tenacity and its cumulativeness – is a function of the successive achievements of theory with respect to meeting transhistorical (but *not* suprahistorical) needs – in particular, those associated with the understanding of the natural world as the abiding resource of human existence, the permanent environment of our species-existence. The context of theory-change is therefore the context of *historical* change of the forms of life of the species. Since I take this historical change itself to be a dynamic of social forms of the organization of the production of the means of existence and of the reproduction of the species, in historical materialist terms, the history of theoretical change is necessarily interwoven with this history: *not* as its reflex, in some simple-minded base-superstructure theory, or in the vulgar Marxism of simple copy-or reflection-theory but rather, insofar as theory itself is a vital and efficacious component of this history.

But what of *scientific* theory? Is all theorizing scientific? I admitted, earlier, a useful demarcation between proto-science and science. I think the demarcation comes, historically, with the division of labor. When theoretical cognitive praxis becomes a distinctive social institution – i.e. when the role of theorizing becomes the function of a *part* of society, then science *as* a social institution begins its truly revolutionary course. In the *Metaphysics*, Aristotle already suggested this division of labor as the beginning of science in his remarks on the origins of rational mathematical theory in Egypt. He suggested that only with a surplus, such that leisure could be provided for a priestly class, could science proper begin. I think he was right.

I also think that this marks science, as an institution with a class-character, and, in such class-societies, with a class role. This complicates its history. It is no longer simply the truth-seeking theoretical activity which represents and also helps to guide successful practice, in some simple sense of an undifferentiated human community providing for its needs. It becomes involved with its role not simply in the acquisition of theoretical knowledge, but the acquisition of this knowledge for differentiated purposes, in historical class societies. Ideology and special interests intrude on science, in the very conditions of division of labor which give it birth. Socially differentiated needs – *historical* needs – both mediate and distort its truth-acquiring function. The simple progress of rationality is replaced by the messier dialectical process of rationality for some purpose. The point is that what we characterize as rationality, *itself* has a history, and it is a social history.

What I am suggesting, sketchily and programmatically here, is a much richer and more complex ontology of science, and therefore a much more difficult project for the philosophy and history of science. But I am further suggesting that such a philosophy and history of science – whatever the necessary division of labor between philosophers and historians, whatever the requirements of specialization and rigor – must nevertheless be an integrated philosophy and history of science – a philosophical-historical theory of science as a changing and developing social human activity. This is what I would propose, therefore, as the missing type of Agreeable Internal Relations between philosophy and history of science.

Boston University

NOTE

* Earlier versions of this paper were given at Queens College of the City University of New York (in December, 1973) and at the London School of Economics (in May 1974). On these occasions, I benefited from the remarks and criticisms of many colleagues and especially Benjamin Nelson and J. W. N. Watkins (the severity of whose criticism was tempered by his graciousness). I also want to thank my colleague Robert S. Cohen, for many helpful criticisms and suggestions, which led to several revisions of earlier drafts. But I especially want to thank Imre Lakatos for all the fun we had together, puzzling about and arguing some of these issues. I miss his responses and his wit very badly, and it is harder to think about these things without him.

WOLFGANG YOURGRAU

COSMOLOGY AND LOGIC – AN INTRACTABLE ISSUE?

Some time ago, Peter Bergmann stressed in a very convincing paper that cosmology has now become an empirical science, since the observational methods at cosmological distances have been improved to such an extent that cosmology should no longer be regarded as "a field for unchecked speculation". This implies automatically that cosmology could now be treated in the same manner as physics, chemistry, biology, geology, or any other scientific discipline of experiential knowledge. He omitted to refer to cosmogony – a domain in which all sorts of conjectures are still offered. In other words, cosmology that treats the structure of the universe, enjoys a scientific status that cosmogony which deals with the origin of the universe, for instance, its age and its evolution up to the present state, does not possess. There is a tendency to-day to blur the demarcation line between cosmology and cosmogony, though a modicum of reasoning shows cogently that cosmology and cosmogony are two different approaches to the universe, notwithstanding the fact that they have a lot in common and are definitely very closely related – yet they are not identical.

The scientific status of cosmology is clearly demonstrated by the simple fact that very often theoretical physicists and astrophysicists investigate some cosmological problems with the same rigor with which they examine questions in their own proper domain.

I fully agree with Bergmann's views that cosmology can now be considered to be a science, but I wish he had also investigated the status of cosmogony which is still open to many misinterpretations of a dubious nature, to wit, Velikovsky's popular fairy tale that had confused, unfortunately, even some cosmogonists who began to question their own scientifically-oriented approach to the subject.

The late cosmogonist George Gamow, who was a trained theoretical physicist and perhaps one of the best produced in our age, did not draw a sharp line between cosmology and cosmogony, as the totality of his work in the areas of cosmology and cosmogony testifies. For him, those two branches of knowledge were overlapping to such an extent that they

R. S. Cohen et al. (eds.), Essays in Memory of Imre Lakatos, 739–753. *All Rights Reserved*
Copyright © 1976 by D. Reidel Publishing Company, Dordrecht–Holland

could hardly be distinguished from each other. According to my opinion, he went too far in his attempts to look at the origin, development, and age of the universe as if they were so inseparable that they had to be investigated as if they were one single science.

It is well known to physicists as well as to philosophers of science that logic has gradually penetrated some of the most essential branches of physics as, for instance, quantum theory and general relativity theory. There exists to-day a substantial collection of articles and books produced by serious, and often renowned physicists, logicians, and philosophers of science who applied, and are still applying, some kind of logic to the most consequential branches of theoretical physics. Alfred Landé, Gustav Bergmann, Josef Jauch *et al.*, have done some interesting work as far as the logic of quanta is concerned and thinkers like the very active Swedish scholar, Törnebohm, are still engaged in establishing a logical metalanguage for general relativity theory.

I find those manifold enterprises of logicalizing quantum and general relativity theory very interesting, especially since the convincibility of the outcome of all those attempts depends on what type of logic one employs: For example, whether or not one works with formalistic, modal, inductive, combinatory logic, or any of the many-valued logics; we cannot deny that all these endeavors may one day permit us to transpose the physico-mathematical quantum and general relativity theories into formal logical systems. However, I think that the situation is entirely different as far as cosmology and cosmogony are concerned. Granted, it may be possible to treat by means of logic some detailed aspect of one or the other of those theories of the universe. But one would have to evaluate even satisfactory results of such attempts as being no more than a logical game that does not give us any deeper insight into the scientific status of the respective subject which was exposed to the specific method of the logician in question. To treat cosmology or cosmogony as a whole by resorting to any kind of logic seems to me a task which is *a priori* doomed to failure. The reasons for this disbelief are the following:

(1) Although one can neither view quantum theory nor general relativity theory as closed theories, e.g. in contrast to geometrical optics, the two theories as such are consistent, coherent, intelligible, communicable, and cast in a strict mathematical form. The fact that new

empirical discoveries may force us to make adjustments to some of the relevant equations, would not represent any hindrance for a competent logician. Since both theories are also subject to empirical revisions, at least in part, the logician who is accustomed to work with frequent changes of the values of his variables, could adjust his formalism without great difficulties to varying equations. In other words, should experiential findings compel the quantum or relativity physicist to "patch up" the respective theory, the logician will easily be able to cope with the new situation and adjust his formalism accordingly.

The situation is entirely different when we concentrate upon cosmology and cosmogony. Here any drastic change of hitherto accepted variables and the numerical changes of constants might easily affect basically any of the theories that reflect systematically the origin, structure and age of the universe. As it happens, we have at least 15 rivaling theories of cosmology and cosmogony and as far as cosmogony is concerned, the number is even higher, if one enumerates all the traditional and novel constructs in that particular field. The expansion of the universe is definitely accepted by most workers in that domain. There exists certain empirical evidence for such an expansion. The irritating Olbers' paradox can be regarded as being resolved and Hubble's law has so far not been refuted, neither theoretically nor empirically. But all the other variables and constants have steadily been revised owing to our much improved observational techniques.

To this day there is no uniform verdict as to whether or not the universe is finite or infinite. At present, we are inclined to postulate an infinite universe, but there are no powerful theoretical or empirical arguments for this belief or assumption – for the simple reason that the hitherto accepted belief in the constancy of laws of nature (even if they may only be statistically valid) may turn out to be what it has always been, namely, a wishful belief into such a constancy. The reason was, in all probability, that one can calculate physical phenomena only if one accepts such an invariance as far as laws of nature are concerned. Yet, as has been shown by Dirac, Jordan *et al.*, it is highly possible that even the so-called constants of nature may be proved in fact to be parameters or even variables, provided one works with the notion of cosmological times rather than with the ordinary concept of time.

(2) As has been pointed out so often, as far as we know, the universe is

unique and not a multiverse. This is certainly of some advantage from the point of view of statistics, but there are difficulties associated with this uniqueness. Thus, Peter Bergmann has correctly emphasized, our present theoretical knowledge implies field singularities at the "infant" stages of an expanding universe. This irrefutable argument indicates that perhaps theoretical concepts which we willingly accept as valid at present, may turn out to be entirely inadequate for our comprehension of highly condensed matter. The logician who has to work with such arguments must find it difficult if a *whole* system or theory deprives him of any reliable substance he has to treat in a logical manner, because we deal here not only with minor changes of values for variables or parameters, but with fundamental issues characteristic for our current insight into cosmology. Moreover, even Peter Bergmann, one of Einstein's staunchest protagonists, is aware of the intrinsic weakness of Einstein's theory of relativity due to the effect which "the cosmological environment" has exerted on various local experiments.

It is still a moot point whether or not cosmogony, named a science, deserves this appellation as was already pointed out before. Cosmogony is still in the very embryonic state and encumbered by too many conjectures which depend more on the particular subjective preference of the author for any one of the hypotheses, rather than upon valid theories supported by solid, reliable evidence. In brief: Current researches support Gamow's contention that there is no strict barrier between cosmology and cosmogony, but if this contention is correct, the task of any logician to approach cosmology by using the tools at his disposal would be automatically doomed.

Although we might consider to-day the separation of cosmology and cosmogony into two actually different domains as an unwarranted approach, the fact for the worker in this field remains that the treatment of the origin of the universe and its present structure cannot be executed without arriving at more or less subjective extrapolations from the present to the past and from a hypothetical past to the present. And although Hubble's law, which indicates an expansion of the universe, appears to be valid, it is now merely considered to be part of an empirically verified theory that, in its present form, seems to indicate a continuous expansion of the universe. Furthermore, the so-called background radiation is a factor which would escape any logical treatment, because we encounter here something which can be neither regarded as a variable *per se*, nor a constant or even a parameter. We seem to have met in this case with a situation which will forever escape the logician's approach.

(3) Let us suppose, in view of the aforementioned remarks, that it may be expedient to treat cosmogony and cosmology as two distinguishable branches of knowledge, notwithstanding that they may actually be only artificially divided in order to facilitate the task of the contemporary cosmologist. Abbé Lemaître, perhaps one of the greatest cosmogonists of our time, calls his notion of the Primeval Atom a hypothesis and not a theory. As a matter of fact, he seems to be skeptical when the question of "proof" is concerned, and now we come to the crux of the matter: How could the logician approach propositions that seem to allow only on *very rare* occasions, namely in modern cosmology, to deal with unambiguous propositions which almost always lead to new questions that have to be answered either theoretically or empirically?

We have not yet arrived at a logic of questions besides those few attempts to show that questions may also be treated as if they were propositions – in contrast to all tradition and all logics with which we work today. The logician must realize – independent of whether we treat cosmology and cosmogony as one singular science or as two different disciplines of knowledge – that he cannot approach problems concerning the whole universe by means of the tools at his disposal. This does not mean that there will not be certain "islands" in cosmology or cosmogony where we may be able to express observational data or certain parts of a theory in a logical manner. This would not give us much satisfaction and we would be hindered to attempt generalizations and extrapolations from present initial situations because, as has been shown by Terletskii, such a process would lead us very soon into unsolvable paradoxes. In other words, to take arbitrarily a certain equation or formula which is expressible in logical terms, would amount to no more than to an unfruitful game which would not even illuminate the actual logical structure of the subject under discussion.

To illustrate our contention, let us take from special relativity theory the well-known observer in S formula:

$$dx^2 + dy^2 + dz^2 - c^2\, dt^2 = 0;$$

analogously, an observer in S' would find that

$$dx'^2 + dy'^2 + dz'^2 - c^2\, dt'^2 = 0,$$

where S and S' can designate as the standard configuration of the two frames of reference S and S'. This is an example from special relativity and would therefore fulfill our claim that without doubt certain propositions, i.e., in special relativity theory, could be transposed into logical formulae. It is in this whole discussion irrelevant what type of logic is employed.

After the aforementioned remarks we may now, if not methodically then at least informally, try to relate aspects of general relativity and cosmology to logic and attempt to examine whether or not the selected concepts or equations (formulae) permit the application of a logic or sometimes even of a metalogic.

First of all, we have also in cosmology, as in many branches of physics and chemistry, so-called constants. Of course, I am referring to universal or fundamental constants. This is slightly different in the case of relativity. For instance, if we do not change by resorting to certain co-ordinates transformations the form of any law in physics or cosmology, i.e., if we obtain the same law in various sets of co-ordinates, such a law is *invariant* or *co-variant* with respect to certain transformations. But these transformations have to be explicitly stated. Here we find indubitably a link with logic and the reader will bear in mind that by logic we mean the most suitable kind of logic applicable to topics mentioned above. Our knowledge of the meaning of the words, 'place', 'distance', 'mass', or 'weight' is still on occasion ambiguous. Bergmann holds that if "the ratio of force to acceleration is constant, different for every individual body, this constant is named the mass of the body". This definition might have satisfied Newton, but certainly not Mach whose definition of mass is definitely superior to that of Newton, as has been shown by Yourgrau and van der Merwe.

Further, whether or not we jettison the principle of relativity and postulate that there exists a universal frame of reference, is relevant to the cosmologist and relativist, but hardly to the logician who is not interested whether the frame of reference is universal or relative and whether or not the speed of light is independent of the direction of propagation. If he happens to be at rest relative to the ground of a body in motion or not, for our logician it would be actually without importance from a *merely* logical point of view. He could, however, attempt to translate the equations of the electromagnetic field. The situation is different when we deal with an effect commonly called the Lorentz contraction. For instance, if we take a moving rod (the material is irrelevant), relative to the observer, it appears contracted by the factor $\sqrt{1-v^2/c^2}$. We can even logically deal with the fact that every rigid body appears to be of maximal length when at rest relative to the observer.

How can we treat logically infinitesimal distances? If we use Cartesian coordinates, the infinitesimal distance is expressed by $ds^2 = dx_k\, dx_k$. But we do not yet know how to define the concept of "distance"! The infinitesimal distance is expressed by equations of the form

$$ds^2 = g_{ij} d^{\xi i} d^{\xi j},$$

where ds^2 is an invariant, g_{ij} is a covariant tensor, and in Riemannian space, g_{ij} are the components of the metric tensor. But the line element in special as well as in general relativity, independent of the transformation we choose, expresses distance and we have not yet found a universally satisfactory definition of "distance". For in cosmology, in contrast to physics proper, the usages of the term 'distance' are in no way identical. The logician would be then only able to work in an *ad hoc* area and could hardly arrive at a universally-valid formula. But we encounter the difficulty in finding a universally-valid definition of distance also in other branches of cosmology. According to Whitrow, Hubble's law *confirms* the linear relationship between redshift and distance. Again, the term 'distance' is here employed without explicit definition and seems to hold merely for this particular aspect of cosmology.

Milne derived from his uniformly expanding universe model a formula that relates the time t, the average density ρ at the time t and the constant of Newtonian gravitation G. Working with Euclidean space and opposing the uniformity of expansion of his universe model, ρ becomes the inverse cube of the temporal epoch in question. He arrived at the equation $4\pi/3G\rho t^2 = 1$. Of course, this implies that in Milne's model G is not a true constant but will grow proportionately with t! It is interesting to note in this context that Hoyle proposed an equation similar to that of Milne by combining the local average density, the reciprocal T of Hubble's constant a and the gravitational constant G: $4\pi/3G\rho T^2 = 1/2$.

The main difference here between Milne and Hoyle lies in the fact that for Hoyle ρ, T and G are *true* physical constants, but for Milne they *all* vary with the advance of time. It is a strange coincidence that Hoyle's relationship resembles very much the Einstein-de Sitter model of the universe; in the latter case, however, G alone is a true constant of nature. Finally, in the model of the universe by Eddington we have $4\pi/3G\rho T^2 = = 1/q^3 - 3q + 2$; q is proportional to the present value of the radius of the universe to its initial value. Eddington holds that in his formula *solely* ρ, T and q vary with time, whereas G alone is a *true* constant of nature.

As North has correctly pointed out, in the model of the universe by Eddington, the present radius of 'our' expanding universe is greater than the radius R of Einstein's model of the universe. This may not be of dramatic relevance to cosmologists who always have a choice to select that type of universe which seems to be most plausible to them; in contradistinction to North I think that the above fact is rather important for the theoretical physicist as it does not correspond to an eigenstate. Since it does not correspond to an eigenstate, it is irrelevant to Schrödinger's wave equation with all its theoretical implications, and any attempt of a unified theory is thereby automatically doomed.

I mentioned before that we do not possess any unique definition of distance in cosmology nor in astrophysics, but we also have no uniform conception of space. Leibniz mentioned the existence of "no real space out of the material universe." He conceived space as an order of co-existences and time as an "order of successions." This strange reasoning of Leibniz was very perspicaciously realized by North who asked the rhetorical question of why Newton was in two minds to regard movement as something relative to physical things, yet was fully aware of our "logical need" to relate the motion of one object to another. The difference between Newton and Leibniz is that Newton was able to answer his question about true motion in a manner which satisfied thinkers for at least three centuries, i.e., before Einstein and Mach investigated that question anew and arrived at a more satisfactory explanation.

All the aforementioned remarks would only permit an occasional treatment by one of our many logics, and perhaps I should stress here that I make a sharp distinction between mathematics (pure and applied) and any of the logics which have been constructed during the nineteenth and twentieth centuries. It is a well known fact to cosmologists and astrophysicists that the displacement of the lines in a stellar spectrum are useful if you wish to discover the radial velocity. Suppose λ is the laboratory wavelength which appears in our stellar spectrum to be of a wavelength $\lambda + d\lambda$; $d\lambda$, of course, is the change in λ. We can then measure the displacement of the lines by $\delta = d\lambda/\lambda$. This expression is independent of the line chosen, provided the cause of the phenomenon is the movement or motion of the star. McVittie realized at once that the magnitude of the radial velocity v is related to the displacement of δ and the velocity of light by the formula $v = c\delta$.

It is obvious from the foregoing statements that not only the term 'distance' but also the term 'space' escape explicit definition and can only be operationally defined according to the subject under discussion. North provided us with virtual synonyms for space: volume, displacement, distance or interval, place, vacant place, totality of geometrical points or coordinate values, metric, etc. Of course, all those terms refer in some way or other to the location of things of objects. For instance, a line-element would fall under the category of distance, yet the non-stationery line-element

$$ds^2 = c^2 dt^2 - \exp(\eta t)(dx^2 + dy^2 + dz^2)$$

can be understood as being necessary for the velocity of light to reduce with time. Hence, the line-element which falls under the category of distance is here employed as a measure of time. For the logician all this is highly confusing because he is forced to jump consistently from mathematical expressions to interpretations, to a metalanguage, and the difficulty is that the same terms used in physics and cosmology are not always synonyms, but represent sometimes different entities or situations altogether. McVittie raises the question as to whether δ varies from line to line in a certain galaxy spectrum or does it indeed remain constant. The constancy of δ in the spectrum of the galaxy would be valid beyond any doubt, provided the redshift could be measured not only at optical but also at radio wavelengths.

The greatest headache for the cosmologist and logician in this discussion is caused by Einstein's cosmological constant λ, a constant of integration. Einstein became aware of this integration constant when he was dealing with the solution of cosmological problems. Hence the name of cosmological constant which was not a very felicitous choice. McVittie is correct when he doubts that the cosmological constant was discovered by the cosmological investigation of the universe. He rightly claims that the λ term is the result of purely mathematical reasoning. It would be valid in the case of the Einstein tensor and be present in Einstein's celebrated equation for mathematical or aesthetical reasons. In the end, Einstein was very satisfied that he could dispense with the λ term, owing to Friedmann's famous paper of 1922.

The logician will have less difficulties in dealing with cosmology when he considers the space-curvature constant k which has the values: $+1$, 0 or -1, subject to space being spherical, Euclidean, or hyperbolic. It seems that these three values of k are entirely arbitrary and depend more or less on the taste of the examiner. Likewise, the scale-factor R is a certain function of time t, but it is entirely uncertain what specific function is at issue.

These remarks are intended merely to show that since cosmology is in no way yet a uniform and definitive science, but hopefully on the way to being one, certain paradoxes must follow. Most cosmologists to-day will accept the expanding universe and thus they will be compelled to maintain that the curvature might change with time. I fully agree with North that to explain the meaning of the preceding viewpoint sometimes it is neither necessary to execute the logically impossible task of having someone outside the universe to observe the expansion, nor is it necessary to specify into which form or space or entity the universe expands. In other words, on either reason it can be contended that it is somewhat meaningless to speak about the expansion of the universe in the glib manner which is so prominent to-day. Any cautious logician would not hesitate to consider a rigid formulation of this concept as being logically without any foundation.

Hubble's parameter (wrongly usually named Hubble's constant) cannot be invoked to represent a true constant of nature. It can only be determined by having recourse to observation. It is customary to regard Hubble's parameter as the reciprocal of time, or rather a time, but this doesn't help us much as long as we do not define "distance" in such a way that the definition holds throughout cosmological discussions. At present we are in the habit of conceiving of the velocity of recession of galaxies as being proportional to the distances of the galaxies. We are further accustomed to assume that whenever a galaxy has a redshift equal to unity, its velocity must equal that of light. Those 'beliefs' are only made possible as long as we have not been able to define unambiguously distance. It looks to-day as if our observables (telescopic, radio and radar) are a more reliable guidance to current cosmology than one of the manyfold cosmological theories. But we know for certain that at each moment *t*, space in our model of the universe must possess constant curvature. We know that Euclidean space is of a curvature zero, and therefore called flat space. Positive curvature of spherical space can be compared to the two-dimensional surface of a sphere in three dimensions. Thus, in a spherical space, we do not encounter any boundary and yet the total volume of space is finite. The third type of space which is considered by cosmologists is hyperbolic, or space of negative curvature. This space too, like the other two spaces, possesses no boundary, but like Euclidean space, it has an infinite volume. It may be tantamount to an interesting game if we would express all this in one of our logics, but I doubt that it could be even done without introducing a plethora of *ad hoc* assumptions so that the actual value of such a logical transformation would be, in the case of space or spaces, not very fruitful.

But the main difficulty of logical transposition of cosmological concepts and relations is still most obvious in the case of Hubble's law. As stated before: $d\lambda/\lambda$, where $d\lambda$ is the displacement of a spectral line, the normal wavelength would be λ. It is claimed by most cosmologists that $d\lambda/\lambda$, the so-called redshift, applies to any celestial body, especially to all the galaxies, (nebulae or clusters). The so-called displacement, $d\lambda$, has to be always positive and the wavelength of a displaced line $\lambda + d\lambda$ must always be positive if we treat Hubble's law as an expansion law. Wavelengths are known to increase by the factor $(\lambda + d\lambda)/\lambda$, i.e. $1 + d\lambda/\lambda$. Hubble correctly reminds us that $E\lambda = \text{constant}$. If this product stays constant, redshifts characterized by increasing wavelengths, will automatically reduce the energy in the quanta. The reason is obvious: It is known to every worker in the field that the energy of any light quantum, when multiplied by the wavelength of that particular quantum, is of constant value. Hence, according to the prevalent interpretation of redshift, there will be a loss of energy even if we should be able to find another convincing interpretation of redshift. At present we regard (conveniently) redshifts as Doppler effects. Or, in cosmological terms, as velocity shifts usually treated as actual motion of recession.

Whitrow is right that if redshifts are not velocity shifts, then they must represent some new principles in physics so far undiscovered. Of course, cosmologists are strongly inclined to treat redshifts as actual motion, notwithstanding the fact that the telescopic observations are still in the process of achieving greater efficiency and are in no way yet to be regarded as definitive and exhaustively reliable. We are confronted with more than 15 models of the universe and most of them are evolutionary. In one model – as McVittie has shown – a uniform motion of contraction of the gas in question is as much possible as an expanding universe. The last arbiter will have to be our appeal to observation. To date we cannot say with unconditional certainty whether all galaxies (nebulae, clusters) move in such a manner that the Hubble constant, or rather parameter, is a true fundamental constant of nature. Theory will not help us in such a case and we must have to have recourse again to observation.

One of the most exciting, though in the end embarrassing, theories of cosmology is that of Bondi and Gold which was later extended by Hoyle. The steady-state theory or creation of matter from *ex nihilo* was suggested by Bondi and Gold in 1948. According to their theory, the universe has neither a beginning nor an end. In contrast to most cosmologies, the BGH theory does not deal at all with gravitation. True, Hoyle tried twice to modify Einstein's field equations which contain the general relativity conception of gravitation, but the new energy tensor introduced by Hoyle becomes incompatible with the law of conservation of energy whereas the Einstein tensor is still compatible with that fundamental law in physics. In parentheses one may mention that in Hoyle's new field equations the Einstein cosmical constant is supposed to be zero. And we know what this assumption implies.

It was Gamow who, among others, attacked the steady-state theory for a variety of reasons. His main argument was that they were in sharp conflict with some of the most fundamental laws known in physics and cosmology. The introduction of the cosmological constant, which may be called universal repulsion in contrast to gravitational attraction, is logically not necessary at all for the steady-state theory which is one of the two well-known non-evolutionary cosmologies. According to BGH, there is no chance at all that the present expansion will ever come to a halt, nor will we ever encounter a recession or a so-called collapse.

In spite of all arguments advanced against the validity of the steady-state theory, it appears as if it experiences a renaissance. Perhaps the various arguments advanced against certain parts of general relativity theory have encouraged some workers in the field to re-examine some of the main claims of the BGH theory.

It looks as if at present we are only left with assumptions without which we cannot even construct a plausible cosmology: Homogeneity, isotropy, that is, uniformity throughout. But uniformity can have different connotations in cosmology and is not quite identical with the principle of general homogeneity which is so closely related to isotropy. Those three terms are rather close in their meaning, for by 'isotropic' we never mean anything else but that the distribution of the arrangement of the galaxies in space is always the same in whatever direction we choose to look: symmetry.

These few remarks on cosmological issues which are of current interest do not claim to be the elementary building blocks of a new cosmological edifice or theory. The reader is bound to be acquainted with the main content of this paper, especially, the cosmologist and astrophysicist. Gradually almost all branches of science have become subject to systematic investigation by logicians who applied one or more of the many kinds of logic to any particular science. According to my knowledge, cosmology has so far not been examined by logicians. Perhaps the reason for this in no way regrettable ommission, lies in the fact that cosmology has only recently become an exact science, and compared to pure mathematics or the axiomatic foundations of logic, it is still far remote from being compared with theoretical physics or with the conceptual and empirical advances in chemistry, biology, etc. There are few fundamental principles which are absolutely necessary to work in the field of cosmology, but the subject is inundated with *ad hoc* assumptions, postulates, constants – to name only a few. Perhaps one day it will be possible to construct a meta-cosmology in which the various mathematical formalisms can on one hand be interpreted empirically and on the other, where the mathematical formulations can become interesting material for a meta-mathematics that would allow the logician to establish a meta-metalogic.

I have worked, without any results, in this direction for the fun of it – not that I believe in the scientific value of such an enterprise, but all my attempts have failed because it looks as if cosmology in its present stage is not a suitable subject for any formal logical treatment.

POSTSCRIPT

I am very indebted to J. D. North for drawing my attention to a point that is usually entirely ignored. It cannot be gainsaid that Lemaître was fully aware of the respective positive and negative qualities in connection with the models of the universe by Einstein and de Sitter. We recall that de Sitter regarded space as well as space-time as being both homogeneous, but separately so. Strangely enough, and apparently contradicting himself, he maintained nevertheless that the partition of space-time into space and time disturbs the homogeneity. It seems evident that as a mathematician, Lemaître evaluated symmetry almost as if it were objective truth. From the point of view of logic, we are facing here a highly undesirable situation, namely, an inconsistency in a matter of utmost relevance. It is obvious from this chapter that we find ourselves still in stark winter as far as the relationship between cosmology and logic is concerned. Our late friend Imre Lakatos, with whom I often discussed this subject, was convinced that soon we would move into a bright spring – an optimism I cannot, at this moment, share.

University of Denver

BIBLIOGRAPHY

Bergmann, P. G., *Found. Phys.* **1** (1970), 1.
Bergmann, P. G., *Introduction to the Theory of Relativity*, Prentice Hall, New York, 1942.
Einstein, A., *Relativity*, Methuen, London, 1921.
Gamow, G., *The Creation of the Universe*, The New American Library of World Literature, New York, 1952.
Hubble, E., *The Realm of the Nebulae*, Dover, New York, 1958.
McVittie, G. C., *General Relativity and Cosmology*, Chapman and Hall, London, 1956.
McVittie, G. C., *Fact and Theory in Cosmology*, Eyre & Spottiswoode, London, 1961.
North, J. D., *The Measure of the Universe*, Clarendon Press, Oxford, 1965.
Terletskii, Ya. P., *Paradoxes in the Theory of Relativity*, Plenum Press, New York, 1968.
Whitrow, G. J., *The Structure of the Universe*, Hutchinson's University Library, No. 29, London.
Whitrow, G. J., *The Structure and Evolution of the Universe*, Harper & Bros., New York, 1959.

Yourgrau, W. and van der Merwe, A., 'Did Ernst Mach "Miss the Target"?' *Synthese* **18** (1968), 234–250.

Curry, H. B., Feys, R., and Craig, W., *Combinatory Logic*, Vol. I, North-Holland, Amsterdam, 1958.

Hintikka, J. and Suppes, P. (eds.), *Aspects of Inductive Logic*, North-Holland, Amsterdam, 1966.

Kneale, W. M., *The Development of Logic*, Clarendon Press, Oxford, 1962.

Prior, A. N., *Formal Logic*, Clarendon Press, Oxford, 1955.

Rosser, J. B. and Turquette, A. R., *Many-Valued Logics*, North-Holland, Amsterdam, 1952.

INDEX OF NAMES

SYNTHESE LIBRARY

Monographs on Epistemology, Logic, Methodology, Philosophy of Science, Sociology of Science and of Knowledge, and on the Mathematical Methods of Social and Behavioral Sciences

Managing Editor:

JAAKKO HINTIKKA (Academy of Finland and Stanford University)

Editors:

ROBERT S. COHEN (Boston University)
DONALD DAVIDSON (The Rockefeller University and Princeton University)
GABRIËL NUCHELMANS (University of Leyden)
WESLEY C. SALMON (University of Arizona)

1. J. M. BOCHEŃSKI, *A Precis of Mathematical Logic*. 1959, X + 100 pp.
2. P. L. GUIRAUD, *Problèmes et méthodes de la statistique linguistique*. 1960, VI + 146 pp.
3. HANS FREUDENTHAL (ed.), *The Concept and the Role of the Model in Mathematics and Natural and Social Sciences, Proceedings of a Colloquium held at Utrecht, The Netherlands, January 1960*. 1961, VI + 194 pp.
4. EVERT W. BETH, *Formal Methods. An Introduction to Symbolic Logic and the Study of effective Operations in Arithmetic and Logic*. 1962, XIV + 170 pp.
5. B. H. KAZEMIER and D. VUYSJE (eds.), *Logic and Language. Studies dedicated to Professor Rudolf Carnap on the Occasion of his Seventieth Birthday*. 1962, VI + 256 pp.
6. MARX W. WARTOFSKY (ed.), *Proceedings of the Boston Colloquium for the Philosophy of Science, 1961–1962*, Boston Studies in the Philosophy of Science (ed. by Robert S. Cohen and Marx W. Wartøfsky), Volume I. 1973, VIII + 212 pp.
7. A. A. ZINOV'EV, *Philosophical Problems of Many-Valued Logic*. 1963. XIV + 155 pp.
8. GEORGES GURVITCH, *The Spectrum of Social Time*. 1964, XXVI + 152 pp.
9. PAUL LORENZEN, *Formal Logic*. 1965, VIII + 123 pp.
10. ROBERT S. COHEN and MARX W. WARTOFSKY (eds.), *In Honor of Philipp Frank*, Boston Studies in het Philosophy of Science (ed. by Robert S. Cohen and Marx W. Wartofsky), Volume II. 1965, XXXIV + 475 pp.
11. EVERT W. BETH, *Mathematical Thought. An Introduction to the Philosopy of Mathematics*. 1965, XII + 208 pp.
12. EVERT W. BETH and JEAN PIAGET, *Mathematical Epistemology and Psychology*. 1966, XII + 326 pp.
13. GUIDO KÜNG, *Ontology and the Logistic Analysis of Language. An Enquiry into the Contemporary Views on Universals*. 1967, XI + 210 pp.
14. ROBERT S. COHEN and MARX W. WARTOFSKY (eds.), *Proceedings of the Boston Colloquium for the Philosophy of Science 1964–1966, in Memory of Norwood Russell Hanson*, Boston Studies in the Philosophy of Science (ed. by Robert S. Cohen and Marx W. Wartofsky), Volume III. 1967, XLIX + 489 pp.

15. C. D. Broad, *Induction, Probability, and Causation. Selected Papers.* 1968, XI + 296 pp.
16. Günther Patzig, *Aristotle's Theory of the Syllogism. A logical-Philosophical Study of Book A of the Prior Analytics.* 1968, XVII + 215 pp.
17. Nicholas Rescher, *Topics in Philosophical Logic.* 1968, XIV + 347 pp.
18. Robert S. Cohen and Marx W. Wartofsky (eds.), *Proceedings of the Boston Colloquium for the Philosophy of Science 1966–1968*, Boston Studies in the Philosophy of Science (ed. by Robert S. Cohen and Marx W. Wartofsky), Volume IV. 1969, VIII + 537 pp.
19. Robert S. Cohen and Marx W. Wartofsky (eds.), *Proceedings of the Boston Colloquium for the Philosophy of Science 1966–1968*, Boston Studies in the Philosophy of Science (ed. by Robert S. Cohen and Marx W. Wartofsky), Volume V. 1969, VIII + 482 pp.
20. J. W. Davis, D. J. Hockney, and W. K. Wilson (eds.), *Philosophical Logic.* 1969, VIII + 277 pp.
21. D. Davidson and J. Hintikka (eds.), *Words and Objections. Essays on the Work of W. V. Quine.* 1969, VIII + 366 pp.
22. Patrick Suppes, *Studies in the Methodology and Foundations of Science. Selected Papers from 1911 to 1969*, XII + 473 pp.
23. Jaakko Hintikka, *Models for Modalities. Selected Essays.* 1969, IX + 220 pp.
24. Nicholas Rescher *et al.* (eds.), *Essays in Honor of Carl G. Hempel. A Tribute on the Occasion of his Sixty-Fifth Birthday.* 1969, VII + 272 pp.
25. P.V. Tavanec (ed.), *Problems of the Logic of Scientific Knowledge.* 1969, VII + 429 pp.
26. Marshall Swain (ed.), *Induction, Acceptance, and Rational Belief.* 1970, VII + 232 pp.
27. Robert S. Cohen and Raymond J. Seeger (eds.), *Ernst Mach: Physicist and Philosopher*, Boston Studies in the Philosophy of Science (ed. by Robert S. Cohen and Marx W. Wartofsky), Volume VI. 1970, VIII + 295 pp.
28. Jaakko Hintikka and Patrick Suppes, *Information and Inference.* 1970, X + 336 pp.
29. Karel Lambert, *Philosophical Problems in Logic. Some Recent Developments.* 1970, VII + 176 pp.
30. Rolf A. Eberle, *Nominalistic Systems.* 1970, IX + 217 pp.
31. Paul Weingartner and Gerhard Zecha (eds.), *Induction, Physics, and Ethics. Proceedings and Discussions of the 1968 Salzburg Colloquium in the Philosophy of Science.* 1970, X + 382 pp.
32. Evert W. Beth, *Aspects of Modern Logic.* 1970, XI + 176 pp.
33. Risto Hilpinen (ed.), *Deontic Logic: Introductory and Systematic Readings.* 1971, VII + 182 pp.
34. Jean-Louis Krivine, *Introduction to Axiomatic Set Theory.* 1971, VII + 98 pp.
35. Joseph D. Sneed, *The Logical Structure of Mathematical Physics.* 1971, XV + 311 pp.
36. Carl R. Kordig, *The Justification of Scientific Change.* 1971, XIV + 119 pp.
37. Milič Čapek, *Bergson and Modern Physics*, Boston Studies in the Philosophy of Science (ed. by Robert S. Cohen and Marx W. Wartofsky), Volume VII. 1971, XV + 414 pp.
38. Norwood Russell Hanson, *What I do Not Believe, and Other Essays* (ed. by Stephen Toulmin and Harry Woolf), 1971, XII + 390 pp.
39. Roger C. Buck and Robert S. Cohen (eds.), *PSA 1970. In Memory of Rudolf Carnap*, Boston Studies in the Philosophy of Science (ed. by Robert S. Cohen and Marx W. Wartofsky), Volume VIII. 1971, LXVI + 615 pp. Also available as paperback.
40. Donald Davidson and Gilbert Harman (eds.), *Semantics of Natural Language.* 1972, X + 769 pp. Also available as paperback.

41. YEHOSHUA BAR-HILLEL (ed.), *Pragmatics of Natural Languages*. 1971, VII + 231 pp.
42. SÖREN STENLUND, *Combinators, λ-Terms and Proof Theory*. 1972, 184 pp.
43. MARTIN STRAUSS, *Modern Physics and Its Philosophy. Selected Papers in the Logic, History, and Philosophy of Science*. 1972, X + 297 pp.
44. MARIO BUNGE, *Method, Model and Matter*. 1973, VII + 196 pp.
45. MARIO BUNGE, *Philosophy of Physics*. 1973, IX + 248 pp.
46. A. A. ZINOV'EV, *Foundations of the Logical Theory of Scientific Knowledge (Complex Logic)*, Boston Studies in the Philosophy of Science (ed. by Robert S. Cohen and Marx W. Wartofsky), Volume IX. Revised and enlarged English edition with an appendix, by G. A. Smirnov, E. A. Sidorenka, A. M. Fedina, and L. A. Bobrova. 1973, XXII + 301 pp. Also available as paperback.
47. LADISLAV TONDL, *Scientific Procedures*, Boston Studies in the Philosophy of Science (ed. by Robert S. Cohen and Marx W. Wartofsky), Volume X. 1973, XII + 268 pp. Also available as paperback.
48. NORWOOD RUSSELL HANSON, *Constellations and Conjectures*, (ed. by Willard C. Humphreys, Jr.), 1973, X + 282 pp.
49. K. J. J. HINTIKKA, J. M. E. MORAVCSIK, and P. SUPPES (eds.), *Approaches to Natural Language. Proceedings of the 1970 Stanford Workshop on Grammar and Semantics*. 1973, VIII + 526 pp. Also available as paperback.
50. MARIO BUNGE (ed.), *Exact Philosophy – Problems, Tools, and Goals*. 1973, X + 214 pp.
51. RADU J. BOGDAN and ILKKA NIINILUOTO (eds.), *Logic, Language, and Probability*. A selection of papers contributed to Sections IV, VI, and XI of the Fourth International Congress for Logic, Methodology, and Philosophy of Science, Bucharest, September 1971. 1973, X + 323 pp.
52. GLENN PEARCE and PATRICK MAYNARD (eds.), *Conceptual Chance*. 1973, XII + 282 pp.
53. ILKKA NIINILUOTO and RAIMO TUOMELA, *Theoretical Concepts and Hypothetico-Inductive Inference*. 1973, VII + 264 pp.
54. ROLAND FRAÏSSÉ, *Course of Mathematical Logic* – Volume 1: *Relation and Logical Formula*. 1973, XVI + 186 pp. Also available as paperback.
55. ADOLF GRÜNBAUM, *Philosophical Problems of Space and Time*. Second, enlarged edition, Boston Studies in the Philosophy of Science (ed. by Robert S. Cohen and Marx W. Wartofsky), Volume XII. 1973, XXIII + 884 pp. Also available as paperback.
56. PATRICK SUPPES (ed.), *Space, Time, and Geometry*. 1973, XI + 424 pp.
57. HANS KELSEN, *Essays in Legal and Moral Philosophy*, selected and introduced by Ota Weinberger. 1973, XXVIII + 300 pp.
58. R. J. SEEGER and ROBERT S. COHEN (eds.), *Philosophical Foundations of Science. Proceedings of an AAAS Program, 1969*. Boston Studies in the Philosophy of Science (ed. by Robert S. Cohen and Marx W. Wartofsky), Volume XI. 1974, X + 545 pp. Also available as paperback.
59. ROBERT S. COHEN and MARX W. WARTOFSKY (eds.), *Logical and Epistemological Studies in Contemporary Physics*, Boston Studies in the Philosophy of Science (ed. by Robert S. Cohen and Marx W. Wartofsky), Volume XIII. 1973, VIII + 462 pp. Also available as paperback.
60. ROBERT S. COHEN and MARX W. WARTOFSKY (eds.), *Methodological and Historical Essays in the Natural and Social Sciences. Proceedings of the Boston Colloquium for the Philosophy of Science, 1969–1972*, Boston Studies in the Philosophy of Science (ed. by Robert S. Cohen and Marx W. Wartofsky), Volume XIV. 1974, VIII + 405 pp. Also available as paperback.
61. ROBERT S. COHEN, J. J. STACHEL and MARX W. WARTOFSKY (eds.), *For Dirk Struik*.

Scientific, Historical and Polical Essays in Honor of Dirk J. Struik, Boston Studies in the Philosophy of Science (ed. by Robert S. Cohen and Marx W. Wartofsky), Volume XV. 1974, XXVII + 652 pp. Also available as paperback.

62. KAZIMIERZ AJDUKIEWICZ, *Pragmatic Logic*, transl. from the Polish by Olgierd Wojtasiewicz. (1974, XV + 460 pp.
63. SÖREN STENLUND (ed.), *Logical Theory and Semantic Analysis. Essays Dedicated to Stig Kanger on His Fiftieth Birthday*. 1974, V + 217 pp.
64. KENNETH F. SCHAFFNER and ROBERT S. COHEN (eds.), *Proceedings of the 1972 Biennial Meeting, Philosophy of Science Association*, Boston Studies in the Philosophy of Science (ed. by Robert S. Cohen and Marx W. Wartofsky), Volume XX. 1974, IX + 444 pp. Also available as paperback.
65. HENRY E. KYBURG, JR., *The Logical Foundations of Statistical Inference*. 1974, IX + 421 pp.
66. MARJORIE GRENE, *The Understanding of Nature: Essays in the Philosophy of Biology*, Boston Studies in the Philosophy of Science (ed. by Robert S. Cohen and Marx W. Wartofsky), Volume XXIII. 1974, XII + 360 pp. Also available as paperback.
67. JAN M. BROEKMAN, *Stucturalism: Moscow, Prague, Paris*. 1974, IX + 117 pp.
68. NORMAN GESCHWIND, *Selected Papers on Language and the Brain*, Boston Studies in the Philosophy of Science (ed. by Robert S. Cohen and Marx W. Wartofsky), Volume XVI. 1974, XII + 549 pp. Also available as paperback.
69. ROLAND FRAÏSSÉ, *Course of Mathematical Logic* – Volume II: *Model Theory*. 1974, XIX + 192 pp.
70. ANDRZEJ GRZEGORCZYK, *An Outline of Methematical Logic. Fundamental Results and Notions Explained with All Details*. 1974, X + 596 pp.
71. FRANZ VON KUTSCHERA, *Philosophy of Language*. 1975, VII + 305 pp.
72. JUHA MANNINEN and RAIMO TUOMELA (eds.), *Essays on Explanation and Understanding. Studies in the Foundations of Humanities and Social Sciences*. 1976, VII + 440 pp.
73. JAAKKO HINTIKKA (ed.), *Rudolf Carnap, Logical Empiricist. Materials and Perspectives*. 1975, LXVIII + 400 pp.
74. MILIČ ČAPEK (ed.), *The Concepts of Space and Time. Their Structure and Their Development*. Boston Studies in the Philosophy of Science (ed. by Robert S. Cohen and Marx W. Wartofsky), Volume XXII. 1976, LVI + 570 pp. Also available as paperback.
75. JAAKKO HINTIKKA and UNTO REMES, *The Method of Analysis. Its Geometrical Origin and Its General Significance*. Boston Studies in the Philosophy of Science (ed. by Robert S. Cohen and Marx W. Wartofsky), Volume XXV. 1974, XVIII + 144 pp. Also available as paperback.
76. JOHN EMERY MURDOCH and EDITH DUDLEY SYLLA, *The Cultural Context of Medieval Learning. Proceedings of the First International Colloquium on Philosophy, Science, and Theology in the Middle Ages – September 1973*. Boston Studies in the Philosophy of Science (ed. by Robert S. Cohen and Marx W. Wartofsky), Volume XXVI. 1975, X + 566 pp. Also available as paperback.
77. STEFAN AMSTERDAMSKI, *Between Experience and Metaphysics. Philosophical Problems of the Evolution of Science*. Boston Studies in the Philosophy of Science (ed. by Robert S. Cohen and Marx W. Wartofsky), Volume XXXV. 1975, XVIII + 193 pp. Also available as paperback.
78. PATRICK SUPPES (ed.), *Logic and Probability in Quantum Mechanics*. 1976, XV + 541 pp.
80. JOSEPH AGASSI, *Science in Flux*. Boston Studies in the Philosophy of Science (ed. by Robert S. Cohen and Marx W. Wartofsky), Volume XXVIII. 1975, XXVI + 553 pp. Also available as paperback.

81. Sandra G. Harding (ed.), *Can Theories Be Refuted? Essays on the Duhem-Quine Thesis.* 1976, XXI + 318 pp. Also available as paperback.
84. Marjorie Grene and Everett Mendelsohn (eds.), *Topics in the Philosophy of Biology.* Boston Studies in the Philosophy of Science (ed. by Robert S. Cohen and Marx W. Wartofsky), Volume XXVII. 1976, XIII + 454 pp. Also available as paperback.
85. E. Fischbein, *The Intuitive Sources of Probabilistic Thinking in Children.* 1975, XIII + 204 pp.
86. Ernest W. Adams, *The Logic of Conditionals. An Application of Probability to Deductive Logic.* 1975, XIII + 156 pp.
89. A. Kasher (ed.), *Language in Focus: Foundations, Methods and Systems. Essays Dedicated to Yehoshua Bar-Hillel.* Boston Studies in the Philosophy of Science (ed. by Robert S. Cohen and Marx W. Wartofsky), Volume XLIII. 1976, XXVIII + 679 pp. Also available as paperback.
90. Jaakko Hintikka, *The Intentions of Intentionality and Other New Models for Modalities.* 1975, XVIII + 262 pp. Also available as paperback.
93. Radu J. Bogdan, *Local Induction.* 1976, XIV + 340 pp.
95. Peter Mittelstaedt, *Philosophical Problems of Modern Physics.* Boston Studies in the Philosophy of Science (ed. by Robert S. Cohen and Marx W. Wartofsky), Volume XVIII. 1976, X + 211 pp. Also available as paperback.
96. Gerald Holton and William Blanpied (eds.), *Science and Its Public: The Changing Relationship.* Boston Studies in the Philosophy of Science (ed. by Robert S. Cohen and Marx W. Wartofsky), Volume XXXIII. 1976, XXV + 289 pp. Also available as paperback.
97. Myles Brand and Douglas Walton (eds.), *Action Theory. Proceedings of the Winnipeg Conference on Human Action, Held at Winnipeg, Manitoba, Canada, 9-11 May 1975.* 1976, VI + 345 pp.

SYNTHESE HISTORICAL LIBRARY

Texts and Studies
in the History of Logic and Philosophy

Editors:

N. KRETZMANN (Cornell University)
G. NUCHELMANS (University of Leyden)
L. M. DE RIJK (University of Leyden)

1. M. T. BEONIO-BROCCHIERI FUMAGALLI, *The Logic of Abelard.* Translated from the Italian. 1969, IX + 101 pp.
2. GOTTFRIED WILHELM LEIBNITZ, *Philosophical Papers and Letters.* A selection translated and edited, with an introduction, by Leroy E. Loemker. 1969, XII + 736 pp.
3. ERNST MALLY, *Logische Schriften,* ed. by Karl Wolf and Paul Weingartner. 1971, X + 340 pp.
4. LEWIS WHITE BECK (ed.), *Proceedings of the Third International Kant Congress.* 1972, XI + 718 pp.
5. BERNARD BOLZANO, *Theory of Science,* ed. by Jan Berg. 1973, XV + 398 pp.
6. J. M. E. MORAVCSIK (ed.), *Patterns in Plato's Thought. Papers arising out of the 1971 West Coast Greek Philosophy Conference.* 1973, VIII + 212 pp.
7. NABIL SHEHABY, *The Propositional Logic of Avicenna: A Translation from al-Shifā: al-Qiyās,* with Introduction, Commentary and Glossary. 1973, XIII + 296 pp.
8. DESMOND PAUL HENRY, *Commentary on De Grammatico: The Historical-Logical Dimensions of a Dialogue of St. Anselm's.* 1974, IX + 345 pp.
9. JOHN CORCORAN, *Ancient Logic and Its Modern Interpretations.* 1974, X + 208 pp.
10. E. M. BARTH, *The Logic of the Articles in Traditional Philosophy.* 1974, XXVII + 533 pp.
11. JAAKKO HINTIKKA, *Knowledge and the Known. Historical Perspectives in Epistemology.* 1974, XII + 243 pp.
12. E. J. ASHWORTH, *Language and Logic in the Post-Medieval Period.* 1974, XIII + 304 pp.
13. ARISTOTLE, *The Nicomachean Ethics.* Translated with Commentaries and Glossary by Hypocrates G. Apostle. 1975, XXI + 372 pp.
14. R. M. DANCY, *Sense and Contradiction: A Study in Aristotle.* 1975, XII + 184 pp.
15. WILBUR RICHARD KNORR, *The Evolution of the Euclidean Elements. A Study of the Theory of Incommensurable Magnitudes and Its Significance for Early Greek Geometry.* 1975, IX + 374 pp.
16. AUGUSTINE, *De Dialectica.* Translated with the Introduction and Notes by B. Darrell Jackson. 1975, XI + 151 pp.

3 5282 00589 6652

Printed in the United States
22939LVS00003BA/2

9 789027 706553